Handbook of Numerical Analysis

Volume 17

Handbook of Numerical Methods for Hyperbolic Problems

Basic and Fundamental Issues

Handbook of Numerical Analysis

Series Editors

Handbook of Numerical Analysis
Volume 17

Handbook of Numerical Methods for Hyperbolic Problems

Basic and Fundamental Issues

Edited by

Rémi Abgrall
Institut für Mathematik & Computational
Science Universität Zürich,
Zürich, Switzerland

Chi-Wang Shu
Division of Applied Mathematics
Brown University
Providence, RI, United States

ELSEVIER
Amsterdam • Boston • Heidelberg • London • New York • Oxford
Paris • San Diego • San Francisco • Singapore • Sydney • Tokyo
North-Holland is an imprint of Elsevier
N·H

North-Holland is an imprint of Elsevier
Radarweg 29, PO Box 211, 1000 AE Amsterdam, The Netherlands
The Boulevard, Langford Lane, Kidlington, Oxford OX5 1GB, United Kingdom

ISBN: 978-0-444-63789-5
ISSN: 1570-8659

For information on all North-Holland publications visit our website at https://www.elsevier.com/

Publisher: Zoe Kruze
Acquisition Editor: Kirsten Shankland
Editorial Project Manager: Hannah Colford
Production Project Manager: Radhakrishnan Lakshmanan
Cover Designer: Matthew Limbert

Typeset by SPi Global, India

Contents

Contributors

Numbers in Parentheses indicate the pages on which the author's contributions begin.

J.L.F. Azevedo (199), Instituto de Aeronautica e Espaço, São Jose dos Campos, SP, Brazil

P. Bochev (289), Center for Computing Research, Sandia National Laboratories, Albuquerque, NM, United States

A. Bonfiglioli (403), Scuola di Ingegneria, Università degli Studi della Basilicata, Potenza, Italy

M.H. Carpenter (495), NASA Langley Research Center, Hampton, VA, United States

B. Cockburn (173), School of Mathematics, University of Minnesota, Minneapolis, MN, United States

C.M. Dafermos (1), Division of Applied Mathematics, Brown University, Providence, RI, United States

B. Després (77), Sorbonne Universités, UPMC Univ Paris 06, UMR 7598, Laboratoire Jacques-Louis Lions, Paris, France

A. Ern (265), Université Paris-Est, CERMICS (ENPC), Marne-la-Vallée cedex 2, France

M. Falcone (603), Università di Roma "La Sapienza", Roma, Italy

R. Ferretti (603), Università Roma Tre, Roma, Italy

T.C. Fisher (495), Sandia National Laboratories, Albuquerque, NM, United States

U.S. Fjordholm (123), NTNU, Trondheim, Norway

J. Glimm (383), Stony Brook University, Stony Brook, NY, United States

S. Gottlieb (549), University of Massachusetts Dartmouth, North Dartmouth, MA, United States

J.-L. Guermond (265), Texas A&M University, College Station, TX, United States

M. Gunzburger (289), Florida State University, Tallahassee, FL, United States

J.S. Hesthaven (441), EPFL-SB-MATHICSE-MCSS, Ecole Polytechnique Fédérale de Lausanne, Lausanne, Switzerland

A. Hsu (383), Stony Brook University, Stony Brook, NY, United States

A. Jameson (227), Stanford University, Stanford, CA, United States

R. Kaufman (383), Stony Brook University, Stony Brook, NY, United States

D.I. Ketcheson (549), CEMSE Division, King Abdullah University of Science and Technology (KAUST), Thuwal, Saudi Arabia

S. Kokh (77), Maison de la Simulation USR 3441, Digiteo Labs, bât. 565, PC 190; DEN/DANS/DM2S/STMF/LMSF, CEA Saclay, Gif-sur-Yvette, France

A. Kurganov (525), Tulane University, New Orleans, LA, United States

C. Lacor (199), Vrije Universiteit, Brussel, Belgium

F. Lagoutière (77), Laboratoire de Mathématiques d'Orsay, Univ. Paris-Sud, CNRS, Université Paris-Saclay, Orsay, France

H. Lim (383), Stony Brook University, Stony Brook, NY, United States

Y. Liu (199), NASA Ames Research Center, CA, United States

R. Loubère (319), Institut de Mathématiques de Toulouse and CNRS, Toulouse Cedex 9, France

P.-H. Maire (319), CEA/CESTA, Le Barp, France

J. Melvin (383), Stony Brook University, Stony Brook, NY, United States

F. Nasuti (403), Dip. di Ingeneria Meccanica e Aerospaziale, Univerità degli studi di Roma "La Sapienza", Rome, Italy

N.C. Nguyen (173), Massachusetts Institute of Technology, Cambridge, MA, United States

E.J. Nielsen (495), NASA Langley Research Center, Hampton, VA, United States

M. Onofri (403), Dip. di Ingeneria Meccanica e Aerospaziale, Univerità degli studi di Roma "La Sapienza", Rome, Italy

R. Paciorri (403), Dip. di Ingeneria Meccanica e Aerospaziale, Univerità degli studi di Roma "La Sapienza", Rome, Italy

M. Parsani (495), King Abdullah University of Science and Technology (KAUST), Extreme Computing Research Center (ECRC), Thuwal, Saudi Arabia

J. Peraire (173), Massachusetts Institute of Technology, Cambridge, MA, United States

J.-M. Qiu (353), University of Houston, Houston, United States

J. Qiu (147), School of Mathematical Sciences, Fujian Provincial Key Laboratory of Mathematical Modeling and High-Performance Scientific Computing, Xiamen University, Xiamen, Fujian, PR China

B. Rebourcet (319), CEA/DAM Ile de France, Arpajon Cedex, France

D. She (383), Stony Brook University, Stony Brook, NY, United States

C.-W. Shu (103), Brown University, Providence, RI, United States

T. Sonar (55), Institut Computational Mathematics, Technische Universität Braunschweig, Braunschweig, Germany

M. Svärd (495), University of Bergen, Bergen, Norway

E. Tadmor (467), Center of Scientific Computation and Mathematical Modeling (CSCAMM), Department of Mathematics, Institute for Physical Science and Technology, University of Maryland, College Park, MD, United States

E.F. Toro (19), Laboratory of Applied Mathematics, DICAM, University of Trento, Trento, Italy

P.E. Vincent (227), Imperial College London, South Kensington, London, United Kingdom

Z.J. Wang (199), University of Kansas, Lawrence, KS, United States

F.D. Witherden (227), Stanford University, Stanford, CA, United States

N. Yamaleev (495), Old Dominion University, Norfolk, VA, United States

Q. Zhang (147), Nanjing University, Nanjing, Jiangsu, PR China

Y.-T. Zhang (103), University of Notre Dame, Notre Dame, IN, United States

H. Zhao (585), University of California, Irvine, CA, United States

Introduction

R. Abgrall* and C.-W. Shu†
*Institut für Mathematik, Universität Zürich, Zürich, Switzerland
†Brown University, Providence, RI, USA

These two volumes represent the 17th and 18th volume of *Handbook of Numerical Analysis*. It is entirely devoted to the numerical methods designed for approximating the solution of hyperbolic equations, or of equations that write as a sum of operators where the most important, in terms of the behaviour of the solution, is the hyperbolic one. An example is the Navier–Stokes equations with high Reynolds number where the solution behaviour is essentially dictated by the hyperbolic operator (here the Euler system), except in boundary layers because of the boundary conditions.

Hyperbolic partial differential equations appear often in applications. The most important application, already mentioned, is fluid dynamics, including specific flows such as multiphase flows, magneto-hydrodynamics and water waves. Other application areas include Maxwell equations, kinetic equations, and traffic flow models and networks. The solutions of hyperbolic partial differential equations often involve discontinuities, making mathematical analysis and numerical simulations difficult. In the past few decades there has been a large amount of literature in the design, analysis and application of various numerical algorithms for solving hyperbolic equations. The current volumes attempt to have experts in different types of algorithms write concise summaries so that the readers can find a variety of algorithms under different situations and become familiar with their relative advantages and limitations.

This is a formidable task. We had to make choices because the field has grown tremendously since the early ages dating back to von Neumann in the United States and researchers from the former Soviet Union such as Rusanov and Godunov. This field has grown up for various reasons. The demand on diverse high-tech areas ranging from airplanes and rockets, to the nuclear and car industries as well as more recently the green industry, to name just a few, necessitates to master better and better tools to improve performance. If it was possible in the early ages to rely on analytical solutions and experimental facilities only, this is no longer the case because of various constraints: economical, technological (weight and so on), energy consumption, etc. This evolution has needed improved algorithms, i.e., more and more accurate as well as more and more robust ones. Hence the research on algorithm has grown up and then exploded since the early 1970s.

In parallel, and also triggered by the same needs, computers have been more and more powerful from scalar, to vectors, then parallel and now massively parallel and hybrid architectures. This evolution of technology has also had a strong impact on the algorithms development.

Because of its success, it is now possible to compute more and more complicated problems, both in terms of geometry and physics.

There is still a lot to do to improve and understand the numerical methods designed for hyperbolic problems. The aim of these two volumes is to give a picture of the current state of the art.

In order to introduce the subject, we have asked Professor Dafermos from Brown University to provide a short summary on the theory of hyperbolic equations. Then, if one looks at the table of content, one would realize that we have tried to cover not only the classical topics, such as the finite volume method and the Riemann solvers that are the building blocks of many of the algorithms, but also less standard methods. Examples include algorithms for computing sharp transition propagated by linearly degenerate waves. Other examples are given by the ENO/WENO family. In that case we have tried to go over the classical description, by giving some analysis of the methods. Other high-order methods are also considered such as the discontinuous Galerkin (DG) ones, the more recent Hybrid DG schemes, high-order finite element methods, front-tracking methods, methods for Lagrangian hydrodynamics and entropy stable schemes. Time discretization is also considered, as well as more specialized problems like the simulation of flows with low Mach numbers, level set techniques and numerical methods for Hamilton–Jacobi equations.

Unfortunately, it is not possible, even in two quite thick volumes, to provide an exhaustive coverage of the state of the art. Even though the table of content seems to be exhaustive, many topics are still missing. For example we have chosen to be quite restrictive on the subject of time stepping: there is no coverage on ADER and IMEX methods. The handling of problems with source terms is touched by two chapters (well-balanced schemes, asymptotic preserving schemes), but there is no direct coverage on stiff source terms. If we have a chapter on methods for Cartesian meshes, there is no direct coverage on the application of immersed boundary methods. Similarly we have chosen to consider the problem of meshing in a specific way; there is no direct coverage on adaptive mesh refinement. The problem on boundary conditions is considered in two chapters (SAT-SPB schemes and inverse Lax–Wendroff procedure), but much more could have been said. It was simply impossible to cover the whole field and we apologize for this.

To end this introduction, we would like to thank all the contributors to these volumes, as well as the referees. Both have been extremely efficient.

ACKNOWLEDGEMENTS

R.A. has been supported in part by SNF grant # 200021_153604. C.-W.S. has been supported in part by NSF grant DMS-1418750.

Chapter 1

Introduction to the Theory of Hyperbolic Conservation Laws

C.M. Dafermos
Division of Applied Mathematics, Brown University, Providence, RI, United States

Chapter Outline

ABSTRACT

This is a brief, informal introduction to nonlinear hyperbolic conservation laws, underscoring their inherent properties (wave breaking, entropy conditions) and sketching the state of the art in their analysis.

Keywords: Hyperbolic conservation laws, Entropy, Viscosity, Shocks, Riemann Problem

AMS Classification Codes: 35L65, 35L67

1 INTRODUCTION

The conservation laws of gas dynamics, elastodynamics, electrodynamics and other branches of classical physics are typically expressed by hyperbolic partial differential equations or systems thereof. In particular, it is hyperbolic systems that provide the proper mathematical setting for a host of wave phenomena.

The salient feature of solutions to nonlinear hyperbolic systems resulting from conservation laws is wave breaking, which triggers the development of jump discontinuities that propagate on as shock waves. This renders the mathematical theory particularly hard, as it must cope with weak solutions. The difficulty is exacerbated by the fact that uniqueness and stability are lost in the realm of weak solutions. As a remedy, one seeks selection criteria, motivated by physical or mathematical considerations, that hopefully weed out all spurious solutions, singling out the admissible one. In consequence of these difficulties, and despite considerable progress achieved over the past 50 years,

Handbook of Numerical Analysis, Vol. 17. http://dx.doi.org/10.1016/bs.hna.2016.08.003

central mathematical issues are still wide open, especially in several spatial dimensions. It is fair to admit that at the present time computation has outpaced the theory.

The aim of this introductory chapter is to provide a description of the basic structure of hyperbolic systems of conservation laws and to survey their inherent properties. We shall not provide specific bibliographic references. The reader who seeks more detailed information should consult one or more of numerous existing texts. The list includes Smoller (1994) (a clear introduction to the basics, albeit somewhat dated), Holden and Risebro (2015) (a readable, successful marriage of theory with numerics), Serre (1999) (an insightful introduction to the basics, supplemented with an interesting selection of more advanced topics), Bressan (2000) (a nice exposition starting out at an introductory elementary level and becoming progressively more focused and technical) and Dafermos (2016) (an encyclopaedic coverage of the field, with voluminous bibliography). The perspective and style of presentation in this chapter are borrowed from Dafermos (2016).

The author is indebted to Zheng Sun for his valuable assistance in drawing the figures.

2 BASIC STRUCTURE OF HYPERBOLIC CONSERVATION LAWS

The canonical form of a system of n conservation laws in k spatial dimensions reads

$$\partial_t U + \sum_{\alpha=1}^{k} \partial_\alpha F_\alpha(U) = 0. \tag{1}$$

The (unknown) n-dimensional *state vector* field U is a function of the k-dimensional spatial variable x and the scalar temporal variable t. For $\alpha = 1, \ldots, k$, the *flux* $F_\alpha(U)$ is a given smooth function from $\mathbb{R}^n$ to $\mathbb{R}^n$ and ∂_α stands for $\partial/\partial x_\alpha$. The terminology, with origins in classical physics, stems from the observation that (1) holds on some domain of $\mathbb{R}^k$ if and only if

$$\frac{d}{dt}\int_\Omega U dx + \oint_{\partial\Omega} \sum_{\alpha=1}^{k} \nu_\alpha F_\alpha(U) dS = 0 \tag{2}$$

for all smooth subdomains Ω, with $\partial\Omega$ denoting the boundary of Ω and ν standing for the exterior unit normal on $\partial\Omega$. Indeed, the 'physical' interpretation of (2) is that the n-vector valued quantity with density U is conserved, in the sense that the rate of change in the amount stored in Ω is balanced by the rate of flux, in or out of Ω, through $\partial\Omega$.

In what follows we will employ matrix notation, identifying $\mathbb{R}^n$ with column vectors $\mathbb{M}^{n\times 1}$. The symbol D will denote the gradient operator in $\mathbb{R}^n$, mapping scalar fields into $\mathbb{M}^{1\times n}$ row vector fields and $\mathbb{M}^{n\times 1}$ column vector fields into $\mathbb{M}^{n\times n}$ matrix fields.

The system of conservation laws (1) will be called *hyperbolic* if for any fixed $U \in \mathbb{R}^n$ and $\nu \in \mathbb{S}^{k-1}$, the $n \times n$ matrix $\sum_{\alpha=1}^{k} \nu_\alpha \mathrm{D}F_\alpha(U)$ has real eigenvalues $\lambda_1(U, \nu), \ldots, \lambda_n(U, \nu)$ and an associated set of n linearly independent eigenvectors $R_1(U, \nu), \ldots, R_n(U, \nu)$.

An important subclass of hyperbolic conservation laws are the *symmetric systems* (1) with the property that for $\alpha = 1, \ldots, k$ and any $U \in \mathbb{R}^n$, the matrices $\mathrm{D}F_\alpha(U)$ are symmetric.

As a consequence of hyperbolicity, for any fixed $U \in \mathbb{R}^n$ and $\nu \in \mathbb{S}^{k-1}$, all functions in the form

$$V(x,t) = u(\nu \cdot x - \lambda_i(U,\nu)t) R_i(U,\nu), \tag{3}$$

depicting waves with amplitude collinear to $R_i(U, \nu)$, travelling in the direction ν with speed $\lambda_i(U, \nu)$, are solutions of the system resulting from linearizing (1) about U:

$$\partial_t V + \sum_{\alpha=1}^{k} \mathrm{D}F_\alpha(U) \partial_\alpha V = 0. \tag{4}$$

The simplest example is provided by the *scalar conservation law*

$$\partial_t u + \sum_{\alpha=1}^{k} \partial_\alpha f_\alpha(u) = 0. \tag{5}$$

However, the primordial, and still most important, example is the system of the Euler equations

$$\begin{cases} \partial_t \rho + \operatorname{div}(\rho v) = 0 \\ \partial_t(\rho v) + \operatorname{div}(\rho v \otimes v) + \operatorname{grad} p(\rho) = 0, \end{cases} \tag{6}$$

which govern the isentropic flow of a gas. In (6), ρ denotes the (mass) density, v stands for velocity and p is the pressure. The gradient and divergence operate with respect to the spatial variable. Eq. $(6)_1$ expresses conservation of mass while $(6)_2$ states conservation of (linear) momentum. The system is hyperbolic so long as $p'(\rho) > 0$.

The notion of entropy plays a very important role in the theory of hyperbolic conservation laws. A scalar function $\eta(U)$ is called an *entropy* for the system (1), associated with the *entropy flux* $q_\alpha(U)$, $\alpha = 1, \ldots, k$, if for $U \in \mathbb{R}^n$

$$\mathrm{D}q_\alpha(U) = \mathrm{D}\eta(U) \mathrm{D}F_\alpha(U), \quad \alpha = 1, \ldots, k. \tag{7}$$

This is equivalent to requiring that any smooth solution U of (1) satisfies automatically the additional conservation law

$$\partial_t \eta(U) + \sum_{\alpha=1}^{k} \partial_\alpha q_\alpha(U) = 0. \tag{8}$$

In the scalar case (5), any function $\eta(u)$ qualifies as entropy, with associated entropy flux $q_\alpha(u) = \int \eta'(u) f'(u) du, \alpha = 1, \ldots, k$. Also, for $k = 1$ and $n = 2$, the system (7) of two equations in two unknowns, yields a rich family of entropy–entropy flux pairs (η, q). In all other cases, however, since $kn > k+1$, (7) is overdetermined so one should not expect the existence of nontrivial entropies for generic systems (1). Nevertheless, it turns out that in virtually all interesting systems arising in physics, the fluxes are judiciously selected so that an entropy exists. Moreover, quite often, though not always, this entropy is a convex function of U.

A case in point is the system (6) of the Euler equations, which is equipped with the entropy–entropy flux pair

$$\eta = \rho\varepsilon(\rho) + \frac{1}{2}\rho|v|^2, \quad q = [\rho\varepsilon(\rho) + \frac{1}{2}\rho|v|^2 + p(\rho)]v, \tag{9}$$

where $\varepsilon(\rho) = \int \rho^{-2} p(\rho) d\rho$ is the internal energy. It turns out that the hyperbolicity condition $p'(\rho) > 0$ renders η convex, as a function of the canonical state vector $(\rho, \rho v)$. The extra conservation law (8) here expresses conservation of mechanical energy.

Any symmetric system (1) is endowed with the entropy–entropy flux pair

$$\eta(U) = \frac{1}{2}|U|^2, \quad q_\alpha = U \cdot F_\alpha(U) - \varphi_\alpha(U), \tag{10}$$

where φ_α is a 'potential' with $D\varphi_\alpha = F_\alpha$, which exists since DF_α is symmetric.

It is easily seen that, conversely, any system of conservation laws (1) possessing a convex entropy $\eta(U)$ is symmetrized by introducing the new state vector $V = D\eta(U)$. In particular, any such system is hyperbolic.

The Cauchy problem is locally well posed for any system of conservation laws (1) endowed with a convex entropy. Indeed, for any function U_0 in the Sobolev space $H^l(\mathbb{R}^k)$, with $l > \frac{1}{2}k + 1$, there exists a unique C^1 solution U of (1) defined on $\mathbb{R}^k \times [0, T_\infty)$ and satisfying the initial condition

$$U(x, 0) = U_0(x), \quad x \in \mathbb{R}^k. \tag{11}$$

The lifespan T_∞ is maximal in the sense that either $T_\infty = \infty$ or else $T_\infty < \infty$, in which case $\max_{\mathbb{R}^k} |\nabla U(\cdot, t)| \to \infty$, as $t \to T_\infty$.

The proof of the above proposition rests on establishing a priori bounds on the $L^2(\mathbb{R}^k)$ norms of U and all of its spatial derivatives up to order l. These are derived by means of 'energy' estimates induced by the presence of the convex entropy. For instance, (8) yields a bound on the $L^2(\mathbb{R}^k)$ norm of U. The restriction $l > \frac{1}{2}k + 1$ is needed in order to keep $\| \nabla U \|_{L^\infty}$ bounded, in which case the family of energy estimates closes.

It turns out that for nonlinear hyperbolic systems of conservation laws the case of finite lifespan for smooth solutions is the rule rather than the exception. This comes as a result of the wave breaking effect: As waves move at different speeds, "compressive" wave profiles get progressively steeper and eventually break. This scenario is easily seen in the setting of the scalar conservation law. Assume $u(x, t)$ is a local smooth solution of the Cauchy problem for (5), with initial values $u(x, 0) = u_0(x)$. Consider characteristics associated with u as trajectories of the ordinary differential equation $\frac{dx_\alpha}{dt} = f'_\alpha(u(x,t)), \alpha = 1,\ldots,k$. Letting an overdot denote the derivative $\partial_t + \sum f'_\alpha(u)\partial_\alpha$ in the characteristic direction, we may write (5) as $\dot{u} = 0$, which shows that characteristics are straight lines along which u is constant. Thus, with any (x, t) is associated $y \in \mathbb{R}^k$ such that

$$y_\alpha = x_\alpha - tf'_\alpha(u(x,t)), \quad \alpha = 1,\ldots,k \tag{12}$$

and $u(x, t) = u_0(y)$. This easily implies

$$\partial_\beta u(x,t) = \frac{\partial_\beta u_0(y)}{1 + t\sum_{\alpha=1}^{k} \partial_\alpha f'_\alpha(u_0(y))}, \quad \beta = 1,\ldots,k. \tag{13}$$

We conclude that if $\sum \partial_\alpha f'_\alpha(u_0(y))$ takes negative values at any points, attaining a negative minimum, say $-\varepsilon$, on $\mathbb{R}^k$, ∇u must blow up at $t = \frac{1}{\varepsilon}$.

Wave breaking is particularly pronounced in one space dimension, where waves are confined and cannot avoid interacting with one another. In several space dimensions, depending on the geometry of the system, wave breaking may be impeded by wave dispersion, which has the opposite effect. As we saw earlier, this is not the case for the scalar conservation law (5). The situation is quite different for the Euler equations (6), in three spatial dimensions. For this system, wave breaking and dispersion are evenly matched and their competition is very keen. Dispersion manages to prolong the lifespan of smooth solutions with initial derivatives of size ε to $O\left(\exp\frac{1}{\varepsilon}\right)$—much longer than the $O\left(\frac{1}{\varepsilon}\right)$ lifespan of smooth solutions to the scalar conservation law. Nevertheless, eventually wave breaking prevails and the derivatives of the solution blow up.

In view of the breakdown of smooth solutions, in order to get solutions in the large, one has to resort to weak, distributional solutions, namely bounded measurable vector fields U on some domain $\mathcal{X}$ of $\mathbb{R}^k \times \mathbb{R}$, which satisfy

$$\int\int_{\mathcal{X}} [\partial_t \phi U + \sum_{\alpha=1}^{k} \partial_\alpha \phi F_\alpha(U)] dxdt = 0, \tag{14}$$

for all smooth test functions ϕ with compact support in $\mathcal{X}$. Notice that it is possible to define weak solutions because conservation laws (1) are in divergence form.

In particular, let us seek weak solutions defined on all of $\mathbb{R}^k \times \mathbb{R}$ in the form

$$U(x,t) = \begin{cases} U_- & \text{for } \nu\cdot x - st < 0 \\ U_+ & \text{for } \nu\cdot x - st > 0, \end{cases} \tag{15}$$

where U_-, U_+ are constant states in $\mathbb{R}^n$, $\nu \in \mathbb{S}^{k-1}$ and s is a scalar. It is a simple exercise to verify that U in (15) will satisfy (14) for all test functions ϕ if and only if

$$\sum_{\alpha=1}^{k} \nu_\alpha [F_\alpha(U_+) - F_\alpha(U_-)] - s[U_+ - U_-] = 0. \tag{16}$$

Recalling the definition of hyperbolicity one sees that for any fixed $U_- \in \mathbb{R}^n$ and $\nu \in \mathbb{S}^{k-1}$ it is possible to find U_+ in the vicinity of U_- such that $U_+ - U_-$ is nearly collinear to $R_i(U_-, \nu)$ and (16) holds for some s near $\lambda_i(U_-, \nu)$. Such a weak solution of (1) is termed a planar *shock wave* with amplitude $U_+ - U_-$ propagating in the direction ν with speed s, and (16) is called the *Rankine–Hugoniot jump condition.*

More generally, there exist piecewise smooth weak solutions to (1) with jump discontinuities across curved shocks. In that situation (16) still holds across the shock, though now ν, U_-, U_+ and s are no longer constant, as they may vary from point to point on the shock.

Familiarity with weak solutions to hyperbolic conservation laws is enhanced by visualizing them as composites of continuous waves and shock waves, with the understanding that these two constituents may be finely blended. Though it is not presently known whether every L^∞ weak solution fits the above description, this is certainly the case at least for solutions U of class BV, for which $\partial_t U$ and $\partial_\alpha U$ are Radon measures. Indeed, the domain of $U \in BV$ is the union of three disjoint sets, namely: (a) the set of points of approximate continuity, in the sense of Lebesgue; (b) the set of points of (approximate) jump discontinuity, which is an at most countable family of disjoint $C^1 k$-dimensional manifolds, across which the jump condition (16) holds; and (c) a 'small' residual set whose k-dimensional Hausdorff measure is nil.

As we shall see in the next section, a serious obstacle for dealing with weak solutions is the loss of uniqueness in the Cauchy problem. Accordingly, additional requirements must be imposed, in order to weed out spurious solutions and single out the unique admissible one. In what follows, we outline two methods in that direction, which are dictated, or at least motivated, by physics.

Assume our system (1) is endowed with an entropy–entropy flux pair (η, q), with $D^2\eta(U)$ positive definite. Recall that under such conditions any smooth solution of (1) satisfies automatically the extra conservation law (8). However, this is no longer the case for weak solutions of (1). We now stipulate that a bounded measurable weak solution U of (1) satisfies the *entropy admissibility condition*, relative to η, on a domain $\mathcal{X}$ of $\mathbb{R}^k \times \mathbb{R}$, if

$$\partial_t \eta(U) + \sum_{\alpha=1}^{k} \partial_\alpha q_\alpha(U) \leq 0 \tag{17}$$

holds in the sense of distributions, that is

$$\int\int_{\mathcal{X}}[\partial_t\psi\eta(U)+\sum_{\alpha=1}^{k}\partial_\alpha\psi q_\alpha(U)]dxdt\geq 0, \tag{18}$$

for all nonnegative smooth test functions ψ, with compact support in $\mathcal{X}$. In particular, all smooth solutions of (1) satisfy this condition and thus are deemed admissible.

In the physical applications, the inequality (17) typically manifests, directly or indirectly, the second law of thermodynamics.

The entropy admissibility condition is particularly effective for scalar conservation laws (5), for which, as we saw earlier, any convex function $\eta(u)$ may serve as entropy. The approach pioneered by Kruzkov is to deem a weak solution u of (5) admissible if it satisfies

$$\partial_t\eta(u)+\sum_{\alpha=1}^{k}\partial_\alpha q_\alpha(u)\leq 0 \tag{19}$$

for all convex functions $\eta(u)$ and $q_\alpha(u)=\int\eta'(u)f'(u)du$. It has been shown that for any $u_0(x)\in L^\infty(\mathbb{R}^k)$ there exists a unique admissible solution $u(x, t)$ of (5) on $\mathbb{R}^k\times[0,\infty)$, with initial value $u_0(x)$. Furthermore, admissible solutions are strongly stable as they have the following L^1 contraction property:

$$\int_{\mathbb{R}^k}|u(x,t)-\bar{u}(x,t)|dx\leq\int_{\mathbb{R}^k}|u_0(x)-\bar{u}_0(x)|dx,\quad 0\leq t<\infty, \tag{20}$$

holds for any pair $(u,\bar{u})$ of admissible solutions with initial values $(u_0,\bar{u}_0)$.

In particular, applying (20) with $\bar{u}_0(x)=u_0(x+e)$ *and thereby* $\bar{u}(x,t)=u(x+e,t)$, where e is an arbitrary k-vector, we deduce that initial data u_0 of class BV generate solutions u to the Cauchy problem for the scalar conservation law (5) that are also of class BV and the variation of $u(\cdot, t)$ over $\mathbb{R}^k$ is a nonincreasing function of t on $[0,\infty)$.

By contrast, systems of conservation laws (1) with $n\geq 2$ typically possess a single convex entropy so that the admissibility condition (17) does not generally suffice for uniqueness of solutions to the Cauchy problem. In particular, it has been shown that for certain initial data the Cauchy problem for the system of the Euler equations (6) admits infinitely many weak solutions satisfying the entropy admissibility condition (17), relative to the entropy (9). Nevertheless, it turns out that even a single inequality (17) for a convex entropy suffices for securing uniqueness and stability of smooth solutions to the Cauchy problem, not only within the class of smooth solutions but even within the broader class of admissible weak solutions (so-called *weak-strong stability*). Specifically, if on $\mathbb{R}^k\times[0,T]$ there exist a smooth solution $\bar{U}(x,t)$, with initial values $\bar{U}_0(x)$, and also a weak solution $U(x, t)$, with initial values $U_0(x)$, satisfying the entropy admissibility condition (17), then

$$\int_{\mathbb{R}^k}|U(x,t)-\bar{U}(x,t)|^2dx\leq ce^{at}\int_{\mathbb{R}^k}|U_0(x)-\bar{U}_0(x)|^2dx,\quad 0\leq t\leq T. \tag{21}$$

The proof of this property is established with the help of the so-called *relative entropy*

$$H(U,\bar{U}) = \eta(U) - \eta(\bar{U}) - \mathrm{D}\eta(\bar{U})[U - \bar{U}]. \tag{22}$$

It is easy to see that for solutions of the form (15) the entropy admissibility condition (18) reduces to the jump condition

$$\sum_{\alpha=1}^{k} \nu_\alpha [q_\alpha(U_+) - q_\alpha(U_-)] - s[\eta(U_+) - \eta(U_-)] \le 0. \tag{23}$$

For arbitrary weak solutions U satisfying the entropy admissibility condition, the left-hand side of (17) is a nonpositive distribution, and thereby a measure. In particular, it turns out that when U is of class BV the above measure is concentrated on the set of points of jump discontinuity and (17) reduces to the requirement that (23) holds across every shock. This has motivated the widely held conjecture that the admissibility of weak solutions hinges exclusively on a localized test, such as (23), to be applied to every point of jump discontinuity, involving just U_-, U_+ and s, interrelated through (16). Even though it may not be universally valid, the above premise enjoys wide applicability, as we shall see in the next section.

We now turn to an alternative admissibility criterion for weak solutions, which is also motivated by physics. The isentropic flow of a viscous gas is governed by the Navier–Stokes equations

$$\begin{cases} \partial_t \rho + \mathrm{div}(\rho v) = 0 \\ \partial_t(\rho v) + \mathrm{div}(\rho v \otimes v) + \mathrm{grad}\, p(\rho) = (\lambda + \mu)\mathrm{grad\ div}\, v + \mu \Delta v, \end{cases} \tag{24}$$

where λ and μ are viscosity coefficients. Any gas in nature has some, perhaps minute, viscosity and hence admissible solutions to the Euler equations (6) should be viewed as asymptotic solutions to the Navier–Stokes equations (24), with viscosity coefficients tending to zero.

One may extend the above argument to general systems (1) as follows. Next to (1), we consider the system

$$\partial_t U + \sum_{\alpha=1}^{k} \partial_\alpha F_\alpha(U) = \mu \Delta U, \tag{25}$$

where μ is a positive 'viscosity' parameter, and postulate that a weak solution U of (1) satisfies the *viscosity admissibility condition* if it is the $\mu \to 0$ limit of smooth solutions U_μ of the parabolic system (25).

Assume (1) is endowed with an entropy–entropy flux pair (η, q), with $\eta(U)$ convex. Suppose $\{U_\mu\}$ is a family of solutions of (25), with U_μ converging, boundedly almost everywhere, as $\mu \to 0$, to a weak solution U of (1). Multiplying (25) by $\mathrm{D}\eta(U_\mu)$ and using (7) yields

$$\partial_t \eta(U_\mu) + \sum_{\alpha=1}^{k} \partial_\alpha q_\alpha(U_\mu) = \mu \Delta \eta(U_\mu) - \mu \sum_{\alpha=1}^{k} \partial_\alpha U_\mu^\top \mathrm{D}^2 \eta(U_\mu) \partial_\alpha U_\mu. \tag{26}$$

As $\mu \to 0$, the left-hand side of (26) converges, in the sense of distributions, to the left-hand side of (17). On the right-hand side of (26), the first term converges to zero while the second term stays nonpositive, since $\eta(U)$ is convex. We thus conclude that U satisfies (17), i.e. the viscosity admissibility condition implies the entropy admissibility condition. However, as we shall see in the next section, the converse is not generally true.

In the scalar case, for consistency with the notation in (5), we write (25) in the form

$$\partial_t u + \sum_{\alpha=1}^{k} \partial_\alpha f_\alpha(u) = \mu \Delta u. \tag{27}$$

Solutions to the Cauchy problem for (27) have the L^1 contraction property (20). This estimate serves as the tool for showing that, as $\mu \to 0$, the solution $u_\mu(x, t)$ of the Cauchy problem for (27), with initial data $u_0(x)$ in L^∞, converges, boundedly almost everywhere, to the unique weak solution $u(x, t)$ of (5), with the same initial values, which satisfies (19) for all convex functions $\eta(u)$. Thus for scalar conservation laws, the viscosity admissibility condition is equivalent to the entropy admissibility condition, *for all* convex entropies.

By contrast, for $n \geq 2$, the convergence of solutions of (25), as $\mu \to 0$, has been established only under quite restrictive hypotheses. Consequently, there is no straightforward way to test whether any particular solution of (1) satisfies the viscosity admissibility condition. In practice, the testing is performed for solutions of the form (15), with the expectation that local admissibility of shocks renders the entire solution admissible.

Accordingly, one seeks to capture the solution (15) as the $\mu \to 0$ limit of solutions U_μ of (25) in the form

$$U_\mu(x,t) = V(\xi), \quad \xi = \frac{\nu \cdot x - st}{\mu}, \tag{28}$$

depicting fronts propagating in the direction ν with speed s, and becoming progressively steeper, as μ decreases. Then V must satisfy the ordinary differential equation

$$\ddot{V} = -s\dot{V} + \sum_{\alpha=1}^{k} \nu_\alpha \dot{F}_\alpha(V) \tag{29}$$

on $(-\infty, \infty)$, with boundary conditions $V(-\infty) = U_-, V(\infty) = U_+$. Integrating (29),

$$\dot{V} = -s[V - U_-] + \sum_{\alpha=1}^{k} \nu_\alpha [F_\alpha(V) - F_\alpha(U_-)]. \tag{30}$$

Notice that the right-hand side of (30) vanishes both at $V=U_-$ and $V=U_+$, the latter by virtue of the jump condition (16). We conclude that the solution (15) of (1) satisfies the viscosity admissibility condition if there exists an orbit of (30) joining the equilibrium points U_- and U_+. The function $V\ (\xi)$ is called the shock profile or the shock structure. It allows us, so to say, to observe the shock under the microscope.

There are many other topics of interest in the theory of hyperbolic systems of conservation laws.

When dealing with solutions of (1) on some domain of $\mathbb{R}^k$ the question of assigning proper boundary conditions requires deep analysis, even in the context of smooth solutions.

One often encounters in the applications so-called systems of balance laws

$$\partial_t U + \sum_{\alpha=1}^{k} \partial_\alpha F_\alpha(U) = G(U), \tag{31}$$

with a *source* term $G(U)$ manifesting relaxation. In that case, the source has a dissipative effect inducing global existence of smooth solutions to the Cauchy problem, when the initial data are smooth and 'small'.

As noted earlier, the Cauchy problem for scalar conservation laws, $n=1$, is well-posed, in the setting of admissible weak solutions, for any $k\geq 1$. On the other hand, virtually nothing is known on the existence and uniqueness of weak solutions to the Cauchy problem for systems, $n\geq 2$, when $k\geq 2$. However, considerable progress has been made for systems in one spatial dimension, $k=1$. The following section will provide an overview.

3 STRICTLY HYPERBOLIC SYSTEMS IN ONE SPATIAL DIMENSION

This section surveys aspects of the theory of hyperbolic systems of conservation laws

$$\partial_t U + \partial_x F(U) = 0 \tag{32}$$

in one spatial dimension. As in Section 2 , the state vector U and the flux $F(U)$ take values in $\mathbb{R}^n$. The system is called *strictly hyperbolic* if for any $U \in \mathbb{R}^n$ the Jacobian matrix $DF(U)$ possesses n real distinct eigenvalues $\lambda_1(U) < \cdots < \lambda_n(U)$, called *characteristic speeds*, and thereby linearly independent eigenvectors $R_1(U),\ldots,R_n(U)$. The theory of hyperbolic systems that are not strictly hyperbolic is still incomplete, even for the simplest case where two characteristic speeds coalesce in just a single point of $\mathbb{R}^n$.

In our discussion we will employ as demonstration models the scalar conservation law

$$\partial_t u + \partial_x f(u) = 0 \tag{33}$$

and the so-called *p-system*

$$\begin{cases} \partial_t u - \partial_x v = 0 \\ \partial_t v + \partial_x p(u) = 0, \end{cases} \tag{34}$$

which is strictly hyperbolic when $p'(u) < 0$, with characteristic speeds $\lambda_1 = -\sqrt{-p'(u)}$ and $\lambda_2 = \sqrt{-p'(u)}$.

The p-system is the Lagrangian version of the Euler equations (6), in one spatial dimension. It governs the rectilinear flow of a gas in a duct, with u standing for specific volume (the inverse of density) and v denoting velocity. The same system governs the one-dimensional motion of elastic solids (longitudinal vibrations of a bar, shearing motion of a slab, oscillation of a string, etc.). In that context one usually replaces in (34) the pressure p with the negative stress $-\sigma$.

The contrast in behaviour between linear and nonlinear hyperbolic systems (32), in one spatial dimension, is particularly pronounced when the latter satisfy the *genuine nonlinearity* condition, namely, after normalizing the eigenvectors R_i,

$$\mathrm{D}\lambda_i(U)R_i(U) = 1, \quad i = 1, \ldots, n, \quad U \in \mathbb{R}^n. \tag{35}$$

In particular, the scalar conservation law (33) is genuinely nonlinear when $f''(u) \neq 0$, for all u. The simplest example is the celebrated Burgers equation

$$\partial_t u + \partial_x \left(\frac{1}{2} u^2 \right) = 0, \tag{36}$$

which is ubiquitous in the area of conservation laws. Despite its apparent simplicity, (36) exemplifies many of the principal features of genuinely nonlinear systems of hyperbolic conservation laws and provides an excellent model for an initial approach to the subject.

The p-system (34) is genuinely nonlinear when $p''(u) \neq 0$, for all u. This is the natural assumption for fluids, where $p''(u) > 0$, while in solids the stress function $\sigma(u)$ may have inflection points.

As we saw in the previous section, smooth solutions to the Cauchy problem for the scalar conservation law, in any number of spatial dimensions, typically break down in finite time because of wave breaking. This process is particularly transparent in the setting of the Burgers equation. Assume u is a smooth solution to the Cauchy problem for (36), defined on $(-\infty, \infty) \times [0, T)$. Consider the characteristics $\frac{dx}{dt} = u(x,t)$ and denote differentiation $\partial_t + u\partial_x$ in the characteristic direction by an overdot. Differentiation of (36) with respect to x yields that the derivative $v = \partial_x u$ satisfies the Bernoulli equation $\dot{v} + v^2 = 0$, along the characteristic. Thus if $v(0) = -\varepsilon < 0$, $\partial_x u$ must blow up at time $t = \frac{1}{\varepsilon}$. The same mechanism is responsible for the blowing up of smooth solutions to the Cauchy problem for general genuinely nonlinear hyperbolic systems of conservation laws in one spatial dimension, but the proof is considerably more complicated. One should expect generic blowing up of smooth solutions even for nonlinear hyperbolic systems of conservation laws that are not genuinely nonlinear, but this needs to be demonstrated individually for each case.

In view of the above, to solve the Cauchy problem in the large one has to resort to weak solutions that contain shocks. The Rankine–Hugoniot jump condition (16), satisfied across shocks, here reduces to

$$F(U_+) - F(U_-) - s[U_+ - U_-] = 0. \tag{37}$$

The standard terminology is that 'U_+ is joined to U_- by a shock of speed s'. In particular, for the scalar conservation law (33),

$$s = \frac{f(u_+) - f(u_-)}{u_+ - u_-}. \tag{38}$$

We now demonstrate, in the context of the Burgers equation, the loss of uniqueness of weak solutions, reported in Section 2. Indeed, the Cauchy problem for (36) with initial data $u(x, 0) = -1$, for $x < 0$, and $u(x, 0) = 1$, for $x > 0$, admits infinitely many weak solutions, including the following two:

$$u(x,t) = \begin{cases} -1 & -\infty < \frac{x}{t} < -1 \\ \frac{x}{t} & -1 \le \frac{x}{t} \le 1 \\ 1 & 1 < \frac{x}{t} < \infty, \end{cases} \tag{39}$$

$$u(x,t) = \begin{cases} -1 & -\infty < \frac{x}{t} < 0 \\ 1 & 0 < \frac{x}{t} < \infty. \end{cases} \tag{40}$$

This pathology is encountered in every nonlinear system (32) of conservation laws.

In order to weed out spurious solutions, so as to restore uniqueness to the Cauchy problem, we appeal to the admissibility criteria introduced in Section 2, as related to shocks. We begin with the case of the scalar conservation law.

Consider a shock for (33) joining the states u_- and u_+. Its speed s is given by (38). Eq. (30) for the structure of the shock here takes the form

$$\dot{v} = -s(v - u_-) + f(v) - f(u_-). \tag{41}$$

The shock will satisfy the viscosity admissibility condition if there exists a solution $v(\tau)$ of (41) on $(-\infty, \infty)$, with $v(-\infty) = u_-$ and $v(\infty) = u_+$. Thus the right-hand side of (41) should not change sign between u_- and u_+, or equivalently

$$f(v) - h(v) \begin{cases} \ge 0 \text{ if } u_- < v < u_+ \\ \le 0 \text{ if } u_+ < v < u_-, \end{cases} \tag{42}$$

where $h(v)$ is the straight line segment (chord), with slope s, connecting the points $(u_-, f(u_-))$ and $(u_+, f(u_+))$ on the graph of f. This is the celebrated *Oleinik E-condition.*

Looking at the same question from the perspective of the entropy admissibility condition, we fix any convex function $\eta(u)$ as entropy and determine the associated entropy flux $q(u) = \int \eta'(u)f'(u)du$. After a simple calculation, (23) here yields

$$-\int_{u_-}^{u_+} \eta''(v)[f(v) - h(v)]dv = q(u_+) - q(u_-) - s[\eta(u_+) - \eta(u_-)] \leq 0. \quad (43)$$

It is clear that the Oleinik E-condition (42) is equivalent to (43), *for all* convex $\eta(u)$. On the other hand, (43) for just one convex $\eta(u)$ does not necessarily imply (42), unless $f''(u) \neq 0$, for all u, i.e. when (33) is genuinely nonlinear. In the genuinely nonlinear case, Eqs. (42) and (43) reduce to the celebrated *Lax E-condition*

$$f'(u_+) < s < f'(u_-). \quad (44)$$

In particular, for the Burgers equation (36), the Lax E-condition (44) reduces to $u_+ < u_-$, so that the solution (40) violates the viscosity and the entropy admissibility conditions. In fact (39) is the unique admissible solution to that Cauchy problem.

We now turn to the question of admissibility of shocks for general strictly hyperbolic systems of conservation laws (32). The first task is to determine the *Hugoniot locus*, associated with a fixed state U_-, which consists of all states U_+ that satisfy (37) for some s, and thereby may be joined to U_- by a shock. The part of the Hugoniot locus contained in a small neighbourhood of U_- has a definite and simple structure. Notice that when (37) holds with $|U_+ - U_-|$ small, then s must be close to one of the characteristic speeds $\lambda_i(U_-)$ and $U_+ - U_-$ must be almost collinear to the associated eigenvector $R_i(U_-)$. Indeed, since (32) is assumed to be strictly hyperbolic, it follows by standard bifurcation theory that for each $i = 1,\ldots, n$ there exists a smooth curve $W_i(\tau)$ in some neighbourhood of U_- and associated scalar function $s_i(\tau)$ such that

$$F(W_i(\tau)) - F(U_-) - s_i(\tau)[W_i(\tau) - U_-] = 0, \quad (45)$$

$$W_i(0) = U_-, \quad (46)$$

$$\dot{W}_i(0) = R_i(U_-), \quad (47)$$

$$s_i(0) = \lambda_i(U_-), \quad (48)$$

$$\dot{s}_i(0) = \frac{1}{2} \mathrm{D}\lambda_i(U_-)R_i(U_-). \quad (49)$$

$W_i(\tau)$ is called the *i-shock curve* of (1) through U_- and the Hugoniot locus near U_- is the union of the n shock curves.

As an example, consider the p-system (34), with Rankine–Hugoniot conditions (37) in the form

$$\begin{cases} v_+ - v_- + s(u_+ - u_-) = 0 \\ p(u_+) - p(u_-) - s(v_+ - v_-) = 0, \end{cases} \tag{50}$$

whence

$$s = \pm\sqrt{-\frac{p(u_+) - p(u_-)}{u_+ - u_-}}. \tag{51}$$

It is here convenient to parametrize the shock curve through (u_-, v_-) by u, in which case

$$v = v_- - s(u - u_-), \tag{52}$$

taking s from (51), with the negative square root for the 1-shock curve and the positive square root for the 2-shock curve.

For the above simple system, (52) describes not only the local but even the global Hugoniot locus. In general, however, the global portrait of the Hugoniot locus may be quite complex, containing detached branches, islands, etc. When the system is not strictly hyperbolic, even the local portrait of the Hugoniot locus may be geometrically varied and complex. Accordingly, here we shall limit discussion to systems (32) that are strictly hyperbolic and for shocks of small amplitude.

The i-shocks must pass the admissibility test. For the viscosity admissibility condition, (30) here reduces to

$$\dot{V} = -s[V - U_-] + F(V) - F(U_-). \tag{53}$$

One seeks a solution $V(\tau)$ of the ordinary differential equation (53) on $(-\infty, \infty)$, with boundary conditions $V(-\infty) = U_-, V(\infty) = U_+$. On the other hand, if (32) is endowed with an entropy–entropy flux pair (η, q), with $\eta(U)$ convex, the entropy admissibility condition (23) for the shock here reduces to

$$q(U_+) - q(U_-) - s[\eta(U_+) - \eta(U_-)] \leq 0. \tag{54}$$

As in the scalar case, it turns out that the viscosity condition always implies the entropy condition. Furthermore, whenever the system (32) is genuinely nonlinear (35), then an i-shock of small amplitude satisfies the viscosity condition if and only if it satisfies the entropy condition, and also if and only if the *Lax E-condition*

$$\lambda_i(U_+) < s < \lambda_i(U_-) \tag{55}$$

holds. Thus, with reference to the i-shock curve and by virtue of (48) and (49), we deduce that, for genuinely nonlinear systems, an i-shock with small amplitude will be admissible when $U_+ = W_i(\tau)$, for $\tau < 0$, and inadmissible when $U_+ = W_i(\tau)$, for $\tau > 0$.

When the system (32) is not genuinely nonlinear, the viscosity admissibility condition still manages to weed out all inadmissible shocks, whereas the entropy condition and the Lax E-condition are less selective, being merely necessary but not sufficient for admissibility. A strengthening of the Lax E-condition that is equivalent to the viscosity condition, even when the system is not genuinely nonlinear, is provided by the *Liu E-condition*, which stipulates that an i-shock joining $U_+ = W_i(\tau)$ to U_- will be admissible if $s_i(\sigma) \geq s_i(\tau)$, for all σ between 0 and τ. Thus, the Liu E-condition generalizes the Oleinik E-condition, from scalar conservation laws to strictly hyperbolic systems of conservation laws. The issue of admissibility of shocks in systems that are not strictly hyperbolic is still unresolved.

An important feature of systems (32) is that they are invariant under uniform stretching of the space-time variables, as a result of which they admit *self-similar solutions* in the form

$$U(x,t) = V(\xi), \quad \xi = \frac{x}{t}. \tag{56}$$

Any constant state, $V(\xi) = \bar{U}$, is a trivial case of a self-similar solution. Another important example is provided by step functions, $V(\xi) = U_-$, for $\xi < s$, and $V(\xi) = U_+$, for $\xi > s$, where U_- and U_+ satisfy (37) and thus may be joined by a shock of speed s.

Next we consider the possibility of Lipschitz continuous self-similar solutions (56), with $V(\xi) = U_-$, for $\xi \leq \xi_-$, $V(\xi) = U_+$, for $\xi \geq \xi_+$, and $V(\xi)$ smooth, for $\xi_- \leq \xi \leq \xi_+$. In that case, one says that 'U_+ is joined to U_- by a *centred rarefaction wave*'—terminology borrowed from gas dynamics. We investigate this question under the assumption that the system (32) is genuinely nonlinear (35). Substituting from (56) into (32) yields

$$[DF(V(\xi)) - \xi I]\dot{V}(\xi) = 0. \tag{57}$$

Thus, assuming $\dot{V}(\xi) \neq 0$,

$$\lambda_i(V(\xi)) - \xi, \tag{58}$$

and $\dot{V}(\xi)$ is collinear to $R_i(V(\xi))$, for some $i = 1, \ldots, n$ (i-rarefaction wave). Differentiating (58) with respect to ξ and using (35) yields

$$\dot{V} = R_i(V). \tag{59}$$

Thus $\xi_- = \lambda_i(U_-)$, $\xi_+ = \lambda_i(U_+)$ and $U_+ = V(\xi_+)$, where $V(\xi)$ is the solution of the ordinary differential equation (59) with initial value $V(\xi_-) = U_-$. We reparametrize $V(\xi)$, replacing ξ with the new parameter $\tau = \xi - \xi_-$, denote it by $V_i(\tau)$ and call it the *i-rarefaction curve* through U_-.

As an example, consider the p-system (34), under the assumption $p''(u) > 0$ of genuine nonlinearity. As with the shock curve (52), it is convenient to parametrize the rarefaction curve through (u_-, v_-) by u, in which case it assumes the form

$$v = v_- \pm \int_{u_-}^{u} \sqrt{-p'(w)}dw, \tag{60}$$

where the minus sign gives the 1-rarefaction curve and the plus sign yields the 2-rarefaction curve.

Fig. 1 provides a sketch of both the i-shock curve and the i-rarefaction curve through U_-. Notice that if we discard the part of the shock curve associated with inadmissible shocks and concatenate the admissible part of the i-shock curve with the i-rarefaction curve, we end up with a C^1 curve called the *i-wave curve* through U_-.

The celebrated *Riemann Problem* is the Cauchy problem for (32) with initial data

$$U(x,0) = \begin{cases} U_l & -\infty < x < 0 \\ U_r & 0 < x < \infty, \end{cases} \tag{61}$$

where U_l and U_r are given constant states. Since the above initial data are invariant under stretching of the x-variable, one expects that the solution to the Riemann Problem will be self-similar, (56). As (32) remains invariant under extreme stretches and extreme contractions of the space-time variables, it is expected that the solution of the Riemann problem should depict both the local and the large time behaviour of admissible weak solutions to the Cauchy problem, under general initial conditions. One might say that the solution of the Riemann problem provides the instrument for observing general solutions 'under the microscope' as well as 'through the telescope'. Furthermore, the solution to the Riemann problem has been used as a building block for constructing general solutions, for theoretical or numerical purposes.

For genuinely nonlinear systems, it is possible to synthesize the solution to the Riemann problem with the help of the wave curves, introduced above. Fig. 2 shows how this is done for the case of the p-system (34). Given the end-states (u_l, v_l) and (u_r, v_r), one has to locate an 'intermediate' state (u_m, v_m) with the property that (u_m, v_m) lies on the 1-wave curve through (u_l, v_l) and, in turn, (u_r, v_r) lies on the 2-wave curve through (u_m, v_m). This is possible because 1- and 2-wave curves intersect transversely, by strict hyperbolicity. The resulting solution to the Riemann Problem consists of the three constant states (u_l, v_l), (u_m, v_m), (u_r, v_r) together with a 1-wave joining (u_m, v_m) to

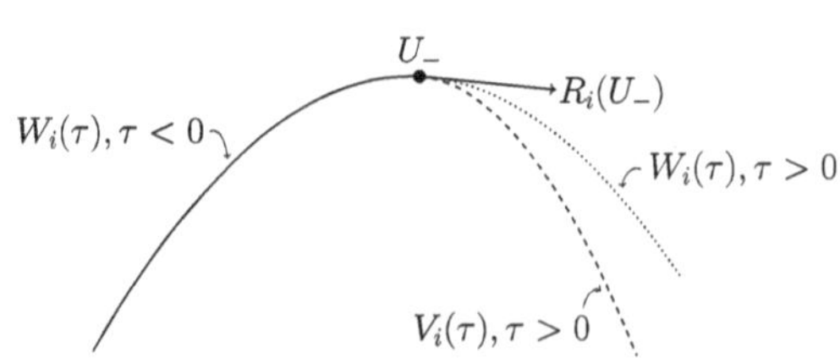

FIG. 1 Shock and rarefaction wave curves.

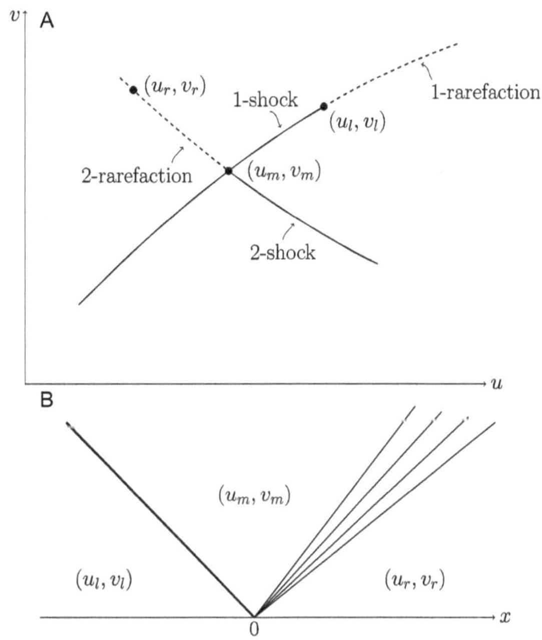

FIG. 2 Solution to the Riemann problem.

(u_l, v_l) and a 2-wave joining (u_r, v_r) to (u_m, v_m). Because of the relative location of (u_l, v_l), (u_r, v_r) in Fig. 2 , the 1-wave turns out to be a shock and the 2-wave a rarefaction. However, any combination is possible.

The situation is similar for general genuinely nonlinear systems (32). By means of the implicit function theorem, one shows that for any given states U_l and U_r, with $|U_r - U_l|$ sufficiently small, there exists a unique solution to the Riemann problem, with initial data (61), which is composed of $n+1$ constant states $U_l=U_0, U_1,\ldots, U_n=U_r$, with U_i joined to U_{i-1} by either an admissible i-shock or by an i-rarefaction wave.

The solution to the Riemann problem for strictly hyperbolic systems (32) that are not genuinely nonlinear has a similar structure, involving as above $n+1$ constant states $U_l=U_0, U_1,\ldots, U_n=U_r$. However, now U_i is joined to U_{i-1} by an i-wave that is no longer necessarily a single i-shock or i-rarefaction wave, but possibly a composite of several such shocks and rarefactions.

For systems that are not strictly hyperbolic, the solution to the Riemann problem may assume a variety of forms and the issue of admissibility and uniqueness is not completely resolved.

We now turn to the general Cauchy problem for strictly hyperbolic systems (32), under initial conditions

$$U(x,0) = U_0(x), \quad -\infty < x < \infty. \tag{62}$$

The state of the art may be summarized as follows: If the total variation of $U_0(x)$ is sufficiently small, there exists a unique admissible weak solution $U(x, t)$ of (32), (62) on the upper half-plane. The function U is in BV with respect to the space-time variables. Furthermore, for each fixed t, the function $U(\cdot, t)$ has bounded variation on $(-\infty, \infty)$ and

$$\mathrm{TV}_{(-\infty,\infty)}U(\cdot,t) \le c\mathrm{TV}_{(-\infty,\infty)}U_0(\cdot), \qquad 0 \le t < \infty. \tag{63}$$

The above solution has been constructed by three methods, namely: (a) The *random choice* method of Glimm, with essential extensions by Tai-Ping Liu. It uses solutions to the Riemann problem as building blocks for approximating the general solution. (b) The *front tracking* algorithm developed by the Italian School headed by Bressan. The strategy in this approach is to construct approximate solutions that contain only constant states and shocks, by replacing rarefaction waves with fans of inadmissible, albeit very weak, rarefaction shocks. Shock interactions are then handled by solving Riemann problems. (c) The *vanishing viscosity* approach of Bianchini and Bressan. In all three methods, the crucial step is to establish the bound (63) on the approximate solutions, for some uniform constant c.

For the above theorem to hold, the restriction that the total variation of the initial data must be small is essential. Indeed, there are cases of systems in which weak solutions under initial data with large total variation break down in finite time. A major challenge to the theory at the present time is to determine whether the systems of hyperbolic conservation laws arising in physics are free from such pathologies. An even greater challenge is the issue of existence of weak solutions in the large, for systems in more than one spatial dimension: This is currently *terra incognita*.

REFERENCES

Bressan, A., 2000. Hyperbolic Systems of Conservation Laws. The One-dimensional Cauchy Problem. Oxford University Press, Oxford.

Dafermos, C.M., 2016. Hyperbolic Conservation Laws in Continuum Physics, fourth ed. Springer, Heidelberg.

Holden, H., Risebro, N.H., 2015. Front Tracking for Hyperbolic Conservation Laws, second ed. Springer, New York.

Serre, D., 1999. Systems of Conservation Laws, vols. 1–2. Cambridge University Press, Cambridge.

Smoller, J.A., 1994. Shock Waves and Reaction-Diffusion Equations, second ed. Springer, New York.

Chapter 2

The Riemann Problem: Solvers and Numerical Fluxes

E.F. Toro
Laboratory of Applied Mathematics, DICAM, University of Trento, Trento, Italy

Chapter Outline

ABSTRACT

Though introduced by Bernhard Riemann more than 150 years ago (Riemann, 1860), the Riemann problem entered the field of modern computational science, with the pioneering work of Godunov (1959), almost a century later. Here, the Riemann problem is first defined and illustrative examples are given, before providing essential background on hyperbolic equations for later use. The exact solution of the Riemann problem for the compressible Euler equation, the canonical hyperbolic system, is then presented in some detail. This sets the bases for studying and critically assessing approximate solution methods, such as the Roe solver, HLL (Harten, Lax, van Leer), HLLC (Harten, Lax, van Leer, Contact) and an Osher–Solomon type solver. Uses of the Riemann

Handbook of Numerical Analysis, Vol. 17. http://dx.doi.org/10.1016/bs.hna.2016.09.015

problem solution are discussed, starting from its role in defining numerical fluxes for finite volume and discontinuous Galerkin finite element methods. Alternative approaches for defining numerical fluxes are also touched upon, including flux vector splitting, centred and multistage type fluxes, such as MUlti-STAge (MUSTA), Krylov type and PVM (polynomial viscosity matrix) schemes. Criteria to judge existing Riemann solvers and related concepts are discussed, along with relevant references for further study. Some possible generalisations of the classical Riemann problem are mentioned, notably multidimensional Riemann solvers and the high-order or generalised Riemann problem.

Keywords: Hyperbolic equations, Riemann problem, Exact solution, Approximate solvers, Complete Riemann solver, Numerical flux, Finite volume methods, DG methods

AMS Classification Codes: 65 (1940–now, Numerical analysis), 76 (1940–now, Fluid mechanics), 35 (1940–now, Partial differential equations)

1 PRELIMINARIES

Here we introduce some basic definitions, give elementary examples and present selected concepts on hyperbolic systems and their numerical approximation.

1.1 Definitions and Simple Examples

Consider the scalar Cauchy problem (or initial value problem, IVP)

$$\left.\begin{array}{lll} \text{PDE}: & \mathbf{L}[q] = s(q(x,t)), & -\infty < x < \infty,\ t > 0, \\ \text{IC}: & q(x,0) = h(x), & -\infty < x < \infty. \end{array}\right\} \tag{1}$$

Here $q(x, t)$ is the sought unknown function, x is distance, t is time, $\mathbf{L}$ is a differential operator associated with a partial differential equation (PDE), $s(q)$ is a prescribed function, called source term and $h(x)$ is the initial condition (IC), also a prescribed function of x, a profile.

Definition: The Riemann problem. The Riemann problem is the special Cauchy problem

$$\left.\begin{array}{ll} \text{PDE}: & \mathbf{L}[q] = 0, \quad -\infty < x < \infty,\ t > 0, \\ \text{IC}: & q(x,0) = h(x) = \begin{cases} q_L & \text{if } x < 0, \\ q_R & \text{if } x > 0, \end{cases} \end{array}\right\} \tag{2}$$

with q_L and q_R two prescribed constant states. The initial discontinuity may be placed at any point $x = x_0$, but one usually takes $x_0 = 0$. Problem (2) is also called the *classical Riemann problem* (CRP) (Riemann, 1860) in that the initial condition $h(x)$ is a piece-wise constant function and the PDE is homogeneous, that is, $s(q) = 0$. The CRP can be generalised in a number of ways, as seen later in this chapter.

Example 1 (The linear advection equation). The simplest example is furnished by the Riemann problem for the linear advection equation with constant characteristic speed λ, namely

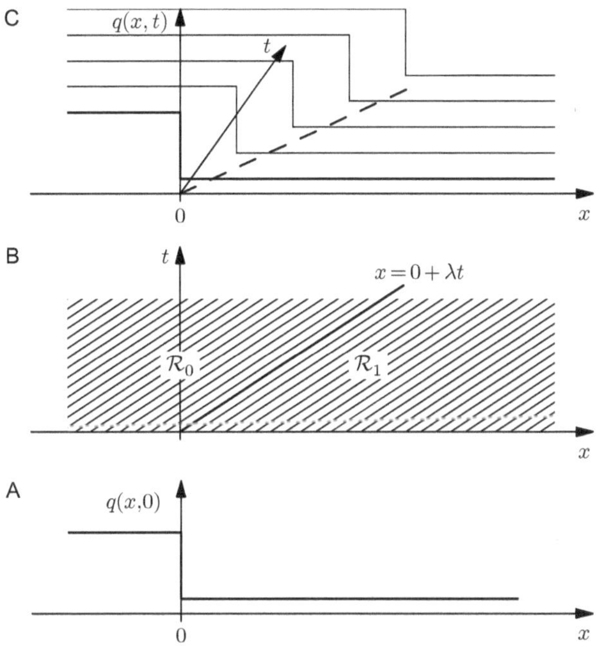

FIG. 1 The Riemann problem for the linear advection equation for $\lambda > 0$.

$$
\left.
\begin{array}{ll}
\text{PDE}: & \mathbf{L}[q] \equiv \partial_t q(x,t) + \lambda \partial_x q(x,t) = 0, \quad -\infty < x < \infty,\ t > 0, \\
\text{IC}: & q(x,0) = h(x) = \begin{cases} q_L & \text{if } x < 0, \\ q_R & \text{if } x > 0. \end{cases}
\end{array}
\right\} \tag{3}
$$

The exact solution for $\lambda > 0$, in terms of the characteristic line $x = \lambda t$ emanating from the origin, is

$$
q(x,t) = \begin{cases} q_L & \text{if } \dfrac{x}{t} < \lambda, \text{ Region} \mathcal{R}_0, \\ q_R & \text{if } \dfrac{x}{t} > \lambda, \text{ Region} \mathcal{R}_1. \end{cases} \tag{4}
$$

Fig. 1 depicts the solution (4). Frame (A) illustrates the piece-wise constant initial condition with discontinuity at $x = 0$. Frame (B) illustrates the complete solution in the x–t plane, with two constant regions, $\mathcal{R}_0$ to the left of the line $x/t = \lambda$ and $\mathcal{R}_1$ to the right $x/t = \lambda$. Frame (C) illustrates the solution $q(x, t)$ as a function of x and t. More generally, for the linear PDE in (3) the solution of the Cauchy problem with arbitrary initial condition $q(x, 0) = h(x)$ is

$$
q(x,t) = h(x - \lambda t), \tag{5}
$$

from which the special case (4) follows. Note that solution (5) is simply the initial profile $h(x)$ translated in space as time evolves.

Example 2 (The inviscid Burgers equation). We now give a nonlinear example, namely the Riemann problem for the inviscid Burgers equation

$$\left.\begin{array}{ll}\text{PDE}: & \mathbf{L}[q] \equiv \partial_t q(x,t) + \partial_x f(q(x,t)) = 0, \quad f(q) = \frac{1}{2}q^2, \quad -\infty < x < \infty,\ t > 0,\\ \text{IC}: & q(x,0) = h(x) = \begin{cases} q_L & \text{if} \quad x < 0, \\ q_R & \text{if} \quad x > 0. \end{cases}\end{array}\right\} \tag{6}$$

The exact solution consists of two cases: a shock (when $q_L > q_R$), or a rarefaction (when $q_L \le q_R$), namely

$$\left.\begin{array}{l} \left.\begin{array}{l} q(x,t) = \begin{cases} q_L & \text{if } \frac{x}{t} < s, \\ q_R & \text{if } \frac{x}{t} > s, \end{cases} \\ s = \frac{1}{2}(q_L + q_R), \end{array}\right\} \text{if } q_L > q_R \text{ (shock)}, \\ \left. q(x,t) = \begin{cases} q_L & \text{if } \frac{x}{t} \le q_L, \\ \frac{x}{t} & \text{if } q_L < \frac{x}{t} < q_R, \\ q_R & \text{if } \frac{x}{t} \ge q_R, \end{cases} \right\} \text{if } q_L \le q_R \text{ (rarefaction)}. \end{array}\right\} \tag{7}$$

Fig. 2A illustrates the case of a shock of speed $s = \frac{1}{2}(q_L + q_R)$, while Fig. 2B illustrates the case of a rarefaction. For the cases shown $q_L \ge 0$, $q_R \ge 0$ but other configurations are possible.

Example 3 (The diffusion, or heat, equation). Currently, the Riemann problem is associated almost exclusively with hyperbolic equations. However, it is also possible to pose the Riemann problem for other types of PDEs. The Riemann problem for the diffusion equation is

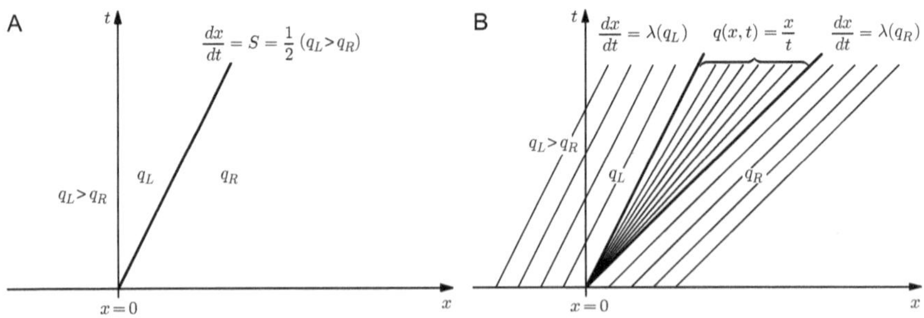

FIG. 2 Solution of the Riemann problem for the Burgers equation. Frame (A): shock wave if $q_L > q_R$. Frame (B): rarefaction wave if $q_L \le q_R$.

$$\left.\begin{array}{ll} \text{PDE}: & \mathbf{L}[q] \equiv \partial_t q(x,t) - \alpha \partial_x^{(2)} q(x,t) = 0, \quad -\infty < x < \infty,\ t > 0, \\ \text{IC}: & q(x,0) = \begin{cases} q_L & \text{if} \quad x < 0, \\ q_R & \text{if} \quad x > 0, \end{cases} \end{array}\right\} \tag{8}$$

where α is the diffusion coefficient, a constant here. The differential equation is parabolic and signal speeds are infinite, unlike the hyperbolic case in which speeds are always finite. The exact solution (Toro and Hidalgo, 2009) is

$$q(x,t) = \frac{1}{2}(q_L + q_R) + \frac{1}{2}(q_R - q_L) erf\left(\frac{x}{2\sqrt{\alpha t}}\right), \tag{9}$$

which is illustrated in Fig. 3. The left frame shows solution profiles at various times while the right frame shows the entire solution in the x–t plane.

1.2 Hyperbolic Systems and Finite Volume Methods

Consider first-order systems of m hyperbolic balance laws written as

$$\partial_t \mathbf{Q}(x,t) + \partial_x \mathbf{F}(\mathbf{Q}(x,t)) = \mathbf{S}(\mathbf{Q}(x,t)), \tag{10}$$

where $\mathbf{Q} = [q_1, q_2, \ldots, q_m]^T$ is the vector of conserved variables, $\mathbf{F}(\mathbf{Q}) = [f_1, f_2, \ldots, f_m]^T$ is the physical flux vector and $\mathbf{S}(\mathbf{Q}) = [s_1, s_2, \ldots, s_m]^T$ is the source term vector, containing no derivatives of $\mathbf{Q}$. The (homogeneous) case in which $\mathbf{S}(\mathbf{Q}) = \mathbf{0}$ defines a system of conservation laws. System (10) can also be written in quasi-linear form as

$$\partial_t \mathbf{Q} + \mathbf{A}(\mathbf{Q}) \partial_x \mathbf{Q} = \mathbf{S}(\mathbf{Q}), \quad \mathbf{A}(\mathbf{Q}) = \frac{\partial \mathbf{F}(\mathbf{Q})}{\partial \mathbf{Q}}, \tag{11}$$

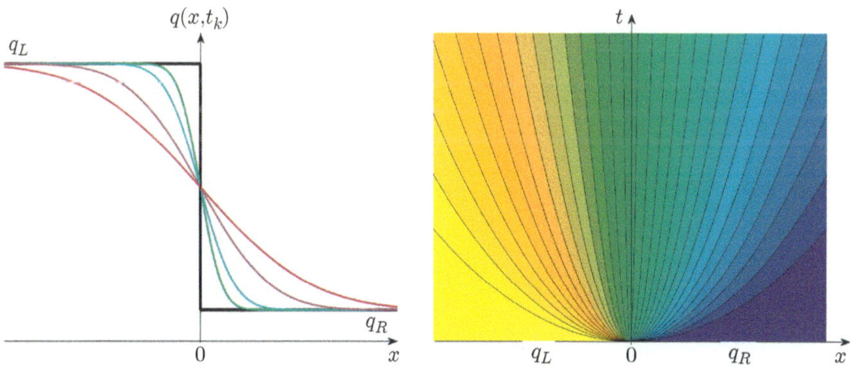

FIG. 3 Solution of the Riemann problem for the linear diffusion equation. *Left frame*: initial condition (*thick black line*) and solution profiles at various times t_k, with $t_0 = 0$ corresponding to the discontinuous initial condition. As time increases, the discontinuous profile diffuses. *Right frame* shows entire solution on x–t plane.

where $\mathbf{A}(\mathbf{Q}) = (a_{ij}) \equiv (\partial f_i / \partial q_j)$ is the Jacobian matrix. Elementary mathematical analysis of system (11) considers the *principal part* of the equations (the differential part) and requires the eigenstructure of the system, that is, eigenvalues and eigenvectors.

Definition: eigenvalues. The eigenvalues of system (11) are the eigenvalues of the Jacobian matrix $\mathbf{A}(\mathbf{Q})$, which in turn are the roots of the characteristic polynomial

$$P(\lambda) = Det(\mathbf{A} - \lambda \mathbf{I}) = 0. \tag{12}$$

Here $\mathbf{I}$ is the $m \times m$ identity matrix. We assume the eigenvalues $\lambda_i(\mathbf{Q})$ exist and are ordered as follows

$$\lambda_1(\mathbf{Q}) \leq \lambda_2(\mathbf{Q}) \leq \cdots \leq \lambda_m(\mathbf{Q}). \tag{13}$$

Definition: right and left eigenvectors. A right eigenvector $\mathbf{R}_i(\mathbf{Q})$ of (11) associated to $\lambda_i(\mathbf{Q})$ satisfies $\mathbf{A}\mathbf{R}_i = \lambda_i \mathbf{R}_i$, while a left eigenvector $\mathbf{L}_j(\mathbf{Q})$ of (11) associated to $\lambda_j(\mathbf{Q})$ satisfies $\mathbf{L}_j \mathbf{A} = \lambda_j \mathbf{L}_j$. We denote the two sets of eigenvectors as

$$\mathcal{R} = \{\mathbf{R}_1(\mathbf{Q}), \mathbf{R}_2(\mathbf{Q}), \ldots, \mathbf{R}_m(\mathbf{Q})\}, \qquad \mathcal{L} = \{\mathbf{L}_1(\mathbf{Q}), \mathbf{L}_2(\mathbf{Q}), \ldots, \mathbf{L}_m(\mathbf{Q})\}. \tag{14}$$

For linear systems with constant coefficients, $\mathbf{A}$ is constant and so are its eigenvalues and eigenvectors. For convenience we usually drop the argument of $\mathbf{A}$, λ_i, $\mathbf{R}_i$ and $\mathbf{L}_i$.

Definition: hyperbolicity. System (11) is hyperbolic if all eigenvalues (13) are real and the corresponding right (or left) eigenvectors in (14) form a complete set of linearly independent eigenvectors.

Definition: characteristic fields. For a hyperbolic system, the *characteristic speed* $\lambda_i(\mathbf{Q})$, or the pair $(\lambda_i, \mathbf{R}_i)$, defines a *characteristic field*, called the λ_i-field, or the $\mathbf{R}_i$-field or simply the i-field.

Characteristic fields are associated to wave families and can be classified. But before doing so we first recall the *gradient* of an eigenvalue $\lambda_i(\mathbf{Q})$

$$\nabla \lambda_i(\mathbf{Q}) = \left[\frac{\partial}{\partial q_1}\lambda_i, \frac{\partial}{\partial q_2}\lambda_i, \ldots, \frac{\partial}{\partial q_m}\lambda_i\right]. \tag{15}$$

Definition: linearly degenerate and genuinely nonlinear fields. A λ_i-characteristic field is said to be *linearly degenerate* if

$$\nabla \lambda_i(\mathbf{Q}) \cdot \mathbf{R}_i(\mathbf{Q}) = 0, \quad \forall \mathbf{Q} \in \mathcal{R}^m, \tag{16}$$

where $\mathcal{R}^m$ is the set of real-valued vectors of m components, called *phase space*. A λ_i-characteristic field is said to be *genuinely nonlinear* if

$$\nabla \lambda_i(\mathbf{Q}) \cdot \mathbf{R}_i(\mathbf{Q}) \neq 0, \quad \forall \mathbf{Q} \in \mathcal{R}^m. \tag{17}$$

Here we consider hyperbolic systems in which genuinely nonlinear characteristic fields are associated with either shock waves or rarefactions waves and

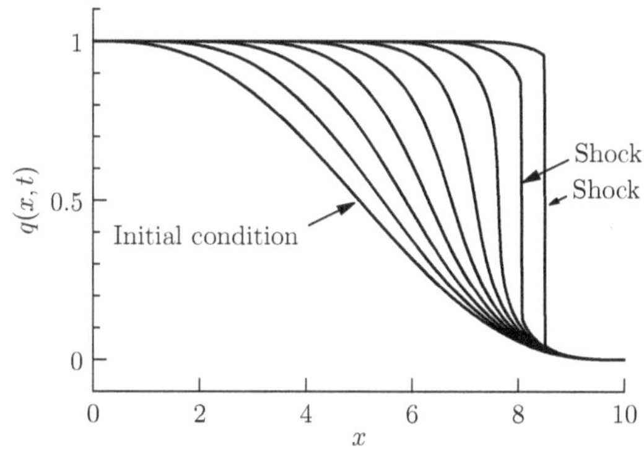

FIG. 4 Illustration of shock wave formation for the inviscid Burgers equation, starting from smooth initial condition at $t = 0$.

linearly degenerate fields are associated to contact discontinuities and shear waves. Note that for a linear hyperbolic system, the Jacobian matrix is constant and so are its eigenvalues and eigenvectors. Therefore, all characteristic fields in a linear system are linearly degenerate.

A distinguishing feature of hyperbolic equations is that they admit discontinuous solutions, even if the initial condition is smooth. Fig. 4 illustrates this situation via the inviscid Burgers equation. For this reason, hyperbolic equations are formulated in *integral form*

$$\left.\begin{aligned}\int_{x_L}^{x_R} \mathbf{Q}(x, t_2)\, dx = &\int_{x_L}^{x_R} \mathbf{Q}(x, t_1)\, dx + \int_{t_1}^{t_2} \mathbf{F}(\mathbf{Q}(x_L, t))\, dt - \int_{t_1}^{t_2} \mathbf{F}(\mathbf{Q}(x_R, t))\, dt \\ &+ \int_{t_1}^{t_2}\int_{x_L}^{x_R} \mathbf{S}(\mathbf{Q}(x, t))\, dxdt,\end{aligned}\right\} \tag{18}$$

so as to admit discontinuous (or weak, or generalised) solutions. Form (18) can be derived by performing exact space–time integration of (10) in the *control volume* $V = [x_L, x_R] \times [t_1, t_2]$.

Finite volume methods, Godunov-type methods in particular, depart from the integral form (18) by redefining the control volume V (called a finite volume) in a computational setting as $V = \left[x_{i-\frac{1}{2}}, x_{i+\frac{1}{2}}\right] \times [t^n, t^{n+1}]$. Then, introducing the *mesh size* $\Delta x = x_{i+\frac{1}{2}} - x_{i-\frac{1}{2}}$ and the *time step* $\Delta t = t^{n+1} - t^n$ we can write (18) as

$$\mathbf{Q}_i^{n+1} = \mathbf{Q}_i^n - \frac{\Delta t}{\Delta x}\left(\mathbf{F}_{i+\frac{1}{2}} - \mathbf{F}_{i-\frac{1}{2}}\right) + \Delta t \mathbf{S}_i, \tag{19}$$

where all vectorial quantities are integral averages, namely

$$\left.\begin{array}{l}\mathbf{Q}_i^n = \dfrac{1}{\Delta x}\displaystyle\int_{x_{i-\frac{1}{2}}}^{x_{i+\frac{1}{2}}} \mathbf{Q}(x, t^n)dx, \\ \mathbf{F}_{i+\frac{1}{2}} = \dfrac{1}{\Delta t}\displaystyle\int_{t^n}^{t^{n+1}} \mathbf{F}\left(\mathbf{Q}\left(x_{i+\frac{1}{2}}, t\right)\right)dt, \quad \mathbf{S}_i = \dfrac{1}{\Delta t \Delta x}\displaystyle\int_{t^n}^{t^{n+1}}\int_{x_{i-\frac{1}{2}}}^{x_{i+\frac{1}{2}}} \mathbf{S}(\mathbf{Q}(x, t))dxdt. \end{array}\right\} \tag{20}$$

Relation (19) is exact, even if it looks like a finite difference approximation. To construct a specific finite volume method of the form (19) one needs to specify suitable approximations to $\mathbf{F}_{i+\frac{1}{2}}$ (the numerical flux) and $\mathbf{S}_i$ (the numerical source) in integrals (20). The finite volume method based on (19) and (20) introduces a set of constant states $\left\{\mathbf{Q}_i^n\right\}_{i=1}^{i=M}$ (integral averages), where M is the total number of cells in the discretised domain. Consequently, the finite volume method introduces, in a very natural manner, a sequence of local Riemann problems. For example, at the interface $x = x_{i+\frac{1}{2}}$, the adjacent states $\mathbf{Q}_i^n$ and $\mathbf{Q}_{i+1}^n$, being integral averages, define two constant states that provide the initial conditions for the given equations, thus generating a local Riemann problem, the special Cauchy problem

$$\left.\begin{array}{ll}\text{PDEs}: & \partial_t \mathbf{Q}(x, t) + \partial_x \mathbf{F}(\mathbf{Q}(x, t)) = \mathbf{0}, \quad x \in \mathcal{R}^1, \quad t > 0, \\ \text{ICs}: & \mathbf{Q}(x, 0) = \begin{cases} \mathbf{Q}_L \equiv \mathbf{Q}_i^n & \text{if } x < x_0 = x_{i+\frac{1}{2}}, \\ \mathbf{Q}_R \equiv \mathbf{Q}_{i+1}^n & \text{if } x > x_0 = x_{i+\frac{1}{2}}. \end{cases} \end{array}\right\} \tag{21}$$

It is convenient to transform global coordinates to local coordinates so that t^n is the initial time and the interface $x_{i+\frac{1}{2}}$ is positioned at $x = 0$. *Godunov-type finite volume methods* for solving the initial-boundary value problem for hyperbolic balance laws (10) interpret the integral form of the conservation laws (19) in an approximate manner in which the numerical flux is

$$\mathbf{F}_{i+\frac{1}{2}} = \frac{1}{\Delta t}\int_0^{\Delta t} \mathbf{F}(\mathbf{Q}_{i+\frac{1}{2}}(0, t))dt = \mathbf{F}\left(\mathbf{Q}_{i+\frac{1}{2}}(0, 0_+)\right), \tag{22}$$

where $\mathbf{Q}_{i+\frac{1}{2}}(x, t)$ is the solution of the local Riemann problem (21) at the interface $x = 0$. The second equality follows from the fact that the similarity solution $\mathbf{Q}_{i+\frac{1}{2}}(x, t)$ is constant along the interface $x = 0$.

In the Godunov method, the solution of the local Riemann problem could be exact or approximate. Hence one speaks of exact Riemann solvers (instead of Riemann problem solvers) and approximate Riemann solvers. Since the Riemann solver is commonly used to construct the numerical flux, one often assigns the name *Riemann solver* to the *numerical flux*, or vice versa. Note that the Riemann problem (21) is unavoidable, it is a consequence of the finite volume integral formulation, independently of the particular approach for defining the numerical flux. Thus, there cannot be *Riemann problem free* finite volume methods. For background on finite volume methods see for example Godlewski and Raviart (1996), Toro (2009), LeVeque (2002) and Vázquez-Cendón (2015).

In the next section we study the exact solver for the Riemann problem for a typical hyperbolic system and derive the corresponding Godunov method.

2 EXACT SOLUTION OF THE RIEMANN PROBLEM FOR THE EULER EQUATIONS

The Euler equations for a compressible fluid are the most typical hyperbolic system in use, and although in computational practice one tends to employ an approximate solution to the Riemann problem to compute the numerical flux, knowing the exact solution is important for a number of very good reasons. First, it constitutes the reference solution that should be used to assess the performance of numerical methods, of any type. Then, detailed knowledge of the exact solver may inspire the derivation of approximate solvers or may provide sound criteria to evaluate the plausibility of proposed approximate solvers, or interpret the failure of these, or may inspire possible corrections to approximate solvers exhibiting obvious shortcomings. There are several exact Riemann solvers for the Euler equations in the literature. They differ amongst themselves mainly in the way the problem is formulated, the variables used and the actual nonlinear equation(s) to be solved. Here we follow the exact solver in Toro (1989a) (see also Colella and Glaz, 1985; Dutt, 1986; Godunov, 1959, 1976; Gottlieb and Groth, 1988; Pike, 1993; Quartapelle et al., 2003; Schleicher, 1993; Smoller, 1994; Toro, 1989a, 2009; van Leer, 1979). Ref. Menikoff and Plohr (1989) is highly recommended. For systems other than the Euler equations see Toro (1992), Schwendeman et al. (2006), Giacomazzo and Rezzolla (2006) and Antón et al. (2010). For historical information see Riemann (1860).

2.1 Equations and Structure of the Solution

The (homogeneous) Euler equations in one space dimension, in differential conservation form, read

$$\partial_t \mathbf{Q}(x,t) + \partial_x \mathbf{F}(\mathbf{Q}(x,t)) = \mathbf{0}. \tag{23}$$

$\mathbf{Q}(x,t) = [\rho, \rho u, E]^T$ is the vector of conserved variables and $\mathbf{F}(\mathbf{Q}) = [\rho u, \rho u^2 + p, u(E+p)]^T$ is the flux vector. Here, ρ is density, u is particle velocity, p is pressure and E is total energy given in terms of kinetic energy $\frac{1}{2}u^2$ and specific internal energy e as

$$E = \rho\left(\frac{1}{2}u^2 + e\right), \quad e(\rho, p) = \frac{p}{\rho(\gamma - 1)}. \tag{24}$$

The function $e = e(\rho, p)$ is called the (caloric) *equation of state*. Here we have taken the ideal gas case, where $1 < \gamma < 3$ is the ratio of specific heats, a constant. For air at moderate pressures and temperatures one uses $\gamma = 1.4$. The eigenvalues of the Euler equations are

$$\lambda_1(\mathbf{Q}) = u - a, \quad \lambda_2(\mathbf{Q}) = u, \quad \lambda_3(\mathbf{Q}) = u + a, \tag{25}$$

where a is the speed of sound, given as

$$a = \sqrt{\frac{\gamma p}{\rho}}. \tag{26}$$

The corresponding right eigenvectors are

$$\mathbf{R}_1(\mathbf{Q}) = \begin{bmatrix} 1 \\ u - a \\ H - ua \end{bmatrix}, \quad \mathbf{R}_2(\mathbf{Q}) = \begin{bmatrix} 1 \\ u \\ \frac{1}{2}u^2 \end{bmatrix}, \quad \mathbf{R}_3(\mathbf{Q}) = \begin{bmatrix} 1 \\ u + a \\ H + ua \end{bmatrix}, \tag{27}$$

where $H = (E + p)/\rho$ is the total specific enthalpy and scaling factors for the eigenvectors have been set to unity. The Riemann problem for the Euler equations (23) is the IVP

$$\left.\begin{array}{ll} \text{PDEs}: & \partial_t \mathbf{Q}(x,t) + \partial_x \mathbf{F}(\mathbf{Q}(x,t)) = \mathbf{0}, \quad x \in \mathcal{R}^1, \quad t > 0, \\ \text{ICs}: & \mathbf{Q}(x,0) = \begin{cases} \mathbf{Q}_L & \text{if} \quad x < 0, \\ \mathbf{Q}_R & \text{if} \quad x > 0. \end{cases} \end{array}\right\} \tag{28}$$

The solution of the Riemann problem (28) consists of three wave families corresponding to the λ_1, λ_2 and λ_3 characteristic fields and separate four constant regions in the half x–t plane. The resulting waves may be of three types: shocks (S), rarefactions (R) and contact discontinuities (C). A priori, it is not known what types of waves will emerge from the solution. Fig. 5 depicts the four possible wave patterns arising from the solution, namely: (A) rarefaction-contact-rarefaction; (B) rarefaction-contact-shock; (C) shock-contact-rarefaction and (D) shock-contact-shock. Fig. 6 depicts the generic structure of the solution in the x–t plane. The solution in the entire x–t half plane is characterised by four constant regions (wedges) $\mathcal{R}_0$, $\mathcal{R}_1$, $\mathcal{R}_2$ and $\mathcal{R}_3$ separated by three wave families. The λ_1 and λ_3 characteristic fields are genuinely nonlinear and are associated with either shocks or rarefactions; the intermediate field λ_2 is linearly degenerate and is associated with a contact discontinuity. Regions $\mathcal{R}_0$ (left data) and $\mathcal{R}_3$ (right data) are known a priori. $\mathcal{R}_1$ and $\mathcal{R}_2$ must be found, as well as the wave types associated to the non-linear fields (shocks or rarefactions).

Finding the solution of the Riemann problem means finding $\mathbf{Q}(x, t)$ for $x \in \mathcal{R}^1$ and $t > 0$. Here we follow the method proposed in Toro (1989b), see also chapter 4 of Toro (2009). The solution strategy proceeds in two steps. First, one computes the pressure and velocity in the *Star Region* depicted in Fig. 6, called the *Star Region*. Then, the wave types are determined and the solution in the full half plane is found.

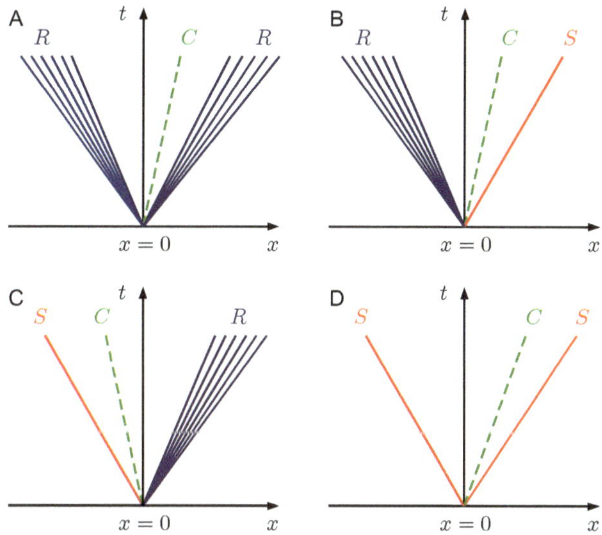

FIG. 5 Four possible wave patterns emerging from the solution of the Riemann problem for the ideal, compressible Euler equations: (A) rarefaction-contact-rarefaction; (B) rarefaction-contact-shock; (C) shock-contact-rarefaction and (D) shock-contact-shock.

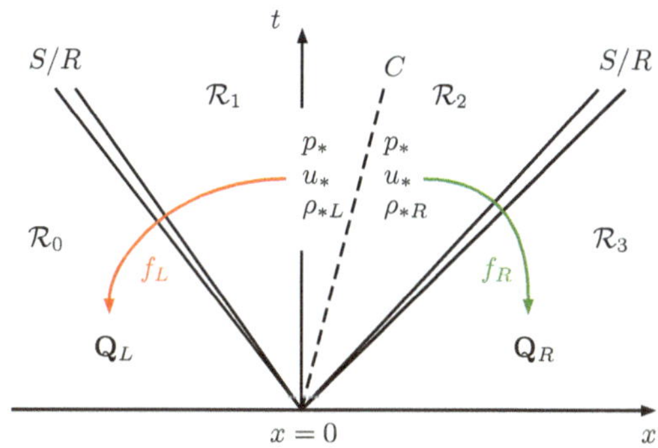

FIG. 6 General wave configuration for the solution of the Riemann problem for the Euler equations. The *Star Region* is connected to left and right data states via functions f_L and f_R, respectively. Pressure and velocity are constant across the contact discontinuity, while density changes discontinuously.

2.2 Pressure and Velocity in the Star Region

Here we establish equations for computing the pressure p_* and the particle velocity u_* in the *Star Region*.

Proposition: solution for p_* and u_*. *The solution for pressure* p_* *in the Riemann problem* (28) *for the ideal Euler equations* (23) *is given by the root of*

$$f(p, \mathbf{Q}_L, \mathbf{Q}_R) \equiv f_L(p, \mathbf{Q}_L) + f_R(p, \mathbf{Q}_R) + \Delta u = 0, \quad \Delta u \equiv u_R - u_L, \tag{29}$$

where

$$f_L(p, \mathbf{Q}_L) = \begin{cases} (p - p_L)\left[\dfrac{A_L}{p + B_L}\right]^{\frac{1}{2}} & \text{if } p > p_\mathrm{L} \text{ (shock)}, \\ \dfrac{2a_L}{(\gamma - 1)}\left[\left(\dfrac{p}{p_L}\right)^{\frac{\gamma-1}{2\gamma}} - 1\right] & \text{if } p \le p_\mathrm{L} \text{ (rarefaction)}, \end{cases} \tag{30}$$

$$f_R(p, \mathbf{Q}_R) = \begin{cases} (p - p_R)\left[\dfrac{A_R}{p + B_R}\right]^{\frac{1}{2}} & \text{if } p > p_\mathrm{R} \text{ (shock)}, \\ \dfrac{2a_R}{(\gamma - 1)}\left[\left(\dfrac{p}{p_R}\right)^{\frac{\gamma-1}{2\gamma}} - 1\right] & \text{if } p \le p_\mathrm{R} \text{ (rarefaction)}, \end{cases} \tag{31}$$

with

$$A_L = \frac{2}{(\gamma + 1)\rho_L}, \quad B_L = \frac{(\gamma - 1)}{(\gamma + 1)}p_L, \quad A_R = \frac{2}{(\gamma + 1)\rho_R}, \quad B_R = \frac{(\gamma - 1)}{(\gamma + 1)}p_R. \tag{32}$$

Once p_* *is determined from solving* (29), *the solution for the velocity* u_* *in the Star Region follows as*

$$u_* = \frac{1}{2}(u_L + u_R) + \frac{1}{2}[f_R(p_*, \mathbf{Q}_R) - f_L(p_*, \mathbf{Q}_L)]. \tag{33}$$

Proof. (omitted). See Toro (2009) for details. □

Fig. 6 shows that f_L governs relations *across* the left nonlinear wave and connects the unknown particle speed u_* to the known state $\mathbf{Q}_\mathrm{L}$ on the left side. Analogously, f_R governs relations *across* the right wave and connects u_* to $\mathbf{Q}_\mathrm{R}$. The form of f_L and f_R depends on whether the corresponding nonlinear wave is a shock or a rarefaction. For shocks one applies the Rankine–Hugoniot conditions and for rarefactions one applies generalised Riemann invariants. The functions f_L and f_R are called Lax curves, or simply wave curves. Analysis of the pressure function $f(p) = f_\mathrm{L} + f_\mathrm{R} + \Delta u$ in (29) reveals that, for physically admissible data, the solution p_* exists and is unique. It can be verified that $f(p)$ is monotone and concave down, by calculating the first and second derivatives of f_K (K=L, R) with respect to p. It can also be seen that $f'_\mathrm{K} \to 0$ as $p \to \infty$ and $f''_\mathrm{K} \to 0$ as $p \to \infty$. Fig. 7 depicts $f(p)$ for fixed initial data for density and pressure ρ_L, p_L and ρ_R, p_R and for three cases of velocity difference $\Delta u = u_\mathrm{R} - u_\mathrm{L}$ denoted as Δu_1, Δu_2 and Δu_3. By decreasing Δu from Δu_1 to Δu_3 $f(p)$ is shifted rightwards, giving rise to roots p_{*1}, p_{*2} and p_{*3} in increasing order. Δu, p_L and p_R are the most important parameters for $f(p)$. With reference to Fig. 7 we define

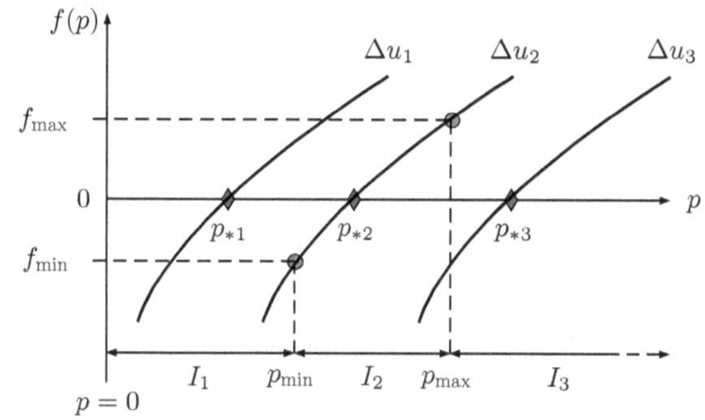

FIG. 7 Behaviour of the pressure function $f(p)$ in the solution of the Riemann problem for the Euler equations assuming $\Delta u_1 > \Delta u_2 > \Delta u_3$.

$$p_{\min} = \min(p_L, p_R), \quad p_{\max} = \max(p_L, p_R), \quad f_{\min} = f(p_{\min}), \quad f_{\max} = f(p_{\max}). \tag{34}$$

Recall that $f(p)$ is the pressure function denoted as $f(p, \mathbf{Q}_L, \mathbf{Q}_R)$ in (29). For given p_L, p_R it is the velocity difference Δu which determines the value of p_* and the specific wave pattern (one out of four) in Fig. 5, namely

$$\left.\begin{array}{lll} p_* \in I_1 = (0, p_{\min}) & \text{if } f_{\min} > 0 \text{ and } f_{\max} > 0 & \text{(RCR)}, \\ p_* \in I_2 = [p_{\min}, p_{\max}] & \text{if } f_{\min} \leq 0 \text{ and } f_{\max} \geq 0 & \text{(RCS or SCR)}, \\ p_* \in I_3 = (p_{\max}, \infty) & \text{if } f_{\min} < 0 \text{ and } f_{\max} < 0 & \text{(SCS)}. \end{array}\right\} \tag{35}$$

The wave pattern can be identified a priori without solving the Riemann problem, by simply noting the signs of $f_{\min}$ and $f_{\max}$. For nonvacuum initial data $\mathbf{Q}_L$, $\mathbf{Q}_R$ there exists a unique positive solution p_* for pressure, provided Δu satisfies the *pressure positivity condition*

$$(\Delta u)_{\text{crit}} \equiv \frac{2a_L}{\gamma - 1} + \frac{2a_R}{\gamma - 1} > u_R - u_L. \tag{36}$$

Otherwise, *vacuum* is generated by the nonlinear waves, usually very strong rarefactions. The structure of the solution in this case is different from that depicted in Fig. 6 and so is the method of solution; see Toro (2009). From the analysis of $f(p)$ it follows that the Newton–Raphson method is a suitable iterative method to solve (29) numerically to the desired accuracy. As a guess value p_0 to start the iteration we suggest to use the *Two-Rarefaction* approximation

$$p_{\text{TR}} = \left[\frac{a_L + a_R - \frac{1}{2}(\gamma - 1)(u_R - u_L)}{a_L / p_L^{\frac{\gamma - 1}{2\gamma}} + a_R / p_R^{\frac{\gamma - 1}{2\gamma}}} \right]^{\frac{2\gamma}{\gamma - 1}}. \tag{37}$$

This is obtained from the exact function (29) for pressure under the assumption that the two nonlinear waves are rarefaction waves. Care is required when resorting to simpler guess values, as the Newton–Raphson iteration may fail by predicting negative pressure iterates.

2.3 The Complete Solution and the 3D Case

Having found the pressure and velocity in the *Star Region*, the rest of the solution follows through a sampling procedure that involves the determination of the type of waves present and their speeds (Toro, 2009). Fig. 8 shows an example of the complete, exact solution of a Riemann problem for the Euler equations. This particular Riemann problem is a modification of the Sod shock tube problem (Sod, 1978), in which initial velocities are zero. Often in the literature, a Riemann problem is referred to as Sod's problem. This is incorrect. The modified Sod test problem here admits a transonic left rarefaction wave, which can be verified by observing that the rarefaction wave straddles the vertical line emerging from $x = 0.3$. This particular Riemann problem is suitable for assessing the entropy-satisfying properties of numerical methods. So far we

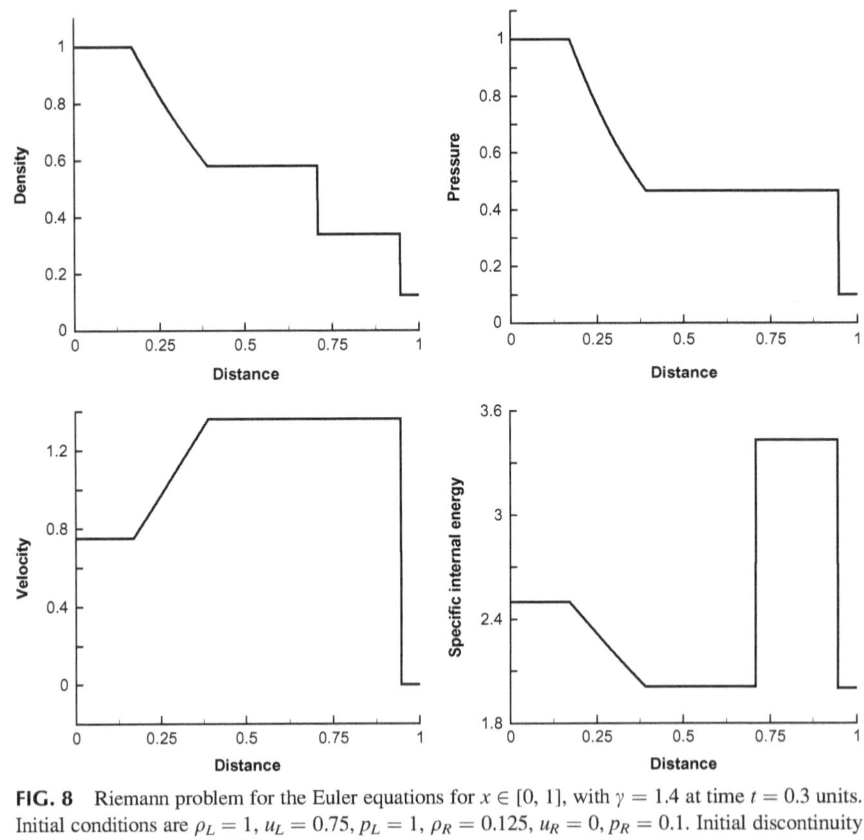

FIG. 8 Riemann problem for the Euler equations for $x \in [0, 1]$, with $\gamma = 1.4$ at time $t = 0.3$ units. Initial conditions are $\rho_L = 1$, $u_L = 0.75$, $p_L = 1$, $\rho_R = 0.125$, $u_R = 0$, $p_R = 0.1$. Initial discontinuity is positioned at $x_0 = 0.3$. The solution contains a (transonic or sonic) left rarefaction wave, a contact discontinuity and a right shock.

have dealt with the one-dimensional case. The three-dimensional Euler equations read

$$\partial_t \mathbf{Q} + \partial_x \mathbf{F}(\mathbf{Q}) + \partial_y \mathbf{G}(\mathbf{Q}) + \partial_z \mathbf{H}(\mathbf{Q}) = \mathbf{0}. \tag{38}$$

Here we consider the equations in the *normal* direction to some specified boundary in 3D space. It is sufficient to consider the *split* problem in the x-direction, for which $\mathbf{Q}$ and the flux $\mathbf{F}(\mathbf{Q})$ are

$$\left.\begin{aligned} \mathbf{Q} &= [\rho,\ \rho u,\ \rho v,\ \rho w,\ E], \\ \mathbf{F}(\mathbf{Q}) &= \left[\rho u,\ \rho u^2 + p,\ \rho uv,\ \rho uw,\ u(E+p)\right]. \end{aligned}\right\} \tag{39}$$

Eigenvalues and corresponding eigenvectors are

$$\lambda_1(\mathbf{Q}) = u - a, \quad \lambda_2(\mathbf{Q}) = \lambda_3(\mathbf{Q}) = \lambda_4(\mathbf{Q}) = u, \quad \lambda_5(\mathbf{Q}) = u + a. \tag{40}$$

The corresponding right eigenvectors are

$$\begin{gathered} \mathbf{R}_1(\mathbf{Q}) = \begin{bmatrix} 1 \\ u-a \\ v \\ w \\ H - ua \end{bmatrix}, \quad \mathbf{R}_2(\mathbf{Q}) = \begin{bmatrix} 1 \\ u \\ v \\ w \\ \frac{1}{2}u^2 \end{bmatrix}, \quad \mathbf{R}_3(\mathbf{Q}) = \begin{bmatrix} 0 \\ 0 \\ 1 \\ 0 \\ v \end{bmatrix}, \\ \mathbf{R}_4(\mathbf{Q}) = \begin{bmatrix} 0 \\ 0 \\ 0 \\ 1 \\ w \end{bmatrix}, \quad \mathbf{R}_5(\mathbf{Q}) = \begin{bmatrix} 1 \\ u+a \\ v \\ w \\ H + ua \end{bmatrix}. \end{gathered} \tag{41}$$

The exact solution of the Riemann problem for the split 3D case is identical to that for the 1D case, regarding ρ, u and p. The solution for the tangential velocity components v and w is

$$v(x,t) = \begin{cases} v_L & \text{if } x/t < u_*, \\ v_R & \text{if } x/t > u_*, \end{cases} \quad w(x,t) = \begin{cases} w_L & \text{if } x/t < u_*, \\ w_R & \text{if } x/t > u_*. \end{cases} \tag{42}$$

That is, $v(x, t)$ defines a shear wave in the y-direction and $w(x, t)$ defines a shear wave in the z-direction, with discontinuous jumps in the data. The three intermediate characteristic fields define three discontinuous waves, an entropy wave across which ρ jumps discontinuously and two shear waves. Numerical resolution of these intermediate discontinuous waves is challenging and depends crucially on whether the solver for the Riemann problem acknowledges the presence of these waves or not.

2.4 Uses of the Exact Solution of the Riemann Problem

An early use of the exact solution of the Riemann problem was, and still is, in shock tube physics, an active discipline for more than a century and relevant to aerospace sciences. The shock tube problem is a Riemann problem with zero initial velocities. This is also the case in hydraulic engineering, where the dam-break problem is the analogue of the shock tube problem. Starting with the work of Godunov (1959), the Riemann problem solution was employed in the construction of numerical methods and in theoretical studies of hyperbolic equations (Glimm, 1965).

Today, a primary use of the exact solution of the Riemann problem is in the assessment of the performance of numerical methods for hyperbolic equations, even if the numerical methods being tested do not explicitly use the Riemann problem, or a Riemann solver. Only then developers can demonstrate that numerical approximations are correct, wave speeds are correct, wave jumps are correct and that discontinuities are entropy satisfying. Once an exact Riemann solver is available, the user may design special Riemann problems to assess potential shortcomings of numerical methods. In Toro (2009) readers can find a carefully selected suite of test problems, that is Riemann problems, and from *eleuteriotoro.com* readers may download a library that includes an exact Riemann solver.

The exact Riemann solver can also be used locally in numerical methods of the Godunov type (Godunov, 1959) and for solving the generalised Riemann problem (GRP) to very high accuracy (Toro and Titarev, 2002). Also, this solver can be used in Glimm's method (Chorin, 1976; Colella, 1982), Front Tracking Methods (Dafermos, 2010; Glimm et al., 1998a,b; Holden and Risebro, 2015; Risebro and Tveito, 1991) and the Shock Fitting Method (Salas, 2009). The Riemann problem solution (exact or approximate) can also be used in smooth particle hydrodynamics (SPH) methods (Ben Moussa, 2001; Ben Moussa and Vila, 2000; Ferrari et al., 2000; Vila, 1999), in discontinuous Galerkin finite element methods (Cockburn and Shu, 1989; Cockburn et al.,1989, 2000; Dumbser et al., 2008a) and in some continuous finite element methods (Guermond and Popov, 2015, 2016). For many applications, including ideal gas dynamics (Toro, 2009), shallow water flows (Toro, 2001) and haemodynamics (Toro, 2016), the exact Riemann solver is not as expensive as is sometimes stated. This becomes particularly evident when going to high order, multiple space dimensions, including source terms and performing grid generation tasks. Then the (relative) cost difference between an exact solver and some other numerical flux diminishes tremendously. However, there are problems in which a Godunov method with the exact solver becomes prohibitively expensive, notably gas dynamics with general equations of state (Toro et al., 2015). In such cases one must resort to approximate Riemann solvers or centred methods that do not explicitly solve the Riemann problem.

Boundary conditions naturally involve the Riemann problem, whether we are aware of it or not. For example, imagine shallow water flow in a 1D horizontal channel with high, impermeable vertical walls at the ends. A reflective boundary condition is obvious, with zero normal velocity through the boundary. This condition can be imposed immediately on the flux. The problem that remains to be determined is the pressure term. Posing a suitable local Riemann problem gives the correct zero velocity and, as a bonus, also the pressure term.

2.5 Approximate Riemann Solvers: Beware

For numerical purposes it is attractive, and often mandatory, to use approximate solutions to the local Riemann problem. The exact Riemann solver, even if not necessarily too expensive in comparison, is complex and rather tedious to construct and implement. However, the experience accumulated in the last 3 decades has revealed that great care is required when adopting or proposing an approximate Riemann solver for practical use to perform serious computations. Below I give a, nonexhaustive, list of points to be considered.

- *Robustness.* A robust solver runs under all, or almost all, circumstances, even at the expense of accuracy or solution correctness. Usually, linearised Riemann solvers lack robustness.
- *Entropy-satisfying shocks.* In addition to theoretical considerations, in a practical setting, transonic rarefactions will reveal whether the scheme is entropy satisfying or not. Unphysical rarefaction shocks may be computed. Linearised Riemann solvers again will be entropy violating, unless appropriate entropy fixes are put in place.
- *Early vacuum.* The Euler equations admit vacuum (Toro, 2009). However, some Riemann solvers, notably linearised solvers, produce spurious vacuum, that is the pressure attains zero or negative values even when the exact solver does not contain vacuum. The pressure positivity condition (37) tells us exactly when vacuum is to be produced. In fact that condition could be implemented in codes to detect admissible vacuum conditions and take appropriate action.
- *Wave model: complete/incomplete Riemann solver.* Some Riemann solvers approximate the structure of the solution by neglecting intermediate characteristic fields and adopting a reduced *wave model.* A typical example is the HLL (Harten, Lax, van Leer) Riemann solver to be described later, which for the 1D Euler equations acknowledges the fastest nonlinear fields and neglects the contact discontinuity. The HLL wave model has 2 waves, instead of 3, while the wave model for the Rusanov solver has 1 wave and nonupwind fluxes, e.g. the Lax–Friedrichs flux, have none. We call a Riemann solver *complete* if its wave model contains all characteristic fields present in the exact Riemann solver. Otherwise we call the solver *incomplete.*
- *The ideal Riemann solver.* A desirable Riemann solver for practical use must be nonlinear and complete. Linearised solvers need entropy fixes, not always obvious, lack robustness for strong waves and fail for near vacuum conditions. Incomplete Riemann solvers add excessive numerical dissipation to neglected characteristic fields, usually intermediate waves, such as contact discontinuities and shear waves. As a matter of fact an incomplete Riemann solver may have no obvious advantage over a nonupwind scheme. The difference will not be obvious for nonlinear waves such as shock waves and even for intermediate fields, a good centred scheme might be comparable to an incomplete Riemann solver. Despite intensive

research in the last two to three decades on this subject, I feel that we have not yet found the *ideal Riemann solver*. The search continues.

- *The wider hyperbolic world.* Rapid advances in mathematical modelling and computational science have brought in a plethora of new, challenging hyperbolic problems. A distinctive area concerns the formulation of parabolic problems as hyperbolic problems with stiff source terms (see e.g. Montecinos and Toro, 2014; Montecinos et al., 2014; Toro and Montecinos, 2014). For hyperbolic formulations of dispersive equations, the reader is referred to Mazaheri et al. (2016). These approaches produce enlarged hyperbolic systems for which even the eigenstructure analysis is challenging, let alone devising complete nonlinear Riemann solvers. A further step in this direction are new mathematical models emerging from the school of Godunov and collaborators (Dumbser et al., 2016; Peshkov and Romenski, 2008).

I conclude this section by acknowledging the rather practical approach taken here. I have avoided challenging difficulties posed by some special hyperbolic systems, such as resonant systems, for example. The Riemann problem for these systems admit multiple solutions; see for example Toro and Siviglia (2013), Han et al. (2015a) and Han et al. (2015b). For theoretical aspects of the Riemann problem the reader is referred to the excellent books by Smoller (1994), Dafermos (2010), Bressan (2000) and the many references therein. The next few sections are devoted to approximate Riemann solvers.

3 THE ROE APPROXIMATE RIEMANN SOLVER

The original Roe approximate Riemann solver was first communicated in Roe (1981). There are by now essentially two approaches to derive the solver, namely the original one (Roe, 1981) and that of Roe and Pike (1984). The latter tends to be preferred and is the one presented here, very succinctly. Detailed derivations for both approaches are found in chapter 11 of Toro (2009). The Roe solver is *linearised*, for which it requires an entropy fix, but *complete*, which means that it represents all characteristic fields present in the equations, at least for the Euler equations. Below I sketch the steps of the Roe–Pike approach in algorithmic fashion.

- *Analytical eigenstructure.* It is assumed that the system of interest is hyperbolic and that analytical expressions for the full eigenstructure are available, namely eigenvalues $\lambda_i(\mathbf{Q})$ and right eigenvectors $\mathbf{R}_i(\mathbf{Q})$.
- *Analytical wave strengths.* Analytical expressions for wave strengths $\alpha_i(\mathbf{Q})$ are found assuming a linearisation of the governing equations based on the assumption that the data states $\mathbf{Q}_L$ and $\mathbf{Q}_R$ in the Riemann problem are close to a reference state $\hat{\mathbf{Q}}$. Note that this is not the Roe linearisation and $\hat{\mathbf{Q}}$ is not the Roe averaged vector.
- *Approximate eigenstructure and wave strengths.* The analytical expressions for $\lambda_i(\mathbf{Q})$, $\mathbf{R}_i(\mathbf{Q})$ and $\alpha_i(\mathbf{Q})$ are evaluated at an unknown averaged state $\tilde{\mathbf{W}}$, the Roe averages, namely

$$\tilde{\lambda}_i = \lambda_i(\tilde{\mathbf{W}}), \quad \tilde{\mathbf{R}}_i = \mathbf{R}_i(\tilde{\mathbf{W}}), \quad \tilde{\alpha}_i = \alpha_i(\tilde{\mathbf{W}}). \tag{43}$$

- *Equations for the sought Roe averages.* To find $\tilde{\mathbf{W}}$ it is assumed

$$\Delta\mathbf{Q} = \mathbf{Q}_R - \mathbf{Q}_L = \sum_{k=1}^{m} \tilde{\alpha}_k \tilde{\mathbf{R}}_k, \quad \Delta\mathbf{F} = \mathbf{F}_R - \mathbf{F}_L = \sum_{k=1}^{m} \tilde{\alpha}_k \tilde{\lambda}_k \tilde{\mathbf{R}}_k. \tag{44}$$

- *Algebraic problem.* Finally, solve the nonlinear algebraic system (44) to find the Roe averages $\tilde{\mathbf{W}}$. Here solutions may be nonunique and care is needed to find the physically meaningful values.
- *The Roe numerical flux.* Once the Roe averages $\tilde{\mathbf{W}}$ are available, the Roe numerical flux follows

$$\mathbf{F}_{i+\frac{1}{2}} = \frac{1}{2}(\mathbf{F}_L + \mathbf{F}_R) - \frac{1}{2}\sum_{k=1}^{m} \tilde{\alpha}_k \, |\tilde{\lambda}_k| \tilde{\mathbf{R}}_k. \tag{45}$$

For the 3D x-split ideal Euler equations the Roe averages are given as

$$\left.\begin{array}{l} \tilde{\rho} = \sqrt{\rho_L \rho_R}, \ \tilde{u} = \dfrac{\sqrt{\rho_L}u_L + \sqrt{\rho_R}u_R}{\sqrt{\rho_L} + \sqrt{\rho_R}}, \ \tilde{v} = \dfrac{\sqrt{\rho_L}v_L + \sqrt{\rho_R}v_R}{\sqrt{\rho_L} + \sqrt{\rho_R}}, \\ \tilde{w} = \dfrac{\sqrt{\rho_L}w_L + \sqrt{\rho_R}w_R}{\sqrt{\rho_L} + \sqrt{\rho_R}}, \ \tilde{H} = \dfrac{\sqrt{\rho_L}H_L + \sqrt{\rho_R}H_R}{\sqrt{\rho_L} + \sqrt{\rho_R}}, \ \tilde{a} = \left((\gamma - 1)\left(\tilde{H} - \dfrac{1}{2}\tilde{\mathbf{V}}^2\right)\right)^{\frac{1}{2}}, \end{array}\right\} \tag{46}$$

where $\tilde{\mathbf{V}}^2 = \tilde{u}^2 + \tilde{v}^2 + \tilde{w}^2$. Now $\tilde{\alpha}_i, \tilde{\lambda}_i$ and $\tilde{\mathbf{R}}_i$ are computed according to (43) and then the Roe intercell flux $\mathbf{F}_{i+\frac{1}{2}}$ follows from formulae (45).

As pointed out earlier, being a linearised solver, the Roe flux needs an *entropy fix*. Otherwise, unphysical, entropy violating shocks are computed. This is most obvious in the presence of transonic rarefaction waves. Fig. 9 shows results for the Roe method for the transonic rarefaction test problem of Fig. 8, for the scheme without and with entropy fix. The result on the left is a convincing argument to implement an entropy fix in the Roe solver. For entropy fixes see chapter 11 of Toro (2009).

There are numerical versions of the Roe scheme, where the derivation of all algebraic expressions is avoided (Castro and Toro, 2014; Dumbser and Toro, 2011a,b). This gives the approach a degree of flexibility than can be exploited in complex applications as well when solving nonconservative hyperbolic systems.

4 THE HLL APPROXIMATE RIEMANN SOLVER

Here we study the approximate Riemann solver first proposed in Harten et al. (1983). An approximate solution of the Riemann problem (28) is sought, with the aim of finding *directly* a numerical flux

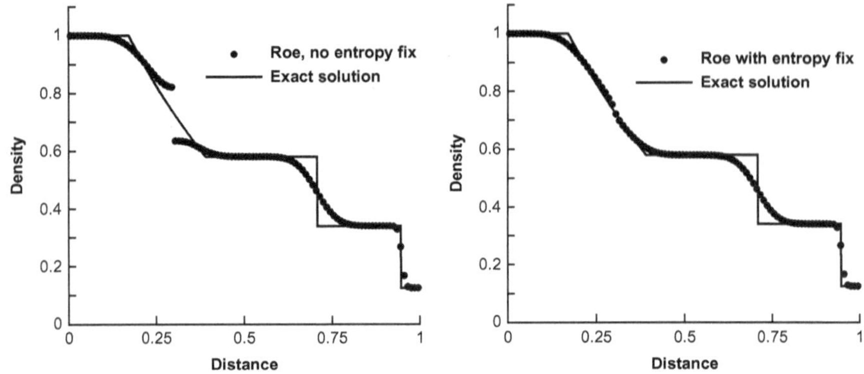

FIG. 9 Numerical (*symbol*) and exact (*line*) solutions to the modified Sod test problem containing a left transonic rarefaction, at time $t = 0.3$, $M = 100$ cells and Courant number $C_{cfl} = 0.9$. *Left frame* shows the Roe method without entropy fix and the *right frame* shows the Roe method with entropy fix.

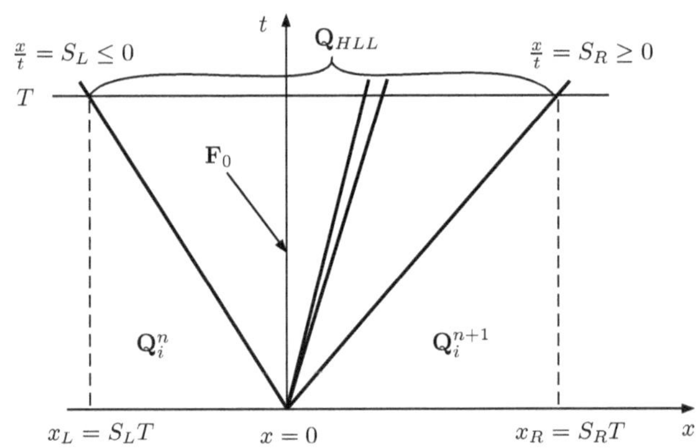

FIG. 10 Wave configuration for the derivation of the HLL flux for a subcritical, or subsonic, wave pattern, $S_L \leq 0$ and $S_R \geq 0$. Intermediate states are averaged into a single state.

$$\mathbf{F}_0 = \frac{1}{T}\int_0^T \mathbf{F}(\mathbf{Q}(0, t))dt \tag{47}$$

for an arbitrary time $T > 0$, where $\mathbf{Q}(0, t)$ is an approximate solution of the Riemann problem along the t-axis. We first establish some useful integral relations in appropriately chosen control volumes. Given initial conditions $\mathbf{Q}_L$ and $\mathbf{Q}_R$, assume S_L and S_R to be the fastest left and right signals emerging from the solution of the Riemann problem, see Fig. 10. For $T > 0$ we define the distances $x_L = TS_L$, $x_R = TS_R$, consider the control volume $[x_L, 0] \times [0, T]$ in x-t space and apply the integral form (18) of the differential conservation laws

$$\partial_t \mathbf{Q} + \partial_x \mathbf{F}(\mathbf{Q}) = \mathbf{0} \tag{48}$$

to obtain

$$\int_{x_L}^{0} \mathbf{Q}(x, T)dx = \int_{x_L}^{0} \mathbf{Q}(x, 0)dx + \int_{0}^{T} \mathbf{F}(\mathbf{Q}(x_L, t))dt - \int_{0}^{T} \mathbf{F}(\mathbf{Q}(0, t))dt. \tag{49}$$

Evaluation of the first and second terms on the right-hand side gives

$$\int_{x_L}^{0} \mathbf{Q}(x, 0)dx = -S_L T \mathbf{Q}_L, \quad \int_{0}^{T} \mathbf{F}(\mathbf{Q}(x_L, t))dt = T\mathbf{F}(\mathbf{Q}_L), \tag{50}$$

which substituted into (49) followed by division through by T yields

$$\mathbf{F}_0 = \frac{1}{T}\int_{0}^{T} \mathbf{F}(\mathbf{Q}(0, t))dt = -S_L \mathbf{Q}_L + \mathbf{F}(\mathbf{Q}_L) - \frac{1}{T}\int_{x_L}^{0} \mathbf{Q}(x, T))dx. \tag{51}$$

To define $\mathbf{F}_0$ approximately it is sufficient to find an approximation to the integral on the right-hand side of (51). This is accomplished by finding an approximate state $\mathbf{Q}(x, T)$. Applying the integral form (18) of the conservation laws (48) in the control volume $[x_L, x_R] \times [0, T]$, see Fig. 10, we obtain

$$\int_{x_L}^{x_R} \mathbf{Q}(x, T)dx = \int_{x_L}^{x_R} \mathbf{Q}(x, 0)dx + \int_{0}^{T} \mathbf{F}(\mathbf{Q}(x_L, t))dt - \int_{0}^{T} \mathbf{F}(\mathbf{Q}(x_R, t))dt. \tag{52}$$

Evaluation of the first term on the right-hand side gives

$$\int_{x_L}^{x_R} \mathbf{Q}(x, 0)dx = -S_L T \mathbf{Q}_L + S_R T \mathbf{Q}_R. \tag{53}$$

Substitution of this into (52) and evaluation of the remaining terms give

$$\int_{x_L}^{x_R} \mathbf{Q}(x, T)dx = T[S_R \mathbf{Q}_R - S_L \mathbf{Q}_L + \mathbf{F}(\mathbf{Q}_L) - \mathbf{F}(\mathbf{Q}_R)], \tag{54}$$

which on division through by $x_R - x_L = T(S_R - S_L)$ produces the sought averaged state

$$\mathbf{Q}_{HLL} = \frac{1}{(x_R - x_L)}\int_{x_L}^{x_R} \mathbf{Q}(x, T)dx = \frac{S_R \mathbf{Q}_R - S_L \mathbf{Q}_L + \mathbf{F}(\mathbf{Q}_L) - \mathbf{F}(\mathbf{Q}_R)}{S_R - S_L}. \tag{55}$$

We now use the state $\mathbf{Q}_{HLL}$ to evaluate the integral on the right-hand side of (51), giving

$$\mathbf{F}_0 = \frac{S_R \mathbf{F}(\mathbf{Q}_L) - S_L \mathbf{F}(\mathbf{Q}_R) + S_L S_R (\mathbf{Q}_R - \mathbf{Q}_L)}{S_R - S_L}. \tag{56}$$

Then the HLL intercell flux for the approximate Godunov method is found by sampling the solution, namely

$$\mathbf{F}_{i+\frac{1}{2}}^{HLL} = \begin{cases} \mathbf{F}_L & \text{if} \quad 0 \le S_L, \\ \dfrac{S_R \mathbf{F}(\mathbf{Q}_L) - S_L \mathbf{F}(\mathbf{Q}_R) + S_L S_R (\mathbf{Q}_R - \mathbf{Q}_L)}{S_R - S_L}, & \text{if} \quad S_L \le 0 \le S_R, \\ \mathbf{F}_R & \text{if} \quad 0 \ge S_R. \end{cases} \tag{57}$$

To complete the HLL scheme it is necessary to provide estimates for S_L and S_R, see next section. Note that the HLL approach takes an integral average of all the states lying between the fastest waves in the system. If this contains just two waves, then the assumed *two-wave model* of HLL is correct. However, if the system has more than two equations, then all intermediate waves get averaged. This results in numerical diffusion for intermediate waves. This issue is addressed in a recent paper (Dumbser and Balsara, 2016) by proposing a revised HLLEM Riemann solver (see also the next section). But before that we consider simpler methods that can be derived from the HLL framework by appropriate choices for the wave speeds. A prominent example is the *the Rusanov flux* (Rusanov, 1961), sometimes called the *Local Lax–Friedrichs* flux, or LLF flux; this can be obtained by choosing

$$S^+ = max\{|S_L|, |S_R|\}, \quad S_R = S^+, \quad S_L = -S^+. \tag{58}$$

Substitution of these speeds into (56) gives the Rusanov flux

$$\mathbf{F}_{i+\frac{1}{2}} = \frac{1}{2}[\mathbf{F}(\mathbf{Q}_L) + \mathbf{F}(\mathbf{Q}_R)] - \frac{1}{2}S^+(\mathbf{Q}_R - \mathbf{Q}_L). \tag{59}$$

The Rusanov scheme adopts a *one-wave model.* It is the simplest upwind, nonlinear method, but incomplete for any system. The (true) *Lax–Friedrichs Flux* can be obtained from Rusanov by choosing $S^+ = \frac{\Delta x}{\Delta t}$, namely

$$\mathbf{F}_{i+\frac{1}{2}} = \frac{1}{2}[\mathbf{F}(\mathbf{Q}_L) + \mathbf{F}(\mathbf{Q}_R)] - \frac{1}{2}\frac{\Delta x}{\Delta t}(\mathbf{Q}_R - \mathbf{Q}_L). \tag{60}$$

This simple numerical flux gives the most diffusive of all stable schemes.

5 THE HLLC APPROXIMATE RIEMANN SOLVER

The HLLC (Harten, Lax, van Leer, Contact) Riemann solver (Toro et al., 1994) is a modified version of the HLL solver to account for the presence of intermediate waves, such as contact discontinuities and shear waves (see also Toro et al. (1992) and Toro and Chakraborty (1994)). For subsequent developments see, for example, Batten et al. (1997a) and Batten et al. (1997b).

5.1 Derivation of the HLLC Flux

Here we review the current version of HLLC as applied to the three-dimensional ideal Euler equations (38), extended to include N species equations for N

concentrations ψ_i. Now the vector $\mathbf{Q}$ of conserved variables and the flux in the x-direction are

$$\left.\begin{aligned}\mathbf{Q} &= [\rho,\, \rho u,\, \rho v,\, \rho w,\, E,\, \rho\psi_1,\ldots,\rho\psi_N],\\ \mathbf{F}(\mathbf{Q}) &= \left[\rho u,\, \rho u^2+p,\, \rho uv,\, \rho uw,\, u(E+p),\, \rho u\psi_1,\ldots,\rho u\psi_N\right].\end{aligned}\right\} \tag{61}$$

For the purpose of determining a numerical flux for the three-dimensional Euler equations that is normal to a cell interface, by virtue of the rotational invariance, it is sufficient to consider the augmented one-dimensional Euler equations aligned in that normal direction. Here, without loss of generality, we assume the normal direction to be the x-direction. First we assume that wave speed estimates S_L, S_* and S_R for the three wave families depicted in Fig. 11 are available. Note, however, that for the x-split 3D multi-component Euler equations the middle eigenvalue has multiplicity $N + 3$, where N is the total number of species equations. Therefore by restoring the contact discontinuity associated to the entropy wave we also restore the shear waves and the N contact discontinuities associated to the M concentrations ψ_k.

Then, by integrating in appropriate control volumes around the waves of speeds S_L and S_R we obtain averaged Rankine–Hugoniot conditions

$$\mathbf{F}_{*L} = \mathbf{F}_L + S_L(\mathbf{Q}_{*L} - \mathbf{Q}_L), \quad \mathbf{F}_{*R} = \mathbf{F}_R + S_R(\mathbf{Q}_{*R} - \mathbf{Q}_R), \tag{62}$$

with

$$\mathbf{Q}_{*K} = [\rho_{*K}, \rho_{*K}u_*, \rho_{*K}v_K, \rho_{*K}w_K, E_{*K}, \rho_{*K}\psi_1,\ldots,\rho_{*K}\psi_N], \text{ for } K=L \text{ or } K=R. \tag{63}$$

This gives rise to a large algebraic system with more unknowns than equations. One way to resolve this difficulty is to introduce a number of assumptions, all consistent with the exact solution, leading to the following expressions for the intermediate state vectors

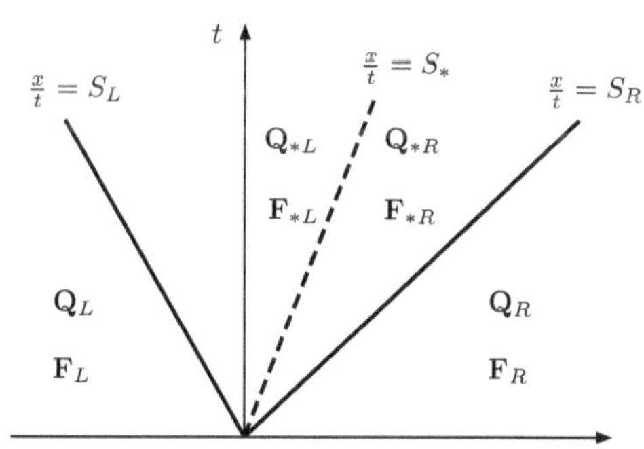

FIG. 11 Wave model for the HLLC approximate Riemann solver for the x-split 3D Euler equations with N species. The middle eigenvalue has multiplicity $N + 3$ defining one entropy wave, two shear waves and N contact discontinuities, all of them recognised by the HLLC solver.

$$\mathbf{Q}_{*K} = \rho_K \left(\frac{S_K - u_K}{S_K - S_*}\right) \begin{bmatrix} 1 \\ S_* \\ v_K \\ w_K \\ \dfrac{E_K}{\rho_K} + (S_* - u_K)\left[S_* + \dfrac{p_K}{\rho_K(S_K - u_K)}\right] \\ (q_1)_K \\ \cdots \\ (q_l)_K \\ \cdots \\ (q_N)_K \end{bmatrix}, \tag{64}$$

for $K = L$ and $K = R$. Then the intermediate fluxes $\mathbf{F}_{*L}$ and $\mathbf{F}_{*R}$ are completely determined and the numerical flux is given as

$$\mathbf{F}_{i+\frac{1}{2}}^{HLLC} = \begin{cases} \mathbf{F}_L & \text{if } 0 \le S_L, \\ \mathbf{F}_{*L} & \text{if } S_L \le 0 \le S_*, \\ \mathbf{F}_{*R} & \text{if } S_* \le 0 \le S_R, \\ \mathbf{F}_R & \text{if } 0 \ge S_R. \end{cases} \tag{65}$$

The missing items for implementing both HLL and HLLC are the wave speed estimates.

5.2 Wave Speed Estimates for HLL and HLLC

For HLL, having a two-wave model, we only need estimates for S_L and S_R. The following choice (Toro, 2009) has proved useful

$$S_L = u_L - a_L q_L, \quad S_R = u_R + a_R q_R, \tag{66}$$

with

$$q_K = \begin{cases} 1 & \text{if } p_* \le p_K \\ \left[1 + \frac{\gamma+1}{2\gamma}(p_*/p_K - 1)\right]^{1/2} & \text{if } p_* > p_K. \end{cases} \tag{67}$$

Here one discriminates between rarefactions and shocks, but an estimate for the pressure p_* is needed. A very reliable choice is the two-rarefaction approximation (37). For the HLLC solver we need additionally the intermediate speed S_*. This can be found in terms of S_L and S_R as follows

$$S_* = \frac{p_R - p_L + \rho_L u_L(S_L - u_L) - \rho_R u_R(S_R - u_R)}{\rho_L(S_L - u_L) - \rho_R(S_R - u_R)}. \tag{68}$$

A crucial difference between HLL and HLLC is seen in the resolution of contact discontinuities and shear waves, with larger differences occurring for slowly moving waves. The extreme case is the stationary wave. To illustrate

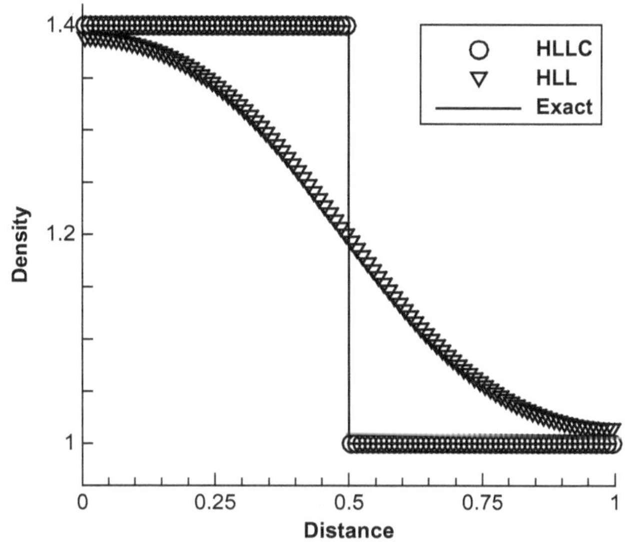

FIG. 12 Computations for an isolated stationary contact in the 1D Euler equations. HLLC (*circles*) and HLL (*triangles*) numerical solutions compared to exact solution (*full line*) at output time $t = 5$.

this point we solve the one-dimensional Euler equations in the domain $0 \leq x \leq 1$ with initial conditions $\rho(x, 0) = 1.4$ for $x \leq 0.5$ and $\rho(x, 0) = 1.0$ for $x > 0.5$, while $u(x, 0) = 0$ and $p(x, 0) = 1$, $\forall x \in [0, 1]$. Fig. 12 shows the numerical results for HLL and HLLC, compared to the exact solution at output time $t = 5.0$ units. The mesh used had $M = 100$ cells and the Courant number was $C_{cfl} = 0.9$. The difference between a complete and incomplete Riemann solver is evident. There are many applications and extensions of the HLLC solver in the literature, see for example Mignone and Bodo (2006) and Tokareva and Toro (2010).

6 A NUMERICAL VERSION OF THE OSHER–SOLOMON RIEMANN SOLVER

Recall that the Osher–Solomon (Osher and Solomon, 1982) numerical flux is constructed from

$$\mathbf{F}_{i+\frac{1}{2}} = \frac{1}{2}(\mathbf{F}(\mathbf{Q}_0) + \mathbf{F}(\mathbf{Q}_1)) - \frac{1}{2}\int_{\mathbf{Q}_0}^{\mathbf{Q}_1} |\mathbf{A}(\mathbf{Q})| d\mathbf{Q}. \tag{69}$$

Here we keep their original notation for the data states, namely $\mathbf{Q}_0 = \mathbf{Q}_L$, $\mathbf{Q}_1 = \mathbf{Q}_R$. It is well known that the Osher–Solomon solver has some limitations, complexity, CPU expense and applicability being three major drawbacks. For full details see chapter 12 of Toro (2009). Recently, a numerical version has been proposed (Dumbser and Toro, 2011a,b), which is presented here. A simplified version of the scheme relies on the choice of the linear path

$$\psi(s, \mathbf{Q}_0, \mathbf{Q}_1) = \mathbf{Q}_0 + s(\mathbf{Q}_1 - \mathbf{Q}_0), \quad s \in [0, 1] \tag{70}$$

to numerically evaluate the integral in (69). The modified scheme is applicable to both conservative and nonconservative hyperbolic systems. Here we deal with the conservative case. Under a change of variables the flux becomes

$$\mathbf{F}_{i+\frac{1}{2}} = \frac{1}{2}(\mathbf{F}(\mathbf{Q}_0) + \mathbf{F}(\mathbf{Q}_1)) - \frac{1}{2}\left(\int_0^1 |\mathbf{A}(\psi(s; \mathbf{Q}_0, \mathbf{Q}_1))| ds\right)\Delta\mathbf{Q}, \tag{71}$$

with $\Delta\mathbf{Q} = \mathbf{Q}_1 - \mathbf{Q}_0$ denoting the jump in initial data. The integration in (71) is performed numerically. For example, using a Gauss–Legendre quadrature rule with G points s_j and associated weights ω_j in the unit interval $I = [0, 1]$ we have

$$\mathbf{F}_{i+\frac{1}{2}} = \frac{1}{2}(\mathbf{F}(\mathbf{Q}_0) + \mathbf{F}(\mathbf{Q}_1)) - \frac{1}{2}\left(\sum_{j=1}^{G} \omega_j |\mathbf{A}(\psi(s_j, \mathbf{Q}_0, \mathbf{Q}_1))|\right)\Delta\mathbf{Q}. \tag{72}$$

For each s_j, $|\mathbf{A}(\psi(s_j, \mathbf{Q}_0, \mathbf{Q}_1))|$ is decomposed using the standard characteristic decomposition

$$|\mathbf{A}(\mathbf{Q})| = \mathbf{R}(\mathbf{Q})|\Lambda(\mathbf{Q})|\mathbf{R}^{-1}(\mathbf{Q}). \tag{73}$$

Some of the features of the new Osher-type Riemann solver include the following: the flux is indeed very simple to implement and is applicable to any hyperbolic system, provided the complete eigenstructure of the system is available, either analytically or numerically. The solver is nonlinear and complete, that is the wave model contains all characteristic fields. Numerical experiments suggest that the scheme is also entropy satisfying. The resulting numerical scheme is very robust requiring no special fixes or tuning of parameters. A drawback of the scheme is computational expense, but experience so far, from complex applications, is indeed very positive and results are superior to those of the original analytical Osher–Solomon scheme, as illustrated through the following test problem. We solve the one-dimensional Euler equations for ideal gases with $\gamma = 1.4$ in the domain $0 \leq x \leq 1$, mesh $M = 100$ cells, Courant number $C_{cfl} = 0.9$, output time $t = 0.012$. The initial conditions are $\rho(x, 0) = 1$ and $u(x, 0) = -19.59745$ for $0 \leq x \leq 1$, $p(x, 0) = 1000$ if $0 \leq x \leq 0.8$ and $p(x, 0) = 0.01$ if $0.8 < x \leq 1$. Fig. 13 shows results compared to the exact solution. The left frame shows the classical Osher–Solomon scheme with P-ordering of integration paths; the scheme fails, it is unable to recognise the shock wave and produces a huge overshoot in density. The right frame shows the numerical Osher-type scheme as compared to the exact solution; its performance is very satisfactory. Recent experience (Müller and Toro, 2013a,b; Müller et al., 2013) suggests that for some particular applications one may need to consider more sophisticated choices for the integration path.

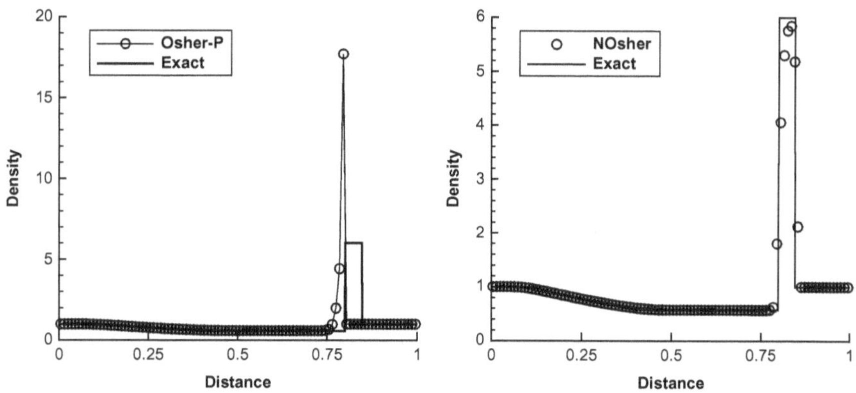

FIG. 13 Classical Osher–Solomon scheme with P-ordering (*left*) and Numerical Osher (*right*) are compared to the exact solution at time $t = 0.012$.

7 OTHER APPROACHES TO CONSTRUCTING NUMERICAL FLUXES

Upwind information for constructing numerical fluxes can also be provided by the flux vector splitting (FVS) approach, generally with less computational effort. However, classical FVS schemes are well known for badly smearing intermediate characteristic fields (van Leer, 1982; Warming and Beam, 1976; Zha and Bilgen, 1993), though more recent FVS-type schemes have overcome this difficulty and today constitute a credible alternative to good (complete and nonlinear) Riemann solvers, at least for classical hyperbolic systems (e.g. Balsara et al., 2016b; Liou and Steffen, 1993; Toro and Vázquez-Cendón, 2012; Toro et al., 2015).

Further down the hierarchy, there are numerical fluxes for finite volume methods that make no explicit use of upwind information contained in the Riemann problem solution. However, it is important to be aware that the Riemann problem is always there by the very definition of finite volume methods via integral averages, that is the governing equations and the piecewise constant initial conditions, are an essential part of the schemes. These schemes are termed *centred* (or centred, or central) or *symmetric* methods. Three classical numerical fluxes in this category are the *the Lax–Friedrichs flux* $\mathbf{F}^{\mathrm{LF}}_{i+\frac{1}{2}}$, *the Godunov centred flux* $\mathbf{F}^{\mathrm{GodC}}_{i+\frac{1}{2}}$ and the *the Lax–Wendroff flux* $\mathbf{F}^{\mathrm{LW}}_{i+\frac{1}{2}}$, given, respectively, as

$$\left.\begin{aligned}
&\mathbf{F}^{\mathrm{LF}}_{i+\frac{1}{2}} = \frac{1}{2}\left[\mathbf{F}(\mathbf{Q}_i^n) + \mathbf{F}(\mathbf{Q}_{i+1}^n)\right] - \frac{1}{2}\frac{\Delta x}{\Delta t}\left[\mathbf{Q}_{i+1}^n - \mathbf{Q}_i^n\right],\\
&\mathbf{F}^{\mathrm{GodC}}_{i+\frac{1}{2}} = \mathbf{F}(\mathbf{Q}^{\mathrm{GodC}}_{i+\frac{1}{2}}), \mathbf{Q}^{\mathrm{GodC}}_{i+\frac{1}{2}} = \frac{1}{2}\left[\mathbf{Q}_i^n + \mathbf{Q}_{i+1}^n\right] - \frac{\Delta t}{\Delta x}\left[\mathbf{F}(\mathbf{Q}_{i+1}^n) - \mathbf{F}(\mathbf{Q}_i^n)\right],\\
&\mathbf{F}^{\mathrm{LW}}_{i+\frac{1}{2}} = \mathbf{F}(\mathbf{Q}^{\mathrm{LW}}_{i+\frac{1}{2}}), \mathbf{Q}^{\mathrm{LW}}_{i+\frac{1}{2}} = \frac{1}{2}\left[\mathbf{Q}_i^n + \mathbf{Q}_{i+1}^n\right] - \frac{1}{2}\frac{\Delta t}{\Delta x}\left[\mathbf{F}(\mathbf{Q}_{i+1}^n) - \mathbf{F}(\mathbf{Q}_i^n)\right].
\end{aligned}\right\} \quad (74)$$

Note the subtle difference between the Godunov centred and the Lax–Wendroff flux. The resulting Lax–Friedrichs scheme is first order and monotone, while the Godunov centred scheme is also first order but non monotone. Moreover the stability limit for the latter is not the usual unity limit but the more restrictive condition $|c| \leq \frac{1}{2}\sqrt{2}$, where c is the Courant number. The Lax–Wendroff scheme is second order but nonmonotone.

A relatively new centred flux is the *the FORCE flux* (Toro and Billett, 1996, 2000), whose construction is inspired by the staggered version of Glimm's method (or random choice method) (Glimm, 1965). By systematically replacing random states by integral averages and invoking the integral form of the conservation laws one arrives at a one-step, nonstaggered, conservative scheme with numerical flux that can, surprisingly, be written as the arithmetic mean of the Lax–Friedrichs and the Lax–Wendroff flux schemes, namely

$$\mathbf{F}_{i+\frac{1}{2}}^{\mathrm{FORCE}} = \frac{1}{2}\left(\mathbf{F}_{i+\frac{1}{2}}^{\mathrm{LF}} + \mathbf{F}_{i+\frac{1}{2}}^{\mathrm{LW}}\right). \tag{75}$$

The resulting scheme is first-order, monotone and stable to Courant number unity. Convergence for some nonlinear hyperbolic system is proved in Chen and Toro (2004). The extension of FORCE to multiple space dimensions on general meshes is quite recent and applies to conservative (Toro et al., 2009) and nonconservative systems (Dumbser et al., 2010). See Kurganov et al. (2001), and references therein, for an alternative approach called *Central-Upwind.*

Fig. 14 shows results for the FORCE scheme, as compared to HLL and the exact solution, for the isolated stationary contact test problem for the 1D Euler equations, at time $t = 5$, mesh $M = 100$ cells and Courant number $C_{cfl} = 0.9$. As expected, the FORCE scheme, being a centred scheme, smears the contact discontinuity, particularly badly for the stationary case. However, and surprisingly, the FORCE scheme is less diffusive than the upwind (incomplete) HLL Riemann solver, at least for Courant numbers close to unity. A conclusion is that Riemann solvers, if not complete, give comparable results to those of centred methods, even for slowly moving intermediate waves.

There is another class of approximate Riemann solvers that aim to improve the resolution of intermediate characteristic fields but without explicitly solving the Riemann problem in the classical sense. Taken to the extreme, the idea is to use *no characteristic information at all*, apart from an estimate for the largest signal speed in the system, so as to choose a stable time step for time marching. The approach is fundamentally numerical, in contrast to traditional approaches that aim for analytical expressions for the Riemann solver. The methodology is indeed very appealing for very large and complex systems, as no eigenstructure (eigenvalues and eigenvectors) would be required. The MUSTA (MUlti-STAge) predictor–corrector

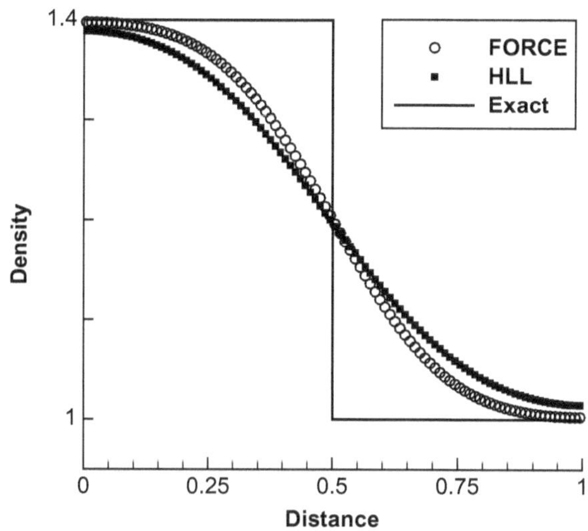

FIG. 14 Computed results for isolated stationary contact in the 1D Euler equations. FORCE scheme (*circles*) is compared to HLL (*squares*) and exact solution (*full line*). Output time is $t = 5$.

approach is probably the first example in this class of methods (Obergaulinger et al., 2009; Titarev and Toro, 2005; Titarev et al., 2008; Toro, 2003; Toro and Titarev, 2004, 2006). The central idea is to *open the Riemann fan* and access the needed upwind information inside the characteristic fan, to compute a numerical flux. There are other approaches that broadly fall within this category of Riemann solvers. They differ in the way the Riemann fan is opened and the amount of characteristic information needed. One variant is the so-called Krylov–Riemann solver (Torrilhon, 2012); another approach is proposed in Castro et al. (2012) and Castro et al. (2014), where polynomial approximations to $|\mathbf{A}|$ are sought, with $\mathbf{A}$ denoting the Jacobian matrix (see also Castro et al., 2016; Cordier et al., 2014; Degond et al., 1999; Ndjinga et al., 2008). These methods have indeed shown improvements to established methods in solving very large and complex hyperbolic systems, for which even obtaining the eigenstructure of the system is challenging or impossible. However such methods have not yet achieved their full potential. Intermediate characteristic fields are still smeared; resolution of stationary fields poses the greatest challenge. Maintaining monotonicity is yet another pending issue. However, particularly in view of the growing number and size of new hyperbolic systems being put forward nowadays, it is quite possible that this class of Riemann solvers may become the only feasible approach to fully resolve all characteristic fields and thus produce accurate Godunov-type methods for large and complex hyperbolic systems in the future.

8 CONCLUDING REMARKS

The Riemann problem has become a broad research theme in computational science, of which a very succinct account has been given in this chapter. To start with, some basic definitions and simple examples of Riemann problems have been given. Then, the exact solution of the Riemann problem for the compressible Euler equations has been presented in some detail, the underlying idea being that a sound understanding of the exact problem is needed in order to address the problem of finding useful approximate solutions. Then, several approaches for finding approximate solutions have been presented and selected examples have been given to emphasise key issues related to the solvers. This chapter has addressed exclusively the first-order schemes associated to the Riemann solvers. These schemes are expected to enjoy the fundamental property of *monotonicity* (for the scalar case) and constitute the building block for constructing high-order accurate Godunov-type finite volume and discontinuous Galerkin (DG) finite element methods.

We have also indicated other uses and generalisations of the Riemann problem, for treating boundary conditions, for example, independently of the particular approach used to construct numerical fluxes. There are several ways of utilising the basic, CRP solver in the construction of high-order methods, starting from the classical second-order TVD methods and going further to semidiscrete type numerical methods based on ENO and WENO reconstruction (Abgrall, 1996; Shu and Osher, 1988, 1989) and discontinuous Galerkin finite element methods (Cockburn and Shu, 1989; Cockburn et al., 1989, 2000; Dumbser et al., 2008a). We have not addressed this theme here. Other related and important topics have been omitted, such as rotated Riemann solvers (Ren, 2003) and multidimensional Riemann solvers; these have received renewed attention in the last few years, see for example Balsara (2014), Balsara et al. (2014), Balsara et al. (2016a) and Boscheri et al. (2014). I have not addressed the *generalised Riemann problem, GRP*, defined as the Cauchy problem for hyperbolic balance laws

$$\left.\begin{array}{ll}\text{PDEs}: & \partial_t \mathbf{Q}(x,t) + \partial_x \mathbf{F}(\mathbf{Q}(x,t)) = \mathbf{S}(\mathbf{Q}(x,t)), \quad x \in R, \quad t > 0 \\ \text{ICs}: & \mathbf{Q}(x,0) = \begin{cases} \mathbf{Q}_L(x) & \text{if} \quad x < 0, \\ \mathbf{Q}_R(x) & \text{if} \quad x > 0. \end{cases}\end{array}\right\} \tag{76}$$

This Cauchy problem is a twofold generalisation of the CRP. First the PDEs admit source terms; in the CRP the source term is zero. Then the initial conditions are general piece-wise smooth vector-valued functions, instead of piece-wise constant, as in the CRP studied in this chapter. In Ben-Artzi and Falcovitz (1984) the special GRP for the homogeneous Euler equations (no source terms) and piece-wise linear initial condition was presented, giving rise to second-order accurate methods. Then, in a less obvious way, in Harten et al. (1987) the homogenous GRP but for more general initial conditions than those in Ben-Artzi and Falcovitz (1984) was posed and solved. Le Floch and

Raviart (1988) posed the general, though homogeneous, GRP problem (76) and developed an existence and uniqueness theory. Toro and Titarev (2002) solved the complete problem (76) to any order of accuracy, generalising the solver proposed by Toro et al. (2001) for linear systems. It also generalises the GRP in Ben-Artzi and Falcovitz (1984) and that in Harten et al. (1987), which did not include source terms. New approximate solvers for the GRP have been presented in Dumbser et al. (2008b), Castro and Toro (2008), Toro and Montecinos (2015), Goetz and Iske (2016) and Goetz and Dumbser (2016). The family of ADER schemes, first communicated in Toro et al. (2001), is based on the approximate solution of the GRP (76), resulting in fully discrete, one-step numerical schemes of arbitrary order of accuracy in both space and time (no theoretical accuracy barrier). For a partial review of ADER schemes see chapters 19 and 20 of Toro (2009). The CRP (piece-wise constant data and no source terms) also plays a crucial role in solving the generalised Riemann problem. Solution procedures put forward so far, reduce the solution procedure for the GRP to that of solving a sequence of CRPs. This is yet another reason for further improving solution methods for the CRP.

ACKNOWLEDGEMENTS

The author gratefully acknowledges the contribution of Christian Contarino (Universitá degli Studi di Trento) and Arturo Hidalgo (Universidad Politécnica de Madrid) in the preparation of this chapter.

REFERENCES

Abgrall, R., 1996. On essentially non-oscillatory schemes on unstructured meshes: analysis and implementation. J. Comput. Phys. 114, 45–58.

Antón, L., Miralles, J.A., Martí, J.M., Ibáñez, J.M., Aloy, M.A., 2010. Relativistic magnetohydrodynamics: renormalized eigenvectors and full decomposition Riemann solver. Astrophys. J. Suppl. Ser. 188, 1–31.

Balsara, D., 2014. Multidimensional Riemann problem with self-similar internal structure. Part I—application to hyperbolic conservation laws on structured meshes. J. Comput. Phys. 277, 163–200.

Balsara, D.S., Dumbser, M., Abgrall, R., 2014. Multidimensional HLLC Riemann solver for unstructured meshes-with application to Euler and MHD flows. J. Comput. Phys. 261, 172–208.

Balsara, D., Vides, J., Gurski, K., Nkonga, B., Dumbser, M., Garain, S., Audit, E., 2016a. A two-dimensional Riemann solver with self-similar sub-structure–alternative formulation based on least squares projection. J. Comput. Phys. 304, 138–161.

Balsara, D.S., Montecinos, G.I., Toro, E.F., 2016b. Exploring various flux vector splittings for the magnetohydrodynamic system. J. Comput. Phys. 311, 1–21.

Batten, P., Clarke, N., Lambert, C., Causon, D., 1997a. On the choice of wave speeds for the HLLC Riemann solver. SIAM J. Sci. Stat. Comp. 18, 1553–1570.

Batten, P., Leschziner, M.A., Goldberg, U.C., 1997b. Average-state Jacobians and implicit methods for compressible viscous and turbulent flows flows. J. Comput. Phys. 137, 38–78.

Ben-Artzi, M., Falcovitz, J., 1984. A second order Godunov-type scheme for compressible fluid dynamics. J. Comput. Phys. 55, 1–32.

Ben Moussa, B., 2001. Meshless particle methods: recent developments for non-linear conservation laws in bounded domains. In: Toro, E.F. (Ed.), Godunov Methods: Theory and Applications. Kluwer Academic/Plenum Publishers, New York/Boston/Dordrecht/London/Moscow, pp. 97–108.

Ben Moussa, B., Vila, J.P., 2000. Convergence of SPH method for scalar nonlinear conservation laws. SIAM J. Numer. Anal. 37 (3), 863–887.

Boscheri, W., Balsara, D.S., Dumbser, M., 2014. Lagrangian ADER-WENO finite volume schemes on unstructured triangular meshes based on genuinely multidimensional HLL Riemann solvers. J. Comput. Phys. 267, 112–138.

Bressan, A., 2000. Hyperbolic Systems of Conservation Laws. The One-Dimensional Cauchy Problem. Oxford University Press, Oxford.

Castro, C.E., Toro, E.F., 2008. Solvers for the high-order Riemann problem for hyperbolic balance laws. J. Comput. Phys. 227, 2481–2513.

Castro, C.E., Toro, E.F., 2014. Roe-type Riemann solver for general hyperbolic systems. Int. J. Numer. Methods Fluids 75, 467–486.

Castro, M.J., Gallardo, J.M., Marquina, A., 2012. A class of computationally fast first order finite volume solvers: PVM methods. SIAM J. Sci. Comput. 34, A2173–A2196.

Castro, M.J., Gallardo, J.M., Marquina, A., 2014. A class of incomplete Riemann solvers based on uniform rational approximations to the absolute value function. J. Sci. Comput. 60 (2), 363–389.

Castro, M.J., Gallardo, J.M., Marquina, A., 2016. Approximate Osher-Solomon schemes for hyperbolic systems. Appl. Math. Comput. 272 (2), 347–368.

Chen, G.Q., Toro, E.F., 2004. Centred schemes for non-linear hyperbolic equations. J. Hyperbol. Differ. Equ. 1 (1), 531–566.

Chorin, A.J., 1976. Random choice solutions of hyperbolic systems. J. Comput. Phys. 22, 517–533.

Cockburn, B., Shu, C.W., 1989. TVB Runge-Kutta local projection discontinuous Galerkin finite element method for conservation laws II: general framework. Math. Comput. 52, 411–435.

Cockburn, B., Li, S.Y., Shu, C.W., 1989. TVB Runge-Kutta local projection discontinuous Galerkin finite element method for conservation laws III: one dimensional systems. J. Comput. Phys. 84, 90–113.

Cockburn, B., Karniadakis, G.E., Shu, C.W., 2000. Discontinuous Galerkin Methods. Lecture Notes in Computational Science and Engineering, Springer-Verlag, Berlin\Heidelberg.

Colella, P., 1982. Glimm's method for gas dynamics. SIAM J. Sci. Stat. Comput. 3 (1), 76–110.

Colella, P., Glaz, H.H., 1985. Efficient solution algorithms for the Riemann problem for real gases. J. Comput. Phys. 59, 264–289.

Cordier, F., Degond, P., Kumbaro, A., 2014. Phase appearance or disappearance in two-phase flows. J. Sci. Comput. 58, 115–148.

Dafermos, C.M., 2010. Hyperbolic Conservation Laws in Continuum Physics, third ed. Springer-Verlag, Berlin\Heidelberg.

Degond, P., Peyrard, P.F., Russo, G., Velledieu, P., 1999. Polynomial upwind schemes for hyperbolic systems. C. R. Acad. Sci. Paris 328, 479–483.

Dumbser, M., Balsara, D., 2016. A new efficient formulation of the HLLEM Riemann solver for general conservative and non-conservative hyperbolic systems. J. Comput. Phys. 304, 275–319.

Dumbser, M., Toro, E.F., 2011a. A simple extension of the Osher Riemann solver to general non-conservative hyperbolic systems. J. Sci. Comput. 48, 70–88.

Dumbser, M., Toro, E.F., 2011b. On universal Osher-type schemes for general nonlinear hyperbolic conservation laws. Commun. Comput. Phys. 10, 635–671.

Dumbser, M., Balsara, D., Toro, E.F., Munz, C.D., 2008a. A unified framework for the construction of one-step finite-volume and discontinuous Galerkin schemes. J. Comput. Phys. 227, 8209–8253.

Dumbser, M., Enaux, C., Toro, E.F., 2008b. Finite volume schemes of very high order of accuracy for stiff hyperbolic balance laws. J. Comput. Phys. 227 (8), 3971–4001.

Dumbser, M., Castro, M.J., Parés, C., Toro, E.F., Hidalgo, A., 2010. FORCE schemes on unstructured meshes II: non-conservative hyperbolic systems. Comput. Methods Appl. Mech. Eng. 199 (9–12), 625–647.

Dumbser, M., Peshkov, I., Romenski, E., Zanotti, O., 2016. High order ADER schemes for a unified first order hyperbolic formulation of continuum mechanics: viscous heat-conducting fluids and elastic solids. J. Comp. Phys. 314, 824–862.

Dutt, P., 1986. A Riemann Solver Based on a Global Existence Proof for the Riemann Problem.

Ferrari, A., Fraccarollo, L., Dumbser, M., Toro, E.F., Armanini, A., 2000. Three-dimensional flow evolution after a dam break. J. Fluid Mech. 663, 456–477.

Giacomazzo, B., Rezzolla, L., 2006. The exact solution of the Riemann problem in relativistic magnetohydrodynamics. J. Fluid Mech. 562, 223–259.

Glimm, J., 1965. Solution in the large for nonlinear hyperbolic systems of equations. Comm. Pure. Appl. Math. 18, 697–715.

Glimm, J., Graham, M.J., Grove, J.W., Li, X.L., Smith, T.M., Tan, D., Tangerman, F., Zhang, Q., 1998a. Front tracking in two and three dimensions. Comput. Math. Appl. 35, 1–11.

Glimm, J., Grove, J.W., Li, X.L., Shyue, K.M., Zhang, Q., Zeng, Y., 1998b. Three dimensional front tracking. SIAM J. Sci. Comput. 19, 703–727.

Godlewski, E., Raviart, P.A., 1996. Numerical Approximation of Hyperbolic Systems of Conservation Laws. Applied Mathematical Sciences, vol. 118. Springer-Verlag, New York.

Godunov, S.K., 1959. A finite difference method for the computation of discontinuous solutions of the equations of fluid dynamics. Mat. Sbornik 47, 357–393.

Godunov, S.K., 1976. Numerical Solution of Multi-Dimensional Problems in Gas Dynamics. Nauka Press, Moscow.

Goetz, C.R., Dumbser, M., 2016. A novel solver for the generalized Riemann problem based on a simplified LeFloch-Raviart expansion and a local space-time discontinuous Galerkin formulation. J. Sci. Comput. 69 (2), 805–840. http://dx.doi.org/10.1007/s10915-016-0218-5.

Goetz, C.R., Iske, A., 2016. Approximate solutions of generalized Riemann problems for nonlinear systems of hyperbolic conservation laws. Math. Comput. 85, 35–62.

Gottlieb, J.J., Groth, C.P.T., 1988. Assessment of Riemann solvers for unsteady one-dimensional inviscid flows of perfect gases. J. Comput. Phys. 78, 437–458.

Guermond, J.L., Popov, B., 2015. Invariant domains and first-order continuous finite element approximation for hyperbolic systems. arXiv preprint arXiv:1509.07461.

Guermond, J.L., Popov, B., 2016. Fast estimation of the maximum wave speed in the Riemann problem for the Euler equations. J. Comput. Phys. 321, 908–916.

Han, E., Warnecke, G., Toro, E.F., Siviglia, A., 2015a. On Riemann Solutions to Weakly Hyperbolic Systems: Part 1. Modelling Subcritical Flows in Arteries. .

Han, E., Warnecke, G., Toro, E.F., Siviglia, A., 2015b. On Riemann Solutions to Weakly Hyperbolic Systems: Part 2. Modelling Supercritical Flows in Arteries. Isaac Newton Institute for Mathematical Sciences, University of Cambridge, UK. Technical Report NI15004-NPA.

Harten, A., Lax, P.D., van Leer, B., 1983. On upstream differencing and Godunov-type schemes for hyperbolic conservation laws. SIAM Rev. 25 (1), 35–61.

Harten, A., Engquist, B., Osher, S., Chakravarthy, S.R., 1987. Uniformly high order accuracy essentially non-oscillatory schemes III. J. Comput. Phys. 71, 231–303.

Holden, H., Risebro, N.H., 2015. Front Tracking for Hyperbolic Conservation Laws, second ed. Springer, Heidelberg\Dordrecht\ London\New York. ISBN: 978-3-662-47506-5.

Kurganov, A., Noelle, S., Petrova, G., 2001. Semi-discrete central-upwind scheme for hyperbolic conservation laws and Hamilton-Jacobi equations. SIAM J. Sci. Comput 23, 707–740.

Le Floch, P., Raviart, P.A., 1988. An asymptotic expansion for the solution of the generalized Riemann problem. Part 1: general theory. Ann. Inst. Henri Poincaré. Analyse non Lineáre 5 (2), 179–207.

LeVeque, R.J., 2002. Finite Volume Methods for Hyperbolic Problems. Cambridge University Press, Cambridge.

Liou, M.S., Steffen, C.J., 1993. A new flux splitting scheme. J. Comput. Phys. 107, 23–39.

Mazaheri, A., Ricchiuto, M., Nishikawa, H., 2016. A first-order hyperbolic system approach for dispersion. J. Comput. Phys. 321, 593–605.

Menikoff, R., Plohr, B.J., 1989. The Riemann problem for fluid flow of real materials. Rev. Mod. Phys. 61, 75–130.

Mignone, A., Bodo, G., 2006. An HLLC Riemann solver for relativistic flows. Mon. Not. R. Astron. Soc. 368 (3), 1040–1054.

Montecinos, G.I., Toro, E.F., 2014. Reformulations for general advection-diffusion-reaction equations and locally implicit ADER schemes. J. Comput. Phys. 275, 415–442.

Montecinos, G.I., Müller, L.O., Toro, E.F., 2014. Hyperbolic reformulation of a 1D viscoelastic blood flow model and ADER finite volume schemes. J. Comput. Phys. 266, 101–123.

Müller, L.O., Toro, E.F., 2013a. A global multi-scale model for the human circulation with emphasis on the venous system. Int. J. Numer. Methods Biomed. Eng. 30 (7), 681–725.

Müller, L.O., Toro, E.F., 2013b. Well-balanced high-order solver for blood flow in networks of vessels with variable properties. Int. J. Numer. Methods Biomed. Eng. 29, 388–1411.

Müller, L.O., Parés, C., Toro, E.F., 2013. Well-balanced high-order numerical schemes for one-dimensional blood flow in vessels with varying mechanical properties. J. Comput. Phys. 242, 53–85.

Ndjinga, M., Kumbaro, A., de Vuyst, F., Laurent-Gengoux, P., 2008. Numerical simulation of hyperbolic two-phase flow models using a Roe-type solver. Nucl. Eng. Des. 238, 2075–2083.

Obergaulinger, M., Cerdá-Durán, P., Müller, E., Aloy, M.A., 2009. Semi-global simulations of the magneto-rotational instability in core collapse supernovae. Astron. Astrophys. 498, 241–271.

Osher, S., Solomon, F., 1982. Upwind difference schemes for hyperbolic conservation laws. Math. Comput. 38 (158), 339–374.

Peshkov, I., Romenski, E., 2008. A hyperbolic model for viscous Newtonian flows. Contin. Mech. Thermodyn. 28 (1), 85–104.

Pike, J., 1993. Riemann solvers for perfect and near-perfect gases. AIAA J. 31 (10), 1801–1808.

Quartapelle, L., Castelletti, L., Guardone, A., Quaranta, G., 2003. Solution of the Riemann problem of classical gasdynamics. J. Comput. Phys. 190, 118–140.

Ren, Y.X., 2003. A robust shock-capturing scheme based on rotated Riemann solvers. Comput. Fluids 32, 1379–1403.

Riemann, B., 1860. Über die Fortpflanzung ebener Luftwellen von endlicher Schwingungsweite. Abhandlungen der Königlichen Gesellschaft der Wissenschaften zu Göttingen 8, 43–65.

Risebro, N.H., Tveito, A., 1991. Front tracking applied to nonstrictly hyperbolic systems of conservation laws. SIAM J. Sci. Stat. Comput. 12, 1401–1419.

Roe, P.L., 1981. Approximate Riemann solvers, parameter vectors, and difference schemes. J. Comput. Phys. 43, 357–372.

Roe, P.L., Pike, J., 1984. Efficient construction and utilisation of approximate Riemann solutions. In: Glowinski, R., Lions, J.L. (Eds.), Computing Methods in Applied Science and Engineering. North-Holland, The Netherlands\Boca Raton\London\New York.

Rusanov, V.V., 1961. Calculation of interaction of non-steady shock waves with obstacles. J. Comput. Math. Phys. USSR 1, 267–279.

Salas, M., 2009. A Shock-Fitting Primer. CRC Press, Boca Raton, FL.

Schleicher, M., 1993. Ein einfaches und effizientes verfahren zur loesung des Riemann-problems. Z. Flugwiss. Weltraumforsch. 17, 265–269.

Schwendeman, D.W., Wahle, C.W., Kapila, A.K., 2006. The Riemann problem and a high-resolution Godunov method for a model of compressible two-phase flow. J. Comput. Phys. 212, 490–526.

Shu, C.W., Osher, S., 1988. Efficient implementation of essentially non-oscillatory shock-capturing schemes. J. Comput. Phys. 77, 439–471.

Shu, C., Osher, S., 1989. Efficient implementation of essentially non-oscillatory shock-capturing schemes II. J. Comput. Phys. 83, 32–78.

Smoller, J., 1994. Shock Waves and Reaction-Diffusion Equations. Springer-Verlag, New York.

Sod, G.A., 1978. A survey of several finite difference methods for systems of nonlinear hyperbolic conservation laws. J. Comput. Phys. 27, 1–31.

Titarev, V.A., Toro, E.F., 2005. MUSTA schemes for multi-dimensional hyperbolic systems: analysis and improvements. Int. J. Numer. Meth. Fluids 49, 117–147.

Titarev, V.A., Romenski, E., Toro, E.F., 2008. MUSTA-type upwind fluxes for non-linear elasticity. Int. J. Numer. Meth. Eng. 73, 897–926.

Tokareva, S.A., Toro, E.F., 2010. HLLC-type Riemann solver for the Baer-Nunziato equations of compressible two-phase flow. J. Comput. Phys. 229, 3573–3604.

Toro, E.F., 1989a. A fast Riemann solver with constant covolume applied to the random choice method. Int. J. Numer. Meth. Fluids 9, 1145–1164.

Toro, E.F., 1989b. A weighted average flux method for hyperbolic conservation laws. Proc. R. Soc. London A423, 401–418.

Toro, E.F., 1992. Riemann problems and the WAF method for solving two-dimensional shallow water equations. Philos. Trans. R. Soc., A A338, 43–68.

Toro, E.F., 2001. Shock-Capturing Methods for Free-Surface Shallow Flows. Wiley and Sons Ltd., Chichester, UK.

Toro, E.F., 2003. Multi-Stage Predictor-Corrector Fluxes for Hyperbolic Equations. Isaac Newton Institute for Mathematical Sciences, University of Cambridge, UK. Technical Report NI03037-NPA.

Toro, E.F., 2009. Riemann Solvers and Numerical Methods for Fluid Dynamics, third ed. Springer-Verlag, Berlin\Heidelberg.

Toro, E.F., 2016. Brain venous haemodynamics, neurological diseases and mathematical modelling. A review. Appl. Math. Comput. 272, 542–579.

Toro, E.F., Billett, S.J., 1996. Centred TVD Schemes for Hyperbolic Conservation Laws. Department of Mathematics and Physics, Manchester Metropolitan University, UK. Technical Report MMU-9603.

Toro, E.F., Billett, S.J., 2000. Centred TVD schemes for hyperbolic conservation laws. IMA J. Numer. Anal. 20, 47–79.

Toro, E.F., Chakraborty, A., 1994. Development of an approximate Riemann solver for the steady supersonic Euler equations. Aeronaut. J. 98, 325–339.

Toro, E.F., Hidalgo, A., 2009. ADER finite volume schemes for diffusion-reaction equations. Appl. Numer. Math. 59, 73–100.

Toro, E.F., Montecinos, G.I., 2014. Advection-diffusion-reaction equations: hyperbolisation and high-order ADER discretizations. SIAM J. Sci. Comput. 36 (5), A2423–A2457.

Toro, E.F., Montecinos, G.I., 2015. Implicit, semi-analytical solution of the generalised Riemann problem for stiff hyperbolic balance laws. J. Comput. Phys. 303, 146–172.

Toro, E.F., Siviglia, A., 2013. Flow in collapsible tubes with discontinuous mechanical properties: mathematical model and exact solutions. Commun. Comput. Phys. 13 (2), 361–385.

Toro, E.F., Titarev, V.A., 2002. Solution of the generalised Riemann problem for advection-reaction equations. Proc. R. Soc. London A 458, 271–281.

Toro, E.F., Titarev, V.A., 2004. MUSTA Schemes for Systems of Conservation Laws. Isaac Newton Institute for Mathematical Sciences, University of Cambridge, UK. Technical Report NI04033-NPA.

Toro, E.F., Titarev, V.A., 2006. MUSTA Schemes for Systems of Conservation Laws. J. Comput. Phys. 216, 403–429.

Toro, E.F., Vázquez-Cendón, M.E., 2012. Flux splitting schemes for the Euler equations. Comput. Fluids 70, 1–12.

Toro, E.F., Spruce, M., Speares, W., 1992. Restoration of the Contact Surface in the HLL-Riemann Solver. Department of Aerospace Science, Collegue of Aeronautics, Cranfield Institute of Technology, UK. Technical Report CoA-9204.

Toro, E.F., Spruce, M., Speares, W., 1994. Restoration of the Contact Surface in the HLL-Riemann Solver. Shock Waves 4, 25–34.

Toro, E.F., Millington, R.C., Nejad, L.A.M., 2001. Towards very high-order Godunov schemes. In: Toro, E.F. (Ed.), Godunov Methods: Theory and Applications. Kluwer Academic/Plenum Publishers, New York, pp. 905–937.

Toro, E.F., Hidalgo, A., Dumbser, M., 2009. FORCE schemes on unstructured meshes I: conservative hyperbolic systems. J. Comput. Phys. 228, 3368–3389.

Toro, E.F., Castro, C.E., Lee, B.J., 2015. A novel numerical flux for the 3D Euler equations with general equation of state. J. Comput. Phys. 303, 80–94.

Torrilhon, M., 2012. Krylov-Riemann solver for large hyperbolic systems of conservation laws. SIAM J. Sci. Comput. 34, A2072–A2091.

van Leer, B., 1979. Towards the ultimate conservative difference scheme V. A second order sequel to Godunov's method. J. Comput. Phys. 32, 101–136.

van Leer, B., 1982. Flux-Vector Splitting for the Euler Equations. NASA Langley Research Center, USA. Technical Report ICASE 82-30.

Vázquez-Cendón, M.E., 2015. Solving Hyperbolic Equations with Finite Volume Methods. Springer, Cham\Heidelberg\New York\Dordrecht\London.

Vila, J.P., 1999. On particle weighted methods and smooth particle hydrodynamics. Math. Models Methods Appl. Sci. 9 (2), 161–209.

Warming, R.F., Beam, R.W., 1976. Upwind second order difference schemes with applications in aerodynamic flows. AIAA J. 24, 1241–1249.

Zha, G.C., Bilgen, E., 1993. Numerical solution of Euler equations by a new flux vector splitting scheme. Int. J. Numer. Meth. Fluids 17, 115–144.

Chapter 3

Classical Finite Volume Methods

T. Sonar
Institut Computational Mathematics, Technische Universität Braunschweig, Braunschweig, Germany

Dedication: Dedicated to Gerald Warnecke on the occasion of his 60th birthday.

Chapter Outline

ABSTRACT

We report on the development of classical finite volume methods from their first occurrence in the 1960s up to the 1980s and 1990s where this class of methods presented themselves as workhorses in CFD. We describe implementations on quadrilateral as well as on triangular grids and exploit the span from central difference schemes augmented with artificial viscosity to upwind and ENO methods.

Keywords: Finite volume methods, Artificial viscosity, Quadrilateral and triangular grids, ENO reconstruction

AMS Classification Codes: 65M08, 65N08, 76M12

1 SOME PHILOSOPHICAL REMARKS

Following Ansorge and Sonar (2009) let us consider the simple linear scalar transport equation

Handbook of Numerical Analysis, Vol. 17. http://dx.doi.org/10.1016/bs.hna.2016.09.005

$$L(u) := \partial_t u + \partial_x u + \partial_y u = \partial_t u + \nabla \cdot F(u) = 0,$$

where $F(u) := (u, u)^T$, in two space dimensions for the sake of simplicity.

Historically, finite difference methods (FDM) enjoy the longest history in numerically solving partial differential equations. The basic idea, philosophically spoken, lies in the *discretisation of the differential operators* involved, e.g. with upwind differences

$$\partial_x u \approx \frac{U^n_{i+1,j} - U^n_{i,j}}{\Delta x},$$

and

$$\partial_y u \approx \frac{U^n_{i,j+1} - U^n_{i,j}}{\Delta x},$$

and forward in time difference

$$\partial_t u \approx \frac{U^{n+1}_{i,j} - U^n_{i,j}}{\Delta x}.$$

The numerical solution is therebye represented by purely discrete values $U^n_{i,j} \approx u(i\Delta x, j\Delta y, n\Delta t)$ at points of a grid with grid sizes Δx, Δy, and Δt, $i, j \in \mathbb{Z}$, $n \in \mathbb{N}$. The discrete equation to be solved then appears as

$$\frac{U^{n+1}_{i,j} - U^n_{i,j}}{\Delta t} + \frac{U^n_{i+1,j} - U^n_{i,j}}{\Delta x} + \frac{U^n_{i,j+1} - U^n_{i,j}}{\Delta x} = 0.$$

Whether weak solutions are approximated or in which sense convergence to weak solutions takes place is a question with long history (see Jovanović and Süli, 2014 for recent research on linear partial differential equations). In case of nonlinear hyperbolic conservation laws a breakthrough was the theorem of Lax and Wendroff (1960), of course, which we will discuss next.

In contrast, the finite element methods (FEM), coming into practical use via variational principles in the 1960s, *approximate the solution* in a finite dimensional vector space, therebye starting from a weak formulation

$$\forall \Phi \in W: \int_{\Omega \times [0,T]} \{u\partial_t \Phi + u\partial_x \Phi + u\partial_y \Phi\}\, dt\, dx\, dy = 0$$

where W denotes a space of compactly supported test functions in space and time. Replacing the test function space W by a finite dimensional space W^h and the solution space by the same space yields

$$\forall \Phi^h \in W^h: \int_{\Omega \times [0,T]} \{U^h \partial_t \Phi^h + U^h \partial_x \Phi^h + U^h \partial_y \Phi^h\} dt\, dx\, dy = 0$$

with $U^h(x, y, t) = \sum_{i=0}^n a_i \Phi^h(x, y, t)$ if $\dim W^h = n$ and if the Φ^h form a basis $\{\Phi_1, \ldots, \Phi_n\}$ of W^h. Most often W^h is chosen to be a space of polynomials for simplicity. One finally arrives at a system of equations

$$\sum_{i=1}^{n}\left(\int_{\Omega\times[0,T]}\Phi_i\{\partial_t\Phi_j+\partial_x\Phi_j+\partial_y\Phi_j\}dt\,dx\,dy\right)a_i=0,\quad j=1,\dots,n$$

for the unknown coefficients a_i defining the approximative solution.

As in the FDM a grid is needed, but now to determine interpolation nodes for the approximating functions, of course. In FEM one uses conforming triangulations as grids in 2D in general while the FDM relies on conforming grids consisting of quadrilaterals. Hence, FEM can be viewed as more flexible if curved boundaries have to be discretised.

FEMs of the above type are often called *space-time FEMs*. It is often convenient to separate the time derivative, treat it in a finite difference manner, and discretise only the space derivatives with finite elements. We have chosen test and ansatz functions from the same space W^h and these methods are called *Bubnov–Galerkin methods*. Further possibilities arise if one allows the ansatz space being different from the test space. This leads to *Petrov–Galerkin methods* (Morton, 1996, p. 153ff.). In our days *discontinuous finite element methods* (or discontinuous Galerkin methods, DGM) enjoy some attention. They exploit polynomial approximation functions on the elements but numerical flux functions from finite differences between two elements (cp. Dolejši and Feistauer, 2015). These types of methods are often felt as being extensions to finite volume methods (FVM) but for classical FVMs this is not the case.

The FVM shares some of its properties with the FEM and the FDM. All FVMs start from the divergence form

$$L(u)=\partial_t u+\nabla\cdot F(u)=0$$

and integrate over a time-fixed spatial element σ,

$$\frac{d}{dt}\int_\sigma u dx\,dy+\oint_{\partial\sigma}\langle F(u),n\rangle ds=0,$$

where Gauss's integral theorem was used to transform the volume integral into a surface integral. The unit outwards pointing normal vector at $\partial\sigma$ is denoted by n and $\langle\cdot,\cdot\rangle$ denotes the Euclidean inner product. Introducing the cell average

$$\bar{u}_\sigma(t):=\frac{1}{|\sigma|}\int_\sigma u(x,y,t)dx\,dy$$

where $|\sigma|$ denotes the measure (length, area, volume) of the element σ, we see that a FVM is an *evolution equation for cell averages*

$$\frac{d}{dt}\bar{u}_\sigma=-\frac{1}{|\sigma|}\oint_{\partial\sigma}\langle F(u),n\rangle ds.$$

Although some authors like to view FVMs as being part of the Petrov–Galerkin family of FEMs we take a different viewpoint here. We follow Heinrich (1987) and see FVMs as a generalisation of FDMs to unstructured grids. Hence, FVMs can be seen as bridging the gap between FDMs and FEMs.

2 ON THE LAX–WENDROFF THEOREM

FVMs seem to be perfectly suited to the conservation or divergence form of partial differential equations since they appear automatically in conservation form, and this form is important per se as we know from the now classical result of Lax and Wendroff (1960). This result stated that if a FDM in conservation form converges at all, then the limit function is indeed a weak solution of the partial differential equation. In its original form the Lax–Wendroff theorem required a uniform grid in one space dimension and other technical assumptions like continuous fluxes and L^∞-boundedness of the sequence of discrete numerical solutions when the mesh parameter tends to zero. It is, however, not clear a priori that this type of theorem extends to FVMs on non-Cartesian or triangular grids. Kröner et al. (1996) have shown that the theorem holds for two-dimensional polygonal meshes under some restrictive conditions on the mesh, cp. their conditions (2.3) and (2.4) in Kröner et al. (1996). Godlewski and Raviart (1996, p. 375 ff.) have relaxed the conditions for a triangular mesh. Even more general meshes are considered by Elling (2007) and his Lax–Wendroff-type theorem is fairly general indeed.

The importance of the Lax–Wendroff theorem is clearly understood if one studies the failure of convergence and strange behaviour if nonconservative numerical discretisation methods are applied to conservation laws (see Hou and LeFloch, 1994).

3 HISTORICAL REMARKS

To the knowledge of the author, the first occurrence of a FVM in the literature is in the famous book by Varga (1962) on iterative solvers for linear systems of equations. Even earlier MacNeal (1953) exploited the idea to formulate a FDM on a distorted grid and used a FV-like approach. The term 'finite volume method' was coined only in the 1970s, however, for example in McDonald (1971) and Rizzi and Inouye (1973).

In the early 1960s Preissmann (1961) was advocating a finite volume method for the St. Venant equations of hydraulic flow for which he coined the name *box scheme*. This now historically famous *Preissmann box scheme* is in modern notation the simplest *cell-vertex schemes* for the equation

$$\partial_t u + \partial_x f(u) = s(x,u),$$

i.e.

$$\begin{aligned}
&\frac{1}{2}(x_{j+1}-x_j)\left[U_{j+1}^{n+1}+U_j^{n+1}-U_{j+1}^n-U_j^n\right]\\
&+\frac{1}{2}(t^{n+1}-t^n)\left[F_{j+1}^{n+1}+F_{j+1}^n-F_j^{n+1}-F_j^n\right]\\
&=\frac{1}{4}(x_{j+1}-x_j)(t^{n+1}-t^n)\left[S_{j+1}^{n+1}+S_j^{n+1}+S_{j+1}^n+S_j^n\right]
\end{aligned} \tag{1}$$

on the 'box grid' of Fig. 1.

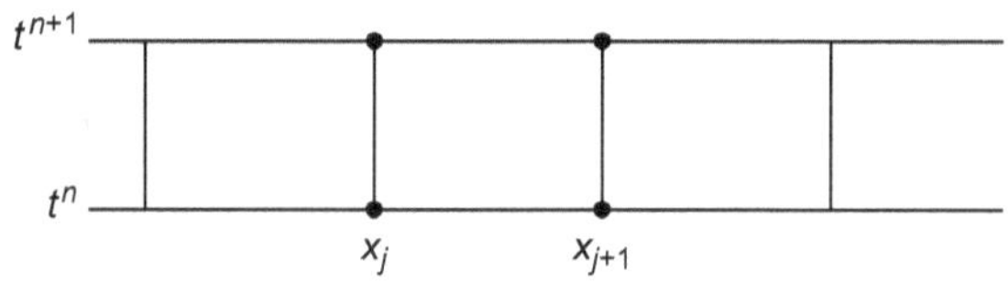

FIG. 1 The cell-vertex box grid according to Preissmann.

The method results from integrating the partial difference equation over the rectangular box shown in (1), employing Gauss's integral theorem to convert the volume integral on the left-hand side to an integral along the boundary as follows. Writing $\nabla_{(t,x)} \cdot G(u) := \partial_t u + \partial_x f(u)$ for $G(u) = (u, f(u))^T$ we get

$$\int_B \nabla_{(t,x)} \cdot G(u) dt\, dx = \int_B s(x,u) dt\, dx$$

where B denotes the box. Gauss's integral theorem then yields

$$\oint_{\partial B} \langle G(u), n\rangle ds = \int_B s(x,u) dt\, dx,$$

i.e.

$$\oint_{\partial B} u n_t + f(u) n_x ds = \int_B s(x,u) dt\, dx,$$

where we used the notation $n = (n_t, n_x)$ for the components of the unit outwards normal vector. The integration along the four boundary parts of the box is

$$\begin{aligned}
\oint_{\partial B} u n_t + f(u) n_x ds &= \int_{x_j}^{x_{j+1}} \begin{pmatrix} u(t^n, x) \\ f(u(t^n, x)) \end{pmatrix} \cdot \begin{pmatrix} -1 \\ 0 \end{pmatrix} dx \\
&+ \int_{t^n}^{t^{n+1}} \begin{pmatrix} u(t, x_{j+1}) \\ f(u(t, x_{j+1})) \end{pmatrix} \cdot \begin{pmatrix} 0 \\ 1 \end{pmatrix} dt \\
&+ \int_{x_{j+1}}^{x_j} \begin{pmatrix} u(t^{n+1}, x) \\ f(u(t^{n+1}, x)) \end{pmatrix} \cdot \begin{pmatrix} 1 \\ 0 \end{pmatrix} dx \\
&+ \int_{t^{n+1}}^{t^n} \begin{pmatrix} u(t, x_j) \\ f(u(t, x_j)) \end{pmatrix} \cdot \begin{pmatrix} 0 \\ -1 \end{pmatrix} dt
\end{aligned}$$

and approximating each of the integrals by means of the trapezoidal rule (we write F as an abbreviation of the approximation of $f(u)$) results in

$$\oint_{\partial B} un_t + f(u)n_x ds \approx -\frac{1}{2}(x_{j+1} - x_j)(U_{j+1}^n + U_j^n)$$
$$+\frac{1}{2}(t^{n+1} - t^n)(F_{j+1}^{n+1} + F_{j+1}^n)$$
$$+\frac{1}{2}(x_{j+1} - x_j)(U_{j+1}^{n+1} + U_j^{n+1})$$
$$-\frac{1}{2}(t^{n+1} - t^n)(F_j^{n+1} + F_j^n)$$

or

$$\oint_{\partial B} un_t + f(u)n_x ds \approx \frac{1}{2}(x_{j+1} - x_j)\left[-U_j^n - U_{j+1}^n + U_{j+1}^{n+1} + U_j^{n+1}\right]$$
$$+\frac{1}{2}(t^{n+1} - t^n)\left[F_{j+1}^{n+1} + F_{j+1}^n - F_j^{n+1} - F_j^n\right],$$

which is the left-hand side of (1). The right-hand side of (1) results simply from the mean of s taken at the four vertices. Hence, the Preissmann box scheme is a true FVM.

In 1959 an ingenious difference scheme for gas dynamics was developed by Godunov, see chapter 2 in Holt (1984) or chapter 12 in Richtmyer and Morton (1967). This scheme gives rise to a *cell-centred* FVM. If we consider

$$U_j^n \approx \frac{1}{x_{j+1/2} - x_{j-1/2}} \int_{x_{j-1/2}}^{x_{j+1/2}} u(x, t^n) dx$$

as an approximation to the cell average then a Godunov-type method for the approximation of

$$\partial_t u + \partial_x f(u) = 0$$

reads as

$$U_j^{n+1} = U_j^n - \frac{\Delta t}{\Delta x}\left[F_{j+1/2}^{n+1/2} - F_{j-1/2}^{n+1/2}\right] \tag{2}$$

where $F_{j+1/2}^{n+1/2}$ is an approximation (average) of the flux between times t^n and t^{n+1} at the cell boundary $x_{j+1/2}$, see Fig. 2.

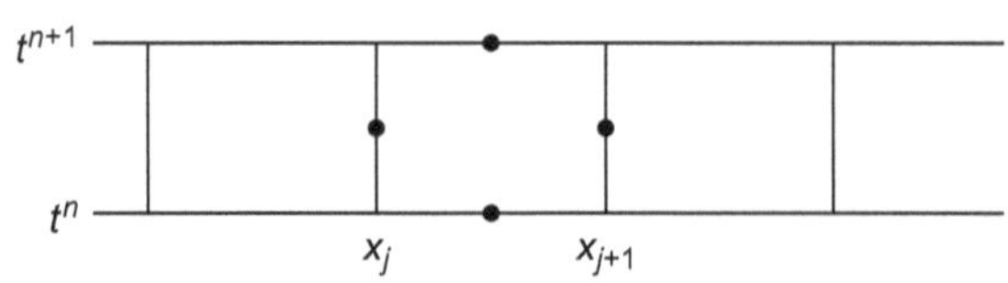

FIG. 2 The cell-vertex box grid used in the Godunov method.

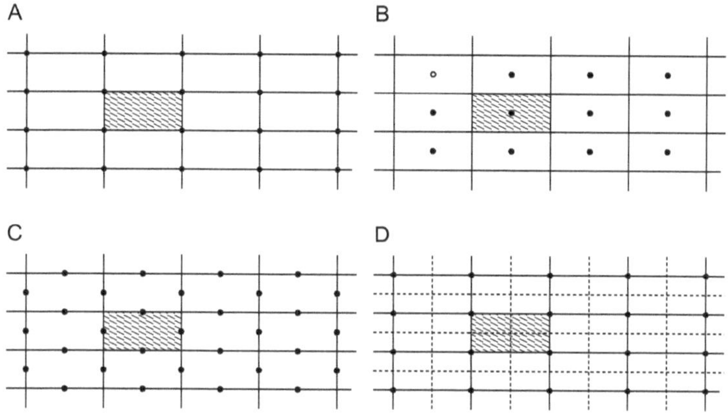

FIG. 3 Four types of finite volume grids. (A) Cell vertex. (B) Cell centre. (C) Cell edge. (D) Vertex-centred.

Here a main advantage of the FVMs can be observed, namely the automatic satisfaction of conservation. Summing (2) over any set of contiguous cells, say $l \leq j \leq r$, results in (cp. Morton and Sonar, 2007, p. 157)

$$\sum_{j=l}^{r} \Delta x(U_j^{n+1} - U_j^n) + \Delta t \left[F_{j+1/2}^{n+1/2} - F_{j-1/2}^{n+1/2} \right] = 0.$$

In Godunov's scheme the numerical flux F is determined by the solution of a Riemann problem at cell boundaries. However, numerous other—simpler—choices can be found in the literature (see Toro, 1999 for an overview).

In general, there are four slightly differing boxes used in FVMs: cell-vertex, cell centre, cell edge, and vertex-centred boxes. These four types are shown in Fig. 3 for the case of quadrilateral grid structures.

We close our historic excursion by remarking that besides 'box methods' the names 'box integration methods', 'balance methods' and 'finite control volume methods' were and are still in use (Heinrich, 1987, p. 10).

4 WEAK SOLUTIONS AND FINITE VOLUME METHODS

Although the notion of weak solutions is well understood the situation in case of FVMs seems to be not so easy due to the (strong) time derivative of the cell average. Considering the system of hyperbolic conservation laws

$$\partial_t u + \sum_{i=0}^{d} \partial x_i f_i(u) = \nabla_{(t,x)} \cdot \begin{pmatrix} u \\ f_1(u) \\ \vdots \\ f_d(u) \end{pmatrix} = 0 \tag{3}$$

for $u = u(x,t) \in \mathbb{R}^m$, $(x,t) \in \mathbb{R}^d \times \mathbb{R}^+$, with initial data $u(x, 0) = u^0(x)$ a weak solution u is defined to be a function in $L^1_{\text{loc}}(\mathbb{R}^d \times \mathbb{R}^+)^m$ so that for all test functions $\varphi \in C_0^1(\mathbb{R}^d \times [0, \infty[)^m$ it holds

$$\int_{\mathbb{R}^d \times \mathbb{R}^+} \left[u \cdot \partial_t \varphi + \sum_{i=0}^{d} f_i(u) \cdot \partial_{x_i} \varphi \right] dx\, dt + \int_{\mathbb{R}^d} u^0(x) \cdot \varphi(x,0) dx = 0. \qquad (4)$$

It is well known that this solution space is too large in the sense that unphysical solutions may be accepted as valid weak solutions and that entropy conditions have to be taken into account, but that is not our concern here. If we introduce a control volume $\Omega \subset \mathbb{R}$ and integrate over a $(d + 1)$-dimensional cylinder $\Omega \times]t, t + \Delta t[$, the we obtain

$$\int_{\Omega} [u(x,t+\Delta t) - u(x,t)] d\Omega + \int_t^{t+\Delta t} \oint_{\partial\Omega} \mathcal{F}(u) \cdot n ds\, dt = 0$$

with the help of the Gaussian integral theorem. Here we have introduced the shorthand notation $\mathcal{F} := (f_1, \ldots, f_d)$. As usual n denotes the unit vector at $\partial\Omega$ pointing outwards Ω.

We generalise our notion slightly to a more general $(d + 1)$-dimensional control volume Ω^b by introducing the generalised variable $x^b := (t, x_1, \ldots, x_d)^T$. Analogously we denote the outer unit normal vector to $\partial\Omega^b$ as n^b and the surface measure as ds^b. Taken the overall divergence form of our Eq. (3) into account we now can write the integral form compactly as

$$\oint_{\partial\Omega^b} \begin{pmatrix} u \\ \mathcal{F}(u) \end{pmatrix} \cdot n^b ds^b = 0. \qquad (5)$$

The question to be answered is how the integral representation (5) relates to the notion of weak solutions (4). Surprisingly, this question was answered already in 1919 by means of Haar's lemma. Haar's lemma was further exploited in the works by Morrey (1960) and Klötzler (1970), so that we may arrive at the following result (see Morton and Sonar, 2007, p. 163).

Theorem 1. *(Haar's Lemma)*

Suppose u and the fluxes f_i, $i = 1, \ldots, d$, are summable over the bounded region $\Omega^b \subset \mathbb{R}^d \times \mathbb{R}^+$. Then

$$\oint_{\partial C} \begin{pmatrix} u \\ \mathcal{F}(u) \end{pmatrix} \cdot n^b ds^b = 0$$

holds for almost all cuboids $C \subset \Omega^b$, if and only if

$$\int_{\Omega^b} \left[u \cdot \partial_t \varphi + \sum_{i=1}^{d} f_i(u) \cdot \partial_{x_i} \varphi \right] dx\, dt = 0$$

holds for every φ vanishing on or near $\partial\Omega^b$ and is uniformly Lipschitz on $\overline{\Omega^b}$.

The same results holds for balls instead of cuboids, and Bruhn (1985) has succeeded to extend the theorem to quite general shapes of control volumes. It is this type of theorem which we exploit if we speak about weak solutions in connection with FVMs.

5 THE CELL-CENTRED SCHEME OF JAMESON, SCHMIDT AND TURKEL

With the advent of powerful computer technology around 1980 the need for reliable and robust FVMs in the field of gas dynamics, and in particular in transonic flows, became felt in the aerospace industry and in academic institutions as well. The seminal paper (Jameson et al., 1981) marked a most important step in this direction since it contained the construction of a robust, second-order accurate, cell-centred FVM on quadrilateral cells. Since the construction of the FVM is not completely independent from the Euler equations of gas dynamics we begin by stating the problem (see Kroll and Jain, 1987).

The system of Euler equations governing compressible inviscid flow in two space dimensions is given by

$$\partial_t u + \partial_{x_1} f_1(u) + \partial_{x_2} f_2(u) = 0,$$

where

$$u = \begin{pmatrix} \rho \\ \rho v_1 \\ \rho v_2 \\ \rho E \end{pmatrix}; \; f_1(u) = \begin{pmatrix} \rho v_1 \\ \rho v_1^2 + p \\ \rho v_1 v_2 \\ \rho H v_1 \end{pmatrix}; \; f_2(u) = \begin{pmatrix} \rho v_2 \\ \rho v_1 v_2 \\ \rho v_2^2 + p \\ \rho H v_2 \end{pmatrix}.$$

Here ρ, $v = (v_1, v_2)^T$, p, E and H are density, velocity, pressure, total energy and total enthalpy, respectively. Assuming a perfect gas the equations

$$p = (\kappa - 1)\rho\left(E - \frac{|v|^2}{2}\right)$$

$$H = E + \frac{p}{\rho}$$

are valid, where κ denotes the ratio of specific heats which, in case of dry air, can be given as $\kappa = 1.4$. The four equations in Euler's system correspond to the conservation of mass, momentum (in the two Cartesian directions) and energy. The system consists of four equations for the four unknowns ρ, v_1, v_2 and E. The system is easily shown to be of hyperbolic type. The main range of applications lay in the computation of stationary flows, and hence the time derivative could be used to converge to a steady solution.

Writing as before

$$\mathcal{F}(u) := (f_1(u), f_2(u))$$

we arrive at the FV formulation

$$\frac{d}{dt}\int_\Omega u dx + \oint_{\partial\Omega} \mathcal{F}(u) \cdot n ds = 0, \tag{6}$$

where n denotes the outwards unit normal vector to $\partial\Omega$, and Ω denotes a fixed spatial domain. Inserting the fluxes into the definition of $\mathcal{F}$ we get

$$\mathcal{F}(u) = \begin{pmatrix} \rho v \\ \rho v_1 v + p e_1 \\ \rho v_2 v + p e_2 \\ \rho H v \end{pmatrix},$$

where $e_1 = (1, 0)^T$ and $e_2 = (0, 1)^T$ denote the Cartesian basis vectors in $\mathbb{R}^2$.

In order to arrive at a cell-centred FVM we look at a part of a grid as in Fig. 4. The point (i, j) is supposed to denote the cell centre of cell $\Omega_{i,j}$. The boundary of $\Omega_{i,j}$ may be denoted by

$$\partial\Omega_{i,j} = S_{i,j-1/2} \cup S_{i+1/2} \cup S_{i,j+1/2} \cup S_{i-1/2,j}$$

where $S_{i,j-1/2}$ denotes the edge between cells $\Omega_{i,j}$ and $\Omega_{i,j-1}$, etc. Computing the area of cell $\Omega_{i,j}$ by

$$|\Omega_{i,j}| = \frac{1}{2}\big[\big(x_{1,(i+1,j+1)} - x_{1,(i,j)}\big)\big(x_{2,(i,j+1)} - x_{2,(i+1,j)}\big) - \big(x_{2,(i+1,j+1)} - x_{2,(i,j)}\big)\big(x_{1,(i,j+1)} - x_{1,(i+1,j)}\big)\big]$$

we can introduce the cell average on $\Omega_{i,j}$ as

$$\overline{U}_{i,j} := \frac{1}{|\Omega_{i,j}|}\int_{\Omega_{i,j}} u dx_1\, dx_2.$$

With a view on (6) we can hence write

$$|\Omega_{i,j}|\frac{d}{dt}\overline{U}_{i,j} = -\oint_{\partial\Omega} \mathcal{F}(u)\cdot n ds$$

which is an evolution equation for cell averages and we have to care for the boundary integral now. The boundary integral is simply a sum over the four edges of cell $\Omega_{i,j}$. Thinking of the flux values as being located at the midpoints of the edges, labelled $(i, j - 1/2)$, $(i + 1/2, j)$, $(i, j + 1/2)$ and

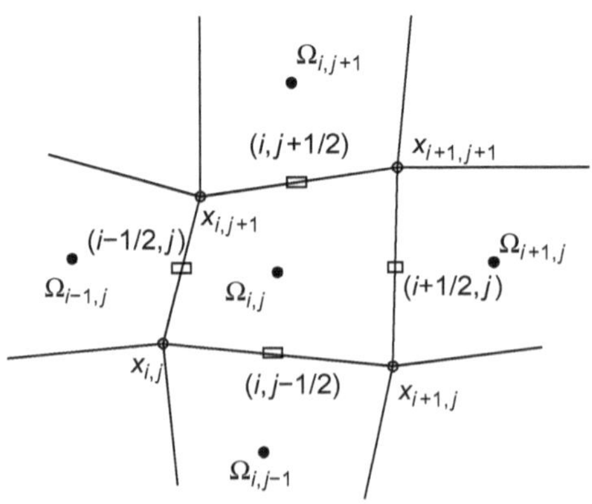

FIG. 4 The cell-centred grid around cell $\Omega_{i,j}$.

$(i - 1/2, j)$ in Fig. 4 and employing a simple midpoint rule for the quadrature over an edge, we can write

$$\oint_{\partial\Omega} \mathcal{F}(u)\cdot n ds \approx \mathcal{F}(u_{i,j-1/2})|S_{i,j-1/2}|n_{i,j-1/2} + \mathcal{F}(u_{i+1/2,j})|S_{i+1/2,j}|n_{i+1/2,j} + \mathcal{F}(u_{i,j+1/2})|S_{i,j+1/2}|n_{i,j+1/2} + \mathcal{F}(u_{i-1/2,j})|S_{i-1/2,j}|n_{i-1/2,j}$$

where $|S_{i,j-1/2}|$ denotes the length of edge $S_{i,j-1/2}$ and $n_{i,j-1/2}$ is the corresponding outwards unit vector. We compute the outwards unit normals by

$$n_{i,j-1/2} = \frac{1}{|S_{i,j-1/2}|}\begin{pmatrix} x_{2,(i+1,j)} - x_{2,(i,j)} \\ x_{1,(i,j)} - x_{1,(i+1,j)} \end{pmatrix}$$

and choose the simple mean value

$$\mathcal{F}(u_{i,j-1/2}) := \mathcal{F}\left(\frac{1}{2}(\overline{U}_{i,j} + \overline{U}_{i,j-1})\right)$$

for the flux located at edge $S_{i,j-1/2}$, we finally arrive at the semidiscrete cell-centred FVM

$$\begin{aligned}\frac{d}{dt}\overline{U}_{i,j} = -\frac{1}{|\Omega_{i,j}|}\Bigg\{ & \mathcal{F}\left(\frac{1}{2}(\overline{U}_{i,j} + \overline{U}_{i,j-1})\right)\begin{pmatrix} x_{2,(i+1,j)} - x_{2,(i,j)} \\ x_{1,(i,j)} - x_{1,(i+1,j)} \end{pmatrix} \\ & + \mathcal{F}\left(\frac{1}{2}(\overline{U}_{i,j} + \overline{U}_{i+1,j})\right)\begin{pmatrix} x_{2,(i+1,j+1)} - x_{2,(i+1,j)} \\ x_{1,(i+1,j)} - x_{1,(i+1,j+1)} \end{pmatrix} \\ & + \mathcal{F}\left(\frac{1}{2}(\overline{U}_{i,j} + \overline{U}_{i,j+1})\right)\begin{pmatrix} x_{2,(i,j+1)} - x_{2,(i+1,j+1)} \\ x_{1,(i+1,j+1)} - x_{1,(i,j+1)} \end{pmatrix} \\ & + \mathcal{F}\left(\frac{1}{2}(\overline{U}_{i,j} + \overline{U}_{i-1,j})\right)\begin{pmatrix} x_{2,(i,j)} - x_{2,(i,j+1)} \\ x_{1,(i,j+1)} - x_{1,(i,j)} \end{pmatrix}\Bigg\}\end{aligned} \tag{7}$$

In Jameson et al. (1981) and Kroll and Jain (1987) a classical four-stage Runge–Kutta method was used for the time stepping. However, the scheme derived thus far is unstable since the spatial disretisation corresponds to central differences. Therefore, the classical idea of artificial viscosity due to von Neumann and Richtmyer (1950) was implemented to stabilise the method.

Denoting the right-hand side of (7) by $-\mathcal{Q}_{i,j}$ the FVM including artificial dissipation is then

$$\frac{d}{dt}\overline{U}_{i,j} = -\mathcal{Q}_{i,j} + \frac{1}{|\Omega_{i,j}|}\mathcal{D}_{i,j} \tag{8}$$

where the artificial dissipation $\mathcal{D}_{i,j}$ is given by

$$\mathcal{D}_{i,j} = D_{i+1/2,j} - D_{i-1/2,j} + D_{i,j+1/2} - D_{i,j-1/2},$$

$$D_{i+1/2,j} = \alpha_{i+1/2,j}\left(\varepsilon^{(2)}_{i+1/2,j}\Delta_i\overline{U}_{i,j} - \varepsilon^{(4)}_{i+1/2,j}\Delta_i^3\overline{U}_{i-1,j}\right)$$

Here $\varepsilon^{(2)}_{i+1/2,j}$ and $\varepsilon^{(4)}_{i+1/2,j}$ are two coefficients to be described below, $\alpha_{i+1/2,j}$ is a scaling factor computed by

$$\alpha_{i+1/2,j} = \frac{1}{2}\left(\frac{|\Omega_{i,j}|}{\Delta t^*_{i,j}} + \frac{|\Omega_{i+1,j}|}{\Delta t^*_{i+1,j}}\right),$$

with $\Delta t^*_{i,j}$ denoting an estimate of the time step for cell $\Omega_{i,j}$ with unit Courant number, and Δ_i being the forward difference in x_1-direction, i.e.

$$\Delta_i\overline{U}_{i,j} = \overline{U}_{i+1,j} - \overline{U}_{i,j}.$$

The symbol Δ_i^3 denotes the third power of the operator Δ_i, i.e.

$$\begin{aligned}\Delta_i^3\overline{U}_{i-1,j} &= \Delta_i^2(\Delta_i\overline{U}_{i-1,j}) = \Delta_i(\Delta_i\overline{U}_{i,j} - \Delta_i\overline{U}_{i-1,j})\\ &= \Delta_i(\overline{U}_{i+1,j} - \overline{U}_{i,j} - \overline{U}_{i,j} + \overline{U}_{i-1,j})\\ &= \overline{U}_{i+2,j} - \overline{U}_{i+1,j} - 2(\overline{U}_{i+1,j} - \overline{U}_{i,j}) + \overline{U}_{i,j} - \overline{U}_{i-1,j}\\ &= \overline{U}_{i+2,j} - 3\overline{U}_{i+1,j} + 3\overline{U}_{i,j} - \overline{U}_{i-1,j}.\end{aligned}$$

The formulae for $D_{i,j+1/2}$, etc., are analogously constructed.

Since first and third differences are used here the overall artificial dissipation corresponds to second and fourth spatial derivatives. Up to this point the FVM was constructed without reference to the system of conservation laws discretised, i.e. it can be used for any other hyperbolic systems. It is in the computation of the dissipation coefficients $\varepsilon^{(2)}_{i+1/2,j}$ and $\varepsilon^{(4)}_{i+1/2,j}$ that reference is made to the Euler system. Since some background dissipation is necessary anyway but additional dissipation is required in the vicinity of shocks the definition of the dissipation coefficients is

$$\varepsilon^{(2)}_{i+1/2,j} = k^{(2)}\max\left\{\nu_{i+2,j}, \nu_{i+1,j}, \nu_{i,j}, \nu_{i-1,j}\right\},$$

$$\varepsilon^{(4)}_{i+1/2,j} = \max\left\{0, k^{(4)} - \varepsilon^{(2)}_{i+1/2,j}\right\}$$

where

$$\nu_{i,j} = \frac{|p_{i+1,j} - 2p_{i,j} + p_{i-1,j}|}{|p_{i+1,j} + 2p_{i,j} + p_{i-1,j}|}$$

and $k^{(2)}$, $k^{(4)}$ are suitably chosen constants. Hence, $\varepsilon^{(2)}_{i+1/2,j}$ is proportional to a normalised second difference of the pressure.

We do not go into the details of implementing boundary conditions here. Details can be found in the literature given.

Concerning the order of the overall scheme one can state the following facts: In case of regular and continuously changing grids and smooth solutions the FVM is approximately second-order accurate. If the grid is regular and the solution contains a shock then the FVM degenerates to first order in the vicinity of the discontinuity due to the dissipation controlled by $\varepsilon^{(2)}_{i+1/2,j}$. But even if the solution is smooth the overall accuracy may drop down to first order or the scheme may even fail to be consistent at all, although reports on such severe failures are not known. The reason for this behaviour lies in the fact that in nonregular grids the location of the fluxes may no longer be close to the intersection point of edge and the connecting line of the cell centres.

The consistency problems of the cell-centred FVM of the type described above was analysed in the PhD thesis (Rossow, 1988), and a cell-vertex scheme was shown to remedy the situation.

6 CELL-VERTEX SCHEMES ON QUADRILATERAL GRIDS

Early cell-vertex FVMs on quadrilateral grids were proposed by Ni (1981) and Hall (1985) for applications in gas dynamics. In 1974 a simple scheme of this type was already analysed by Gushchin and Shchennikov (1974) for model equations. Later, a rigorous mathematical analysis followed from the group of Morton in Oxford (see Morton, 1996, p. 258ff.).

The flux variables are now associated with the four edge points of each cell. Hence $U_{i,j}$ now is the approximate value of the conservative variables *in the cell vertex* (i, j) and *not* within the cell $\Omega_{i,j}$. These four cell-vertex values can be used to give values at the midpoints of the cell via

$$U_{i,j-1/2} = \frac{1}{2}\left(U_{i,j} + U_{i+1,j}\right),$$

$$U_{i+1/2,j} = \frac{1}{2}\left(U_{i+1,j} + U_{i+1,j+1}\right),$$

$$U_{i,j+1/2} = \frac{1}{2}\left(U_{i+1,j+1} + U_{i,j+1}\right),$$

$$U_{i-1/2,j} = \frac{1}{2}\left(U_{i,j+1} + U_{i,j}\right),$$

and fluxes can be computed at the cell interfaces by means of

$$\mathcal{F}(u_{i,j-1/2}) \approx \mathcal{F}(U_{i,j-1/2}),$$

etc. Analogously to (7) we get the balance of the fluxes

$$\begin{aligned}\Phi_{i,j} = -\frac{1}{|\Omega_{i,j}|}\Bigg\{ & \mathcal{F}\left(\frac{1}{2}(U_{i,j}+U_{i,j-1})\right)\begin{pmatrix} x_{2,(i+1,j)} - x_{2,(i,j)} \\ x_{1,(i,j)} - x_{1,(i+1,j)} \end{pmatrix} \\ & + \mathcal{F}\left(\frac{1}{2}(U_{i,j}+U_{i+1,j})\right)\begin{pmatrix} x_{2,(i+1,j+1)} - x_{2,(i+1,j)} \\ x_{1,(i+1,j)} - x_{1,(i+1,j+1)} \end{pmatrix} \\ & + \mathcal{F}\left(\frac{1}{2}(U_{i,j}+U_{i,j+1})\right)\begin{pmatrix} x_{2,(i,j+1)} - x_{2,(i+1,j+1)} \\ x_{1,(i+1,j+1)} - x_{1,(i,j+1)} \end{pmatrix} \\ & + \mathcal{F}\left(\frac{1}{2}(U_{i,j}+U_{i-1,j})\right)\begin{pmatrix} x_{2,(i,j)} - x_{2,(i,j+1)} \\ x_{1,(i,j+1)} - x_{1,(i,j)} \end{pmatrix}\Bigg\}\end{aligned} \tag{9}$$

for every cell $\Omega_{i,j}$, and the flux balances of the four cells $\Omega_{i,j}$, $\Omega_{i,j-1}$, $\Omega_{i-1,\,j-1}$ and $\Omega_{i-1,j}$ sharing vertex $(i,\,j)$ (cp. Fig. 5) need now to be projected to this vertex. It is this step in which classical cell-vertex FVMs differ. Following Ni (1981) a simple arithmetic mean is used to give

$$\frac{d}{dt}U_{i,j} = \frac{1}{4}\{\Phi_{i,j} + \Phi_{i,j-1} + \Phi_{i-1,j-1} + \Phi_{i-1,j}\} \tag{10}$$

Hall (1985), however, used a mean with a weighting according to the volume of the contributing cells and even upwind techniques could be implemented (cp. Morton, 1996, p. 274ff.).

We shall not go into more detail concerning classical cell-vertex FVMs here but only remark that for the time stepping in (10) usually a classical Runge–Kutta scheme is used. Additionally, since the basic discretisation is based on central differencing, an artificial viscosity is needed as in the case of the cell-centred FVM. For more details on implementation, in particular, boundary condition, see Ni (1981) or Rossow (1988).

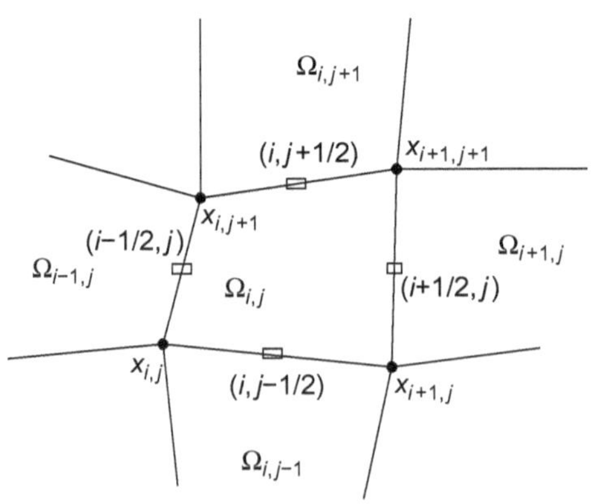

FIG. 5 The cell-vertex grid around cell $\Omega_{i,j}$.

7 FINITE VOLUME METHODS ON UNSTRUCTURED GRIDS

The 1980s saw a tremendous development in the numerical analysis of discretisation methods for hyperbolic conservation laws. Moving away from the classical idea of using central differences augmented by artificial dissipation researchers like Harten, Osher, Tadmor and many others came up with numerical techniques in which no 'fudge factors' like $k^{(2)}$ and $k^{(4)}$ in the construction of the artificial dissipation terms were needed at all (cp. Godlewski and Raviart, 1991 for an overview).

Very soon these techniques, namely the use of robust approximate Riemann solvers in connection with TVD- or ENO-techniques, were taken from their native finite difference background and incorporated into FVMs on more flexible triangular grids. We follow the representation given in Morton and Sonar (2007).

For the sake of simplicity, we consider gas dynamics in a bounded open domain $\Omega \subset \mathbb{R}^2$ where the boundary $\partial\Omega = \overline{\Omega}\backslash\Omega$ is already polygonal. A *triangulation* $\mathcal{T}^h$ of $\overline{\Omega}$ is a set of triangles $T_i \subset \overline{\Omega}$, $i = 1, \ldots, \#T$, such that

- $\overline{\Omega} = \cup_{i\in\{1,\ldots,\#T\}} T_1$,
- every $T_i \in \mathcal{T}^h$ is closed and nonempty,
- for any two $T_i, T_j \in \mathcal{T}^h$ with $i \neq j$ it holds that $\mathring{T}_i \cap \mathring{T}_i = \varnothing$.

We shall not consider triangulations in full generality but confine ourselves to more regular triangular tessellations. We call a triangulation *conforming* if the following additional condition is satisfied:

- every edge of any $T_i \in \mathcal{T}^h$ is either a subset of $\partial\Omega$ or the edge of another T_j, $j \neq i$.

Conformity ensures that the overall grid does not contain so-called hanging nodes, i.e. vertices of one triangle lying within an edge of another triangle.

7.1 Cell-Centred Finite Volume Methods

We start with the description of a very basic cell-centred FVM (Fig. 6).

Considering the hyperbolic system

$$\partial_t u + \partial_{x_1} f_1(u) + \partial_{x_2} f_2(u) = \partial_t u + \nabla \cdot \mathcal{F}(u) = 0$$

and its weak finite volume form stemming from integrating over a triangle T_i,

$$\frac{d}{dt}\overline{U}_{T_i} = -\frac{1}{|T_i|}\oint_{\partial T_i} f_1(u)n_1 + f_2(u)n_2\, ds. \tag{11}$$

Here, $\overline{U}_{T_i}$ denotes the cell average $\overline{U}_{T_i} = \frac{1}{|T_i|}\int_{T_i} u dx_1\, dx_2$, $n = (n_1, n_2)$ is the outwards unit normal vector to ∂T_i, and $|T_i|$ denotes the area of triangle T_i. The question now remains is the way of approximating the right-hand side of (11).

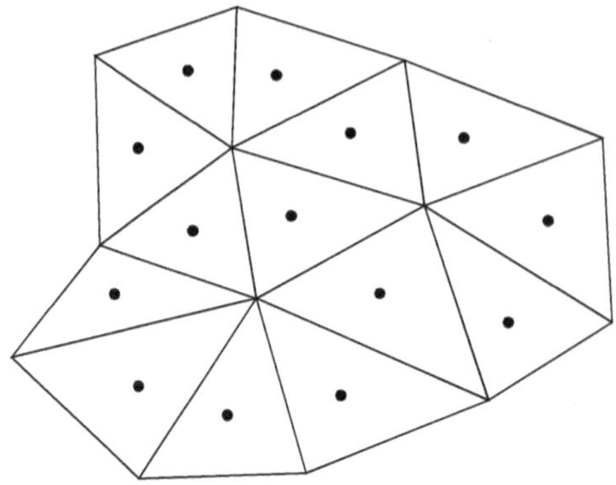

FIG. 6 The cell-centred triangular grid.

To this end, consider the index set

$$N(i) := \{j \in \mathbb{N} \mid T_i \cap T_j \text{ is an edge of } T_i\}$$

of the triangles neighbouring T_i, so that (11) can be written as

$$\frac{d}{dt}\overline{U}_{T_i} = -\frac{1}{|T_i|}\sum_{j\in N(i)}\int_{\partial T_i\cap\partial T_j}\sum_{k=1}^{2} f_k(u)n_{ij,k}\,ds, \tag{12}$$

$n_{ij} = (n_{ij,1}, n_{ij,2})$ being the unit normal vector to $\partial T_i \cap \partial T_j$ and pointing outwards with respect to T_i.

The crucial step now is to choose a *numerical Riemann solver* or numerical flux function

$$(u_i, u_j; n) \mapsto H(u_i, u_j; n)$$

which is consistent with the given fluxes in the sense that

$$H(u, u; n) = \sum_{k=1}^{2} f_k(u)n_k = \mathcal{F}(u)\cdot n.$$

Choosing central differences $H(u_i, u_j; n) = \mathcal{F}\left(\frac{1}{2}(u_i + u_j)\right)\cdot n$ would lead us directly to the triangular analogue of the cell-centred FVM (7), and we would come up with an unstable FVM needing artificial viscosity. Instead, we choose a robust first-order approximate Riemann solver like the numerical flux functions of Lax–Friedrichs, Harten–Lax–van Leer, Roe, Osher–Solomon or any other of those available in the literature (cp. Toro, 1999). A very simple, robust example can be found in the so-called local Lax–Friedrichs flux for the Euler system of gas dynamism

$$H^{\mathrm{lLF}}(u_i, u_j; n) := \frac{1}{2}((f(u_i) + f(u_j))\cdot n - \phi(u_j - u_i))$$

where ϕ is a locally varying scalar of the form

$$\phi = \phi(u_i, u_j) = \max_{w \in \left\{u_i, u_j, \frac{u_i + u_j}{2}\right\}} |v(w) \cdot n| + a(w),$$

with $v = (v_1, v_2)^T$ being the flow velocity and $a = \sqrt{\kappa \frac{p}{\rho}}$ denoting the speed of sound. It is also possible to use certain matrices for ϕ but we will not go into any detail here.

Approximating the line integral in (12) in the simplest possible manner we hence arrive at

$$\frac{d}{dt}\overline{U}_{T_i} = -\frac{1}{|T_i|}\sum_{j \in N(i)} |\partial T_i \cap \partial T_j| H(\overline{U}_{T_i}, \overline{U}_{T_j}; n_{ij}), \tag{13}$$

where n_{ij} denotes the unit normal vector at edge $\partial T_i \cap \partial T_j$, pointing outwards with respect to T_i. The FVM (13) is a very basic, easy to implement, first-order accurate, cell-centred FVM for which the initial data is given by

$$\overline{U}_{T_i}(0) = \frac{1}{|T_i|}\int_{T_i} u(x,0)dx = \frac{1}{|T_i|}\int_{T_i} u_0(x)dx.$$

Had we introduced Gauss points on the edges $\partial T_i \cap \partial T_j$ we could have achieved much higher accuracy of our cell-centred FVM. To this end consider the transformation of the edge $\partial T_i \cap \partial T_j$ with end points x_i and x_j to $[-1, 1]$ and introduce the parametrisation

$$[-1,1] \ni s \mapsto x_{ij}(s) = \frac{x_i + x_j + s(x_j - x_i)}{2}$$

of the edge. Then (12) yields

$$\frac{d}{dt}\overline{U}_{T_i} = -\frac{1}{|T_i|}\sum_{j \in N(i)} \frac{|\partial T_i \cap T_j|}{2} \int_{-1}^{1} \sum_{k=1}^{2} f_k(u(x_{ij}(s),t)) n_{ij,k}\, ds.$$

Denoting the number of Gauss points on each edge by n_G, let the position of these points be $x_{ij}(s_\nu), \nu = 1, \ldots, n_G$ and the quadrature weights by ω_ν, it follows

$$\frac{d}{dt}\overline{U}_{T_i} = -\frac{1}{|T_i|}\sum_{j \in N(i)} \frac{|\partial T_i \cap T_j|}{2} \left\{ \sum_{\nu=1}^{n_G} \omega_\nu \sum_{k=1}^{2} f_k(u(x_{ij}(s_\nu),t)) n_{ij,k} + \mathcal{O}(h^{2n_G}) \right\}, \tag{14}$$

h denoting a typical grid cell measure. The problem with our formulation now, however, is the assumption that we know the exact solution u at the Gauss points, which we do not know, of course. Here another important ingredient of FVMs come into play—the recovery process (Morton and Sonar, 2007).

Sonar (1997, Morton and Sonar, 2007, p. 207 ff.) was the first to apply the theory of optimal recovery to the recovery process in FVMs and to exploit the mathematical ideas behind the recovery of polynomials and radial splines from given cell averages. We will not go into detail here but conclude our outline of higher-order accurate FVMs by the following remarks. Imagine we had recovered polynomials p_i and p_j on T_i and T_j, respectively, from the cell averages of surrounding triangles. Then we can employ an approximate Riemann solver H to result in the cell-centred FVM

$$\frac{d}{dt}\overline{U}_{T_i} = -\frac{1}{|T_i|}\sum_{j\in N(i)} \frac{|\partial T_i \cap T_j|}{2}\left\{\sum_{\nu=1}^{n_G} \omega_\nu H(p_i(x_{ij}(s_\nu),t), p_j(x_{ij}(s_\nu),t); n_{ij})\right\}. \quad (15)$$

It is easily seen that if our recovery polynomials fulfil

$$H(p_i(x_{ij}(s_\nu),t), p_j(x_{ij}(s_\nu),t); n_{ij}) = \sum_{k=1}^{s} f_k(u(x_{ij}(s_\nu,t))n_{ij,k} + \mathcal{O}(h^r),$$

then the resulting FVM has spatial accuracy order of $\min\{r, 2n_G\}$.

7.2 Vertex-Centred Finite Volume Methods

Many different names are in use for the vertex-centred FVMs which we will describe now: box methods, secondary grid methods, etc.

The main idea is to use a conforming triangular grid but consider boxes around each of the nodes of the triangulation as control volumes. There are some different geometrical constructions of boxes in use; we use here boxes which piecewise consist of the straight lines connecting the barycentre of a triangle with the midpoint of its edges. Hence, the boundary between neighbouring boxes B_i and B_j consists of the two straight line pieces l_{ij}^1 and l_{ij}^2 with two unit normal vectors n_{ij}^1 and n_{ij}^2 as shown in Fig. 8.

Introducing the index set

$$N(i) := \{j \in \mathbb{N} \mid B_i \cap B_j \text{ is an edge of } B_i\}$$

and employing the notation of Fig. 7 we get the analogue of Eq. (12) in the form

$$\frac{d}{dt}\overline{U}_{B_i} = -\frac{1}{|B_i|}\sum_{j\in N(i)}\sum_{m=1}^{2}\int_{l_{ij}^m}\sum_{k=1}^{2} f_k(u)n_{ij,k}^m\, ds, \quad (16)$$

where the cell average is taken over a volume B_i, of course. Introducing an approximate Riemann solver as before leads to the basic vertex-centred FVM

$$\frac{d}{dt}\overline{U}_{B_i} = -\frac{1}{|B_i|}\sum_{j\in N(i)}\sum_{m=1}^{2}|l_{ij}^m| H(\overline{U}_{B_i}(t), \overline{U}_{B_j}(t); n_{ij}^m), \quad (17)$$

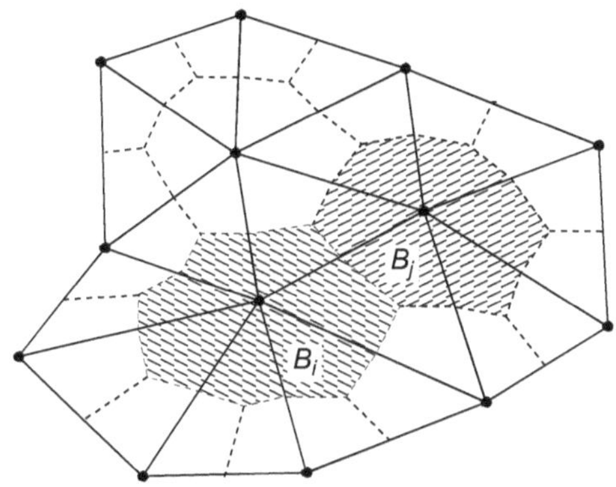

FIG. 7 The vertex-centred triangular grid.

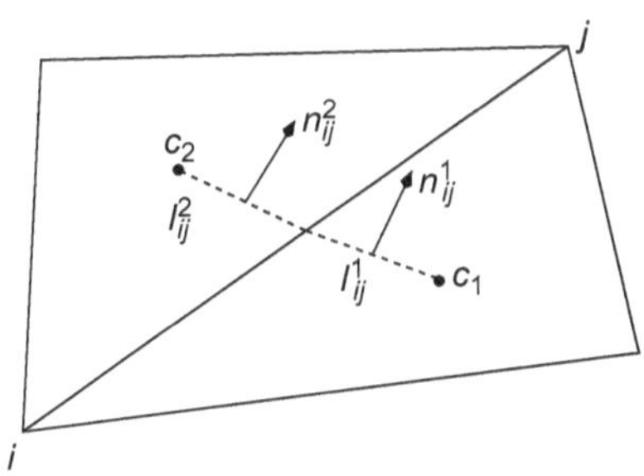

FIG. 8 The boundary between the boxes B_i and B_j.

which is a robust, first-order scheme. Introducing a parametrisation on the edge segment l_{ij}^m via

$$[-1,2] \ni s \mapsto x_{ij}^m(s) = \frac{1}{2}\left(c_m + \frac{x_i + x_j}{2}\right) + \frac{s}{2}\left(\frac{x_i + x_j}{2} - c_m\right),$$

where c_1 and c_2 denote the barycentres according to Fig. 8. If we now apply to (16) a Gaussian quadrature rule on the edge l_{ij}^m we get

$$\frac{d}{dt}\overline{U}_{B_i} = -\frac{1}{|B_i|}\sum_{j \in N(i)}\sum_{m=1}^{2}\frac{|l_{ij}^m|}{2}\left\{\sum_{v=1}^{n_G}\sum_{k=1}^{2}\omega_v^m f_k(u(x_{ij}^m(s_v),t))n_{ij,k}^m + \mathcal{O}(h^{2n_G})\right\},$$

where, as in the case of the cell-centred FVM, ω_v^m denote the weights of the quadrature rule and s_v the parameter values of the Gauss points.

If we succeed in finding recovery polynomials p_k on the boxes and if we employ an approximate Riemann solver, we find a vertex-centred FVM to be

$$\frac{d}{dt}\overline{U}_{B_i} = -\frac{1}{|B_i|}\sum_{j \in N(i)}\sum_{m=1}^{2}\frac{|l_{ij}^m|}{2}\omega_v^m H(p_i(x_{ij}^m(s_v),t), p_j(x_{ij}^m(s_v),t; n_{ij}^m). \qquad (18)$$

The FVM is of spatial order $\min\{r, 2n_G\}$ if

$$H(p_i(x_{ij}^m(s_\nu),t), p_j(x_{ij}^m(s_\nu),t; n_{ij}^m) = \sum_{k=1}^{2} f_k(u(x_{ij}^m(s_\nu),t))n_{ij,k}^m + \mathcal{O}(h^r).$$

7.3 Remarks on Recovery

We will not go into the details of the theory of (optimal) recovery from cell averages, which were described in Abgrall and Sonar (1997), Iske and Sonar (1996) and Sonar (1997, 1998), but give an instructive example. In the cell-centred method (15) we consider a neighbourhood of triangle T_i as shown in Fig. 9. Durlofsky et al. (1992) proposed a linear recovery in which the three stencils

$$K_1(T_i) := \{T_i, T_{i_1}, T_{i_2}\}, \quad K_2(T_i) := \{T_i, T_{i_1}, T_{i_3}\}, \quad K_1(T_i) := \{T_i, T_{i_2}, T_{i_2}\},$$

give rise to three recovery polynomials

$$p_i^{(k)} = a_{00}^{(k)} + a_{10}^{(k)}(x_1 - c_{i,1}) + a_{01}^{(k)}(x_2 - c_{i,2}), \quad k = 1,2,3$$

for triangle T_i, where $c_i = (c_{i,1}, c_{i,2})$ denotes the coordinates of the barycentre of triangle T_i. The coefficients of the three polynomials are computed via three linear system

$$\frac{1}{|T_j|} p_i^{(k)} = \overline{U}_{T_j},$$

for $(k, j) \in \{(1, i), (2, i_1), (3, i_2)\}$, $(k, j) \in \{(1, i), (2, i_1), (3, i_3)\}$ and $(k, j) \in \{(1, i), (2, i_2), (3, i_3)\}$, respectively. Having computed three possible linear polynomials we now have to choose the one which will serve as a recovery polynomial. This is done in Durlofsky et al. (1992) in a TVD-like manner as follows.

If the cell average on T_i is maximal in comparison with the three cell averages on T_{i_k}, $k = 1, 2, 3$, then no polynomial is chosen at all and the

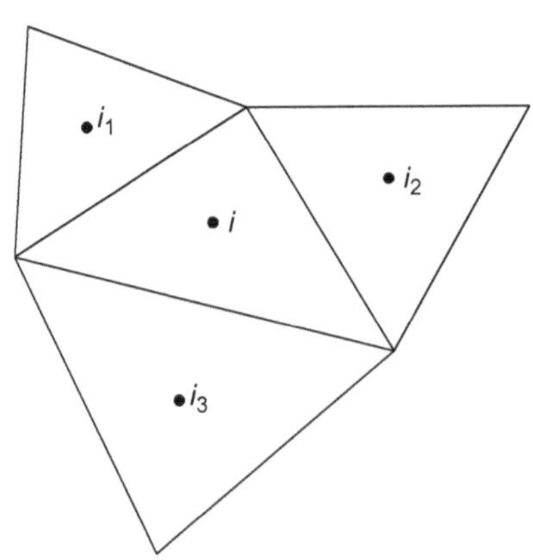

FIG. 9 A neighbourhood of triangle T_i.

computation proceed with cell averages; i.e. the FVM is first-order accurate in space. Otherwise, the most steepest polynomial $\left(\max_k |\nabla p_i^{(k)}|\right)$ is chosen and then tested, whether the values of this polynomial at the midpoints of the three edges of T_i lie between the cell averages on T_i and its three neighbours. If this is the case, the recovery polynomial is found. If the test fails, the next steepest polynomial is chosen from the set of three and tested again. If the test fails again, the remaining polynomial is chosen and tested, and if our test fails again the computation proceeds with the piecewise constant cell averages.

The Durlofsky–Engquist–Osher recovery hence tries to find the steepest recovery polynomial available but reduces the spatial order to one at extrema, as every TVD method does.

A simple ENO recovery can be described by choosing the polynomial $p_i \in \{p_i^{(1)}, p_i^{(2)}, p_i^{(3)}\}$ which is the less steepest, i.e. which satisfies $|\nabla p_i| = \min_k |\nabla p_i^{(k)}|$. The remaining FVM might show some small oscillations but turns out to be a very robust and reliable second-order scheme.

Similar algorithms for quadratic or even higher degree polynomials can be constructed along the lines described, and this is of course possible also for vertex-centred FVMs. The complication then arise from the algorithms to choose different stencils in the neighbourhood of a triangle or box. Besides TVD and ENO FVMs also WENO-type methods can be considered, of course. Since very wide stencils introduce stability problems other techniques like the subcell reconstruction may be applied but this would leave the realm of *classical* FVMs.

REFERENCES

Abgrall, R., Sonar, T., 1997. On the use of Mühlbach expansions in the recovery step of ENO methods. Num. Math. 76, 1–25.

Ansorge, R., Sonar, T., 2009. Mathematical Models of Fluid Dynamics—Modelling,Theory, Basic Numerical Facts. An Introduction. Wiley-VCH, Weinheim.

Bruhn, G., 1985. Erhaltungssätze und schwache Lösungen in der Gasdynamik. Math. Methods Appl. Sci. 7, 470–479.

Dolejši, V., Feistauer, M., 2015. Discontinuous Galerkin Method—Analysis and Applications to Compressible Flow. Springer, Cham, Heidelberg, New York.

Durlofsky, L.J., Engquist, B., Osher, S., 1992. Triangle based adaptive stencils for the solution of hyperbolic conservation laws. J. Comp. Phys. 98, 64–73.

Elling, V., 2007. A Lax-Wendroff type theorem for unstructured quasi-uniform grids. Math. Comp. 76 (257), 251–272.

Godlewski, E., Raviart, P.-A., 1991. Hyperbolic Systems of Conservation Laws. Mathematiques & Applications, Publication trimestrielle numéro 3/4, Ellipses, Paris.

Godlewski, E., Raviart, P.-A., 1996. Numerical Approximation of Hyperbolic Systems of Conservation Laws. Springer, New York, Berlin, Heidelberg.

Gushchin, V.A., Shchennikov, V.V., 1974. A monotonic difference scheme of second-order accuracy. U.S.S.R. Comput. Math. Math. Phys. 14, 252–256.

Hall, M.G., 1985. Cell-vertex multigrid scheme for the solution of the Euler equations. In: Proceedings of the Conference on Numerical Methods for Fluid Dynamics, Reading. Clarendon Press, Oxford.
Heinrich, B., 1987. Finite Difference Methods on Irregular Networks—A Generalized Approach to Second Order Elliptic Problems. Birkhäuser Verlag, Basel, Boston, Stuttgart.
Holt, M., 1984. Numerical Methods in Fluid Dynamics. Springer-Verlag, Berlin, Heidelberg.
Hou, T.Y., LeFloch, P.G., 1994. Why nonconservative schemes converge to wrong solutions: error analysis. Math. Comp. 62, 497–530.
Iske, A., Sonar, T., 1996. On the structure of function spaces in optimal recovery of point data for ENO-schemes by radial basis functions. Num. Math. 74, 177–201.
Jameson, A., Schmidt, W., Turkel, E., 1981. Numerical solutions of the Euler equations by finite volume methods using Runge Kutta time stepping schemes. AIAA paper 81-1259.
Jovanović, B.S., Süli, E., 2014. Analysis of Finite Difference Schemes—For Linear Partial Differential Equations with Generalized Solutions. Springer-Verlag, London.
Klötzler, R., 1970. Mehrdimensionale Variationsrechnung. Birkhäuser Verlag, Basel.
Kroll, N., Jain, R.K., 1987. Solution of Two-Dimensional Euler Equations—Experiences with a Finite Volume Code. DFVLR, Germany. DFVLR-Forschungsbericht 87-41.
Kröner, D., Rokyta, M., Wierse, M., 1996. A Lax-Wendroff type theorem for upwind finite volume schemes in 2-D. East-West J. Numer. Math. 4, 279–292.
Lax, P.D., Wendroff, B., 1960. Systems of conservation laws. Comm. Pure Appl. Math. 13 (2), 217–237.
MacNeal, R.H., 1953. An asymmetrical finite difference network. Q. Appl. Math. 11, 295–310.
McDonald, P.W., 1971. The Computation of Transonic Flow Through Two-Dimensional Gas Turbine Cascades. ASME Proc. Paper 71-GT-89, ASME, New York.
Morrey, C.B., 1960. Multiple integral problems in the calculus of variations and related topics. Ann. Scuola Norm Pisa (III) 14, 1–61.
Morton, K.W., 1996. Numerical Solution of Convection-Diffusion Problems. Chapman & Hall, London, Glasgow.
Morton, K.W., Sonar, T., 2007. Finite volume methods for hyperbolic conservation laws. Act. Num. 16, 155–238.
Ni, R.H., 1981. Multiple grid scheme for solving the Euler equations. AIAA Paper 81-1025.
Preissmann, A., 1961. Propagation des intumescences dans les canaux et rivières. In: 1st Congrès de l'Assoc. Française de Calc, AFCAL, Grenoble, pp. 433–442.
Richtmyer, R.D., Morton, K.W., 1967. Difference Methods for Initial-Value Problems. Interscience Publ., New York, Chichester.
Rizzi, A.W., Inouye, M., 1973. Time split finite volume method for three dimensional blunt-body flows. AIAA J. 11, 1478–1485.
Rossow, C.-C., 1988. Berechnung von Strömungsfeldern durch Lösung der Euler-Gleichungen mit einer erweiterten finite-volumen diskretisierungsmethode. Fakultät für Maschinebau und Elektrotechnik, Technische Universität Braunschweig (dissertation).
Sonar, T., 1997. Mehrdimensionale ENO-verfahren. B.G. Teubner, Stuttgart.
Sonar, T., 1998. On families of pointwise optimal finite volume ENO approximations. SIAM J. Num. Anal. 35 (6), 2350–2369.
Toro, E.F., 1999. Riemann Solvers and Numerical Methods for Fluid Dynamics—An Introduction. Springer, Berlin, Heidelberg.
Varga, R.S., 1962. Matrix Iterative Analysis. Prentice-Hall Int, London.
von Neumann, J., Richtmyer, R.D., 1950. A method for the numerical calculations of hydrodynamical shocks. J. Appl. Phys. 21, 232–237.

Chapter 4

Sharpening Methods for Finite Volume Schemes

B. Després*, S. Kokh[†,‡] and F. Lagoutière[§]
*Sorbonne Universités, UPMC Univ Paris 06, UMR 7598, Laboratoire Jacques-Louis Lions, Paris, France
[†]Maison de la Simulation USR 3441, Digiteo Labs, bât. 565, PC 190, CEA Saclay, Gif-sur-Yvette, France
[‡]DEN/DANS/DM2S/STMF/LMSF, CEA Saclay, Gif-sur-Yvette, France
[§]Laboratoire de Mathématiques d'Orsay, Univ. Paris-Sud, CNRS, Université Paris-Saclay, Orsay, France

Chapter Outline

ABSTRACT

We review sharpening methods for finite volume schemes, with an emphasis on the basic structure of sharpening methods. It covers high-order methods and nonlinear techniques for linear advection, Glimm's method, antidiffusion techniques, and the interaction of these techniques with the PDE structures. Additional approaches like level sets, interface reconstruction, and Vofire are also discussed. We also present the algorithmic structure of the downwind method for a simple two components problem.

Handbook of Numerical Analysis, Vol. 17. http://dx.doi.org/10.1016/bs.hna.2016.09.016

Keywords: Sharpening methods, Finite volume schemes, Antidiffusion, Interface tracking

AMS Classification Codes: 65-02, 65-M08

1 INTRODUCTION

The present chapter deals with sharpening methods for finite volume (FV) schemes understood as discretization strategies for the enhancement of sharp profiles in numerical simulations. We restrict the scope to finite volume schemes since they are the numerical method of choice for compressible computation fluid dynamics where the exact or approximate solutions may exhibit strong gradients that account for shocks or contact discontinuities.

We will more specifically focus on the calculations of interfaces associated with linearly degenerate fields (contact discontinuities), material contact discontinuities, or free boundaries that are tracked across the computational domain. Although we shall consider numerical methods that are compliant with shock capturing, we shall not discuss the approximations of shocks in this chapter.

Interface tracking has motivated a considerable amount of contributions since the early days of scientific computing and numerical analysis. Therefore reviewing exhaustively all the methods that have been published to date seems quite unrealistic and we apologize in advance to the community for all the works that will not be mentioned in the sequel. We propose to sketch a map of these methods by relying on mathematical and algorithmic arguments that can be used to analyze the efficiency. We hope that this effort may also help classifying the methods that will not be discussed in this document and help understanding the sharpening mechanisms at play within the numerical schemes that are available in the literature.

The chapter is organized as follows. Most of the common material (that is high-order and nonlinear techniques, the Glimm's scheme, the notion of antidiffusion, level sets, multidimensional issues) is presented for linear equations in Section 2. The introduction of sharpening methods in nonlinear systems is evoked in Section 3. References are provided inside the text.

2 SHARPENING METHODS FOR LINEAR EQUATIONS

Sharpening methods for linear equations use two important generic ideas: the first one is to use high-order schemes, and it may seem paradoxical at first sight; the second idea is based on compression with nonlinear techniques; other strategies rely on the Glimm's scheme, on PDEs to represent the interface, or reconstruct locally as in the volume of fluid (VOF) method. Most of the ideas can be presented on the advection equation with velocity $u \in \mathbb{R}$, which serves as a model problem. It writes

$$\partial_t c(t,x) + u\partial_x c(t,x) = 0, \quad x \in \mathbb{R}, \quad t > 0,$$

together with a Cauchy datum $c(0, x) = c^{\text{ini}}(x)$.

2.1 High-Order Methods

References to high-order discretization of nonlinear equations are Toro (2009), Russo et al. (2008), Titarev and Toro (2007), and Toro and Titarev (2005). The fact that high-order methods have the ability to sharpen discontinuities is kind of a paradox. Indeed, local Taylor expansions show bad convergence behaviour for profiles involving discontinuities or strong gradients.

We give hereafter a simple explanation of the corresponding sharpening based on the theory of linear Strang's stencils. Let $\Delta t > 0$ and Δx be respectively the time and space steps. We consider a series of instants $t^n = n\Delta t$ and the classical discretization of the real line into intervals $[x_{j-1/2}, x_{j+1/2}]$, whereby $x_j = j\Delta x$ and $x_{j+1/2} = (j + 1/2)\Delta x$. We note c_j^n an approximation of c at instant t^n within the cell $[x_{j-1/2}, x_{j+1/2}]$ and set $c^n = (c_j^n)$. The initial numerical datum can be taken as $c_j^0 = c^{ini}(x_j)$ (this is especially done when dealing with smooth solutions and high-order methods) or $c_j^0 = \int_{x_{j-1/2}}^{x_{j+1/2}} c^{ini}(x)\, dx/\Delta x$ (usually when dealing with nonsmooth data).

The analysis is here limited to explicit and compact schemes with a stencil of $p + 1$ contiguous cells. In a simplified finite difference form on a Cartesian grid, the family of linear schemes may read

$$c_j^{n+1} = \sum_{r=k-p}^{k} \alpha_r c_{j+r}^n, \qquad \alpha_r = \alpha_r(\nu). \tag{1}$$

The coefficients α_r are functions of the Courant–Friedrichs–Lewy (CFL) number $\nu - u\Delta t/\Delta x$. It is possible to write a scheme with order p in time and space using (1). Once p has been chosen, k determines the shift of the scheme. Basic examples are the well-known upwind scheme $c_j^{n+1} = (1-\nu)c_j^n + \nu c_{j-1}^n$, when $(p, k) = (1, 0)$, $\alpha_{-1} = \nu$ and $\alpha_0 = 1 - \nu$, the Lax–Wendroff scheme (Lax and Wendroff, 1960) $c_j^{n+1} = H^{\text{LW}}(c^n)_j = (1-\nu^2)c_j^n + \frac{\nu+\nu^2}{2}c_{j-1}^n + \frac{\nu^2-\nu}{2}c_{j+1}^n$, with $(p, k) = (2, 1)$, $\alpha_{-1} = (\nu^2 + \nu)/2$, $\alpha_0 = 1 - \nu^2$ and $\alpha_1 = (\nu^2 + \nu)/2$, and the Beam–Warming scheme (Warming and Beam, 1976): $c_j^{n+1} = H^{\text{BW}}(c^n)_j = \left(1 - \frac{3}{2}\nu + \frac{1}{2}\nu^2\right)c_j^n + (2\nu - \nu^2)c_{j-1}^n + \frac{\nu^2-\nu}{2}c_{j-2}^n$, with $(p, k) = (2, 0)$, $\alpha_{-2} = (\nu^2 - \nu)/2$, $\alpha_{-1} = 2\nu - \nu^2$ and $\alpha_0 = 1 - 3\nu/2 + \nu^2/2$. Under the hypothesis that $\sum_r \alpha_r = 1$, which is a natural assumption that ensures the conservativity of the algorithm, these schemes may be rewritten also as finite volume methods in their classical form

$$\frac{c_j^{n+1} - c_j^n}{\Delta t} + u\frac{c_{j+\frac{1}{2}}^n - c_{j-\frac{1}{2}}^n}{\Delta x} = 0. \tag{2}$$

The conversion between the two forms is let to the reader because it does not have impact on the following discussion. A third order in time and space $O3$ scheme $(p, k) = (3, 1)$ is defined by a convex combination (Després, 2008) of the Lax–Wendroff scheme and the Beam–Warming scheme: $c_j^{n+1} = (1-\alpha)H^{\mathrm{LW}}(c^n)_j + \alpha H^{\mathrm{BW}}(c^n)_j$ with $\alpha = \frac{1+\nu}{3}$. The seminal works of Iserles and Strang (Strang, 1968; Iserles and Strang, 1983) show that the order in time and space, p, can be arbitrary large. Nevertheless, *the only pairs (p, k) for which there exists schemes such that the l^2 norm is non-increasing at any iterate for all $\nu \leq 1$ are $p = 2k + 1$, $p = 2k$* and $p = 2k + 2$. In the following, such schemes are called IS-schemes.

The stability in L^1 of *IS*-schemes has been given in Després (2009): *Assume moreover the order is odd, that is $p = 2k + 1$. Then the scheme is stable in all L^q: there exists a constant $D_p > 0$ such that $\|c^n\|_{L^q} \leq D_p \|c^0\|_{L^q} \forall n, \forall \nu \in [0, 1], \forall c^0$ and $\forall q \in [1, \infty]$.*

Equipped with these fundamental results, a convergence result that provides a sharp convergence estimate for an initial datum with bounded variation (BV datum) can be stated (Després, 2008). The proof is done by regularization of the BV profile and use of the L^1 stability. In this result, c^n is to be understood as the constant by cell function that takes the value c_j^n in the cell number j, namely $[(j - 1/2)\Delta x, (j + 1/2)\Delta x]$.

Theorem 1. *Assume $c^{\mathrm{ini}} \in L^\infty \cap BV$ (in space dimension 1, this is just the* BV *space). Consider an IS-scheme, with $p = 2k + 1$ odd. Assume $\nu \leq 1$. Then*

$$\|c^n - c(n\Delta t)\|_{L^1} \leq C_p |c^{\mathrm{ini}}|_{BV} \left(\Delta x^a T^b + \Delta x\right) \tag{3}$$

with $a = \frac{p}{p+1}$ and $b = \frac{1}{p+1}$.

Here, as the estimate is for nonsmooth data and thus is of order less than 1, the initial numerical datum can be chosen both as point values or mean values.

Using very high-order schemes means choosing p very large. In this case $\frac{p}{p+1}$ is very close to 1. This is optimal because an error of order 1 is what we get by a 1 cell translation of the Heavyside function. In a nutshell: very high-odd-order advection schemes have nearly optimal order of convergence in L^1 even for discontinuous initial data. It means that the very high-order feature of such schemes is able to sharpen discrete profiles with strong gradients. Perhaps even more important for applications is the very small dependence with respect to the time T since $\frac{1}{p+1}$ is close to zero for large p. This means that the difference between the true solution and the numerical solution does not evolve significantly in time. That is the sharpening effect is time independent. This theoretical behaviour is the solution of the apparent paradox explained at the beginning of the section.

Nevertheless the drawbacks of these high-orders (linear) FV methods are that they do not satisfy the maximum principle, according to a well-known theorem by Godunov.

2.2 Compression Within a BV Setting

It is known since (Harten, 1984; Yee et al., 1984; le Roux, 1977) that the bounded variation (BV) setting is a convenient framework of the construction of numerical nonlinear FV schemes with good sharpening properties. For one-dimensional problems the BV setting is strongly related to the preservation of the maximum principle (Roe, 1985; Sweby, 1984, 1985). In some cases the sharpening effect is so pronounced that it is called squaring.

The general situation can be explained a follows. Consider the FV formulation

$$\frac{c_j^{n+1} - c_j^n}{\Delta t} + u\frac{c_{j+\frac{1}{2}}^n - c_{j-\frac{1}{2}}^n}{\Delta x} = 0, \quad u > 0. \tag{4}$$

Scheme (4) can be recast into

$$c_j^{n+1} = c_j^n - \nu\left(c_{j+\frac{1}{2}}^n - c_{j-\frac{1}{2}}^n\right), \quad \nu = u\frac{\Delta t}{\Delta x}. \tag{5}$$

The numerical fluxes $c_{j+\frac{1}{2}}^n$ are yet to be defined at this point of the construction. The design principle is to impose the maximum principle under the form

$$\min(c_j^n, c_{j-1}^n) \equiv m_{j-1/2}^n \leq c_j^{n+1} \leq M_{j-1/2}^n \equiv \max(c_j^n, c_{j-1}^n), \tag{6}$$

which is legitimate for advection to the right. If the advection is to the left ($u < 0$) one takes $\min(c_j^n, c_{j+1}^n) \equiv m_{j+1/2}^n \leq c_j^{n+1} \leq M_{j+1/2}^n \equiv \max(c_j^n, c_{j+1}^n)$. We consider the classical formula (see Sweby, 1984)

$$c_{j+\frac{1}{2}}^n = c_j^n + \frac{1}{2}(1-\nu)(c_{j+1}^n - c_j^n)\varphi_{j+\frac{1}{2}}^n, \quad \forall j. \tag{7}$$

where the correction factor $\varphi_{j+\frac{1}{2}}^n$ is a limiter or slope limiter. It is usually defined as a function of the local slope ratio

$$\varphi_{j+\frac{1}{2}}^n = \varphi\left(r_{j+\frac{1}{2}}^n\right), \quad r_{j+\frac{1}{2}}^n = \frac{c_j^n - c_{j-1}^n}{c_{j+1}^n - c_j^n}.$$

There are natural additional constraints for the definition of the slope limiter. A first one writes $\varphi(1) = 1$: it gives back the Lax–Wendroff flux in case $r = 1$, and, generally, the second order when the datum is smooth. A second constraint can be $\varphi(r) = 0$ for any $r \leq 0$: this is a way to enforce a local preservation of the maximum principle, this is explained in the classical textbook (Toro, 2009). Another idea could be to add $\varphi(r) = r\varphi\left(\frac{1}{r}\right)$ (Toro, 2009). There are variants where these conditions are relaxed, see for example Dubois and Mehlman (1996) and Schmidtmann et al. (2016).

Most of the formulas published in the literature make use of the minmod function. Its value is given as follows: if $ab \leq 0$ then minmod(a, b) = 0; if $a > 0$ and $b > 0$, then $\text{minmod}(a,b) = \min(a,b)$; if $a < 0$ and $b < 0$, then $\text{minmod}(a,b) = \max(a,b)$. Then the multidimensional function $\text{minmod} : \mathbb{R}^p \to \mathbb{R}$ is defined recursively for $p \geq 2$ independently of the ordering by

$$\text{minmod}(\mathbf{a}) = \text{minmod}(\text{minmod}(\mathbf{b}), c) \quad \text{for} \quad \mathbf{a} = (\mathbf{b}, c) \in \mathbb{R}^p, \quad \mathbf{b} \in \mathbb{R}^{p-1}, \quad c \in \mathbb{R}.$$

A first classical result is that if the slope limiter satisfies

$$0 \leq \varphi(r) \leq 2\text{minmod}(1, r) \tag{8}$$

then the scheme (4) with the flux (7) satisfies the maximum principle. Even if this is a very classical result, we provide a proof since it will explain how to modify (8) for deriving schemes with even stronger sharpening effect. One has

$$\begin{aligned} c_j^{n+1} &= c_j^n - \nu\left(c_j^n + \frac{1}{2}(1-\nu)(c_{j+1}^n - c_j^n)\varphi_{j+\frac{1}{2}}^n \right. \\ &\quad \left. - c_{j-1}^n - \frac{1}{2}(1-\nu)(c_j^n - c_{j-1}^n)\varphi_{j-\frac{1}{2}}^n\right) \\ &= c_j^n - \nu\left(1 + \frac{1}{2}(1-\nu)\left(\frac{\varphi_{j+\frac{1}{2}}^n}{r_{j+\frac{1}{2}}^n} - \varphi_{j-\frac{1}{2}}^n\right)\right)(c_j^n - c_{j-1}^n), \end{aligned}$$

that is $c_j^{n+1} = (1 - L_j^n)c_j^n + L_j^n c_{j-1}^n$, $L_j^n = \nu + \frac{\nu(1-\nu)}{2}\left(\frac{\varphi_{j+\frac{1}{2}}^n}{r_{j+\frac{1}{2}}^n} - \varphi_{j-\frac{1}{2}}^n\right)$. The maximum principle is satisfied provided $0 \leq L_j^n \leq 1$, that is

$$0 \leq \nu + \frac{\nu(1-\nu)}{2}\left(\frac{\varphi_{j+\frac{1}{2}}^n}{r_{j+\frac{1}{2}}^n} - \varphi_{j-\frac{1}{2}}^n\right) \leq 1.$$

Assume (8) holds. Then $0 \leq \varphi_{j-\frac{1}{2}}^n \leq 2$ and $1 - \frac{1-\nu}{2}\varphi_{j-\frac{1}{2}}^n \geq 1 - (1-\nu) \geq 0$, thus $0 \leq C_j^n$. One notices that (8) also yields $0 \leq \varphi_{j+\frac{1}{2}}^n \leq 2r_{j+\frac{1}{2}}^n$. Therefore $1 + \frac{1-\nu}{2}\frac{\varphi_{j+\frac{1}{2}}^n}{r_{j+\frac{1}{2}}^n} \leq 1 + (1-\nu) = 2 - \nu$. Finally

$$\nu+\frac{\nu(1-\nu)}{2}\frac{\varphi^n_{j+\frac{1}{2}}}{r^n_{j+\frac{1}{2}}}\le 2\nu-\nu^2\le 1,\quad \forall \nu\in[0,1],$$

which ends the proof.

A huge number of formulas has been proposed in the literature. We just review the most usual ones. The Minmod flux writes

$$\varphi(r)=\text{minmod}(1,r). \tag{9}$$

The Superbee flux writes

$$\varphi(r)=\max(0,\min(1,2r),\min(2,r)). \tag{10}$$

Remark 1. (Squaring/sharpening behaviour of Superbee).
The notion of sharpening is not present at this stage of the discussion. It is introduced by noticing that the SuperBee limiter is squaring. This has been reported in the literature in Toro (2009) and many other texts. Squaring means that if an initial smooth profile is chosen, for example in the form of a Gaussian, then the numerical solution has the tendency to converge to a mass preserving square profile for $t\to\infty$. This behaviour necessarily increases the L^2 norm of the profile.
Squaring is usually considered as a consequence of the strong nonlinearity of SuperBee. Even if it is a well-documented behaviour, we know of no definitive proof. But on the contrary, it is easy to understand that the minmod limiter cannot sharpen. To this end we consider the semidiscrete (that is continuous in time) version of the scheme

$$\frac{\mathrm{d}}{\mathrm{d}t}c_j(t)+u\frac{c_{j+\frac{1}{2}}(t)-c_{j-\frac{1}{2}}(t)}{\Delta x}=0,\quad j\in\mathbb{Z}. \tag{11}$$

Since Δt vanishes, the flux (7) is simplified taking $\nu=0$.
Lemma 1. *The semidiscrete scheme (11) with the flux $c_{j+\frac{1}{2}}=c_j+\frac{1}{2}(c_{j+1}-c_j)\varphi_{j+\frac{1}{2}}$ and the Minmod limiter (9) satisfies the a priori estimate*

$$\frac{d}{dt}\left(\sum_{j\in\mathbb{Z}}|c_j(t)|^2\right)\le 0. \tag{12}$$

So, as a corollary of Remark 1, this scheme cannot sharpen.
The same property holds for similar schemes with a limiter $0\le\varphi(r)\le 1$ for all r. The proof proceeds as follows. One has

$$\frac{\Delta x}{2}\frac{\mathrm{d}}{\mathrm{d}t}\left(\sum_{j\in\mathbb{Z}}|c_j|^2\right)=\Delta x\sum_j c_j\frac{\mathrm{d}}{\mathrm{d}t}c_j=-u\sum_j c_j\left(c_{j+\frac{1}{2}}-c_{j-\frac{1}{2}}\right)$$

$$=-u\sum_j c_j(c_j-c_{j-1})-u\sum_j c_j\left(c_{j+\frac{1}{2}}-c_j\right)+u\sum_j c_j\left(c_{j-\frac{1}{2}}-c_{j-1}\right).$$

It is easy to check the identities

$$\begin{cases} \sum_j c_j(c_j - c_{j-1}) = \frac{1}{2}\sum_j |c_j - c_{j-1}|^2, \\ \sum_j c_j\left(c_{j+\frac{1}{2}} - c_j\right) = \frac{1}{2}\sum_j c_j(c_{j+1} - c_j)\varphi_{j+\frac{1}{2}}, \\ \qquad = \frac{1}{2}\sum_j c_{j-1}(c_j - c_{j-1})\varphi_{j-\frac{1}{2}}, \\ \sum_j c_j\left(c_{j-\frac{1}{2}} - c_{j-1}\right) = \frac{1}{2}\sum_j c_j(c_j - c_{j-1})\varphi_{j-\frac{1}{2}}. \end{cases}$$

Therefore by summation and rearrangements

$$\frac{\Delta x}{2}\frac{\mathrm{d}}{\mathrm{d}t}\left(\sum_{j\in\mathbb{Z}} |c_j|^2\right) = -\frac{u}{2}\sum_j |c_j - c_{j-1}|^2\left(1 - \varphi_{j-\frac{1}{2}}\right) \le 0,$$

which shows that the L^2 norm decreases. It makes squaring impossible. The proof is ended. It can be generalized to the fully discrete scheme with the same conclusion. A corollary is as follows.

Lemma 2 (Necessary condition for sharpening).
A slope limiter that sharpens is necessarily such that $\varphi(r) > 1$ for some $r \in \mathbb{R}$. This condition is satisfied by the SuperBee formula (10), for which $\lim_{r\to\infty}\varphi(r) = 2$.

2.3 Inequality and Antidiffusion

This sharpening strategy is more radical. It is naturally introduced in the context of BV schemes (Després and Lagoutière, 2001a), see also Toro (2009), and has been adapted to ENO techniques (Shu, 2009) in Xu and Shu (2006, 2005). We refer to Shyue and Xiao (2014), Chen et al. (2012), Rongsan and De-kang (2011), Kokh and Lagoutière (2010), Shu (2009), Jaouen and Lagoutière (2007), Xu and Shu (2006), Xu and Shu (2005), Billaud Friess and Kokh (2014), and Goudon et al. (2013) for the use of such methods for different problems.

We shall note $m_{j+1/2} = \min(c_j^n, c_{j+1}^n)$, $M_{j+1/2} = \max(c_j^n, c_{j+1}^n)$

$$\lambda_{j+1/2} = \frac{\Delta x}{u\Delta t}(c_j^n - M_{j-1/2}) + M_{j-1/2}^n \text{ and } \Lambda_{j+1/2} = \frac{\Delta x}{u\Delta t}(c_j^n - m_{j-1/2}) + m_{j-1/2}^n. \tag{13}$$

We observe that $\lambda_{j+1/2} \le \Lambda_{j+1/2}$ if the CFL condition $u\Delta t \le \Delta x$ is satisfied. A basic property writes as follows.

Lemma 3. *Under CFL, the upwind flux choice* $c^n_{j+1/2} = c^n_j$ *belongs to the interval* $[\lambda_{j+1/2}, \Lambda_{j+1/2}] \cap [m_{j+1/2}, M_{j+1/2}]$, *which ultimately ensures the maximum principle (Després and Lagoutière, 2001a; Xu and Shu, 2006, 2005).*

In this context, one introduces compression, or sharpening, or antidiffusion, by using the most extreme formulated choice. Let $\omega_{j+1/2}$ and $\Omega_{j+1/2}$ such that $[\omega_{j+1/2}, \Omega_{j+1/2}] = [\lambda_{j+1/2}, \Lambda_{j+1/2}] \cap [m_{j+1/2}, M_{j+1/2}]$. One obtains

$$c^n_{j+1/2} = \begin{cases} \Omega_{j+1/2}, & \text{if } \Omega_{j+1/2} \leq c^n_{i+1}, \\ c^n_{j+1}, & \text{if } \omega_{j+1/2} \leq c^n_{i+1} \leq \Omega_{j+1/2}, \\ \omega_{j+1/2}, & \text{if } c^n_{i+1} \leq \omega_{j+1/2}. \end{cases}$$

An equivalent definition (still for the case $u > 0$) is given in the following lemma.

Lemma 4. *The limited downwind flux defined above is equivalent to the so-called Ultra-Bee flux limiter flux (see Toro, 2009) defined as*

$$c^{n+1}_j = c^n_j - \nu(c^n_j - c^n_{j-1}) - \frac{\nu(1-\nu)}{2}(\varphi^n_{j+1/2}(c^n_{j+1} - c^n_j) - \varphi^n_{j-1/2}(c^n_j - c^n_{j-1}))$$

with $\varphi^n_{j+1/2} = \varphi(r^n_{j+1/2}, \nu)$ *and* $\varphi(r, \nu) = minmod\left(\frac{2r}{\nu}, \frac{2}{1-\nu}\right)$.

The limiter is now function of the slope r and of the Courant number ν. The scheme is called *limited downwind* in the following.

Lemma 5. *This limited downwind scheme is exact for step initial conditions (Després and Lagoutière, 2001a).*

Confirmation is by starting from an initial data which is not a step function, but a (discretized) smooth function. One observes (under a surprising technical condition CFL≠1/2) that the smooth profile is replaced a step function close by a step function with an approximation error is $O(\Delta x)$. After that first stage the step function is perfectly transported. So in some sense the Ultra-Bee limiter is a perfect sharpener. The sharpening effect is so pronounced that it may resemble an instability, but it is not.

This technique was incorporated in FV algorithms for the simulation of two-component fluid flows, for the mass fraction, volume fraction, or colour function of components, in, e.g., Despres and Lagoutiere (2007), Kokh and Lagoutière (2010), and extended to *multicomponent* in Jaouen and Lagoutière (2007) and Billaud Friess and Kokh (2014).

This was also modified to apply to nonlinear discontinuities such as classical shocks (Aguillon and Chalons, 2016) and nonclassical shocks in the scalar context (Boutin et al., 2008) and in the context of systems in Aguillon (2016).

2.4 Glimm's Method

At this stage of the discussion the problem is the following: either one accepts to violate the maximum principle although this can be very critical, for example when the transported unknown is the mass or volume fraction of a fluid in a multicomponent flow, or one has to use linear first order or a nonlinear scheme (see Section 2.2). Yet there exists an alternative, that was first proposed by Glimm, in Glimm (1965) for theoretical analysis purposes. This method avoids the numerical diffusion of first order stable schemes because it does not involve any "projection" on the mesh, and it does not create new values of the solution a each time step in the case of linear transport.

To describe it briefly, let us consider once again the upwind scheme written as $c_j^{n+1} = (1-\nu)c_j^n + \nu c_{j-1}^n$ with ν the CFL number. The smearing of the profiles comes from the (strictly) convex combination that appears in the formula. This scheme can be interpreted as a two-step scheme: exact transport of the profile for a time Δt, and then projection on the mesh (the upwind scheme is the Godunov scheme). Glimm proposes to avoid the projection by taking one of the two values that are present in cell j after one time step: c_{j-1}^n or c_j^n. The choice is performed randomly: c_{j-1}^n is chosen with probability ν, and c_j^n is chosen with probability $1 - \nu$. This interpretation is correct since $0 < 1 \leq \nu$ (resp. $0 \leq 1 - \nu < 1$) under CFL. In the more general context of nonlinear problems, the algorithm is based on the resolution of the Riemann problems at each interface and on the choice of a random variable δ^n(different from one time step to the other), chosen according to the uniform law between 0 and Δx. Then the updated value of the unknown in the cell j is defined by taking the value of the solution of the Riemann problem at time Δt at position $(j - 1/2)\Delta x + \delta^n$. This was shown by Glimm to converge, with probability 1, and it is clear that it does not *smear* profiles, at least when the profile is a step and in the linear context. Let us note that this random procedure has the drawback that the scheme is nonconservative; however, this does not prevent the scheme to converge to the entropy solution for nonlinear problems. Note also that the randomness is not mandatory: the only property that is required for $(\delta^n)_n$ is that it is an equidistributed (with low discrepancy) sequence. The Van der Corput sequence, which is such a deterministic sequence, is shown to give qualitatively very good (better than a random sequence) results in Colella (1982).

One can notice that, for the linear transport equation (1), the upwind scheme is the *expectation* of Glimm's scheme. This observation was used to prove error estimates for the upwind scheme on general meshes, using central-limit type estimates, in Delarue and Lagoutière (2011).

In space dimension 1 and in the context of linearly degenerate fields (which correspond to material discontinuities) an FV algorithm based on a Lagrange–Remap (formulated as Lagrange-transport) strategy with a random sampling technique for the transport part, for the simulation of two-component compressible fluid flows is derived in Chalons and Coquel (2012), with very good efficiency. See also Chalons (2007), Chalons and Goatin (2008), and Bachman et al. (2013).

In the more particular context of *nonlinear* material discontinuities that are present in some viscous-dispersive limits of systems with fields that are neither genuinely nonlinear nor linearly degenerate, with so-called *nonclassical shocks* (see LeFloch, 2002), the random choice method was shown in Chalons and LeFloch (2003) to give very good (and convergent) results, which is very difficult in this context.

The tentation to use such a scheme in higher dimension is great, but it is known since Chorin (1976) that it is not satisfactory for genuinely nonlinear conservation laws. Colella (1982) proposed a modification of the random choice algorithm that involves the Godunov method and that seem to be convergent. Unfortunately this modification of Glimm's algorithm does not allow to preserve sharp fronts. Nevertheless, for linear or linearly degenerate fields, this random choice procedure shows great efficiency, at least on Cartesian grids. This has been investigated and analyzed, for example in Helluy and Jung (2014) and Helluy and Jung (2013).

2.5 PDE Models and Sharpening Methods

Level sets methods are discussed in Osher and Fedkiw (2003), Osher and Sethian (1988), and Sethian (1996). This is a very popular set of numerical methods for interface modelling that has been applied to many problems. In the present context, the idea is to rely on a partial differential equation to transport a colour function (our definition of a colour function f is that it takes value in [0, 1], so that: if a point x is such that $f(x) = 0$ has colour equal to 0; if $f(x) = 1$ then x has a colour equal to 1; and finally $0 < f(x) < 1$ corresponds to intermediate colours). No colours below 0 and above 1 are considered in this presentation, but it is not mandatory. A typical elementary question with the level set approach is about the influence of the numerical parameters on the level set. In certain cases the answer is that the method can be insensitive to this parameters.

To understand this property we consider the simplest colour function at initial time $c^{\text{ini}}(x) = H(x)$, that is $c^{\text{ini}}(x) = 0$ for $x < 0$ and $c_0(x) = 1$ for $x > 1$. Instead of manipulating the upwind first order scheme, we use its modified equation (that is to say, the PDE it is consistent with at the *second* order in time and space). We thus consider the function c_μ solution of the advection equation with viscosity

$$\partial_t c_\mu + u\partial_x c_\mu = \mu \partial_{xx} c_\mu, \quad \mu = \frac{\Delta x}{2}(1 - \nu),$$

where $0 \leq \nu \leq 1$ is the CFL number. The modified equation is a second order approximation of the upwind scheme. The interface is recovered at any time t by as the 1/2 level set $\Gamma_\mu(t) = x$ such that $c_\mu(x, t) = 1/2$. It is easy to prove that x exists and is unique for $t > 0$ and $0 \leq \nu < 1$: this is a consequence of well-known integral representation formula detailed below. One has more.

Lemma 6. *For all $t > 0$ and $0 \leq \nu < 1$, the 1/2 level set is exact: that is $\Gamma_\mu(t) = ut$.*

One has with the fundamental solution of the heat equation

$$c_\mu(x,t) = \frac{1}{\sqrt{4\pi\mu t}} \int_{\mathbb{R}} \exp\left(-(x-y-ut)^2/(4\mu t)\right) H(y) dy.$$

So

$$c_\mu(ut,t) = \frac{1}{\sqrt{4\pi\mu t}} \int_0^\infty \exp\left(-y^2/(4\mu t)\right) dy = \frac{1}{\sqrt{2\pi}} \int_0^\infty \exp\left(-y^2/2\right) dy = \frac{1}{2}.$$

Even if this argument is very elementary, it explains that level set methods have the ability to predict the interface with great accuracy, even if the underlying scheme for the transport of the colour function is low order. In the context of this review chapter, it is perfect sharpening.

2.6 Nature of the Grid/Mesh

The discussion so far was restricted to one-dimensional grids. The extension of the previous FV algorithms to general multidimensional grids poses two fundamental difficulties.

A first one is that sharpening techniques are highly nonlinear methods. A good sharpening technique is in practice equipped with a method which controls the oscillations due to strong nonlinear interactions. In dimension $d = 1$, this principle is mostly based on the BV setting. The issue is that this bound on the total variation is lost in dimension $D = 2$ and greater. This has been proved in a famous article (Goodman and LeVeque, 1985) on a Cartesian grid. This unfortunate situation has the consequence that the preservation of the maximum principle does not yield a control of some special oscillations which develop mostly tangentially to the isolines of the exact profiles: an important reference in this direction is the series (Turek and Kuzmin, 2005; Kuzmin and Möller, 2005b,a). See also Després et al. (2010).

A confirmation of this behaviour is the 2D algorithm in Després and Lagoutière (2001a). It is shown that the extension of the Ultra-Bee scheme with directional splitting is exact for squares. But unfortunately this algorithm is not equipped with a control of 2D variations. It can be interpreted as a distant consequence of the Goodman and LeVeque (1985) theorem. In consequence this algorithm is useless for calculations of profiles with values which are not exactly 0 and or 1. Even if the initial data is an indicatrix function, its boundary is not necessarily a 2D step function: in this situation one observes oscillations at the boundary between 0 and 1: these oscillations are perfectly bounded in L^∞ norm because directional splitting preserves the maximum principle; but they are not bounded in the BV seminorm because the BV seminorm is a global quantity destroyed by directional splitting. An attempt is been made in Després and Lagoutière (2001b) to overcome this failure, but the numerical results are deceptive (not published), probably due to the curse explained by the Goodman and LeVeque (1985) theorem.

2.7 Interface Reconstruction and VOF

The simple line interface calculation (SLIC) (Noh and Woodward, 1976) is an extremely popular method that presents a nearly all purpose methodology for FV interface sharpening. The design principle of SLIC is to reconstruct parallel and/or anti parallel perfect interfaces in Cartesian cells from the knowledge of volume fractions. In dimension 1, for a step initial condition, it is equivalent to the limited downwind scheme (that can be seen as a reconstruction algorithm, where the reconstructed solution is a step function in every cell). Even if it is an extremely simple method, the results are quite good when comparing with the implementation cost and run time. This is probably the reason why it is still a reference. With respect to SLIC, the volume of fluid (VOF) (Hirt and Nichols, 1981) has the huge advantage to reconstruct interface with any direction. Even without discussing the simplicity of the method, it is clear that this information is a kind of first order interface reconstruction while SLIC can be considered as a zeroth order interface reconstruction. Another feature of VOF is that the normal direction of the interface is computed from the discrete gradient of some volume fractions. It is possible to optimize the performance of VOF by changing the parameters of the discrete gradient operator and of the method used to evolve the volume fractions.

It must be noted that SLIC is not PDE based and VOF is only partially PDE based. In consequence it is not really possible to perform a convergence analysis of the algorithms, but only on parts of them. The Youngs algorithm (Youngs, 1984) has a similar nature.

2.8 Vofire

We give some details of the Vofire method, which is a multidimensional nonlinear FV scheme. The geometrical idea relies on the following observation: in dimension greater than 2, the numerical diffusion can be decomposed into two different diffusions: the longitudinal diffusion, along the velocity field, which is typically one dimensional, and the transverse diffusion, which is really due to the fact that the mesh is multidimensional. This distinction between the two phenomena could appear arbitrary, but is in accordance with basic numerical tests. Consider for example an initial condition which is the characteristic function of the square $(0.25, 0.75) \times (0.25, 0.75)$. This profile

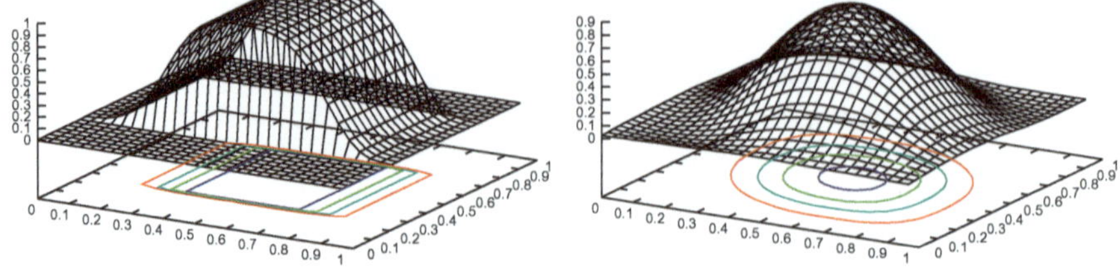

FIG. 1 Upwind scheme. The initial condition is the characteristic function of a square. Final time $t = 1$. Periodic boundary conditions. On the left: the velocity $\mathbf{u} = (1, 0)^T$ is aligned with the mesh; the result displays only longitudinal diffusion. On the right: the velocity $\mathbf{u} = (1, 1)^T$ is not aligned with the mesh. The consequence is that there is both longitudinal and transverse diffusion.

is advected with the upwind scheme. The velocity direction $\mathbf{u}$ has a great influence on the result. It is illustrated in Fig. 1.

We here propose to restrict to triangular meshes, on which it is simpler to expose the Vofire technique. Thus we consider the following type of mesh structure:

The idea to "fight" against these two diffusion phenomena is to use, as for the limited downwind seen as a (discontinuous) reconstruction scheme, a reconstruction procedure in every cell. This reconstruction will be here two-fold: it will consist in a first reconstruction that will be transverse, and in a second one that will be done along the velocity field. The velocity field $\mathbf{u}$ is constant for simplicity, but this assumption can be removed. Note that the transverse diffusion actually depends more on the shape of the cells than on the velocity, so that this assumption of constant velocity is not so much restrictive here. For a given cell T_j, $N^-(j)$ denotes the set of cells that are adjacent to T_j and *upwind*: $N^-(j) = \{T_m \text{ such that } \overline{T_j} \cap \overline{T_m} \text{ is of nonzero one} - \text{dimensional Lebesgue measure and } u \cdot n_{j,m} < 0\}$, and $N^+(j)$ denotes the set of downwind cells to T_j. As in dimension 1, the fundamental requirement of the scheme is that it satisfies an upwind maximum principle:

$$\min\left(c_j^n, \min_{k \in N^-(j)} c_k^n\right) \le c_j^{n+1} \le \max\left(c_j^n, \max_{k \in N^-(j)} c_k^n\right) \quad \text{for any } j.$$

The most important part of the procedure, regarding the multidimensional properties of the scheme, is the first one that concerns the transverse reconstruction. As we will see, after this reconstruction, the algorithm will be one dimensional, and one-dimensional techniques (such as the limited downwind scheme for instance) will be applied.

Recall that, for expository purposes, the mesh is assumed to be made with triangles, in dimension 2. The transverse reconstruction consists in breaking a cell in two parts by a segment parallel to the velocity, and modifying the value of the unknown in each of these two subcells. Each triangle T_j has at least one downwind neighbour and at most two. If it has only one downwind neighbour, we do not perform the transverse reconstruction (we do not cut the cell). This can be explained by the fact that when there is only one downwind neighbour, the "information" contained in the cell is not spread transversally by any scheme (with small stencil). Let us thus assume that T_j has two downwind neighbours, T_k and T_l. It has then one upwind neighbour, T_m. We consider the intersection point of the two edges relative to the downwind neighbours and cut T_j along the line passing on this intersection point and parallel to $\mathbf{u}$. The two subcells are denoted $T_{j,k}$ and $T_{j,l}$: $T_{j,k}$ has T_k as (unique) downwind neighbour, and $T_{j,l}$ has T_l as (unique) downwind neighbour. The partioning is illustrated in Fig. 2.

The essential property of this cutting is that now every subcell has exactly one downwind and one upwind neighbour, as will be used below: this is due to the fact that the new normal vector $\mathbf{n}_j$ (see Fig. 3) is orthogonal to the velocity, so that there will be no flux through the new interface. We use

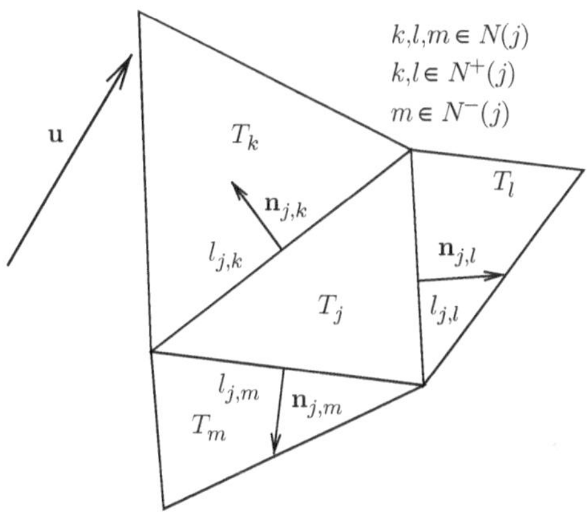

FIG. 2 Mesh and notations (for the sake of brevity, these notations will not be explained further in the text, as they are very classical).

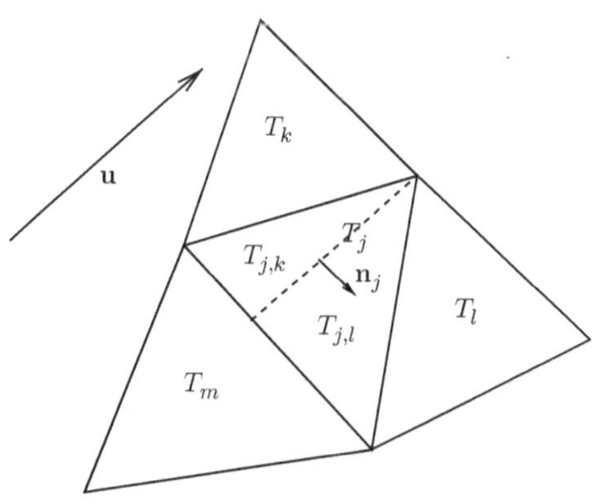

FIG. 3 Transverse reconstruction.

symbols $s_{j,k}$ and $s_{j,l}$ to denote the areas of subcells $T_{j,k}$ and $T_{j,l}$ respectively. Clearly $s_{j,k} + s_{j,l} = s_j$ and $s_{j,k} > 0$ and $s_{j,l} > 0$. The aim is to define a reconstructed value $c^R_{j,k}$ in $T_{j,k}$ and a reconstructed value $c^R_{j,l}$ in $T_{j,l}$. We impose

$$s_{j,k}c^R_{j,k} + s_{j,l}c^R_{j,l} = s_j c^n_j \tag{14}$$

to guarantee the local conservativity. Let us write

$$\begin{cases} c^R_{j,k} = c^n_j + \lambda_{j,k}\left(c^n_k - c^n_j\right), & 0 \le \lambda_{j,k} \le 1, \\ c^R_{j,l} = c^n_j + \lambda_{j,l}\left(c^n_l - c^n_j\right), & 0 \le \lambda_{j,l} \le 1, \end{cases} \tag{15}$$

which means that $c^R_{j,k}$ and $c^R_{j,l}$ must satisfy a condition of local consistency. We introduce the idea of antidissipative schemes which will serve to find a unique value of $\lambda_{j,k}$ and $\lambda_{j,l}$.

Remark 2. Our objective is to choose $\lambda_{j,k}$ and $\lambda_{j,l}$ in order to obtain an antidissipative scheme, with a very low level of numerical diffusion. This is the reason why we seek the largest possible $\lambda_{j,k}$ and $\lambda_{j,l}$ in the interval [0, 1]. This is the same principle as in Section 2.3. But of course we cannot take $\lambda_{j,k} = \lambda_{j,l} = 1$ directly because we ask the reconstruction to be conservative: see Eq. (14).

So our goal is to have the largest $\lambda_{j,k}$ and $\lambda_{j,l}$ in the interval [0, 1], but still satisfying the maximum principle. Eq. (14) can be rewritten as $s_{j,k}c^R_{j,k} + s_{j,l}c^R_{j,l} - s_j c^n_j = s_{j,k}\left(c^R_{j,k} - c^n_j\right) + s_{j,l}\left(c^R_{j,l} - c^n_j\right) = 0$, that is

$$\left[s_{j,k}\left(c^n_k - c^n_j\right)\right]\lambda_{j,k} + \left[s_{j,l}\left(c^n_l - c^n_j\right)\right]\lambda_{j,l} = 0.$$

As we will see, either the data c^n_j is a local transverse maximum or minimum and then (14) implies $\lambda_{j,l} = \lambda_{j,k} = 0$ (it means there is actually no reconstruction), or the data is transversally monotone, and then at least $\lambda_{j,l} = 1$ or $\lambda_{j,k} = 1$. The solution is computed as follows.

(1) *If* $s_{j,k}\left(c^n_k - c^n_j\right)s_{j,l}\left(c^n_l - c^n_j\right) \ge 0$, c^n_j is a local extremum in the transverse direction. Then we do not reconstruct, which means $\lambda_{j,k} = \lambda_{j,l} = 0$ and

$$c^R_{j,l} = c^R_{j,k} = c^n_j. \tag{16}$$

(2) *If* $-\dfrac{s_{j,k}\left(c^n_k - c^n_j\right)}{s_{j,l}\left(c^n_l - c^n_j\right)} > 1$, the solution is obtained by taking $\lambda_{j,l} = 1$,

$$c^R_{j,l} = c^n_l, \quad c^R_{j,k} = c^n_j - \frac{s_{j,l}}{s_{j,k}}\left(c^n_l - c^n_j\right) = \left(s_j c^n_j - s_{j,l}c^n_l\right)/s_{j,k}. \tag{17}$$

(3) *If* $-\dfrac{s_{j,k}\left(c^n_k - c^n_j\right)}{s_{j,l}\left(c^n_l - c^n_j\right)} < 1$, the solution is obtained by taking $\lambda_{j,k} = 1$,

$$c_{j,k}^R = c_k^n, \quad c_{j,l}^R = c_j^n - \frac{s_{j,k}}{s_{j,l}}\left(c_k^n - c_j^n\right) = \left(s_j c_j^n - s_{j,k} c_k^n\right)/s_{j,l}. \tag{18}$$

As the situation is now one dimensional for each subcell T_{jk} and T_{jl}, one can analyze the scheme where the first stage is this reconstruction followed by a second stage which is the upwind scheme. It is obvious that this scheme will provide the maximum principle, as the reconstruction does. What is not so obvious is that the CFL stability condition for the upwind scheme on this new (and finer) mesh *is the same as for the initial mesh.* A simple proof is as follows.

Proof. The reconstructed quantities (15) respect the maximum principle. By construction the scheme is equal to a two-steps algorithm: first step, use the upwind scheme for a mesh which is locally cut in smaller cells, as it is described in Fig. 3, and with cell quantities equal to the reconstructed quantities; second step, project onto the original coarse mesh. Therefore it is sufficient to check that the CFL condition is the same for the original mesh (four cells in Fig. 3) and for the new mesh (five cells in Fig. 3).

Since $\mathbf{u}$ is constant, then $\sum_{k\in N^+(j)} l_{j,k}(\mathbf{u}^T\mathbf{n}_{j,k}) = -\sum_{k\in N^-(j)} l_{j,k}(\mathbf{u}^T\mathbf{n}_{j,k})$. The standard CFL condition for the upwind scheme for the cell T_j thus has the form $\frac{\Delta t}{s_j}\sum_{k\in N^+(j)} l_{j,k}(\mathbf{u}^T\mathbf{n}_{j,k}) \le 1$, that is

$$\frac{\Delta t}{s_j}\left(l_{j,k}(\mathbf{u}^T\mathbf{n}_{j,k}) + l_{j,l}(\mathbf{u}^T\mathbf{n}_{j,l})\right) \le 1. \tag{19}$$

The CFL condition for the subcells $T_{j,k}$ and $T_{j,l}$ are respectively

$$\frac{\Delta t}{s_{j,k}} l_{j,k}(\mathbf{u}^T\mathbf{n}_{j,k}) \le 1 \quad \text{and} \quad \frac{\Delta t}{s_{j,l}} l_{j,l}(\mathbf{u}^T\mathbf{n}_{j,l}) \le 1. \tag{20}$$

Let $l_j = \text{length}\left(\overline{T_{j,k}} \cap \overline{T_{j,l}}\right)$ be the length of the segment separating $T_{j,k}$ and $T_{j,l}$. One has $s_{j,k} = \frac{l_j}{2|\mathbf{u}|} l_{j,k}(\mathbf{u}^T\mathbf{n}_{j,k})$ and $s_{j,l} = \frac{l_j}{2|\mathbf{u}|} l_{j,l}(\mathbf{u}^T\mathbf{n}_{j,l})$ and $s_j = s_{j,k} + s_{j,l} = \frac{l_j}{2|\mathbf{u}|}\left(l_{j,k}(\mathbf{u}^T\mathbf{n}_{j,k}) + l_{j,l}(\mathbf{u}^T\mathbf{n}_{j,l})\right)$. The two inequalities of (20) and inequality (19) thus rewrite $|\mathbf{u}|\frac{2\Delta t}{l_j} \le 1$. So they are equivalent and the proof is ended. □

Some modifications and improvements of the Vofire technique have been proposed in Michel et al. (2010) and Bohbot et al. (2010) for example.

3 COUPLING WITH HYPERBOLIC NONLINEAR EQUATIONS

An issue is to use the previous techniques in complex computational fluid dynamics FV codes with a respect of the key properties necessary for a correct simulation. Of course the notion of a correct simulation and the identification of the key ingredients of a correct coupling are far to be evident. We restrict the discussion to hyperbolic models for compressible fluids for which conservative issues are critical. Indeed the nonlinearity of the equations induces discontinuous solutions such as shocks and contact discontinuities: it is well accepted that the violation of the conservation is only at the cost of a strong deviation with respect to the solution of the Riemann problem (see Zhong et al., 1996; Hou et al., 1999 for a justification). These questions are fiercely debated when dealing with multifluid models.

There are common guidelines for incorporating sharpening techniques into discretization strategies of complex models. Usually one singles out transport effects in the system and update a set of key fluid parameters thanks to a numerical scheme that transports discontinuities as sharply as possible. A delicate matter is generally to preserve good stability and consistency properties of the overall numerical scheme.

3.1 An Example of Discretization for Compressible Flows With Two Components Separated by a Sharp Interface

For the sake of illustrating these ideas, we consider, in space dimension 1, a simple model of compressible flows that involves two perfect gases that was studied in Abgrall (1988) and Larrouturou and Fézoui (1989). The specific heat at constant volume and ratio of specific heat of the fluid $k = 1, 2$ are respectively $c_{vk} > 0$ and $\gamma_k > 1$. The density of the two-phase medium is noted ρ and $Y_1 = Y$ (resp. $Y_2 = 1 - Y$) is the mass fraction of the fluid $k = 1$ (resp. $k = 2$). We suppose that there is a thermal equilibrium between the gases and that the pressure P verifies Dalton's law, then we have

$$P = \frac{\sum_{k=1,2} Y_k(\gamma_k - 1)c_{vk}}{\sum_{k=1,2} Y_k c_{vk}} \rho e, \tag{21}$$

where e is the specific internal energy of the medium. One supposes that the components have the same velocity u and that no mass transfer occurs between the species. If one notes $\rho\mathbf{W} = [\rho Y, \rho u, \rho(e+u^2/2)]^T$, $\mathbf{T}(\mathbf{W}) = [0, 0, P, Pu]^T$ then the flow is governed by

$$\partial_t \rho + \partial_x(\rho u) = 0, \quad \partial_t(\rho\mathbf{W}) + \partial_x(\rho\mathbf{W}u + \mathbf{T}) = 0. \tag{22}$$

System (22) is hyperbolic provided that $\gamma_k > 1$ and it is equipped with jump relations that enable the definition of weak solutions that verifies the transport equation

$$\partial_t Y + u\partial_x Y = 0. \tag{23}$$

Although this model is equipped with a mixture model (based on oversimplified assumptions), if one chooses an initial condition such that $Y(x, t = 0) \in \{0, 1\}$, then no physical mixing should occur in the domain as (23) guarantees that $Y(x, t) \in \{0, 1\}$ for $t > 0$. In this sense, (22) can be used as a model for a flows involving two compressible fluids separated by a sharp interface.

The decoupling between transport and other phenomenon can be achieved thanks to a Lagrange–Remap method. Let us note ξ the Lagrangian space coordinate defined by $d\xi(t; x_0)/dt = u(\xi(t; x_0), t)$ with $\xi(t = 0; x_0) = x_0$. If $(x, t) \mapsto b$ is any fluid parameter, we note $(\xi, t) \mapsto b^{\text{Lag}}$ the Lagrangian field associated with b by $b^{\text{Lag}}(\xi(t; x), t) = b(x, t)$. System (22) can be expressed in the so-called Lagrangian reference frame as follows

$$\rho^{\text{Lag}}(\xi,0)\partial_t(1/\rho^{\text{Lag}}) - \partial_\xi u^{\text{Lag}} = 0, \quad \rho^{\text{Lag}}(\xi,0)\partial_t \mathbf{W}^{\text{Lag}} + \partial_\xi \mathbf{T}^{\text{Lag}} = 0. \tag{24}$$

Given a set of discrete values $(\rho, \rho\mathbf{W})_j^n$ that represent an approximation of the fluid state at instant $t = t^n$ within the cell i, the Lagrange–Remap method is a two-step algorithm (Godlewski and Raviart, 1996; Després, 2010): first, we update the discrete unknowns to a value $(\rho, \rho\mathbf{W})_j^{\text{Lag}}$ by approximating the solution of (24) over $[t^n, t^n + \Delta t]$. Let us remark that the evolution equation for Y expressed in (24) boils down to $\partial_t Y = 0$, therefore it is reasonable to expect that $Y_i^{\text{Lag}} = Y_i^n$. The second step updates the fluid parameter to their values $(\rho, \rho\mathbf{W})_j^{n+1}$ by remapping the Lagrange values $(\rho, \rho\mathbf{W})_j^{\text{Lag}}$ onto the Eulerian mesh. It can read as follows

$$\rho_j^{n+1} - \rho_j^n + \frac{\Delta t}{\Delta x}(\rho_{j+1/2}^{\text{Lag}} u_{j+1/2}^n - \rho_{j-1/2}^{\text{Lag}} u_{j-1/2}^n) = 0, \tag{25a}$$

$$(\rho\mathbf{W})_j^{n+1} - \rho_j^n \mathbf{W}_j^{\text{Lag}} + \frac{\Delta t}{\Delta x}((\rho\mathbf{W})_{j+1/2}^{\text{Lag}} u_{j+1/2}^n - (\rho\mathbf{W})_{j-1/2}^{\text{Lag}} u_{j-1/2}^n) = 0. \tag{25b}$$

The values $u_{j-1/2}^n$ are approximations of the material velocity of the fluid at the cell interface $x = x_{j+1/2}$ that can be estimated with the discretization of (24). One can therefore consider that $u_{j-1/2}^n$ is known when performing (25). The only missing ingredient for achieving the remap procedure is thus the definition of the variable flux $(\rho, \rho\mathbf{W})_{j+1/2}^{\text{Lag}} = [\rho, \rho Y, \rho u, \rho(e + u^2/2)]_{j+1/2}^{\text{Lag}}$. For this problem it is clear that the antidiffusive mechanism should concern the variable Y whose discontinuity carries the location of material interface between the fluids. Suppose given a definition for the fluxes $b_{j+1/2}^{\text{Lag}}$ that is consistent for $b \in \{\rho, \rho u, \rho(e + u^2/2)\}$ and that $\rho_j^n > 0$ and $\rho_{j+1/2}^{\text{Lag}} > 0$. Let us note $\mathfrak{m}_{j+1/2}^n = \min(Y_j^n, Y_{j+1}^n)$, $\mathfrak{M}_{j+1/2}^n = \max(Y_j^n, Y_{j+1}^n)$. Following the ideas introduced in Section 2.3 in the case of pure transport problem, we aim at defining a flux $Y_{j+1/2}^{\text{Lag}}$ that fulfils two requirements.

(i) $Y_{j+1/2}^{\text{Lag}} \in [\mathfrak{m}_{j+1/2}^n, \mathfrak{M}_{j+1/2}^n]$;

(ii) the choice of $Y_{j+1/2}^{\text{Lag}}$ and (25b) should ensure a discrete maximum principle for Y in the cell i (resp. $i + 1$) if $u_{j+1/2}^n > 0$ and $u_{j-1/2}^n > 0$ (resp. $u_{j+1/2}^n < 0$ and $u_{j+3/2}^n < 0$).

For the sake of simplicity, we suppose that $\rho_{j+1/2}^{\text{Lag}}$ is defined by the upwind flux, *i.e.* $\rho_{j+1/2}^{\text{Lag}} u_{j+1/2}^n = \rho_j^{\text{Lag}} (u_{j+1/2}^n)^+ + \rho_{j+1}^{\text{Lag}} (u_{j+1/2}^n)^-$, then we can define the real interval $[d_{j+1/2}, D_{j+1/2}]$ as follows.

- If $u_{j+1/2}^n > 0$ and $u_{j+1/2}^n > 0$ (resp. $u_{j+1/2}^n < 0$), we set

$$d_{j+1/2} = Y_j^n + (\mathfrak{M}_{j-1/2} - Y_j^n)\left[1 - \frac{\Delta x}{u_{j+1/2}^n \Delta t}\right] \quad (\text{resp. } d_{j+1/2} = Y_j^n),$$

$$D_{j+1/2} = Y_j^n + (\mathfrak{m}_{j-1/2} - Y_j^n)\left[1 - \frac{\Delta x}{u_{j+1/2}^n \Delta t}\right] \quad (\text{resp. } D_{j+1/2} = Y_j^n).$$

- If $u_{j+1/2}^n < 0$ and $u_{j+3/2}^n < 0$ (resp. $u_{j+3/2}^n > 0$), we set

$$d_{j+1/2} = Y_{j+1}^n + (\mathfrak{M}_{j+3/2} - Y_{j+1}^n)\left[1 + \frac{\Delta x}{u_{j+1/2}^n \Delta t}\right] \quad (\text{resp. } d_{j+1/2} = Y_{j+1}^n),$$

$$D_{j+1/2} = Y_{j+1}^n + (\mathfrak{m}_{j+3/2} - Y_{j+1}^n)\left[1 - \frac{\Delta x}{u_{j+1/2}^n \Delta t}\right] \quad (\text{resp. } D_{j+1/2} = Y_{j+1}^n).$$

Let us note $[\omega_{j+1/2}, \Omega_{j+1/2}] = [\mathfrak{m}_{j+1/2}^n, \mathfrak{M}_{j+1/2}^n] \cap [d_{j+1/2}, D_{j+1/2}]$. Under the CFL condition

$$|u_{j+1/2}^n| \Delta t / \Delta x < 1, \tag{26}$$

one can check that $[\omega_{j+1/2}, \Omega_{j+1/2}] \neq \emptyset$ as Y_j^n (resp. Y_{j+1}^n) belongs to $[\omega_{j+1/2}, \Omega_{j+1/2}]$ if $u_{j+1/2}^n > 0$ (resp. $u_{j+1/2}^n < 0$). Choosing $Y_{j+1/2}^{\text{Lag}} \in [\omega_{j+1/2}, \Omega_{j+1/2}]$ ensures that (i) and (ii) are verified under the condition (26). In order to enable a sharp transport of Y, one just need to use the limited downwind choice within the interval $[\omega_{j+1/2}, \Omega_{j+1/2}]$, which boils down to set

$$Y_{j+1/2}^{\text{Lag}} = \min(\max(\omega_{j+1/2}, Y_{\text{down}}^{\text{Lag}}), \Omega_{j+1/2}), \tag{27}$$

where $Y_{\text{down}}^{\text{Lag}} = Y_{j+1}^{\text{Lag}}$ (resp. $Y_{\text{down}}^{\text{Lag}} = Y_j^{\text{Lag}}$) if $u_{j+1/2}^n > 0$ (resp. $u_{j+1/2}^n < 0$).

A numerical scheme based on a finite volume approximation of (24) and (25) with the limited downwind choice (27) was studied in Lagoutière (2000) for the model described in this section. It is worth mentioning that up to a careful discretization choice for (24) the overall algorithm is conservative with respect to (ρ, $\rho\mathbf{W}$). Let us also emphasize that the algorithm presented in this section is difficult to use in practice: spurious pressure and velocity oscillations at the material interface may occurs, which is a common issue for this type of problems (Abgrall, 1996). The same method was applied to similar two-phase models with an alternate mixture law in Lagoutière (2000) and Despres and Lagoutiere (2007) that guarantees that constant pressure and velocity profiles are preserved.

3.2 Example of Other Evolution Equation Involving Sharp Interfaces

It is not possible to give an exhaustive list of all possible sharpening techniques implementation, we will try to give hereafter an overview of the works that have been achieved the past years that is inevitably incomplete.

The approach of Section 3.1 has been successfully extended to other systems like the five-equation model of Massoni et al. (2002) and Allaire et al. (2002) in Kokh and Lagoutière (2010) and also for compressibles flows involving an arbitrary number of components separated by interfaces (Jaouen and Lagoutière, 2007; Billaud Friess and Kokh, 2014). Other techniques may be used to sharpen front in systems with interface. For example, considering again system (22), one can discretize directly the transport equation (23) with the limited downwind scheme of Section 2.3 and use a classical finite volume discretization for ρ, ρu and $\rho(e + u^2/2)$, at the cost of deriving a nonconservative numerical scheme. Other sharpening techniques can also be used for compressible two-phase flows with interface similar to (22): the THINC method that was first developed for incompressible flows (Xiao et al., 2005) has been adapted in Shyue and Xiao (2014) to the five-equation model studied in Allaire et al. (2002). This method relies on controlling the spreading of the material interface thanks to an hyperbolic tangent profile. As mentioned in Section 2.4, Glimm's method has also been used for discretizing sharply the evolution of an interface. Indeed, it is possible to sharply let evolve contact discontinuities in a system by providing a dedicated treatment based on a Glimm type random choice method (Chalons, 2007; Chalons and Goatin, 2008). In Bachman et al. (2013) a random choice method is within a Lagrange–Remap strategy to perform the Remap step while preserving sharp profiles. The limited downwind strategy has been implemented to describe interface that are not solely passively advected like problems of reacting gas flows (Kunkun et al., 2014). A VOF-type reconstruction that relies on a level set description of the interface is proposed in Hu et al. (2006) for the simulation of two-component compressible flows.

3.3 Cut-Cells and CFL Condition

Taking as a principle that sharpening techniques have the ability to reconstruct interfaces, it appears that an interface which moves dynamically in a Cartesian mesh may cut cells into smaller cells. Of course it is most of the time, only a geometrical interpretation. However, it has the unfortunate consequence that these small cut cells may have a dramatic influence on the CFL conditions through a complex nonlinear interaction of the parts of the global algorithm (note nevertheless that it is not the case with the Vofire algorithm). This feature is difficult to analyze rigorously in the context of sharpening methods. In practice one observes a posterior the stability or the instability of the scheme/code. More material is to be found in another chapter in this volume. See also Berger (1984).

REFERENCES

Abgrall, R., 1988. Generalization of the roe scheme for the computation of mixture of perfect gases. Rech. Aérospatiale 6, 31–43 (English edition).

Abgrall, R., 1996. How to prevent pressure oscillations in multicomponent flow calculations: a quasi-conservative approach. J. Comput. Phys. 125 (1), 150–160. ISSN 0021-9991. http://dx.doi.org/10.1006/jcph.1996.0085.

Aguillon, N., 2016. Capturing nonclassical shocks in nonlinear elastodynamic with a conservative finite volume scheme. Interfaces Free Bound. 18, 137–159. http://dx.doi.org/10.4171/IFB/360.

Aguillon, N., Chalons, C., 2016. Nondiffusive conservative schemes based on approximate Riemann solvers for Lagrangian gas dynamics. *ESAIM: M2AN* 50 (6), 1887–1916.

Allaire, G., Clerc, S., Kokh, S., 2002. A five-equation model for the simulation of interfaces between compressible fluids. J. Comput. Phys. 181 (2), 577–616. ISSN 0021-9991. http://dx.doi.org/10.1006/jcph.2002.7143.

Bachman, M., Helluy, P., Jung, J., Mathis, H., Müller, S., 2013. Random sampling remap for compressible two-phase flows. Comput. Fluids 86, 275–283. ISSN 0045-7930. http://dx.doi.org/10.1016/j.compfluid.2013.07.010. http://www.sciencedirect.com/science/article/pii/S0045793013002818.

Berger, M., 1984. Computing Methods in Applied Sciences and Engineering, VI (Versailles, 1983). North-Holland, Amsterdam, 491–492.

Billaud Friess, M., Kokh, S., 2014. Simulation of sharp interface multi-material flows involving an arbitrary number of components through an extended five-equation model. J. Comput. Phys. 273, 488–519. ISSN 0021-9991. http://dx.doi.org/10.1016/j.jcp.2014.05.012.

Bohbot, J., Tran, Q.H., Velghe, A., Gillet, N., 2010. A multi-dimensional spatial scheme for massively parallel compressible turbulent combustion simulation. In: Pereira, J.C.F., Sequeira, A. (Eds.), V European Conference on Computational Fluid Dynamics ECCOMAS CFD 2010.

Boutin, B., Chalons, C., Lagoutière, F., LeFloch, P.-G., 2008. Convergent and conservative schemes for nonclassical solutions based on kinetic relations. I. Interfaces Free Bound. 10 (3), 399–421. ISSN 1463-9963. http://dx.doi.org/10.4171/IFB/195.

Chalons, C., 2007. Numerical approximation of a macroscopic model of pedestrian flows. SIAM J. Sci. Comput. 29 (2), 539–555. http://dx.doi.org/10.1137/050641211.

Chalons, C., Coquel, F., 2012. Computing material fronts with a Lagrange-projection approach. In: Hyperbolic Problems–Theory, Numerics and Applications. Volume 1. Ser. Contemp. Appl. Math. CAM, vol. 17. World Sci. Publishing, Singapore, pp. 346–356.

Chalons, C., Goatin, P., 2008. Godunov scheme and sampling technique for computing phase transitions in traffic flow modeling. Interfaces Free Bound. 10 (2), 197–221. ISSN 1463-9963. http://dx.doi.org/10.4171/IFB/186.

Chalons, C., LeFloch, P.-G., 2003. Computing undercompressive waves with the random choice scheme. Nonclassical shock waves. Interfaces Free Bound. 5 (2), 129–158. ISSN 1463-9963. http://dx.doi.org/10.4171/IFB/75.

Chen, Y.G., Price, W.G., Temarel, P., 2012. An anti-diffusive volume of fluid method for interfacial fluid flows. Int. J. Numer. Methods Fluids 68 (3), 341–359. ISSN 0271-2091. http://dx.doi.org/10.1002/fld.2509.

Chorin, A.-J., 1976. Random choice solution of hyperbolic systems. J. Comput. Phys. 22 (4), 517–533. ISSN 0021-9991.

Colella, P., 1982. Glimm's method for gas dynamics. SIAM J. Sci. Statist. Comput. 3 (1), 76–110. ISSN 0196-5204. http://dx.doi.org/10.1137/0903007.

Delarue, F., Lagoutière, F., 2011. Probabilistic analysis of the upwind scheme for transport equations. Arch. Ration. Mech. Anal. 199 (1), 229–268. ISSN 0003-9527. http://dx.doi.org/10.1007/s00205-010-0322-x.

Després, B., 2008. Finite volume transport schemes. Numer. Math. 108 (4), 529–556. ISSN 0029-599X. http://dx.doi.org/10.1007/s00211-007-0128-4.

Després, B., 2009. Uniform asymptotic stability of Strang's explicit compact schemes for linear advection. SIAM J. Numer. Anal. 47 (5), 3956–3976. ISSN 0036-1429. http://dx.doi.org/10.1137/080734571.

Després, B., 2010. Lois de Conservations Eulériennes, Lagrangiennes et Méthodes Numériques. Mathématiques & Applications, vol. 68. Springer-Verlag, Berlin. ISBN 978-3-642-11656-8, x+284. http://dx.doi.org/10.1007/978-3-642-11657-5.

Després, B., Lagoutière, F., 2001a. Contact discontinuity capturing schemes for linear advection and compressible gas dynamics. J. Sci. Comput. 16 (4), 479–524. ISSN 0885-7474; 1573-7691/e. http://dx.doi.org/10.1023/A:1013298408777.

Després, B., Lagoutière, F., 2001b. Generalized Harten formalism and longitudinal variation diminishing schemes for linear advection on arbitrary grids. M2AN Math. Model. Numer. Anal. 35 (6), 1159–1183. ISSN 0764-583X. http://dx.doi.org/10.1051/m2an:2001152.

Despres, B., Lagoutiere, F., 2007. Numerical resolution of a two-component compressible fluid model with interfaces. Prog. Comput. Fluid Dyn. Int. J. 7 (6), 295–310. http://dx.doi.org/10.1504/PCFD.2007.014680.

Després, B., Lagoutière, F., Labourasse, E., Marmajou, I., 2010. An antidissipative transport scheme on unstructured meshes for multicomponent flows. Int. J. Finite Vol. 7 (1), 36. ISSN 1634-0655.

Dubois, F., Mehlman, G., 1996. A non-parameterized entropy correction for Roe's approximate Riemann solver. Numer. Math. 73 (2), 169–208. ISSN 0029-599X. http://dx.doi.org/10.1007/s002110050190.

Glimm, J., 1965. Solutions in the large for nonlinear hyperbolic systems of equations. Commun. Pure Appl. Math. 18, 697–715. ISSN 0010-3640.

Godlewski, E., Raviart, P.A., 1996. Numerical Approximation of Hyperbolic Systems of Conservation Laws. In: Applied Mathematical Sciences. 118, Springer-Verlag, New York.

Goodman, J.-B., LeVeque, R.-J., 1985. On the accuracy of stable schemes for 2D scalar conservation laws. Math. Comput. 45 (171), 15–21. ISSN 0025-5718. http://dx.doi.org/10.2307/2008046.

Goudon, T., Lagoutière, F., Tine, L.-M., 2013. Simulations of the Lifshitz-Slyozov equations: the role of coagulation terms in the asymptotic behavior. Math. Models Methods Appl. Sci. 23 (7), 1177–1215. ISSN 0218-2025. http://dx.doi.org/10.1142/S0218202513500061.

Harten, A., 1984. On a class of high resolution total-variation-stable finite-difference schemes. SIAM J. Numer. Anal. 21 (1), 1–23. ISSN 0036-1429. http://dx.doi.org/10.1137/0721001. With an appendix by Peter D. Lax.

Helluy, P., Jung, J., 2013. Opencl simulations of two-fluid compressible flows with a random choice method. Int. J. Finite Vol. 10, 1–38.

Helluy, P., Jung, J., 2014. Two-fluid compressible simulations on GPU cluster. In: Congrès SMAI 2013. ESAIM Proc. Surveys, vol. 45. EDP Sci., Les Ulis, pp. 349–358. http://dx.doi.org/10.1051/proc/201445036.

Hirt, C.-W., Nichols, B.-D., 1981. Volume of fluid (VOF) method for the dynamics of free boundaries. J. Comput. Phys. 39, 201–225.

Hou, T.Y., Rosakis, P., LeFloch, P.G., 1999. A level-set approach to the computation of twinning and phase-transition dynamics. J. Comput. Phys. 150 (2), 302–331. ISSN 0021-9991. http://dx.doi.org/10.1006/jcph.1998.6179.

Hu, X.-Y., Khoo, B.C., Adams, N.A., Huang, F.L., 2006. A conservative interface method for compressible flows. J. Comput. Phys. 219 (2), 553–578. ISSN 0021-9991. http://dx.doi.org/10.1016/j.jcp.2006.04.001. http://www.sciencedirect.com/science/article/pii/S0021999106001926.

Iserles, A., Strang, G., 1983. The optimal accuracy of difference schemes. Trans. of the AMS, 277 (198(2)), 779–803.

Jaouen, S., Lagoutière, F., Amsterdam, 2007. Numerical transport of an arbitrary number of components. Comput. Methods Appl. Mech. Eng. 196 (33-34), 3127–3140. ISSN 0045-7825. http://dx.doi.org/10.1016/j.cma.2007.02.007.

Kokh, S., Lagoutière, F., 2010. An anti-diffusive numerical scheme for the simulation of interfaces between compressible fluids by means of a five-equation model. J. Comput. Phys. 229 (8), 2773–2809. ISSN 0021-9991. http://dx.doi.org/10.1016/j.jcp.2009.12.003.

Kunkun, T., Beccantini, A., Corre, C., 2014. Combining discrete equations method and upwind downwind-controlled splitting for non-reacting and reacting two-fluid computations: two dimensional case. Comput. Fluids 103, 132–155. ISSN 0045-7930. http://dx.doi.org/10.1016/j.compfluid.2014.07.019. http://www.sciencedirect.com/science/article/pii/S0045793014003041.

Kuzmin, D., Möller, M., 2005a. Algebraic flux correction. I. Scalar conservation laws. In: Flux-Corrected Transport, Sci. Comput. Springer, Berlin, pp. 155–206. http://dx.doi.org/10.1007/3-540-27206-2_6.

Kuzmin, D., Möller, M., 2005b. Algebraic flux correction. II. Compressible Euler equations. In: Flux-Corrected Transport, Sci. Comput. Springer, Berlin, pp. 207–250. http://dx.doi.org/10.1007/3-540-27206-2_7.

Lagoutière, F., 2000. Modélisation mathématique et résolution numérique de problèmes de fluides compressibles à plusieurs constituants. Ph.D. thesis, Université Paris VI.

Larrouturou, B., Fézoui, L., 1989. On the equations of multi-component perfect or real gas inviscid flow. In: Nonlinear Hyperbolic Problems (Bordeaux, 1988). Lecture Notes in Math., vol. 1402. Springer, Berlin, pp. 69–98. http://dx.doi.org/10.1007/BFb0083869.

Lax, P.-D., Wendroff, B., 1960. Systems of conservation laws. Commun. Pure. Appl. Math. 13 (2), 217–237. ISSN 1097-0312. http://dx.doi.org/10.1002/cpa.3160130205.

LeFloch, P.-G., 2002. Hyperbolic Systems of Conservation Laws. The Theory of Classical and Nonclassical Shock Waves, Lectures in Mathematics ETH Zürich Birkhäuser Verlag, Basel. ISBN 3-7643-6687-7, x+294. http://dx.doi.org/10.1007/978-3-0348-8150-0.

le Roux, A.Y., 1977. A numerical conception of entropy for quasi-linear equations. Math. Comput. 31 (140), 848–872. ISSN 0025-5718.

Massoni, J., Saurel, R., Nkonga, B., Abgrall, A., 2002. Some models and Eulerian methods for interface problems between compressible fluids with heat transfer. Int. J. Heat Mass Tran. 45 (6), 1287–1307.

Michel, A., Tran, Q.H., Favennec, G., 2010. A genuinely one-dimensional upwind scheme with accuracy enhancement for multidimensional advection problems. In: ECMOR XII-12th European Conference on the Mathematics of Oil Recovery. European Association of Geoscientists and Engineers.

Noh, W.F., Woodward, P., 1976. SLIC (simple line interface calculation). Commun. Math. Sci. 59, 57–70.

Osher, S., Fedkiw, R., 2003. Level Set Methods and Dynamic Implicit Surfaces. Springer-Verlag, New York.

Osher, S., Sethian, J.-A., 1988. Fronts propagating with curvature-dependent speed: algorithms based on Hamilton-Jacobi formulations. J. Comput. Phys. 79 (1), 12–49. ISSN 0021-9991. http://dx.doi.org/10.1016/0021-9991(88)90002-2.

Roe, P.-L., 1985. Some contributions to the modelling of discontinuous flows. In: Large-Scale Computations in Fluid Mechanics, Part 2 (La Jolla, Calif., 1983), Lectures in Appl. Math., vol. 22. Amer. Math. Soc., Providence, RI, pp. 163–193.

Rongsan, C., De-kang, M., 2011. Entropy-TVD scheme for nonlinear scalar conservation laws. J. Sci. Comput. 47 (2), 150–169. ISSN 0885-7474. http://dx.doi.org/10.1007/s10915-010-9431-9.

Russo, G., Toro, E.F., Titarev, V.A., 2008. ADER-Runge-Kutta schemes for conservation laws in one space dimension. In: Hyperbolic Problems: Theory, Numerics, Applications. Springer, Berlin, pp. 929–936. http://dx.doi.org/10.1007/978-3-540-75712-2_97.

Schmidtmann, B., Seibold, B., Torrilhon, M., 2016. Relations between WENO3 and third-order limiting in finite volume methods. J. Sci. Comput. 68 (2), 624–652.

Sethian, J.-A., 1996. Level Set Methods: Evolving Interfaces in Geometry, Fluid Mechanics, Computer Vision and Materials Sciences. In: Cambridge Monograph on Applied and Computational Mathematics. Cambridge University Press.

Shu, C.-W., 2009. High order weighted essentially nonoscillatory schemes for convection dominated problems. SIAM Rev. 51 (1), 82–126. ISSN 0036-1445. http://dx.doi.org/10.1137/070679065.

Shyue, K.-M., Xiao, F., 2014. An Eulerian interface sharpening algorithm for compressible two-phase flow: the algebraic THINC approach. J. Comput. Phys. 268, 326–354. ISSN 0021-9991. http://dx.doi.org/10.1016/j.jcp.2014.03.010.

Strang, G., 1968. On the construction and comparison of difference schemes. SIAM J. Num. Anal. 5, 506–517.

Sweby, P.-K., 1984. High resolution schemes using flux limiters for hyperbolic conservation laws. SIAM J. Numer. Anal. 21 (5), 995–1011. ISSN 0036-1429. http://dx.doi.org/10.1137/0721062.

Sweby, P.-K., 1985. High resolution TVD schemes using flux limiters. In: Large-Scale Computations in Fluid Mechanics, Part 2 (La Jolla, Calif., 1983). Lectures in Appl. Math., vol. 22. Amer. Math. Soc., Providence, RI, pp. 289–309.

Titarev, V.A., Toro, E.F., 2007. Analysis of ADER and ADER-WAF schemes. IMA J. Numer. Anal. 27 (3), 616–630. ISSN 0272-4979. http://dx.doi.org/10.1093/imanum/drl033.

Toro, E.F., 2009. Riemann Solvers and Numerical Methods for Fluid Dynamics: A Practical Introduction, third ed. Springer-Verlag Berlin Heidelberg.

Toro, E.-F., Titarev, V.-A., 2005. TVD fluxes for the high-order ADER schemes. J. Sci. Comput. 24 (3), 285–309. ISSN 0885-7474. http://dx.doi.org/10.1007/s10915-004-4790-8.

Turek, S., Kuzmin, D., 2005. Algebraic flux correction. III. Incompressible flow problems. In: Flux-Corrected Transport, Sci. Comput. Springer, Berlin, pp. 251–296. http://dx.doi.org/10.1007/3-540-27206-2_8.

Warming, R.-F., Beam, R.-M., 1976. Upwind second-order difference schemes and applications in aerodynamic flows. AIAA J. 14 (9), 1241–1249. ISSN 0001-1452.

Xiao, F., Honma, Y., Kono, T., 2005. A simple algebraic interface capturing scheme using hyperbolic tangent function. Int. J. Numer. Meth. Fluids, 48 (9), 1023–1040. ISSN 1097-0363. http://dx.doi.org/10.1002/fld.975.

Xu, Z., Shu, C.-W., 2005. Anti-diffusive high order WENO schemes for Hamilton-Jacobi equations. Methods Appl. Anal. 12 (2), 169–190. ISSN 1073-2772. http://dx.doi.org/10.4310/MAA. 2005.v12.n2.a6.

Xu, Z., Shu, C.-W., 2006. Anti-diffusive finite difference WENO methods for shallow water with transport of pollutant. J. Comput. Math. 24 (3), 239–251. ISSN 0254-9409.

Yee, H.C., Warming, R.F., Harten, A., 1984. On a class of TVD schemes for gas dynamic calculations. In: Computing Methods in Applied Sciences and Engineering, VI (Versailles, 1983). North-Holland, Amsterdam, pp. 491–492.

Youngs, D.L., 1984. An interface tracking method for a 3D Eulerian hydrodynamics code. AWRE Technical Report 44/92/35. AWRE.

Zhong, X., Hou, T.Y., LeFloch, P.G., 1996. Computational methods for propagating phase boundaries. J. Comput. Phys. 124 (1), 192–216. ISSN 0021-9991. http://dx.doi.org/10.1006/jcph.1996.0053.

Chapter 5

ENO and WENO Schemes

Y.-T. Zhang* and C.-W. Shu†
*University of Notre Dame, Notre Dame, IN, United States
†Brown University, Providence, RI, United States

Chapter Outline

ABSTRACT

The weighted essentially nonoscillatory (WENO) schemes, based on the successful essentially nonoscillatory (ENO) schemes with additional advantages, are a popular class of high-order accurate numerical methods for hyperbolic partial differential equations (PDEs) and other convection-dominated problems. The main advantage of such schemes is their capability to achieve arbitrarily high-order formal accuracy in smooth regions while maintaining stable, nonoscillatory and sharp discontinuity transitions. The schemes are thus especially suitable for problems containing both strong discontinuities and complex smooth solution structures. In this chapter, we review the basic formulation of ENO and WENO schemes, outline the main ideas in constructing the schemes and discuss several of recent developments in using the schemes to solve hyperbolic type PDE problems.

Keywords: Essentially nonoscillatory (ENO) schemes, Weighted essentially nonoscillatory (WENO) schemes, High-order accuracy, Hyperbolic partial differential equations, Convection-dominated problems, Finite volume schemes, Finite difference schemes

AMS Classification Code: 65M99

Handbook of Numerical Analysis, Vol. 17. http://dx.doi.org/10.1016/bs.hna.2016.09.009

1 INTRODUCTION

High-order accuracy numerical methods are especially efficient for solving partial differential equations (PDEs) which contain complex solution structures. Here we refer to high-order accurate numerical methods by those with an order of accuracy at least three, measured by local truncation errors when the solution is smooth. High-order numerical schemes have been applied extensively in computational fluid dynamics for solving convection-dominated problems with both discontinuities/sharp gradient regions and complicated smooth structures, for example, the Rayleigh–Taylor instability simulations (Remacle et al., 2003; Shi et al., 2003; Zhang et al., 2003, 2006b), the shock vortex interactions (Grasso and Pirozzoli, 2001; Zhang et al., 2005, 2006a, 2009) and direct simulation of compressible turbulence (Taylor et al., 2007). Its resolution power over the lower-order schemes was verified in these applications.

For hyperbolic PDEs or convection-dominated problems, their solutions can develop singularities such as discontinuities, sharp gradients, discontinuous derivatives, etc. For problems containing both singularities and complicated smooth solution structures, schemes with uniform high order of accuracy in smooth regions of the solution which can also resolve singularities in an accurate and essentially nonoscillatory (ENO) fashion are desirable, since a straightforward high-order approximation for the nonsmooth region of a solution will generate instability called Gibbs phenomena. A popular class of such schemes is the class of weighted essentially nonoscillatory (WENO) schemes. WENO schemes are designed based on the successful ENO schemes (Harten et al., 1987; Shu and Osher, 1988, 1989) with additional advantages. The first WENO scheme was constructed by Liu, Osher and Chan in their pioneering paper (Liu et al., 1994) for a third-order finite volume version. Jiang and Shu (1996) constructed arbitrary-order accurate finite difference WENO schemes for efficiently computing multidimensional problems, with a general framework for the design of the smoothness indicators and nonlinear weights. The fifth-order finite difference WENO scheme in Jiang and Shu (1996) has been used in most applications.

The main idea of the WENO schemes is to form a weighted combination of several local reconstructions based on different stencils (usually referred to as small stencils) and use it as the final WENO reconstruction. The combination coefficients (also called nonlinear weights) depend on the linear weights, often chosen to increase the order of accuracy over that on each small stencil, and on the smoothness indicators which measure the smoothness of the reconstructed function in the relevant small stencils. Hence an adaptive approximation or reconstruction procedure is actually the essential part of the WENO schemes.

In this article, we review the basic formulation of ENO and WENO schemes, describe the main ideas in constructing the finite volume and finite difference versions of the schemes and emphasize several of recent developments in using the schemes to solve hyperbolic type PDE problems. The organization of this paper is as follows. ENO and WENO approximation or reconstruction procedure is explained in Section 2. In Section 3, we describe the finite volume and finite difference ENO/WENO schemes for solving hyperbolic conservation laws. Several recent developments are discussed in Section 4.

2 ENO AND WENO APPROXIMATIONS

The essential part of the ENO and WENO schemes is an adaptive approximation or reconstruction procedure. In this section, we describe the basic idea of this procedure using the third-order ENO and the fifth-order WENO approximations as examples.

2.1 Reconstruction

We first explain the reconstruction procedure which is the building block of all ENO and WENO approximations. For simplicity of the explanation, a uniform mesh $\cdots < x_0 < x_1 < x_2 < \cdots$ is used and the mesh size $\Delta x = x_{i+1} - x_i$ is a constant. The half-grid points $x_{i+1/2} = \frac{1}{2}(x_i + x_{i+1})$, and the domain is partitioned into computational cells $I_i = (x_{i-1/2}, x_{i+1/2})$, $i = \ldots, 0, 1, 2, \ldots$. We would like to emphasize that the uniform mesh assumption is not necessary for the reconstruction procedure here, although it may be needed for specific cases (for example, a uniform mesh or a smoothly varying mesh should be used in constructing a high-order conservative finite difference ENO or WENO scheme). Given the cell average values

$$\bar{u}_i = \frac{1}{\Delta x}\int_{x_{i-1/2}}^{x_{i+1/2}} u(x)dx \tag{1}$$

of a function $u(x)$ over the cells I_i for all i, we would like to find an approximation of $u(x)$ at a given point, for example, at the half-grid points $x_{i+1/2}$.

Lagrange interpolation technique can be applied here. Define the primitive function of $u(x)$ by

$$U(x) = \int_{x_{-1/2}}^{x} u(\xi)d\xi, \tag{2}$$

where the lower limit $x_{-1/2}$ is irrelevant and can be replaced by any other half-grid point, then the point values of the primitive function $U(x_{i+1/2})$ at all half-grid points can be obtained from the cell average values as the following

$$U(x_{i+1/2}) = \int_{x_{-1/2}}^{x_{i+1/2}} u(\xi)d\xi = \sum_{l=0}^{i} \Delta x \bar{u}_l. \tag{3}$$

Hence we can construct interpolation polynomials for $U(x)$, and approximate $u(x)$ by directly taking the derivative of the interpolation polynomials. Different stencils will lead to different approximations. For example, if we would like to find a polynomial $p_1(x)$ of degree at most two which reconstructs $u(x)$ on the stencil $S_1 = \{I_{i-2}, I_{i-1}, I_i\}$, namely,

$$(\bar{p}_1)_j = \frac{1}{\Delta x}\int_{x_{j-1/2}}^{x_{j+1/2}} p_1(x)dx = \bar{u}_j, \quad j = i-2, i-1, i, \tag{4}$$

a polynomial $P_1(x)$ of degree at most three will be constructed which interpolates the function $U(x)$ at the four half-grid points $x_{j+1/2}$, $j = i-3, i-2, i-1, i$ and let $p_1(x) = P_1'(x)$. The condition (4) can be easily verified for such $p_1(x)$. Hence $u(x_{i+1/2})$ is approximated by $p_1(x_{i+1/2})$. Denoting the approximation by $u_{i+1/2}^{(1)} \triangleq p_1(x_{i+1/2})$, we have an explicit formula for it:

$$u_{i+1/2}^{(1)} = \frac{1}{3}\bar{u}_{i-2} - \frac{7}{6}\bar{u}_{i-1} + \frac{11}{6}\bar{u}_i. \tag{5}$$

This is a third-order accuracy approximation

$$u_{i+1/2}^{(1)} - u(x_{i+1/2}) = O(\Delta x^3), \tag{6}$$

if the function $u(x)$ is smooth in the stencil S_1. Similarly, if a different stencil $S_2 = \{I_{i-1}, I_i, I_{i+1}\}$ is chosen, we could find a different reconstruction polynomial $p_2(x)$ of degree at most two to satisfy $(\bar{p}_2)_j = \bar{u}_j$ for $j = i-1, i, i+1$. Then a different third-order accuracy approximation $u_{i+1/2}^{(2)} \triangleq p_2(x_{i+1/2})$ can be obtained if $u(x)$ is smooth in the stencil S_2. The formula is

$$u_{i+1/2}^{(2)} = -\frac{1}{6}\bar{u}_{i-1} + \frac{5}{6}\bar{u}_i + \frac{1}{3}\bar{u}_{i+1}. \tag{7}$$

The third choice of a approximation stencil to include the "target" cell I_i is $S_3 = \{I_i, I_{i+1}, I_{i+2}\}$. The third reconstruction polynomial $p_3(x)$ of degree at most two to satisfy $(\bar{p}_3)_j = \bar{u}_j$ for $j = i, i+1, i+2$ is constructed and gives another approximation $u_{i+1/2}^{(3)} \triangleq p_3(x_{i+1/2})$. Again the approximation has third-order accuracy if $u(x)$ is smooth in the stencil S_3. The explicit formula of this approximation is

$$u_{i+1/2}^{(3)} = \frac{1}{3}\bar{u}_i + \frac{5}{6}\bar{u}_{i+1} - \frac{1}{6}\bar{u}_{i+2}. \tag{8}$$

Remark 1. Another method of reconstruction is to directly find the polynomial whose averages on the stencil's cells agree with the given values by solving the resulting linear system. This method is convenient to be applied in reconstructions on unstructured meshes. Techniques such as using a closer to orthogonal basis and least square methods were developed to improve the condition numbers of the linear system for reconstructions on high dimensional unstructured meshes. These two approaches result in the same reconstruction and hence the same error. For details, see Ciarlet and Raviart (1972), Abgrall et al. (1999), Abgrall and Sonar (1997) and Zhang and Shu (2003).

2.2 ENO Approximation

For hyperbolic PDEs or convection-dominated problems, solutions often have discontinuities (or sharp gradients). For such solutions, a fixed stencil approximation may not be adequate near discontinuities (or sharp gradient locations). Oscillations happen when the stencils contain the discontinuities (or sharp gradients).

The basic idea of ENO approximation is to adaptively avoid including the discontinuous cell (i.e., the cell on which the solution is discontinuous) in the stencil, if possible (Harten et al., 1987; Shu and Osher, 1988). For the reconstructions in Section 2.1, the ENO approximation is to choose one of the three approximations $u^{(1)}_{i+1/2}$, $u^{(2)}_{i+1/2}$ and $u^{(3)}_{i+1/2}$ given by (5), (7) and (8), respectively, based on the three stencils S_1, S_2 and S_3. The selection criterion is to compare the local smoothness of the reconstruction polynomials, measured by divided differences. Here we describe the procedure to construct a third-order ENO approximation. The job is to find a stencil which must include $x_{i-1/2}$ and $x_{i+1/2}$, such that the primitive function $U(x)$ (hence the corresponding reconstruction polynomial) is the "smoothest" in this stencil comparing with other possible stencils. The divided differences of $U(x)$ are used. We emphasize here that in the implementation of the procedure, the divided differences of $U(x)$ can be expressed completely by the divided differences of the given cell averages $\bar{u}$, without any need to reference $U(x)$ (Harten et al., 1987; Shu and Osher, 1988, 1989). Thus in cell I_i we start with a two point stencil $\widetilde{S}_2(i) = \{x_{i-1/2}, x_{i+1/2}\}$ for $U(x)$, which is equivalent to a one cell stencil $\bar{S}_1(i) = \{I_i\}$ for $\bar{v}$. Next we have two choices to expand the stencil by adding either the left neighbour $x_{i-3/2}$ or the right neighbour $x_{i+3/2}$. This is decided by comparing the absolute values of two relevant divided differences $U[x_{i-3/2}, x_{i-1/2}, x_{i+1/2}]$ and $U[x_{i-1/2}, x_{i+1/2}, x_{i+3/2}]$, and a smaller one implies that the function is "smoother" in that stencil. So, if

$$|U[x_{i-3/2}, x_{i-1/2}, x_{i+1/2}]| < |U[x_{i-1/2}, x_{i+1/2}, x_{i+3/2}]|, \tag{9}$$

the three-point stencil will be taken as

$$\widetilde{S}_3(i) = \{x_{i-3/2}, x_{i-1/2}, x_{i+1/2}\}; \tag{10}$$

otherwise, we will take the stencil

$$\widetilde{S}_3(i) = \{x_{i-1/2}, x_{i+1/2}, x_{i+3/2}\}. \tag{11}$$

This procedure can be repeated to add the next grid point to the stencil, according to the smaller of the absolute values of two relevant divided differences. For a third-order approximation, with one more grid point we will obtain the desired stencil, and one of the approximations $u^{(1)}_{i+1/2}, u^{(2)}_{i+1/2}$ or $u^{(3)}_{i+1/2}$ will be the final ENO approximation. Of course we can continue this procedure to add more grid points to the stencil for a higher-order accuracy ENO approximation.

2.3 WENO Approximation

WENO approximation is based on ENO, with additional advantages. For example, WENO approximation has higher-order accuracy than ENO approximation on the same stencils used in forming the reconstructions. WENO approximation results in more smooth numerical flux than ENO one when it is applied in solving a hyperbolic PDE.

The basic idea of WENO approximation is the following: instead of using only one of the candidate stencils to form the reconstruction, one uses a convex combination of all of them. If all three stencils S_1, S_2 and S_3 of a third-order ENO approximation are combined to form a large stencil $S = \{I_{i-2}, I_{i-1}, I_i, I_{i+1}, I_{i+2}\}$, a reconstruction polynomial $p(x)$ of degree at most four is obtained. $p(x)$ satisfies $\bar{p}_j = \bar{u}_j$, for $j = i-2, i-1, i, i+1, i+2$ and gives an approximation $u_{i+1/2} \triangleq p(x_{i+1/2})$. The explicit formula is

$$u_{i+1/2} = \frac{1}{30}\bar{u}_{i-2} - \frac{13}{60}\bar{u}_{i-1} + \frac{47}{60}\bar{u}_i + \frac{9}{20}\bar{u}_{i+1} - \frac{1}{20}\bar{u}_{i+2}. \tag{12}$$

Notice that this is a fifth-order accurate approximation if the function $u(x)$ is smooth in the large stencil S. Further investigation on the fifth-order approximation $u_{i+1/2}$ in (12) and the three third-order approximations $u^{(1)}_{i+1/2}, u^{(2)}_{i+1/2}$ and $u^{(3)}_{i+1/2}$, defined by (5), (7) and (8) reveals the following linear combination relationship:

$$u_{i+1/2} = \gamma_1 u^{(1)}_{i+1/2} + \gamma_2 u^{(2)}_{i+1/2} + \gamma_3 u^{(3)}_{i+1/2}, \tag{13}$$

where the constants γ_1, γ_2 and γ_3, satisfying $\gamma_1 + \gamma_2 + \gamma_3 = 1$, are called the linear weights. In this case they take values

$$\gamma_1 = \frac{1}{10}, \quad \gamma_2 = \frac{3}{5}, \quad \gamma_3 = \frac{3}{10}. \tag{14}$$

To deal with the situation that $u(x)$ is not smooth, WENO approximation uses the "nonlinear weights" technique to adaptively avoid including the discontinuous cell in the stencil. It chooses the final approximation as a convex combination of the three third-order approximations $u^{(1)}_{i+1/2}, u^{(2)}_{i+1/2}$ and $u^{(3)}_{i+1/2}$:

$$u_{i+1/2} = w_1 u^{(1)}_{i+1/2} + w_2 u^{(2)}_{i+1/2} + w_3 u^{(3)}_{i+1/2}, \tag{15}$$

where $w_j \geq 0$, and $w_1 + w_2 + w_3 = 1$. The nonlinear weight w_j is determined by the "smoothness indicator" β_j, which measures the relative smoothness of the function $u(x)$ in the stencil S_j. A larger β_j indicates that the function $u(x)$ is less smooth in the stencil S_j. In most of the WENO papers, the smoothness indicator β_j is chosen as in Jiang and Shu (1996),

$$\beta_j = \sum_{l=1}^{k} \Delta x^{2l-1} \int_{x_{i-1/2}}^{x_{i+1/2}} \left(\frac{d^l}{dx^l} p_j(x) \right)^2 dx, \tag{16}$$

where k is the polynomial degree of $p_j(x)$ (here, $k = 2$). This is a scaled sum of the square L^2 norms of all the derivatives of the relevant reconstruction polynomial $p_j(x)$ in the relevant cell I_i, with the scaling factor Δx^{2l-1} to make the final explicit formulas for the smoothness indicators independent on the grid size Δx. In this example, the explicit formulas of the smoothness indicators are the following quadratic functions of the cell average values of $u(x)$ in the relevant stencils:

$$\begin{aligned}
\beta_1 &= \frac{13}{12}(\bar{u}_{i-2} - 2\bar{u}_{i-1} + \bar{u}_i)^2 + \frac{1}{4}(\bar{u}_{i-2} - 4\bar{u}_{i-1} + 3\bar{u}_i)^2, \\
\beta_2 &= \frac{13}{12}(\bar{u}_{i-1} - 2\bar{u}_i + \bar{u}_{i+1})^2 + \frac{1}{4}(\bar{u}_{i-1} - \bar{u}_{i+1})^2, \\
\beta_3 &= \frac{13}{12}(\bar{u}_i - 2\bar{u}_{i+1} + \bar{u}_{i+2})^2 + \frac{1}{4}(3\bar{u}_i - 4\bar{u}_{i+1} + \bar{u}_{i+2})^2.
\end{aligned} \tag{17}$$

With these smoothness indicators, the nonlinear weights are defined as

$$w_j = \frac{\alpha_j}{\alpha_1 + \alpha_2 + \alpha_3}, \quad \alpha_j = \frac{\gamma_j}{(\epsilon + \beta_j)^2}, \quad j = 1,2,3. \tag{18}$$

Here ϵ is a small positive number used to avoid the denominator becoming zero and is typically chosen to be $\epsilon = 10^{-6}$ in computations. It is verified in Jiang and Shu (1996) that with such nonlinear weights, the WENO approximations (15) is fifth-order accurate if the function $u(x)$ is smooth in the large stencil S. If $u(x)$ is not smooth in a stencil S_j but is smooth in at least one of the other two stencils, the WENO approximations would guarantee a nonoscillatory result since the contribution from any stencil containing the discontinuity of $u(x)$ has an essentially zero weight.

Remark 2. There may be situations in which all small stencils contain the discontinuity. For example, there may be a discontinuity point in the cell I_i. It turns out that this seemingly difficult case is actually not problematic in ENO or WENO approximations, because the reconstruction polynomials are all essentially monotone in I_i (Harten et al., 1986).

3 ENO AND WENO SCHEMES FOR HYPERBOLIC CONSERVATION LAWS

In this section, we describe the finite volume and finite difference ENO and WENO schemes for solving hyperbolic conservation laws. First the simple one-dimensional scalar equation

$$u_t + f(u)_x = 0 \tag{19}$$

is used to show the ideas of constructing the schemes.

3.1 Finite Volume Schemes

In finite volume schemes, the integral form of the conservation law (19)

$$\frac{d\bar{u}_i(t)}{dt} + \frac{1}{\Delta x_i}\left(f(u_{i+1/2}) - f(u_{i-1/2})\right) = 0 \tag{20}$$

is considered. Here $\bar{u}_i = \frac{1}{\Delta x_i}\int_{I_i} u(x,t)dx$ is the spatial cell average of the solution $u(x, t)$ in the cell I_i. For linear stability of the schemes, upwinding property (i.e. numerical schemes should propagate solution information in the same characteristic direction as that of a hyperbolic PDE) is required. We replace $f(u_{i+1/2})$ by $\hat{f}(u^-_{i+1/2}, u^+_{i+1/2})$, where $\hat{f}(a,b)$ is a monotone numerical flux satisfying the following conditions: (1) $\hat{f}(a,b)$ is nondecreasing in its first argument and nonincreasing in its second argument; (2) $\hat{f}(a,b)$ is consistent with the physical flux $f(u)$, i.e., $\hat{f}(u,u) = f(u)$; (3) $\hat{f}(a,b)$ is Lipschitz continuous with respect to both arguments. Some examples of monotone fluxes include the Godunov flux

$$\hat{f}(a,b) = \begin{cases} \min_{a \le x \le b} f(u) & \text{if } a \le b, \\ \max_{b \le x \le a} f(u) & \text{if } a > b; \end{cases} \tag{21}$$

the Engquist–Osher flux

$$\hat{f}(a,b) = \int_0^a \max(f'(u), 0)du + \int_0^b \min(f'(u), 0)du + f(0); \tag{22}$$

and the Lax–Friedrichs flux

$$\hat{f}(a,b) = \frac{1}{2}[f(a) + f(b) - \alpha(b-a)], \tag{23}$$

where $\alpha = \max_u |f'(u)|$ is a constant and the maximum is taken over the relevant range of u. $u^-_{i+1/2}$ and $u^+_{i+1/2}$ are ENO/WENO approximations based on cell average values in stencils one cell biased to the left and one cell biased to the right, respectively. For a third-order ENO or a fifth-order WENO scheme, the approximation $u^-_{i+1/2}$ uses cell average values in cells $\{I_{i-2}, I_{i-1}, I_i, I_{i+1}, I_{i+2}\}$, while the approximation $u^+_{i+1/2}$ uses those in cells $\{I_{i-1}, I_i, I_{i+1}, I_{i+2}, I_{i+3}\}$. Section 2 gives the detailed procedure for these ENO/WENO reconstructions.

With a monotone numerical flux and ENO/WENO reconstructions, the integral form (20) can be written as a method-of-lines ordinary differential equation (ODE) system

$$\frac{d\bar{u}_i(t)}{dt} = -\frac{1}{\Delta x_i}\left[\hat{f}(u^-_{i+1/2}, u^+_{i+1/2}) - \hat{f}(u^-_{i-1/2}, u^+_{i-1/2})\right]. \tag{24}$$

The ODE system can be discretized by a high-order total variation diminishing (TVD) Runge–Kutta time discretization method, also known as the strong stability preserving (SSP) method (Gottlieb et al., 2001; Shu and Osher, 1988). For example, the most popular TVD Runge–Kutta method is the third-order accurate one given in Shu and Osher (1988). Other time discretization methods may also be applied. For example, the method which uses a Lax–Wendroff procedure to convert all time derivatives into spatial derivatives and discretizes all the spatial derivatives to the correct order of accuracy, e.g., see Harten et al. (1987), Titarev and Toro (2005) and Qiu and Shu (2003).

3.2 Finite Difference Schemes

Finite difference schemes use point values $\{u_i\}$ of the numerical solution to approximate the PDE directly. A finite difference scheme for hyperbolic conservation laws is required to be in conservation form. For Eq. (19), it is

$$\frac{du_i(t)}{dt} + \frac{1}{\Delta x}\left(\hat{f}_{i+1/2} - \hat{f}_{i-1/2}\right) = 0, \tag{25}$$

where $\hat{f}$ is the numerical flux. $\hat{f}_{i+1/2} = \hat{f}(u_{i-p}, \ldots, u_{i+q})$, and it is consistent with the physical flux $\hat{f}(u, \ldots, u) = f(u)$ and is Lipschitz continuous with respect to all its arguments. The scheme is r-th order accurate if

$$\frac{1}{\Delta x}\left(\hat{f}_{i+1/2} - \hat{f}_{i-1/2}\right) = f(u)_x|_{x=x_i} + O(\Delta x^r), \tag{26}$$

when $u(x)$ is smooth in the stencil.

It was found in Shu and Osher (1989) that one can directly use the same ENO/WENO reconstruction procedure in a finite volume scheme to compute the numerical flux $\hat{f}_{i+1/2}$. By defining the cell averages $\bar{h}_i$ of a function $h(x)$ to be $\bar{h}_i \triangleq f(u_i)$, we apply the high-order accuracy ENO/WENO approximation in Section 2 to compute the point values $h(x_{i+1/2})$. Then the numerical flux is obtained

$$\hat{f}_{i+1/2} = h(x_{i+1/2}). \tag{27}$$

For the purpose of linear stability (upwinding), a flux splitting is performed, i.e.

$$f(u) = f^+(u) + f^-(u), \tag{28}$$

where $f^+(u)$ and $f^-(u)$ satisfy $\frac{d}{du}f^+(u) \geq 0, \frac{d}{du}f^-(u) \leq 0$. Upwinding requires that the approximation $\hat{f}^+_{i+1/2}$ for $f^+(u_{i+1/2})$ uses a biased stencil with one more point to the left, and $\hat{f}^-_{i+1/2}$ for $f^-(u_{i+1/2})$ uses a biased stencil with one more point to the right. A very popular flux splitting is the Lax–Friedrichs splitting

$$f^+(u) = \frac{1}{2}(f(u) + \alpha u), \quad f^-(u) = \frac{1}{2}(f(u) - \alpha u), \tag{29}$$

where $\alpha = \max_u |f'(u)|$. Other flux splittings can also be used (Jiang and Shu, 1996). Using the high-order accuracy ENO/WENO approximation to obtain $\hat{f}^+_{i+1/2}$ and $\hat{f}^-_{i+1/2}$, we have the final numerical flux

$$\hat{f}_{i+1/2} = \hat{f}^+_{i+1/2} + \hat{f}^-_{i+1/2}. \tag{30}$$

The resulting ODE system (25) can again be evolved by a high-order time discretization scheme, for example, the third-order accurate TVD Runge–Kutta method.

3.3 Remarks on Multidimensional Problems and Systems

In this section, we make several remarks about using ENO/WENO schemes in solving multidimensional problems and PDE systems. A high-order finite difference scheme for solving a multidimensional problem can be performed dimension by dimension directly on a uniform Cartesian or a smooth curvilinear mesh. Its computational cost is exactly the same as in the one-dimensional case per point per direction. However, for nonuniform and unstructured meshes, high-order finite difference scheme can *not* be applied and a finite volume scheme has to be used. A high-order finite volume scheme is generally more expensive than a finite difference scheme of the same order of accuracy if the same mesh and the same reconstruction procedure are used, since even on a Cartesian mesh, the direct dimension by dimension ENO/WENO reconstruction can *not* be performed for a nonlinear hyperbolic conservation law. For example, in two dimensions, a finite volume scheme with accuracy order higher than two is two to five times as expensive as a finite difference one, depending on the specific coding and computer type. This discrepancy in cost is even bigger for higher dimension problems. A detailed comparison of finite volume and finite difference schemes for solving multidimensional problems in the context of ENO approximations can be found in Casper et al. (1994).

For systems of hyperbolic conservation laws, the ENO/WENO schemes have the same structure as those for the scalar cases. A monotone flux is replaced by an exact or approximate Riemann solver (Toro, 2009). The ENO/WENO reconstruction can be performed either componentwise or in local characteristic directions. Usually, componentwise reconstruction produces satisfactory results for schemes up to third-order accuracy, while characteristic reconstruction produces better nonoscillatory results for higher-order accuracy, albeit with an increased computational cost. Details about the local characteristic decomposition procedure can be found in, e.g., Harten et al. (1987), Shu and Osher (1988) and Shu et al. (1992).

4 SELECTED TOPICS OF RECENT DEVELOPMENTS

In this section, we discuss a few selected topics of recent developments in using ENO/WENO schemes to solve hyperbolic or convection–diffusion problems.

4.1 Unstructured Meshes

While ENO/WENO schemes on structured (either Cartesian or smooth curvilinear) meshes are quite mature, the development of simple and robust ENO/WENO schemes on unstructured meshes (e.g. arbitrary triangular or tetrahedral meshes) for dealing with complex domain geometries is less advanced. The finite volume approach must be used to design ENO/WENO schemes on unstructured meshes for solving hyperbolic conservation laws. We use the two-dimensional conservation law

$$\frac{\partial u}{\partial t}+\frac{\partial f(u)}{\partial x}+\frac{\partial g(u)}{\partial y}=0 \tag{31}$$

as an example, and the computational control volumes are triangles $\{\triangle_i\}$. The semidiscrete finite volume scheme of (31) is formulated as

$$\frac{d\bar{u}_i(t)}{dt}+\frac{1}{|\triangle_i|}\int_{\partial\triangle_i}F\cdot ndS=0 \tag{32}$$

where the cell average $\bar{u}_i(t)=\frac{1}{|\triangle_i|}\int_{\triangle_i}udxdy$, $F=(f,g)^T$, and n is the outward unit normal of the triangle boundary $\partial\triangle_i$. In (32), the line integral is discretized by a q-point Gaussian quadrature formula,

$$\int_{\partial\triangle_i}F\cdot nds\approx\sum_{k=1}^{3}S_k\sum_{j=1}^{q}\widetilde{w}_jF\left(u\left(G_j^{(k)},t\right)\right)\cdot n_k \tag{33}$$

where S_k is the length of the k-th side of $\partial\triangle_i$, $G_j^{(k)}$ and $\widetilde{w}_j$ are the Gaussian quadrature points and weights, respectively, and $F\left(u\left(G_j^{(k)},t\right)\right)\cdot n_k$ is approximated by a numerical flux. For example, if the Lax–Friedrichs flux is used, then we have

$$F\left(u\left(G_j^{(k)},t\right)\right)\cdot n_k \approx \frac{1}{2}\left[\left(F\left(u^-\left(G_j^{(k)},t\right)\right)+F\left(u^+\left(G_j^{(k)},t\right)\right)\right)\right.$$
$$\left.\cdot n_k - \alpha\left(u^+\left(G_j^{(k)},t\right)-u^-\left(G_j^{(k)},t\right)\right)\right], \tag{34}$$

where α is taken as an upper bound for the magnitude of the eigenvalues of the Jacobian in the n_k direction, and u^- and u^+ are the values of u inside the triangle and outside the triangle (inside the neighbouring triangle) at the Gaussian point. q is determined by the order of accuracy of the schemes. For example, if a third-order finite volume scheme is designed, then the two-point Gaussian quadrature $q = 2$ is used. For the line with endpoints P_1 and P_2, the Gaussian quadrature points are $G_1 = cP_1 + (1 - c)P_2$, $G_2 = cP_2 + (1 - c)P_1$, where $c = \frac{1}{2} + \frac{\sqrt{3}}{6}$; and the Gaussian quadrature weights are $\widetilde{w}_1 = \widetilde{w}_2 = \frac{1}{2}$.

The next key step is to build a high-order ENO/WENO reconstruction for the point values at the Gaussian quadrature points. About the development of high-order ENO reconstructions on unstructured meshes, the reader is referred to Abgrall (1994), Abgrall and Sonar (1997) and Augoula and Abgrall (2000). For WENO reconstructions, the big stencil S consisting of triangles is a union of small stencils $\{S_m : m = 1, 2, \ldots, N\}$. Cell average values of u in S are used to construct a polynomial $p(x, y)$, which will have the same cell average as u on the target cell Δ_0 (i.e., the control volume cell). WENO reconstructions need to obtain a linear combination of reconstructions on small stencils. The reconstruction values at the Gaussian points should satisfy

$$p\left(x^G,y^G\right) = \sum_{m=1}^{N} \gamma_m p_m\left(x^G,y^G\right), \tag{35}$$

where (x^G, y^G) is a Gaussian point, p_m is a reconstruction polynomial on a small stencil S_m, and γ_m is the linear weight. Based on (35), nonlinear WENO reconstruction values at the Gaussian points are

$$p_{\text{weno}}\left(x^G,y^G\right) = \sum_{m=1}^{N} \omega_m p_m\left(x^G,y^G\right), \tag{36}$$

where ω_m is a nonlinear WENO weight defined as

$$\omega_m = \frac{\widetilde{\omega}_m}{\sum_{m=1}^{N}\widetilde{\omega}_m}, \qquad \widetilde{\omega}_m = \frac{\gamma_m}{(\epsilon + IS_m)^2}. \tag{37}$$

As that for the WENO reconstructions on structured meshes, IS_m is the smoothness indicator for the m-th reconstruction polynomial $p_m(x, y)$ associated with the m-th small stencil S_m, and ϵ is a small positive number to avoid the denominator to become 0. The smoothness indicator measures how smooth the function p_m is on the target cell Δ_0: the smaller the smoothness indicator, the smoother the function p_m is on Δ_0. For two-dimensional problems, it is defined as

$$IS_m = \sum_{1\le|\alpha|\le k} \int_{\Delta_0} |\Delta_0|^{|\alpha|-1} (D^{\alpha} p_m(x,y))^2 dxdy, \tag{38}$$

where k is the degree of polynomial p_m, α is a multiindex and D is the derivative operator. For the definition of the smoothness indicator for three-dimensional problems, the reader is referred to Zhang and Shu (2009).

There are two types of WENO reconstructions on unstructured meshes in the literature. The major difference between them is the different method to construct small stencils and find linear weights. The first type (type I) reconstruction has an order of accuracy not higher than that of the reconstruction on each small stencil. This is similar as ENO schemes. For this type of WENO reconstructions, the nonlinear weights do not contribute towards the increase of the order of accuracy, and they are designed purely for the purpose of nonlinear stability, or to avoid spurious oscillations. Because type I WENO schemes just need to choose the linear weights as arbitrary positive numbers for better linear stability (e.g. the centred small stencil is assigned a larger linear weight than the others), they are easier to construct than the type II WENO schemes discussed in the following paragraph. For Type I WENO reconstructions, see e.g., Friedrichs (1998) and Titarev et al. (2010) for two-dimensional triangulations and Dumbser and Käser (2007) and Dumbser et al. (2007) for three-dimensional triangulations.

The second type (type II) consists of WENO schemes whose order of accuracy is higher than that of the reconstruction on each small stencil. For example, the third-order WENO scheme on two-dimensional triangular meshes in Hu and Shu (1999) is based on second-order accuracy linear polynomial reconstructions on small stencils, and the fourth-order WENO scheme in Hu and Shu (1999) is based on third-order accuracy quadratic polynomial reconstructions on small stencils. For similar WENO schemes on two-dimensional triangular meshes for solving Hamilton–Jacobi equations, we refer to see Zhang and Shu (2003) and Levy et al. (2006) and for WENO reconstructions on three-dimensional tetrahedral meshes, which belong to type II as well, we refer to see Zhang and Shu (2009). Type II WENO schemes are more difficult to construct, however they have a much more compact stencil than type I WENO schemes of the same accuracy, which is an advantage in applications, such as when the WENO methodology is used as limiters for the discontinuous Galerkin methods, e.g., Qiu and Shu (2005)

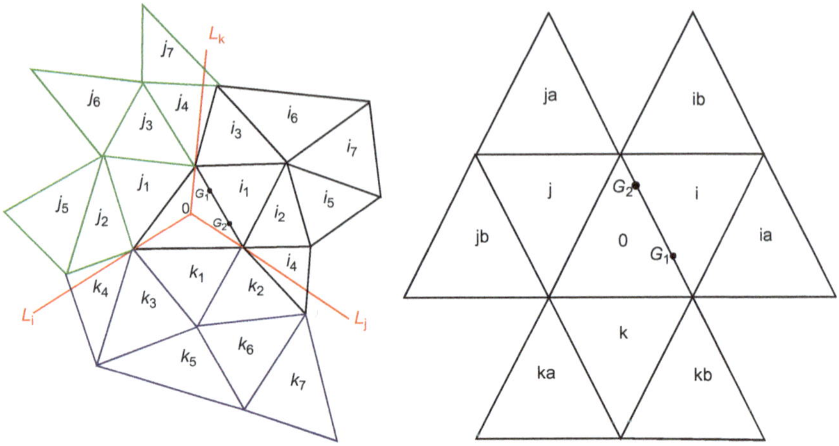

FIG. 1 A big stencil S for a third-order WENO reconstruction. *Left*: the type I; *right*: the type II. *Pictures are reproduced from Liu, Y., Zhang, Y.T., 2013. A robust reconstruction for unstructured WENO schemes. J. Sci. Comput. 54, 603–621, with permission of Springer.*

and Zhu et al. (2008). For example, see the big stencil S for a third-order WENO reconstruction of type I and II in Fig. 1. The type I reconstruction needs more layers of neighbouring triangles of the target cell Δ_0 (i.e., the cell "0" in Fig. 1) than that for the type II reconstruction. Actually the big stencil of the type II reconstruction (the right picture in Fig. 1) is just the central small stencil in the type I reconstruction.

A crucial step in building a type II WENO scheme on unstructured meshes is to construct lower-order polynomials whose weighted average will give the same result as the high-order reconstruction at each Gaussian quadrature point for the flux integral on the element boundary. This step is actually the most difficult step in designing a robust second type high-order WENO schemes on unstructured meshes, since we can not guarantee the quality of the unstructured meshes when the domain geometry is very complicated. Especially, when the spatial domain has higher dimensions (e.g. three-dimensional problems) and complex geometry, the quality of the unstructured meshes is hard to control. Distorted local mesh geometries can be easily generated. The local linear system for finding linear weights could have very large condition number or is even singular at the places where mesh quality is bad (e.g. there are very obtuse triangles). This may lead to negative and very large linear weights in type II WENO schemes, or even the linear weights do *not* exist. For mild negative linear weights, the splitting technique developed in Shi et al. (2002) can be applied effectively. For the degenerate cases in various mesh geometries that linear weights are negative and very large or do *not* exist, a more robust approach is needed. In a recent work (Liu and Zhang, 2013), we hybrid the approaches of type II and type I WENO schemes, and avoid the appearance of negative and very large linear weights no matter how bad the quality of the unstructured meshes is. The idea is to switch to the approach of assigning linear weights of type I WENO schemes at the places where the linear weight system of type II WENO scheme is ill-posed or singular, i.e., the linear weights are negative and very large (larger than a preset threshold value) or do *not* exist. The trade-off is that the compactness of the type II WENO scheme will be lost at these places. But we obtain a robust reconstruction with respect to the quality of unstructured meshes and the complexity of the domain. Furthermore, since for a general triangulation distorted mesh geometries only occur in minor parts of the whole domain, the overall percentage of the places where the type I WENO approach is applied is quite small. We refer to Liu and Zhang (2013) for more details.

4.2 Steady State Problems

Steady state problems for hyperbolic PDEs are common mathematical models appearing in many applications, such as fluid mechanics, optimal control, differential games, image processing and computer vision, geometric optics, etc. Solution information of these boundary value problems propagates along characteristics starting from the boundary. A large nonlinear system is obtained after spatial discretization of a steady state hyperbolic PDE by a high-order WENO scheme. It is still a challenging problem how to solve the large nonlinear system resulting from WENO discretization.

There are at least two factors which may affect efficiency and robustness of computation. One is that a high-order accurate shock capturing scheme such as a WENO scheme often suffers from difficulties in its convergence towards steady state solutions. In Zhang and Shu (2007), a systematic study was carried out and discovered that slight postshock oscillations actually cause this problem. A new smoothness indicator (Zhang and Shu, 2007) and upwind-biased interpolation technique (Zhang et al., 2011) have been developed to improve the convergence of fifth-order WENO scheme for solving steady state of Euler systems. The other factor affecting the performance of computation is the iterative scheme designed for the nonlinear system.

For a highly nonlinear system derived from high-order WENO spatial discretization, one way is to solve it directly with Newton iterations (e.g. Hu et al., 2011), or a more robust method such as the homotopy method (Hao et al., 2013). A major advantage of solving the nonlinear system by Newton iterations or the homotopy method is that the resulting methods are free of the CFL condition, hence have linear computational complexity in solving these boundary value problems. Another way is to solve the large WENO system by fixed-point iterative schemes of Jacobi type or Gauss–Seidel type. The popular time marching approach for solving steady state problems is essentially a Jacobi type fixed-point iterative method. Starting from an initial condition, the numerical solution evolves into a steady state by using a time stepping scheme (e.g. Abgrall and Mezine, 2004; Abgrall and Roe, 2003; Chou and Shu, 2006; Jiang and Shu, 1996). A big advantage of the time marching method is that the computed steady state is stable and usually carries physical properties of the system and the initial condition. However, the computational efficiency of time marching method for obtaining a steady state solution is restricted by the CFL condition. This can be improved by the "fast sweeping" technique. Fast sweeping methods utilize alternating sweeping strategy to cover a family of characteristics in a certain direction simultaneously in each sweeping order. Coupled with the Gauss–Seidel iterations, these methods can achieve a fast convergence speed for computations of steady state solutions of hyperbolic PDEs by high-order WENO schemes (Xiong et al., 2010; Zhang et al., 2006c,d).

Furthermore, to compute steady state of hyperbolic conservation laws, the forward Euler time marching is preferred since only one stage and one step is used, as time direction accuracy has no effects on the numerical accuracy of steady state solutions. However, a high-order WENO scheme (e.g. the fifth-order WENO scheme) coupled with the first-order forward Euler time discretization is linearly unstable (Wang and Spiteri, 2007). Hence a high-order time discretization needs to be coupled with a high-order WENO scheme for steady state problems, which increases the number of iterations for the Jacobi type fixed-point scheme to converge. In a recent work (Wu et al., 2016), based on fifth-order WENO schemes which improve the convergence of the classical WENO schemes by removing slight postshock oscillations (Zhang and Shu, 2007; Zhang et al., 2011), we designed fifth-order fixed-point sweeping WENO methods for steady state of hyperbolic conservation laws. It is discovered that the fast sweeping technique can largely improve the stability of high-order spatial scheme with the forward Euler time marching. Extensive numerical experiments are performed in Wu et al. (2016) to compare four different iterative schemes including the regular forward Euler and Runge–Kutta time marching methods, and the ones coupled with fast sweeping technique. All numerical examples show that the forward Euler time discretization with fast sweeping technique is the most efficient approach for fifth-order WENO computations of steady state of hyperbolic conservation laws.

4.3 Time Discretizations for Convection–Diffusion Problems

High-order WENO schemes are often used to discretize nonlinear convection terms for convection–diffusion PDEs, to deal with the convection-dominated cases or a spatial mixture of convection-dominated and diffusion-dominated cases. A general convection–diffusion problem may contain significant diffusion in some regions and couple nonlinear stiff reaction terms, with dominated convection in other regions. Computational efficiency by using high-order WENO schemes to solve such problems depends heavily on robust time discretizations which permit large time step sizes, since the regular explicit time schemes require very small time step sizes. A fully implicit discretization by using implicit Runge–Kutta or backward difference formula (BDF) methods (see, e.g. Hairer and Wanner, 1991) has large linear stability regions but typically requires the solution of large nonlinear coupled system of algebraic equations. Especially high-order WENO schemes have high level of nonlinearity in the nonlinear weights. Certain iterative schemes such as Newton's method do not seem to be robust near strong shocks for a large time step, and special treatment is required (Gottlieb et al., 2006). Another approach is to avoid solving completely coupled nonlinear systems, for example, to use implicit–explicit (IMEX) Runge–Kutta methods (see, e.g. Kennedy and Carpenter, 2003). To deal with stiffness in a convection–diffusion–reaction problem, exponential integrator is an efficient tool. Recently, implicit integration factor (IIF) WENO methods were developed for solving stiff nonlinear convection–diffusion–reaction equations (Jiang and Zhang, 2013). The methods can be designed for arbitrary order of accuracy and no large nonlinear coupled algebraic system needs to be solved. The stiffness of the system is resolved well and the methods are stable by using time step sizes which are mainly determined by the nonstiff hyperbolic part of the system. To efficiently calculate large matrix exponentials, Krylov subspace approximation is applied in the methods. The time discretizations in Jiang and Zhang (2013) are multistep methods. In Jiang and Zhang (2016), single-step IIF-WENO methods were developed for solving stiff convection–diffusion–reaction equations. The methods are designed carefully to avoid generating positive exponentials in the matrix exponentials, which is necessary for the stability of the exponential integrator schemes.

4.4 Accuracy Enhancement

Efforts have been made to improve the accuracy in high-order WENO schemes. Strategies include modifying the linear or nonlinear weights, modifying the smoothness indicators, or improving the dissipation and/or dispersion properties of WENO schemes. For example, in Henrick et al. (2005), a mapping function was designed to modify the nonlinear weights in Jiang and Shu (1996). The resulting nonlinear weights improve accuracy of the WENO schemes at smooth extrema. In Borges et al. (2008) and Castro et al. (2011), the classical smoothness indicators in Jiang and Shu (1996) were combined to form new smoothness indicators, which also improve accuracy and resolution of the WENO schemes without the mapping. For high frequency wave computations, the resolution can be enhanced via optimizing the dissipation and/or dispersion of the WENO schemes (for example, see Hu et al., 2015; Martin et al., 2006; Ponziani et al., 2003; Wang and Chen, 2001).

ACKNOWLEDGEMENTS

This research was supported by NSF grants DMS-1620108 and DMS-1418750.

REFERENCES

Abgrall, R., 1994. On essentially non-oscillatory schemes on unstructured meshes: analysis and implementation. J. Comput. Phys. 114, 45–58.

Abgrall, R., Mezine, M., 2004. Construction of second order accurate monotone and stable residual distribution schemes for steady problems. J. Comput. Phys. 195, 474–507.

Abgrall, R., Roe, P.L., 2003. High order fluctuation scheme on triangular meshes. J. Sci. Comput. 19, 3–36.

Abgrall, R., Sonar, T., 1997. On the use of Muehlbach expansions in the recovery step of ENO methods. Numer. Math. 76, 1–25.

Abgrall, R., Lantéri, S., Sonar, T., 1999. ENO approximations for compressible fluid dynamics. ZAMM - J. Appl. Math. Mech. 79, 3–28.

Augoula, S., Abgrall, R., 2000. High order numerical discretization for Hamilton-Jacobi equations on triangular meshes. J. Sci. Comput. 15, 197–229.

Borges, R., Carmona, M., Costa, B., Don, W.S., 2008. An improved weighted essentially non-oscillatory scheme for hyperbolic conservation laws. J. Comput. Phys. 227, 3191–3211.

Casper, J., Shu, C.W., Atkins, H.L., 1994. Comparison of two formulations for high-order accurate essentially nonoscillatory schemes. AIAA J. 32, 1970–1977.

Castro, M., Costa, B., Don, W.S., 2011. High order weighted essentially non-oscillatory WENO-Z schemes for hyperbolic conservation laws. J. Comput. Phys. 230, 1766–1792.

Chou, C.S., Shu, C.W., 2006. High order residual distribution conservative finite difference WENO schemes for steady state problems on non-smooth meshes. J. Comput. Phys. 214, 698–724.

Ciarlet, P.G., Raviart, P.A., 1972. General Lagrange and Hermite interpolation in $\mathbf{R}^n$ with application to finite element methods. Arch. Ration. Mech. Anal. 42, 177–199.

Dumbser, M., Käser, M., 2007. Arbitrary high order non-oscillatory finite volume schemes on unstructured meshes for linear hyperbolic systems. J. Comput. Phys. 221, 693–723.

Dumbser, M., Käser, M., Titarev, V.A., Toro, E.F., 2007. Quadrature-free non-oscillatory finite volume schemes on unstructured meshes for nonlinear hyperbolic systems. J. Comput. Phys. 226, 204–243.

Friedrichs, O., 1998. Weighted essentially non-oscillatory schemes for the interpolation of mean values on unstructured grids. J. Comput. Phys. 144, 194–212.

Gottlieb, S., Shu, C.W., Tadmor, E., 2001. Strong stability-preserving high-order time discretization methods. SIAM Rev. 43, 89–112.

Gottlieb, S., Mullen, J.S., Ruuth, S.J., 2006. A fifth order flux implicit WENO method. J. Sci. Comput. 27, 271–287.

Grasso, F., Pirozzoli, S., 2001. Simulations and analysis of the coupling process of compressible vortex pairs: free evolution and shock induced coupling. Phys. Fluids 13, 1343–1366.

Hairer, E., Wanner, G., 1991. Solving Ordinary Differential Equations II, Stiff and Differential-Algebraic Problems. Springer, Berlin.

Hao, W., Hauenstein, J.D., Shu, C.W., Sommese, A.J., Xu, Z., Zhang, Y.T., 2013. A homotopy method based on WENO schemes for solving steady state problems of hyperbolic conservation laws. J. Comput. Phys. 250, 332–346.

Harten, A., Osher, S., Engquist, B., Chakravarthy, S., 1986. Some results on uniformly high order accurate essentially non-oscillatory schemes. Appl. Numer. Math. 2, 347–377.

Harten, A., Engquist, B., Osher, S., Chakravarthy, S., 1987. Uniformly high order essentially non-oscillatory schemes III. J. Comput. Phys. 71, 231–303.

Henrick, A.K., Aslam, T.D., Powers, J.M., 2005. Mapped weighted essentially non-oscillatory schemes: achieving optimal order near critical points. J. Comput. Phys. 207, 542–567.

Hu, C., Shu, C.W., 1999. Weighted essentially non-oscillatory schemes on triangular meshes. J. Comput. Phys. 150, 97–127.

Hu, G.H., Li, R., Tang, T., 2011. A robust WENO type finite volume solver for steady Euler equations on unstructured grids. Commun. Comput. Phys. 9, 627–648.

Hu, X.Y., Wang, B., Adams, N.A., 2015. An efficient low-dissipation hybrid weighted essentially non-oscillatory scheme. J. Comput. Phys. 301, 415–424.

Jiang, G., Shu, C.W., 1996. Efficient implementation of weighted ENO schemes. J. Comput. Phys. 126, 202–228.

Jiang, T., Zhang, Y.T., 2013. Krylov implicit integration factor WENO methods for semilinear and fully nonlinear advection-diffusion-reaction equations. J. Comput. Phys. 253, 368–388.

Jiang, T., Zhang, Y.T., 2016. Krylov single-step implicit integration factor WENO methods for advection-diffusion-reaction equations. J. Comput. Phys. 311, 22–44.

Kennedy, C.A., Carpenter, M.H., 2003. Additive Runge-Kutta schemes for convection-diffusion-reaction equations. Appl. Numer. Math. 44, 139–181.

Levy, D., Nayak, S., Shu, C.W., Zhang, Y.T., 2006. Central WENO schemes for Hamilton-Jacobi equations on triangular meshes. SIAM J. Sci. Comput. 28, 2229–2247.

Liu, Y., Zhang, Y.T., 2013. A robust reconstruction for unstructured WENO schemes. J. Sci. Comput. 54, 603–621.

Liu, X.D., Osher, S., Chan, T., 1994. Weighted essentially non-oscillatory schemes. J. Comput. Phys. 115, 200–212.

Martin, M.P., Taylor, E.M., Wu, M., Weirs, V.G., 2006. A bandwidth-optimized WENO scheme for the effective direct numerical simulation of compressible turbulence. J. Comput. Phys. 220, 270–289.

Ponziani, D., Pirozzoli, S., Grasso, F., 2003. Development of optimized weighted-ENO schemes for multiscale compressible flows. Int. J. Numer. Methods Fluids 42, 953–977.

Qiu, J., Shu, C.W., 2003. Finite difference WENO schemes with Lax-Wendroff-type time discretizations. SIAM J. Sci. Comput. 24, 2185–2198.

Qiu, J., Shu, C.W., 2005. Runge-Kutta discontinuous Galerkin method using WENO limiters. SIAM J. Sci. Comput. 26, 907–929.

Remacle, J.F., Flaherty, J.E., Shephard, M.S., 2003. An adaptive discontinuous Galerkin technique with an orthogonal basis applied to compressible flow problems. SIAM Rev. 45, 53–72.

Shi, J., Hu, C., Shu, C.W., 2002. A technique of treating negative weights in WENO schemes. J. Comput. Phys. 175, 108–127.

Shi, J., Zhang, Y.T., Shu, C.W., 2003. Resolution of high order WENO schemes for complicated flow structures. J. Comput. Phys. 186, 690–696.

Shu, C.W., Osher, S., 1988. Efficient implementation of essentially non-oscillatory shock capturing schemes. J. Comput. Phys. 77, 439–471.

Shu, C.W., Osher, S., 1989. Efficient implementation of essentially non-oscillatory shock capturing schemes II. J. Comput. Phys. 83, 32–78.

Shu, C.W., Zang, T.A., Erlebacher, G., Whitaker, D., Osher, S., 1992. High-order ENO schemes applied to two- and three-dimensional compressible flow. Appl. Numer. Math. 9, 45–71.

Taylor, E.M., Wu, M.W., Martin, M.P., 2007. Optimization of nonlinear error for weighted essentially non-oscillatory methods in direct numerical simulations of compressible turbulence. J. Comput. Phys. 223, 384–397.

Titarev, V.A., Toro, E.F., 2005. ADER schemes for three-dimensional non-linear hyperbolic systems. J. Comput. Phys. 204, 715–736.

Titarev, V.A., Tsoutsanis, P., Drikakis, D., 2010. WENO schemes for mixed-element unstructured meshes. Commun. Comput. Phys. 8, 585–609.

Toro, E.F., 2009. Riemann Solvers and Numerical Methods for Fluid Dynamics: A Practical Introduction. Springer-Verlag, Berlin, Heidelberg.

Wang, Z.J., Chen, R.F., 2001. Optimized weighted essentially nonoscillatory schemes for linear waves with discontinuity. J. Comput. Phys. 174, 381–404.

Wang, R., Spiteri, R.J., 2007. Linear instability of the fifth-order WENO method. SIAM J. Numer. Anal. 45, 1871–1901.

Wu, L., Zhang, Y.T., Zhang, S., Shu, C.W., 2016. High order fixed-point sweeping WENO methods for steady state of hyperbolic conservation laws and its convergence study. Commun. Comput. Phys. 20, 835–869.

Xiong, T., Zhang, M., Zhang, Y.T., Shu, C.W., 2010. Fast sweeping fifth order WENO scheme for static Hamilton-Jacobi equations with accurate boundary treatment. J. Sci. Comput. 45, 514–536.

Zhang, Y.T., Shu, C.W., 2003. High order WENO schemes for Hamilton-Jacobi equations on triangular meshes. SIAM J. Sci. Comput. 24, 1005–1030.

Zhang, S., Shu, C.W., 2007. A new smoothness indicator for the WENO schemes and its effect on the convergence to steady state solutions. J. Sci. Comput. 31, 273–305.

Zhang, Y.T., Shu, C.W., 2009. Third order WENO scheme on three dimensional tetrahedral meshes. Commun. Comput. Phys. 5, 836–848.

Zhang, Y.T., Shi, J., Shu, C.W., Zhou, Y., 2003. Numerical viscosity and resolution of high-order weighted essentially nonoscillatory schemes for compressible flows with high Reynolds numbers. Phys. Rev. E 68, 046709.

Zhang, S., Zhang, Y.T., Shu, C.W., 2005. Multistage interaction of a shock wave and a strong vortex. Phys. Fluids 17, 116101.

Zhang, S., Zhang, Y.T., Shu, C.W., 2006. Interaction of an oblique shock wave with a pair of parallel vortices: shock dynamics and mechanism of sound generation. Phys. Fluids 18, 126101.

Zhang, Y.T., Shu, C.W., Zhou, Y., 2006. Effects of shock waves on Rayleigh-Taylor instability. Phys. Plasmas 13, 062705.

Zhang, Y.T., Zhao, H.K., Chen, S., 2006. Fixed-point iterative sweeping methods for static Hamilton-Jacobi equations. Methods Appl. Anal. 13, 299–320.

Zhang, Y.T., Zhao, H.K., Qian, J., 2006. High order fast sweeping methods for static Hamilton-Jacobi equations. J. Sci. Comput. 29, 25–56.

Zhang, S., Jiang, S., Zhang, Y.T., Shu, C.W., 2009. The mechanism of sound generation in the interaction between a shock wave and two counter rotating vortices. Phys. Fluids 21, 076101.

Zhang, S., Jiang, S., Shu, C.W., 2011. Improvement of convergence to steady state solutions of Euler equations with the WENO schemes. J. Sci. Comput. 47, 216–238.

Zhu, J., Qiu, J., Shu, C.W., Dumbser, M., 2008. Runge-Kutta discontinuous Galerkin method using WENO limiters II: unstructured meshes. J. Comput. Phys. 227, 4330–4353.

Chapter 6

Stability Properties of the ENO Method

U.S. Fjordholm
NTNU, Trondheim, Norway

Chapter Outline

ABSTRACT

We review the currently available stability properties of the essentially nonoscillatory reconstruction procedure, such as its monotonicity and nonoscillatory properties, the sign property, upper bounds on cell interface jumps, and a total variation-type bound. We also outline how these properties can be applied to derive stability and convergence of high-order accurate schemes for conservation laws.

Keywords: ENO reconstruction, High-order accuracy, Stability, Sign property

AMS Classification Codes: 65D05, 65M12

1 INTRODUCTION

The ENO (essentially nonoscillatory) reconstruction method is a method of recovering—to a high degree of accuracy—a function v, given only discrete information such as a finite number of point values $v_i = v(x_i)$ or local averages

$$\bar{v}_i = \fint_{I_i} v(x)dx, \qquad i \in \mathbb{Z}$$

Handbook of Numerical Analysis, Vol. 17. http://dx.doi.org/10.1016/bs.hna.2016.09.004

over cells $I_i = [x_{i-1/2}, x_{i+1/2})$. (Here and below we will denote $\fint_I = \frac{1}{|I|}\int$.) The method was first developed as a means of increasing the order of accuracy of numerical methods for hyperbolic conservation laws. Solutions of these types of partial differential equations (PDEs) are at best *piecewise* smooth and can have large jump discontinuities. The ENO method accomplishes the feat of approximating v to a high degree of accuracy in smooth parts, while avoiding "Gibbs-like" oscillations near the discontinuities. The purpose of this chapter is to review the currently known stability properties of the ENO method and the application of these to numerical methods for hyperbolic conservation laws.

The ENO reconstruction method was originally developed by Harten (1986) and further developed and analyzed by Harten, Osher, Engquist, and Chakravarthy in a series of papers (Harten and Osher, 1987; Harten et al., 1986, 1987). Since then, it has been generalized and applied to a number of areas. In this paper, we will concentrate on the one-dimensional version of the ENO reconstruction method and its application to approximate one-dimensional scalar conservation laws. Thus, we leave out a large body of work on multidimensional generalizations of ENO, related "ENO-type" reconstruction methods, and applications of ENO to systems of conservation laws, as well as other fields such as data compression/representation and image analysis/reconstruction. Multidimensional ENO methods were introduced by Shu and Osher (1988) on Cartesian (tensor-product) meshes and generalized to unstructured (triangular) meshes by Harten and Chakravarthy (1991) and by Abgrall and Lafon (1993) (see also Abgrall, 1994). Related "ENO-type" methods include the highly successful Weighted ENO method (Liu et al., 1994; Jiang and Shu, 1996), biased ENO (Shu, 1990), ENO-SR (subcell resolution) (Harten, 1989) and its multidimensional generalization GENO (geometric ENO) (Siddiqi et al., 1997), and ENO-EA (edge adapted) (Arandiga et al., 2008). For applications of ENO apart from conservation laws, we mention in particular Harten's work on multiresolution methods (Harten, 1996; see also Aràndiga and Donat, 2000).

Here follows an outline of the rest of the chapter. In Section 2, we motivate and describe the ENO reconstruction method. In Section 3, we briefly describe the application of the ENO method to (scalar) conservation laws; we show that the resulting second-order accurate scheme is convergent; and we derive a list of *a priori* bounds that imply convergence of (one class of) higher-order ENO-type schemes. Section 4 is the main section of the chapter. We start by listing some immediate stability properties of the ENO method and move on to describing some of the more nontrivial properties such as the sign property, upper bounds on jumps, and the "ENO" property.

We have attempted to make this paper as self-contained as possible. In particular, Sections 2 and 4 should be accessible also to readers without a background in PDEs.

2 THE ENO RECONSTRUCTION METHOD

For the sake of completeness, we describe here the ENO reconstruction method. We refer to the review article by Zhang and Shu (2016) for further details.

Let us fix a partition $(I_i)_{i\in\mathbb{Z}}$ of the real line, where each *cell* I_i is an interval $I_i = [x_{i-1/2}, x_{i+1/2})$ of length $\Delta x_i = x_{i+1/2} - x_{i-1/2}$, bounded from above by $\Delta x = \max_i \Delta x_i$. Let $(\bar{v}_i)_{i\in\mathbb{Z}} \subset \mathbb{R}$ be a given collection of numbers, which we interpret as the cell averages of some unknown function v,

$$\bar{v}_i = \fint_{I_i} v(x)dx. \tag{1}$$

The ENO reconstruction method (Harten, 1986; Harten et al., 1987) aims to *reconstruct* v by producing a collection of $(k-1)$th-order polynomials $p_i = p_i(x)$ which approximate v to kth order:

$$p_i(x) = v(x) + e(x)\Delta x_i^k \qquad \forall x \in I_i, \tag{2}$$

where $e(x)$ denotes the leading-order error term. The reconstruction is required to conserve mass, in the sense that

$$\fint_{I_i} p_i(x)dx = \bar{v}_i \qquad \forall i \in \mathbb{Z}, \tag{3}$$

and is required to be as "nonoscillatory" as possible.

The properties of accuracy and mass conservation are automatically satisfied if p_i interpolates the cell average values $\bar{v}_j$ over any of the k *stencils*

$$\{s,\ldots,s+k-1\}, \quad i-k+1 \le s \le i.$$

Thus, p_i is defined as the unique $(k-1)$th-order polynomial which satisfies

$$\fint_{I_j} p_i(x)dx = \bar{v}_j \quad \text{for} \quad j = s,\ldots,s+k-1, \tag{4}$$

for some integer $s = s_i \in \{i-k+1,\ldots,i\}$ called the *stencil index*. The problem of finding p_i satisfying (4) is a somewhat nonstandard interpolation problem, and Harten (1986) suggested two approaches.

In the *reconstruction via deconvolution* (RD) approach, it is observed that (1) is a convolution of v with the indicator function over I_i. Taylor expanding v and comparing with (4) results in an upper triangular linear system for p_i.

In the *reconstruction via primitive function* (RP) approach, we define the primitive of v,

$$V(x) = \int_{-\infty}^{x} v(x)dx \tag{5}$$

(the lower limit of this integral is irrelevant) and observe that V is precisely known at every cell interface,

$$V(x_{i+1/2}) = \sum_{j\le i} \Delta x_j \bar{v}_j.$$

If we let P_i be the unique kth-order polynomial which interpolates V over the points $\{x_{s-1/2},\ldots,x_{s+k-1/2}\}$, then the $(k-1)$th-order polynomial $p_i(x)=\frac{d}{dx}P_i(x)$ satisfies (4).

The RD approach requires a uniform mesh (i.e. $\Delta x\equiv$ const.), whereas the RP approach works for any (one-dimensional) mesh. Even on a uniform mesh, the RD and RP approaches are *not* equivalent; i.e., they yield distinct reconstructions p_i. We are unaware of any further work on the RD methodology beyond the original papers by Harten and Osher (1987) and Harten et al. (1986, 1987), and we will concentrate on RP for the remainder of this chapter (see also Remark 4).

2.1 Choosing the Stencil Index

The algorithm to select the stencil index s_i is what characterizes the ENO reconstruction procedure. A naive choice of the stencil index could be the all-upwind or all-downwind stencils $s_i\equiv i-k+1$ or $s_i\equiv i$; however, given the possible nonsmoothness or discontinuity of v, these choices would lead to "Gibbs phenomena"—large oscillations in nonsmooth regions.

Harten (1986) proposed an iterative, data-dependent algorithm to compute s_i. The algorithm is based upon the divided differences of V, defined as

$$\begin{cases} V[x_{i+1/2}]=V(x_{i+1/2}) \\ V[x_{i-1/2},\ldots,x_{j+1/2}]=\dfrac{V[x_{i+1/2},\ldots,x_{j+1/2}]-V[x_{i-1/2},\ldots,x_{j-1/2}]}{x_{j+1/2}-x_{i-1/2}} \end{cases} \quad \forall\ i<j.$$

Starting with the stencil $\{x_{i-1/2}, x_{i+1/2}\}$, the ENO stencil selection procedure adds either the left or the right point $x_{i-3/2}$ or $x_{i+3/2}$, depending on which of the divided differences $V[x_{i-3/2}, x_{i-1/2}, x_{i+1/2}]$ or $V[x_{i-1/2}, x_{i+1/2}, x_{i+3/2}]$ is the smallest. This process is then iterated, ending up with an interpolation stencil $\{x_{s_i-1/2},\ldots,x_{s_i+k+1/2}\}$ for some $s_i\in\{i-k+1,\ldots,i\}$. Recalling that the kth divided difference of V is an approximation of the $(k-1)$th derivative of v,

$$V[x_{i-1/2},\ldots,x_{i+k-1/2}]=\frac{V^{(k)}(\xi)}{k!}=\frac{v^{(k-1)}(\xi)}{k!}, \qquad \xi\in[x_{i-1/2},x_{i+k-1/2}],$$

we see that the ENO procedure iteratively adds a new point to the interpolation stencil "in the direction of smoothness".

Since $V[x_{i-1/2},x_{i+1/2}]=\bar{v}_i$, we can write

$$V[x_{i-1/2},\ldots,x_{j+1/2}]=[\bar{v}_i,\ldots,\bar{v}_j]\quad \forall i\leq j$$

where the "cell average divided differences" are defined as

$$\begin{cases} [\bar{v}_i]=\bar{v}_i \\ [\bar{v}_i,\ldots,\bar{v}_j]=\dfrac{[\bar{v}_{i+1},\ldots,\bar{v}_j]-[\bar{v}_i,\ldots,\bar{v}_{j-1}]}{x_{j+1/2}-x_{i-1/2}} \end{cases} \quad \forall\ i<j. \tag{6}$$

We summarize the ENO procedure using this notation.

ALGORITHM 1 (ENO Stencil Selection Procedure)

$s_i^1 = 0$
for $\ell = 1, \ldots, k-1$ **do**
 if $\left|\left[\bar{v}_{s_i^\ell - 1}, \ldots, \bar{v}_{s_i^\ell + \ell - 1}\right]\right| < \left|\left[\bar{v}_{s_i^\ell}, \ldots, \bar{v}_{s_i^\ell + \ell}\right]\right|$ **then**
  $s_i^{\ell+1} = s_i^\ell - 1$
 else
  $s_i^{\ell+1} = s_i^\ell$
 end if
end for
$s_i = s_i^k$
Let P_i interpolate V over $\{x_{s_i - 1/2}, \ldots, x_{s_i + k - 1/2}\}$
Define $p_i(x) = \frac{d}{dx} P_i(x)$

The implications of the ENO stencil selection procedure are easiest to see with the Newton form of the interpolating polynomial P_i. It is straightforward to show by induction that the Newton form of P_i can be expressed as

$$P_i(x) = \sum_{\ell=0}^{k} V[x_{s_i^\ell - 1/2}, \ldots, x_{s_i^\ell + \ell - 1/2}] \prod_{m=0}^{\ell-1} (x - x_{s_i^{\ell-1} + m - 1/2}),$$

where we have defined $s_i^{-1} = s_i^0 = i$. After differentiating and using the notation (6), we get

$$\begin{aligned} p_i(x) &= \sum_{\ell=1}^{k} V[x_{s_i^\ell - 1/2}, \ldots, x_{s_i^\ell + \ell - 1/2}] \sum_{n=0}^{\ell-1} \prod_{\substack{m=0 \\ m \neq n}}^{\ell-1} (x - x_{s_i^{\ell-1} + m - 1/2}) \\ &= \sum_{\ell=1}^{k} [\bar{v}_{s_i^\ell}, \ldots, \bar{v}_{s_i^\ell + \ell - 1}] \sum_{n=0}^{\ell-1} \prod_{\substack{m=0 \\ m \neq n}}^{\ell-1} (x - x_{s_i^{\ell-1} + m - 1/2}) \end{aligned} \tag{7}$$

(see also Harten, 1986, p. 81). Thus, the ENO stencil selection procedure chooses each index s_i^ℓ so that the above coefficients $[\bar{v}_{s_i^\ell}, \ldots, \bar{v}_{s_i^\ell + \ell - 1}]$ are as small as possible, thereby obtaining the least oscillatory polynomial possible.

Note that both the ENO stencil selection procedure and the formula for p_i can be written completely in terms of the divided differences of $\bar{v}$. Thus, it is not necessary to compute the primitive V or its divided differences.

Remark 1. There is also a point-value version of the ENO reconstruction method. Given the point values $v_i = v(x_i)$ of some function v, this method employs a similar algorithm to obtain a reconstruction $p_i(x) = v(x) + O(\Delta x_i^k)$. The reconstruction p_i is given by the $(k-1)$th-order polynomial interpolating $(v_j)_{j \in \mathbb{Z}}$ over the points $x_{s_i}, \ldots, x_{s_i + k - 1}$, where s_i is obtained by replacing every occurrence of $\bar{v}_j$ in Algorithm 1 by v_j. See Shu and Osher (1988) for further details and Fjordholm (2013) and Fjordholm et al. (2012b) for a stability analysis.

3 APPLICATION TO CONSERVATION LAWS

The ENO method was originally developed as a means of increasing the order of accuracy of finite volume schemes for hyperbolic conservation laws. We consider here only one-dimensional, scalar conservation laws

$$\begin{aligned} \partial_t u + \partial_x f(u) &= 0 \\ u(x,0) &= u_0(x). \end{aligned} \tag{1}$$

To establish the notation and some useful identities, we briefly review this setting in Section 3.1. We refer to the article by Zhang and Shu (2016) for further details. In Section 3.2, we see that the second-order ENO method results in a TVD (total variation diminishing), convergent finite volume scheme for scalar conservation laws. In Section 3.3, we review one approach to obtaining convergent higher-order accurate schemes.

Below we use the notation from Section 2. Furthermore, we denote

$$[[\bar{v}]]_{i+1/2} = \bar{v}_{i+1} - \bar{v}_i, \quad \{\bar{v}\}_{i+1/2} = \frac{\bar{v}_i + \bar{v}_{i+1}}{2}. \tag{2}$$

3.1 Finite Volume Methods

A (semidiscrete) finite volume method for (1) aims to compute an approximation of the cell averages

$$\bar{v}_i(t) \approx \fint_{I_i} u(x,t)dx \ \ \forall \ t \geq 0$$

of the exact (entropy) solution of (1). A consistent, conservative finite volume method for (1) is then of the form

$$\frac{d}{dt}\bar{v}_i(t) = -\frac{1}{\Delta x_i}\left(F_{i+1/2} - F_{i-1/2}\right) \tag{3}$$

for some $F_{i+1/2} = F(\bar{v}_{i-m+1}, \ldots, \bar{v}_{i+m})$, and F is a numerical flux function such as the Godunov, Lax–Friedrichs, and Engquist–Osher fluxes. One class of (formally) high-order accurate schemes is obtained by letting

$$F_{i+1/2} = F\left(v^-_{i+1/2}, v^+_{i+1/2}\right) \tag{4}$$

for some monotone flux F. Here, $v^{\pm}_{i+1/2}$ are the reconstructed cell interface values

$$v^-_{i+1/2} = p_i(x_{i+1/2}, t), \quad v^+_{i+1/2} = p_{i+1}(x_{i+1/2}, t), \quad p(x,t) = \mathcal{R}(\bar{v}(t))(x) \tag{5}$$

for some reconstruction operator $\mathcal{R}$ such as ENO.

To obtain a fully discrete method, we discretize the temporal domain $t \in [0, \infty)$ into discrete points $t^n = n\Delta t$ for some $\Delta t > 0$ (which we for simplicity assume is constant), and the aim is to approximate

$$\bar{v}_i^n \approx \fint_{I_i} u(x,t^n)dx \;\; \forall \;\; i \in \mathbb{Z}.$$

An explicit, fully discrete finite volume method for (1) is then of the form

$$\bar{v}_i^{n+1} = \bar{v}_i^n - \frac{\Delta t}{\Delta x_i}\left(F_{i+1/2}^n - F_{i-1/2}^n\right) \tag{6}$$

for some $F_{i+1/2}^n = F\left(\bar{v}_{i-m+1}^n, \ldots, \bar{v}_{i+m}^n\right)$. This scheme is TVD if

$$\mathrm{TV}(\bar{v}^{n+1}) \leq \mathrm{TV}(\bar{v}^n), \tag{7}$$

so-called after Harten (1983).

The scheme (6) can be viewed as a (first-order accurate) forward Euler discretization of (3) (see Godlewski and Raviart, 1991, Section II.3.3 for a rigorous derivation; cf. also Harten et al., 1986, p. 352). Higher-order accurate methods can be obtained using multistep methods or strong stability preserving Runge–Kutta methods (Gottlieb et al., 2001), which consist of convex combinations of (6).

3.2 TVD ENO Schemes

Consider now the (formally) second-order accurate scheme (6) with a flux (4) using a second-order reconstruction method. Any second-order reconstruction $(p_i)_{i\in\mathbb{Z}}$ of cell averages $(\bar{v}_i)_{i\in\mathbb{Z}}$ must necessarily be of the form

$$p_i(x) = \bar{v}_i + \sigma_i(x - x_i) \tag{8}$$

where $\sigma_i \in \mathbb{R}$ is the slope of p_i. This slope is commonly written in the *slope limited* form

$$\sigma_i = \varphi(\theta_i^+)[[\bar{v}]]_{i+1/2}, \quad \theta_i^+ = \frac{[[\bar{v}]]_{i-1/2}}{[[\bar{v}]]_{i+1/2}} \tag{9}$$

for some $\varphi : \mathbb{R} \to \mathbb{R}$ called a *slope limiter*. Using Harten's work (1983), Sweby (1984) showed that if the slope limiter satisfies

$$\left|\varphi(\theta_1) - \frac{\varphi(\theta_2)}{\theta_2}\right| \leq 2 \;\; \forall \;\; \theta_1, \theta_2 \in \mathbb{R}, \tag{10}$$

then the explicit discretization (6) is both TVD and uniformly bounded, so the computed solution satisfies

$$\mathrm{TV}(\bar{v}^n) \leq \mathrm{TV}(\bar{v}^0), \quad \|\bar{v}^n\|_{L^\infty} \leq \|\bar{v}^0\|_{L^\infty} \;\; \forall \;\; n \in \mathbb{N}.$$

As a consequence, there is a subsequence Δt_m, $\Delta x_m \to 0$ for which the computed solutions converge towards a weak solution.

It is not hard to see that the second-order ENO reconstruction can be written as (8), (9) with the slope limiter

$$\varphi(\theta)=\begin{cases}\theta & \text{if } |\theta|<1\\ 1 & \text{if } |\theta|\geq 1.\end{cases} \tag{11}$$

Although this limiter does not lie in the "TVD region" introduced by Sweby Sweby (1984), it *does* satisfy (10). Therefore, the scheme (6), (4) using second-order ENO reconstruction is both TVD and uniformly bounded and hence converges (subsequentially) towards a weak solution.

3.3 Convergence of High-Order Schemes

A uniform bound on the total variation of a sequence of approximate solutions—such as the bound (7) provided by TVD schemes—prevents the buildup of high-frequency oscillations, a necessary requirement for the strong convergence of the sequence. However, it is well known that any TVD scheme for (1) is at most second-order accurate when measured in L^1. Thus, any proof of stability or convergence of higher (than second) order accurate schemes must necessarily relax the TVD requirement, while still preventing high-frequency oscillations.

We present here one class of convergent, high-order accurate schemes, the so-called TECNO schemes (Fjordholm, 2013; Fjordholm et al., 2012a). As a motivation, we first derive the necessary a priori bounds for a parabolic regularization of (1), which can be thought of as the effective (modified) equation of the numerical scheme. We then perform the analogous computations for the TECNO schemes.

3.3.1 Motivation

Consider the following regularization of (1):

$$\begin{aligned}\partial_t v^\varepsilon+\partial_x f(v^\varepsilon) &=\varepsilon\partial_{xx}v^\varepsilon\\ v^\varepsilon(x,0) &=v_0^\varepsilon(x),\end{aligned} \tag{12}$$

where v_0^ε converges to u_0 as $\varepsilon\to 0$. The term $\varepsilon\partial_{xx}v^\varepsilon$ can be thought of as the numerical viscosity of a numerical scheme, and $\varepsilon\sim\Delta x^k$, where k is the order of accuracy of the method. Multiplying (12) by $2v^\varepsilon$ we obtain

$$\partial_t(v^\varepsilon)^2+\partial_x q(v^\varepsilon)=\varepsilon\partial_{xx}(v^\varepsilon)^2-2\varepsilon(\partial_x v^\varepsilon)^2, \tag{13}$$

where q satisfies $q'(u)=2uf'(u)$ for all $u\in\mathbb{R}$. Integrating (13) over $x\in\mathbb{R}$, $t\in[0,T]$ gives

$$\int_{\mathbb{R}} v^\varepsilon(x,T)^2\;dx=\int_{\mathbb{R}} u_0^\varepsilon(x)^2\;dx-2\varepsilon\int_0^T\int_{\mathbb{R}}(\partial_x v^\varepsilon)^2\;dxdt. \tag{14}$$

Thus, we have the two bounds

$$\| v^{\varepsilon}(T) \|_{L^2(\mathbb{R})} \le \| u_0 \|_{L^2(\mathbb{R})} \tag{15a}$$

$$2\varepsilon \int_0^T \int_{\mathbb{R}} (\partial_x v^{\varepsilon})^2 \ dxdt \le \| u_0 \|^2_{L^2(\mathbb{R})} \tag{15b}$$

for all $\varepsilon > 0$, i.e., a uniform L^2 bound and a "weak TV bound". From these, compensated compactness techniques can be used to show that a subsequence $v^{\varepsilon'}$ converges to a weak solution of (1) as $\varepsilon' \to 0$. Since the second term on the right-hand side of (13) is nonpositive, we find that any strong limit $u = \lim_{\varepsilon' \to 0} v^{\varepsilon'}$ satisfies the entropy condition

$$\partial_t u^2 + \partial_x q(u) \le 0. \tag{16}$$

We conclude that the *whole* sequence $(v^{\varepsilon})_{\varepsilon > 0}$ converges strongly to the (unique) entropy solution of (1).

3.3.2 TECNO Schemes

We consider now the semidiscrete finite volume method (3) with a numerical flux function of the form

$$F_{i+1/2} = F^*_{i+1/2} - c_{i+1/2} \langle\langle v \rangle\rangle_{i+1/2}. \tag{17}$$

Here, $\langle\langle v \rangle\rangle_{i+1/2} = v^+_{i+1/2} - v^-_{i+1/2}$ is the cell interface jump in the reconstructed values (cf. (5)) for some reconstruction operator $\mathcal{R}$, to be determined. The diffusion constant $c_{i+1/2}$ is some number satisfying $c_{\max} \ge c_{i+1/2} \ge c_{\min} > 0$, and F^* is a Lipschitz continuous numerical flux, to be determined. Note that if the reconstructed values satisfy, say,

$$|\langle\langle v \rangle\rangle_{i+1/2}| \le C |[[\bar{v}]]_{i+1/2}| \tag{18}$$

for some $C > 0$ independent of v, then F is Lipschitz continuous—a natural assumption in the convergence analysis of finite volume schemes.

Multiplying (3) by $2\bar{v}_i(t)$ we obtain

$$\begin{aligned} \frac{d}{dt}\bar{v}_i^2 + 2\bar{v}_i \frac{F^*_{i+1/2} - F^*_{i-1/2}}{\Delta x_i} &= 2\bar{v}_i \frac{c_{i+1/2}\langle\langle v \rangle\rangle_{i+1/2} - c_{i-1/2}\langle\langle v \rangle\rangle_{i-1/2}}{\Delta x_i} \\ &= 2\frac{c_{i+1/2}\{v\}_{i+1/2}\langle\langle v \rangle\rangle_{i+1/2} - c_{i-1/2}\{v\}_{i-1/2}\langle\langle v \rangle\rangle_{i-1/2}}{\Delta x_i} \\ &\quad - \frac{c_{i+1/2}[[\bar{v}]]_{i+1/2}\langle\langle v \rangle\rangle_{i+1/2} + c_{i-1/2}[[\bar{v}]]_{i-1/2}\langle\langle v \rangle\rangle_{i-1/2}}{\Delta x_i}. \end{aligned}$$

Assuming that we can write $2\bar{v}_i(F^*_{i+1/2} - F^*_{i-1/2}) = (Q^*_{i+1/2} - Q^*_{i-1/2})$ (as in the step from (12) to (13)) for some "numerical entropy flux Q^*", we can define $Q_{i+1/2} = Q^*_{i+1/2} - 2c_{i+1/2}\{\bar{v}\}_{i+1/2}\langle\langle v \rangle\rangle_{i+1/2}$ and obtain

$$\frac{d}{dt}\bar{v}_i^2 + \frac{Q_{i+1/2} - Q_{i-1/2}}{\Delta x_i} = -\frac{c_{i+1/2}[[\bar{v}]]_{i+1/2}\langle\langle v \rangle\rangle_{i+1/2} + c_{i-1/2}[[\bar{v}]]_{i-1/2}\langle\langle v \rangle\rangle_{i-1/2}}{\Delta x_i}. \tag{19}$$

Summing over $i \in \mathbb{Z}$ and integrating over $t \in [0, T]$, we get

$$\sum_{i\in\mathbb{Z}} \bar{v}_i(T)^2 \Delta x_i = \sum_{i\in\mathbb{Z}} \bar{v}_i(0)^2 \Delta x_i - 2\int_0^T \sum_{i\in\mathbb{Z}} c_{i+1/2}[[\bar{v}]]_{i+1/2}\langle\langle v\rangle\rangle_{i+1/2} dt \tag{20}$$

(compare with (14)). Assuming now that

$$[[\bar{v}]]_{i+1/2}\langle\langle v\rangle\rangle_{i+1/2} \geq 0 \quad \forall \ i \in \mathbb{Z}, \tag{21}$$

i.e., that the jumps $\bar{v}_{i+1} - \bar{v}_i$ and $v_{i+1/2}^+ - v_{i+1/2}^-$ have the same sign, we can conclude that

$$\| v_{\Delta x}(T)\|_{L^2(\mathbb{R})} \leq \| v_{\Delta x}(0)\|_{L^2(\mathbb{R})}, \tag{22a}$$

$$2\int_0^T \sum_{i\in\mathbb{Z}} c_{i+1/2}[[\bar{v}]]_{i+1/2}\langle\langle v\rangle\rangle_{i+1/2} dt \leq \| v_{\Delta x}(0) \|_{L^2(\mathbb{R})}^2 \tag{22b}$$

(compare with (15a)). The property (21) also ensures that the right-hand side of (19) is nonpositive, so that

$$\frac{d}{dt}\bar{v}_i^2 + \frac{Q_{i+1/2} - Q_{i-1/2}}{\Delta x_i} \leq 0$$

(compare with (16)), i.e., a discrete entropy inequality is satisfied.

The bound (22b) is not quite a weak TV bound like (15b)—for this we would need a bound of the form

$$\int_0^T \sum_{i\in\mathbb{Z}} |[[\bar{v}]]_{i+1/2}|^p dt \leq C \tag{23}$$

for some $p \geq 1$ and $C > 0$ independent of Δx.

We have thus arrived at a list of properties which enable a convergence proof of the finite volume method (6): The upper bound on reconstructed jumps (18), the *sign property* (21), and the "weak TV bound" (23).

The sign property and the upper bound have been proven for the ENO reconstruction method and are discussed in Sections 4.2 and 4.3, respectively. For $k = 2$, it has been proven—and conjectured for $k > 2$—that the "reconstructed TV bound" (22b) implies the "weak TV bound" (23). This is discussed in Section 4.4. We refer to this conjecture as the *ENO TV conjecture*.

In Fjordholm (2013) and Fjordholm et al. (2012a) the authors constructed schemes of the form (3), (17) which uses the ENO reconstruction method—the so-called TECNO schemes. We summarize the main convergence theorem here and refer to Fjordholm (2013) for the proof.

Theorem 1. *For every* k *for which the ENO TV conjecture holds, we have the following. If the approximate solution computed by the kth-order TECNO method is L^∞-bounded, then the sequence of approximate solutions converges to the entropy solution of (1) as $\Delta x \to 0$.*

Remark 2. With some extra effort, the above computation can be generalized from the square entropy v^2 to arbitrary entropies $\eta(v)$. See the review article by Tadmor (2016) (see also Fjordholm et al., 2012a; Tadmor, 2003) for more information on so-called entropy stable methods.

4 ENO STABILITY PROPERTIES

In this section, we review the currently known stability properties of the ENO reconstruction method. In Section 4.1, we summarize some immediate (but nevertheless useful) properties of the ENO reconstruction. In Section 4.2, we prove the *sign property* of the ENO method, and in Section 4.3, we prove an upper bound on the jump $\langle\langle v\rangle\rangle = v^+_{i+1/2} - v^-_{i+1/2}$. We discuss the ENO TV conjecture in Section 4.4. Recall from Section 3.3 that all of these properties are essential for the convergence of the high-order TECNO schemes.

In Section 4.5, we prove some well-known *mesh-dependent* properties of ENO. As it turns out, the sign property is a necessary ingredient in a rigorous proof of some of these properties. We conclude in Section 4.6 by mentioning some deficiencies of ENO.

4.1 Immediate Properties

4.1.1 Mesh Invariance and Linearity

Under the mapping $x \to a + bx$ for any $a \in \mathbb{R}$ and $b > 0$, the reconstructed polynomial is $p_i\left(\frac{x-a}{b}\right)$. If $(\bar{v}_i)_{i\in\mathbb{Z}}$ is replaced by $(\alpha\bar{v}_i + \beta)_{i\in\mathbb{Z}}$ for any $\alpha, \beta \in \mathbb{R}$, then the ENO reconstruction $p_i(x)$ is replaced by $\alpha p_i(x) + \beta$.

4.1.2 Discontinuity Across Cell Edges

As a rule of thumb, the ENO reconstruction $p = \mathcal{R}(\bar{v})$ is discontinuous *at least* at every kth cell interface $x_{i+1/2}$. To see this, note that neighbouring cells with the same stencil index $s_i = s_{i+1}$ have the same reconstruction $p_i = p_{i+1}$ (and are thus continuous at $x_{i+1/2}$), whereas if $s_i < s_{i+1}$ then $p_i \neq p_{i+1}$, and hence $p_i(x_{i+1/2}) \neq p_{i+1}(x_{i+1/2})$ (except in very rare cases, such as when v is itself a $(k-1)$th-order polynomial). Since s_i must change at least at every kth index i, this yields a discontinuity in p.

At points of discontinuity $x_{i+1/2}$, the size of the jump $p_{i+1}(x_{i+1/2}) - p_i(x_{i+1/2})$ is $O(\Delta x^k)$ (see Section 4.2). Note that the cell interface jump $p_{i+1}(x_{i+1/2}) - p_i(x_{i+1/2})$ can—and often will—be zero even when $\bar{v}_{i+1} - \bar{v}_i \neq 0$.

4.1.3 Uniform kth-Order Accuracy

Let $v \in C^\infty(\mathbb{R})$ with primitive $V(x)$ defined in (5). Through a Taylor expansion of V, it is easy to see that the ENO reconstruction $p = \mathcal{R}(\bar{v})$ of $(\bar{v}_i)_{i\in\mathbb{Z}}$ is a kth-order approximation of v. More specifically, p_i satisfies the relation (2) with an error term $|e(x)| \leq C \|\frac{d^k v}{dx^k}\|_{L^\infty}$ for some $C = C_k$. In each cell I_i, the error term $e(x)$ is continuous (but not Lipschitz continuous) with at least one zero. It is discontinuous only at those cell interfaces $x_{i+1/2}$ where p is discontinuous (see Section 4.1.2).

4.2 The Sign Property

Consider a reconstruction procedure $\mathcal{R}$, mapping a collection of cell averages $(\bar{v}_i)_{i\in\mathbb{Z}}$ to a piecewise polynomial function $\sum_i p_i 1_{I_i}$. As before, define the cell interface values $v^-_{i+1/2} = p_i(x_{i+1/2})$ and $v^+_{i+1/2} = p_{i+1}(x_{i+1/2})$ and the jump $\langle\langle v\rangle\rangle_{i+1/2} = v^+_{i+1/2} - v^-_{i+1/2}$. We say that $\mathcal{R}$ satisfies the *sign property* if for every $i\in\mathbb{Z}$,

$$\begin{array}{lll} \text{if } [[\bar{v}]]_{i+1/2} > 0, & \text{then } \langle\langle v\rangle\rangle_{i+1/2} \geq 0 \\ \text{if } [[\bar{v}]]_{i+1/2} < 0, & \text{then } \langle\langle v\rangle\rangle_{i+1/2} \leq 0 \\ \text{if } [[\bar{v}]]_{i+1/2} = 0, & \text{then } \langle\langle v\rangle\rangle_{i+1/2} = 0. \end{array} \tag{1}$$

As we have seen in Section 3.3, the sign property implies that the diffusion coefficient in finite volume schemes of the form (6), (17) has the right sign.

The sign property is illustrated in Fig. 1, which shows a third-, fourth-, and fifth-order ENO reconstruction of randomly chosen cell averages. Even though the reconstructed polynomial may have large variations within each cell, its jumps at cell interfaces always have the same sign as the jumps of the cell averages.

In Fjordholm (2013) and Fjordholm et al. (2012b), it was shown that the kth-order ENO reconstruction satisfies the sign property, for any $k\in\mathbb{N}$ and for any mesh $(x_{i+1/2})_{i\in\mathbb{Z}}$. We provide here a sketch of the proof.

Proof of ENO sign property (sketch). The first step is to derive the following expression for the jump in reconstructed values:

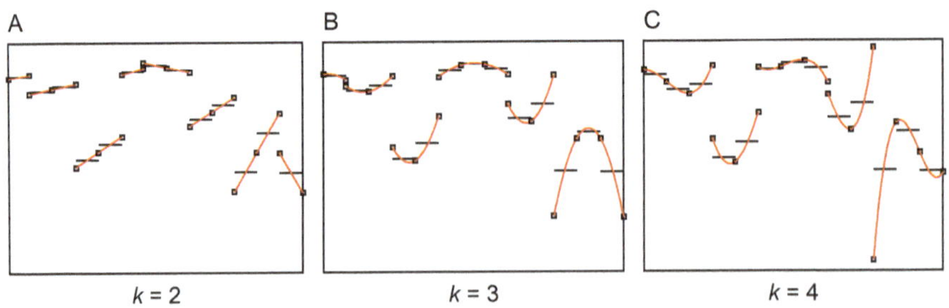

FIG. 1 ENO reconstruction of randomly chosen cell averages. *Black lines*: cell averages. *Red (gray* in the print version) *curves*: reconstruction. *Squares*: cell interface values. (A) $k = 2$, (B) $k = 3$, and (C) $k = 4$.

$$\langle\langle v\rangle\rangle_{i+1/2} = \sum_{s=s_i}^{s_{i+1}-1} [\bar{v}_s,\ldots,\bar{v}_{s+k}] X_{i,s} \tag{2}$$

where

$$X_{i,s} := (x_{s+k+1/2} - x_{s-1/2}) \prod_{\substack{m=0 \\ m\neq i-s}}^{k-1} (x_{i+1/2} - x_{s+m+1/2}).$$

When $s_i = s_{i+1}$, i.e., the neighbouring stencils are the same, then (2) yields $\langle\langle v\rangle\rangle_{i+1/2} = 0$ and the reconstruction is continuous across $x_{i+1/2}$. Observe that (2) expresses $\langle\langle v\rangle\rangle_{i+1/2}$ in terms of only kth-order divided differences of $\bar{v}$, instead of divided differences of order $1,\ldots,k-1$, as one might expect from (7). In particular, when $k = 1$, we get $\langle\langle v\rangle\rangle_{i+1/2} = [\bar{v}_i, \bar{v}_{i+1}] (x_{i+3/2} - x_{i-1/2}) = [[v]]_{i+1/2}$, as expected.

The proof of (2) amounts to a simple manipulation of Newton polynomials, but the idea is quite clear: Both $v_{i+1/2}^-$ and $v_{i+1/2}^+$ are kth-order approximations of $v(x_{i+1/2})$, with truncation terms of the form $[\bar{v}_s,\ldots,\bar{v}_{s+k}] = \frac{1}{k!}\frac{d^k v}{dx^k}(\xi)$.

The next step is to show that each summand in (2) has the same sign as $[[\bar{v}]]_{i+1/2}$. Because $\operatorname{sgn}(X_{i,s}) = (-1)^{s+k+1}$, we need only to show that

$$[[\bar{v}]]_{i+1/2} [\bar{v}_s,\ldots,\bar{v}_{s+k}] (-1)^{s+k+1} \geq 0 \;\; \forall \;\; s = s_i^k,\ldots,s_{i+1}^k - 1. \tag{3}$$

The proof of (3) is obvious for $k = 1$. Assume that (3) holds for some $k \geq 1$. It suffices to consider the case $[[\bar{v}]]_{i+1/2} > 0$, so we have

$$[\bar{v}_s,\ldots,\bar{v}_{s+k}] (-1)^{s+k+1} \geq 0 \;\; \text{for} \;\; s = s_i^k,\ldots,s_{i+1}^k - 1.$$

The fact that $[\bar{v}_s,\ldots,\bar{v}_{s+k+1}] (-1)^{s+k+2} \geq 0$ for $s = s_i^{k+1},\ldots,s_{i+1}^{k+1} - 1$ then follows by writing out the definition of these $(k+1)$th divided differences in terms of kth divided difference and using the induction hypothesis and the ENO choice of s^{k+1}. We refer to Fjordholm (2009) and Fjordholm et al. (2012b) for the full proof. □

We emphasize that the sign property is mesh independent, in the sense that it holds for *any* mesh $(x_{i+1/2})_{i\in\mathbb{Z}}$, regardless of the mesh width Δx_i.

Remark 3. The "point-value version" of ENO (see Remark 1) also satisfies the sign property (1) (see Fjordholm, 2013; Fjordholm et al., 2012b).

Remark 4. It is easy to confirm by numerical experiments that the "RD" (reconstruction with deconvolution) ENO method does *not* satisfy the sign property. Indeed, fig. 3b of Harten et al. (1987), which shows a fourth-order RD ENO reconstruction, clearly violates the sign property at the fifth cell interface from the left.

4.3 Upper Bound on Jumps

In Fjordholm (2013) and Fjordholm et al. (2012b), it was shown that the ENO reconstruction procedure satisfies—in addition to the sign property—an upper bound on the jumps in the reconstructed polynomial. More precisely, for every $k \in \mathbb{N}$, the kth-order ENO reconstruction satisfies

$$0 \leq \frac{\langle\langle v \rangle\rangle_{i+1/2}}{[[\bar{v}]]_{i+1/2}} \leq C_{k,i} \quad \forall \; i \in \mathbb{Z}, \tag{4}$$

where $C_{k,i}$ depends only on k and on the ratios $|I_j|/|I_\ell|$ of neighbouring cell sizes. (Note that the first inequality in (4) is merely a restatement of the sign property (1)). Recall from Section 3.3.2 that his bound ensures Lipschitz continuity of the numerical flux (17).

In the case of a uniform mesh, $|I_i| \equiv$ const., the constant $C_{k,\,i} \equiv C_k$ can be computed explicitly (see Table 1). By way of an example, it was also found that the upper bound (4) is sharp. Indeed, if

$$\bar{v}_i = \begin{cases} 0 & \text{if } i \text{ is odd} \\ 1 & \text{if } i \text{ is even and } i \leq 4 \\ 1-\varepsilon & \text{if } i \text{ is even and } i > 4. \end{cases}$$

for any $\varepsilon > 0$, then the upper bound in (4) is attained in the limit $\varepsilon \to 0$. Fig. 2 shows these worst-case scenarios for $k = 2, 3, 4, 5$ and $\varepsilon = 10^{-10}$.

4.4 The ENO TV Conjecture

Any compactness argument for numerical approximations of the conservation law (1) requires some "weak TV bound" of the form (23). To conclude such a bound on the basis of the available "weak reconstructed TV bound" (22b), it would seem that a lower bound of the form $|\langle\langle v \rangle\rangle_{i+1/2}| \geq |[[\bar{v}]]_{i+1/2}|$ for all i is

TABLE 1 The Upper Bound in (4) for a Uniform Mesh

k	Upper Bound C_k
1	1
2	2
3	10/3 = 3.333...
4	16/3 = 5.333...
5	128/15 = 8.533...
6	208/15 = 13.866...

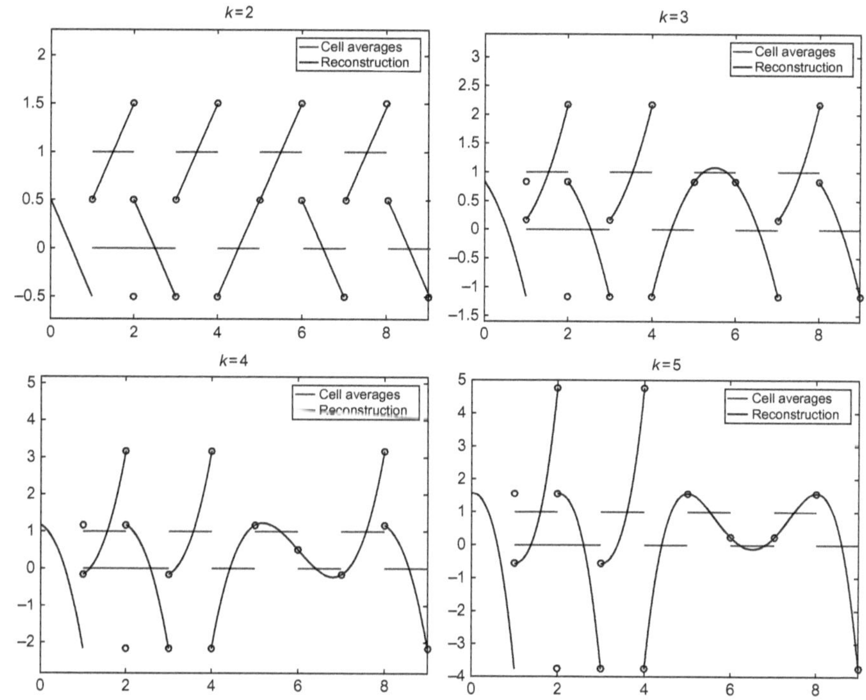

FIG. 2 Worst-case cell interface jumps for $k = 2, 3, 4, 5$.

required. However, such a bound is impossible due to the possibility that $\langle\langle v\rangle\rangle_{i+1/2} = 0$ even when $[[\bar{v}]]_{i+1/2} \neq 0$ (see Section 4.1.2).

In Fjordholm (2013) the following inequality was conjectured for the kth-order ENO reconstruction method:

$$\sum_{i\in\mathbb{Z}} |[[\bar{v}]]_{i+1/2}|^{k+1} \leq C \| \bar{v} \|_{L^\infty}^{k-1} \sum_{i\in\mathbb{Z}} [[\bar{v}]]_{i+1/2} \langle\langle v\rangle\rangle_{i+1/2} \tag{5}$$

for some $C > 0$ independent of $\bar{v}$ and Δx. Clearly, if this were to hold, then the "weak reconstructed TV bound" (22b), together with an L^∞ bound on $\bar{v}$, would imply (23). The only case for which (5) has been proven is for $k = 2$, and we include the proof here. For the sake of simplicity, we assume that the mesh is uniform.

Proof of (5) for $k = 2$. Denote $\Delta\bar{v}_i = \bar{v}_{i+1} - \bar{v}_i$, and iteratively $\Delta^k \bar{v}_i = \Delta^{k-1}(\Delta\bar{v})_i$. The formula (2) yields

$$\begin{aligned}\sum_{i\in\mathbb{Z}} [[\bar{v}]]_{i+1/2} \langle\langle v\rangle\rangle_{i+1/2} &= \sum_{i\in\mathbb{Z}} |\Delta\bar{v}_i| \sum_{j=s_i^2}^{s_{i+1}^2-1} a_i |\Delta^2 \bar{v}_j| \\ &\geq a \sum_{i\in\mathbb{Z}} |\Delta\bar{v}_i| \sum_{j=s_i^2}^{s_{i+1}^2-1} |\Delta^2 \bar{v}_j|\end{aligned}$$

for constants $a_i \geq a > 0$ only dependent on i and s_i^2. For every $j \in \mathbb{Z}$, there is precisely one index $i \in \mathbb{Z}$ such that $j \in \{s_i^2, \ldots, s_{i+1}^2 - 1\}$, and we denote this index i by $i = \iota_j^2$. Thus, we can write

$$\sum_{i\in\mathbb{Z}} |\Delta\bar{v}_i| \sum_{j=s_i^2}^{s_{i+1}^2-1} |\Delta^2\bar{v}_j| = \sum_{j\in\mathbb{Z}} |\Delta\bar{v}_{\iota_j^2}||\Delta^2\bar{v}_j|.$$

It is straightforward to show that for $k = 2$, the index ι is given by

$$\iota_j^2 = \begin{cases} j & \text{if } |\Delta\bar{v}_j| > |\Delta\bar{v}_{j+1}| \\ j+1 & \text{if } |\Delta\bar{v}_j| \leq |\Delta\bar{v}_{j+1}|, \end{cases} \tag{6}$$

and as a consequence,

$$|\Delta\bar{v}_{\iota_j^2}| = \max\left(|\Delta\bar{v}_j|, |\Delta\bar{v}_{j+1}|\right). \tag{7}$$

Starting with the left-hand side of (5) with $k = 2$, we get

$$\begin{aligned}
\sum_{i\in\mathbb{Z}} |\Delta\bar{v}_i|^3 &= \sum_{i\in\mathbb{Z}} |\Delta\bar{v}_i|\Delta\bar{v}_i\Delta\bar{v}_i \\
(\textit{summation-by-parts}) \quad &= -\sum_{i\in\mathbb{Z}} \bar{v}_{i+1}\Delta(|\Delta\bar{v}_i|\Delta\bar{v}_i) \\
&= -\sum_{i\in\mathbb{Z}} \bar{v}_{i+1}\left((\Delta|\Delta\bar{v}_i|)\Delta\bar{v}_i + |\Delta\bar{v}_i|\Delta^2\bar{v}_i\right) \\
&\leq 2\sum_{i\in\mathbb{Z}} |\bar{v}_{i+1}||\Delta^2\bar{v}_i||\Delta\bar{v}_i| \\
(\textit{relabeling } i \mapsto j \textit{ and using } (7)) \quad &\leq 2\,\|\bar{v}\|_{L^\infty} \sum_{j\in\mathbb{Z}} |\Delta^2\bar{v}_j||\Delta\bar{v}_{\iota_j^2}|
\end{aligned}$$

This completes the proof. □

4.5 Mesh-Dependent Properties

The "mesh-dependent properties" of ENO are those properties which are satisfied asymptotically as $\Delta x \to 0$. In other words, for a *fixed* underlying function $v(x)$, these are properties of ENO that are satisfied on *sufficiently fine* meshes. Although these properties function as a proof of concept of the ENO reconstruction method, they are of limited value in applications to numerical methods for conservation laws (1) because for such applications, the cell averages in question will themselves depend (nonlinearly) on the mesh. As such, these properties cannot be used in a proof of stability or convergence of numerical schemes for (1).

Below, we use the term "shock" to refer to any jump discontinuity of the underlying function v. For simplicity, we will assume that $\Delta x_i \equiv \text{const}$.

4.5.1 Uniform kth-Order Accuracy up to Discontinuities

If v is a piecewise C^∞ function with finitely many jump discontinuities ("shocks"), then for *sufficiently small* Δx, the ENO reconstruction is a

kth-order approximation of v in all cells not containing a shock (Harten et al., 1986). Indeed, if Δx is sufficiently small, then there are at least k cells in-between the shocked cells. Moreover, the ℓth divided difference $[\bar{v}_s,\ldots,\bar{v}_{s+\ell}]$ over any stencil containing a shocked cell behaves as $O(\Delta x^{-\ell})$. Thus, if Δx is small enough then in every nonshocked cell, the ENO stencil selection procedure can, and will, select an ENO stencil $\{s_i,\ldots, s_i + k - 1\}$ *not* containing a shock. The property of uniform kth-order accuracy then follows as in Section 4.1.3.

4.5.2 Monotonicity in Shocked Cells

Harten et al. (1986) proved that the primitive P_i of the ENO reconstruction p_i will be *monotone* in every cell containing a discontinuity of V. This property is of limited value since (a) the primitive V is always continuous, and (b) we are primarily interested in p_i, not P_i. However, it turns out that the same property in fact holds for the ENO reconstruction p_i (see Fig. 3).

Proposition 1. *Let* v *be a piecewise* C^∞ *function with finitely many shocks. Then for sufficiently small* Δx*, the ENO reconstruction of* v *is monotone in every cell containing a jump discontinuity—more precisely, it is strictly increasing at positive jumps and strictly decreasing at negative jumps.*

Proof. The proof is similar in spirit to the proof of Harten et al. (1986, Theorem 4.1).

By choosing Δx sufficiently small, we may assume that shocked cells are at least k cells from one another, and hence it suffices to consider the case $v = w + H$, where w is Lipschitz continuous and H is piecewise constant with a single jump discontinuity at $x=\bar{x}\in(x_{i-1/2},x_{i+1/2})$, for some index $i\in\mathbb{Z}$. By the linearity of the ENO method (see Section 4.1.1), we may assume that

$$H(x)=\begin{cases}0 & \text{if } x<\bar{x}\\ 1 & \text{if } x>\bar{x}.\end{cases}$$

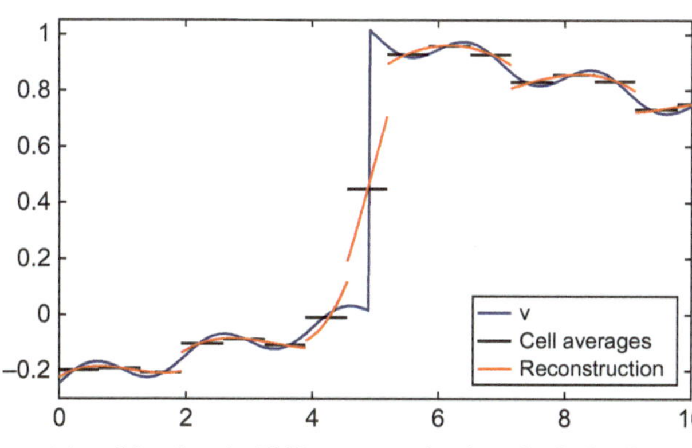

FIG. 3 Monotonicity of fourth-order ENO reconstruction in a shocked cell.

Moreover, we may assume that $k \geq 3$ since the cases $k = 1$ (piecewise constant reconstruction) and $k = 2$ (piecewise linear reconstruction) are immediate.

Let $S = \{s_i^k, \ldots, s_i^k + k - 1\}$ denote the ENO reconstruction stencil in cell i and let $I = \bigcup_{j \in S} I_j$. We can write $p_i = q + G$, where q and G are $(k-1)$th-order polynomials which interpolate $(\bar{w}_j)_{j \in S}$ and $(\bar{H}_j)_{j \in S}$, respectively. Since w is Lipschitz continuous, we have $|[\bar{w}_j, \ldots, \bar{w}_{j+\ell}]| \leq C \Delta x^{-\ell}$ for all j, ℓ, so from (7) we get

$$\left\| \frac{dq}{dx} \right\|_{L^\infty(I)} \leq C \tag{8}$$

for some C independent of Δx.

Since G interpolates $(\bar{H}_j)_{j \in S}$, there is at least one point $y_j \in I_j$ for every $j \in S, j \neq i$ such that

$$G(y_j) = H(y_j) = \begin{cases} 0 & \text{if } j < i \\ 1 & \text{if } j > i. \end{cases}$$

If there is more than one such root in cell I_j, we select the root y_j which is closest to $\bar{x}$. By Rolle's theorem, the function $\frac{dG}{dx}$ has a zero in every interval of the form

$$\begin{array}{ll} (y_{j-1}, y_j) & \text{for } s_i < j < i \\ (y_j, y_{j+1}) & \text{for } i < j < s_i + k - 1. \end{array} \tag{9}$$

Note that cell I_i intersects none of the above intervals. We will show that $\frac{dG}{dx}$ cannot have a zero in $(y_{i-1}, y_{i+1}) \supset I_i$. Choosing Δx small enough and using (8), we can then conclude that also $p_i = q + G$ must be monotone in I_i.

We divide into two cases:

Case 1: $s_i \in \{i, i - k + 1\}$, i.e., there are no cells in the stencil either to the left or to the right of I_i. In this case, there are exactly $k - 2$ intervals of the form (9). Since the $(k-2)$th-order polynomial $\frac{dG}{dx}$ can have at most $k - 2$ zeros, it cannot have another zero in I_i.

Case 2: $s_i \notin \{i, i - k + 1\}$. In this case, there are exactly $k - 3$ intervals of the form (9). From the jump expression (2) and the sign property (see Section 4.2), we get

$$\begin{aligned} (p_{i+1} - p_i)(x_{i+1/2}) &= \sum_{s=s_i}^{s_{i+1}-1} [\bar{v}_s, \ldots, \bar{v}_{s+k}] X_{i,s} \\ &\geq \sum_{s=s_i}^{s_{i+1}-1} |[\bar{v}_s, \ldots, \bar{v}_{s+k}]| \Delta x^k \\ &\geq b_{i+1/2} \end{aligned}$$

for some $b_{i+1/2} > 0$ independent of Δx. Here, we have used the fact that $[\bar{v}_s, \ldots, \bar{v}_{s+k}] \sim \Delta x^{-k}$ for all $s \in \{i - k, \ldots, i\}$. Similarly, we get

$$(p_i - p_{i-1})(x_{i-1/2}) \geq b_{i-1/2}$$

for some $b_{i-1/2} > 0$ independent of Δx. Thus,

$$\begin{cases} G(x_{i-1/2}) = (p_i - q)(x_{i-1/2}) \geq b_{i-1/2} + (p_{i-1} - q)(x_{i-1/2}) = b_{i-1/2} + O(\Delta x), \\ G(x_{i+1/2}) = (p_i - q)(x_{i+1/2}) \leq -b_{i+1/2} + (p_{i+1} - q)(x_{i+1/2}) = 1 - b_{i+1/2} + O(\Delta x). \end{cases}$$

Choosing Δx small enough that the "$O(\Delta x)$" terms are smaller than $b_{i\pm 1/2}$, we find that

$$G(y_{i-1}) = 0, \quad G(x_{i-1/2}) > 0, \quad G(x_{i+1/2}) < 1, \quad G(y_{i+1}) = 1,$$

and hence,

$$\frac{dG}{dx}(y_{i-1}) \geq 0, \quad \frac{dG}{dx}(y_{i+1}) \geq 0.$$

Thus, if $\frac{dG}{dx}$ has a zero in (y_{i-1}, y_{i+1}), there must be at least two of them (or one zero with multiplicity at least 2). But the $(k-2)$th-order polynomial $\frac{dG}{dx}$ already has $k - 3$ zeros in the intervals (9), so it cannot any zeros in (y_{i-1}, y_{i+1}). □

4.5.3 Essentially Nonoscillatory

The "ENO" property, from which ENO derives its name, can be roughly stated as follows: Up to a term of order Δx^k, the total variation of the ENO reconstruction p is less than that of v. As with the monotonicity property, Harten et al. (1986) proved this only for the primitives P, V, not for the reconstruction p itself. However, with Proposition 1 in place we can establish this result also for p.

Theorem 2. *Assume that v is piecewise C^∞ with finitely many jump discontinuities. Then for sufficiently small Δx, there exists a function $z = z(x)$ such that*

$$z(x) = p(x) + O(\Delta x^k) \ \ \forall \ x, \quad TV(z) \leq TV(v),$$

where $p = \mathcal{R}(\bar{v})$ is the ENO reconstruction of v.

Proof. Let Δx be sufficiently small that $p(x) = v(x) + O(\Delta x^k)$ in all nonshocked cells (see Section 4.5.1). Decrease Δx further such that p is monotone in all shocked cells (see Section 4.5.2). We choose $z(x) = v(x)$ in nonshocked cells, and $z(x) = p(x)$ in shocked cells. After an $O(\Delta x^k)$ modification near the interfaces $x_{i+1/2}$ between shocked and nonshocked cells, the sign property implies that the total variation does not increase at these points. □

Remark 5. Although the above theorem says nothing about $\text{TV}(p)$, it may be shown that $\text{TV}(p) \leq \text{TV}(v) + O(\Delta x^k)$ for sufficiently small Δx.

4.6 ENO Deficiencies

Despite satisfying numerous stability properties, the ENO reconstruction method suffers from some deficiencies which makes it less attractive for certain applications such as numerical methods for linear conservation laws.

4.6.1 $\mathcal{R}$ Is Discontinuous

The ENO reconstruction $\mathcal{R}:(\bar{v}_i)_{i\in\mathbb{Z}} \to p$ is discontinuous, in the sense that a small change in $\bar{v}_i$ (such as round-off errors) can change the switch in the ENO stencil selection procedure, thus producing a different reconstruction p_j. Although this stencil switching might not be a problem in practice, the discontinuous nature of ENO-based methods makes their analysis significantly more difficult.

4.6.2 Inefficient Use of Information

Although the final ENO reconstruction p_i in a cell only relies on k values, the ENO stencil selection procedure depends on all $2k-1$ neighbouring points. This is an inefficient use of information; using all $2k-1$ points would potentially give up to $(2k-1)$th-order accuracy in smooth parts of the solution. This situation is exacerbated in multiple dimensions.

The WENO method uses a much more compact interpolation stencil and might therefore be more suitable for multidimensional problems.

4.6.3 Instabilities in Linear Problems

Rogerson and Meiburg (1990) reported on a series of numerical experiments with an ENO-based fourth-order finite difference schemes for the periodic linear advection equation

$$\partial_t u + \partial_x u = 0, \quad x \in [-\pi, \pi).$$

They observed the expected fourth-order convergence rate with $u_0(x) = \sin(x)$, but with $u_0(x) = \sin(x)^4$ they observed a decay in the convergence rate at moderately high values of N, the number of meshpoints.

We approximate the above initial value problem using a Godunov-type finite volume scheme with fourth-order ENO reconstruction and a fourth-order Runge–Kutta time integrator. (Rogerson and Meiburg computed with the so-called ENO-Roe method (Shu and Osher, 1989), but the problem persists in other variants of ENO method and hence seems to be inherent to the ENO reconstruction procedure.) Fig. 4 (top row) shows the fourth-order divided difference $[\bar{v}_i, \ldots, \bar{v}_{i+4}]$ at various times. High-frequency oscillations appear quickly at the critical points $x = 0$, $x = \pm\frac{\pi}{2}$, and $x = \pm\pi$, and over time, these oscillations propagate into the lower-order divided differences, finally polluting the solution $\bar{v}_i$. The oscillations near $x = \pm\frac{\pi}{2}$ stay bounded, whereas the oscillations near $x = 0$, $x = \pm\pi$ grow unboundedly.

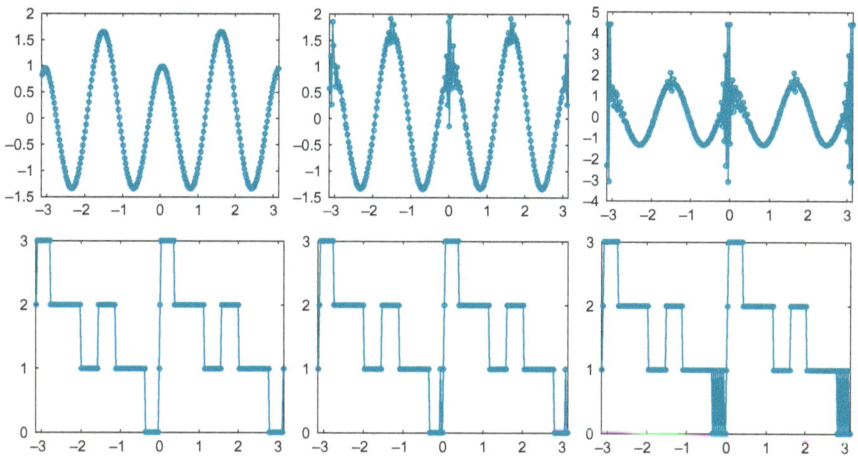

FIG. 4 Fourth-order divided differences (*top row*) and stencil offset r_i (*bottom row*) at $t = 0$ (*left*), $t \approx 0.02$ (*middle*), and $t \approx 0.04$ (*right*).

Fig. 4 (bottom row) shows the *stencil offset* $r_i^k = i - s_i^k \in \{0, \ldots, k-1\}$. (Recall from Section 4.1.2 that every interface $x_{i+1/2}$ where $r_{i+1} \geq r_i$ will have a discontinuity in the reconstruction, which might lead to larger truncation errors.) Near the oscillatory points $x = 0$, $x = \pm\pi$, the ENO method selects the stencils $r_i^3 = 0$ and $r_i^3 = 3$. Rogerson and Meiburg (1990) call these stencils *linearly unstable*: setting $r_i^3 \equiv 0$ or $\equiv 3$ for all i will give an unconditionally unstable, divergent scheme, whereas $r_i \equiv 1$ or $\equiv 2$ gives a stable, convergent scheme. Although this heuristic explanation might very well be the root of the problem, the nonlinear nature of ENO makes this problem very hard to analyze rigorously. Further discussion can be found in Abgrall and Lafon (1993, Section 5), Harten (1987), and Shu (1990). We mention in closing that the WENO method does not exhibit these instabilities for this particular problem (Shu, 1998).

5 SUMMARY

The ENO method has been enormously influential in the numerics community for hyperbolic conservation laws. Despite its highly nonlinear (even discontinuous) nature, it yields expressions and formulas which are rather easy to analyze, and enjoys several surprising properties such as the nonoscillatory property, the sign property and upper bounds on discontinuities. As discussed in Section 4.6, certain ENO-based finite volume methods suffer from instabilities which prevent convergence. A rigorous analysis of this problem would be highly interesting (not to mention difficult) and might lead to provably stable ENO-type methods.

ACKNOWLEDGEMENTS

Research supported in part by the grant *Waves and Nonlinear Phenomena* (WaNP) from the Research Council of Norway.

REFERENCES

Abgrall, R., 1994. On essentially non-oscillatory schemes on unstructured meshes: analysis and implementation. J. Comput. Phys 114, 45–58. http://dx.doi.org/10.1006/jcph.1994.1148.

Abgrall, R., Lafon, F.C., 1993. ENO schemes on unstructured meshes. INRIA. Technical report RR-2099.

Aràndiga, F., Donat, R., 2000. Nonlinear multiscale decompositions: the approach of A. Harten. Numer. Algorithms 1572-926523 (2), 175–216. http://dx.doi.org/10.1023/A:1019104118012.

Arandiga, F., Cohen, A., Donat, R., Dyn, N., Matei, B., 2008. Approximation of piecewise smooth functions and images by edge-adapted (ENO-EA) nonlinear multiresolution techniques. Appl. Comput. Har. Anal. 24, 225–250.

Fjordholm, U.S., 2009. Structure preserving finite volume methods for the shallow water equations. Master's thesis, University of Oslo.

Fjordholm, U.S., 2013. High-order accurate entropy stable numerical schemes for hyperbolic conservation laws. PhD thesis, ETH Zürich. No. 21025.

Fjordholm, U.S., Mishra, S., Tadmor, E., 2012a. Arbitrarily high order accurate entropy stable essentially non-oscillatory schemes for systems of conservation laws. SIAM J. Numer. Anal. 50, 544–573.

Fjordholm, U.S., Mishra, S., Tadmor, E., 2012b. ENO reconstruction and ENO interpolation are stable. Found. Comput. Math. 1615-3375. 13 (2), 139–159. http://dx.doi.org/10.1007/s10208-012-9117-9.

Godlewski, E., Raviart, P.-A., 1991. Hyperbolic Systems of Conservation Laws. Ellipses, Paris.

Gottlieb, S., Shu, C.-W., Tadmor, E., 2001. High order time discretization methods with the strong stability property. SIAM Rev. 43, 89–112.

Harten, A., 1983. On the symmetric form of systems of conservation laws with entropy. J. Comput. Phys. 0021-9991. 49 (1), 151–164. http://dx.doi.org/10.1016/0021-9991(83)90118-3. http://www.sciencedirect.com/science/article/pii/0021999183901183.

Harten, A., 1986. On high-order accurate interpolation for non-oscillatory shock capturing schemes. In: Dafermos, C., Ericksen, J.L., Kinderlehrer, D., Slemrod, M. (Eds.), Oscillation Theory, Computation, and Methods of Compensated Compactness, vol. 2, The IMA Volumes in Mathematics and Its Applications. Springer, New York, pp. 71–105.

Harten, A., 1987. On the nonlinearity of modern shock-capturing schemes. In: Chorin, A.J., Majda, A.J. (Eds.), Wave Motion: Theory, Modelling, and Computation, vol. 7, Mathematical Sciences Research Institute Publications. Springer, USA, pp. 147–201.

Harten, A., 1989. ENO schemes with subcell resolution. J. Comput. Phys. 0021-9991. 83 (1), 148–184. http://dx.doi.org/10.1016/0021-9991(89)90226-X.

Harten, A., 1996. Multiresolution representation of data: a general framework. SIAM J. Numer. Anal. 33 (3), 1205–1256. http://dx.doi.org/10.1137/0733060.

Harten, A., Chakravarthy, S.R., 1991. Multi-dimensional ENO schemes for general geometries. ICASE. 91–76.

Harten, A., Osher, S., 1987. Uniformly high-order accurate nonoscillatory schemes, I. SIAM J. Numer. Anal. 24 (2), 279–309.

Harten, A., Engquist, B., Osher, S., Chakravarthy, S.R., 1986. Some results on uniformly high-order accurate essentially nonoscillatory schemes. Appl. Numer. Math. 2, 347–377.

Harten, A., Engquist, B., Osher, S., Chakravarthy, S.R., 1987. Uniformly high order accurate essentially non-oscillatory schemes, III. J. Comput. Phys. 0021-9991. 71 (2), 231–303. http://dx.doi.org/10.1016/0021-9991(87)90031-3.

Jiang, G.-S., Shu, C.-W., 1996. Efficient implementation of weighted ENO schemes. J. Comput. Phys. 0021-9991. 126 (1), 202–228. http://dx.doi.org/10.1006/jcph.1996.0130.

Liu, X.-D., Osher, S., Chan, T., 1994. Weighted essentially non-oscillatory schemes. J. Comput. Phys. 0021-9991. 115 (1), 200–212. http://dx.doi.org/10.1006/jcph.1994.1187. http://www.sciencedirect.com/science/article/pii/S0021999184711879.

Rogerson, A.M., Meiburg, E., 1990. A numerical study of the convergence properties of ENO schemes. J. Sci. Comput. 1573-7691. 5 (2), 151–167. http://dx.doi.org/10.1007/BF01065582.

Shu, C.-W., 1990. Numerical experiments on the accuracy of ENO and modified ENO schemes. J. Sci. Comput. 0885-7474. 5 (2), 127–149. http://dx.doi.org/10.1007/BF01065581.

Shu, C.-W., 1998. Essentially non-oscillatory and weighted essentially non-oscillatory schemes for hyperbolic conservation laws. In: Quarteroni, A. (Ed.), Advanced Numerical Approximation of Nonlinear Hyperbolic Equations. Lecture Notes in Mathematics, vol. 1697. Springer-Verlag, Berlin, pp. 325–432. (Cockburn, B., Johnson, C., Shu, C.-W., Tadmor, E.).

Shu, C.-W., Osher, S., 1988. Efficient implementation of essentially non-oscillatory shock-capturing schemes. J. Comput. Phys. 0021-9991. 77 (2), 439–471. http://dx.doi.org/10.1016/0021-9991(88)90177-5.

Shu, C.-W., Osher, S., 1989. Efficient implementation of essentially non-oscillatory shock-capturing schemes, II. J. Comput. Phys. 0021-9991. 83 (1), 32–78. http://dx.doi.org/10.1016/0021-9991(89)90222-2.

Siddiqi, K., Kimia, B.B., Shu, C.-W., 1997. Geometric shock-capturing ENO schemes for subpixel interpolation, computation and curve evolution. Graph. Model Im. Proc. 59 (5), 278–301.

Sweby, P.K., 1984. High resolution schemes using flux limiters for hyperbolic conservation laws. SIAM J. Numer. Anal. 21 (5), 995–1011. http://dx.doi.org/10.1137/0721062. http://link.aip.org/link/?SNA/21/995/1.

Tadmor, E., 2003. Entropy stability theory for difference approximations of nonlinear conservation laws and related time dependent problems. Acta Numer. 12, 451–512.

Tadmor, E., 2016. Entropy stable schemes. In: Abgrall, R., Shu, C.-W. (Eds.), Handbook of Numerical Methods for Hyperbolic Problems, vol. 17. Elsevier, Amsterdam, pp. 467–493.

Zhang, Y.-T., Shu, C.-W., 2016. ENO and WENO Schemes. In: Abgrall, R., Shu, C.-W. (Eds.), Handbook of Numerical Methods for Hyperbolic Problems, vol. 17. Elsevier, Amsterdam, pp. 103–122.

Chapter 7

Stability, Error Estimate and Limiters of Discontinuous Galerkin Methods

J. Qiu* and Q. Zhang†

*School of Mathematical Sciences, Fujian Provincial Key Laboratory of Mathematical Modeling and High-Performance Scientific Computing, Xiamen University, Xiamen, Fujian, PR China

†Nanjing University, Nanjing, Jiangsu, PR China

Chapter Outline

ABSTRACT

In this chapter, we review the development of discontinuous Galerkin (RKDG) methods for conservation laws and focus on the stability, error estimates and limiters for RKDG methods. The stability and error estimates are core properties of RKDG, and limiter is an important component of RKDG methods for solving conservation laws with strong shocks in the solutions, which is applied to detect discontinuities and control spurious oscillations near such discontinuities.

Handbook of Numerical Analysis, Vol. 17. http://dx.doi.org/10.1016/bs.hna.2016.06.001

Keywords: RKDG method, Strong stability preserving, Runge–Kutta, Stability, Error estimate, Limiter

AMS Classification Codes: 65M12, 65N12

1 INTRODUCTION

In this chapter, we review some development of discontinuous Galerkin (DG) method and focus mainly on the stability, error estimate and limiters for the d-dimensional conservation law

$$\partial_t u + \nabla \cdot f(u) = 0, \tag{1}$$

with the initial solution $u(x, 0) = u_0(x)$, where $f(u)$ is the given smooth flux function. For simplicity, we mainly consider the case when the exact solution is periodic or compactly supported.

After the first version of DG method, which was introduced by Reed and Hill (1973), in the framework of neutron linear transport, the DG method has been paid more and more attention, because of its many advantages. For example, this method has strong stability and optimal accuracy to capture discontinuous jump sharply and combines the advantages of finite element method and finite difference method. An important development in the DG method is in the late 1980s, when Cockburn and Shu (1991, 1989, 1998) and Cockburn et al. (1989, 1990) combine the Runge–Kutta time discretization and the DG spatial discretization, with exact or approximate Riemann solvers as interface fluxes and total variation bounded (TVB) limiter (Shu, 1987) to achieve nonoscillatory properties for strong shocks, to easily solve *nonlinear* time-dependent hyperbolic conservation laws (1). These schemes are termed Runge–Kutta discontinuous Galerkin (RKDG) methods. Till now there have been many published papers in this field, for example, see the review paper and books (Cockburn, 1998; Cockburn and Shu, 2001) and the others.

This chapter shall focus on two issues. One is the theoretical analysis for this kind of method, and the other is the design and application of limiters.

The error estimates of DG method have been paid attention to by many authors. Lasaint and Raviart (1974) proved first the suboptimal order for general triangulations, and optimal order for Cartesian grids. Later, Johnson and Pitkäranta (1986) proved a quasi-optimal rate of convergence for general triangulation and Peterson (1991) confirmed its sharpness. Note that optimal error estimate can be achieved for some meshes with special structure (Richter, 1988; Cockburn et al., 2008). However, the above discussions are carried out for either the steady problem or the space-time DG method and semidiscrete DG method for unsteady problems. In this chapter, we mainly discuss the error estimates to the fully discrete RKDG methods for the conservation law (1), whether the exact solution has sufficient smoothness or not.

One of the main difficulties in using RKDG methods to solve (1) with possibly strong shocks or contact discontinuities is that the numerical solution might cause spurious oscillations. These spurious oscillations might lead to nonlinear instability. One common strategy to control these oscillations is to apply nonlinear limiters to RKDG methods. Many limiters have been studied in the literature for RKDG methods, such as the *minmod*-type TVB limiter (Cockburn and Shu, 1989, 1998; Cockburn et al., 1989, 1990), the moment-based limiter (Biswas et al., 1994) and an improved moment limiter (Burbeau et al., 2001). These limiters belong to the slope-type limiters and they do control oscillations very well at the price of possibly degrading the accuracy of the numerical solution at smooth extrema. Another type of limiters is the WENO-type limiters, which are based on the weighted essentially nonoscillatory (WENO) methodology (Jiang and Shu, 1996; Liu et al., 1994) and can achieve both high-order accuracy and nonoscillatory property near discontinuities. This type of limiters includes the WENO limiter (Qiu and Shu, 2005c; Zhu et al., 2008) and the HWENO limiter (Luo et al., 2007; Qiu and Shu, 2004, 2005b), which use the classical WENO finite volume methodology for reconstruction and thus require a wide stencil, especially for higher order methods. More recently, the new WENO limiting procedures were developed for RKDG methods (Zhong and Shu, 2013; Zhu et al., 2013, 2016, submitted for publication) on both the structure and unstructured meshes, and the idea is to reconstruct the entire polynomial on the target cell by a convex combination of polynomials on this cell and its immediate neighbouring cells, with suitable adjustments for conservation and with the nonlinear weights of the convex combination following the classical WENO procedure. The subcell limiting procedures were developed in Dumbser et al. (2014) and Zanotti et al. (2015a,b), which breaks the DG cell into subcells and then uses WENO ideas for limiting.

The chapter is organized as follows. In Section 2, we present the framework of RKDG method. Then we present some stability results in Section 3, and some error estimates in Section 4. In Section 5, we introduce some good limiter used in the RKDG method. Finally, some concluding remarks are given in Section 6.

2 IMPLEMENTATION OF DG METHODS

We start with the description in the bounded interval $I = (0, 1)$; however, it works also in multi-dimensions. Divide it into N cells with boundary points $0 = x_{\frac{1}{2}} < x_{\frac{3}{2}} < \cdots < x_{N+\frac{1}{2}} = 1$ and denote the cell size of $I_i = [x_{i-1/2}, x_{i+1/2}]$ by $h_i = x_{i+1/2} - x_{i-1/2}$. The maximum cell size is denoted by $h = \max_i h_i$. For simplicity of presentation, we would like to assume that the used mesh is quasi-uniform; namely, there exists a positive constant C independent of h, such that $Ch_i \geq h$ for every $i = 1, 2, \ldots, N$, as h goes to zero.

The numerical solution and the test function are both considered in the following discontinuous finite element space:

$$V_h = V_h^k = \{ v \in L^2(I) : v|_{I_i} \in \mathbb{P}^k(I_i), i = 1, \ldots, N \}, \tag{2}$$

where $\mathbb{P}^k(I_i)$ is the space of polynomials of degree at most $k \geq 0$ on the cell I_i. Note that the functions in V_h are allowed to have discontinuities across element interfaces. For any function $v_h \in V_h$, there are two limits along different directions at each element boundary point, namely, the left-value v_h^- and the right-value v_h^+. Further, the jump and mean, respectively, are denoted by

$$[\![v_h]\!] = v_h^+ - v_h^- \text{ and } \{\!\{v_h\}\!\} = \frac{1}{2}(v_h^+ + v_h^-). \tag{3}$$

2.1 Semidiscrete Version

First we define the semidiscrete DG(k) method as follows. We would like to find the numerical solution $u_h(t) \in V_h$ for any time $t > 0$, such that

$$(u_{h,t}, v_h) = \mathcal{H}(u_h, v_h), \quad \forall\, v_h \in V_h, \tag{4}$$

where the global DG spatial discretization is defined by

$$\mathcal{H}(u_h, v_h) = \sum_{1 \leq i \leq N} \left[\hat{f}(u_h)_{i+1/2} [\![v_h]\!]_{i+1/2} + \int_{I_i} f(u_h) v_{h,x} \mathrm{d}x \right]. \tag{5}$$

Here $(\cdot, \cdot)$ is the usual inner product in $L^2(I)$, and

$$\hat{f}(u_h) = \hat{f}(u_h^-, u_h^+) \tag{6}$$

is the numerical flux defined on each element boundary point. For example, the Lax–Friedrich numerical flux

$$\hat{f}(u_h^-, u_h^+) = \frac{1}{2}\left[f(u_h^-) + f(u_h^+) - C[\![u_h]\!]\right] \tag{7}$$

is used widely in practice, where $C = \max|f'(u)|$. Obviously, it is an E-flux or monotone flux, since it is not decreasing for the first argument and not increasing for the second argument. More numerical flux can be found in Qiu et al. (2006).

The initial solution is usually given as the approximation of the given solution $u_0(x)$. For example, $u_h^0 = \pi_h u_0(x)$ is the local L^2-projection of $u_0(x)$, such that

$$\int_0^1 (\pi_h u_0(x) - u_0(x)) v_h(x) \mathrm{d}x = 0, \quad \forall v_h \in V_h. \tag{8}$$

Below we would not mention the setting of initial solution, since it only affects the numerical error, but not the numerical stability.

Remark 1. Eq. (5) is obtained by simple summary of DG formulation on each element, as the periodic boundary condition has been used in the above process. The other boundary condition, for example, the inflow boundary condition (Zhang, 2011), can be treated with in a similar way.

2.2 SSPRK Algorithms

The second feature of RKDG method is the time marching. One of the famous and successful treatment is adopting the explicit total variation diminishing Runge–Kutta time marching; please refer to the series papers of Cockburn and Shu (1991, 1989, 1998) and Cockburn et al. (1989, 1990). Now this kind of time marching has been considered in the term *strong stability-preserving* (SSP) algorithms (Gottlieb et al., 2001).

It is to say that we would like to seek the solution u_h^n at the time level $t^n = n\tau$ step by step, where τ is the time step. The time step could actually change from step to step; for simplicity, in this chapter we take it as a constant. According to the Osher–Shu representation (Shu and Osher, 1988), the general construction of RKDG(s, r, k) method is given as follows, where s and r are the stages and the order of the used Runge–Kutta time marching, respectively, and k is the degree of piecewise polynomials. Assume that the numerical solution $u_h^n \in V_h$ has been obtained, we will solve successively each stage solution $u_h^{n,\ell+1} \in V_h$ for $\ell = 0, 1, 2, \ldots, s-1$, by virtue of the variation form

$$(u_h^{n,\ell+1}, v_h) = \sum_{0 \le \kappa \le \ell} a_{\ell\kappa}(u_h^{n,\kappa}, v_h) + b_\ell \mathcal{H}(u_h^{n,\ell}, v_h)\tau, \quad \forall v_h \in V_h, \tag{9}$$

where $u_h^{n,0} = u_h^n$ and $u_h^{n+1} = u_h^{n,s}$. Note that the coefficients, $a_{\ell\kappa}$ and b_ℓ, are given for the used time marching, with the distinguish property that they are all nonnegative. The parameters in the RKDG(3, 3, k) and RKDG(2, 2, k) are given in Table 1.

The above fully discrete schemes are actually implemented explicitly because the mass matrix is easy to be inverted due to the block diagonal structure. The mass matrix will be diagonal when a local orthogonal basis is chosen for polynomials on each element.

TABLE 1 Parameters: Left: RKDG(2,2,k); Right: RKDG(3,3,k)

			1			1
1		1	$\frac{3}{4}$	$\frac{1}{4}$		$\frac{1}{4}$
$\frac{1}{2}$	$\frac{1}{2}$	$\frac{1}{2}$	$\frac{1}{3}$	0	$\frac{2}{3}$	$\frac{2}{3}$

2.3 Limiters

The method described in Sections 2.1 and 2.2 can compute solutions to (1) which are either smooth or have weak shocks and other discontinuities without further modification. If the discontinuities are strong, however, the scheme will generate significant oscillations and even nonlinear instability. To avoid such difficulties, a nonlinear limiter procedure is used after each Runge–Kutta inner stage (or after the complete Runge–Kutta time step) to control the numerical solution. There are many limiters which exist in the literature, for example, the minmod-type limiters, the moment-based limiters, the improved moment-based limiters and the WENO-type limiter. We will describe these limiters in details in Section 5.

3 STABILITY

In this section, we collect some stability results on the DG methods. To show this, let us recall some inverse properties of finite element space V_h. Specially, for any function $v_h \in V_h$, there hold[a]

$$\| (v_h)_x \| \leq \mu_1 h^{-1} \| v_h \|, \quad \| v_h \|_{\Gamma_h,\pm} \leq \mu_2 h^{-1/2} \| v_h \|, \quad \| v_h \|_{\Gamma_h,*} \leq \mu_3 h^{-1/2} \| v_h \|, \tag{10}$$

where the inverse constants, μ_1, μ_2 and μ_3, are independent of v_h and solely depend on the degree k of the piecewise polynomials (Zhang and Shu, 2009, in preparation). The sharp values are listed in Table 2 for $k \leq 4$. Here Γ_h denotes all element boundary points,

$$\| v \|_{\Gamma_h,\pm} = \left[\sum_{i=1}^{N} |v^{\pm}_{i+1/2}|^2 \right]^{1/2} \quad \text{and} \quad \| v \|_{\Gamma_h,*} = \left[\| v \|^2_{\Gamma_h,-} + \| v \|^2_{\Gamma_h,+} \right]^{1/2}. \tag{11}$$

TABLE 2 Inverse Constants on the Uniform Mesh (Zhang and Shu, 2009, in preparation): $k \leq 4$

k	0	1	2	3	4
μ_1	0	$\sqrt{12} \approx 3.46$	$\sqrt{60} \approx 7.75$	$\sqrt{2(45+\sqrt{1605})}$ ≈ 13.04	$\sqrt{2(105+3\sqrt{805})}$ ≈ 19.50
μ_2	1	2	3	4	5
μ_3	$\sqrt{2}$	$\sqrt{6}$	$\sqrt{12}$	$\sqrt{20}$	$\sqrt{30}$

[a]Note that the above L^2-norm of derivative should be understood element by element.

For the linear case, the stability in L^2-norm is clear, for both semidiscrete version and fully discrete version. However, for the nonlinear case, the answer is not completely clear till now.

3.1 Linear Stability in L^2-Norm

Assume $f(u) = \beta u$ with β being a given constant. In this case, $\hat{f}(u_h) \equiv \hat{f}(u_h^-, u_h^+)$ is the so-called upwind numerical flux

$$\hat{f}(u_h) = \hat{f}(u_h^-, u_h^+) = \begin{cases} \beta u_h^-, & \text{if } \beta > 0, \\ \beta u_h^+, & \text{if } \beta < 0; \end{cases} \tag{12}$$

thus the global DG spatial discretization in (4) is defined explicitly by

$$\mathcal{H}(u_h, v_h) = \sum_{1 \le i \le N} \left[\hat{f}(u_h)_{i+1/2} [\![v_h]\!]_{i+1/2} + \int_{I_i} \beta u_h v_{h,x} \mathrm{d}x \right]. \tag{13}$$

The following three properties (Zhang and Shu, 2009, 2010) about the DG space discretization provide an important distribution to the success of DG method, which can achieved after a simple application of integration by parts, the inverse properties together with Cauchy–Schwarz inequality.

Lemma 1. *The bilinear functional has the approximating antisymmetric property*

$$\mathcal{H}(\psi, \phi) + \mathcal{H}(\phi, \psi) = -\sum_{1 \le j \le N} |\beta| [\![\phi]\!]_{j+\frac{1}{2}} \cdot [\![\psi]\!]_{j+\frac{1}{2}}, \quad \forall \phi, \psi \in V_h. \tag{14}$$

As a direct conclusion, the bilinear functional has the negative semidefined property

$$\mathcal{H}(\phi, \phi) = -\frac{1}{2} \sum_{1 \le j \le N} |\beta| [\![\phi]\!]_{j+1/2}^2 = -\frac{1}{2} |\beta| \, \| \, [\![\phi]\!] \, \|_{\Gamma}^2, \quad \forall \phi \in V_h. \tag{15}$$

Furthermore, the bilinear functional is continuous and bounded in $V_h \times V_h$, in the sense

$$|\mathcal{H}(\phi, \psi)| \le |\beta| \, \| \psi \| \left[\mu_1 h^{-1} \| \phi \| + \mu_2 h^{-1/2} \| \, [\![\phi]\!] \, \|_{\Gamma} \right], \quad \forall \psi, \phi \in V_h. \tag{16}$$

Now the L^2-norm stability of semidiscrete version is easy. It is followed from the negative semidefined property that

$$\frac{1}{2} \| u_h(T) \|^2 + \frac{|\beta|}{2} \int_0^T \| \, [\![u_h(t)]\!] \, \|_{\Gamma}^2 \, \mathrm{d}t \le \frac{1}{2} \| u_h(0) \|^2, \tag{17}$$

which reflects the subtle built-in dissipation mechanism of the DG method and allows more accurate than the standard Galerkin methods. However, this stability mechanism from the square of jumps is very weak; hence the time marching must be treated carefully, if the time step only satisfies the standard

CFL condition that the ratio of the time step over the mesh's size is fixed in a constant. Although RKDG(r, s, k) method has been used successfully in numerical practice, the stability analysis is nontrivial under the SSP framework (Gottlieb et al., 2001), because Euler-forward time marching of DG method is linearly unstable under the standard CFL (Courant–Friedrichs–Lewy) condition. Thus we have to find another way to prove the L^2-norm stability in theory.

Roughly speaking, there are two main strategies to do this. The trivial analysis is the so-called Fourier's technique, which can give the sharp CFL condition or the maximum CFL number. But, this technique demands too much assumptions that the used mesh is uniform and the boundary condition is given in periodical status. It is also hard to develop into the linear varying-coefficient problems, the nonlinear problems, the general boundary condition and multidimensional problems. The second strategy is energy analysis to overcome the above difficulties. This motivation comes from the optimal error estimate for two RKDG methods to solve the nonlinear conservation law (Zhang and Shu, 2004, 2010), which is obtained by virtue of the suitable projection and the stability analysis for the linear case.

Restricted by the page limitation, we would like to focus on the energy analysis and present only the sketch. The following material is partially taken from Zhang and Shu (2009). The important idea is to introduce some differences of stage solutions

$$\mathbb{D}_\ell u_h^n = \sum_{0\le\kappa\le\ell} \sigma_{\ell\kappa} u_h^{n,\kappa}, \quad \ell = 1,\ldots,s, \tag{18}$$

such that

$$(\mathbb{D}_{\ell+1} u_h^n, v_h) = \frac{\tau}{\ell+1}\mathcal{H}(\mathbb{D}_\ell u_h^n, v_h), \quad \forall v_h \in V_h, \tag{19}$$

holds for $\ell = 0, 1, \ldots, s-1$. Here and after we denote $\mathbb{D}_0 u_h^n = u_h^n$ for simplicity. The combination coefficients $\gamma_{\ell\kappa}$ are given constants independent on the numerical solution and satisfy the consistent condition $\sum_{0\le\kappa\le\ell}\sigma_{\ell\kappa} = 0$. This purpose is easily achieved by a series of linear combinations of (9). It depends on the used time marching. For example, in RKDG(2,2,k) they are defined as

$$\mathbb{D}_1 u_h^n = u_h^{n,1} - u_h^n, \quad \mathbb{D}_2 u_h^n = u_h^{n+1} - u_h^{n,1}, \tag{20}$$

and in RKDG(3,3,k) they are defined as

$$\mathbb{D}_1 u_h^n = u_h^{n,1} - u_h^n, \quad \mathbb{D}_2 = 2u_h^{n,2} - u_h^{n,1} - u_h^n, \quad \mathbb{D}_2 = u_h^{n+1} - 2u_h^{n,2} + u_h^n. \tag{21}$$

These differences of stage solutions can be looked upon as the approximation of different order time derivatives, due to (19) and the following observation

$$\| \mathbb{D}_{\ell+1} u_h^n \| \le \frac{1}{\ell+1} K\lambda \| \mathbb{D}_\ell u_h^n \|, \quad \ell = 0, 1, \ldots, s-1, \tag{22}$$

where K is a bounding constant depending solely on the inverse constants.

The next procedure is to establish the energy equation, such as

$$\| u_h^{n+1} \|^2 - \| u_h^n \|^2 = \sum_{0\le \ell \le s-1} \gamma_\ell \mathcal{H}(u_h^{n,\ell}, u_h^{n,\ell})\tau + \sum_{1\le i\le j\le s} \omega_{ij}(\mathbb{D}_i u_h^n, \mathbb{D}_j u_h^n), \tag{23}$$

where γ_ℓ and ω_{ij} are some given constants and can be sought by a trivial algebraic manipulation. Assume furthermore

$$\gamma_\ell \ge 0, \quad \ell = 0, 1, \ldots, s-1, \tag{24}$$

which can be verified at least for both RKDG(2,2,k) and RKDG(3,3,k) methods. Two terms on the right-hand side of (23) have distinguishing meanings. The first one represents the elemental stability owing to the DG spatial discretization, and the second one represents the interactional function owing to the time marching.

The main and difficult work in the whole energy analysis is how to control sharply the second term on the right-hand side of (23). This work depends strongly on the properties of DG spatial discretization, as well as the deep investigation of different stability mechanisms. For example, an additional numerical stability in RKDG(3,3,k) method shows up explicitly in the term $-\| \mathbb{D}_2^n \|^2$. It is to say that the dissipative nature of the RKDG(3,3,k) method comes from not only DG space discretization but also time marching. However, the stability mechanisms in RKDG(2,2,k) method are totally different. To obtain strong stability in L^2-norm, we have to balance the dissipation of DG spatial discretization and the antidissipation of the used time marching, since the used time marching does not belong to the so-called $A(\pi)$-stable algorithm in the terminology of the ODE solvers. This aim to obtain stability in L^2-norm can be completed by using a special property holding only for $k = 1$, namely,

$$\| \mathbb{D}_2 u_h^n \| \le \frac{1}{2}\mu_2 (|\beta|\lambda\tau)^{1/2} \left[\| [\![\mathbb{D}_1 u_h^n]\!] \|_\Gamma + \lambda\mu_1 \| [\![\mathbb{D}_0 u_h^n]\!] \|_\Gamma \right]. \tag{25}$$

This inequality can be obtained by filtering the average of numerical solution in each element, which can be extended to the generalized slope function for high-order piecewise polynomials (Cheng and Shu, 2010). For more detailed analysis, see Zhang and Shu (2009).

Finally, the strong L^2-norm stability results for two types of RKDG method can be established under the standard CFL condition.

Theorem 1. *Let* $\mathrm{u_h}$ *be the numerical solution of the RKDG(s, r, k) scheme. If the CFL number* $\lambda = |\beta|\tau h^{-1}$ *satisfies*

$$\mu_2^2 \lambda (\sqrt{2} + \mu_1 \lambda)^2 \le 2, \tag{26}$$

for $\mathrm{s} = \mathrm{r} = 2$ *and* $k \le 1$, *or satisfies*

$$\lambda \le \frac{12}{3\mu_3^2 + \sqrt{9\mu_3^4 + 48(\mu_1 + \sqrt{2}\mu_2\mu_3)^2}}, \tag{27}$$

for $\mathrm{s} = \mathrm{r} = 3$ *and arbitrary* $\mathrm{k} \ge 0$, *then* $\| u_h^{n+1} \| \le \| u_h^n \|$ *for any* n.

Remark 2. For the uniform mesh, it follows from the inverse constant listed in Table 2 and Theorem 1 that the maximum of CFL number for RKDG(2,2,1) method and RKDG(3,3,2) method, respectively, is approximated to be 0.1391 and 0.0904. There exists a gap to the sharp CFL number, 0.33 and 0.20 for above two schemes, respectively, which has been given by the Fourier technique and the numerical experiments.
Remark 3. Along the same line, we are able to use energy technique and obtain L^2-norm stability of other RKDG methods (Zhang and Shu, in preparation) with either second-order or third-order SSPRK time marching, for example, RKDG(3,2,1), RKDG(4,2,1), and RKDG(4,3,k) methods. However, how to extend the above work to the RKDG method with higher order Runge–Kutta time marching is an open problem till now.

3.2 Nonlinear Stability

There are a few stability results for the nonlinear case. The L^2-norm stability of semidiscrete DG method has been proved for arbitrary degree $k \geq 0$ by Jiang and Shu (1994), due to the famous *cell entropy inequality*

$$\frac{\mathrm{d}}{\mathrm{d}t}\| u_h \|_{I_i}^2 + \hat{F}_{i+\frac{1}{2}} - \hat{F}_{i-\frac{1}{2}} \leq 0, \tag{28}$$

where

$$\hat{F} = \hat{f}(u_h^-, u_h^+)u_h^- - \int_0^{u_h^-} f(s)\mathrm{d}s$$

is consistent with the entropy flux for the square entropy. The above cell entropy inequality can be proved for semidiscrete and Euler-forward (thus SSP-type Runge–Kutta) time marching, if the piecewise constant is considered. However, the nonlinear stability for the fully discrete RKDG method is an open problem till now, for high-order piecewise polynomials.

Given the function of limiter in RKDG methods, there are some more results in stability. For example, it has been proved in Cockburn (1998) that the numerical solution of RKDG method satisfies the total variation diminishing of means property

$$|\bar{u}_h^n|_{TV(0,1)} \leq |\bar{u}_h^0|_{TV(0,1)}, \quad \forall n \geq 0, \tag{29}$$

under a suitable CFL condition, when some slope limiter (such as MUSCL limiter) in each stage updating is used. Here $\bar{u}_h^n$ is the piecewise constant, made up of the average of u_h in each element. Similarly, the TVBM properties have been proved for some limiters, if the limiter can overcome the order reduction on the so-called sonic point. The detailed contents about limiter's implementation will be given in Section 5.

4 ERROR ESTIMATES

To show the numerical advantage of RKDG method, the error estimate is necessary. In this section, we mainly present several optimal (and/or quasi-optimal) error estimates in L^2-norm and so on.

4.1 Scalar Equation with Smooth Solution

We start from the scalar conservation law with sufficiently smooth solution. The material presented in this section is mainly taken from Zhang and Shu (2004, 2010). They introduce a quantity at each element boundary point

$$\alpha(\hat{f};w_h)=\begin{cases}\{f(\{\!\{w_h\}\!\})-\hat{f}(w_h^-,w_h^+)\}/[\![w_h]\!], & \text{if } [\![w_h]\!]\neq 0,\\ \frac{1}{2}f'(\{\!\{w_h\}\!\}), & \text{if } [\![w_h]\!]=0,\end{cases} \tag{30}$$

to describe the numerical viscosity coming from the numerical flux $\hat{f}$. It is easy to see that $\alpha(\hat{f};w_h)=|\beta|/2$ for the upwind flux (12). If $\hat{f}$ is an E-flux, a trivial analysis shows the following properties:

- Obviously there holds $\alpha(\hat{f};w_h)\geq 0$. Therefore, the numerical stability of semidiscrete DG method can be expressed approximately in an explicit form $\alpha(\hat{f};u_h)\,[\![u_h]\!]^2$ at every element endpoints. Note that we do not assume $\alpha(\hat{f};w_h)$ to have a uniform lower-bounded above zero, which leads to a major difficulty in analysis.
- It follows from the above definition that

$$\frac{1}{2}|f'(\{\!\{w_h\}\!\})|\leq\alpha(\hat{f};w_h)+C_*|\,[\![w_h]\!]\,|, \tag{31}$$

where C_* is a bounding constant depending solely on the nonlinearity of given flux $f(\cdot)$. Furthermore, it is reasonable to assume that $\alpha(\hat{f};w_h)$ is Lipschitz continuous with each argument. Hence, we can see that $\alpha(\hat{f};w_h)\approx\frac{1}{2}|f'(\{\!\{w_h\}\!\})|$ in first order of jump.

Due to the explicit expression on the numerical viscosity, we can obtain the optimal error estimate for semidiscrete version (Xu and Shu, 2007) and fully discrete version (Zhang and Shu, 2004, 2010). The next theorem shows the optimal error estimates for the fully discrete version.

Theorem 2. *Assume $T=M\tau$ for simplicity. For the RKDG(s, r, k) method, there holds the following optimal (or quasi-optimal) error estimate:*

$$\|\,u_h^M-u(\cdot,T)\,\|\leq C(h^{k+\sigma}+\tau^r), \tag{32}$$

under the standard CFL condition $\tau\leq\gamma h$, where γ is a suitable CFL number. Here $s=r=2$ and $k=1$ for RKDG(2,2,1) method, and $s=r=3$ and arbitrary $k\geq 1$ for RKDG(3,3,k) method. In general, $\sigma=1/2$ when the E -flux is used, and moreover $\sigma=1$ when the upwind numerical flux is used.

The main tools used in analysis are the projection's techniques and the detailed investigation on the numerical flux. To deal with those errors resulting from the time marching, those discussion and techniques used in the stability analysis are important and useful, too. For more details, see Zhang and Shu (2004, 2010).

4.2 Symmetrizable System with Smooth Solution

The above error estimate can be extended to the symmetrizable system (1), including Euler equations. Namely, there exists a mapping $u(v):\mathbb{R}^m \to \mathbb{R}^m$ so that when transformed into

$$u_v v_t + f_v v_x = 0,$$

the matrix $\mathbb{Q}(v) \equiv u_v$ is symmetric positive definite matrix and $f_v = f_u u_v$ is also symmetric matrix. The main development is the abstract definition of generalized E-flux and how to describe the corresponding numerical viscosity. The following material is taken from Luo et al. (2015).

Definition 1. The numerical flux $\hat{f}(w_h^-, w_h^+)$ is local Lipschitz continuous with each argument and consistent with $f(w)$. It is called a generalized E-flux, if there exist a rotation position s_κ and an adjusting matrix $\mathbb{Y}_\kappa$, such that

$$[\![w_h]\!]^\top \mathbb{Q}(s_\kappa)\{f(r_\kappa) - \hat{f}(w_h^-, w_h^+)\} + [\![w_h]\!]^\top \mathbb{Y}_\kappa [\![w_h]\!] \geq 0, \quad \kappa = 1,2, \tag{33}$$

for both $r_1 = w_h^-$ and $r_2 = w_h^+$, where $s_\kappa = s_\kappa(w_h^-, w_h^+)$ lies in the standard super-rectangle with two vertices w_h^+ and w_h^-, and each element in $\mathbb{Y}_\kappa = \mathbb{Y}_\kappa(w_h^-, w_h^+)$ has first order of jump, with the common bound depending only on the local Lipschitz constant of $\mathbb{Q}(\cdot)$ in the above super-rectangle.

This definition covers many numerical fluxes for the system, for example, Lax–Friedrich flux. Now we can introduce the numerical viscosity matrix at each element boundary point

$$\mathcal{A}(\hat{f}; w_h) \equiv \frac{1}{2}\mathbb{Q}(s_1)\mathcal{D}\hat{f}^{(1)}[w_h^-, w_h^+] - \frac{1}{2}\mathbb{Q}(s_2)\mathcal{D}\hat{f}^{(2)}[w_h^-, w_h^+], \tag{34}$$

where $\hat{f}^{(1)}(r) = \hat{f}(r, w_h^+)$ and $\hat{f}^{(2)}(r) = \hat{f}(w_h^-, r)$ are single-value functions with respect to the vectored-variable r. Here we have used the concept of the generalized Newton difference quotient $\mathcal{D}g[a,b]$, for any given function $g = (g_1, g_2, \ldots, g_m)^\top : \mathbb{R}^m \to \mathbb{R}^m$ between two points a and b in m-dimensional space. Specifically, each element in $\mathcal{D}g[a,b]$ is defined by

$$(\mathcal{D}g[a,b])_{ij} = \frac{g_i(a^{(j-1)}) - g_i(a^{(j)})}{a_j - b_j}, \tag{35}$$

where $a^{(j-1)}$ is an m-dimensional vector defined as

$$a^{(j)} = (b_1, \ldots, b_{j-1}, b_j, a_{j+1}, \ldots, a_m)^\top, \quad j = 1, 2, \ldots, m-1, \tag{36}$$

together with $a^{(0)} = a$ and $a^{(m)} = b$. If the denominator is equal to zero, the term on the right-hand side of (35) should be understood as the limit when the denominator goes to zero.

Now we can have almost the same error estimate as Theorem 2, by using energy technique with more careful treatment on the rotation in the middle point and boundary point of each element. The discussion is more complex than scalar case, because the numerical viscosity matrix (34) is not less than zero in an approximation status. For more details, see Zhang and Shu (2006) and Luo et al. (2015).

4.3 Scalar Equation with Discontinuous Initial Solution

It is well known that the numerical oscillation will happen when the initial solution contains a discontinuous point and the piecewise polynomials $k \geq 1$. Consider the linear constant hyperbolic equation, namely $f(u) = \beta u$ in (1). The detailed analysis shows that the pollution region around the characteristic line across the discontinuous point is only restricted in a narrow zone, and the optimal error estimate outside the pollution region is also preserved. The next theorem (Zhang and Shu, 2014) stated this double-optimal result for RKDG(3,3,k) method with arbitrary $k \geq 0$.

Theorem 3. *Assume $T = M\tau$ for simplicity, and $k \geq 1$. Under the standard CFL condition, namely, $\tau \leq \lambda h$ with λ being suitably small, there is the optimal error estimate for RKDG(3,3,k) method to solve the linear constant hyperbolic equation*

$$\| u(\cdot, T) - u_h^M \|_{L^2(\mathbb{R} \setminus \mathcal{R}_T)} \leq C_1 (h^{k+1} + \tau^3),$$

out of the pollution region

$$\mathcal{R}_T = \left(\beta T - C_2 \sqrt{T\beta} h^{1/2} \log \frac{1}{h}, \beta T + C_3 \sqrt{T\beta} h^{1/2} \log \frac{1}{h} \right).$$

The above bounding constants are all independent of h, τ and λ^{-1}.

This theorem is proved by energy analysis with two special weight functions, in order to detect the left side and right side of pollution region. The analysis is long and complex, which involves many technical points, for example, the superconvergence results, the generalized slope function and the highest frequency component, as well as the complex treatment of those errors coming from the Runge–Kutta time marching.

The above result is independent on λ^{-1} and hence also holds for the semi-discrete DG method. The numerical results given in Zhang and Shu (2014) verify the sharpness of the above results.

Remark 4. Similar works have been carried out by many authors. For example, Johnson and Pitkäranta (1986) considered the space-time DG method and proved that the pollution region at the crosswind direction has the width of $\mathcal{O}(h^{1/2}\log(1/h))$, and Cockburn and Guzman (2008) considered RKDG (2,2,1) method and proved similar results.

4.4 Other Error Estimates

There are three kinds of studies on the superconvergence of DG methods. The first kind of study is that between the numerical solution and a special projection of exact solution, such as

$$\| \mathcal{P}_h u - u_h \| \leq Ch^{k+1+\gamma},$$

where $\mathcal{P}_h u$ is the local Gauss–Radau projection, which is defined as an example, by

$$(\mathcal{P}_h u - u, v_h)_{I_i} = 0, \ \forall v_h \in \mathbb{P}^{k-1}(I_i); \ \ (\mathcal{P}_h u - u)^-_{i+\frac{1}{2}} = 0, \ \ i = 1,2,\ldots,N, \quad (37)$$

for the linear constant hyperbolic with the flowing is from left to right. Here $\gamma > 0$ is the so-called superconvergence order. It is proved to be $\gamma = 1/2$ by Cheng and Shu (2010), and then it was developed to $\gamma = 1$ by Yang and Shu (2012). A nonlinear conservation law has been considered by Meng et al. (2012). The second one is that between the numerical solution and the exact solution on some special point in each element. For example, there hold $(k + 2)$-order accuracy at the Radau points and moreover $(2k + 1)$-order accuracy at the downwind endpoint. Detailed statement and technical proof can be found in Cao et al. (2014), where the modification polynomials based on Legendre polynomials play a very important role. The last one is the postprocessing of numerical solutions, such that

$$\| G_h {}^* u_h - u \| \leq Ch^{2k+1},$$

where G_h is a suitable kernel in the convolution manipulation. The main technique is the negative-norm analysis and the control on the difference quotient of the solution on the staggered mesh. For more details, refer to Cockburn et al. (2003) and Mirzaee et al. (2012) and references included therein.

The studies on the posterior estimate are also carried for a long time, although there are a few development, compared with the other error estimates. For example, Giles and Süli (2002) and Adjerid and Baccouch (2009) used the dual technique and construct an estimation indicator. It is worthy to point out that the superconvergence results can play very important role in this field.

5 LIMITERS FOR DISCONTINUOUS GALERKIN METHODS

An important component of RKDG methods for solving conservation laws, with strong shocks in the solutions, is a nonlinear limiter, which is applied to detect discontinuities and control spurious oscillations near such discontinuities. There are many limiters which exist in the literature, for example, the minmod-type limiters, the moment-based limiters and the improved moment-based limiters. In this section, we will review these limiters and describe a robust limiter, the WENO-type limiter, which was developed in recent years.

Below we would like to use the notations in Cockburn and Shu (1989) to describe this procedure; however, we emphasize that the procedure actually does not depend on the specific basis chosen for the polynomials. In order to implement the DG methods for computation, we adopt a local orthogonal basis over I_i, $\{v_l^{(i)}(x),\ l=0,1,\ldots,k\}$, namely the scaled Legendre polynomials:

$$v_0^{(i)}(x)=1,\quad v_1^{(i)}(x)=\frac{x-x_i}{h_i/2},\quad v_2^{(i)}(x)=\left(\frac{x-x_i}{h_i/2}\right)^2-\frac{1}{3},\ldots,$$

where the points x_i are the centres of the cells $I_i = [x_{i-1/2}, x_{i+1/2}]$. Then the numerical solution $u_h(x, t)$ in the space V_h^k can be written as

$$u_h(x,t)=\sum_{l=0}^{k}u_i^{(l)}(t)v_l^{(i)}(x),\ \text{for } x\in I_i \tag{38}$$

and the degrees of freedom $u_i^{(l)}(t)$ are the moments defined by

$$u_i^{(l)}(t)=\frac{1}{a_l}\int_{I_i}u_h(x,t)v_l^{(i)}(x)dx,\quad l=0,1,\ldots,k,$$

where $a_l=\int_{I_i}(v_l^{(i)}(x))^2dx$ are the normalization constants since the basis is not orthonormal. In order to determine the approximate solution, we evolve the degrees of freedom $u_i^{(l)}$:

$$\begin{aligned}\frac{d}{dt}u_i^{(l)}+\frac{1}{a_l}\Bigg(-\int_{I_i}f(u_h(x,t))\frac{d}{dx}v_l^{(i)}(x)dx+\hat{f}(u_{i+1/2}^-,u_{i+1/2}^+)v_l^{(i)}(x_{i+1/2})\\ -\hat{f}(u_{i-1/2}^-,u_{i-1/2}^+)v_l^{(i)}(x_{i-1/2})\Bigg)=0,\quad l=0,1,\ldots,k,\end{aligned} \tag{39}$$

where $u_{i+1/2}^{\pm}=u_h(x_{i+1/2}^{\pm},t)$ are the left and right limits of the discontinuous solution u_h at the cell interface $x_{i+1/2}$, and $\hat{f}(u^-,u^+)$ is a monotone flux (nondecreasing in the first argument and nonincreasing in the second argument) for the scalar case and an exact or approximate Riemann solver for the system case. The integral term in (39) can be computed either exactly or by a suitable numerical quadrature accurate to at least $O(h^{k+l+2})$.

The semidiscrete scheme (39), written as $U_t = L(U)$, is then discretized in time by a nonlinearly stable Runge–Kutta time discretization which is described in Section 2.2. A limiter will be employed after each stage of time marching. For convenience of notations, below we would like to omit the time variable.

5.1 Traditional Limiters

Now we list a few traditional limiters, such as TVB limiter, moment limiter and modified moment limiter.

- The minmod-based TVB limiter (Cockburn and Shu, 1989). Denote:

$$u_{i+1/2}^- = u_i^{(0)} + \tilde{u}_i, \quad u_{i-1/2}^+ = u_i^{(0)} - \tilde{\tilde{u}}_i.$$

From (38), we can see that

$$\tilde{u}_i = \sum_{l=1}^{k} u_i^{(l)} v_l^{(i)}(x_{i+1/2}), \quad \tilde{\tilde{u}}_i = -\sum_{l=1}^{k} u_i^{(l)} v_l^{(i)}(x_{i-1/2}).$$

These are modified by either the standard minmod limiter (Harten, 1983)

$$\tilde{u}_i^{(mod)} = m(\tilde{u}_i, \Delta_+ u_i^{(0)}, \Delta_- u_i^{(0)}), \quad \tilde{\tilde{u}}_i^{(mod)} = m(\tilde{\tilde{u}}_i, \Delta_+ u_i^{(0)}, \Delta_- u_i^{(0)}), \tag{40}$$

where $\Delta_+ u_i^{(0)} = u_{i+1}^{(0)} - u_i^{(0)}$, $\Delta_- u_i^{(0)} = u_i^{(0)} - u_{i-1}^{(0)}$, and the minmod function m is given by

$$m(a_1, a_2, \ldots, a_n) = \begin{cases} s \cdot \min_{1 \le j \le n} |a_j| & \text{if } \operatorname{sign}(a_1) = \operatorname{sign}(a_2) = \cdots = \operatorname{sign}(a_n) = s, \\ 0 & \text{otherwise,} \end{cases} \tag{41}$$

or the TVB modified minmod function (Shu, 1987)

$$\tilde{m}(a_1, a_2, \ldots, a_n) = \begin{cases} a_1 & \text{if } |a_1| \le Mh^2, \\ m(a_1, a_2, \ldots, a_n) & \text{otherwise,} \end{cases} \tag{42}$$

where $M > 0$ is a constant. Then we reconstruct the new moment from $\tilde{u}_i^{(mod)}$, $\tilde{\tilde{u}}_i^{(mod)}$ and $u_i^{(0)}$. The choice of M depends on the solution of the problem. For scalar problems, it is possible to estimate M by the initial condition as in Cockburn and Shu (1989) (M is proportional to the second derivative of the initial condition at smooth extrema); however, it is more difficult to estimate M for the system case.

- Moment limiter of Biswas et al. (1994). The moment-based limiter in Biswas et al. (1994) is given by

$$u_i^{(l),mod} = \frac{1}{2l-1} m\left((2l-1)u_i^{(l)}, u_{i+1}^{(l-1)} - u_i^{(l-1)}, u_i^{(l-1)} - u_{i-1}^{(l-1)}\right), \tag{43}$$

where m is again the TVD minmod function (41). This limiter is applied adaptively. First, the highest-order moment $u^{(k)}$ is limited. Then the limiter is applied to successively lower-order moments when the next higher order moment on the interval has been changed by the limiting. In this way, the limiting is applied only where it is needed, and accuracy is retained in smooth regions.

- A modification of the moment limiter by Burbeau et al. (2001). If (43) is enacted, that is, $u_i^{(l),mod} \neq u_i^{(l)}$, then

$$\hat{u}_i^{(l),mod} = \frac{1}{2l-1} m\left((2l-1)u_i^{(l)}, u_{i+1/2}^{(l-1)+} - u_i^{(l-1)}, u_i^{(l-1)} - u_{i-1/2}^{(l-1)-}\right), \tag{44}$$

where

$$u_{i+1/2}^{(l-1)+} = u_{i+1}^{(l-1)} - (2l-1)u_{i+1}^{(l)}, \quad u_{i-1/2}^{(l-1)-} = u_{i-1}^{(l-1)} + (2l-1)u_{i-1}^{(l)}.$$

Again this limiter is applied adaptively as moment limiter.

These limiters tend to degrade accuracy when mistakenly used in smooth regions of the solution. In order to overcome the drawback of these limiters, from 2003, Qiu and colleagues have studied using WENO as limiter for RKDG methods, with the goal of obtaining a robust and high-order limiting procedure to simultaneously obtain uniform high-order accuracy and sharp, nonoscillatory shock transition for RKDG methods.

5.2 WENO Reconstruction as a Limiter for the RKDG Method

In this section, we will describe the procedure of WENO reconstruction as a limiter for the RKDG method (Qiu and Shu, 2005c; Zhu et al., 2008). The WENO-type limiter procedure is separated into two parts:

1. Identify the "troubled cells", namely those cells which might need the limiting procedure;
2. Reconstruct polynomials in "troubled cells" using WENO reconstruction which only maintain the original cell averages (conservation).

For the first part, there are many troubled cell indicators which can be used. In Qiu and Shu (2005a), we have systematically studied and compared a few different troubled cell indicators for the RKDG methods using WENO methodology as limiters. Extensive one- and two-dimensional simulations on the hyperbolic systems of Euler equations indicate that the minmod-based TVB indicator (when the TVB constant M is suitably chosen) and the KXRCF indicator by Krivodonova et al. (2004) are better than other choices in all the test cases. Recently, the troubled cell indicators based on wavelets and outlier detectors were presented by Vuik and Ryan (2014):

- TVB minmod troubled cell indicator, if, in (40), we have $\tilde{u}_i^{(mod)} \neq \tilde{u}_i$ or $\tilde{\tilde{u}}_i^{(mod)} \neq \tilde{\tilde{u}}_i$, then the cell is identified as a troubled cell, and be marked for further reconstruction.
- A shock detection technique based on a strong superconvergence at the outflow boundary of each element in smooth regions for the discontinuous Galerkin method by Krivodonova et al. (2004). We will denote the troubled cell indicator as the KXRCF indicator. The KXRCF indicator can be described as follows.

  Partition the boundary of a cell I_i into two portions ∂I_i^- (inflow, $\vec{v} \cdot \vec{n} < 0$) and ∂I_i^+ (outflow, $\vec{v} \cdot \vec{n} > 0$). The cell I_i is identified as a troubled cell marked for further reconstruction, if

$$\frac{\left|\int_{\partial I_i^-}(u^h|_{I_i}-u^h|_{I_{n_i}})ds\right|}{h_i^{\frac{k+1}{2}}|\partial I_i^-|\,||u^h|_{I_i}||}>1,$$

where h_i is the radius of the circumscribed circle in the element I_i. I_{n_i} is the neighbour of I_i on the side of ∂I_i^- and the norm is based on an element average in one-dimensional case.

Let I_i be a troubled cell which is identified by one of the troubled cell indicators which are described above, we will reconstruct the degrees of freedom, or the moments, $u_i^{(l)}$ for the troubled cell I_i for $l = 1, \ldots, k$ and retain only the cell average $u_i^{(0)}$.

- We identify k+1 small stencils S_j, $j = 0, \ldots, k$, such that I_i belongs to each of them. Here we set $S_j = \cup_{l=0}^{k} I_{i+j-l}$. We denote by $\mathcal{T} = \cup_{j=0}^{k} S_j$ the larger stencil which contains all the cells from the k+1 smaller stencils.

  We have a kth degree polynomial reconstruction denoted by $q_j(x)$, associated with each of the stencils S_j, $j = 0, \ldots, k$, such that the cell average of $q_j(x)$ in each of the cells in the stencil S_j agrees with the given cell average of u, i.e.

$$\frac{1}{\Delta x_{i+j-l}}\int_{I_{i+j-l}} q_j(x)dx = u_{i+j-l}^{(0)}, l=0,\ldots,k.$$

We also have a higher order (2k)th degree polynomial reconstruction denoted by $Q(x)$, associated with the larger stencil $\mathcal{T}$, such that

$$\frac{1}{\Delta x_{i+l}}\int_{I_{i+l}} Q(x)dx = u_{i+l}^{(0)}, l=-k,\ldots,k.$$

The detail of the construction of $q_j(x)$ and $Q(x)$ can be found in Shu (1998).

- We find the combination coefficients, also called linear weights γ_j, $j = 0, 1, \ldots, k$ satisfying:

$$A: \int_{I_i} Q(x)v_l^{(i)}(x)dx = \sum_{j=0}^{k}\gamma_j\int_{I_i} q_j(x)v_l^{(i)}(x)dx, \quad l=1,\ldots,k;$$

$$B: Q(x_G) = \sum_{j=0}^{k}\gamma_j q_j(x_G).$$

- We compute the smoothness indicator, denoted as β_j for each stencil S_j, which measures how smooth the function $q_j(x)$ on cell I_i,

$$\beta_j = \sum_{l=1}^{k}\int_{x_{i-1/2}}^{x_{i+1/2}}(\Delta x)^{2l-1}(q_j^{(l)})^2dx,$$

  where $q_j^{(l)}$ is the lth derivative of $q_j(x)$.

- We compute the nonlinear weight ω_j based on the smoothness indicator

$$\omega_j = \frac{\alpha_j}{\sum_{l=0}^{k}\alpha_l}, \text{ with } \alpha_j = \frac{\gamma_j}{(\varepsilon + \beta_j)^2}, j = 0, 1, \ldots, k,$$

where $\varepsilon > 0$ is a small number to avoid the denominator to become 0.
- The final WENO approximation is then given by:

$$A : u_i^{(l)} = \frac{1}{a_l}\sum_{j=0}^{k}\omega_j \int_{I_i} q_j(x) v_l^{(i)}(x)dx, \quad l = 1, \ldots, k;$$

$$B : u(x_G) = \sum_{j=0}^{k}\omega_j q_j(x_G).$$

- Reconstruction of moments based on the reconstructed point values for procedure B:

$$u_i^{(l)} = \frac{\Delta x}{a_l}\sum_G w_G u(x_G) v_l^{(i)}(x_G), \quad l = 1, \ldots, k.$$

Remark 5. For procedure A, there are not the linear weights for $\mathbb{P}^3$ case. For procedure B, we use the two-point and four-point Gauss quadrature in the $\mathbb{P}^1$ and $\mathbb{P}^3$ cases, respectively. For the $\mathbb{P}^2$ case, we use either the four-point Gauss–Lobatto quadrature points or three-point Gauss quadrature points. But there are negative linear weights when three-point Gauss quadrature points are used.
Remark 6. For the system cases, in order to achieve better qualities at the price of more complicated computations, the WENO reconstruction limiter is always used with a local characteristic field decomposition (see, e.g., Shu, 1998 for details).

5.3 Hermite WENO Reconstruction

WENO limiters work well in all our numerical test cases, including 1D, 2D and 3D, structure and unstructured meshes (Qiu and Shu, 2005c; Zhu et al., 2008; Zhu and Qiu, 2012), but for $\mathbb{P}^2$ and $\mathbb{P}^3$ cases, the compactness of DG is destroyed. In order to maintain the compactness of DG methods, we developed the following Hermite WENO (HWENO) limiter (Qiu and Shu, 2004, 2005b). For $\mathbb{P}^2$ case, we summarize the procedure to reconstruct the first and second moments $u_i^{(1)}$ and $u_i^{(2)}$ for a troubled cell I_i using HWENO. First, we reconstruct the following polynomials:

$$\int_{I_{i+j}} q_0(x)dx = u_{i+j}^{(0)}a_0, \quad j = -1, 0; \qquad \int_{I_{i-1}} q_0(x)v_1^{(i-1)}(x)dx = u_{i-1}^{(1)}a_1;$$

$$\int_{I_{i+j}} q_1(x)dx = u_{i+j}^{(0)}a_0, \quad j = 0, 1; \qquad \int_{I_{i+1}} q_1(x)v_1^{(i+1)}(x)dx = u_{i+1}^{(1)}a_1;$$

$$\int_{I_{i+j}} q_2(x)dx = u_{i+j}^{(0)} a_0, \quad j = -1, 0, 1;$$

$$\int_{I_{i+j}} Q(x)dx = u_{i+j}^{(0)} a_0, \quad j = -1, 0, 1; \quad \int_{I_{i+j}} Q(x) v_1^{(i+j)}(x)dx = u_{i+j}^{(1)} a_1, \quad j = -1, 1,$$

Then, following the routine *A* of WENO reconstruction, we can obtain the new moment $u_i^{(1)}$. To reconstruct $u_i^{(2)}$, we have the following polynomials:

$$\int_{I_{i+j}} q_0(x)dx = u_{i+j}^{(0)} a_0, \quad \int_{I_{i+j}} q_0(x) v_1^{(i+j)}(x)dx = u_{i+j}^{(1)} a_1, \quad j = -1, 0;$$

$$\int_{I_{i+j}} q_1(x)dx = u_{i+j}^{(0)} a_0, \quad \int_{I_{i+j}} q_1(x) v_1^{(i+j)}(x)dx = u_{i+j}^{(1)} a_1, \quad j = 0, 1;$$

$$\int_{I_{i+j}} q_2(x)dx = u_{i+j}^{(0)} a_0, \quad j = -1, 0, 1; \quad \int_{I_i} q_2(x) v_1^{(i)} dx = u_i^{(1)} a_1;$$

$$\int_{I_{i+j}} Q(x)dx = u_{i+j}^{(0)} a_0, \quad \int_{I_{i+j}} Q(x) v_1^{(i+j)}(x)dx = u_{i+j}^{(1)} a_1, \quad j = -1, 0, 1.$$

Then following the routine *A* of WENO reconstruction, we can obtain the new moment $u_i^{(2)}$.

5.4 A Simple WENO-Type Limiter

In 2013, Zhong and Shu developed a simple WENO-type limiter for DG (Zhong and Shu, 2013). Let I_i be a troubled cell, we use stencil $S = \{I_{i-1}, I_i, I_{i+1}\}$. Denote the solutions of the DG method on these three cells as polynomials $q_0(x)$, $q_1(x)$ and $q_2(x)$, respectively. We would like to modify $q_1(x)$ to $q_1^{new}(x)$. In order to make sure that the reconstructed polynomial maintains the original cell average of q_1 in the target cell I_i, the following modifications are taken:

$$\tilde{q}_0 = q_0 - \bar{\bar{q}}_0 + \bar{\bar{q}}_1, \quad \tilde{q}_1 = q_1, \quad \tilde{q}_2 = q_0 - \bar{\bar{q}}_2 + \bar{\bar{q}}_1,$$

$$\bar{\bar{q}}_0 = \frac{1}{\Delta x_i} \int_{I_i} q_0(x)dx, \quad \bar{\bar{q}}_1 = \frac{1}{\Delta x_i} \int_{I_i} q_1(x)dx, \quad \bar{\bar{q}}_2 = \frac{1}{\Delta x_i} \int_{I_i} q_2(x)dx.$$

The final nonlinear WENO reconstruction polynomial $q_1^{new}(x)$ is now defined by a convex combination of these modified polynomials:

$$q_1^{new}(x) = \omega_0 \tilde{q}_0(x) + \omega_1 \tilde{q}_1(x) + \omega_2 \tilde{q}_2(x).$$

If $\omega_0 + \omega_1 + \omega_2 = 1$, then q_1^{new} has the same cell average and order of accuracy as q_1.

Computational formulae of ω_0, ω_1 and ω_2 are the same as in WENO reconstruction. The linear weights can be chosen to be any set of positive numbers adding up to one. Since for smooth solutions the central cell is usually the best one, a larger linear weight is put on the central cell than on the neighbouring cells, i.e.

$$\gamma_0 < \gamma_1 \text{ and } \gamma_1 > \gamma_2.$$

In Zhong and Shu (2013), they take:

$$\gamma_0 = 0.001, \quad \gamma_1 = 0.998, \quad \gamma_2 = 0.001,$$

which can maintain the original high order in smooth regions and can keep essentially nonoscillatory shock transitions in all their numerical examples.

5.5 A Simple and Compact HWENO Limiter

This new HWENO limiter (Zhu et al., 2016) is an modification of the simple WENO limiter proposed recently by Zhong and Shu (2013). Both limiters use information of the DG solutions only from the target cell and its immediate neighbouring cells, thus maintaining the original compactness of the DG scheme. The goal of both limiters is to obtain high-order accuracy and nonoscillatory properties simultaneously. The main novelty of the new HWENO limiter in this chapter is to reconstruct the polynomial on the target cell in a least square fashion, while the simple WENO limiter (Zhong and Shu, 2013) is to use the entire polynomial of the original DG solutions in the neighbouring cells with an addition of a constant for conservation. The modification improves the robustness in the computation of problems with strong shocks or contact discontinuities and can get better resolutions for some examples for the P^3 case without the help of positivity-preserving limiters, and without changing the compact stencil of the DG scheme.

In order to make sure that the reconstructed polynomial maintains the original cell average of q_1 in the troubled cell I_i, the following modifications are taken:

$$\int_{I_{i-1}} (\tilde{q}_0(x) - q_0(x))^2 dx = \min \int_{I_{i-1}} (\phi(x) - q_0(x))^2 dx,$$

$$\int_{I_{i+1}} (\tilde{q}_2(x) - q_2(x))^2 dx = \min \int_{I_{i+1}} (\phi(x) - q_2(x))^2 dx$$

for $\forall \phi(x) \in \mathbb{P}^k$ with $\int_{I_i} \phi(x) dx = \int_{I_i} q_1(x) dx$.

For notational consistency, we also denote $\tilde{q}_1(x) = q_1(x)$. Then we follow the routine of Zhong and Shu (2013) and obtain the final nonlinear WENO reconstruction polynomial $q_1^{new}(x)$.

For two-dimensional cases, the reconstruction procedure of the limiters refers to Cockburn et al. (1990), Cockburn and Shu (1998), Biswas et al. (1994), Burbeau et al. (2001), Qiu and Shu (2005c), Qiu and Shu (2005b),

Zhu et al. (2008), Zhu and Qiu (2009), Zhu and Qiu (2011), Zhu and Qiu (2012), Zhong and Shu (2013), Zhu et al. (2013), and Zhu et al. (2016, submitted for publication).

6 CONCLUDING AND REMARKS

In this chapter, we reviewed the stability, error estimates and limiters for RKDG methods. For the stability, we focused on nonlinear stability by the energy analysis strategy; for error estimates, the priori estimate and posterior estimate are reviewed, and the traditional limiters and WENO-type limiters are shown in Section 5.

REFERENCES

Adjerid, S., Baccouch, M., 2009. The discontinuous Galerkin method for two-dimensional hyperbolic problems. II. A posteriori error estimation. J. Sci. Comput. 38, 15–49.

Biswas, R., Devine, K.D., Flaherty, J., 1994. Parallel, adaptive finite element methods for conservation laws. Appl. Numer. Math. 14, 255–283.

Burbeau, A., Sagaut, P., Bruneau, C.H., 2001. A problem-independent limiter for high-order Runge-Kutta discontinuous Galerkin methods. J. Comput. Phys. 169, 111–150.

Cao, W.X., Zhang, Z.M., Zou, Q.S., 2014. Superconvergence of discontinuous Galerkin methods for linear hyperbolic equations. SIAM J. Numer. Anal. 52, 2555–2573.

Cheng, Y.D., Shu, C.-W., 2010. Superconvergence of discontinuous Galerkin and local discontinuous Galerkin schemes for linear hyperbolic and convection-diffusion equations in one space dimension. SIAM J. Numer. Anal. 47, 4044–4072.

Cockburn, B., 1998. An introduction to the discontinuous Galerkin method for convection-dominated problem. In: Cockburn, B., Johnson, C., Shu, C.-W., Tadmor, E. (Eds.), Advanced Numerical Approximation of Nonlinear Hyperbolic Equations, Lecture Notes in Mathematics. vol. 1697. Springer, pp. 325–432.

Cockburn, B., Guzman, J., 2008. Error estimates for the Runge-Kutta discontinuous Galerkin method for the transport equation with discontinuous initial data. SIAM J. Numer. Anal. 46, 1364–1398.

Cockburn, B., Shu, C.-W., 1989. TVB Runge-Kutta local projection discontinuous Galerkin finite element method for scalar conservation laws II: general framework. Math. Comp. 52, 411–435.

Cockburn, B., Shu, C.-W., 1991. The Runge-Kutta local projection $\mathbb{P}^1$-discontinuous Galerkin method for scalar conservation laws. RAIRO Modél. Math. Anal. Numér. 25, 337–361.

Cockburn, B., Shu, C.-W., 1998. The Runge-Kutta discontinuous Galerkin finite element method for conservation laws V: multidimensional systems. J. Comput. Phys. 141, 199–224.

Cockburn, B., Shu, C.-W., 2001. Runge-Kutta discontinuous Galerkin methods for convection-dominated problems. J. Sci. Comput. 16, 173–261.

Cockburn, B., Lin, S.Y., Shu, C.-W., 1989. TVB Runge-Kutta local projection discontinuous Galerkin finite element method for conservation laws III: one dimensional systems. J. Comput. Phys. 84, 90–113.

Cockburn, B., Hou, S., Shu, C.-W., 1990. TVB Runge-Kutta local projection discontinuous Galerkin finite element method for conservation laws IV: the multidimensional case. Math. Comp. 54, 545–581.

Cockburn, B., Luskin, M., Shu, C.-W., Süli, E., 2003. Enhanced accuracy by post-processing for finite element methods for hyperbolic equations. Math. Comp. 72, 577–606.

Cockburn, B., Dong, B., Guzmán, J., 2008. Optimal convergence of the original DG method for the transport-reaction equation on special meshes. SIAM J. Numer. Anal. 46, 1250–1265.

Dumbser, M., Zanotti, O., Loubere, R., Diot, S., 2014. A posteriori subcell limiting of the discontinuous Galerkin finite element method for hyperbolic conservation laws. J. Comput. Phys. 278, 47–75.

Giles, M.B., Süli, E., 2002. Adjoint methods for PDEs: a posteriori error analysis and postprocessing by duality. Acta Numer. 11, 145–236.

Gottlieb, S., Shu, C.-W., Tadmor, E., 2001. Strong stability preserving high order time discretization methods. SIAM Rev. 43, 89–112.

Harten, A., 1983. High resolution schemes for hyperbolic conservation laws. J. Comput. Phys. 49 (1983), 357–393.

Jiang, G., Shu, C.-W., 1994. On cell entropy inequality for discontinuous Galerkin methods. Math. Comp. 62, 531–538.

Jiang, G., Shu, C.-W., 1996. Efficient implementation of weighted ENO schemes. J. Comput. Phys. 126, 202–228.

Johnson, C., Pitkäranta, J., 1986. An analysis of the discontinuous Galerkin method for a scalar hyperbolic equation. Math. Comp. 46, 1–26.

Krivodonova, L., Xin, J., Remacle, J., Chevaugeon, N., Flaherty, J.E., 2004. Shock detection and limiting with discontinuous Galerkin methods for hyperbolic conservation laws. Appl. Numer. Math. 48, 323–338.

Lasaint, P., Raviart, P.-A., 1974. On a finite element method for solving the neutron transport equation. Mathematical aspects of finite elements in partial differential equations. Proc. Sympos., Math. Res. Center, Univ. Wisconsin, Madison, Wis., 1974. Publication No. 33, Math. Res. Center, Univ. Wisconsin-Madison, Academic Press, New York, pp. 89–123.

Liu, X., Osher, S., Chan, T., 1994. Weighted essentially non-oscillatory schemes. J. Comput. Phys. 115, 200–212.

Luo, H., Baum, J.D., Lohner, R., 2007. A Hermite WENO-based limiter for discontinuous Galerkin method on unstructured grids. J. Comput. Phys. 225, 686–713.

Luo, J., Shu, C.-W., Zhang, Q., 2015. A priori estimates to smooth solutions of the third order Runge-Kutta discontinuous Galerkin method for symmetrizable systems of conservation laws. Math. Model. Numer. Anal. 49, 991–1018.

Meng, X., Shu, C.-W., Zhang, Q., Wu, B.Y., 2012. Superconvergence of discontinuous Galerkin methods for scaler nonlinear conservation law in one space dimension. SIAM J. Numer. Anal. 50, 2336–2356.

Mirzaee, H., Ryan, J.K., Kirby, R.M., 2012. Efficient implementation of smoothness-increasing accuracy-conserving (SIAC) filter for discontinuous Galerkin solutions. J. Sci. Comput. 52, 85–112.

Peterson, T., 1991. A note on the convergence of the discontinuous Galerkin method for a scalar hyperbolic equation. SIAM J. Numer. Anal. 28, 133–140.

Qiu, J., Shu, C.-W., 2004. Hermite WENO schemes and their application as limiters for Runge-Kutta discontinuous Galerkin method: one dimensional case. J. Comput. Phys. 193, 115–135.

Qiu, J., Shu, C.-W., 2005a. A comparison of trouble cell indicators for Runge-Kutta discontinuous Galerkin method using WENO limiters. SIAM J. Sci. Comput. 27, 995–1013.

Qiu, J., Shu, C.-W., 2005b. Hermite WENO schemes and their application as limiters for Runge-Kutta discontinuous Galerkin method II: two dimensional case. Comput. Fluids 34, 642–663.

Qiu, J., Shu, C.-W., 2005c. Runge-Kutta discontinuous Galerkin method using WENO limiters. SIAM J. Sci. Comput. 26, 907–929.

Qiu, J., Khoo, B.C., Shu, C.-W., 2006. A numerical study for the performance of the Runge-Kutta discontinuous Galerkin method based on different numerical fluxes. J. Comput. Phys. 212, 540–565.

Reed, W.H., Hill, T.R., 1973. Triangular mesh methods for the neutron transport equation, Los Alamos Scientific Laboratory Report LA-UR-73-479. Los Alamos, NM.

Richter, G.R., 1988. An optimal-order error estimate for the discontinuous Galerkin method. Math. Comp. 50, 75–88.

Shu, C.-W., 1987. TVB uniformly high-order schemes for conservation laws. Math. Comp. 49, 105–121.

Shu, C.-W., 1998. Essentially non-oscillatory and weighted essentially non-oscillatory schemes for hyperbolic conservation laws. In: Cockburn, B., Johnson, C., Shu, C.-W., Tadmor, E. (Eds.), Advanced Numerical Approximation of Nonlinear Hyperbolic Equations, Lecture Notes in Mathematics, vol. 1697. Springer, Berlin, pp. 325–432.

Shu, C.-W., Osher, S., 1988. Efficient implementation of essentially non-oscillatory shock capturing schemes. J. Comput. Phys. 77, 439–471.

Vuik, M., Ryan, J., 2014. Multiwavelet troubled cell indicator for discontinuity detection of discontinuous Galerkin schemes. J. Comput. Phys. 270, 138–160.

Xu, Y., Shu, C.-W., 2007. Error estimates of the semi-discrete local discontinuous Galerkin method for nonlinear convection-diffusion and KdV equations. Comput. Methods Appl. Mech. Eng. 196, 3805–3822.

Yang, Y., Shu, C.-W., 2012. Analysis of optimal superconvergence of discontinuous Galerkin method for linear hyperbolic equation. SIAM J. Numer. Anal. 50, 3110–3133.

Zanotti, O., Fambri, F., Dumbser, M., 2015a. Solving the relativistic magnetohydrodynamics equations with ADER discontinuous Galerkin methods, a posteriori subcell limiting and adaptive mesh refinement. Mon. Not. R. Astron. Soc. 452, 3010–3029.

Zanotti, O., Fambri, F., Dumbser, M., Hidalgo, A., 2015b. Space-time adaptive ADER discontinuous Galerkin finite element schemes with a posteriori sub-cell finite volume limiting. Comput. Fluids 118, 204–224.

Zhang, Q., 2011. Third order explicit Runge-Kutta discontinuous Galerkin method for linear conservation law with inflow boundary condition. J. Sci. Comput. 46 (2), 294–313.

Zhang, Q., Shu, C.-W., 2004. Error estimates to smooth solution of Runge-Kutta discontinuous Galerkin methods for scalar conservation laws. SIAM J. Numer. Anal. 42, 641–666.

Zhang, Q., Shu, C.-W., 2006. Error estimates to smooth solution of Runge-Kutta discontinuous Galerkin method for symmetrizable system of conservation laws. SIAM J. Numer. Anal. 44, 1702–1720.

Zhang, Q., Shu, C.-W., 2009. Stability analysis and a priori error estimates to the third order explicit Runge-Kutta discontinuous Galerkin method for scalar conservation laws. Technical Report 2009-28. http://www.dam.brown.edu/scicomp/reports/2009-28/.

Zhang, Q., Shu, C.-W., 2010. Stability analysis and a priori error estimates to the third order explicit Runge-Kutta discontinuous Galerkin method for scalar conservation laws. SIAM J. Numer. Anal. 48, 1038–1064.

Zhong, X., Shu, C.-W., 2013. A simple weighted essentially nonoscillatory limiter for Runge-Kutta discontinuous Galerkin methods. J. Comput. Phys. 232, 397–415.

Zhang, Q., Shu, C.-W., 2014. Error estimate for the third order explicit Runge-Kutta discontinuous Galerkin method for linear hyperbolic equation with discontinuous initial solution. Num. Math. 126, 703–740.

Zhang, Q., Shu, C.-W., in preparation. Stability analysis of Runge-Kutta discontinuous Galerkin methods for linear hyperbolic equation, in preparation.

Zhu, J., Qiu, J., 2009. Hermite WENO schemes and their application as limiters for Runge-Kutta discontinuous Galerkin method III: unstructured meshes. J. Sci. Comput. 39, 293–321.

Zhu, J., Qiu, J., 2011. Local Runge-Kutta discontinuous Glaerkin method using WENO type limiters for convection-diffusion equations. J. Comput. Phys. 230, 4353–4375.

Zhu, J., Qiu, J., 2012. Runge-Kutta discontinuous Galerkin method using WENO type limiters: three dimensional unstructured meshes. Commun. Comput. Phys. 11, 985–1005.

Zhu, J., Qiu, J., Shu, C.-W., Dumbser, M., 2008. Runge-Kutta discontinuous Galerkin method using WENO limiters II: unstructured meshes. J. Comput. Phys. 227, 4330–4353.

Zhu, J., Zhong, X., Shu, C.-W., Qiu, J., 2013. Runge-Kutta discontinuous Galerkin method using a new type of WENO limiters on unstructured meshes. J. Comput. Phys. 248, 200–220.

Zhu, J., Zhong, X., Shu, C.-W., Qiu, J., 2016. Runge-Kutta discontinuous Galerkin method with a simple and compact Hermite WENO limiter. Commun. Comput. Phys. 19, 944–969.

Zhu, J., Zhong, X., Shu, C.-W., Qiu, J., submitted for publication. Runge-Kutta discontinuous Galerkin method with a simple and compact Hermite WENO limiter on unstructured meshes. Commun. Comput. Phys., submitted for publication.

Chapter 8

HDG Methods for Hyperbolic Problems

B. Cockburn*, N.C. Nguyen[†] and J. Peraire[†]
*School of Mathematics, University of Minnesota, Minneapolis, MN, United States
[†]Massachusetts Institute of Technology, Cambridge, MA, United States

Chapter Outline

ABSTRACT
We give a short overview of the development of the so-called hybridizable discontinuous Galerkin methods for hyperbolic problems. We describe the methods, discuss their main features and display numerical results which illustrate their performance. We do this in the framework of wave propagation problems. In particular, we show that these methods are amenable to static condensation, and hence to efficient implementation, both for time-dependent (with implicit time-marching schemes) and for time-harmonic problems; we also show that they can be used with explicit time-marching schemes. We discuss an unexpected, recently uncovered superconvergence property and introduce a new postprocessing for time-harmonic Maxwell's equations. We end by providing bibliographical notes.

Keywords: Discontinuous Galerkin methods, Hybridization, Hyperbolic problems

AMS Classification Codes: 65N60, 35L04, 35Q61

1 INTRODUCTION

We give a short overview of the development of the so-called hybridizable discontinuous (HDG) methods for hyperbolic problems. The HDG methods are discontinuous Galerkin methods which were originally devised for numerically approximating steady-state problems and implicit time-marching schemes for time-dependent problems. Their distinctive feature is that they are amenable to *static condensation* and hence to efficient implementation. They turned out to be more accurate that other DG methods, as will be shown later.

The HDG methods were introduced by Cockburn et al. (2009a) in the framework of steady-state diffusion as part of the effort that started at the end of last century to devise efficient DG methods for second-order elliptic problems. The development of the HDG methods was then spearheaded by the authors who extended them to a variety of problems in computational fluid dynamics including convection–diffusion (Nguyen et al., 2009a,b), the incompressible Navier–Stokes equations (Nguyen et al., 2010b, 2011d) and the compressible Euler and Navier–Stokes equations (Nguyen et al., 2015a; Peraire et al., 2010); to partial differential equations in continuum mechanics, see Nguyen and Peraire (2012) and the references therein; to wave propagation problems in the time-domain (Nguyen et al., 2011a; Stanglmeier et al., 2016); as well as to the frequency domain (Nguyen et al., 2011c, 2015b).

In this chapter, we describe the HDG methods, highlight some of their main features and provide numerical experiments displaying their performance. In particular, we show that they can be efficiently implemented, that they can be used with either implicit or explicit time-marching schemes and that they do possess recently uncovered superconvergence properties. We do this for the acoustic wave equation in Section 2, for the elastic wave equation in Section 3 and for the time-harmonic Maxwell's equation in Section 4. In Section 5, we end with a few bibliographic notes.

2 THE ACOUSTICS WAVE EQUATION

In this section, we describe HDG methods for the numerical solution of the acoustic wave equation

$$\rho\frac{\partial^2 u}{\partial t^2} - \nabla\cdot(\kappa\nabla u) = f, \quad \text{in } \Omega\times(0,T]. \tag{1}$$

By introducing the velocity $v = u_t$ and the flux $\boldsymbol{q} = -\kappa\nabla u$, we can write (1) as the following system of first-order equations:

$$\begin{aligned} \kappa^{-1}\frac{\partial \boldsymbol{q}}{\partial t} + \nabla v &= 0, \quad \text{in } \Omega\times(0,T], \\ \rho\frac{\partial v}{\partial t} + \nabla\cdot\boldsymbol{q} &= f, \quad \text{in } \Omega\times(0,T]. \end{aligned} \tag{2a}$$

The exact solution $(v, \boldsymbol{q})$ satisfies the following initial conditions

$$\begin{aligned} v(\boldsymbol{x},t=0) &= v_0(\boldsymbol{x}), \\ \boldsymbol{q}(\boldsymbol{x},t=0) &= \boldsymbol{q}_0(\boldsymbol{x}), \end{aligned} \tag{2b}$$

and a Robin boundary condition

$$-\boldsymbol{q}\cdot\boldsymbol{n}+\alpha v = g, \quad \text{on } \partial\Omega\times(0,T]. \tag{2c}$$

The coefficient α varies on the boundary $\partial\Omega$ and represents different types of boundary conditions. Specifically, the Neumann boundary condition corresponds to $\alpha = 0$, the Dirichlet boundary condition to $1/\alpha = 0$ and the first-order absorbing boundary condition to $\alpha = \sqrt{\kappa\rho}$. We assume that $\kappa(\boldsymbol{x})$, $\rho(\boldsymbol{x})$ and $\alpha(\boldsymbol{x})$ are scalar positive functions.

We begin with the spatial discretization of the wave equation (2) and the temporal integration of the semidiscrete form using both explicit and implicit time-stepping methods. We end by presenting numerical experiments to demonstrate their performance.

2.1 Spatial Discretization

Let $\mathcal{T}_h$ be a collection of disjoint elements that partition Ω. We denote by $\partial\mathcal{T}_h$ the set $\{\partial K : K\in\mathcal{T}_h\}$. For an element K of the collection $\mathcal{T}_h$, $F=\partial K\cap\partial\Omega$ is the boundary face if the $d-1$ Lebesgue measure of F is nonzero. For two elements K^+ and K^- of the collection $\mathcal{T}_h$, $F=\partial K^+\cap\partial K^-$ is the interior face between K^+ and K^- if the $d-1$ Lebesgue measure of F is nonzero. Let $\mathcal{E}_h^o$ and $\mathcal{E}_h^\partial$ denote the set of interior and boundary faces, respectively. We denote by $\mathcal{E}_h$ the union of $\mathcal{E}_h^o$ and $\mathcal{E}_h^\partial$.

Let $\mathcal{P}_k(D)$ denote the set of polynomials of degree at most k on a domain D. We are going to use the following discontinuous finite element spaces:

$$\begin{aligned} W_h &= \{w\in L^2(\Omega) \quad : w|_K\in W(K),\ \forall K\in\mathcal{T}_h\}, \\ \boldsymbol{V}_h &= \{\boldsymbol{p}\in(L^2(\Omega))^d : \boldsymbol{p}|_K\in\boldsymbol{V}(K),\ \forall K\in\mathcal{T}_h\}. \end{aligned}$$

Some appropriate choices for the local space $W(K)\times\boldsymbol{V}(K)$ on K include

$$W(K)\times\boldsymbol{V}(K)\equiv\begin{cases} \mathcal{P}_k(K)\times(\mathcal{P}_k(K))^d, \\ \mathcal{P}_{k-1}(K)\times(\mathcal{P}_k(K))^d, \\ \mathcal{P}_k(K)\times\left((\mathcal{P}_k(K))^d+\boldsymbol{x}\mathcal{P}_k(K)\right). \end{cases}$$

These spaces correspond to the equal-order elements, the BDM elements (Arnold and Brezzi, 1985) and the RT elements (Raviart and Thomas, 1977), respectively. In addition, we introduce a traced finite element space

$$M_h = \{\mu\in L^2(\mathcal{E}_h) \ : \ \mu|_F\in\mathcal{P}_k(F),\ \ \forall F\in\mathcal{E}_h\}.$$

For functions $\boldsymbol{w}$ and $\boldsymbol{v}$ in $(L^2(D))^d$, we denote $(\boldsymbol{w},\boldsymbol{v})_D = \int_D \boldsymbol{w}\cdot\boldsymbol{v}$. For functions w and v in $L^2(D)$, we denote $(w,v)_D = \int_D wv$ if D is a domain in $\mathbb{R}^d$ and $\langle w,v\rangle_D = \int_D wv$ if D is a domain in $\mathbb{R}^{d-1}$. We then introduce

$$(w,v)_{\mathcal{T}_h} = \sum_{K\in\mathcal{T}_h} (w,v)_K, \quad \langle \mu,\eta\rangle_{\partial\mathcal{T}_h} = \sum_{K\in\mathcal{T}_h} \langle \mu,\eta\rangle_{\partial K},$$

for w, v defined on $\mathcal{T}_h$ and μ, η defined on $\partial\mathcal{T}_h$.

The HDG methods for the wave equation (2) seek to define $(\boldsymbol{q}_h, v_h, \widehat{v}_h)(t) \in \boldsymbol{V}_h \times W_h \times M_h$, for $t \in [0, T]$, as a solution of the following system

$$\left(\kappa^{-1}\frac{\partial \boldsymbol{q}_h}{\partial t},\boldsymbol{r}\right)_{\mathcal{T}_h} - (v_h, \nabla\cdot\boldsymbol{r})_{\mathcal{T}_h} + \langle \widehat{v}_h, \boldsymbol{r}\cdot\boldsymbol{n}\rangle_{\partial\mathcal{T}_h} = 0, \tag{3a}$$

$$\left(\rho\frac{\partial v_h}{\partial t},w\right)_{\mathcal{T}_h} - (\boldsymbol{q}_h, \nabla w)_{\mathcal{T}_h} + \langle \widehat{\boldsymbol{q}}_h\cdot\boldsymbol{n}, w\rangle_{\partial\mathcal{T}_h} = (f,w)_{\mathcal{T}_h}, \tag{3b}$$

$$-\langle \widehat{\boldsymbol{q}}_h\cdot\boldsymbol{n},\mu\rangle_{\partial\mathcal{T}_h\backslash\partial\Omega} + \langle -\widehat{\boldsymbol{q}}_h\cdot\boldsymbol{n} + \alpha\widehat{v}_h, \mu\rangle_{\partial\Omega} = \langle g,\mu\rangle_{\partial\Omega}, \tag{3c}$$

for all $(\boldsymbol{r}, w, \mu) \in \boldsymbol{V}_h \times W_h \times M_h$ and all $t \in (0, T]$, where the numerical flux is defined as

$$\widehat{\boldsymbol{q}}_h\cdot\boldsymbol{n} = \boldsymbol{q}_h\cdot\boldsymbol{n} + \tau(v_h - \widehat{v}_h), \quad \text{on } \partial\mathcal{T}_h. \tag{3d}$$

If the stabilization function is taken as $\tau = \sqrt{\kappa\rho}$, we obtain the well-known upwinding flux.

Note that the equations (2a) require v and the normal component of $\boldsymbol{q}$ to be continuous across the set of interior faces $\mathcal{E}_h^o l \times (0,T)$. The HDG method takes into account these *transmission conditions* by requiring that the corresponding numerical traces $\widehat{v}_h$ and the normal component of $\widehat{\boldsymbol{q}}_h$ be single valued on that set. The first condition is satisfied by taking $\widehat{v}_h(t)$ in M_h and the second by imposing equation (3c) for any $t \in [0, T]$. For other ways of defining HDG methods, we refer to Cockburn (2016b) and the references therein.

This semidiscretization gives rise to a system of ODEs to be solved by using some time-marching methods. As we are going to see in the next subsection, the form presented here is useful when using implicit time-marching methods because it takes advantage of the fact that the HDG methods are amenable to static condensation. When using explicit time-marching methods, a better way of presenting the method is the following: find $(\boldsymbol{q}_h, v_h) \in \boldsymbol{V}_h \times W_h$ such that for all $K \in \mathcal{T}_h$,

$$\left(\kappa^{-1}\frac{\partial \boldsymbol{q}_h}{\partial t},\boldsymbol{r}\right)_K - (v_h, \nabla\cdot\boldsymbol{r})_K + \langle \widehat{v}_h, \boldsymbol{r}\cdot\boldsymbol{n}\rangle_{\partial K} = 0, \quad \forall\, \boldsymbol{r} \in \boldsymbol{V}(K), \tag{4a}$$

$$\left(\rho\frac{\partial v_h}{\partial t},w\right)_K - (\boldsymbol{q}_h, \nabla w)_K + \langle \widehat{\boldsymbol{q}}_h\cdot\boldsymbol{n}, w\rangle_{\partial K} = (f,w)_K, \quad \forall\, w \in W(K), \tag{4b}$$

where, for any given face $F \in \partial K$,

$$\widehat{v}_h = \begin{cases} \dfrac{\tau^+ v_h^+ + \tau^- v_h^-}{\tau^+ + \tau^-} + \dfrac{1}{\tau^+ + \tau^-}(\boldsymbol{q}_h^+ \cdot \boldsymbol{n}^+ + \boldsymbol{q}_h^- \cdot \boldsymbol{n}^-), & \text{if } F \in \mathcal{E}_h^o, \\ \dfrac{\tau}{\tau+\alpha} v_h + \dfrac{1}{\tau+\alpha}(\mathsf{P}g + \boldsymbol{q}_h \cdot \boldsymbol{n}), & \text{if } F \in \partial\Omega, \end{cases} \tag{4c}$$

and

$$\widehat{\boldsymbol{q}}_h \cdot \boldsymbol{n} = \boldsymbol{q}_h \cdot \boldsymbol{n} + \tau(v_h - \widehat{v}_h) \quad \text{on } \partial K. \tag{4d}$$

Here $\mathsf{P}g$ denotes the L^2 projection of g onto the space M_h, and

$$v_h^{\pm}|_F = v_h|_{F \in \partial K^{\pm}}, \quad \text{and} \quad \boldsymbol{q}_h^{\pm}|_F = \boldsymbol{q}_h|_{F \in \partial K^{\pm}},$$

where K^+ and K^- are two elements sharing the face F. Hence, v_h^- and $\boldsymbol{q}_h^-$ (respectively, v_h^+ and $\boldsymbol{q}_h^+$) are nothing but the value of v_h and $\boldsymbol{q}_h$ on the face F from the element K^- (respectively, K^+). We can easily show that when the stabilization function is taken to be a constant on each face lying on $\partial\mathcal{T}_h$, the system (4) is equivalent to the original formulation (3) (Nguyen et al., 2011a; Stanglmeier et al., 2016).

2.2 Temporal Discretization

We now show how to obtain a fully discrete scheme by discretizing the above system of ODEs by several different time-marching methods, two being implicit and the other two explicit.

2.2.1 BDF Methods

We will only discuss the backward-Euler method since higher-order BDF methods follow a similar way. Using the backward-Euler scheme for the discretization of the time derivative in (3), we find that the approximate solution $(\boldsymbol{q}_h^n, v_h^n, \widehat{v}_h^{\,n}) \in \boldsymbol{V}_h \times W_h \times M_h$ at time step n satisfies the following equations

$$\left(\frac{\boldsymbol{q}_h^n}{\kappa\Delta t}, \boldsymbol{r}\right)_{\mathcal{T}_h} - (v_h^n, \nabla \cdot \boldsymbol{r})_{\mathcal{T}_h} + \langle \widehat{v}_h^{\,n}, \boldsymbol{r} \cdot \boldsymbol{n} \rangle_{\partial\mathcal{T}_h} = \left(\frac{\boldsymbol{q}_h^{n-1}}{\kappa\Delta t}, \boldsymbol{r}\right)_{\mathcal{T}_h}, \tag{5a}$$

$$\left(\rho\frac{v_h^n}{\Delta t}, w\right)_{\mathcal{T}_h} - (\boldsymbol{q}_h^n, \nabla w)_{\mathcal{T}_h} + \langle \widehat{\boldsymbol{q}}_h^{\,n} \cdot \boldsymbol{n}, w \rangle_{\partial\mathcal{T}_h} = \left(f^n + \rho\frac{v_h^{n-1}}{\Delta t}, w\right)_{\mathcal{T}_h}, \tag{5b}$$

$$-\langle \widehat{\boldsymbol{q}}_h^{\,n} \cdot \boldsymbol{n}, \mu \rangle_{\partial\mathcal{T}_h \backslash \partial\Omega} + \langle -\widehat{\boldsymbol{q}}_h^{\,n} \cdot \boldsymbol{n} + \alpha\widehat{v}_h^{\,n}, \mu \rangle_{\partial\Omega} = \langle g^n, \mu \rangle_{\partial\Omega}, \tag{5c}$$

for all $(\boldsymbol{r}, w, \mu) \in \boldsymbol{V}_h \times W_h \times M_h$, where

$$\widehat{\boldsymbol{q}}_h^{\,n} \cdot \boldsymbol{n} = \boldsymbol{q}_h^n \cdot \boldsymbol{n} + \tau(v_h^n - \widehat{v}_h^{\,n}), \quad \text{on } \partial\mathcal{T}_h. \tag{5d}$$

Here $(\boldsymbol{q}_h^n, v_h^n, \widehat{v}_h^{\,n})$ represents the numerical approximation to the exact solution $(\boldsymbol{q}(t^n), u(t^n), \widehat{v}(t^n))$ at time t^n. We then find $u_h^n \in W_h$ such that

$$\frac{1}{\Delta t}(u_h^n,w)_{\mathcal{T}_h}=(v_h^n,w)_{\mathcal{T}_h}+\frac{1}{\Delta t}(u_h^{n-1},w)_{\mathcal{T}_h},\quad \forall w\in W_h. \tag{6}$$

The fully discrete system (5) can be efficiently solved by locally eliminating $(\boldsymbol{q}_h, \boldsymbol{u}_h)$ to obtain a linear system in terms of the globally coupled degrees of freedom of $\widehat{v}_h$. We refer to Nguyen et al. (2011a) for a detailed discussion.

2.2.2 DIRK Methods

Next, we apply the DIRK formula represented by the coefficients (a_{ij}, b_i, c_i), $1 \le i \le q$, $1 \le j \le i$, to the semidiscrete system (3). We denote by $(\boldsymbol{q}_h^{n,i}, v_h^{n,i}, \widehat{v}_h^{n,i})$ the numerical approximation to the exact solution $(\boldsymbol{q}(t^{n,i})|_{\mathcal{T}_h}, v(t^{n,i})|_{\mathcal{T}_h}, v(t^{n,i})|_{\mathcal{E}_h})$, where $t^{n,i} = t^{n-1} + c_i\Delta t$, $1 \le i \le q$. Given the approximate solution at time t^{n-1}, $(\boldsymbol{q}_h^{n-1}, v_h^{n-1}, \widehat{v}_h^{n-1})$, we find the intermediate solutions $(\boldsymbol{q}_h^{n,i}, v_h^{n,i}, \widehat{v}_h^{n,i}) \in \boldsymbol{V}_h \times W_h \times M_h$ satisfying

$$\left(\frac{\boldsymbol{q}_h^{n,i}}{\kappa\Delta t},\boldsymbol{v}\right)_{\mathcal{T}_h}-(v_h^{n,i},\nabla\cdot\boldsymbol{v})_{\mathcal{T}_h}+\left\langle\widehat{v}_h^{n,i},\boldsymbol{v}\cdot\boldsymbol{n}\right\rangle_{\partial\mathcal{T}_h}=\left(\frac{\boldsymbol{p}_h^{n,i}}{\kappa},\boldsymbol{v}\right)_{\mathcal{T}_h}, \tag{7a}$$

$$\left(\frac{\rho v_h^{n,i}}{a_{ii}\Delta t},w\right)_{\mathcal{T}_h}-(\boldsymbol{q}_h^{n,i},\nabla w)_{\mathcal{T}_h}+\left\langle\widehat{\boldsymbol{q}}_h^{n,i}\cdot\boldsymbol{n},w\right\rangle_{\partial\mathcal{T}_h}=(f^{n,i}+\rho s_h^{n,i},w)_{\mathcal{T}_h}, \tag{7b}$$

$$-\left\langle\widehat{\boldsymbol{q}}_h^{n,i}\cdot\boldsymbol{n},\mu\right\rangle_{\partial\mathcal{T}_h\backslash\partial\Omega}+\left\langle-\widehat{\boldsymbol{q}}_h^{n,i}\cdot\boldsymbol{n}+\alpha\widehat{v}_h^{n,i},\mu\right\rangle_{\partial\Omega}=\langle g^{n,i},\mu\rangle_{\partial\Omega}, \tag{7c}$$

for all $(\boldsymbol{v}, w, \mu) \in \boldsymbol{V}_h \times W_h \times M_h$, where

$$\widehat{\boldsymbol{q}}_h^{n,i}\cdot\boldsymbol{n}=\boldsymbol{q}_h^{n,i}\cdot\boldsymbol{n}+\tau(v_h^{n,i}-\widehat{v}_h^{n,i}),\quad\text{on }\partial\mathcal{T}_h. \tag{7d}$$

The terms $s_h^{n,i}$ and $\boldsymbol{p}_h^{n,i}$ on the right-hand side of (7) are given by

$$s_h^{n,i}=\frac{v_h^{n-1}}{a_{ii}\Delta t}+\sum_{j=1}^{i-1}\frac{a_{ij}}{a_{ii}}\left(\frac{v_h^{n,j}}{a_{jj}\Delta t}-s_h^{n,j}\right),\quad i=1,\ldots,q,$$

$$\boldsymbol{p}_h^{n,i}=\frac{\boldsymbol{q}_h^{n-1}}{a_{ii}\Delta t}+\sum_{j=1}^{i-1}\frac{a_{ij}}{a_{ii}}\left(\frac{\boldsymbol{q}_h^{n,j}}{a_{jj}\Delta t}-\boldsymbol{p}_h^{n,j}\right),\quad i=1,\ldots,q.$$

The discrete systems (7) must be solved sequentially from $i = 1, 2, \ldots, q$. Hence, a q-stage DIRK scheme requires us to solve q discrete systems which are very similar to the system (5) resulting from the backward-Euler method.

Once the intermediate solutions have been computed, the numerical solution $(\boldsymbol{q}_h^n, v_h^n)$ at time t^n is given by

$$(\boldsymbol{q}_h^n,v_h^n)=(\boldsymbol{q}_h^{n-1},v_h^{n-1})+\Delta t\sum_{i=1}^{q}b_i(\boldsymbol{y}_h^{n,i},z_h^{n,i}), \tag{8}$$

where

$$\boldsymbol{y}_h^{n,i} = \frac{\boldsymbol{q}_h^{n,i} - \boldsymbol{q}_h^{n-1}}{a_{ii}\Delta t} - \sum_{j=1}^{i-1} \frac{a_{ij}}{a_{ii}} \boldsymbol{y}_h^{n,j}, \quad i = 1, \ldots, q,$$

$$z_h^{n,i} = \frac{v_h^{n,i} - v_h^{n-1}}{a_{ii}\Delta t} - \sum_{j=1}^{i-1} \frac{a_{ij}}{a_{ii}} z_h^{n,j}, \quad i = 1, \ldots, q.$$

Finally, we compute u_h^n by using the same DIRK scheme to discretize the ODE $\partial u_h/\partial t = v_h$.

2.2.3 Adams–Bashforth Methods

The Adams–Bashforth (AB) methods are linear multistep explicit methods. The forward-Euler method is a first-order AB method. Here we discuss the forward-Euler method since higher-order AB methods can be constructed in a similar way. Given the solution at the previous time step $(\boldsymbol{q}_h^n, v_h^n, u_h^n)$, we first compute the approximate traces as

$$\widehat{v}_h^{\,n} = \begin{cases} \dfrac{\tau^+ v_h^{+n} + \tau^- v_h^{-n}}{\tau^+ + \tau^-} - \dfrac{1}{\tau^+ + \tau^-}(\boldsymbol{q}_h^{+n} \cdot \boldsymbol{n}^+ + \boldsymbol{q}_h^{-n} \cdot \boldsymbol{n}^-), & \text{if } F \in \mathcal{E}_h \backslash \partial\Omega, \\ \dfrac{\tau}{\tau + \alpha} v_h^n + \dfrac{1}{\tau + \alpha}(\mathsf{P}g^n + \alpha \boldsymbol{q}_h^n \cdot \boldsymbol{n}), & \text{if } F \in \partial\Omega, \end{cases} \tag{9}$$

and $\widehat{\boldsymbol{q}}_h^{\,n} \cdot \boldsymbol{n} = \boldsymbol{q}_h^n \cdot \boldsymbol{n} + \tau(v_h^n - \widehat{v}_h^{\,n})$ for all faces F of $\mathcal{E}_h$. We then determine the numerical solution $(\boldsymbol{q}_h^{n+1}, v_h^{n+1}, u_h^{n+1}) \in \boldsymbol{V}(K) \times W(K) \times W(K)$ at the next time step by solving

$$\begin{aligned} \left(\frac{1}{\kappa}\frac{\boldsymbol{q}_h^{n+1} - \boldsymbol{q}_h^n}{\Delta t^n}, \boldsymbol{r}\right)_K - (v_h^n, \nabla \cdot \boldsymbol{r})_K + \langle \widehat{v}_h^{\,n}, \boldsymbol{r} \cdot \boldsymbol{n} \rangle_{\partial K} &= 0, \\ \left(\rho \frac{v_h^{n+1} - v_h^n}{\Delta t^n}, w\right)_K - (\boldsymbol{q}_h^n, \nabla w)_K + \langle \widehat{\boldsymbol{q}}_h^{\,n} \cdot \boldsymbol{n}, w \rangle_{\partial K} &= (f^n, w)_K, \\ \left(\frac{u_h^{n+1} - u_h^n}{\Delta t^n}, z\right)_K - (v_h^n, z)_K &= 0, \end{aligned} \tag{10}$$

for all $(\boldsymbol{r}, w, z) \in \boldsymbol{V}(K) \times W(K) \times W(K)$ and for all elements $K \in \mathcal{T}_h$.

It is clear that we compute the numerical solution at any time step in an element-by-element fashion. Therefore, explicit HDG methods have the same computational complexity as other explicit DG methods. Higher-order AB methods can be used as well, provided that the numerical solutions at the earlier time steps are available.

2.3 SSP-RK Methods

Lastly, we describe the SSP-RK(q, q) scheme (Chen et al., 2005; Gottlieb et al., 2000) to integrate the semidiscrete system (4) in time. For $i = 0, \ldots, q - 1$, we compute

$$\widehat{v}_h^{n,i-1} = \begin{cases} \dfrac{\tau^+ v_h^{+n,i-1} + \tau^- v_h^{-n,i-1}}{\tau^+ + \tau^-} \\ + \dfrac{1}{\tau^+ + \tau^-}(\boldsymbol{q}_h^{+n,i-1} \cdot \boldsymbol{n}^+ + \boldsymbol{q}_h^{-n,i-1} \cdot \boldsymbol{n}^-), & \text{if } F \in \mathcal{E}_h \backslash \partial\Omega, \\ \dfrac{\tau}{\tau+\alpha} v_h^{n,i-1} + \dfrac{1}{\tau+\alpha}(\mathsf{P}g^{n,i-1} + \alpha \boldsymbol{q}_h^{n,i-1} \cdot \boldsymbol{n}), & \text{if } F \in \partial\Omega, \end{cases} \tag{11}$$

and $\widehat{\boldsymbol{q}}_h^{n,i-1} \cdot \boldsymbol{n} = \boldsymbol{q}_h^{n,i-1} \cdot \boldsymbol{n} + \tau(v_h^{n,i-1} - \widehat{v}_h^{n,i-1})$ for all faces F of $\mathcal{E}_h$; we then determine $(\boldsymbol{q}_h^{n,i}, v_h^{n,i}, v_h^{n,i}) \in \boldsymbol{V}(K) \times W(K) \times W(K)$ as the solution of

$$\begin{aligned} \left(\frac{1}{\kappa}\frac{\boldsymbol{q}_h^{n,i} - \boldsymbol{q}_h^{n,i-1}}{\Delta t}, \boldsymbol{r}\right)_K - (v_h^{n,i-1}, \nabla \cdot \boldsymbol{r})_K + \left\langle \widehat{v}_h^{n,i-1}, \boldsymbol{r} \cdot \boldsymbol{n} \right\rangle_{\partial K} &= 0, \\ \left(\rho \frac{v_h^{n,i} - v_h^{n,i-1}}{\Delta t}, w\right)_K - (\boldsymbol{q}_h^{n,i-1}, \nabla w)_K + \left\langle \widehat{\boldsymbol{q}}_h^{n,i-1} \cdot \boldsymbol{n}, w \right\rangle_{\partial K} &= (f^{n,i-1}, w)_K, \\ \left(\frac{v_h^{n,i} - v_h^{n,i-1}}{\Delta t}, z\right)_K - (v_h^{n,i-1}, z)_K &= 0, \end{aligned} \tag{12}$$

for all ($\boldsymbol{r}$, w, z) $\in$ $\boldsymbol{V}(K) \times W(K) \times W(K)$ and for all elements $K \in \mathcal{T}_h$. We finally set

$$(\boldsymbol{q}_h^n, v_h^n, v_h^n) = \sum_{i=0}^{s} \alpha_{q,i} (\boldsymbol{q}_h^{n,i}, v_h^{n,i}, v_h^{n,i}), \tag{13}$$

where the coefficients $\alpha_{q,\, i}$ are precisely those corresponding to the SSP-RK scheme (q, q) (Chen et al., 2005; Gottlieb et al., 2000), namely

$$\begin{aligned} &\alpha_{1,0} = 1, \quad \alpha_{q,i} = \frac{1}{i}\alpha_{q-1,i-1}, \quad i = 1, \ldots, q-2, \\ &\alpha_{q,q} = \frac{1}{q!}, \quad \alpha_{q,q-1} = 0, \quad \alpha_{q,0} = 1 - \sum_{i=1}^{q-1} \alpha_{q,i}. \end{aligned} \tag{14}$$

The SSP-RK(q, q) scheme has q stages and q orders of accuracy. Each stage of the SSP-RK(q, q) scheme is exactly the forward-Euler method described earlier.

2.4 Postprocessing

The numerical results we present in the next subsection are going to involve two elementwise postprocessings defined as follows. The first is a new approximation to u: on every simplex $K \in \mathcal{T}_h$, we take $u_h^{n*} \in \mathcal{P}_{k+1}(K)$, such that

$$\begin{aligned} (\nabla u_h^{n*}, \nabla w)_K &= (\boldsymbol{q}_h^n, \nabla w)_K, \quad \forall\, w \in \mathcal{P}_{k+1}(K), \\ (u_h^{n*}, 1)_K &= (u_h^n, 1)_K. \end{aligned} \tag{15}$$

The second is a new approximation to u_t: On every simplex $K \in \mathcal{T}_h$, we take $v_h^{n*} \in \mathcal{P}_{k+1}(K)$, such that

$$\begin{aligned} \left(\nabla v_h^{n*}, \nabla w\right)_K &= -\left(v_h^n, \Delta w\right)_K + \langle \widehat{v}_h^{\,n}, \nabla w \cdot \boldsymbol{n} \rangle_{\partial K}, \quad \forall\, w \in \mathcal{P}_{k+1}(K), \\ \left(v_h^{n*}, 1\right)_K &= \left(v_h^n, 1\right)_K. \end{aligned} \tag{16}$$

As we are going to see, it turns out that both postprocessings u_h^* and v_h^* have better orders of convergence that the original approximations u_h and v_h, respectively. Note that this local postprocessing can be performed at suitable time steps, where these more accurate approximations are needed.

2.5 Numerical Results

We consider the wave equation on a unit square $\Omega = (0, 1) \times (0, 1)$ with boundary condition $v = 0$ on $\partial\Omega$ and initial condition $u_0 = 0$ and $v_0 = \sin(\pi x)\sin(\pi y)$. For $\rho = \kappa = 1$ and $f = 0$, the problem has the following exact solution

$$u = \frac{1}{\sqrt{2}\pi}\sin(\pi x)\sin(\pi y)\sin(\sqrt{2}\pi t), \qquad v = \sin(\pi x)\sin(\pi y)\cos(\sqrt{2}\pi t).$$

We use triangular meshes obtained by splitting a regular $n \times n$ Cartesian grid into a total of $2n^2$ triangles, giving uniform element sizes of $h = 1/n$.

We present the L^2-errors and orders of convergence for the numerical approximations in Table 1 for the DIRK schemes and Table 2 for the SSP-RK schemes. We observe that the approximate field variables converge with the optimal order $k + 1$, while the postprocessed displacement and velocity converge with order $k + 2$. The HDG methods yield optimal convergence for the approximate gradient, while many other DG methods provide suboptimal convergence of order k. Furthermore, the postprocessed displacement and velocity converge one order higher than the original approximations.

These convergence properties were first reported by Nguyen et al. (2011a) and later proven (for the semidiscrete case) by Cockburn and Quenneville-Bélair (2014). A priori error estimates for $v - v_h^*$ remain an open problem though.

3 THE ELASTIC WAVE EQUATIONS

The elastic wave equations are different from the scalar acoustic wave equation in that they are vectorial and have two different wave speeds, namely, pressure (primary) wave speed and shear (secondary) wave speed. Although there are several different formulations of the elastic wave equations, we will focus on HDG methods for the displacement gradient-velocity–pressure formulation.

TABLE 1 History of Convergence Results Using DIRK(k + 1, k + 2) Schemes

		$\|u-u_h\|_{\mathcal{T}_h}$		$\|v-v_h\|_{\mathcal{T}_h}$		$\|q-q_h\|_{\mathcal{T}_h}$		$\|u-u_h^*\|_{\mathcal{T}_h}$		$\|v-v_h^*\|_{\mathcal{T}_h}$	
k	$1/h$	*Error*	*Order*	*Error*	*Order*	*Error*	*Order*	*Error*	*Order*	*Error*	*Order*
2	2	7.29e−3	–	1.72e−2	–	3.01e−2	–	6.16e−3	–	1.71e−2	–
	4	4.80e−4	3.92	2.16e−3	2.99	2.00e−3	3.91	2.77e−4	4.48	1.99e−3	3.11
	8	4.47e−5	3.42	1.86e−4	3.54	1.84e−4	3.44	7.02e−6	5.30	1.40e−4	3.83
	16	5.24e−6	3.09	1.81e−5	3.36	2.15e−5	3.10	2.54e−7	4.79	8.73e−6	4.00
	32	6.36e−7	3.04	2.08e−6	3.12	2.61e−6	3.04	1.44e−8	4.14	5.36e−7	4.03
3	2	5.80e−4	–	1.60e−3	–	2.67e−3	–	1.97e−4	–	1.59e−3	–
	4	3.12e−5	4.22	8.22e−5	4.29	1.38e−4	4.27	4.92e−6	5.33	8.05e−5	4.30
	8	1.78e−6	4.13	5.20e−6	3.98	7.74e−6	4.16	1.37e−7	5.17	3.78e−6	4.41
	16	1.06e−7	4.07	3.32e−7	3.97	4.56e−7	4.08	4.05e−9	5.08	1.14e−7	5.05
	32	6.46e−9	4.04	2.09e−8	3.99	2.77e−8	4.04	1.24e−10	5.03	1.50e−9	6.24

TABLE 2 History of Convergence Results Using SSP-RK(k + 2, k + 2) Schemes

		$\|\|u-u_h\|\|_{\mathcal{T}_h}$		$\|\|v-v_h\|\|_{\mathcal{T}_h}$		$\|\|q-q_h\|\|_{\mathcal{T}_h}$		$\|\|u-u_h^*\|\|_{\mathcal{T}_h}$		$\|\|v-v_h^*\|\|_{\mathcal{T}_h}$	
k	$1/h$	*Error*	*Order*	*Error*	*Order*	*Error*	*Order*	*Error*	*Order*	*Error*	*Order*
2	2	4.13e−3	–	9.84e−3	–	1.65e−2	–	2.13e−3	–	8.64e−3	–
	4	4.01e−4	3.37	1.06e−3	3.22	1.65e−3	3.32	1.02e−4	4.38	5.19e−4	4.06
	8	4.44e−5	3.17	1.27e−4	3.06	1.83e−4	3.18	4.82e−6	4.40	2.80e−5	4.21
	16	5.24e−6	3.08	1.60e−5	2.99	2.15e−5	3.09	2.59e−7	4.22	1.61e−6	4.12
	32	6.36e−7	3.04	2.02e−6	2.99	2.61e−6	3.04	1.53e−8	4.08	9.81e−8	4.04
3	2	5.75e−4	–	1.62e−3	–	2.66e−3	–	1.82e−4	–	1.33e−3	–
	4	3.12e−5	4.21	8.22e−5	4.30	1.38e−4	4.27	4.63e−6	5.29	3.59e−5	5.21
	8	1.78e−6	4.13	5.21e−6	3.98	7.74e−6	4.15	1.31e−7	5.15	1.03e−6	5.13
	16	1.06e−7	4.07	3.32e−7	3.97	4.56e−7	4.08	3.88e−9	5.07	3.05e−8	5.07
	32	6.46e−9	4.04	2.09e−8	3.99	2.77e−8	4.04	1.19e−10	5.03	8.97e−10	5.09

Let $\boldsymbol{u}$ represent the displacement field, λ and μ the Lamé moduli, ρ the density of the elastic isotropic material, and $\boldsymbol{b}$ a time-dependent body force. Let Ω be an open bounded domain in $\mathbb{R}^d$ and T a fixed final time. The motion of the elastic isotropic body Ω is governed by:

$$\rho\frac{\partial^2 \boldsymbol{u}}{\partial t^2} - \nabla\cdot[\mu\nabla\boldsymbol{u} + (\mu+\lambda)(\nabla\cdot\boldsymbol{u})\mathbf{I}] = \boldsymbol{b}, \quad \text{in } \Omega\times(0,T]. \tag{17}$$

We introduce the velocity field $\boldsymbol{v} = \partial\boldsymbol{u}/\partial t$, the displacement gradient tensor $\mathbf{H} = \nabla\boldsymbol{u}$ and the hydrostatic pressure $p = (\mu + \lambda)(\nabla \cdot \boldsymbol{u})$. We then rewrite (17) as the first-order system

$$\begin{aligned} \frac{\partial \mathbf{H}}{\partial t} - \nabla\boldsymbol{v} &= 0, \quad \text{in } \Omega\times(0,T], \\ \rho\frac{\partial \boldsymbol{v}}{\partial t} - \nabla\cdot(\mu\mathbf{H} + p\mathbf{I}) &= \boldsymbol{b}, \quad \text{in } \Omega\times(0,T], \\ \epsilon\frac{\partial p}{\partial t} - \nabla\cdot\boldsymbol{v} &= 0, \quad \text{in } \Omega\times(0,T], \end{aligned} \tag{18}$$

where $\epsilon = 1/(\mu + \lambda)$, and $\mathbf{I}$ is the second-order identity tensor. Associated with this system is the boundary condition

$$(\mu\boldsymbol{H} + p\boldsymbol{I})\cdot\boldsymbol{n} + \alpha\boldsymbol{v} = \boldsymbol{g}, \quad \text{on } \partial\Omega\times(0,T]$$

and initial condition

$$\boldsymbol{v} = \boldsymbol{v}_0, \quad \mathbf{H} = \mathbf{H}_0, \quad p = p_0, \quad \text{on } \Omega\times\{t=0\}.$$

For simplicity of exposition, we assume that $\epsilon > 0$, which in essence means that the elastic solid is either compressible or nearly incompressible. The incompressible limit $\epsilon = 0$ requires the average pressure condition and can be treated by the augmented Lagrangian method (Nguyen et al., 2010a, 2011b).

3.1 Spatial Discretization

In addition to the finite element spaces defined in Section 2.2, we introduce the following new finite element spaces:

$$\begin{aligned} \mathbf{G}_h &= \{\mathbf{N}\in(L^2(\mathcal{T}_h))^{d\times d} : \mathbf{N}|_K \in (W(K))^{d\times d}, \ \forall K\in\mathcal{T}_h\}, \\ \boldsymbol{M}_h &= \{\boldsymbol{\mu}\in(L^2(\mathcal{E}_h))^d \quad : \boldsymbol{\mu}|_F \in (\mathcal{P}_k(F))^d, \ \forall F\in\mathcal{E}_h\}. \end{aligned}$$

We then define volume and boundary inner products associated with $\mathbf{G}_h$ as

$$(\mathbf{N},\mathbf{L})_{\mathcal{T}_h} = \sum_{K\in\mathcal{T}_h}(\mathbf{N},\mathbf{L})_K, \quad \langle\mathbf{N},\mathbf{L}\rangle_{\partial\mathcal{T}_h} = \sum_{K\in\mathcal{T}_h}\langle\mathbf{N},\mathbf{L}\rangle_{\partial K},$$

for $\mathbf{N},\mathbf{L}\in(L^2(\mathcal{T}_h))^{d\times d}$. Note that $(\mathbf{N}, \mathbf{L})_D$ denotes the integral of $\mathrm{tr}(\mathbf{N}^T\mathbf{L})$ over D, where tr is the trace operator.

The HDG methods then find an approximation $(\mathbf{H}_h, \boldsymbol{v}_h, p_h, \widehat{\boldsymbol{v}}_h) \in \mathbf{G}_h \times \boldsymbol{V}_h \times W_h \times \boldsymbol{M}_h$ at time t such that

$$\left(\frac{\partial \mathbf{H}_h}{\partial t}, \mathbf{N}\right)_{\mathcal{T}_h} + (\boldsymbol{v}_h, \nabla \cdot \mathbf{N})_{\mathcal{T}_h} - \langle \widehat{\boldsymbol{v}}_h, \mathbf{N} \cdot \boldsymbol{n} \rangle_{\partial \mathcal{T}_h} = 0, \tag{19a}$$

$$\left(\rho \frac{\partial \boldsymbol{v}_h}{\partial t}, \boldsymbol{w}\right)_{\mathcal{T}_h} + (\mu \mathbf{H}_h + p_h \mathbf{I}, \nabla \boldsymbol{w})_{\mathcal{T}_h} - \left\langle (\mu \widehat{\mathbf{H}}_h + \widehat{p}_h \mathbf{I}) \cdot \boldsymbol{n}, \boldsymbol{w} \right\rangle_{\partial \mathcal{T}_h} = (\boldsymbol{b}, \boldsymbol{w})_{\mathcal{T}_h}, \tag{19b}$$

$$\left(\epsilon \frac{\partial p_h}{\partial t}, q\right)_{\mathcal{T}_h} + (\boldsymbol{v}_h, \nabla q)_{\mathcal{T}_h} - \langle \widehat{\boldsymbol{v}}_h \cdot \boldsymbol{n}, q \rangle_{\partial \mathcal{T}_h} = 0, \tag{19c}$$

$$\left\langle (\mu \widehat{\mathbf{H}}_h + \widehat{p}_h \mathbf{I}) \cdot \boldsymbol{n}, \boldsymbol{\mu} \right\rangle_{\partial \mathcal{T}_h} + \langle \alpha \widehat{\boldsymbol{v}}_h, \boldsymbol{\mu} \rangle_{\partial \Omega} = \langle \boldsymbol{g}, \boldsymbol{\mu} \rangle_{\partial \Omega}, \tag{19d}$$

for all $(\mathbf{N}, \boldsymbol{w}, q, \boldsymbol{\mu}) \in \mathbf{G}_h \times \boldsymbol{V}_h \times W_h \times \boldsymbol{M}_h$, where

$$(\mu \widehat{\mathbf{H}}_h + \widehat{p}_h \mathbf{I}) \cdot \boldsymbol{n} = (\mu \mathbf{H}_h + p_h \mathbf{I}) \cdot \boldsymbol{n} - \mathbf{S}(\boldsymbol{v}_h - \widehat{\boldsymbol{v}}_h). \tag{19e}$$

Here $\mathbf{S}$ is a second-order tensor consisting of stabilization parameters which can be set to $\sqrt{(\mu + \lambda)\rho} \boldsymbol{I}$.

The semidiscrete form (19) can be equivalently reformulated into finding $(\mathbf{H}_h, \boldsymbol{v}_h, p_h)$ such that for all $K \in \mathcal{T}_h$,

$$\left(\frac{\partial \mathbf{H}_h}{\partial t}, \mathbf{N}\right)_K + (\boldsymbol{v}_h, \nabla \cdot \mathbf{N})_K - \langle \widehat{\boldsymbol{v}}_h, \mathbf{N} \cdot \boldsymbol{n} \rangle_{\partial K} = 0, \tag{20a}$$

$$\left(\rho \frac{\partial \boldsymbol{v}_h}{\partial t}, \boldsymbol{w}\right)_K + (\mu \mathbf{H}_h + p_h \mathbf{I}, \nabla \boldsymbol{w})_K - \left\langle (\mu \widehat{\mathbf{H}}_h + \widehat{p}_h \mathbf{I}) \cdot \boldsymbol{n}, \boldsymbol{w} \right\rangle_{\partial K} = (\boldsymbol{b}, \boldsymbol{w})_K, \tag{20b}$$

$$\left(\epsilon \frac{\partial p_h}{\partial t}, q\right)_K + (\boldsymbol{v}_h, \nabla q)_K - \langle \widehat{\boldsymbol{v}}_h \cdot \boldsymbol{n}, q \rangle_{\partial K} = 0, \tag{20c}$$

where, for any given face $F \in \partial K$,

$$\widehat{\boldsymbol{v}}_h = \begin{cases} \dfrac{\tau^+ \boldsymbol{v}_h^+ + \tau^- \boldsymbol{v}_h^-}{\tau^+ + \tau^-} \\ -\dfrac{1}{\tau^+ + \tau^-}((\mu \mathbf{H}_h^+ + p_h^+ \mathbf{I}) \cdot \boldsymbol{n}^+ + (\mu \mathbf{H}_h^- + p_h^- \mathbf{I}) \cdot \boldsymbol{n}^-), & \text{if } F \in \mathcal{E}_h^o, \\ \dfrac{\tau}{\tau + \alpha} \boldsymbol{v}_h + \dfrac{1}{\tau + \alpha}(\mathsf{P}\boldsymbol{g} - (\mu \mathbf{H}_h + p_h \mathbf{I}) \cdot \boldsymbol{n}), & \text{if } F \in \partial \Omega, \end{cases} \tag{20d}$$

and

$$(\mu \widehat{\mathbf{H}}_h + \widehat{p}_h \mathbf{I}) \cdot \boldsymbol{n} = (\mu \mathbf{H}_h + p_h \mathbf{I}) \cdot \boldsymbol{n} - \mathbf{S}(\boldsymbol{v}_h - \widehat{\boldsymbol{v}}_h) \quad \text{on } \partial K. \tag{20e}$$

In this formulation, both $\widehat{\boldsymbol{v}}_h$ and $(\mu \widehat{\mathbf{H}}_h + \widehat{p}_h \mathbf{I}) \cdot \boldsymbol{n}$ are explicitly determined from the numerical solution $(\mathbf{H}_h, \boldsymbol{v}_h, p_h)$.

While the first formulation (19) is useful for implicit time-stepping methods, the second formulation (20) is convenient for explicit time-stepping methods. Since the temporal discretization in this case is very similar to that in the scalar wave equation, it will not be discussed here.

3.2 Local Postprocessing

As with the acoustic wave equation, we can define two new approximations which will converge faster than the corresponding original approximations. The postprocessed displacement field $\boldsymbol{u}_h^{n*} \in (\mathcal{P}_{k+1}(K))^d$ satisfies, on every simplex $K \in \mathcal{T}_h$,

$$\begin{aligned}(\nabla \boldsymbol{u}_h^{n\,*}, \nabla \boldsymbol{w})_K &= (\mathbf{H}_h^n, \nabla \boldsymbol{w})_K, \quad \forall\, \boldsymbol{w} \in (\mathcal{P}_{k+1}(K))^d,\\ (\boldsymbol{u}_h^{n\,*}, 1)_K &= (\boldsymbol{u}_h^n, 1)_K.\end{aligned} \tag{21}$$

The postprocessed velocity field $\boldsymbol{v}_h^{n\,*} \in (\mathcal{P}_{k+1}(K))^d$ is obtained by locally solving

$$\begin{aligned}(\nabla \boldsymbol{v}_h^{n\,*}, \nabla \boldsymbol{w})_K &= -(\boldsymbol{v}_h^n, \Delta \boldsymbol{w})_K + \langle \widehat{\boldsymbol{v}}_h^n, \nabla \boldsymbol{w} \cdot \boldsymbol{n} \rangle_{\partial K} \quad \forall\, \boldsymbol{w} \in (\mathcal{P}_{k+1}(K))^d,\\ (\boldsymbol{v}_h^{n\,*}, 1)_K &= (\boldsymbol{v}_h^n, 1)_K.\end{aligned} \tag{22}$$

Since the local postprocessing can be carried out at any particular timestep and performed at the element level, the postprocessed displacement and velocity are very inexpensive to compute. Note that the postprocessing is effective only if the temporal accuracy is of order $k + 2$.

3.3 Numerical Results

We consider the elastic wave equations on a unit square $\Omega = (0, 1) \times (0, 1)$ with $\mu = 1$ and $\rho = 1$. The exact solution is given by

$$\begin{aligned}u_1 &= -x^2 y(2y-1)(x-1)^2(y-1)\sin(\pi t),\\ u_2 &= xy^2(2x-1)(x-1)(y-1)^2\sin(\pi t).\end{aligned}$$

The source term $\boldsymbol{b}$ is determined from the above solution. The Dirichlet boundary data are determined as the restriction of the exact solution on the boundary. Likewise the initial data are taken as an instantiation of the exact solution at time $t = 0$. Our triangular meshes are constructed upon regular $n \times n$ Cartesian grids ($h = 1/n$). The stabilization parameter is set to $\tau = 1$.

We present the L^2-errors and orders of convergence of the numerical approximations at the final time $T = 0.5$ in Table 3 for $\lambda = 1$ (compressible case) and in Table 4 for $\lambda = 1000$ (nearly incompressible case). These results are obtained using the DIRK(2,3) scheme for $k = 1$ and the DIRK(3,4) scheme for $k = 2$, and a fixed ratio $h/\Delta t = 4$. We observe that the approximate field variables converge with the optimal order $k + 1$ even for the nearly incompressible case.

TABLE 3 History of Convergence Results for a Compressible Material ($\lambda = 1$)

		$\|u-u_h\|_{\mathcal{T}_h}$		$\|v-v_h\|_{\mathcal{T}_h}$		$\|\sigma-\sigma_h\|_{\mathcal{T}_h}$		$\|u-u_h^*\|_{\mathcal{T}_h}$		$\|v-v_h^*\|_{\mathcal{T}_h}$	
k	$1/h$	*Error*	*Order*	*Error*	*Order*	*Error*	*Order*	*Error*	*Order*	*Error*	*Order*
1	4	3.79e−4	–	1.94e−3	–	2.08e−3	–	1.74e−4	–	1.28e−3	–
	8	1.12e−4	1.76	4.51e−4	2.11	5.07e−4	2.04	2.53e−5	2.78	1.74e−4	2.88
	16	3.04e−5	1.88	1.06e−4	2.09	1.26e−4	2.01	3.27e−6	2.95	2.18e−5	2.99
	32	7.90e−6	1.94	2.60e−5	2.03	3.16e−5	2.00	4.12e−7	2.99	2.96e−6	2.89
	64	2.01e−6	1.97	6.45e−6	2.01	7.93e−6	2.00	5.16e−8	3.00	3.99e−7	2.89
2	4	5.14e−5	–	2.26e−4	–	3.27e−4	–	1.78e−5	–	2.41e−4	–
	8	8.01e−6	2.68	2.90e−5	2.96	4.21e−5	2.96	1.20e−6	3.89	7.10e−6	5.08
	16	1.10e−6	2.87	3.67e−6	2.98	5.25e−6	3.00	7.39e−8	4.02	4.53e−7	3.97
	32	1.43e−7	2.94	4.60e−7	3.00	6.54e−7	3.01	4.52e−9	4.03	2.70e−8	4.07
	64	1.82e−8	2.97	5.75e−8	3.00	8.14e−8	3.00	2.78e−10	4.02	1.68e−9	4.01

TABLE 4 History of Convergence Results for a Nearly Incompressible Material ($\lambda = 1000$)

		$\|u-u_h\|_{\mathcal{T}_h}$		$\|v-v_h\|_{\mathcal{T}_h}$		$\|\sigma-\sigma_h\|_{\mathcal{T}_h}$		$\|u-u_h^*\|_{\mathcal{T}_h}$		$\|v-v_h^*\|_{\mathcal{T}_h}$	
k	1/*h*	*Error*	*Order*	*Error*	*Order*	*Error*	*Order*	*Error*	*Order*	*Error*	*Order*
1	4	3.75e−4	–	1.94e−3	–	2.2e−3	–	1.72e−4	–	1.26e−3	–
	8	1.12e−4	1.75	4.49e−4	2.11	5.41e−4	2.02	2.57e−5	2.74	1.71e−4	2.89
	16	3.04e−5	1.88	1.06e−4	2.08	1.33e−4	2.02	3.37e−6	2.93	2.13e−5	3.00
	32	7.90e−6	1.94	2.60e−5	2.03	3.33e−5	2.00	4.26e−7	2.98	2.87e−6	2.89
	64	2.01e−6	1.97	6.45e−6	2.01	8.33e−6	2.00	5.34e−8	2.99	3.85e−7	2.90
2	4	5.11e−5	–	2.24e−4	–	3.67e−4	–	1.80e−5	–	2.40e−4	–
	8	7.98e−6	2.68	2.88e−5	2.96	4.82e−5	2.93	1.22e−6	3.89	6.91e−6	5.12
	16	1.09e−6	2.87	3.66e−6	2.98	6.12e−6	2.98	7.44e−8	4.03	4.20e−7	4.04
	32	1.43e−7	2.94	4.59e−7	2.99	7.89e−7	2.96	4.52e−9	4.04	2.48e−8	4.08
	64	1.82e−8	2.97	5.75e−8	3.00	9.95e−8	2.99	2.78e−10	4.02	1.48e−9	4.07

Furthermore, we observe that both the postprocessed displacement and velocity converge with order $k+2$, which are one order higher than the original approximations. Since the local postprocessing is inexpensive, the HDG methods provide better convergence and accuracy than other DG methods.

These convergence properties were first reported by Nguyen et al. (2011a). For the semidiscrete case, they can be obtained by an analysis similar to that for the acoustic wave equation in Cockburn and Quenneville-Bélair (2014). Again, a priori error estimates for $\boldsymbol{v}-\boldsymbol{v}_h^*$ remain an open problem.

4 THE ELECTROMAGNETIC WAVE EQUATIONS

In this section, we restrict our attention to devising HDG methods for the Maxwell's equations in frequency domain. Numerical treatment of the Maxwell's equations in time domain follows from the discussion in this section and in Section 2.

Let us consider the time-harmonic Maxwell's equations in a connected and bounded domain $\Omega \in \mathbb{R}^3$ with zero charge density and zero conductivity:

$$\nabla\times\boldsymbol{E}+i\mu\omega\boldsymbol{H}=0,\qquad \nabla\times\boldsymbol{H}-i\epsilon\omega\boldsymbol{E}=\boldsymbol{J},\qquad \text{in }\Omega\subset\mathbb{R}^3, \tag{23}$$

where $\boldsymbol{E}$, $\boldsymbol{H}$ and $\boldsymbol{J}$ are the electric field, magnetic field and current source, respectively. In addition, ω is the frequency, ϵ the permittivity and μ the permeability of the medium. We assume that the electromagnetic field satisfies the following impedance condition

$$-\boldsymbol{n}\times\boldsymbol{H}+\alpha\boldsymbol{n}\times\boldsymbol{E}\times\boldsymbol{n}=\boldsymbol{g},\quad \text{on }\partial\Omega, \tag{24}$$

for some given scalar function α and vectorial function $\boldsymbol{g}$.

4.1 Numerical Discretization

To define the numerical approximation of the Maxwell's equations (23), we introduce the following approximation spaces

$$\begin{aligned}\boldsymbol{V}_h &:= \{\boldsymbol{v}\in[L^2(\mathcal{T}_h)]^3 \;:\; \boldsymbol{v}|_K\in[\mathcal{C}_k(K)]^3,\ \forall K\in\mathcal{T}_h\},\\ \boldsymbol{M}_h^t &:= \{\boldsymbol{\eta}\in[L^2(\mathcal{E}_h)]^3 \;:\; \boldsymbol{\eta}|_F\in[\mathcal{C}_k(F)]^3,\ (\boldsymbol{\eta}\cdot\boldsymbol{n})|_F=0,\ \forall F\in\mathcal{E}_h\}.\end{aligned} \tag{25}$$

Here $\mathcal{C}_k(D)$ denote the space of *complex-valued* polynomials of degree at most k on D. We then define the inner products for our approximation spaces as

$$\begin{aligned}(w,v)_{\mathcal{T}_h} &:= \sum_{K\in\mathcal{T}_h}\int_K w\bar{v}, &\quad (\boldsymbol{w},\boldsymbol{v})_{\mathcal{T}_h} &:= \sum_{j=1}^{3}(w_j,v_j)_{\mathcal{T}_h},\\ \langle w,v\rangle_{\partial\mathcal{T}_h} &:= \sum_{K\in\mathcal{T}_h}\int_{\partial K} w\bar{v}, &\quad \langle\boldsymbol{w},\boldsymbol{v}\rangle_{\partial\mathcal{T}_h} &:= \sum_{j=1}^{3}\langle w_j,v_j\rangle_{\partial\mathcal{T}_h}.\end{aligned} \tag{26}$$

Here the bar denotes a complex conjugate which is applied only to the second argument of the inner products.

Note that $\boldsymbol{M}_h^t$ consists of vector-valued functions whose normal component is zero on any face $F \in \mathcal{E}_h$. Hence, a member of $\boldsymbol{M}_h^t$ can be characterized by two tangential vectors on the faces: if $\boldsymbol{t}_1^F$ and $\boldsymbol{t}_2^F$ denote independent tangent vectors on F, we can write the restriction of $\boldsymbol{\eta} \in \boldsymbol{M}_h^t$ on F as

$$\boldsymbol{\eta}|_F = \eta_1^F \boldsymbol{t}_1^F + \eta_2^F \boldsymbol{t}_2^F, \tag{27}$$

where $\eta_1^F \in \mathcal{C}_k(F)$ and $\eta_2^F \in \mathcal{C}_k(F)$ are complex-valued polynomials of degree at most k on F. Hence, the vector-valued function $\boldsymbol{\eta} \in \boldsymbol{M}_h^t$ is characterized by two scalar functions η_1 and η_2.

The HDG method seeks $(\boldsymbol{E}_h, \boldsymbol{H}_h, \widehat{\boldsymbol{E}}_h^t) \in \boldsymbol{V}_h \times \boldsymbol{V}_h \times \boldsymbol{M}_h^t$ such that

$$(i\omega\mu \boldsymbol{H}_h, \boldsymbol{R})_{\mathcal{T}_h} + (\boldsymbol{E}_h, \nabla \times \boldsymbol{R})_{\mathcal{T}_h} + \left\langle \widehat{\boldsymbol{E}}_h^t, \boldsymbol{R} \times \boldsymbol{n} \right\rangle_{\partial \mathcal{T}_h} = 0, \tag{28a}$$

$$(\boldsymbol{H}_h, \nabla \times \boldsymbol{W})_{\mathcal{T}_h} + \left\langle \widehat{\boldsymbol{H}}_h, \boldsymbol{W} \times \boldsymbol{n} \right\rangle_{\partial \mathcal{T}_h} - (i\epsilon\omega \boldsymbol{E}_h, \boldsymbol{W})_{\mathcal{T}_h} = (\boldsymbol{J}, \boldsymbol{W})_{\mathcal{T}_h}, \tag{28b}$$

$$-\left\langle \boldsymbol{n} \times \widehat{\boldsymbol{H}}_h, \boldsymbol{\eta} \right\rangle_{\partial \mathcal{T}_h} + \left\langle \alpha \widehat{\boldsymbol{E}}_h^t, \boldsymbol{\eta} \right\rangle_{\partial \Omega} = \langle \boldsymbol{g}, \boldsymbol{\eta} \rangle_{\partial \Omega}, \tag{28c}$$

for all $(\boldsymbol{R}, \boldsymbol{W}, \boldsymbol{\eta}) \in \boldsymbol{V}_h \times \boldsymbol{V}_h \times \boldsymbol{M}_h^t$, where

$$\widehat{\boldsymbol{H}}_h = \boldsymbol{H}_h + \tau(\boldsymbol{E}_h - \widehat{\boldsymbol{E}}_h^t) \times \boldsymbol{n}. \tag{28d}$$

Here the stabilization parameter is chosen as $\tau = \sqrt{\dfrac{\epsilon\omega^2}{\mu}}$. This completes the HDG method for solving the time-harmonic Maxwell's equations.

The structure of the HDG method makes itself amenable to an efficient implementation. Note that the first two equations in (28) can be thought as characterizing $(\boldsymbol{E}_h, \boldsymbol{H}_h)$ in terms of $\widehat{\boldsymbol{E}}_h$. Eq. (28c) is then the equation that determines the actual values of the unknown $\widehat{\boldsymbol{E}}_h$. In this manner, the only globally coupled degrees of freedom are those of $\widehat{\boldsymbol{E}}_h$. As a result, the HDG method can provide more accurate solutions at much lower computational cost than standard frequency-domain DG method.

4.2 Local Postprocessing

We propose a new local postprocessing to obtain new approximations of the electric and magnetic fields, which converges with an additional order in the $\mathcal{H}^{\text{curl}}(\mathcal{T}_h)$-norm. A remarkable feature of this new local postprocessing is that it is effective even in three dimensions, whereas the local postprocessing introduced in our previous work (Nguyen et al., 2011c) is only applicable in two dimensions.

We find the new approximate electric field $\boldsymbol{E}_h^*$ as the element of $[\mathcal{C}_{k+1}(K)]^3$ such that for all $K \in \mathcal{T}_h$,

$$(\nabla \times \boldsymbol{E}_h^*, \boldsymbol{W})_K = -(i\mu\omega \boldsymbol{H}_h, \boldsymbol{W})_K, \quad \forall\, \boldsymbol{W} \in \nabla \times [\mathcal{C}_{k+1}(K)]^3, \tag{29a}$$

$$(\boldsymbol{E}_h^*, \boldsymbol{Y})_K = (\boldsymbol{E}_h, \boldsymbol{Y})_K, \qquad \forall\, \boldsymbol{Y} \in \nabla \mathcal{C}_{k+2}(K). \tag{29b}$$

Similarly, we find the new approximate magnetic field $\boldsymbol{H}_h^*$ as the element of $[\mathcal{C}_{k+1}(K)]^3$ such that for all $K \in \mathcal{T}_h$,

$$(\nabla \times \boldsymbol{H}_h^*, \boldsymbol{W})_K = (i\epsilon\omega \boldsymbol{E}_h + \boldsymbol{J}, \boldsymbol{W})_K, \quad \forall\, \boldsymbol{W} \in \nabla \times [\mathcal{C}_{k+1}(K)]^3, \tag{30a}$$

$$(\boldsymbol{H}_h^*, \boldsymbol{Y})_K = (\boldsymbol{H}_h, \boldsymbol{Y})_K, \qquad \forall\, \boldsymbol{Y} \in \nabla \mathcal{C}_{k+2}(K). \tag{30b}$$

It is obvious that $\nabla \times \boldsymbol{E}_h^*$ and $\nabla \times \boldsymbol{H}_h^*$ are nothing but the projection of $i\mu\omega \boldsymbol{H}_h$ and $i\epsilon\omega \boldsymbol{E}_h + \boldsymbol{J}$, respectively, onto the space of divergence-free functions in $[\mathcal{P}_{k+1}(K)]^3$. Therefore, we expect that both $\boldsymbol{E}_h^*$ and $\boldsymbol{H}_h^*$ converge with order $k+1$ in the $\mathcal{H}^{\text{curl}}(\mathcal{T}_h)$-norm, whereas $\boldsymbol{E}_h$ and $\boldsymbol{H}_h$ converge with order k in the $\mathcal{H}^{\text{curl}}(\mathcal{T}_h)$-norm.

4.3 Numerical Results

We consider the time-harmonic Maxwell's equations on a unit cube $\Omega = (0, 1)^3$ with $\mu = 1$, $\epsilon = 2$, $\alpha = 0$ and $\omega = 1$. For $\boldsymbol{J} = 0$ the problem has the exact solution

$$\begin{aligned}
E_x &= \sin(\omega y)\sin(\omega z), & H_x &= i\sin(\omega x)(\cos(\omega y) - \cos(\omega z)),\\
E_y &= \sin(\omega x)\sin(\omega z), & H_y &= i\sin(\omega y)(\cos(\omega z) - \cos(\omega x)),\\
E_z &= \sin(\omega y)\sin(\omega x), & H_z &= i\sin(\omega z)(\cos(\omega x) - \cos(\omega y)),
\end{aligned}$$

The boundary data $\boldsymbol{g}$ is determined from the exact solution. The tetrahedral meshes are constructed upon regular $n \times n \times n$ Cartesian grids ($h = 1/n$) by splitting each cube into six tetrahedral.

We present the L^2-errors and orders of convergence of the numerical approximations in Table 5 and the postprocessed quantities in Table 6. We observe that the approximate electric and magnetic fields converge with order $k+1$ in the L^2-norm, but only order k in the $\mathcal{H}^{\text{curl}}(\mathcal{T}_h)$-norm. Furthermore, we observe that the postprocessed electric and magnetic fields converge with order $k+1$ in the $\mathcal{H}^{\text{curl}}(\mathcal{T}_h)$-norm, which are one order higher than the original approximations. The theoretical justification of these results is still an open problem.

5 BIBLIOGRAPHIC NOTES

5.1 Time-Dependent Wave Propagation

The devising of HDG methods for the acoustic wave equation was carried out as an almost immediate consequence of the introduction of HDG methods for steady-state diffusion. After all, both equations share the same second-order strongly elliptic operator. However, not all convergence properties which hold

TABLE 5 History of Convergence Results for the Approximate Solution

		$\|E-E_h\|_{\mathcal{T}_h}$		$\|E-E_h\|_{\mathcal{H}^{curl}(\mathcal{T}_h)}$		$\|H-H_h\|_{\mathcal{T}_h}$		$\|H-H_h\|_{\mathcal{H}^{curl}(\mathcal{T}_h)}$	
k	1/*h*	*Error*	*Order*	*Error*	*Order*	*Error*	*Order*	*Error*	*Order*
1	2	2.94e−2	–	9.90e−2	–	8.41e−3	–	2.20e−1	–
	4	7.77e−3	1.92	4.46e−2	1.15	2.18e−3	1.95	1.10e−1	1.00
	6	1.94e−3	2.00	2.14e−2	1.06	5.85e−4	1.90	5.52e−2	1.00
	8	4.81e−4	2.01	1.05e−2	1.02	1.54e−4	1.93	2.76e−2	1.00
2	2	9.49e−4	–	1.32e−2	–	6.56e−4	–	3.28e−2	–
	4	1.33e−4	2.84	3.37e−3	1.97	8.74e−5	2.91	8.15e−3	2.01
	6	1.90e−5	2.81	8.47e−4	1.99	1.12e−5	2.96	2.03e−3	2.00
	8	2.87e−6	2.73	2.12e−4	2.00	1.42e−6	2.98	5.09e−4	2.00
3	2	8.72e−5	–	1.40e−3	–	5.51e−5	–	1.74e−3	–
	4	5.59e−6	3.96	1.73e−4	3.02	3.51e−6	3.97	2.28e−4	2.93
	6	3.53e−7	3.99	2.15e−5	3.01	2.23e−7	3.98	2.92e−5	2.97
	8	2.22e−8	3.99	2.67e−6	3.00	1.41e−8	3.99	3.69e−6	2.98

TABLE 6 History of Convergence Results for the Postprocessed Solution

		$\|E - E_h^*\|_{\mathcal{T}_h}$		$\|E - E_h^*\|_{\mathcal{H}^{curl}(\mathcal{T}_h)}$		$\|H - H^*{}_h\|_{\mathcal{T}_h}$		$\|H - H_h^*\|_{\mathcal{H}^{curl}(\mathcal{T}_h)}$	
k	$1/h$	*Error*	*Order*	*Error*	*Order*	*Error*	*Order*	*Error*	*Order*
1	2	3.19e−2	–	3.44e−2	–	1.05e−2	–	6.26e−2	–
	4	8.42e−3	1.92	9.05e−3	1.93	2.69e−3	1.97	1.67e−2	1.90
	6	2.10e−3	2.00	2.27e−3	1.99	7.05e−4	1.93	4.21e−3	1.99
	8	5.23e−4	2.01	5.68e−4	2.00	1.83e−4	1.95	1.05e−3	2.00
2	2	9.56e−4	–	1.58e−3	–	8.34e−4	–	2.06e−3	–
	4	1.34e−4	2.84	2.07e−4	2.93	1.08e−4	2.95	2.82e−4	2.87
	6	1.91e−5	2.81	2.76e−5	2.91	1.38e−5	2.97	3.85e−5	2.87
	8	2.88e−6	2.73	3.81e−6	2.86	1.74e−6	2.99	5.46e−6	2.82
3	2	8.36e−5	–	1.03e−4	–	4.88e−5	–	1.75e−4	–
	4	5.43e−6	3.95	6.71e−6	3.95	3.20e−6	3.93	1.13e−5	3.94
	6	3.44e−7	3.98	4.26e−7	3.98	2.05e−7	3.97	7.20e−7	3.98
	8	2.17e−8	3.99	2.69e−8	3.99	1.29e−8	3.98	4.54e−8	3.99

for HDG methods for steady-state diffusion problems (Chen and Cockburn, 2012, 2014; Cockburn et al., 2008, 2009b, 2010, 2012a,b) can be immediately obtained for time-dependent wave equations. In particular, the wave equation does not have a smoothing effect which could generate superconvergence results, as happens for the heat equation, see Chabaud and Cockburn (2012). However, in Cockburn and Quenneville-Bélair (2014), it was shown how to obtain the superconvergence results we have illustrated in Section 2; a comparison with other mixed and DG methods can also be found there. Although therein we only used simplexes and spaces of polynomials of degree k, similar convergence and superconvergence results do hold for meshes made of elements of arbitrary shape. This can be obtained by using the so-called theory of M-decompositions developed by Cockburn et al. (2016b), Cockburn and Fu (2016a) and Cockburn and Fu (2016b). In a similar way, HDG methods for the elastic wave equation can be easily obtained once HDG methods for linear elasticity (Cockburn and Shi, 2013; Fu et al., 2015; Nguyen and Peraire, 2012; Soon et al., 2009) are obtained.

The first HDG methods for wave propagation were proposed by Nguyen et al. (2011a), where implicit time-marching methods were used, and in Stanglmeier et al. (2016), where explicit time-marching methods were used. In both papers, the superconvergence properties of the semidiscrete method uncovered in Cockburn and Quenneville-Bélair (2014) were shown to hold for the corresponding implicit and explicit time-marching schemes, respectively.

The HDG methods we have presented here can be also used with PML absorbing boundary conditions, as shown in Nguyen et al. (2011a). HDG methods, which are not dissipative, have similar superconvergence properties and have been developed by Cockburn et al. (2016a).

5.2 Time-Harmonic Wave Propagation

HDG methods for time-harmonic hyperbolic equations are also strongly related to the HDG methods for steady-state diffusion problems. The first HDG method for the Helmholtz equation was introduced by Griesmaier and Monk (2011). The same optimal convergence and superconvergence properties of the associated steady-state diffusion were proven. In Feng and Xing (2013), a wide family of discontinuous Galerkin methods, which included the HDG methods, were proven to be stable regardless of the wave number. The methods used piecewise linear approximations. In Cui and Zhang (2014), an analysis of the HDG methods for the Helmholtz equations was carried which shows that the method is stable for any wave number, mesh and polynomial degree and which recovers the orders of convergence and superconvergence obtained previously by Griesmaier and Monk (2011). A method for arbitrarily large wave numbers is proposed by Nguyen et al. (2015b).

The first HDG for the time-harmonic Maxwell's equations was proposed by Nguyen et al. (2011c) in two-space dimensions. The extension of the method to three-dimensions was done in Li et al. (2013). A variation was introduced by Li et al. (2015). HDG method for the time-harmonic equations of elastodynamics can be found in Hungria et al. (2016).

5.3 Further Reading Material

A systematic way of defining HDG methods for Friedrichs' systems has been developed by Bui-Thanh (2015). A general construction of DG methods for these problems can be found in Ern and Guermond (2006a,b, 2008). An overview of the development of DG (and in particular, HDG) methods for fluid dynamics can be found in Cockburn (2016a). An overview of the HDG methods for steady-state diffusion can be found in Cockburn (2016b). Therein, the relation between static condensation, hybridization and the devising of HDG methods is carefully explored.

ACKNOWLEDGEMENTS

The authors would like to thank Rémi Abgrall and Chi-Wang Shu for the invitation to write this chapter. B.C. was supported in part by the National Science Foundation (Grant DMS-1522657) and by the University of Minnesota Supercomputing Institute. N.C.N. and J.P. were supported in part by the Singapore-MIT Alliance.

REFERENCES

Arnold, D.N., Brezzi, F., 1985. Mixed and nonconforming finite element methods: implementation, postprocessing and error estimates. RAIRO Modél. Math. Anal. Numér. 19, 7–32.

Bui-Thanh, T., 2015. From Godunov to a unified hybridized discontinuous Galerkin framework for partial differential equations. J. Comput. Phys. 295, 114–146.

Chabaud, B., Cockburn, B., 2012. Uniform-in-time superconvergence of HDG methods for the heat equation. Math. Comp. 81, 107–129.

Chen, Y., Cockburn, B., 2012. Analysis of variable-degree HDG methods for convection-diffusion equations. Part I: general nonconforming meshes. IMA J. Numer. Anal. 32, 1267–1293.

Chen, Y., Cockburn, B., 2014. Analysis of variable-degree HDG methods for convection-diffusion equations. Part II: semimatching nonconforming meshes. Math. Comp. 83, 87–111.

Chen, M.-H., Cockburn, B., Reitich, F., 2005. High-order RKDG methods for computational electromagnetics. J. Sci. Comput. 22/23, 205–226.

Cockburn, B., 2016. Discontinuous Galerkin methods for computational fluid dynamics. In: Stein, E., de Borst, R., Hughes, T.J.R. (Eds.), Encyclopedia of Computational Mechanics, second ed. vol. 3. John Wiley & Sons, Ltd., England. 111 pp. (to appear).

Cockburn, B., 2016. Static condensation, hybridization, and the devising of the HDG methods. In: Barrenechea, G.R., Brezzi, F., Cagniani, A., Georgoulis, E.H. (Eds.), Building Bridges: Connections and Challenges in Modern Approaches to Numerical Partial Differential Equations. Lect. Notes Comput. Sci. Eng., vol. 114. Springer-Verlag, Berlin. 51 pp. LMS Durham Symposia funded by the London Mathematical Society, Durham, UK, July 8–16, 2014 (to appear).

Cockburn, B., Fu, G., 2016. Superconvergence by M-decompositions. Part II: construction of two-dimensional finite elements. Modél. Math. Anal. Numér. (to appear).

Cockburn, B., Fu, G., 2016. Superconvergence by M-decompositions. Part III: construction of three-dimensional finite elements. Modél. Math. Anal. Numér. (to appear).

Cockburn, B., Quenneville-Bélair, V., 2014. Uniform-in-time superconvergence of HDG methods for the acoustic wave equation. Math. Comp. 83, 65–85.

Cockburn, B., Shi, K., 2013. Superconvergent HDG methods for linear elasticity with weakly symmetric stresses. IMA J. Numer. Anal. 33, 747–770.

Cockburn, B., Dong, B., Guzmán, J., 2008. A superconvergent LDG-hybridizable Galerkin method for second-order elliptic problems. Math. Comp. 77, 1887–1916.

Cockburn, B., Gopalakrishnan, J., Lazarov, R., 2009. Unified hybridization of discontinuous Galerkin, mixed and continuous Galerkin methods for second order elliptic problems. SIAM J. Numer. Anal. 47, 1319–1365.

Cockburn, B., Guzmán, J., Wang, H., 2009. Superconvergent discontinuous Galerkin methods for second-order elliptic problems. Math. Comp. 78, 1–24.

Cockburn, B., Gopalakrishnan, J., Sayas, F.-J., 2010. A projection-based error analysis of HDG methods. Math. Comp. 79, 1351–1367.

Cockburn, B., Qiu, W., Shi, K., 2012. Conditions for superconvergence of HDG methods for second-order elliptic problems. Math. Comp. 81, 1327–1353.

Cockburn, B., Qui, W., Shi, K., 2012. Conditions for superconvergence of HDG methods on curvilinear elements for second-order eliptic problems. SIAM J. Numer. Anal. 50, 1417–1432.

Cockburn, B., Fu, X., Hungria, A., Ji, L., Sayas, F.-J., 2016. Stormer-Numerov HDG methods for the acoustic wave equation. (submitted for publication).

Cockburn, B., Fu, G., Sayas, F.-J., 2016. Superconvergence by M-decompositions. Part I: general theory for HDG methods for diffusion. Math. Comp. (to appear).

Cui, J., Zhang, W., 2014. An analysis of HDG methods for the Helmholtz equation. IMA J. Numer. Anal. 34 (1), 279–295.

Ern, A., Guermond, J.-L., 2006. Discontinuous Galerkin methods for Friedrichs' systems. Part I. General theory. SIAM J. Numer. Anal. 44, 753–778.

Ern, A., Guermond, J.-L., 2006. Discontinuous Galerkin methods for Friedrichs' systems. Part II. Second-order elliptic PDEs. SIAM J. Numer. Anal. 44 (6), 2363–2388.

Ern, A., Guermond, J.-L., 2008. Discontinuous Galerkin methods for Friedrichs' systems. III. Multifield theories with partial coercivity. SIAM J. Numer. Anal. 46 (2), 776–804.

Feng, X., Xing, Y., 2013. Absolutely stable local discontinuous Galerkin methods for the Helmholtz equation with large wave number. Math. Comp. 82, 1269–1296.

Fu, G., Cockburn, B., Stolarski, H., 2015. Analysis of an HDG method for linear elasticity. Int. J. Numer. Methods Eng. 102 (3–4), 551–575.

Gottlieb, S., Shu, C.-W., Tadmor, E., 2000. Strong stability preserving high order time discretization methods. SIAM Rev. 43, 89–112.

Griesmaier, R., Monk, P., 2011. Error analysis for a hybridizable discontinuous Galerkin method for the Helmholtz equation. J. Sci. Comput. 49 (3), 291–310.

Hungria, A., Prada, D., Sayas, F.-J., 2016. HDG methods for elastodynamics. (submitted for publication).

Li, L., Lanteri, S., Perrussel, R., 2013. A hybridizable discontinuous Galerkin method for solving 3D time-harmonic Maxwell's equations. Numerical Mathematics and Advanced Applications 2011, Springer, Heidelberg, pp. 119–128.

Li, L., Lanteri, S., Perrussel, R., 2015. A class of locally well-posed hybridizable discontinuous Galerkin methods for the solution of time-harmonic Maxwell's equations. Comput. Phys. Commun. 192, 23–31.

Nguyen, N.C., Peraire, J., 2012. Hybridizable discontinuous Galerkin methods for partial differential equations in continuum mechanics. J. Comput. Phys. 231, 5955–5988.

Nguyen, N.C., Peraire, J., Cockburn, B., 2009. An implicit high-order hybridizable discontinuous Galerkin method for linear convection-diffusion equations. J. Comput. Phys. 228, 3232–3254.

Nguyen, N.C., Peraire, J., Cockburn, B., 2009. An implicit high-order hybridizable discontinuous Galerkin method for nonlinear convection-diffusion equations. J. Comput. Phys. 228, 8841–8855.

Nguyen, N.C., Peraire, J., Cockburn, B., 2010. A hybridizable discontinuous Galerkin method for Stokes flow. Comput. Methods Appl. Mech. Eng. 199, 582–597.

Nguyen, N.C., Peraire, J., Cockburn, B., 2010. A hybridizable discontinuous Galerkin method for the incompressible Navier-Stokes equations (AIAA Paper 2010-362). In: Proceedings of the 48th AIAA Aerospace Sciences Meeting and Exhibit, Orlando, FL.

Nguyen, N.C., Peraire, J., Cockburn, B., 2011. High-order implicit hybridizable discontinuous Galerkin methods for acoustics and elastodynamics. J. Comput. Phys. 230, 3695–3718.

Nguyen, N.C., Peraire, J., Cockburn, B., 2011. Hybridizable discontinuous Galerkin methods. In: Hesthaven, J., Ronquist, E. (Eds.), Spectral and High Order Methods for Partial Differential Equations. Lect. Notes Comput. Sci. Eng., vol. 76. Springer-Verlag, Berlin, Heidelberg, pp. 63–84.

Nguyen, N.C., Peraire, J., Cockburn, B., 2011. Hybridizable discontinuous Galerkin methods for the time-harmonic Maxwell's equations. J. Comput. Phys. 230, 7151–7175.

Nguyen, N.C., Peraire, J., Cockburn, B., 2011. An implicit high-order hybridizable discontinuous Galerkin method for the incompressible Navier-Stokes equations. J. Comput. Phys. 230, 1147–1170.

Nguyen, N.C., Peraire, J., Cockburn, B., 2015. A class of embedded discontinuous Galerkin methods for computational fluid dynamics. J. Comput. Phys. 302, 674–692.

Nguyen, N.C., Peraire, J., Reitich, F., Cockburn, B., 2015. A phase-based hybridizable discontinuous Galerkin method for the numerical solution of the Helmholtz equation. J. Comput. Phys. 290, 318–335.

Peraire, J., Nguyen, N.C., Cockburn, B., 2010. A hybridizable discontinuous Galerkin method for the compressible Euler and Navier-Stokes equations (AIAA Paper 2010-363). In: Proceedings of the 48th AIAA Aerospace Sciences Meeting and Exhibit, Orlando, FL.

Raviart, P.A., Thomas, J.M., 1977. A mixed finite element method for second order elliptic problems. In: Galligani, I., Magenes, E. (Eds.), Mathematical Aspects of Finite Element Method, Lecture Notes in Math. 606. Springer-Verlag, New York, pp. 292–315.

Soon, S.-C., Cockburn, B., Stolarski, H.K., 2009. A hybridizable discontinuous Galerkin method for linear elasticity. Int. J. Numer. Methods Eng. 80 (8), 1058–1092.

Stanglmeier, M., Nguyen, N.C., Peraire, J., Cockburn, B., 2016. An explicit hybridizable discontinuous Galerkin method for the acoustic wave equation. Comput. Methods Appl. Mech. Eng. 300, 748–769.

Chapter 9

Spectral Volume and Spectral Difference Methods

Z.J. Wang*, Y. Liu†, C. Lacor‡ and J.L.F. Azevedo§
**University of Kansas, Lawrence, KS, United States*
†*NASA Ames Research Center, CA, United States*
‡*Vrije Universiteit, Brussel, Belgium*
§*Instituto de Aeronautica e Espaço, São Jose dos Campos, SP, Brazil*

Chapter Outline

ABSTRACT
This chapter describes two related methods, the spectral volume and spectral difference methods for hyperbolic conservation laws. Similar to the discontinuous Galerkin method, both are inspired by the finite element and finite volume methods in that multiple degrees of freedom are defined in each element, and the cell-averaged mean obeys the conservation principle. We review the history, the connection, and recent developments of both methods and highlight their pros and cons with analysis and numerical demonstrations.

Keywords: Spectral volume method, Spectral difference method, High order, Unstructured grids, Discontinuous Galerkin, Finite volume

AMS Classification Codes: 65M08, 65M60, 75N15

1 INTRODUCTION

The 1980s saw the major development of so-called high-resolution methods for hyperbolic conservation laws, such as MUSCL (Van Leer, 1979), the total variation diminishing (TVD) scheme (Harten, 1983), and essentially nonoscillatory (ENO) methods (Harten et al., 1987). Many of these were originally formulated in a finite difference (FD) context (e.g. Godunov, 1959), but can be extended to a finite volume (FV) framework. The basic FV idea was successfully extended to unstructured meshes in the late 1980s, e.g., in Fezoui and Stoufflet (1989), and such methods have been further developed to become the core algorithms in many production and commercial CFD codes.

Partly due to the tremendous success enjoyed by FV methods in the 1990s, the CFD community paid little attention to a new method called discontinuous Galerkin (DG) (Reed and Hill, 1973; Cockburn and Shu, 1989; Bassi and Rebay, 1997; Cockburn et al., 2000), which can be viewed a hybrid of finite element (FE) and FV methods. It uses the FE concept by allowing multiple degrees of freedom (DOFs) in a single element, unlike the FV method in which only the cell-averaged variable is associated with an element. Another reason that prevented the DG method from wider acceptance was the myth that it was much more expensive than the FV method. The myth may not be true when one compares the cost to generate computational results of equal accuracy (see Wang et al., 2013).

By the 2000s, many in the CFD community have "rediscovered" the DG method, especially for applications with complex geometries and high accuracy requirements. In an attempt to formulate a FV version of the DG method, the spectral volume (SV) method was developed in a series of papers (Wang, 2002; Wang and Liu, 2002, 2004, 2006; Wang et al., 2004; Liu et al., 2006a; Sun et al., 2006). At that time, it was realized that extending the FV method directly to higher order accuracy led to robustness and efficiency issues because of the large reconstruction stencils (Delanaye and Liu, 1999). The basic idea of the SV method is to allow each element to have multiple DOFs in the form of subcell averages. The partition of an element into subcells is a critical problem determining the stability and accuracy of the method. In the original SV paper, the Lebesgue constant of the partition was used as a generic criterion to determine the partition quality, and numerical tests were then performed to show accuracy and stability (Wang and Liu, 2002). Partitions with smaller Lebesgue constants were developed for triangles in Liu et al. (2003), which also presented the first high-order partitions of a 3D simplex. Chen (2006a,b) obtained partitions with very small Lebesgue constants for both 2D and 3D simplexes through an optimization approach. Later Van den Abeele and Lacor (2007) used Fourier analysis to analyse the partition quality, minimize dissipation and dispersion errors, and determine stability for triangular elements. They also discovered a weak instability in some earlier partitions, and developed new low dissipation and dispersion error partitions while confirming that the 2D partitions found by Chen (2006b) are stable. In addition, it was discovered that low Lebesgue constants do not necessarily guarantee stability. Harris and Wang (2009a,b) developed a stable fifth-order partition for a triangle. Although several efforts have been made to find stable high-order partitions for the 3D simplex, e.g., by Van den Abeele et al. (2009), no success has been reported so far. An analysis of the SV and SD methods was also provided by Lacor and Van den Abeele (2011).

Soon after the first SV paper appeared, Choi et al. (2004) successfully extended the method to handle the shallow water equation. Zhang and Shu (2005) compared the DG and SV methods in terms of accuracy and stability. Sun et al. (2006) were the first to extend the method to 2D laminar Navier–Stokes. Haga et al. (2006) implemented the SV method on Japan's Earth Simulator, the fastest computer in the world at the time. Liu et al. (2006a) applied the method to 3D computational electromagnetics problems. Cheruvu et al. (2007) developed a SV method for a transport equation on the cubed sphere. Harris et al. (2008) developed a more efficient quadrature-free implementation of the SV method. Van den Abeele et al. (2007a) analysed dissipation and dispersion errors, and Kannan and Wang (2009) studied viscous flux formulas in the context of a p-multigrid approach. Haga et al. (2009) was the first to employ the SV method to solve the Reynolds averaged Navier–Stokes equations. Xu et al. (2009) developed a hierarchical reconstruction approach for the SV method on unstructured grids for shock capturing. Breviglieri et al. (2010) developed an implicit SV method capable of handling shock waves with an excellent iterative convergence property. Cozzolino et al. (2012) developed a well-balanced SV method for the shallow water equation. Cimorelli et al. (2012) studied how to recover the solution derivative at interfaces for the evaluation of diffusive fluxes. Meister et al. (2012) compared the DG and SV formulations on triangulations using Proriol–Koornwinder–Dubiner polynomials. Raghavendra (2011) and Kannan (2012) extended the SV method to handle derivatives higher than second order.

The extension of the SV method to three dimensions becomes challenging for several reasons. First, the partition problem is complicated and can generate subcells which are complex polyhedrals. Second, it is difficult to generate partitions that result in stable numerical schemes. Third, many quadrature points are introduced on subcell interfaces, which make the SV method expensive. In an effort to remedy the difficulties faced by the SV method, Liu et al. (2004, 2006b) conceived the spectral difference (SD) method to mimic an FD method for efficiency. In this regard, we can say that the SD method is an FD version of the DG method.

The original development of the SD method was carried out on a triangle to handle complex geometries (Liu et al., 2004, 2006b). Two sets of points, the solution points (SPs) and flux points (FPs), are defined. Some of the FPs must be located at element interfaces to allow Riemann fluxes to be computed there to couple the elements, achieve conservation, and provide dissipation for stability. We wish to emphasize that the SD method is independent of where the SPs are located, as explained in Liu et al. (2004) and shown in Van den Abeele et al. (2008). It is, therefore, not necessary to have staggered SPs and FPs in the SD method. In fact, some of the SPs can be located at the FPs to maximize efficiency. For example, Huang et al. (2005) employed the FPs as the SPs and adapted the order of the polynomial through an artificial viscosity approach for shock capturing. In 1D, the SD method is identical to the staggered-grid multidomain method developed by Kopriva and Kolias (1996).

Wang and Liu (2005) and Wang et al. (2007) extended the SD method to handle the Euler equations, and May and Jameson (2006) extended the SD method for both the Euler and Navier–Stokes equations. Van den Abeele et al. (2007b) established an unexpected connection between the SV and SD methods and discovered that they are identical if the SV partitioning points coincide with the FPs in the SD method. Sun et al. (2007) developed a Navier–Stokes solver for unstructured hexahedral meshes and implemented an implicit nonlinear LU-SGS approach for the high-order SD solver (Sun et al., 2009). Parsani et al. (2010b, 2011a) theoretically analysed the stability properties of this implicit LU-SGS when used as a smoother for a p-multigrid approach combined with SV schemes of second up to fourth order. Results are shown for different 2D compressible laminar Navier–Stokes problems. Van den Abeele et al. (2008) performed an extensive analysis of stability and accuracy of the SD method and found that the use of Gauss–Lobatto points as the FPs results in a weak instability. Liang et al. (2009a,b) developed an SD method for mixed unstructured grids. Cagnone and Nadarajah (2009) developed a frequency domain SD method for periodic inviscid flows. Yang and Wang (2009) developed a parameter-free generalized moment limiter for the SD method. Li et al. (2010) compared h- and p-adaptations for the SD method, and Liang et al. (2011a) extended the p-multigrid SD method for incompressible flow problems. Jameson (2010) proved the stability of the SD methods for all orders of accuracy with the Gauss points as interior FPs. Mohammad et al. (2010) and Parsani et al. (2010a) performed large eddy simulations of turbulent flow using high-order SD solvers. Zhou and Wang (2010) developed absorbing boundary conditions for the SD method for computational aeroacoustics problems.

The SD method was extended to dynamic meshes by Yu et al. (2011) and Liang et al. (2011b). Ou and Jameson (2011) simulated flapping wing aerodynamics of realistic configurations using an SD solver. May (2011) established an interesting connection between the SD and the DG methods. Parsani et al. (2011a,b) applied and validated a high-order solver for induced noise simulations. Gao et al. (2012) developed an optimized SD method for computational aeroacoustics problems. Balan et al. (2012) discovered a stable high-order SD method on a 2D simplex by using a special space, the Raviart–Thomas space, for the flux functions. Meister et al. (2012) compared the SD and DG methods on triangulations using PKD polynomials. Parsani et al. (2013) developed optimized explicit Runge–Kutta schemes for the SD method for wave propagation problems. Lodato et al. (2014) developed a wall-modelled large eddy simulation approach for the SD method on unstructured meshes. Premasuthan et al. (2014) implemented an artificial viscosity approach for the SD method for shock capturing. Wang et al. (2015) developed an SD code to handle stratified convection in rotating spherical shells. Xie et al. (2015) employed hierarchical polynomial bases for the SD method. Choi (2015) developed a hybrid SD/FV method for conservation laws. Zhang and Liang (2015) developed a sliding mesh SD solver to couple rotating and stationary domains. Lamouroux et al. (2016) proposed a compact limiter for the SV and SD methods.

The reminder of the chapter is organized as follows. In Section 2, we present the basic 1D SV and SD methods and prove that they are identical. Then we describe the SV and SD methods on triangles, briefly discuss their efficiency on Cartesian elements, and highlight their pros and cons in Section 3. Some sample computational results are given in Section 4. Finally conclusions are summarized in Section 5.

2 ONE-DIMENSIONAL FORMULATIONS

Consider the following 1D conservation law

$$\frac{\partial u(x,t)}{\partial t}+\frac{\partial f(u)}{\partial x}=0, \tag{1}$$

where u is the state variable and f is the flux. The computational domain $[a, b]$ is partitioned into N elements, and the ith element is $V_i \equiv \left[x_{i-1/2}, x_{i+1/2}\right]$, with $x_{1/2}=a$, and $x_{N+1/2}=b$, and the element size is $\Delta x_i=x_{i+1/2}-x_{i-1/2}$. Each element is mapped into a standard element $[-1, 1]$ with the following linear transformation

$$\xi=2\left(x-x_{i-1/2}\right)/\Delta x_i-1. \tag{2}$$

The conservation law can be transformed from the physical element to the standard element

$$\frac{\partial u(x,t)}{\partial t}+\frac{\partial f}{\partial \xi}\frac{\partial \xi}{\partial x}=\frac{\partial u(\xi,t)}{\partial t}+\frac{2}{\Delta x_i}\frac{\partial f}{\partial \xi}=0. \tag{3}$$

Either (1) or (3) can be used depending on the context. On each element, the approximate solution is a degree k polynomial with no continuity assumed at element interfaces.

2.1 SV Method

The standard element is partitioned into $(k+1)$ control volumes (CVs) using $k+2$ points $\{\xi_{j-1/2}, j=1,k+2\}$, with $\xi_{1/2}=-1$, and $\xi_{k+1+1/2}=1$, as shown

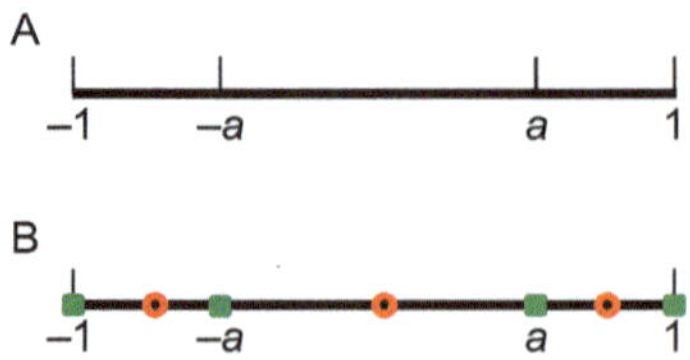

FIG. 1 A SV partition and the placement of solution and FPs in an SD scheme. (A) $k=2$ SV partition and (B) solution (●) and flux (■) points in a $k=2$ SD element.

in Fig. 1A for $k=2$. The jth CV on the standard element is defined by $[\xi_{j-1/2,} \xi_{j+1/2,}]$. The corresponding CV in the ith element is denoted by $V_{i,j} \equiv [x_{i,j-1/2}, x_{i,j+1/2}]$, and its size is $\Delta x_{i,j} = x_{i,j+1/2} - x_{i,j-1/2}$. Integrating (1) over each CV, we obtain

$$\int_{V_{i,j}} \left(\frac{\partial u}{\partial t} + \frac{\partial f}{\partial x}\right) dx = \frac{d\bar{u}_{i,j}}{dt} \Delta x_{i,j} + f\left(u_{i,j+1/2}\right) - f\left(u_{i,j-1/2}\right) = 0, \tag{4}$$

where $\bar{u}_{i,j}$ are the CV-averaged solutions, and they are the DOFs. In order to compute the fluxes at CV interfaces, a degree k solution polynomial $U_i^{\text{SV}}(\xi)$ is reconstructed from the DOFs within element i to approximate the true solution, i.e.,

$$u(x,t) \approx U_i^{\text{SV}}(\xi,t) = \sum_{j=1}^{k+1} \bar{u}_{i,j}(t) L_j^{\text{SV}}(\xi), \tag{5}$$

where $L_j^{\text{SV}}(\xi)$ are degree k polynomials called shape functions satisfying the following equations

$$\frac{1}{\xi_{k+1/2} - \xi_{k-1/2}} \int_{\xi_{k-1/2}}^{\xi_{k+1/2}} L_j^{\text{SV}}(\xi) d\xi = \delta_{jk}, \tag{6}$$

with δ_{jk} the Kronecker delta. It is obvious that the solution is continuous within each element, but discontinuous across element boundaries. Therefore, at internal CV interfaces, the analytical flux is computed as follows:

$$f\left(u_{i,j+1/2}\right) \approx F_{i,j+1/2} = f\left(U_i^{\text{SV}}\left(\xi_{j+1/2}\right)\right). \tag{7}$$

At the element interface, a Riemann flux is computed since the numerical solution is discontinuous

$$F_{i+1/2} = f_{\text{Riem}}\left(U_{i-1}^{\text{SV}}(1), U_{i+1}^{\text{SV}}(-1)\right). \tag{8}$$

This Riemann flux is used to update the CVs sharing an element interface, i.e.,

$$F_{i,k+1+1/2} = F_{i+1,1/2} = F_{i+1/2}. \tag{9}$$

Finally the semidiscrete SV scheme becomes

$$\frac{d\bar{u}_{i,j}}{dt} = -\frac{1}{\Delta x_{i,j}}\left(F_{i,j+1/2} - F_{i,j-1/2}\right). \tag{10}$$

Any time integration schemes can be used for (10). Popular choices include the SSP Runge–Kutta (Gottlieb and Shu, 1998) or classical Runge–Kutta schemes.

From the very beginning, it was realized that the stability of the SV schemes hinges on how the element is partitioned into CVs (Wang, 2002; Wang and Liu, 2002). Based on approximation theory, the Lebesgue constant was used as a criterion to measure the partition quality. However, it was also shown using Fourier analysis that the Lebesgue constant cannot accurately predict either accuracy or stability (Van den Abeele et al., 2007a,b). Because of the connection of the SV and SD methods, the stability issue will be discussed after the presentation of the SD method.

2.2 SD Method

Two sets of points are defined on the standard element. One set is called the SPs and the other is named the FPs, as shown in Fig. 1B. The DOFs are the solutions at the $k+1$ SPs, denoted $\left\{\xi_j^{SP}, j = 1, k+1\right\}$. The DOFs are then used to reconstruct a degree k polynomial on element i

$$U_i^{SD}(\xi, t) = \sum_{j=1}^{k+1} u_{i,j}(t) L_j^{SP}(\xi), \tag{11}$$

where $u_{i,j}$ are the solutions at the SPs and $L_j^{SP}(\xi)$ are the degree k Lagrange polynomials based on the SPs.

In order to mimic an FD method, we reconstruct a flux polynomial, which is one order higher than the solution polynomial. Therefore, $k+2$ FPs are defined on the standard element, denoted $\left\{\xi_j^{FP}, j = 1, k+2\right\}$. The first and last FPs coincide with the element boundary, i.e., $\xi_1^{FP} = -1$, $\xi_{k+2}^{FP} = 1$, in order to couple neighbouring elements, and achieve conservation. The fluxes at the FPs are computed in exactly the same way as the fluxes at the CV interfaces in the SV method, except that the solution polynomial in the SV method, $U_i^{SV}(\xi)$, is replaced by its counterpart in the SD method, $U_i^{SD}(\xi)$. Hence we use the same symbol to denote the fluxes at the FPs, $\left\{F_{i,j}, j = 1, k+2\right\}$. Then a degree $k+1$ flux polynomial is reconstructed from the fluxes at the FPs

$$f(u) \approx F_i^{SD}(\xi) = \sum_{j=1}^{k+2} F_{i,j} L_j^{FP}(\xi), \tag{12}$$

where $L_j^{FP}(\xi)$ are the degree $k+1$ Lagrange polynomials based on the FPs. Finally the DOFs are updated like an FD scheme

$$\frac{du_{i,j}}{dt} = -\frac{2}{\Delta x_i}\frac{\partial F_i^{\mathrm{SD}}}{\partial \xi}\left(\xi_j^{\mathrm{SP}}\right) = -\frac{2}{\Delta x_i}\sum_{j=1}^{k+2} F_{i,j}\frac{\partial L_j^{\mathrm{FP}}}{\partial \xi}\left(\xi_j^{\mathrm{SP}}\right). \tag{13}$$

Note that the $k+2$ coefficients $\frac{\partial L_j^{\mathrm{FP}}}{\partial \xi}\left(\xi_j^{\mathrm{SP}}\right)$ are universal for all elements. The solution and flux reconstructions need to be performed only once for the standard element and the coefficients are precomputed.

2.3 Equivalence of the SV and SD Methods and Their Stability

From appearance, the DOFs of the SV and SD methods are different, and they are updated using FV and FD-like approaches. It is well known that 1D FV and FD methods are equivalent on uniform meshes, but the equivalence does not extend to nonuniform meshes. Given the fact that SV and SD methods can handle nonuniform meshes without any complication, we did not expect them to be equivalent. In addition, different initial conditions are often applied for the SV and SD methods. For example, if $u(x,0) = \sin(x)$. The natural initialization approach for an SV method is to compute the CV-averaged solution exactly, i.e.,

$$\bar{u}_{i,j} = \int_{V_{i,j}} \sin(x)\mathrm{d}x. \tag{14}$$

Similarly, the nodal values of the exact initial condition can be used to initialize the SPs in the SD method, i.e.,

$$u_{i,j} = \sin\left(x_{i,j}\right). \tag{15}$$

The initializations used in (14) and (15) seem "exact". However because the solutions in the SV and SD methods are always degree k polynomials, the "true" initial conditions are actually the following polynomials:

$$U_i^{\mathrm{SV}}(\xi) = \sum_{j=1}^{k+1} \bar{u}_{i,j} L_j^{\mathrm{SV}}(\xi), \tag{16}$$

$$U_i^{\mathrm{SD}}(\xi) = \sum_{j=1}^{k+1} u_{i,j} L_j^{\mathrm{SP}}(\xi). \tag{17}$$

Clearly, these two polynomials are not the same. Therefore the SV and SD methods start from different initial conditions, even if the analytical initial condition is the same. This is perhaps one of the main reasons that the equivalence of the SV and SD methods was not discovered until 2007, when Van den Abeele's Fourier analysis revealed that they are identical. In hindsight, we should have thought of it earlier. In order to prove the equivalence, the

SV and SD methods must start from the same initial condition, i.e., the same solution polynomial. This is possible because they share the same solution space, the piece-wise continuous degree k polynomials. In Van den Abeele et al. (2007a,b), the following results were proven.

Theorem. *In 1D, an SD method is equivalent to an SV method, and thus* $U_i^{SV}(x,t)=U_i^{SD}(x,t)$, *provided that*

- *The numerical solution is initialized by locally projecting the exact initial solution on the basis polynomials in each cell.*
- *The CV boundary positions for the SV method are the same as the FP positions for the SD method.*
- *The same Riemann solver is used to compute the flux at element interfaces.*

A direct result of this theorem is that the SD method is totally independent of the locations of the SPs. It is unnecessary to stagger the SPs and FPs, and in fact, some of the SPs can be exactly the same as the FPs to maximize efficiency.

Using Fourier analysis, Van den Abeele et al. (2008) also discovered that the use of the Chebyshev–Gauss–Lobatto nodes as the SPs results in a weak instability for $k>1$. Stable nodal sets were described in the same paper. Huynh (2007) showed that the use of Gauss quadrature points as the interior FPs results in stable SD schemes. This result is proved by Jameson (2010) for all orders of accuracy. Due to the equivalence of the SV and SD methods, Gauss quadrature nodes should also be used in an SV scheme to partition an element into CVs.

We wish to point out that the 1D SV and SD methods have similar complexity, which includes the following cost:

- To compute the solutions at the CV interfaces or the FPs.
- To compute the fluxes.
- To update the DOFs.

We want to emphasize that the SD and SV methods are equivalent only in 1D.

3 TWO-DIMENSIONAL FORMULATION ON THE SIMPLEX

In order to present the basic ideas, let us first consider the following 2D scalar conservation law:

$$\frac{\partial u(\boldsymbol{x},t)}{\partial t}+\frac{\partial f(u)}{\partial x}+\frac{\partial g(u)}{\partial y}=0, \tag{18}$$

where u is the state variable, and f and g are fluxes in the x and y directions, respectively, and $\boldsymbol{x}=(x,\ y)$. We further assume that proper initial and boundary conditions are available to make the problem well posed. The computational domain Ω is discretized into N simplex elements (triangles), and the ith element is denoted V_i. Each triangle can be transformed into a standard element (an equilateral triangle) using a linear transformation. The approximate solution on each element is a degree k polynomial, which needs a minimum number of $m=(k+1)(k+2)/2$ DOFs.

3.1 SV Method

The standard element V^{S} is partitioned into m subcells called CVs, and the jth CV of V^{S} is denoted by V_j^{S}. The partition of the standard element can be mapped back to each physical element, V_i, and the corresponding jth CV is $V_{i,j}$. Some sample partitions for the $k=1$–3 SV schemes are displayed in Fig. 2. The cell-averaged conserved variables at time t for CV $V_{i,j}$ are defined as

$$\bar{u}_{i,j}(t)=\frac{\int_{V_{i,j}} u(\boldsymbol{x},t)d\boldsymbol{x}}{|V_{i,j}|}, \tag{19}$$

where $|V_{i,j}|$ is the volume of $V_{i,j}$. Given the cell-averaged state variables for all the CVs, a degree k polynomial $U_i^{\mathrm{SV}}(\boldsymbol{x})$ can be reconstructed such that it is a $k+1$th-order accurate approximation to the solution $u(\boldsymbol{x})$ inside V_i:

$$U_i^{\mathrm{SV}}(\boldsymbol{x})=u(\boldsymbol{x})+O\left(h^{k+1}\right),\boldsymbol{x}\in V_i, \tag{20}$$

where h is the maximum edge length of V_i. This reconstruction can be solved analytically by satisfying the following conditions:

$$\frac{\int_{V_{i,j}} U_i^{\mathrm{SV}}(\boldsymbol{x})d\boldsymbol{x}}{|V_{i,j}|}=\frac{\int_{V_j^{\mathrm{S}}} U_i^{\mathrm{SV}}(\boldsymbol{\xi})d\boldsymbol{\xi}}{\left|V_j^{\mathrm{S}}\right|}=\bar{u}_{i,j},\ j=1,\ldots,m. \tag{21}$$

This polynomial $U_i^{\mathrm{SV}}(\boldsymbol{x})$ is the $k+1$th-order approximation we are looking for as long as the function $u(\boldsymbol{x})$ is smooth in the region covered by V_i. The reconstruction can be more conveniently expressed as

$$U_i^{\mathrm{SV}}(\boldsymbol{\xi})=\sum_{j=1}^{m}\bar{u}_{i,j}L_j^{\mathrm{SV}}(\boldsymbol{\xi}), \tag{22}$$

where $L_j^{\mathrm{SV}}(\boldsymbol{\xi})$ are degree k polynomials called "shape" functions which satisfy

$$\frac{\int_{V_k^{\mathrm{S}}} L_j^{\mathrm{SV}}(\boldsymbol{\xi})d\boldsymbol{\xi}}{\left|V_k^{\mathrm{S}}\right|}=\delta_{jk}. \tag{23}$$

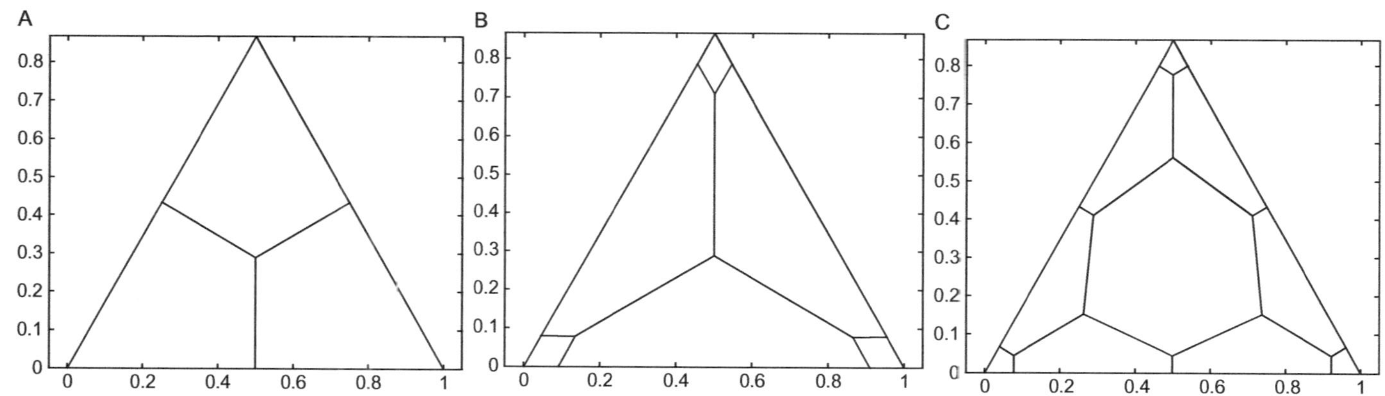

FIG. 2 Partitions of various orders in a triangular spectral volume, the third- and fourth-order partitions are found by Van den Abeele and Lacor (2007). (A) $k=1$ (second order), (B) $k=2$ (third order), and (C) $k=3$ (fourth order).

The high-order reconstruction is then used to generate high-order updates for the CV-averaged state variables. Integrating (18) in $V_{i,j}$, we obtain the following integral equation for the CV-averaged mean:

$$\frac{d\bar{u}_{i,j}}{dt}\left|V_{i,j}\right| + \int_{\partial V_{i,j}} \boldsymbol{f} \cdot \boldsymbol{n} dS = 0. \tag{24}$$

The surface integral can be further broken down into integrals over linear faces. If the fluxes are nonlinear, the integral over each face can be carried out with a Gauss quadrature formula of enough precision

$$\int_{\partial V_{i,j}} \boldsymbol{f} \cdot \boldsymbol{n} dS = \sum_{k=1}^{n_f} \int_{S_k} \boldsymbol{f} \cdot \boldsymbol{n} dS \approx \sum_{k=1}^{n_f} \sum_{q=1}^{n_q} w_q \boldsymbol{f}\left(\boldsymbol{x}_{k,q}\right) \cdot \boldsymbol{n}_k \left|S_k\right|, \tag{25}$$

where n_q is the number of Gauss quadrature points, which should be *integer* $[(k+2)/2]$ to maintain the formal accuracy, w_q is the quadrature weight, and $\boldsymbol{x}_{k,q}$ are the coordinates of the quadrature point. Analytical fluxes are computed at the interior CV boundaries since the solution is assumed continuous

$$\boldsymbol{f}\left(\boldsymbol{x}_{k,q}\right) \cdot \boldsymbol{n}_k \approx F_{k,q}^n = \boldsymbol{f}\left(U_i^{\mathrm{SV}}\left(\boldsymbol{x}_{k,q}\right)\right) \cdot \boldsymbol{n}_k. \tag{26}$$

Riemann fluxes are evaluated at element interface quadrature points, e.g.,

$$\boldsymbol{f}\left(\boldsymbol{x}_{k,q}\right) \cdot \boldsymbol{n}_k \approx F_{k,q}^n = f_{\mathrm{Riem}}\left(u_{\mathrm{L}}\left(\boldsymbol{x}_{k,q}\right), u_{\mathrm{R}}\left(\boldsymbol{x}_{k,q}\right), \boldsymbol{n}_k\right), \tag{27}$$

where u_{L} and u_{R} are the solutions just to the left and right of a face. Any approximate Riemann solvers can be used (Roe, 1981). Finally the update scheme for the DOFs in an SV scheme is

$$\frac{d\bar{u}_{i,j}}{dt} = -\frac{1}{\left|V_{i,j}\right|} \sum_{k=1}^{n_f} \sum_{q=1}^{n_q} w_q F_{k.q}^n \left|S_k\right|. \tag{28}$$

3.2 SD Method

Within each cell or element, we again define two sets of points, i.e., the SPs and FPs. Fig. 3 displays possible placements of SPs and FPs for $k=0$ (first order) to $k=2$ (third order) SD schemes. The SD method on the 2D simplex is also independent of the placement of the SPs. As a result, more efficient placements of SPs are possible by coinciding some of the SPs with the FPs, as shown in Fig. 3D–F. It should be noted that this property is not generally valid for nonsimplex elements such as quadrilaterals in 2D and hexahedra in 3D, though it is satisfied in many practical circumstances (see Van den Abeele et al., 2008). Let the position vector of the jth SP at cell i be denoted by $\boldsymbol{x}_{i,j}$, and the kth FP at cell i be denoted by $\boldsymbol{x}_{i,k}$. Denote $u_{i,j}$ the solution at $\boldsymbol{x}_{i,j}$. Given the solutions at the SPs, an element-wise degree k polynomial can be constructed using a Lagrange-type polynomial basis, i.e.,

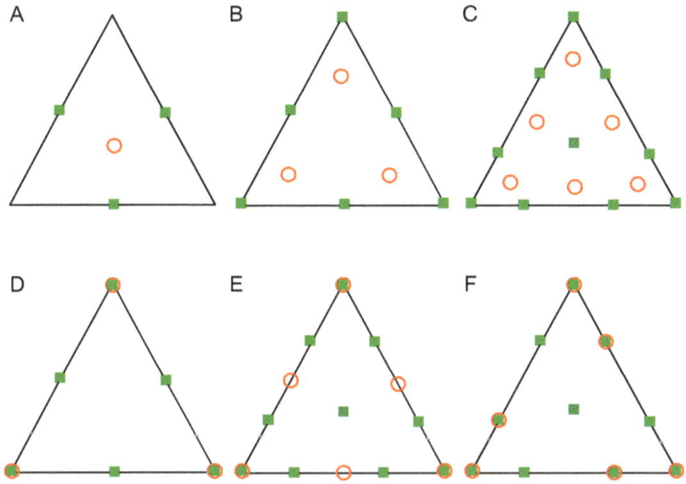

FIG. 3 Placement of solution (*circle*) and flux (*square*) points for a triangular element. (A) $k=0$ (first order), (B) $k=1$ (second order), (C) $k=2$ (third order), (D) second order, (E) third order, and (F) third order.

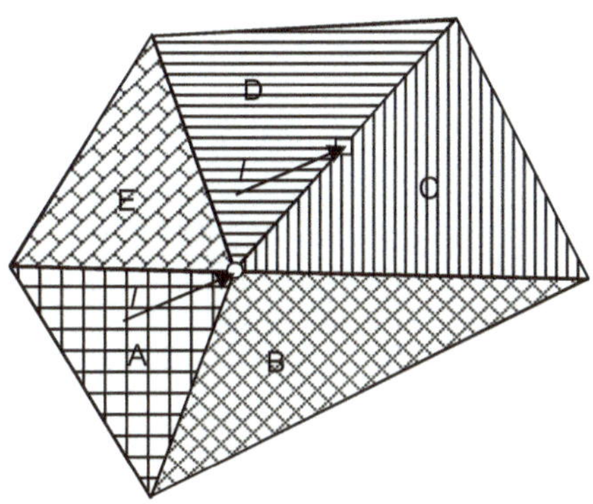

FIG. 4 Illustration of multidimensional Riemann problems at the corner and face point.

$$u_i(\boldsymbol{x}, t) = \sum_{j=1}^{N_{sp}} u_{i,j}(t) L_{i,j}^{SP}(\boldsymbol{x}), \tag{29}$$

where $L_{i,j}^{SP}(\boldsymbol{x})$ are the cardinal basis functions and N_{sp} is the number of basis functions required to support a degree k polynomial reconstruction. Obviously, the locations of the SPs uniquely determine the cardinal basis functions. With the polynomial distribution given in (29), the solutions at the FPs can be computed easily.

Since the solution is discontinuous across element boundaries, the fluxes at the element interfaces are not uniquely defined, for example, at the corner and face points shown in Fig. 4. At the corner point, five solutions exist from all the cells (A–E) sharing the point. At the face point between cells C and D, two solutions exist. The naïve approach is to compute an averaged solution from these multiple solutions and then determine the flux based on the averaged solution. However, it is well known that this naïve approach is equivalent to central differencing and is not stable. An alternative approach is to find the physical solution at the corner and face points at time $t=0^+$ with the discontinuous solution as the initial condition at $t=0$. This idea is of course due to Godunov (1959). For scalar conservation laws, this can be done based on the local wave direction $\boldsymbol{l}=\left(\frac{\partial f}{\partial u}, \frac{\partial g}{\partial u}\right)$, as shown in Fig. 4. Obviously, the solution computed from element D should be used to compute the face flux, while the solution computed from element A should be used to compute the corner flux.

Unfortunately, for systems of equations such as the Euler equations, this "multidimensional Riemann problem" shown in Fig. 4 is very difficult to solve, either analytically or numerically. Simpler approximate Riemann solvers must be found to determine these fluxes. We again turn to the FV method to look for inspiration. Obviously, in order to ensure conservation, the normal component of the flux vector on each face should be identical for the two cells sharing the face. To ensure conservation, a one-dimensional Riemann solver is employed in the face normal direction to compute the common normal flux. Consider the face FP shown in Fig. 5 and denote the outgoing normal from cell C to cell 1 $\boldsymbol{n}_1$. For this interface point, u_{L} is computed from cell C and u_{R} is computed from cell 1. Then the common normal component of the flux can be computed with any Riemann solver.

Since the tangential component of the flux does not affect the conservation property, we have complete freedom in determining it at the face point. In fact, it is not strictly necessary to have a unique tangential component

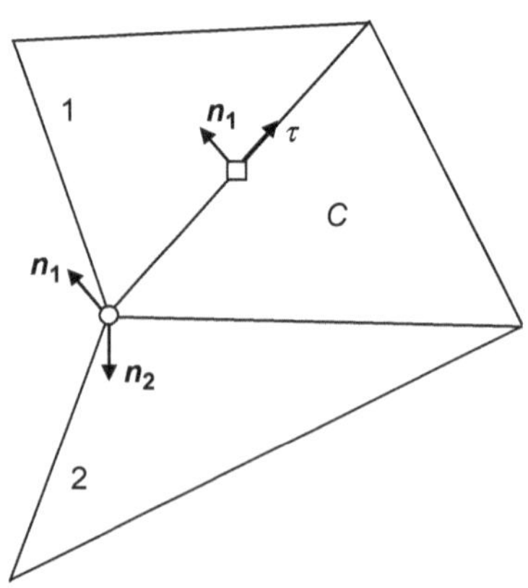

FIG. 5 Flux computation for a corner (○) and a face (□) point using one-dimensional Riemann solvers.

physically at a face point (e.g., think of a contact discontinuity in which density is discontinuous). Let the unit vector in the tangential direction be $\boldsymbol{\tau}$. Here we offer two possibilities. One is to use a unique tangential component by averaging the two tangential components from both sides of the face, i.e.,

$$f^{\tau} = \frac{1}{2}\{\boldsymbol{f}(u_{\mathrm{L}}) + \boldsymbol{f}(u_{\mathrm{R}})\} \cdot \boldsymbol{\tau}. \tag{30}$$

The other possibility is to use its own tangential component from the current cell, allowing the tangential component to be discontinuous. Therefore, the tangential component of the flux on either side is not modified. For the left cell, the tangential and normal components are $\left(\boldsymbol{f}(u_{\mathrm{L}}) \cdot \boldsymbol{\tau}, f^{n}_{\mathrm{Riem}}\right)$, and for the right cell, they become $\left(\boldsymbol{f}(u_{\mathrm{R}}) \cdot \boldsymbol{\tau}, f^{n}_{\mathrm{Riem}}\right)$.

For a corner FP in cell C, two faces (from cell C) share the corner point, as shown in Fig. 5. Let the unit normals of the two faces be $\mathbf{n}_1$ and $\mathbf{n}_2$. Once again, the normal components of flux f^n_1 and f^n_2 in the $\mathbf{n}_1$ and $\mathbf{n}_2$ directions are computed with a one-dimensional Riemann solver in the normal directions. The full flux vector can then be uniquely determined from the two normal flux components:

$$\boldsymbol{f} \cdot \boldsymbol{n}_1 = f^n_1 \tag{31}$$

$$\boldsymbol{f} \cdot \boldsymbol{n}_2 = f^n_2. \tag{32}$$

It is important to emphasize here that fluxes at cell corner points do not have unique values for all the cells sharing the corner. In spite of that, local conservation is guaranteed because neighbouring cells do share a common normal flux at all the FPs. Once the fluxes at all the FPs are recomputed, they are used to form a degree $k+1$ polynomial, i.e.,

$$\boldsymbol{f}_i(\boldsymbol{x}) = \sum_{k=1}^{N_{\mathrm{fp}}} L^{\mathrm{FP}}_{i,k}(\boldsymbol{x}) \boldsymbol{f}_{i,k}, \tag{33}$$

where $L^{\mathrm{FP}}_{i,k}(\boldsymbol{x})$ are the set of cardinal basis functions defined by the FPs, and $\boldsymbol{f}_{i,k}$ are the fluxes at the FPs. Obviously, the divergence of the flux at any point within the cell can be computed using

$$\nabla \cdot \boldsymbol{f}_i(\boldsymbol{x}) = \sum_{k=1}^{N_{\mathrm{fp}}} \nabla L^{\mathrm{FP}}_{i,k}(\boldsymbol{x}) \cdot \boldsymbol{f}_{i,k}. \tag{34}$$

Finally the semidiscrete scheme to update the solution unknowns can be written as

$$\frac{du_{i,j}}{dt} = -\nabla \cdot \boldsymbol{f}_i(\boldsymbol{x}_{i,j}) = -\sum_{k=1}^{N_{\mathrm{fp}}} \nabla L^{\mathrm{FP}}_{i,k}(\boldsymbol{x}_{i,j}) \cdot \boldsymbol{f}_{i,k}. \tag{35}$$

3.3 Efficiency and Stability

As discussed earlier, the SV and SD methods are identical in accuracy and stability and are similar in efficiency in 1D. On a 2D simplex, they become distinctly different methods with a significant difference in efficiency. Let us compare the $k=2$ SV and SD schemes in cost, which can be broken down into three main types:

- cost to compute the solutions at the Gauss quadrature points, or FPs;
- cost to compute the analytical and Riemann fluxes;
- cost to assemble the flux quadratures or flux divergence to form the residual.

In the $k=2$ SV partition shown in Fig. 2B, there are a total of 36 quadrature points, where the solutions need to be computed, and 18 analytical and 18 Riemann flux evaluations. In contrast, the $k=2$ SD scheme requires reconstructions for 10 FPs, 8 analytical, and 12 Riemann flux computations corresponding the FPs as shown in Fig. 3C, E, and F, respectively. Although the cost to form the residual from the fluxes is more tedious to estimate, we expect it to be smaller than the cost of reconstruction and flux computations. Therefore, it is obvious that the SV scheme is much more expensive than the SD scheme. Furthermore, the implementation of the SV schemes at higher orders becomes more complex as the CVs can be arbitrary polygons. This complexity was precisely the motivation for the development of the SD method.

On a 2D Cartesian element, the SD method is also much more efficient than the SV method, as shown in Fig. 6 for $k=2$ SV and SD schemes. For the SV scheme, there are two quadrature points for each CV face, resulting in a total of 48 quadrature points, 24 analytical, and 24 Riemann flux computations. For the SD scheme, there are a total of 24 FPs, with 12 analytical and 12 Riemann flux computations. For the same accuracy, the SV scheme is twice as expensive as the SD scheme. Because of this reason, we strongly

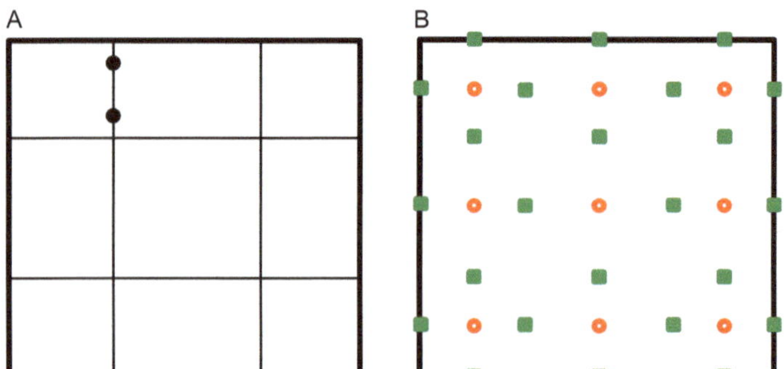

FIG. 6 Schematic of $k=2$ SV and SD schemes on a Cartesian element. (A) Partition for a $k=2$ SV scheme and (B) SPs and FPs for a $k=2$ SD scheme.

favour the SD method over the SV method for quadrilateral and hexahedral elements as the SD method is essentially one dimensional in each coordinate direction.

On a 2D simplex, stable partitions for SV schemes for k up to 4 have been discovered (Van den Abeele et al., 2007a,b; Harris and Wang, 2009a,b). However, an effort to find stable SD schemes for $k>1$ failed (Van den Abeele et al., 2008). Furthermore, a search for stable SV schemes for $k>1$ has not succeeded thus far for the 3D simplex (Van den Abeele et al., 2009).

A breakthrough in the SD method for the 2D simplex was achieved in Balan et al. (2012). Stable SD schemes for $k>1$ were discovered with the use of the Raviart–Thomas (RT) space to approximate the flux vector. The space RT_k is the smallest space (consisting of vector elements) having divergence in P_k (the space of degree k polynomials). Because of this property, it reduces the number of DOFs needed for flux interpolation, and hence the computational cost is reduced compared to that for the traditional SD method. In 2D, the number of DOFs to represent a vector-valued function in the RT_k space is given by

$$N_k^{\mathrm{RT}} = (k+1)(k+3). \tag{36}$$

The flux vector can be approximate using

$$\boldsymbol{f}_i(\boldsymbol{\xi}) = \sum_{k=1}^{N_k^{\mathrm{RT}}} f_k \boldsymbol{\psi}_k(\boldsymbol{\xi}), \tag{37}$$

where $\boldsymbol{\psi}_k(\boldsymbol{\xi})$ are vector interpolation functions which form a basis set in the RT space of degree k, and f_k are the (scalar) flux components, as shown in Fig. 7 for $k=2$. The arrows indicate the scalar components that serve as the DOFs for the RT_2 space. Then the divergence is computed as follows

$$\nabla \cdot \boldsymbol{f}_i(\boldsymbol{\xi}) = \sum_{k=1}^{N_k^{\mathrm{RT}}} f_k \nabla \cdot \boldsymbol{\psi}_k(\boldsymbol{\xi}). \tag{38}$$

Both Fourier analysis and numerical tests showed that this SD-RT method is stable and capable of achieving the expected order of accuracy.

4 NUMERICAL EXAMPLES

For the sake of space, we refer the reader to the references outlined in Section 1 for shock capturing, viscous flows, time integration, dynamic grids, and curved and 3D elements. In this section, we present several sample simulations to demonstrate the capability of the SV and SD methods.

4.1 Double Mach Reflection

This problem is a standard test case for high-resolution schemes and has been studied extensively by many researchers. The computational domain for this

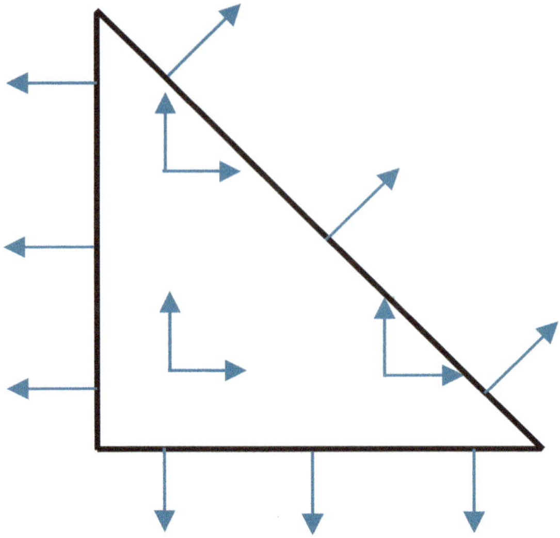

FIG. 7 The degrees of freedom in the RT_2 space.

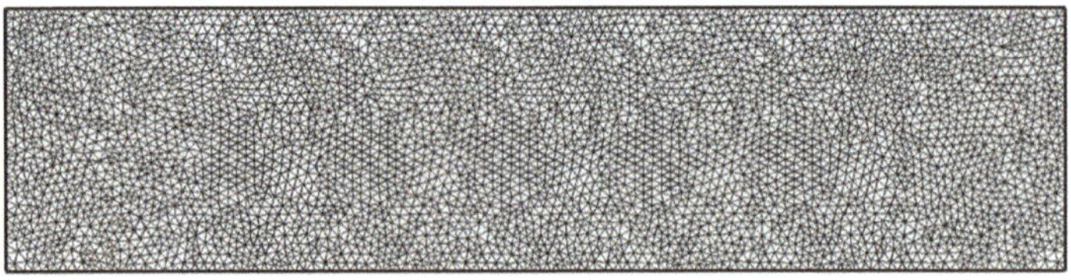

FIG. 8 A coarse mesh for the double Mach reflection problem with size 1/30.

problem is chosen to be $[0, 4] \times [0, 1]$. The reflecting wall lies at the bottom of the computational domain starting from $x = 1/6$. Initially a right-moving Mach 10 shock is positioned at $x = 1/6$, $y = 0$ and makes a 60-degree angle with the x-axis. For the bottom boundary, the exact postshock condition is imposed for the region from $x = 0$ to 1/6 and a solid wall boundary condition is used for the rest. For the top boundary of the computational domain, the solution is set to describe the exact motion of the Mach 10 shock. The left boundary is set as the exact postshock condition, while the right boundary is set as outflow boundary. Three unstructured triangular grids were generated with approximate mesh sizes of 1/30, 1/60, and 1/120. These meshes have 8200, 32,936, and 130,828 triangular cells, respectively, corresponding to 49,200, 197,616, and 784,968 DOFs per equation for a third-order ($k = 2$) SV scheme. The coarsest mesh is displayed in Fig. 8. All the simulations were carried until $t = 0.2$ using the Rusanov flux and a TVD limiter. Fig. 9 shows the density contours on the coarse, medium, and fine grids. The density contours computed with the third-order scheme on the medium mesh display finer

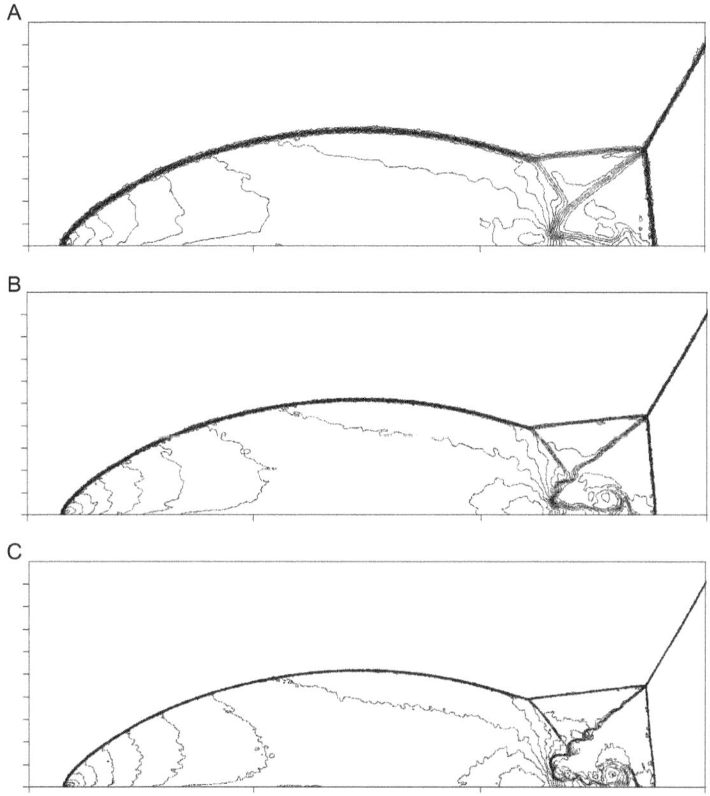

FIG. 9 Density contours computed using a third-order SV scheme with the Rusanov flux and TVD limiter, 30 evenly spaced contours between 1.25 and 21.5. (A) $h=1/30$ (49,200 DOFs), (B) $h=1/60$ (197,616 DOFs), and (C) $h=1/120$ (784,968 DOFs).

structures than those computed with a second-order SV scheme on the fine mesh, as shown in Fig. 10.

4.2 Rayleigh–Taylor Instability Problem With Solution-Based Grid Adaptation

The Rayleigh–Taylor instability (RTI) problem involves a cold fluid overlying a warm fluid. The results presented here are taken from Harris and Wang (2009a,b). Two inviscid fluids are initially taken to be in hydrostatic equilibrium in an isolated chamber. The chamber is 1 unit height and 0.25 units wide. The upper half of the chamber contains a fluid of density two, while the lower half of the chamber contains a fluid of unit density. The initial pressure field is chosen to ensure hydrostatic equilibrium, and an initial perturbation of the velocity field triggers the instability.

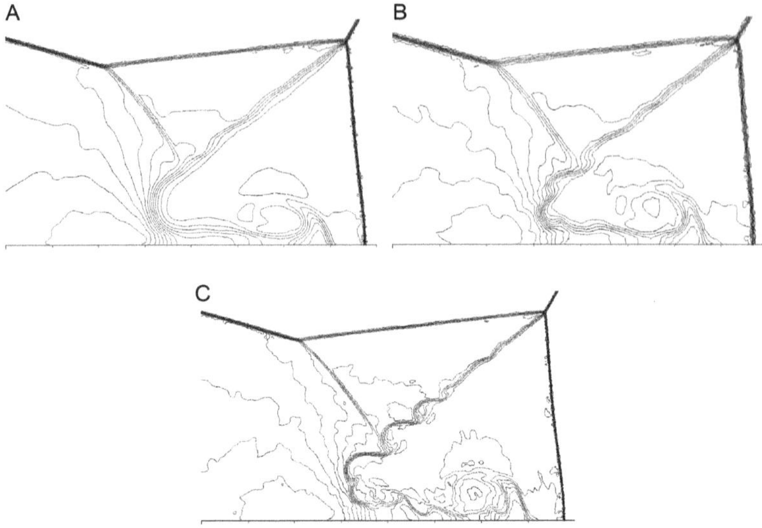

FIG. 10 Close-up view of the density contours near the double Mach stem. (A) Second-order SV scheme, $h=1/120$, (B) third-order SV scheme, $h=1/60$, and (C) third-order SV scheme, $h=1/120$.

TABLE 1 Initial Conditions for RTI Problem

Parameter	Upper Part	Lower Part
ρ	2	1
u	$\varepsilon_x \sin(8\pi x)\cos[\pi(y+1/2)]\sin^{\tau-1}[\pi(y+1/2)]$	Same as the upper part
v	$-\varepsilon_y \cos(8\pi x)\sin^{\tau}[\pi(y+1/2)]$	Same as the upper part
p	$2-2y$	$2-y$

The Geometric Center of the Chamber is taken to be the origin of the coordinate system.

The flow is governed by the Euler equations with the addition of source terms in the y-momentum and energy equations that correspond to unit gravity in the downward direction. The initial data are summarized in Table 1, where $M_0=0.1$, $\tau=6$, $\gamma=1.4$, $\varepsilon_y=M_0\sqrt{\gamma/2}$, and $\varepsilon_x=-\varepsilon_y\tau/16$.

A perturbation is selected which gives rise to a single mode instability, and inviscid wall boundary conditions are used for the chamber walls. While there are no shock waves in this problem, there is a contact discontinuity between the two fluids. Solution-based grid adaptation was applied in this case, and an edge-based criterion was utilized. A TVD limiter was employed.

It is immediately apparent that the behaviour of the RTI problem is heavily dependent on the grid used. Namely, if the initial grid is symmetric,

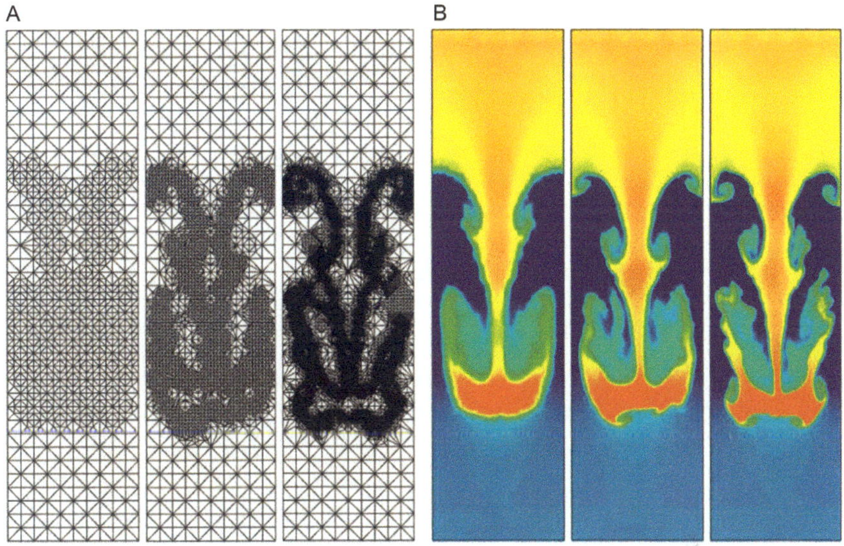

FIG. 11 Third-order results for RTI problem at time = 1.9 obtained using one to three levels of adaptation; (A) levels 1–3 adaptive symmetric grids (*left* to *right*, 1770, 4788, and 13,513 *triangles*) and (B) density contours for symmetric grids (*left* to *right*, 10,620, 28,728, and 81,078 DOFs).

the solution tends to stay symmetric (for the most part), otherwise the solution is completely asymmetric. In all cases, the typical mushroom-cap behaviour is observed, with increasingly complicated flow structure downstream as the number of adaptation levels is increased. Here only the results on the symmetric mesh are presented. Fig. 11 shows the adaptive grids and the computed density contours computed with a third-order SV scheme. Uniform grid refinements were also used for this case. It was shown that local adaptive h-refinement is far more effective than global refinement at resolving the flow features for this problem. In fact, a much more highly resolved solution is obtained using local h-refinement with far fewer DOFs than is necessary for a global refinement strategy to produce similar results.

4.3 Aerodynamic Performance of Flapping Wing

Yu et al. (2011) extended a 3D SD Navier–Stokes solver to handle dynamic meshes and used this solver to study bioinspired aerodynamics of moving bodies (Yu et al., 2012). Here we present a study of thrust generation by a rectangular wing with two different types of wing motions: flapping, and combined flapping and pitching motions. The flow fields for the flapping motion at four different phases, namely 0, 90, 180, and 270 degrees, are displayed in Fig. 12A, B, C, and D, respectively. Herein, the vortex structures are indicated by the Q-criterion coloured by the streamwise velocity. It is found that a large amount of elaborate vortex structures are generated

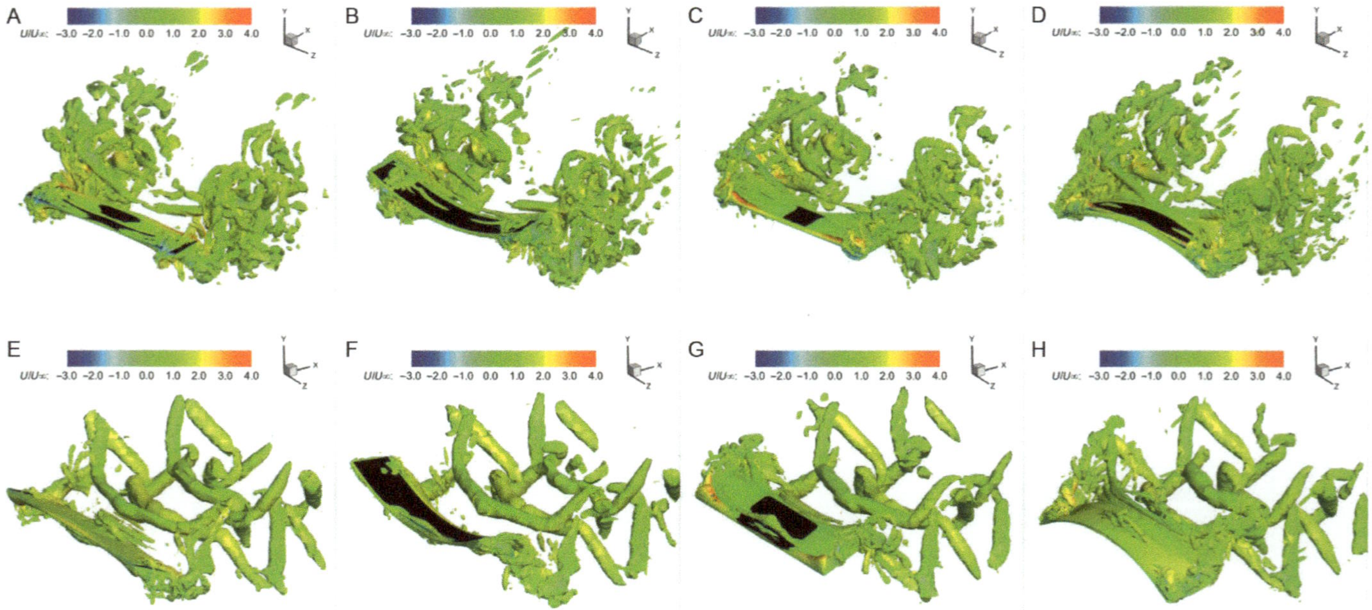

FIG. 12 Comparison of the vortex topologies for a rectangular wing at four phases (0, 90, 180, and 270 degrees) for a flapping motion (A–D) and for a combined flapping and pitching motion (E–H). (A) $\phi=0$ degrees, (B) $\phi=90$ degrees, (C) $\phi=180$ degrees, (D) $\phi=270$ degrees, (E) $\phi=0$ degrees, (F) $\phi=90$ degrees, (G) $\phi=180$ degrees, and (H) $\phi=270$ degrees.

around the flapping wings especially in the wingtip region. It can be inferred from this phenomenon that much flapping energy is wasted when a pure flapping motion is used because the generated small vortices are difficult to efficiently collect to generate thrust. The vortex structures for the combined motion are displayed in Fig. 12E–H at the same four phase angles. The flow field seems much less chaotic. It is clear that because of the effective angle of attack adjustment due to the pitching motion, the breakdown of vortices under the combined flapping and pitching motion is less severe. This indicates that less kinetic energy is dissipated under the combined motion than under the pure flapping motion. In fact, the thrust generated by the combined motion is 30 times larger than that generated by the pure flapping motion.

5 CONCLUSIONS

In this chapter, we reviewed the history and recent development of the SV and SD methods. Here is a list of the main results:

- In 1D, the SV and SD methods are identical if the CV boundaries in the SV method coincide with the FPs in the SD method. The SV and SD methods allow larger time steps than the DG method, but are less accurate.
- The use of Gauss quadrature points as the interior FPs guarantees stability for the SD method.
- The SD method is independent of the placement of the SPs.
- In 2D, the SD method is much more efficient than the SV method on both simplex and quadrilateral elements.
- In 2D, SD schemes are not stable on the simplex for $k>1$. The use of the RT space to approximate the flux vector results in stable SD schemes on the simplex.
- The search for stable SV partitions of the 3D simplex for $k>1$ has not been successful.

The spirit of the SD method has been inherited by the flux reconstruction/correction procedure via reconstruction framework. See the corresponding chapter in this volume for details.

ACKNOWLEDGEMENTS

Wang's research on high-order methods was/has been funded by AFOSR, NASA, DOE, U.S. Navy, NSF, DARPA, as well as Michigan State University, Iowa State University, and the University of Kansas. Azevedo's work has been sponsored by the Brazilian National Scientific Council (CNPq) and by the Sao Paulo Research Foundation (FAPESP).

REFERENCES

Balan, A., May, G., Schöberl, J., 2012. A stable high-order spectral difference method for hyperbolic conservation laws on triangular elements. J. Comput. Phys. 231, 2359–2375.

Bassi, F., Rebay, S., 1997. A high-order accurate discontinuous finite element method for the numerical solution of the compressible Navier-Stokes equations. J. Comput. Phys. 131, 267–279.

Breviglieri, C., Azevedo, J.L.F., Basso, E., Souza, M.A.F., 2010. Implicit high-order spectral finite volume method for inviscid compressible flows. AIAA J. 48 (10), 2365–2376.

Cagnone, J., Nadarajah, S.K., 2009. Implicit nonlinear frequency-domain spectral-difference scheme for periodic Euler flow. AIAA J. 47, 361–372.

Chen, Q.-Y., 2006a. Partitions for spectral (finite) volume reconstruction in the tetrahedron. J. Sci. Comput. 29 (3), 299–319.

Chen, Q.-Y., 2006b. Partitions of a simplex leading to accurate spectral (finite) volume reconstructions. SIAM J. Sci. Comput. 27 (4), 1458–1470.

Cheruvu, V., Nair, R.D., Tufo, H.M., 2007. A spectral finite volume transport scheme on the cubed-sphere. Appl. Numer. Math. 57 (9), 1021–1032.

Choi, J.J., 2015. Hybrid spectral difference/embedded finite volume method for conservation laws. J. Comput. Phys. 295, 285–306.

Choi, B.J., Iskandarani, M., Levin, J., Haidvogel, D.B., 2004. A spectral finite-volume method for the shallow water equations. Mon. Weather Rev. 132 (7), 1777–1791.

Cimorelli, L., Covelli, C., Cozzolino, L., Morte, R.D., Pianese, D., 2012. A derivative recovery spectral volume model for the analysis of constituents transport in one-dimensional flows. J. Math. Syst. Sci. 2, 334–340.

Cockburn, B., Shu, C.-W., 1989. TVB Runge-Kutta local projection discontinuous Galerkin finite element method for conservation laws II: general framework. Math. Comput. 52, 411–435.

Cockburn, B., Karniadakis, G.E., Shu, C.-W. (Eds.), 2000. Discontinuous Galerkin methods. Theory, computation and applications. In: Lecture Notes in Computational Science and Engineering, vol. 11. Springer-Verlag, Berlin.

Cozzolino, L., Della Morte, R., Del Giudice, G., 2012. A well-balanced spectral volume scheme with the wetting–drying property for the shallow-water equations. J. Hydroinform. 14 (3), 745–760. http://dx.doi.org/10.2166/hydro.2012.035.

Delanaye, M., Liu, Yen, 1999. Quadratic reconstruction finite volume schemes on 3D arbitrary unstructured polyhedral grids. AIAA Paper No. 99-3259-CP.

Fezoui, L., Stoufflet, B., 1989. A class of implicit upwind schemes for Euler simulations with unstructured meshes. J. Comput. Phys. 84 (1), 174–206.

Gao, J., Yang, Z., Li, X., 2012. An optimized spectral difference scheme for CAA problems. J. Comput. Phys. 231 (14), 4848–4866.

Godunov, S.K., 1959. A finite-difference method for the numerical computation of discontinuous solutions of the equations of fluid dynamics. Mat. Sb. 47, 271.

Gottlieb, S., Shu, C.-W., 1998. Total variation diminishing Runge–Kutta schemes. Math. Comput. 67, 73–85.

Haga, T., Ohnishi, N., Sawada, K., Masunaga, K., 2006. Spectral volume computation of flowfield in aerospace application using earth simulator. AIAA Paper No. 2006-2823.

Haga, T., Furndate, M., Sawada, K., 2009. RANS simulation using high-order spectral volume method on unstructured tetrahedral grids. AIAA paper 2009–404.

Harris, R., Wang, Z.J., 2009a. High-order adaptive quadrature-free spectral volume method on unstructured grids. Comput. Fluids 38, 2006–2025.

Harris, R., Wang, Z.J., 2009b. Partition design and optimization for high-order spectral volume schemes. AIAA-2009-1333.

Harris, R., Wang, Z.J., Liu, Y., 2008. Efficient quadrature-free high-order spectral volume method on unstructured grids: theory and 2D implementation. J. Comput. Phys. 227 (3), 1620–1642.

Harten, A., 1983. High resolution schemes for hyperbolic conservation laws. J. Comput. Phys. 49, 357–393.

Harten, A., Engquist, B., Osher, S., Chakravarthy, S., 1987. Uniformly high order essentially non-oscillatory schemes III. J. Comput. Phys. 71, 231.

Huang, P.G., Wang, Z.J., Liu, Y., 2005. An implicit space-time spectral difference method for discontinuity capturing using adaptive polynomials. AIAA-2005-5255.

Huynh, H.T., 2007. A flux reconstruction approach to high-order schemes including discontinuous Galerkin methods. AIAA Paper 2007-4079.

Jameson, A., 2010. A proof of the stability of the spectral difference method for all orders of accuracy. J. Sci. Comput. 45, 348–358.

Kannan, R., 2012. A high order spectral volume formulation for solving equations containing higher spatial derivative terms II: improving the third derivative spatial discretization using the LDG2 method. Commun. Comput. Phys. 12 (3), 767–788.

Kannan, R., Wang, Z.J., 2009. A study of viscous flux formulations for a p-multigrid spectral volume Navier Stokes solver. J. Sci. Comput. 41 (2), 165–199.

Kopriva, D.A., Kolias, J.H., 1996. A conservative staggered-grid Chebyshev multidomain method for compressible flows. J. Comput. Phys. 125, 244.

Lacor, C., Van den Abeele, K., 2011. Stability and accuracy analysis of space discretizations. In: Wang, Z.J. (Ed.), Adaptive High-Order Methods in Computational Fluid Dynamics. World Scientific, Singapore.

Lamouroux, R., Gressier, J., Grondin, G., 2016. A high-order compact limiter based on spatially weighted projections for the spectral volume and the spectral differences method. J. Sci. Comput. 67, 375–403. http://dx.doi.org/10.1007/s10915-015-0084-6.

Li, Y., Premasuthan, S., Jameson, A., 2010. Comparison of h-and p-adaptations for spectral difference methods. AIAA paper 2010–4435.

Liang, C., Kannan, R., Wang, Z.J., 2009a. A p-multigrid spectral difference method with explicit and implicit smoothers on unstructured triangular grids. Comput. Fluids 38, 254–265.

Liang, C., Jameson, A., Wang, Z.J., 2009b. Spectral difference method for two-dimensional compressible flow on unstructured grids with mixed elements. J. Comput. Phys. 228, 2847–2858.

Liang, C., Chan, A.S., Jameson, A., 2011a. A p-multigrid spectral difference method for incompressible Navier-Stokes equations. Comput. Fluids 51, 127–135.

Liang, C., Ou, K., Premasuthan, S., Jameson, A., Wang, Z.J., 2011b. High-order accurate simulations of unsteady flow past plunging and pitching airfoils. Comput. Fluids 40, 236–248.

Liu, Y., Vinokur, M., Wang, Z.J., 2003. Three-dimensional high-order spectral finite volume method for unstructured grids. AIAA-2003-3837.

Liu, Y., Vinokur, M., Wang, Z.J., 2004. Discontinuous spectral difference method for conservation laws on unstructured grids. In: Proceedings of the 3rd International Conference on Computational Fluid Dynamics, Toronto, Canada, July 12–16.

Liu, Y., Vinokur, M., Wang, Z.J., 2006a. Spectral (finite) volume method for conservation laws on unstructured grids V: extension to three-dimensional systems. J. Comput. Phys. 212, 454–472.

Liu, Y., Vinokur, M., Wang, Z.J., 2006b. Spectral difference method for unstructured grids I: basic formulation. J. Comput. Phys. 216, 780–801.

Lodato, G., Castonguay, P., Jameson, A., 2014. Structural wall-modeled LES using a high-order spectral difference scheme for unstructured meshes. Flow Turbul. Combust. 92, 579–606.

May, G., 2011. On the connection between the spectral difference method and the discontinuous Galerkin method. Commun. Comput. Phys. 9, 1071–1080. http://dx.doi.org/10.4208/cicp.090210.040610a.

May, G., Jameson, A., 2006. A spectral difference method for the Euler and Navier-Stokes equations. AIAA paper No. 2006-304.

Meister, A., Ortleb, S., Sonar, T., Wirz, M., 2012. A comparison of the discontinuous-Galerkin- and spectral-difference-method on triangulations using PKD polynomials. J. Comput. Phys. 231, 7722–7729.

Mohammad, A.H., Wang, Z.J., Liang, C., 2010. Large eddy simulation of turbulent flow past a cylinder using spectral difference method. Adv. Appl. Math. Mech. 2, 451–466.

Ou, K., Jameson, A., 2011. Towards computational flapping wing aerodynamics of realistic configurations using spectral difference method. AIAA paper 2011–3068.

Parsani, M., Ghorbaniasl, G., Lacor, C., Turkel, E., 2010a. An implicit high-order spectral difference approach for large eddy simulation. J. Comput. Phys. 229 (14), 5373–5393.

Parsani, M., Van den Abeele, K., Lacor, C., Turkel, E., 2010b. Implicit LU-SGS algorithm for high-order methods on unstructured grid with p-multigrid strategy for solving the steady Navier-Stokes equations. J. Comput. Phys. 229, 828–850.

Parsani, M., Ghorbaniasl, G., Lacor, C., 2011a. Analysis of the implicit LU-SGS algorithm for 3rd- and 4th-order spectral volume scheme for solving the steady Navier-Stokes equations. J. Comput. Phys. 230 (19), 7073–7085.

Parsani, M., Ghorbaniasl, G., Lacor, C., 2011b. Validation and application of an high-order spectral difference method for flow induced noise simulation. J. Comput. Acoust. 19, 241–268.

Parsani, M., Ketcheson, D.I., Deconinck, W., 2013. Optimized explicit Runge–Kutta schemes for the spectral difference method applied to wave propagation problems. SIAM J. Sci. Comput. 35 (2), A957–A986.

Premasuthan, S., Liang, C., Jameson, A., 2014. Computation of flows with shocks using the spectral difference method with artificial viscosity, I: basic formulation and application. Comput. Fluids 98, 111–121.

Raghavendra, R., 2011. A high order spectral volume method for equations containing third spatial derivative using an interior penalty formulation. CFD Lett. 3, 74–88.

Reed, W.H., Hill, T.R., 1973. Triangular mesh methods for the neutron transport equation: Los Alamos Scientific Laboratory Report, LA-UR-73-479.

Roe, P.L., 1981. Approximate Riemann solvers, parameter vectors, and difference schemes. J. Comput. Phys. 43, 357–372.

Sun, Y., Wang, Z.J., Liu, Y., 2006. Spectral (finite) volume method for conservation laws on unstructured grids VI: extension to viscous flow. J. Comput. Phys. 215 (1), 41–58.

Sun, Yuzhi, Wang, Z.J., Liu, Yen, 2007. High-order multidomain spectral difference method for the Navier-Stokes equations on unstructured hexahedral grids. Commun. Comput. Phys. 2, 310–333.

Sun, Y., Wang, Z.J., Liu, Y., 2009. Efficient implicit non-linear LU-SGS approach for compressible flow computation using high-order spectral difference method. Commun. Comput. Phys. 5 (2–4), 760–778.

Van den Abeele, K., Lacor, C., 2007. An accuracy and stability study of the 2D spectral volume method. J. Comput. Phys. 226 (1), 1007–1026.

Van den Abeele, K., Broeckhoven, T., Lacor, C., 2007a. Dispersion and dissipation properties of the 1D spectral volume method and application to a p-multigrid algorithm. J. Comput. Phys. 224 (2), 616–636.

Van den Abeele, K., Lacor, C., Wang, Z.J., 2007b. On the connection between the spectral volume and the spectral difference method. J. Comput. Phys. 227 (2), 877–885.

Van den Abeele, K., Lacor, C., Wang, Z.J., 2008. On the stability and accuracy of the spectral difference method. J. Sci. Comput. 37 (2), 162–188.

Van den Abeele, K., Ghorbaniasl, G., Parsani, M., Lacor, C., 2009. A stability analysis for the spectral volume method on tetrahedral grids. J. Comput. Phys. 228 (2), 257–265.

Van Leer, B., 1979. Towards the ultimate conservative difference scheme. V. A second order sequel to Godunov's method. J. Comput. Phys. 32, 101–136.

Wang, Z.J., 2002. Spectral (finite) volume method for conservation laws on unstructured grids: basic formulation. J. Comput. Phys. 178, 210.

Wang, Z.J., Liu, Y., 2002. Spectral (finite) volume method for conservation laws on unstructured grids II: extension to two-dimensional scalar equation. J. Comput. Phys. 179, 665–697.

Wang, Z.J., Liu, Y., 2004. Spectral (finite) volume method for conservation laws on unstructured grids III: extension to one-dimensional systems. J. Sci. Comput. 20, 137.

Wang, Z.J., Liu, Y., 2005. The spectral difference method for the 2D Euler equations on unstructured grids. AIAA Paper No. 2005-5112.

Wang, Z.J., Liu, Y., 2006. Extension of the spectral volume method to high-order boundary representation. J. Comput. Phys. 211, 154–178.

Wang, Z.J., Zhang, L., Liu, Y., 2004. Spectral (finite) volume method for conservation laws on unstructured grids IV: extension to two-dimensional Euler equations. J. Comput. Phys. 194, 716.

Wang, Z.J., Liu, Y., May, G., Jameson, A., 2007. Spectral difference method for unstructured grids II: extension to the Euler equations. J. Sci. Comput. 32 (1), 45–71.

Wang, Z.J., Fidkowski, K.J., Abgrall, R., Bassi, F., Caraeni, D., Cary, A., Deconinck, H., Hartmann, R., Hillewaert, K., Huynh, H.T., Kroll, N., May, G., Persson, P.-O., van Leer, B., Visbal, M., 2013. High-order CFD methods: current status and perspective. Int. J. Numer. Methods Fluids 72, 811–845.

Wang, J., Liang, C., Miesch, M.S., 2015. A compressible high-order unstructured spectral difference code for stratified convection in rotating spherical shells. J. Comput. Phys. 290, 90–111.

Xie, L., Xu, M., Zhang, B., Qiu, Z., 2015. A new spectral difference method using hierarchical polynomial bases for hyperbolic conservation laws. J. Comput. Phys. 284, 431–461.

Xu, Z., Liu, Y., Shu, C.-W., 2009. Hierarchical reconstruction for spectral volume method on unstructured grids. J. Comput. Phys. 228 (16), 5787–5802.

Yang, M., Wang, Z.J., 2009. A parameter-free generalized moment limiter for high-order methods in unstructured grids. Adv. Appl. Math. Mech. 1 (4), 451–480.

Yu, M., Wang, Z.J., Hu, H., 2011. A high-order spectral difference method for unstructured dynamic grids. Comput. Fluids 48, 84–97.

Yu, M., Wang, Z.J., Hu, H., 2012. High-fidelity flapping-wing aerodynamics simulations with a dynamic unstructured grid based spectral difference method. In: Proceedings of the 7th International Conference on Computational Fluid Dynamics, ICCFD7-4104.

Zhang, B., Liang, C., 2015. A simple, efficient, and high-order accurate curved sliding-mesh interface approach to spectral difference method on coupled rotating and stationary domains. J. Comput. Phys. 295, 147–160.
Zhang, M., Shu, C.-W., 2005. An analysis and a comparison between the discontinuous Galerkin method and the spectral finite volume methods. Comput. Fluids 34 (4–5), 581–592.
Zhou, Y., Wang, Z.J., 2010. Absorbing boundary conditions for the Euler and Navier–Stokes equations with the spectral difference method. J. Comput. Phys. 229, 8733–8749.

Chapter 10

High-Order Flux Reconstruction Schemes

F.D. Witherden*, P.E. Vincent† and A. Jameson*
*Stanford University, Stanford, CA, United States
†Imperial College London, South Kensington, London, United Kingdom

Chapter Outline

ABSTRACT
There is an increasing desire among industrial practitioners of computational fluid dynamics to undertake high-fidelity scale-resolving simulations of unsteady flows within the vicinity of complex geometries. Such simulations require numerical methods that can operate on unstructured meshes with low numerical dissipation. The flux reconstruction (FR) approach describes one such family of numerical methods, which includes a particular type of collocation-based nodal discontinuous Galerkin method, and spectral difference methods, as special cases. In this chapter we describe the current state-of-the-art surrounding research into FR methods. To begin, FR is described in one dimension for both advection and advection–diffusion problems. This is followed by a description of its extension to multidimensional tensor product and simplex elements. Stability and accuracy issues are then discussed, including an overview of energy-stability proofs, von Neumann analysis results, and stability characteristics when the flux function of the governing system is nonlinear. Finally, implementation aspects

Handbook of Numerical Analysis, Vol. 17. http://dx.doi.org/10.1016/bs.hna.2016.09.010

are outlined in the context of modern hardware platforms, and three example applications of FR are presented, demonstrating the potential utility of FR schemes for scale resolving simulation of unsteady flow problems.

Keywords: High-order methods, Discontinuous Galerkin method, Flux reconstruction

AMS Classification Codes: 65M60, 65M70

1 INTRODUCTION

There is an increasing desire among industrial practitioners of computational fluid dynamics (CFD) to undertake high-fidelity scale-resolving simulations of transient compressible flows within the vicinity of complex geometries. For example, to improve the design of next generation aircraft, there exists a need to perform simulations—at Reynolds numbers 10^4–10^7 and Mach numbers $M \sim 0.1$–1.0—of highly separated flow over deployed spoilers/airbrakes; separated flow within serpentine intake ducts and flow over entire vehicle configurations at off-design conditions. In order to perform these simulations it is necessary to solve the compressible Navier–Stokes equations. These take the form of a nonlinear conservation law.

When solving the Navier–Stokes equations numerically it is customary to independently discretise space and time. Although there exist a variety of spatial discretisations the three most popular are (Vincent and Jameson, 2011) the finite difference (FD) method in which the governing system is discretised onto a structured grid of points, the finite volume (FV) method in which the computational domain is decomposed into cells and an integral form of the governing system is solved within each cell, and the finite element (FE) method where the computational domain is decomposed into elements inside of which sits a polynomial that is required to satisfy a variational form of the governing system. All of these methods have been used successfully to solve fluid flow problems throughout both industry and academia.

An important consideration when choosing a discretisation is the order of accuracy. This dictates how the error in the solution will respond to a change in the resolution of the grid. Implementations of the above methods are usually first- or second-order accurate in space. A consequence of this is a large degree of numerical dissipation. Such schemes therefore encounter significant difficulties when attempting to simulate fundamentally unsteady phenomena (Jameson and Ou, 2011). This has led to an interest in high-order methods, the promise of which is increased accuracy with a decreased computational cost. One such example of a high-order spatial discretisation are classical spectral methods (Canuto et al., 2006). These methods involve decomposing the solution into modes (in frequency space), which are defined globally within the domain of interest. Spectral methods, however, typically lack the geometrical flexibility since it is often impossible to define continuous global modes within a complex geometry.

Alternatively, it is also possible to construct higher order formulations of the aforementioned FD, FV, and FE schemes. The order of accuracy of an FD scheme can be readily increased by simply expanding the size of the stencil. For FV methods the procedure is somewhat more involved. The most popular high-order FV type schemes are the essentially nonoscillatory (ENO) of Harten et al. (1987) and the weighted ENO (WENO) schemes of Liu et al. (1994). These schemes use an adaptive stencil through an unstructured grid in order to achieve a high-order reconstruction. The adaptive nature of the stencil allows both ENO and WENO schemes to automatically achieve high-order accuracy in the vicinity of shocks and other discontinuities. High-order FE schemes can be constructed by increasing the degree of the polynomial inside of each element. Such schemes are normally termed continuous Galerkin (CG) methods with elements being coupled by requiring that the approximate solution to be piecewise continuous between elements. Further details can be found in the books by Karniadakis and Sherwin (2005) and Solin et al. (2003). A popular alternative to CG is the discontinuous Galerkin (DG) finite element method, first introduced by Reed and Hill in 1973 to solve the neutron transport equation. In DG the solution is not required to be continuous between elements, instead coupling is achieved through the calculation of common fluxes at interfaces. This is similar to the coupling that occurs between cells in FV schemes.

Beyond CG and DG another more recent class of high-order schemes for unstructured grids are spectral difference (SD) methods. Originally proposed under the moniker 'staggered-grid Chebyshev multidomain methods' by Kopriva and Kolias in 1996 their use in CFD was popularised by Sun et al. (2007). In 2007 Huynh proposed the flux reconstruction (FR) approach; a unifying framework that encompasses both under integrated collocation-based nodal DG schemes and, at least for a linear flux function, any SD scheme. Following on from this, in 2009 Gao and Wang introduced a closely related set of methods which they referred to as lifting collocation penalty (LCP) schemes (Gao and Wang, 2009; Wang and Gao, 2009). Subsequently, in 2013 Yu and Wang showed in one dimension (1D) that the LCP schemes are identical to the FR approach. As such several, authors have adopted the name corrections procedure via reconstruction (CPR) to refer to both FR and LCP. Furthermore, Allaneau and Jameson (2011) have showed that it is possible to cast some FR schemes as a filtered nodal DG scheme.

The remainder of this chapter is organised as follows. In Section 2 we present the FR approach in 1D for advection and advection–diffusion type hyperbolic conservation laws. Section 3 extends this methodology to quadrilaterals through a tensor product construction and simplex elements. The stability and accuracy of FR schemes are discussed in Section 4. It is shown how through the use of an energy method a family of one-parameter schemes can be obtained. Details surrounding the implementation of FR on modern hardware platforms—including GPUs—are presented in Section 5. In Section 6 we show how FR can be used to solve the compressible Navier–Stokes equations in three dimensions and present results for a variety of benchmark flow problems. Finally, in Section 7, conclusions are drawn.

2 FR IN 1D

2.1 Advection Problems

Consider using the FR approach to solve

$$\frac{\partial u}{\partial t}+\frac{\partial f}{\partial x}=0 \tag{1}$$

within an arbitrary 1D domain $\mathbf{\Omega}$, where x is a spatial coordinate, t is time, $u = u(x, t)$ is a conserved scalar quantity, and $f = f(u)$ is the flux of u in the x direction.

The first stage of the FR approach involves partitioning $\mathbf{\Omega}$ into N distinct elements, each denoted $\mathbf{\Omega}_n = \{x|x_n < x < x_{n+1}\}$, such that

$$\mathbf{\Omega}=\bigcup_{n=0}^{N-1}\mathbf{\Omega}_n, \qquad \bigcap_{n=0}^{N-1}\mathbf{\Omega}_n=\varnothing. \tag{2}$$

The solution u in Eq. (1) is approximated in each $\mathbf{\Omega}_n$ by $u_n^{\delta D}=u_n^{\delta D}(x,t)$, which is a polynomial of degree p within $\mathbf{\Omega}_n$, and the flux f in Eq. (1) is approximated in each $\mathbf{\Omega}_n$ by $f_n^{\delta}=f_n^{\delta}(x,t)$, which is a polynomial of degree $p + 1$ within $\mathbf{\Omega}_n$. Consequently, a total approximate solution $u^{\delta D} = u^{\delta D}(x, t)$ and a total approximate flux $f^{\delta} = f^{\delta}(x, t)$ can be defined within $\mathbf{\Omega}$ as

$$u^{\delta D}=\sum_{n=0}^{N-1}u_n^{\delta D}\approx u, \quad f^{\delta}=\sum_{n=0}^{N-1}f_n^{\delta}\approx f, \tag{3}$$

where no level of inter-element continuity in $u^{\delta D}$ is explicitly enforced. However, f^{δ} is required to be C_0 continuous at element interfaces.

The second stage of the FR approach involves transforming each $\mathbf{\Omega}_n$ to a standard element $\mathbf{\Omega}_S = \{\xi|-1 \le \xi \le 1\}$ via the mapping

$$\xi=\Gamma_n(x)=2\left(\frac{x-x_n}{x_{n+1}-x_n}\right)-1, \tag{4}$$

which has the inverse

$$x=\Gamma_n^{-1}(\xi)=\left(\frac{1-\xi}{2}\right)x_n+\left(\frac{1+\xi}{2}\right)x_{n+1}. \tag{5}$$

Having performed such a transformation, the evolution of u_n^{δ} within any individual $\mathbf{\Omega}_n$ (and thus the evolution of u^{δ} within $\mathbf{\Omega}$) can be determined by solving the following transformed equation within the standard element $\mathbf{\Omega}_S$

$$\frac{\partial \hat{u}^{\delta D}}{\partial t} + \frac{\partial \hat{f}^{\delta}}{\partial \xi} = 0, \tag{6}$$

where

$$\hat{u}^{\delta D} = \hat{u}^{\delta D}(\xi,t) = J_n u_n^{\delta D}(\Gamma_n^{-1}(\xi),t) \tag{7}$$

is a polynomial of degree p,

$$\hat{f}^{\delta} = \hat{f}^{\delta}(\xi,t) = f_n^{\delta}(\Gamma_n^{-1}(\xi),t), \tag{8}$$

is a polynomial of degree $p + 1$, and $J_n = (x_{n+1} - x_n)/2$.

The third stage of the FR approach involves defining the degree p polynomial $\hat{u}^{\delta D}$ in terms of a nodal basis as follows

$$\hat{u}^{\delta D} = \sum_{i=0}^{p} \hat{u}_i^{\delta D} \, l_i, \tag{9}$$

where l_i are Lagrange polynomials defined as

$$l_i = \prod_{j=0, j \neq i}^{p} \left(\frac{\xi - \xi_j}{\xi_i - \xi_j} \right), \tag{10}$$

ξ_i ($i = 0$ to p) are $p + 1$ distinct solution points within $\mathbf{\Omega}_S$, and $\hat{u}_i^{\delta D} = \hat{u}_i^{\delta D}(t)$ ($i = 0$ to p) are values of $\hat{u}^{\delta D}$ at the solution points ξ_i.

The fourth stage of the FR approach involves constructing a degree p polynomial $\hat{f}^{\delta D} = \hat{f}^{\delta D}(\xi,t)$, defined as the approximate transformed discontinuous flux within $\mathbf{\Omega}_S$. Specifically, $\hat{f}^{\delta D}$ is obtained via a collocation projection at the $p + 1$ solution points, and can hence be expressed as

$$\hat{f}^{\delta D} = \sum_{i=0}^{p} \hat{f}_i^{\delta D} \, l_i, \tag{11}$$

where the coefficients $\hat{f}_i^{\delta D} = \hat{f}_i^{\delta D}(t)$ are simply values of the transformed flux at each solution point ξ_i (evaluated directly from the approximate solution). The flux $\hat{f}^{\delta D}$ is termed discontinuous since it is calculated directly from the approximate solution, which is in general discontinuous between elements.

The fifth stage of the FR approach involves evaluating $\hat{u}^{\delta D}$ at either end of the standard element $\mathbf{\Omega}_S$ (i.e. at $\xi = \pm 1$). These values, in conjunction with analogous information from adjoining elements, are then used to calculate numerical interface fluxes. The exact methodology for calculating such numerical interface fluxes will depend on the nature of the equations being solved. For example, when solving the Euler equations one may use a Roe-type approximate Riemann solver (Roe, 1981), or any other two-point flux formula that provides for an upwind bias. In what follows the numerical interface fluxes associated with the left- and right-hand ends of $\mathbf{\Omega}_S$ (and transformed appropriately for use in $\mathbf{\Omega}_S$) will be denoted $\hat{f}_L^{\delta I}$ and $\hat{f}_R^{\delta I}$, respectively.

The penultimate stage of the FR approach involves constructing the degree $p+1$ polynomial $\hat{f}^{\delta}$, by adding a correction flux $\hat{f}^{\delta C} = \hat{f}^{\delta C}(\xi,t)$ of degree $p+1$ to $\hat{f}^{\delta D}$, such that their sum equals the transformed numerical interface flux at $\xi = \pm 1$, yet in some sense follows $\hat{f}^{\delta D}$ within the interior of $\mathbf{\Omega}_S$. In order to define $\hat{f}^{\delta C}$ such that it satisfies the above requirements, consider first defining degree $p+1$ correction functions $g_L = g_L(\xi)$ and $g_R = g_R(\xi)$ that approximate zero (in some sense) within $\mathbf{\Omega}_S$, as well as satisfying

$$g_L(-1) = 1, \quad g_L(1) = 0, \tag{12}$$

$$g_R(-1) = 0, \quad g_R(1) = 1, \tag{13}$$

and

$$g_L(\xi) = g_R(-\xi). \tag{14}$$

The exact form of g_L and g_R can be varied, determining various stability and accuracy properties; in this sense the FR approach can be considered a family of schemes. In particular, it has been shown that if g_L and g_R are the right and left Radau polynomials, respectively, then an under-integrated collocated-based nodal DG scheme is recovered (De Grazia et al., 2014; Huynh, 2007; Mengaldo et al., 2016), and if g_L and g_R are set to zero at a set of p points within $\mathbf{\Omega}_S$, located symmetrically about the origin, then SD schemes are recovered for a linear flux. A suitable expression for $\hat{f}^{\delta C}$ can now be written in terms of g_L and g_R as

$$\hat{f}^{\delta C} = (\hat{f}_L^{\delta I} - \hat{f}_L^{\delta D})g_L + (\hat{f}_R^{\delta I} - \hat{f}_R^{\delta D})g_R, \tag{15}$$

where $\hat{f}_L^{\delta D} = \hat{f}^{\delta D}(-1,t)$ and $\hat{f}_R^{\delta D} = \hat{f}^{\delta D}(1,t)$. Using this expression, the degree $p+1$ approximate transformed total flux $\hat{f}^{\delta}$ within $\mathbf{\Omega}_S$ can be constructed from the discontinuous and correction fluxes as follows

$$\hat{f}^{\delta} = \hat{f}^{\delta D} + \hat{f}^{\delta C} = \hat{f}^{\delta D} + (\hat{f}_L^{\delta I} - \hat{f}_L^{\delta D})g_L + (\hat{f}_R^{\delta I} - \hat{f}_R^{\delta D})g_R. \tag{16}$$

The final stage of the FR approach involves evaluating the divergence of $\hat{f}^{\delta}$ at each solution point ξ_i using the expression

$$\frac{\partial \hat{f}^{\delta}}{\partial \xi}(\xi_i) = \sum_{j=0}^{k} \hat{f}_j^{\delta D} \frac{\mathrm{d}l_j}{\mathrm{d}\xi}(\xi_i) + (\hat{f}_L^{\delta I} - \hat{f}_L^{\delta D})\frac{\mathrm{d}g_L}{\mathrm{d}\xi}(\xi_i) + (\hat{f}_R^{\delta I} - \hat{f}_R^{\delta D})\frac{\mathrm{d}g_R}{\mathrm{d}\xi}(\xi_i). \tag{17}$$

These values can then be used to advance $\hat{u}^{\delta D}$ in time via a suitable temporal discretisation of the following semi-discrete expression

$$\frac{\mathrm{d}\hat{u}_i^{\delta D}}{\mathrm{d}t} = -\frac{\partial \hat{f}^{\delta}}{\partial \xi}(\xi_i). \tag{18}$$

2.2 Advection Diffusion

Consider using the FR approach to solve Eq. (1) within an arbitrary 1D domain $\mathbf{\Omega}$, where x is a spatial coordinate, t is time, $u = u(x, t)$ is a conserved scalar quantity, and $f = f(u, q)$ is the flux of u in the x direction, where

$$q = \frac{\partial u}{\partial x}. \tag{19}$$

The first stage of the FR approach involves partitioning $\mathbf{\Omega}$ into N distinct elements as per Eq. (2), and constructing approximations of the solution u in Eq. (1) and the flux f in Eq. (1) as per Eq. (3). Additionally, the so-called auxiliary variable q in Eq. (19) is approximated in each $\mathbf{\Omega}_n$ by $q_n^{\delta D} = q_n^{\delta D}(x,t)$, which is a polynomial of degree p within $\mathbf{\Omega}_n$, and the solution u in Eq. (19) is approximated in each $\mathbf{\Omega}_n$ by $u_n^{\delta} = u_n^{\delta}(x,t)$, which is a polynomial of degree $p + 1$ within $\mathbf{\Omega}_n$. Consequently, a total approximate auxiliary variable $q^{\delta D} = q^{\delta D}(x, t)$ and a total approximate solution $u^{\delta} = u^{\delta}(x, t)$ can be defined within $\mathbf{\Omega}$ as

$$q^{\delta D} = \sum_{n=0}^{N-1} q_n^{\delta D} \approx q, \quad u^{\delta} = \sum_{n=0}^{N-1} u_n^{\delta} \approx u, \tag{20}$$

where no level of inter-element continuity in $q^{\delta D}$ is explicitly enforced. However, u^{δ} is required to be C_0 continuous at element interfaces.

The second stage of the FR approach involves transforming each $\mathbf{\Omega}_n$ to a standard element $\mathbf{\Omega}_S$ using the mapping defined by Eq. (4). This leads to Eqs. (21), (7), (8) as well as

$$\hat{q}^{\delta D} = \frac{\partial \hat{u}^{\delta}}{\partial x}, \tag{21}$$

where

$$\hat{q}^{\delta D} = \hat{q}^{\delta D}(\xi,t) = J_n^2 q_n^{\delta D}(\Gamma_n^{-1}(\xi),t) \tag{22}$$

is a polynomial of degree p, and

$$\hat{u}^{\delta} = \hat{u}^{\delta}(\xi,t) = J_n u_n^{\delta}(\Gamma_n^{-1}(\xi),t), \tag{23}$$

is a polynomial of degree $p + 1$.

The third stage of the FR approach involves defining the degree p polynomial $\hat{u}^{\delta D}$ in terms of a nodal basis as per Eq. (9).

The forth stage of the FR approach involves calculating $\hat{u}^{\delta D}$ at the left- and right-hand ends of $\mathbf{\Omega}_S$, denoted $\hat{u}_L^D$ and $\hat{u}_R^D$, respectively. These are subsequently used in conjunction with analogous information from adjoining elements to calculate common solution values at the left- and right-hand ends of $\mathbf{\Omega}_S$, denoted $\hat{u}_L^I$ and $\hat{u}_R^I$, respectively. There are a number of approaches for determining common solution values, including the central flux (CF)

(Hesthaven and Warburton, 2008), local discontinuous Galerkin (LDG) (Cockburn and Shu, 1998), compact discontinuous Galerkin (CDG) (Peraire and Persson, 2008), internal penalty (IP) (Arnold, 1982), and the BR2 (Bassi and Rebay, 1997) approaches.

The fifth stage of the FR approach involves constructing the degree $p + 1$ polynomial $\hat{u}^{\delta}$, by adding a correction flux $\hat{u}^{\delta C} = \hat{u}^{\delta C}(\xi, t)$ of degree $p + 1$ to $\hat{u}^{\delta D}$, such that their sum equals the transformed common solutions at $\xi = \pm 1$, yet in some sense follows $\hat{u}^{\delta D}$ within the interior of $\mathbf{\Omega}_S$. A suitable expression for $\hat{f}^{\delta C}$ can be written in terms of h_L and h_R, which have the same properties as g_L and g_R, respectively, as

$$\hat{u}^{\delta C} = (\hat{u}_L^{\delta I} - \hat{u}_L^{\delta D}) g_L + (\hat{u}_R^{\delta I} - \hat{u}_R^{\delta D}) g_R. \tag{24}$$

Using this expression the degree $p + 1$ approximate transformed total solution $\hat{u}^{\delta}$ within $\mathbf{\Omega}_S$ can be constructed from the discontinuous and correction solutions as follows

$$\hat{u}^{\delta} = \hat{u}^{\delta D} + \hat{u}^{\delta C} = \hat{u}^{\delta D} + (\hat{u}_L^{\delta I} - \hat{u}_L^{\delta D}) g_L + (\hat{u}_R^{\delta I} - \hat{u}_R^{\delta D}) g_R. \tag{25}$$

The sixth stage of the FR approach involves constructing the degree p polynomial $\hat{q}^{\delta D}$. Using Eqs. (19), (25), (9) one can obtain

$$\hat{q}^{\delta D} = \sum_{i=0}^{p} \hat{u}_i^{\delta D} \frac{\mathrm{d}l_i}{\mathrm{d}\xi} + (\hat{u}_L^{\delta I} - \hat{u}_L^{\delta D}) \frac{\mathrm{d}g_L}{\mathrm{d}\xi} + (\hat{u}_R^{\delta I} - \hat{u}_R^{\delta D}) \frac{\mathrm{d}g_R}{\mathrm{d}\xi}. \tag{26}$$

The seventh stage of the FR approach involves constructing a degree p polynomial $\hat{f}^{\delta D} = \hat{f}^{\delta D}(\xi, t)$, defined as the approximate transformed discontinuous flux within $\mathbf{\Omega}_S$ as per Eq. (11), where $\hat{f}_i^{\delta D} = \hat{f}_i^{\delta D}(t)$ at each solution point ξ_i now depend on both $\hat{u}^{\delta D}$ and $\hat{q}^{\delta D}$.

The eighth stage of the FR approach involves evaluating $\hat{u}^{\delta D}$ and $\hat{q}^{\delta D}$ at either end of the standard element $\mathbf{\Omega}_S$ (i.e. at $\xi = \pm 1$). These values, in conjunction with analogous information from adjoining elements, are then used to calculate numerical interface fluxes. The exact methodology for calculating such numerical interface fluxes will depend on the nature of the equations being solved. For example, when solving the Navier–Stokes equations one may use a Roe type approximate Riemann solver (Roe, 1981) for the inviscid component, and one of the aforementioned CF, LDG, CDG, IP, or BR2 approaches for the viscous component. In what follows the numerical interface fluxes associated with the left- and right-hand ends of $\mathbf{\Omega}_S$ (and transformed appropriately for use in $\mathbf{\Omega}_S$) will be denoted $\hat{f}_L^{\delta I}$ and $\hat{f}_R^{\delta I}$, respectively.

The penultimate stage of the FR approach involves constructing the degree $p + 1$ polynomial $\hat{f}^{\delta}$ as per Eq. (16), and the final stage of the FR approach involves evaluating the divergence of $\hat{f}^{\delta}$ at each solution point ξ_i as per Eq. (17). These values can then be used to advance $\hat{u}^{\delta D}$ in time via a suitable temporal discretisation of Eq. (18).

3 FR IN MULTIDIMENSIONS

3.1 Overview

In this section we will describe extension of the FR approach into two dimensions (2D) for both quadrilateral and triangular elements. The approaches presented further generalise into three dimensions and have been used successfully to obtain stable schemes inside of both hexahedra and tetrahedra. All discussions will be cast in terms of solving the following 2D scalar conservation law

$$\frac{\partial u}{\partial t} + \nabla_{xy} \cdot \mathbf{f} = 0 \tag{27}$$

within an arbitrary domain $\boldsymbol{\Omega}$, where x and y are spatial coordinates, t is time, $u = u(x, t)$ is a conserved scalar, $\mathbf{f} = (f, g)$ where $f = f(u)$ and $g - g(u)$ are the fluxes of u in the x and y directions, respectively.

3.2 Tensor Product Elements

Consider partitioning the domain $\boldsymbol{\Omega}$ into N nonoverlapping, conforming quadrilateral elements $\boldsymbol{\Omega}_n$. To facilitate the implementation, each quadrilateral element $\boldsymbol{\Omega}_n$ in the physical domain (x, y) is mapped to a reference element $\boldsymbol{\Omega}_S = (\xi, \eta)| - 1 \leq \xi, \eta \leq 1$ in the transformed space (ξ, η).

Using this mapping, the governing equation in the physical domain can be transformed to an equivalent governing equation in the computational domain which takes the form

$$\frac{\partial \hat{u}}{\partial t} + \nabla_{\xi\eta} \cdot \hat{\mathbf{f}} = 0 \tag{28}$$

where

$$\hat{u} = Ju, \tag{29}$$

$$\hat{\mathbf{f}} = (\hat{f}, \hat{g}) = \left(\frac{\partial y}{\partial \eta} f - \frac{\partial x}{\partial \eta} g, -\frac{\partial y}{\partial \xi} f + \frac{\partial x}{\partial \xi} g \right), \tag{30}$$

$$J = \frac{\partial x}{\partial \xi}\frac{\partial y}{\partial \eta} - \frac{\partial x}{\partial \eta}\frac{\partial y}{\partial \xi}. \tag{31}$$

The approximate solution $\hat{u}^{\delta D}$ within the reference element $\boldsymbol{\Omega}_S$ is represented by a multidimensional polynomial of degree p, defined by its values at a set of $N_p = (p+1)^2$ solution points. These solution points are generated by taking the tensor product of a set of 1D solution points. The approximate solution in the reference element takes the form

$$\hat{u}^{\delta D}(\mathbf{r}, t) = \sum_{i=1}^{p+1} \sum_{j=1}^{p+1} \hat{u}_{i,j}^{\delta D} l_i(\xi) l_j(\eta), \tag{32}$$

where $\{l_i\}$ are the set of 1D Lagrange polynomials associated with the set of solution points and $\hat{u}^{\delta D}_{i,j}$ is the value of the solution at the solution point located at (ξ_i, η_j). Furthermore, a set of $p + 1$ flux points are located along each edge of the quadrilateral for a total of $4(p + 1)$ such points.

Following the 1D FR approach, the total transformed approximate flux $\hat{\mathbf{f}}^\delta = (\hat{f}^\delta, \hat{g}^\delta)$ is written as the sum of a discontinuous component $\hat{\mathbf{f}}^{\delta D}$ and a correction component $\hat{\mathbf{f}}^{\delta C}$,

$$\hat{\mathbf{f}}^\delta = \hat{\mathbf{f}}^{\delta D} + \hat{\mathbf{f}}^{\delta C}. \tag{33}$$

The approximate discontinuous flux $\hat{\mathbf{f}}^{\delta D} = (\hat{f}^{\delta D}, \hat{g}^{\delta D})$ is computed by constructing a degree p polynomial for each of its components as

$$\hat{\mathbf{f}}^{\delta D} = \sum_{i=1}^{p+1} \sum_{j=1}^{p+1} \hat{\mathbf{f}}^{\delta D}_{i,j} l_i(\xi) l_j(\eta), \tag{34}$$

where $\hat{\mathbf{f}}^{\delta D}_{i,j}$ is simply the value of the transformed flux evaluated directly at the solution point (ξ_i, η_j).

The divergence of the discontinuous flux is therefore

$$\nabla_{\xi\eta} \cdot \hat{\mathbf{f}}^{\delta D} = \sum_{i=1}^{p+1} \sum_{j=1}^{p+1} \hat{f}^{\delta D}_{i,j} l'_i(\xi) l_j(\eta) + \sum_{i=1}^{p+1} \sum_{j=1}^{p+1} \hat{g}^{\delta D}_{i,j} l_i(\xi) l'_j(\eta), \tag{35}$$

where l'_i is the first derivative of l_i. The divergence of the transformed correction flux $\nabla_{\xi\eta} \cdot \hat{\mathbf{f}}^{\delta C} = \frac{\partial \hat{f}^{\delta C}}{\partial \eta} + \frac{\partial \hat{g}^{\delta C}}{\partial \xi}$ at solution point (ξ_i, η_i) is computed following the 1D methodology using

$$\frac{\partial \hat{f}^{\delta C}}{\partial \xi}(\xi_i, \eta_i) = \left((\hat{\mathbf{f}} \cdot \hat{n})^{\delta I}_L - \hat{f}^{\delta D}_L \right) \frac{\mathrm{d}h_L}{\mathrm{d}\xi}(\xi_i) + \left((\hat{\mathbf{f}} \cdot \hat{n})^{\delta I}_R - \hat{f}^{\delta D}_R \right) \frac{\mathrm{d}h_R}{\mathrm{d}\xi}(\xi_i), \tag{36}$$

$$\frac{\partial \hat{g}^{\delta C}}{\partial \eta}(\xi_i, \eta_i) = \left((\hat{\mathbf{f}} \cdot \hat{n})^{\delta I}_B - \hat{g}^{\delta D}_B \right) \frac{\mathrm{d}h_B}{\mathrm{d}\eta}(\eta_j) + \left((\hat{\mathbf{f}} \cdot \hat{n})^{\delta I}_T - \hat{g}^{\delta D}_T \right) \frac{\mathrm{d}h_T}{\mathrm{d}\eta}(\eta_j), \tag{37}$$

where $\hat{f}^{\delta D}_L$, $\hat{f}^{\delta D}_R$, $\hat{g}^{\delta D}_L$, and $\hat{g}^{\delta D}_T$ are the transformed discontinuous flux values at the flux points L, R, B, and T located at $(\xi, \eta) = (-1, \eta_j)$, $(1, \eta_j)$, $(\xi_i, -1)$, $(\xi_i, 1)$, respectively. The quantities $(\hat{\mathbf{f}} \cdot \hat{n})^{\delta I}_L$, $(\hat{\mathbf{f}} \cdot \hat{n})^{\delta I}_R$, $(\hat{\mathbf{f}} \cdot \hat{n})^{\delta I}_B$, and $(\hat{\mathbf{f}} \cdot \hat{n})^{\delta I}_T$ are the transformed common interface flux values at the flux points L, R, B, and T, respectively.

3.3 Simplex Elements

Consider partitioning the domain $\boldsymbol{\Omega}$ into N nonoverlapping, conforming triangular elements $\boldsymbol{\Omega}_n$. To facilitate the implementation, each element $\boldsymbol{\Omega}_n$ in

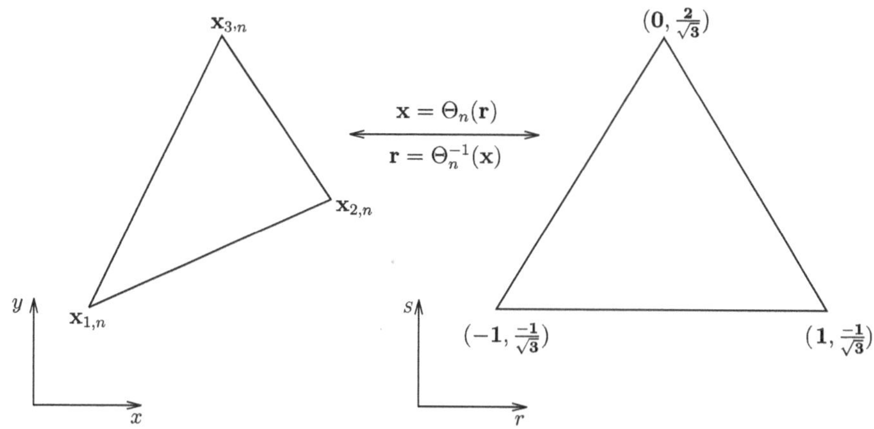

FIG. 1 Mapping between the physical space (x, y) and the computational space (r, s). *From the Ph.D. thesis of P. Castonguay and copyright P. Castonguay, reused with permission.*

physical space is mapped to a reference equilateral triangle $\mathbf{\Omega}_S$, using a mapping $\mathbf{\Theta}_n$, as shown in Fig. 1.

For a linear triangular element, the mapping is

$$\mathbf{x} = \mathbf{\Theta}_n(\mathbf{r}) = \frac{(-3r+2-\sqrt{3}s)}{6}\mathbf{x}_1 + \frac{(2+3r-\sqrt{3}s)}{6}\mathbf{x}_2 + \frac{(2+2\sqrt{3}s)}{6}\mathbf{x}_3 \quad (38)$$

where $\mathbf{x}_1$, $\mathbf{x}_2$, and $\mathbf{x}_3$ are the coordinates of the vertices of the triangular element $\mathbf{\Omega}_n$ in physical space. Using this mapping, Eq. (27) in the physical domain can be transformed to an equivalent governing equation in the computational domain which takes the form

$$\frac{\partial \hat{u}}{\partial t} + \nabla_{rs} \cdot \hat{\mathbf{f}} = 0 \quad (39)$$

where

$$\hat{u} = Ju, \quad (40)$$

$$\hat{\mathbf{f}} = (\hat{f}, \hat{g}) = \left(\frac{\partial y}{\partial s}f - \frac{\partial x}{\partial s}g, -\frac{\partial y}{\partial r}f + \frac{\partial x}{\partial r}g\right), \quad (41)$$

$$J = \frac{\partial x}{\partial r}\frac{\partial y}{\partial s} - \frac{\partial x}{\partial s}\frac{\partial y}{\partial r}. \quad (42)$$

The approximate solution $\hat{u}^{\delta D}$ within the reference element $\mathbf{\Omega}_S$ is represented by a multidimensional polynomial of degree p, defined by its values at a set of $N_p = (p + 1)(p + 2)/2$ solution points (represented by hollow circles in Fig. 2).

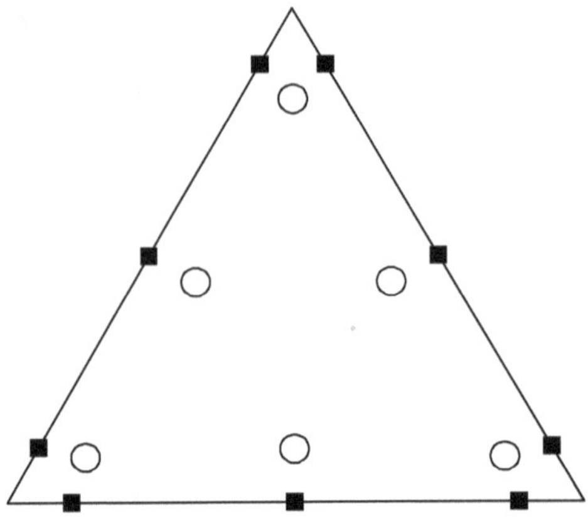

FIG. 2 Solution points (*circles*) and flux points (*squares*) in the reference element for $p = 2$. *From the Ph.D. thesis of P. Castonguay and copyright P. Castonguay, reused with permission.*

The approximate solution in the reference element takes the form

$$\hat{u}^{\delta D}(\mathbf{r},t) = \sum_{i=1}^{Np} \hat{u}_i^{\delta D} l_i(\mathbf{r}) \tag{43}$$

where $\hat{u}_i^{\delta D} = J_n \cdot u^{\delta D}(\Theta_n^{-1}(\mathbf{r}_i),t)$ is the value of $\hat{u}$ at the solution point i of element n, and $l_i(\mathbf{r})$ is the multidimensional Lagrange polynomial associated with the solution point i in the reference equilateral triangle $\boldsymbol{\Omega}_S$. The approximate solution $\hat{u}^{\delta D}$ lies in the space $P_p(\boldsymbol{\Omega}_S)$, defined as the space of polynomials of degree $\leq p$ on $\boldsymbol{\Omega}_S$. Following the 1D FR approach, the total transformed approximate flux $\hat{\mathbf{f}}^{\delta} = (\hat{f}^{\delta}, \hat{g}^{\delta})$ is written as the sum of a discontinuous component $\hat{\mathbf{f}}^{\delta D}$ and a correction component $\hat{\mathbf{f}}^{\delta C}$,

$$\hat{\mathbf{f}}^{\delta} = \hat{\mathbf{f}}^{\delta D} + \hat{\mathbf{f}}^{\delta C}. \tag{44}$$

The approximate discontinuous flux $\hat{\mathbf{f}}^{\delta D} = (\hat{f}^{\delta D}, \hat{g}^{\delta D})$ is computed by constructing a degree p polynomial for each of its components as

$$\hat{f}^{\delta D} = \sum_{i=1}^{N_p} \hat{f}_i^{\delta D} l_i, \quad \hat{g}^{\delta D} = \sum_{i=1}^{N_p} \hat{g}_i^{\delta D} l_i \tag{45}$$

where the coefficients $\hat{f}_i^{\delta D}$ and $\hat{g}_i^{\delta D}$ are the values of the transformed flux at the solution point i evaluated directly from the approximate solution $\hat{u}_i(\hat{f}_i^{\delta D} = \hat{f}(\hat{u}_i)$ and $\hat{g}_i^{\delta D} = \hat{g}(\hat{u}_i))$. The divergence of the discontinuous flux is therefore

$$\nabla_{rs} \cdot \hat{\mathbf{f}}^{\delta D} = \sum_{i=1}^{N_p} \hat{f}_i^{\delta D} \frac{\partial l_i}{\partial r} + \sum_{i=1}^{N_p} \hat{g}_i^{\delta D} \frac{\partial l_i}{\partial s}. \tag{46}$$

On each edge of the element, a set of $N_{fp} = (p + 1)$ flux points are defined (illustrated by squares in Fig. 2) and used to couple the solution between neighbouring elements. The correction flux $\hat{\mathbf{f}}^{\delta C}$ is constructed as follows

$$\hat{\mathbf{f}}^{\delta C}(\mathbf{r}) = \sum_{f=1}^{3}\sum_{j=1}^{N_{fp}}\left[(\hat{\mathbf{f}}\cdot\hat{\mathbf{n}})^{\delta I}_{f,j} - (\hat{\mathbf{f}}^{\delta D}\cdot\hat{\mathbf{n}})_{f,j}\right]\mathbf{h}_{f,j}(\mathbf{r}) \tag{47}$$

$$= \sum_{f=1}^{3}\sum_{j=1}^{N_{fp}}\Delta_{f,j}\mathbf{h}_{f,j}(\mathbf{r}). \tag{48}$$

Regarding Eqs. (47) and (48), various points should be noted. First, expressions subscripted by the indices f, j correspond to a quantity at the flux point j of face f, where $1 \leq f \leq 3$ and $1 \leq j \leq N_{fp}$. The convention used to number the faces and flux points is illustrated in Fig. 3.

For example, $(\hat{\mathbf{f}}^{\delta D}\cdot\hat{\mathbf{n}})_{f,j}$ is the normal component of the discontinuous transformed flux $\hat{\mathbf{f}}^{\delta D}$ at the flux point f, j. In Eq. (47), $(\hat{\mathbf{f}}\cdot\hat{\mathbf{n}})^{\delta I}_{f,j}$ is a normal transformed numerical flux computed at flux point f, j. As in 1D FR, it is computed by first evaluating pairs of $u^{\delta D}$ at each flux point using Eq. (43). At each flux point, we define $u^{\delta D}_{-}$ to be the value of $u^{\delta D}$ computed using the information local to the current element and $u^{\delta D}_{+}$ to be the value of $u^{\delta D}$ computed using information from the neighbouring element sharing the same flux point. Once both approximate solution values ($u^{\delta D}_{-}$ and $u^{\delta D}_{+}$) are evaluated at each flux point, a system specific solver is used to compute a common interface flux based on $u^{\delta D}_{+}$, $u^{\delta D}_{-}$ and the local normal vector $\mathbf{n}_{f,j}$. From this common numerical flux, the transformed normal numerical flux denoted by $(\hat{\mathbf{f}}\cdot\hat{\mathbf{n}})^{\delta I}_{f,j}$ can be obtained. In Eq. (48), $\Delta_{f,j}$ is defined as the difference between the normal transformed numerical flux and the normal transformed discontinuous

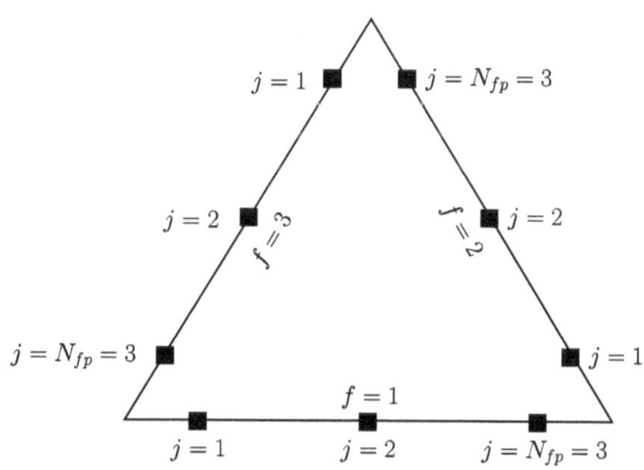

FIG. 3 Numbering convention for the faces and flux points on the reference triangular element. Example shown corresponds to $p = 2$. *From the Ph.D. thesis of P. Castonguay and copyright P. Castonguay, reused with permission.*

flux at the flux point f, j. Finally, $\mathbf{h}_{f,j}(\mathbf{r})$ is a vector correction function associated with flux point f, j. Each vector correction function $\mathbf{h}_{f,j}(\mathbf{r})$ is restricted to lie in the Raviart–Thomas space (Raviart and Thomas, 1977) of order p, denoted by $RT_p(\mathbf{\Omega}_S)$. Because of this property,

$$\begin{aligned} \nabla_{rs} \cdot \mathbf{h}_{f,j} &\in P_p(\mathbf{\Omega}_S) \\ \mathbf{h}_{f,j} \cdot \mathbf{n}|_{\mathbf{\Gamma}_S} &\in R_p(\mathbf{\Gamma}_S) \end{aligned} \tag{49}$$

where $P_p(\mathbf{\Omega}_S)$ is the space of polynomials of degree $\leq p$ on the reference element $\mathbf{\Omega}_S$ and $R_p(\mathbf{\Gamma}_S)$ is the space of polynomials of degree $\leq p$ on the boundary of $\mathbf{\Gamma}_S$. In other words, the divergence of each correction function ($\nabla_{rs} \cdot \mathbf{h}_{f,j}$) is a polynomial of degree p and the normal trace $\mathbf{h}_{f,j} \cdot \hat{\mathbf{n}}$ is also a polynomial of degree p along each edge. Furthermore, it is required that the correction functions $\mathbf{h}_{f,j}$ satisfy

$$\mathbf{h}_{f,j}(\mathbf{r}_{f_2,j_2}) \cdot \mathbf{n}_{f_2,j_2} = \begin{cases} 1 & \text{if } f = f_2 \text{ and } j = j_2 \\ 0 & \text{otherwise} \end{cases} \tag{50}$$

Hence, the following holds at each flux point f, j

$$\hat{\mathbf{f}}^{\delta C}(\mathbf{r}_{f,j}) \cdot \hat{\mathbf{n}}_{f,j} = \left[(\hat{\mathbf{f}} \cdot \hat{\mathbf{n}})^{\delta I}_{f,j} - (\hat{\mathbf{f}}^{\delta D} \cdot \hat{\mathbf{n}})_{f,j} \right] = \Delta_{f,j}. \tag{51}$$

An example of a vector correction function $\mathbf{h}_{f,j}$ is shown in Fig. 4 for the case $p = 2$. To simplify the notation in the following sections, the correction field $\phi_{f,j}(\mathbf{r})$ defined as the divergence of the correction function $\mathbf{h}_{f,j}(\mathbf{r})$ is introduced, i.e.

$$\phi_{f,j}(\mathbf{r}) = \nabla_{rs} \cdot \mathbf{h}_{f,j}(\mathbf{r}). \tag{52}$$

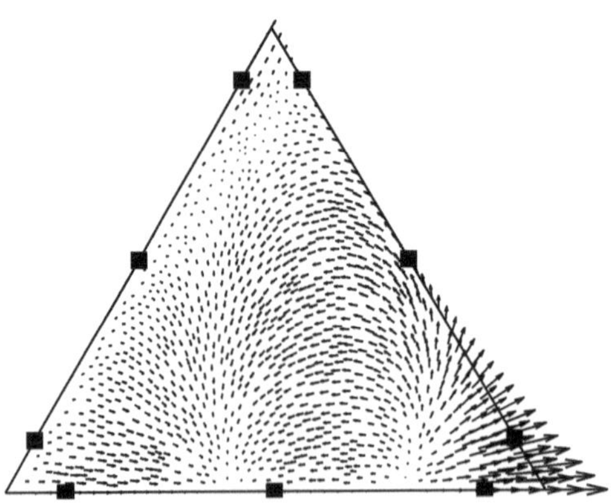

FIG. 4 A correction function associated with flux point $f = 2, j = 1$ for $p = 2$. *From the Ph.D. thesis of P. Castonguay and copyright P. Castonguay, reused with permission.*

Finally, combining Eqs. (39), (44), (46), and (48), the approximate solution values at the solution points can be updated from

$$
\begin{aligned}
\frac{d\hat{u}_i^\delta}{dt} &= -\left(\nabla_{rs}\cdot\hat{\mathbf{f}}^\delta\right)\big|_{\mathbf{r}_i} \\
&= -\left(\nabla_{rs}\cdot\hat{\mathbf{f}}^{\delta D}\right)\big|_{\mathbf{r}_i} - \left(\nabla_{rs}\cdot\hat{\mathbf{f}}^{\delta C}\right)\big|_{\mathbf{r}_i} \\
&= -\sum_{k=1}^{N_p}\hat{f}_k^{\delta D}\frac{\partial l_k}{\partial r}\big|_{\mathbf{r}_i} - \sum_{k=1}^{N_p}\hat{g}_k^{\delta D}\frac{\partial l_k}{\partial s}\big|_{\mathbf{r}_i} - \sum_{f=1}^{3}\sum_{j=1}^{N_{fp}}\Delta_{f,j}\phi_{f,j}(\mathbf{r}_i).
\end{aligned}
\tag{53}
$$

On triangular elements, the nature of a particular FR scheme depends on the location of the solution points, the location of the flux points, the methodology for calculating the transformed numerical interface flux $(\hat{\mathbf{f}}\cdot\hat{\mathbf{n}})_{f,j}^{\delta I}$, and finally, the form of the divergence $\phi_{f,\ j}$ of the correction functions $\mathbf{h}_{f,\ j}$. It has been shown in Castonguay et al. (2012) that a collocation-based (under integrated) nodal DG scheme is recovered on triangular elements if every correction function $\mathbf{h}_{f,\ j}$ is chosen such that

$$\int_{\mathbf{\Omega}_S}\mathbf{h}_{f,j}\varphi d\mathbf{\Omega}_S = 0 \tag{54}$$

for any polynomial $\varphi \in P_{p-1}(\mathbf{\Omega}_S)$, i.e., the space of polynomials of degree $p-1$. For further details, see Castonguay et al. (2012).

4 STABILITY AND ACCURACY OF FR SCHEMES

4.1 Energy Stability

Jameson (2010) proved that a particular SD scheme (recovered via FR) is energy-stable for linear advection problems in 1D. Vincent et al. (2011b) subsequently extended this result, and identified a family of stable FR schemes for linear advection problems in 1D for all orders of accuracy. Specifically, it was proven that if the left and right flux correction functions are defined as

$$g_L = \frac{(-1)^p}{2}\left[L_p - \left(\frac{\eta_p L_{p-1} + L_{p+1}}{1+\eta_p}\right)\right], \tag{55}$$

$$g_R = \frac{1}{2}\left[L_p + \left(\frac{\eta_p L_{p-1} + L_{p+1}}{1+\eta_p}\right)\right]. \tag{56}$$

where p is the degree of the solution polynomial within each element, L_{p-1}, L_p, and L_{p+1} are Legendre polynomials of the denoted degree (normalised such that $|L_p(\pm 1)| = 1$ for all p), and

$$\eta_p = \frac{c(2p+1)(a_p p!)^2}{2}, \tag{57}$$

with

$$a_p = \frac{(2p)!}{2^p (p!)^2}, \tag{58}$$

and c a free parameter in the range

$$\frac{-2}{(2p+1)(a_p p!)^2} < c < \infty, \tag{59}$$

then a broken Sobolev type norm of the discrete solution $||u^{\delta D}||_{p,2}$ with the from

$$||u^{\delta}||_{p,2} = \left[\sum_{n=0}^{N-1} \int_{x_n}^{x_{n+1}} (u_n^{\delta D})^2 + \frac{c}{2}(J_n)^{2p} \left(\frac{\partial^p u_n^{\delta D}}{\partial x^p} \right)^2 \mathrm{d}x \right]^{1/2} \tag{60}$$

is guaranteed to be nonincreasing, and thus bounded. Consequently, by equivalence of norms in the finite-dimensional solution space, any norm of the solution is guaranteed to remain bounded, and thus the method is guaranteed to be stable. The resulting one parameter family of FR schemes, defined in terms of the free parameter c, have been referred to as Energy Stable Flux Reconstruction (ESFR) schemes. It can be noted that judicious choice of the parameter c leads to recovery of various known FR schemes. Specifically, if $c = 0$, then a particular nodal DG scheme is recovered, if

$$c = \frac{2p}{(2p+1)(p+1)(a_p p!)^2} \tag{61}$$

then a particular SD scheme is recovered (the scheme is, in fact, the particular SD scheme that (Huynh, 2007) showed to be Fourier stable, and Jameson (2010) proved to be energy stable), and if

$$c = \frac{2(p+1)}{(2p+1)p(a_p p!)^2} \tag{62}$$

then the g_2 FR scheme (Huynh, 2007) is recovered.

ESFR schemes for linear advection problems in 1D have been extended by Castonguay et al. (2012) to solve linear advection problems on 2D triangular grids. As in the 1D case, a one-parameter family of correction functions were identified that guarantee a particular norm of the solution is nonincreasing. However, unlike in the 1D case, an explicit expression for these correction functions was not presented (instead the divergence of each correction function was defined implicitly via a matrix system). Interestingly, the one-parameter family of schemes did not appear to include an SD scheme as a special case, despite the fact that Balan et al. (2011) were able to identify stable SD schemes on triangular grids for several orders or accuracy. The 1D ESFR schemes have also been extended by Sheshadri and Jameson (2016) to solve linear advection problems on tensor-product quadrilateral elements.

Williams et al. (2013), Castonguay et al. (2013), and Williams and Jameson (2013) have extended ESFR schemes for linear advection problems to develop a range of ESFR schemes for linear advection–diffusion problems on multiple element types. Their approach involves use of ESFR correction functions to construct a continuous polynomial representation of the solution (in addition to a continuous representation of the flux) within each element. Development of an energy-stable treatment for diffusive terms is important, since it is a prerequisite for effective solution of the Navier–Stokes equations.

4.2 von Neumann Analysis

Energy-based stability proofs are powerful since they apply for all orders of accuracy and on nonuniform grids. However, they do not offer insight into all the stability properties of a numerical scheme. Huynh (2007) and Vincent et al. (2011a) presented comprehensive von Neumann analyses of FR methods in order to elucidate further stability properties of the schemes. Their results indicate that the form of the flux correction function has a significant impact on the CFL stability limit associated with a given FR scheme. In the context of 1D ESFR schemes for linear advection, it has been shown that increasing the free parameter c (from zero) can increase the CFL limit by over a factor of two in certain cases. However, this is at the cost of a reduction in the overall accuracy of the scheme.

It has been observed that FR schemes can achieve the expected $p + 1$ order of accuracy for a variety of problems, much like nodal DG schemes (Hesthaven and Warburton, 2008). Additionally, von Neumann analysis has shown that FR schemes can in theory achieve so-called super-accuracy for dispersion and dissipation errors in the asymptotic 'well-resolved' limit, up to order $2p + 1$ (Huynh, 2007; Vincent et al., 2011a). Such super-accuracy has been demonstrated for the solution on 2D structured grids (Vincent et al., 2011a). However, the extent to which it extends to unstructured meshes is yet to be determined. Moreover, it has also been found by von Neumann analysis that steady state solutions are limited to accuracy of order $p + 1$ (Asthana et al., 2016).

More recently Li and Wang (2013) have used von Neumann analysis to develop FR methods optimised for aeroacoustics, and Asthana and Jameson (2015) have used von Neumann analysis to derive a set of spectrally optimal FR schemes, designed to minimise wave propagation error across all resolvable wavenumbers. These schemes were found to yield an improvement over collocation-based nodal DG.

4.3 Nonlinear Stability

Jameson et al. (2011) and Mengaldo et al. (2015) showed that FR methods can be afflicted by an aliasing driven instability if the flux function is nonlinear. Such instabilities are a consequence of aliasing errors (that occur when a polynomial representation of the nonlinear flux is constructed via a collocation projection at the solution points). Jameson et al. (2011) also demonstrated that the location of the solution points plays a critical role in determining the extent of any aliasing driven instabilities. Specifically, they suggest that the solution points should be located at the abscissa of a strong quadrature rule in order to minimise aliasing driven instabilities. This finding is supported by the numerical experiments of Castonguay et al. (2011), who used the FR approach to solve the Euler equations on 2D triangular grids. They found that if the solution points were located at the so-called alpha-optimised points of Hesthaven and Warburton (2008) then the simulations blew up. However, if the solution points were located at the abscissa of a high-strength quadrature rule derived by Taylor et al. (2005), then the simulations remained stable. More recent works by Witherden and Vincent (2014) and Witherden et al. (2016) have further validated this hypothesis. However, it is unclear if, for a given element type, there exists an optimal set of solution points for a range of flow problems.

Other efforts to improve nonlinear stability have included development of limiting strategies for dealing with shocks (Park and Kim, 2016), as well as derivation of summation-by-parts operators for FR (Ranocha et al., 2016); an approach which has previously yielded successful results in a DG context (Gassner, 2013).

5 IMPLEMENTATION

5.1 Overview

In addition to offering high-order accuracy on unstructured mixed grids, FR schemes are also compact in space, and thus when combined with explicit time marching offer a significant degree of element locality. As such, explicit high-order FR schemes are characterised by a large degree of structured computation, even on unstructured grids. Over the past two decades improvements in the arithmetic capabilities of processors have significantly outpaced advances in random access memory. Algorithms which have traditionally been compute bound—such as dense matrix-vector products—are now limited instead by the bandwidth to/from memory. Whereas the CPUss of two decades ago had FLOPS-per-byte of 0.2 more recent chips have ratios upwards of 4. This disparity is not limited to just conventional CPUs. Massively parallel accelerators and coprocessors such as the Nvidia K20X and Intel Xeon Phi 5110P have ratios of 5.24 and 3.16, respectively.

A concomitant of this disparity is that modern hardware architectures are highly dependent on a combination of high-speed caches and/or shared memory to maintain throughput. However, for an algorithm to utilise these efficiently its memory access pattern must exhibit a degree of either spatial or temporal locality. To a first-order approximation the spatial locality of a method is inversely proportional to the amount of memory indirection. On an unstructured grid indirection arises whenever there is coupling between elements. This is potentially a problem for discretisations whose stencil is not compact. Coupling also arises in the context of implicit time stepping schemes. Implementations are therefore very often bound by memory bandwidth. As a secondary trend we note that the manner in which FLOPS are realised has also changed. In the early 1990s commodity CPUs were predominantly scalar with a single core of execution. However, in 2015 processors with eight or more cores are not uncommon. Moreover, the cores on modern processors almost always contain vector processing units. Vector lengths of 512-bits, which permit up to eight double precision values to be operated on at once, will soon be commonplace. It is therefore imperative that compute-bound algorithms are amenable to both multithreading and vectorisation. A versatile means of accomplishing this is by breaking the computation down into multiple, necessarily independent, streams. By virtue of their independence these streams can be readily divided up between cores and vector lanes. This leads directly to the concept of stream processing. A corollary of the above discussion is that compute intensive discretisations which can be formulated within the stream processing paradigm are well suited to acceleration on current and likely future hardware platforms. The FR approach combined with explicit time-stepping is an archetypical of this.

5.2 Salient Aspects of an FR Implementation

The majority of operations within an FR time-step can be cast in terms of matrix–matrix multiplications, in which a fixed, small, approximately square, and sometimes sparse operator matrix multiplies a dynamic, large, 'short-fat', dense state matrix (such multiplies are often referred to as block-by-panel). All remaining operations (e.g. flux evaluations) are point-wise, concerning themselves with either a single solution point inside of an element or two collocating flux points at an interface. Hence, in broad terms, there are five salient aspects of an FR implementation, specifically (i) definition of the constant operator matrices, (ii) specification of the state matrices, (iii) implementation of matrix multiply kernels, (iv) implementation of point-wise kernels, and finally (v) handling of distributed memory parallelism and scheduling of kernel invocations. For full details of a particular cross-platform implementation called PyFR, which has exhibited performance at around 50% of machine peak in benchmark calculations on massively parallel GPU-based clusters see Witherden et al. (2014) and Vincent et al. (2015).

It is often convenient to defer matrix multiplications to the GEMM family of subroutines from a suitable Basic Linear Algebra Subprograms (BLAS) library. BLAS is available for virtually all platforms and optimised versions are often maintained by the hardware vendors themselves (e.g., cuBLAS for Nvidia GPUs). This approach greatly facilitates development of efficient and platform portable code. We note, however, that the matrix sizes encountered in FR are not necessarily optimal from a GEMM perspective. Specifically, GEMM is optimised for the multiplication of large square matrices, whereas the constant operator matrixes in FR are 'small and square' with 10–100 rows/columns, and the state matrices are 'short and fat' with 10–100 rows and 10,000–100,000 columns. Moreover, we note that the constant operator matrices are know a priori and do not change in time. This a priori knowledge can be leveraged to design bespoke matrix multiply kernels that are more efficient than GEMM. The GiMMiK project is an example of such an approach (Wozniak et al., 2016) and has been found to improve performance by factors of two or more.

6 APPLICATIONS

In this section we will demonstrate some applications of the FR approach to solving nonlinear systems in three dimensions. Specifically, we will showcase how FR has been used to solve the compressible Euler and Navier–Stokes equations for three benchmark problems: flow over a circular cylinder, flow around an SD7003 wing, and flow through a T106c low-pressure turbine cascade.

6.1 Solving the Euler and Navier–Stokes Equations

The Navier–Stokes equations govern the dynamics for compressible flow. In the case where the flow is inviscid they reduce to the Euler equations. Both sets of equations take the form of nonlinear conservation laws. The FR approach can be readily extended to solve such systems by applying the approaches prescribed above for scalar FR to each component in the system.

In conservative form the Euler equations can be expressed as

$$\frac{\partial u}{\partial t} + \nabla \cdot \mathrm{f}(u) = 0, \tag{63}$$

where

$$u = \begin{Bmatrix} \rho \\ \rho v_x \\ \rho v_y \\ \rho v_z \\ E \end{Bmatrix}, \quad \mathrm{f} = \mathrm{f}^{(\mathrm{inv})} = \begin{Bmatrix} \rho v_x & \rho v_y & \rho v_z \\ \rho v_x^2 + p & \rho v_y v_x & \rho v_z v_x \\ \rho v_x v_y & \rho v_y^2 + p & \rho v_z v_y \\ \rho v_x v_z & \rho v_y v_z & \rho v_z^2 + p \\ v_x(E+p) & v_y(E+p) & v_z(E+p) \end{Bmatrix}, \tag{64}$$

here ρ is the mass density of the fluid, $\mathrm{v} = (v_x, v_y, v_z)^T$ is the fluid velocity vector, E is the total energy per unit volume and p is the pressure. For a perfect gas the pressure and total energy can be related by the ideal gas law

$$E = \frac{p}{\gamma - 1} + \frac{1}{2}\rho \, \| \mathrm{v} \|^2, \tag{65}$$

where $\gamma = c_p/c_v$ is the ratio of specific heats.

The more general Navier–Stokes equations can be written as an extension of the Euler equations through the inclusion of viscous terms. Within the presentation outlined above the flux now takes the form of $\mathrm{f}=\mathrm{f}^{(\mathrm{inv})} - \mathrm{f}^{(\mathrm{vis})}$ where

$$\mathrm{f}^{(\mathrm{vis})} = \left\{ \begin{array}{ccc} 0 & 0 & 0 \\ \mathcal{T}_{xx} & \mathcal{T}_{yx} & \mathcal{T}_{zx} \\ \mathcal{T}_{xy} & \mathcal{T}_{yy} & \mathcal{T}_{zy} \\ \mathcal{T}_{xz} & \mathcal{T}_{yz} & \mathcal{T}_{zz} \\ \sum_i v_i \mathcal{T}_{ix} + \Delta \partial_x T & \sum_i v_i \mathcal{T}_{iy} + \Delta \partial_y T & \sum_i v_i \mathcal{T}_{iz} + \Delta \partial_z T \end{array} \right\}. \quad (66)$$

In the above we have defined $\Delta = \mu c_p/P_r$ where μ is the dynamic viscosity and P_r is the Prandtl number. The components of the stress-energy tensor are given by

$$\mathcal{T}_{ij} = \mu(\partial_i v_j + \partial_j v_i) - \frac{2}{3}\mu \delta_{ij} \nabla \cdot \mathrm{v}, \quad (67)$$

where δ_{ij} is the Kronecker delta. Using the ideal gas law the temperature can be expressed as

$$T = \frac{1}{c_v} \frac{1}{\gamma - 1} \frac{p}{\rho}, \quad (68)$$

with partial derivatives thereof being given according to the quotient rule.

6.2 Flow Over a Circular Cylinder

Flow over a circular cylinder has been the focus of various experimental and numerical studies (Breuer, 1998; Kravchenko and Moin, 2000; Ma et al., 2000; Norberg, 1998; Parnaudeau et al., 2008; Vermeire and Nadarajah, 2013, 2015). Characteristics of the flow are known to be highly dependent on the Reynolds number Re, defined as

$$Re = \frac{u_\infty D}{\nu}, \quad (69)$$

where u_∞ is the free-stream fluid speed, D is the cylinder diameter, and ν is the fluid kinematic viscosity. Roshko (1953) identified a stable range between $Re =$ 40 and 150 that is characterised by the shedding of regular laminar vortices, as well as a transitional range between $Re = 150$ and 300, and a turbulent range beyond $Re = 300$. These results were subsequently confirmed by Bloor (1964), who identified a similar set of regimes. Later, Williamson (1988) identified two modes of transition from two- to three-dimensional flow. The first, known as Mode-A instability, occurs at $Re \approx 190$ and the second, known as Mode-B instability, occurs at $Re \approx 260$. The turbulent range beyond

$Re = 300$ can be further subclassified into the shear-layer transition, critical, and supercritical regimes as discussed in the review by Williamson (1996).

Previous studies of Witherden et al. have used the FR approach to simulate flow over a circular cylinder at $Re = 3900$, and an effectively incompressible Mach number of 0.2. This case sits in the shear-layer transition regime identified by Williamson (1996), and contains several complex flow features, including separated shear layers, turbulent transition, and a fully turbulent wake. In addition to the recent study of Witherden et al. this test case has also been the focus of a number of other works, both experimental and numerical (Breuer, 1998; Kravchenko and Moin, 2000; Ma et al., 2000; Norberg, 1998; Parnaudeau et al., 2008). In 2013, Lehmkuhl et al. demonstrated that the wake profile for this test case can be classified as one of two modes, a low-energy mode (Mode-L) and a high-energy mode (Mode-H). Specifically, via analysis of a very long period simulation (over 2000 convective times), they showed that the wake fluctuates between these two modes.

6.2.1 Domain and Mesh

In their study Witherden et al. (2015) used a computational domain of dimensions $[-9D, 25D]$; $[-9D, 9D]$; and $[0, \pi D]$ in the stream-, cross-, and span-wise directions, respectively. The cylinder is centred at (0, 0, 0). The stream-wise and cross-wise dimensions are comparable to the experimental and numerical values used by Parnaudeau et al. (2008). The overall domain dimensions are also comparable to those used for DNS studies by Lehmkuhl et al. (2013). The domain is periodic in the span-wise direction, with a no-slip isothermal wall boundary condition applied at the surface of the cylinder, and Riemann invariant boundary conditions applied at the far-field.

In the study, the domain was meshed in two ways. The first mesh consisted of entirely structured hexahedral elements, while the second was unstructured, consisting of prismatic elements in the near wall boundary layer region, and tetrahedral elements in the wake and far-field. Both meshes employed quadratically curved elements and were designed to fully resolve the near wall boundary layer region when $p = 4$. Specifically, the maximum skin friction coefficient was estimated a priori as $C_f \approx 0.075$ based on the LES results of Breuer (1998). The height of the first element was then specified such that when $p = 4$ the first solution point from the wall sits at $y+ \approx 1$, where nondimensional wall units are calculated in the usual fashion as $y+ = u_\tau y/v$ with $u_\tau = \sqrt{C_f/2}u_\infty$.

The hexahedral mesh had 104 elements in the circumferential direction, and 16 elements in the span-wise direction, which when $p = 4$ achieves span-wise resolution comparable to that used in previous studies. The prism/tetrahedral mesh has 116 elements in the circumferential direction, and 20 elements in the span-wise direction, these numbers being chosen to help reduce face aspect ratios at the edges of the prismatic layer; which

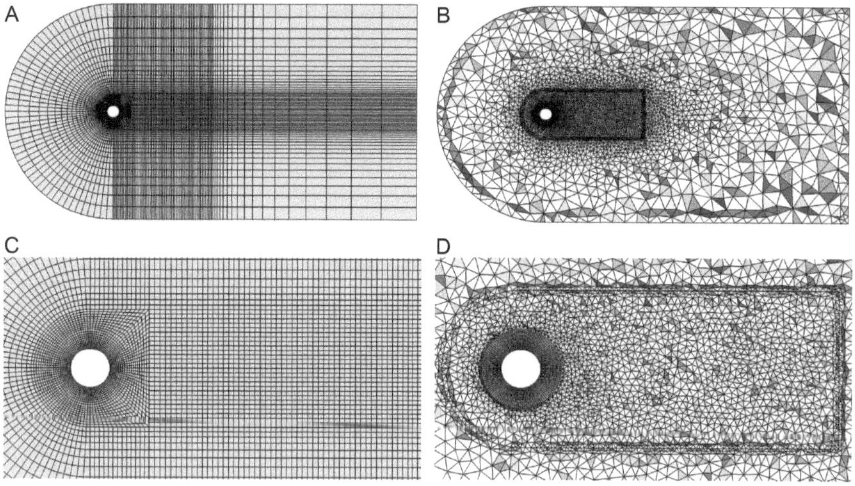

FIG. 5 Cutaways through the two meshes. (A) Hexahedral, far-field. (B) Prism/tetrahedral, far-field. (C) Hexahedral, wake. (D) Prism/tetrahedral, wake. *From Witherden, F.D., Vermeire, B.C., Vincent, P.E., 2015. Heterogeneous computing on mixed unstructured grids with PyFR. Comput. Fluids 120, 173–186, and copyright F.D. Witherden, B.C. Vermeire, P.E. Vincent, reused with permission.*

facilitates transition to the fully unstructured tetrahedral elements in the far-field. In total the hexahedral mesh contained 119,776 elements, and the prism/tetrahedral mesh contained 79,344 prismatic elements and 227,298 tetrahedral elements. Both meshes are shown in Fig. 5.

6.2.2 *Methodology*

The compressible Navier–Stokes equations, with constant viscosity, were solved on each of the two meshes shown in Fig. 5. A DG scheme was used for the spatial discretisation, a Rusanov Riemann solver was used to calculate the inviscid fluxes at element interfaces, and the explicit RK45[2R+] scheme of Kennedy et al. (2000) was used to advance the solution in time. No subgrid model was employed, hence the approach should be considered ILES/DNS (Vermeire et al., 2013, 2014), as opposed to classical LES.

6.2.3 *Accuracy*

In this section we present the time-span-averaged results obtained by Witherden et al. using a cluster of 12 NVIDIA K20c GPUs at $p = 4$, the design resolution for both meshes. Both simulations were run for 1000 convective times, allowing the flow to fluctuate between Mode-H and Mode-L as identified by Lehmkuhl et al. (2013) and Ma et al. (2000). A moving window time-average with a width of 100 convective times is used to extract both modes from the long-period simulation. This yields four datasets including both Mode-H and Mode-L for both the hexahedral and prism/tetrahedral meshes. Both modes are then compared with results from previous experimental and numerical studies, where either one or both of the modes were observed (Lehmkuhl et al., 2013; Ma et al., 2000; Norberg, 1998; Parnaudeau et al., 2008).

Instantaneous surfaces of iso-density are shown in Fig. 6 for both simulations at similar phases of the shedding cycle. We observe laminar flow at the leading edge of the cylinder for both test cases, turbulent transition near the separation points, and fully turbulent flow in the wake region. These are the characteristic features of the shear-layer transition regime, as described by Williamson (1996). The wake is composed of large vortices, alternately shedding off of the upper and lower surfaces of the cylinder, and smaller scale turbulent structures.

Plots of the averaged stream-wise wake profiles are shown in Figs. 7 and 8 for Mode-H and Mode-L, respectively. Both the hexahedral and prism/tetrahedral meshes show excellent agreement with the numerical results of Lehmkuhl et al. (2013) for both modes and with the experimental results of Parnaudeau et al. (2008), which is available for Mode-L. The Mode-H cases exhibit relatively shorter separation bubbles and the Mode-L cases have characteristic inflection points in the wake profile near $x/D \approx 1$.

Plots of the averaged pressure coefficient $\overline{C_p}$ on the surface of the cylinder are shown in Figs. 9 and 10 for both extracted modes and both meshes. The Mode-H results are shown alongside the Mode-H numerical results of Lehmkuhl et al. (2013) and the results from Case I of Ma et al. (2000). The Mode-L results are shown alongside the Mode-L numerical results of Lehmkuhl et al. (2013) and the experimental results of Norberg (1998) at a similar $Re = 4020$, which were extracted from Kravchenko and Moin (2000). Both modes have similar pressure coefficient distributions at the leading face of the cylinder, while the Mode-H case has stronger suction on the trailing face adjacent to the separation bubble. Both modes extracted using both meshes show excellent agreement with their corresponding reference data sets (Table 1).

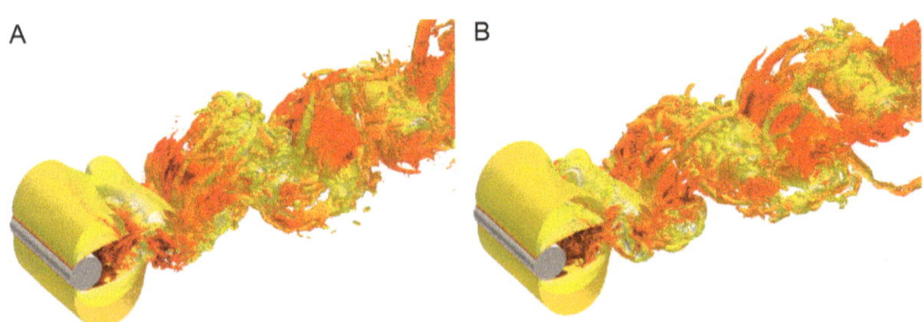

FIG. 6 Instantaneous surfaces of iso-density coloured by velocity magnitude. (A) Structured hexahedral. (B) Unstructured prism/tetrahedral. *From Witherden, F.D., Vermeire, B.C., Vincent, P.E., 2015. Heterogeneous computing on mixed unstructured grids with PyFR. Comput. Fluids 120, 173–186, and copyright F.D. Witherden, B.C. Vermeire, P.E. Vincent, reused with permission.*

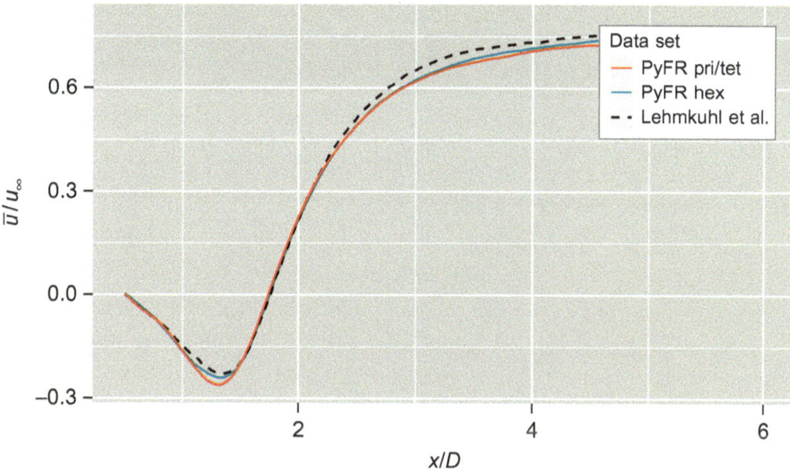

FIG. 7 Averaged wake profiles for Mode-H compared with the numerical results of Lehmkuhl et al. (2013). *From Witherden, F.D., Vermeire, B.C., Vincent, P.E., 2015. Heterogeneous computing on mixed unstructured grids with PyFR. Comput. Fluids 120, 173–186, and copyright F.D. Witherden, B.C. Vermeire, P.E. Vincent, reused with permission.*

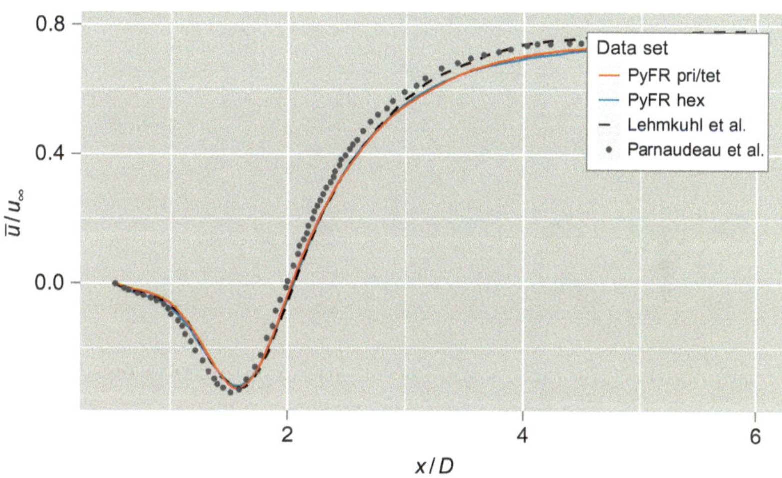

FIG. 8 Averaged wake profiles for Mode-L compared with the numerical results of Lehmkuhl et al. (2013) and experimental results of Parnaudeau et al. (2008). *From Witherden, F.D., Vermeire, B.C., Vincent, P.E., 2015. Heterogeneous computing on mixed unstructured grids with PyFR. Comput. Fluids 120, 173–186, and copyright F.D. Witherden, B.C. Vermeire, P.E. Vincent, reused with permission.*

The averaged pressure coefficient at the base of the cylinder $\overline{C_{pb}}$, and the averaged separation angle θ_s measured from the leading stagnation point are tabulated in Fig. 1 for both modes and meshes. These are shown along with measurements from the experimental results of Norberg (1998), experimental data from Parnaudeau et al. (2008), and DNS data from Lehmkuhl et al. (2013) for

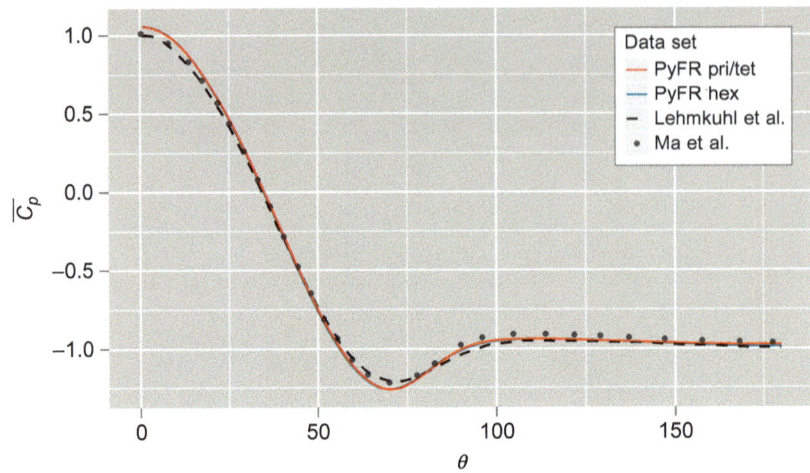

FIG. 9 Averaged pressure coefficient for Mode-H compared with the numerical results of Ma et al. (2000) and Lehmkuhl et al. (2013). *From Witherden, F.D., Vermeire, B.C., Vincent, P.E., 2015. Heterogeneous computing on mixed unstructured grids with PyFR. Comput. Fluids 120, 173–186, and copyright F.D. Witherden, B.C. Vermeire, P.E. Vincent, reused with permission.*

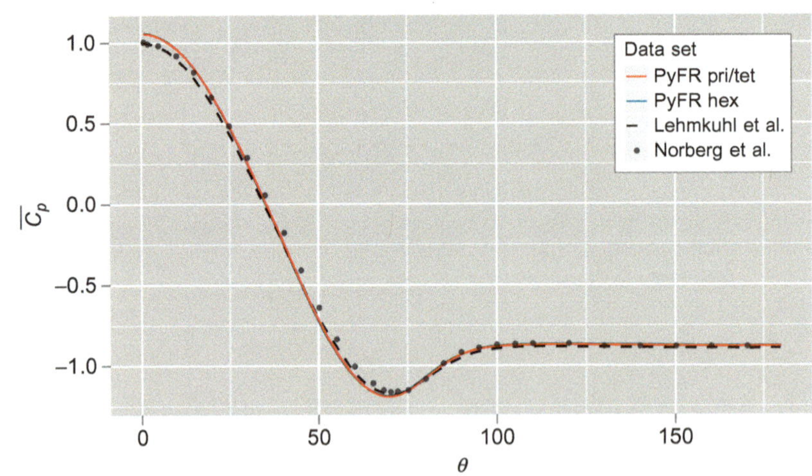

FIG. 10 Averaged pressure coefficient for Mode-L compared with the numerical results of Lehmkuhl et al. (2013) and experimental results of Norberg (1998). *From Kravchenko, A.G., Moin, P., 2000. Numerical studies of flow over a circular cylinder at* $\mathrm{re_d} = 3\ 900$*. Phys. Fluids 12, 403–417 and Witherden, F.D., Vermeire, B.C., Vincent, P.E., 2015. Heterogeneous computing on mixed unstructured grids with PyFR. Comput. Fluids 120, 173–186, and copyright F.D. Witherden, B.C. Vermeire, P.E. Vincent, reused with permission.*

TABLE 1 Comparison of Quantitative Values With Experimental and DNS Results

	Mode-H		Mode-L	
	$-\overline{C_{pb}}$	θ_s/degrees	$-\overline{C_{pb}}$	θ_s/degrees
PyFR Pri/tet	0.974	87.13	0.882	86.90
PyFR Hex	0.987	88.28	0.880	87.71
Parnaudeau et al. (2008)				88.00
Lehmkuhl et al. (2013)	0.980	88.25	0.877	87.80
Norberg (1998) (extracted from Kravchenko and Moin, 2000)			0.880	

both modes. Both measured quantities agree well with the reference data sets for both modes and meshes. The difference in separation angle is less than ~ 1 degree between the current and reference results. The pressure coefficient at base of the cylinder shows that the high-energy Mode-H case has stronger recirculation in the wake, characterised by greater suction at the wall adjacent to the recirculation bubble.

6.3 Flow Over an SD7003 Wing

Both Vermeire et al. (2016) and Williams (2013) have used FR schemes to simulate flow over an SD7003 infinite wing at 4 degrees angle of attack. The SD7003 geometry is that of a low Reynolds number Selig/Donovan airfoil that has a maximum thickness of 9.2% at 30.9% chord, and a maximum camber of 1.2% at 38.3% chord.

The numerical experiments of Williams on the SD7003 geometry were performed on a cylindrical domain formed by extruding a circular domain (of radius $20c$) by $0.2c$ in the z direction where c is the chord. Periodic boundary conditions were prescribed on the front and back of the cylindrical domain, characteristic boundary conditions were prescribed on the sides of the domain, and adiabatic wall boundary conditions were prescribed on the surface of the wing-section. The domain was then meshed with 711,332 quadratically curved tetrahedral elements.

The Reynolds number was taken to be $Re = 10,000$ with the incoming flow having a velocity of Mach $M = 0.2$. The angle of entry for the incoming flow was set as 4 degree in order to simulate a 4 degree angle of attack. The simulations were run using two different tetrahedral ESFR schemes: one with

$c = 0$ designed to recover a collocated nodal DG scheme and another with $c = c^+$ where c^+ is chosen to maximise the permissible explicit time step. The solution was marched forward in time using the RK54 approach and, at each time-step, the inviscid and viscous numerical fluxes were computed using a Rusanov-type Riemann solver. Results were evaluated after the lift and drag reached a pseudo-periodic state.

Lift and drag results for the two simulations of Williams along with the numerical results of Uranga et al. (2011) can be seen in Table 2. Looking at the table it can be seen that both ESFR schemes show good agreement with the coefficients obtained by Uranga et al. The time history of the lift C_L and drag C_D coefficients for the $c = c^+$ simulation are shown in Fig. 11. Instantaneous isosurfaces of density and vorticity for this simulation are shown in Fig. 12.

TABLE 2 Time Averaged Lift $\overline{C_L}$ and Drag $\overline{C_D}$ Coefficients of Williams for the SD7003 Wing at $Re = 10{,}000$ and $M = 0.2$ Using $p = 3$ Tetrahedral ESFR Schemes

Simulation	$\overline{C_L}$	$\overline{C_D}$
Uranga et al. (2011)	0.3743	0.04967
Williams $c = 0$	0.3466	0.04908
Williams $c = c^+$	0.3454	0.04903

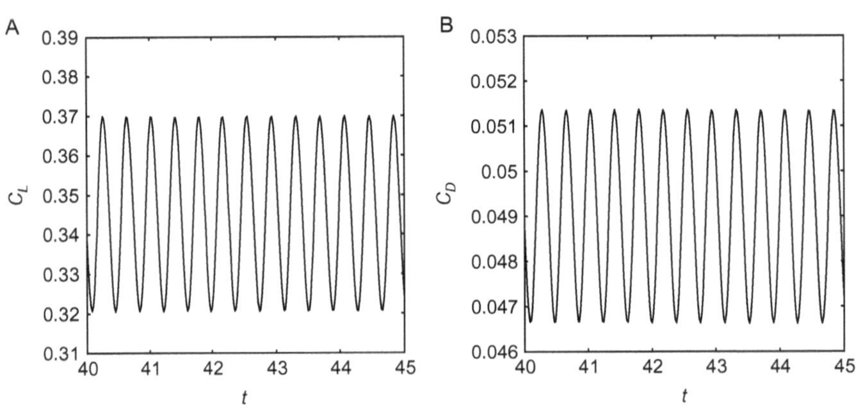

FIG. 11 Temporal variation of the lift and drag coefficients for the SD7003 wing at $Re = 10{,}000$ and $M = 0.2$ using $p = 3$ tetrahedral ESFR scheme with $c = c^+$. (A) Lift coefficient. (B) Drag coefficient. *From Williams, D.M., 2013. Energy stable high-order methods for simulating unsteady, viscous, compressible flows on unstructured grids (Ph.D. thesis), Stanford University, and copyright D.M. Williams, reused with permission.*

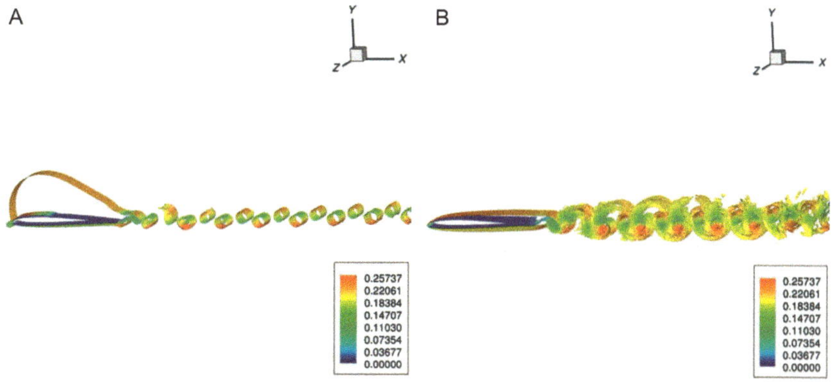

FIG. 12 Instantaneous isosurfaces of density and vorticity, coloured by Mach number, for the SD7003 wing at $Re = 10,000$ and $M = 0.2$ using $p = 3$ tetrahedral ESFR scheme with $c = c^{+}$. (A) Density isosurfaces. (B) Vorticity isosurfaces. *From Williams, D.M., 2013. Energy stable high-order methods for simulating unsteady, viscous, compressible flows on unstructured grids (Ph.D. thesis), Stanford University, and copyright D.M. Williams, reused with permission.*

6.4 T106c Low-Pressure Turbine Cascade

FR schemes have also shown promise in simulating flows over turbomachinery. The T106 low-pressure turbine cascade (Wood et al., 1990) is a popular test case for evaluating CFD solvers. The T106c cascade is defined by imposing a pitch-to-cord ratio of $s/c = 0.95$ and is supported by a wide body of numerical and experimental data (Garai et al., 2016; Hillewaert et al., 2014; Michlek et al., 2012; Pacciani et al., 2011). From the specification of the T106c cascade the chord is $c = 0.09301$ m and the pitch-wise inlet flow angle is 32.7 degree. In this study we will consider the case at Reynolds number $Re = 80{,}000$, with an outlet Mach number of $M = 0.65$. These conditions are of interest as they represent a fully compressible simulation at a meaningful Reynolds number. The working fluid for the case is taken to be air with a total temperature at the inlet of $T_i = 298.15$ K. The ratio of specific heats is $\gamma = 1.4$ with the specific gas constant being given by $R \simeq 287.1\ \mathrm{J\ kg^{-1}\ K^{-1}}$.

Additional characteristics of the flow can be determined using the isentropic flow equations. The exit temperature can be determined as

$$T_e = \frac{T_i}{1 + \frac{\gamma+1}{2} M^2} \simeq 274.92\,\mathrm{K}, \tag{70}$$

with the exit velocity being given by

$$v_e = M\sqrt{\gamma R T_e} \simeq 216.07\,\mathrm{m\,s^{-1}}. \tag{71}$$

Using Sutherland's law the viscosity at the exit is given by

$$\mu_e = \mu_0 \left(\frac{T_e}{T_0}\right)^{\frac{3}{2}} \frac{T_0 + S}{T_e + S}, \tag{72}$$

where $\mu_0 \simeq 1.716 \times 10^{-5}$ Pa s is the viscosity for air at the reference temperature, $T_0 \simeq 273.15$ K, and $S \simeq 110.4$ K is Sutherland's temperature. Hence, $\mu_e \simeq 1.7248 \times 10^{-5}$ Pa s with the density at the outlet being given by

$$\rho_e = \frac{Re\mu_e}{v_e c} \simeq 0.0687\,\text{kg m}^{-3}. \tag{73}$$

Finally, the pressure at the outlet and total pressure can be determined as

$$p_e = \rho_e R T_e \simeq 5419.3\,\text{Pa}, \tag{74}$$

$$p_i = p_e\left(1 + \frac{\gamma - 1}{2} M^2\right)^{\frac{\gamma}{\gamma - 1}} \simeq 7198.5\,\text{Pa}, \tag{75}$$

which serve to fully specify the problem.

The linear T106c cascade used in the experiments of Michlek et al. (2012) consisted of six blades each with a span $h = 2.4c \simeq 0.22$ m. For the case under consideration the inlet turbulence was approximately 0.9%. To simplify the problem setup somewhat we will simulate just a single blade with a reduced span of $0.2c \simeq 0.0186$ m. The domain is periodic in the cross- and span-wise directions. On the surface of the blade an adiabatic wall condition is applied. A fixed total pressure condition is enforced on the inlet of the domain. This condition also enforces the desired pitch-wise angle for the incoming flow. At the exit a Riemann invariant boundary condition is applied. We note here that since Riemann invariant conditions are designed to minimise reflections at the boundary that they do not guarantee a strong enforcement of the exit pressure p_e. Hence, some tuning is required around the pressure enforced by the boundary condition in order to obtain the desired (averaged) exit pressure.

To mesh the domain unstructured hexahedral elements were employed. The elements on the boundary of the blade were quadratically curved and designed to fully resolved the near wall boundary layer region at $p = 2$. Specifically, the height of the first element was specified such that when $p = 2$ that the first solution point from the wall sits at $y+ \approx 1$ with

$$y+ = \frac{u_\tau \rho_e y}{\mu_e} = \frac{\sqrt{C_f} v_e \rho_e y}{\sqrt{2}\mu_e} \approx \frac{(2\log_{10}Re - 0.65)^{-1.15} v_e \rho_e y}{\sqrt{2}\mu_e}, \tag{76}$$

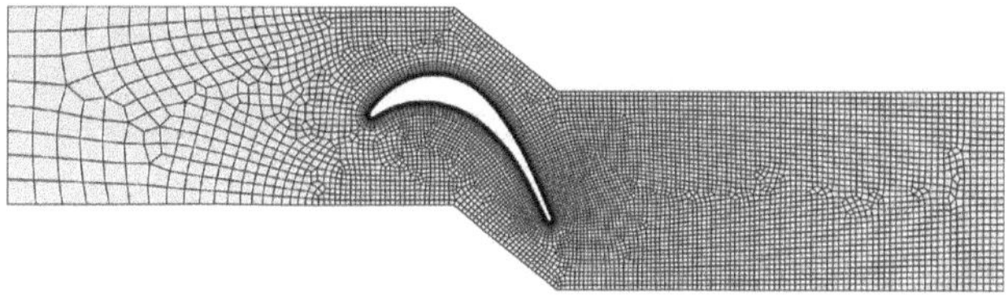

FIG. 13 Cutaway in the x–y plane of the unstructured hexahedral mesh used for the T106c test case. *Figure copyright F.D. Witherden and A. Jameson, reused with permission.*

where in the third step we have used the Schlichting skin friction formula to estimate C_f based off of the Reynolds number. The resulting mesh has a total of eight layers in the span-wise direction. In total the mesh contains 59,936 elements. A cut-away of the mesh can be seen in Fig. 13.

For the simulations a DG scheme was used for the spatial discretisation, a Rusanov Riemann solver was used to calculate the inviscid fluxes at element interfaces, and the explicit low-storage RK45[2R+] Runga–Kutta scheme of Kennedy et al. (2000) was used to advance the solution in time. Local temporal error was managed by utilising a PI type step size controller with absolute and relative error tolerances set to 10^{-5}. Solution points and flux points were placed at a tensor product construction of Gauss–Legendre quadrature points. The upwind and penalty parameters for the LDG scheme were set to $\beta = 1/2$ and $\tau = 1/10$, respectively. The Prandtl number was taken to be $Pr = 0.71$.

The simulations were run at $p = 1$ with a viscosity of $\mu = 3.4 \times 10^{-5}$ until $t = 2.5 \times 10^{-2}$. At this point the simulations were restarted with the correct viscosity of 1.7248×10^{-5} at order $p = 2$ and advanced to 3.5×10^{-2}. Time averaging of the pressure field was enabled at $t = 3.25 \times 10^{-2}$ with the average being accumulated every 50 time steps. This corresponds to approximately two passes over the chord.

A plot of isosurfaces of Q-criterion coloured by velocity magnitude for the fully developed flow are shown in Fig. 14. To compare with experimental data we consider the isentropic Mach number, defined as

$$M^{\text{isentropic}} = \sqrt{\frac{2}{\gamma - 1}\left(\left(\frac{p_e}{p}\right)^{\Gamma} - 1\right)}, \tag{77}$$

where $\Gamma = (\gamma - 1)/\gamma$. A plot of the chord-wise distribution of this can be seen in Fig. 15. Looking at the figure also see that there is a strong level of agreement between PyFR and the experimental data of Michlek et al. (2012). However, there are some discrepancies around peak suction $x/c \sim 0.6$. We note however that such over-prediction has also been observed by both the studies of Hillewaert et al. (2014) and Garai et al. (2016).

FIG. 14 Isosurfaces of Q-criterion coloured by velocity magnitude for the T106c test case with $p = 2$. *Figure copyright F.D. Witherden and A. Jameson, reused with permission.*

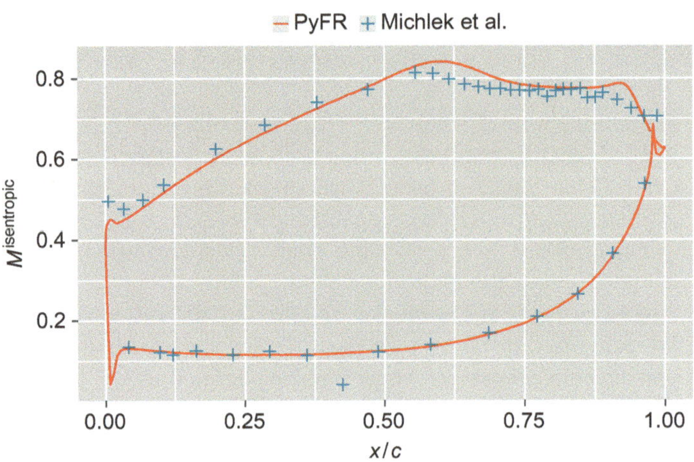

FIG. 15 Chord-wise distribution of the time- and span-averaged isentropic Mach number for the T106c test case compared against the experimental data of Michlek et al. (2012). *Copyright F.D. Witherden and A. Jameson, reused with permission.*

7 SUMMARY

This chapter has provided a review of the FR approach to the formulation of high-order methods that was first proposed by Huynh. This approach provides a unifying framework for the design of high-order discontinuous finite element methods for unstructured grids. As described in Section 2 the solution in each element is represented by a polynomial u^D of degree p via a Lagrange type interpolating polynomial at $p + 1$ collocation points, typically the Gauss–Legendre points. A corresponding piecewise discontinuous flux polynomial f^D is similarly constructed via Lagrange interpolation. Interface fluxes at the element boundaries are computed based on an approximate Riemann solver from the left and right states at each interface, and a continuous flux f^C of degree $p + 1$ is then constructed by propagating the difference between the Riemann solution and the value of f^D into interior using correction polynomials of degree $p + 1$. The solution is then advanced by taking the divergence of the continuous flux f^C. The FR method can be extended to advection–diffusion problems by writing the equation as a first order system in a manner similar to DG methods.

The nature of a particular FR scheme to solve a hyperbolic conservation law depends on three factors: the location of the solution points, the choice of approximate Riemann solver for the interface fluxes, and the choice of the correction functions. It was shown by Huynh that the correction functions can be chosen to recover the nodal DG and spectral difference methods.

Section 4 outlines the ways in which the FR approach can be extended to multidimensional problems both for quadrilateral and hexahedral elements using a tensor product formulation, and to simplex elements by using vector correction functions restricted to lie in the Raviart–Thomas space of order p. The stability and accuracy of FR schemes in addressed in Section 5, highlighting in particular the form of energy stable FR schemes (proposed by Vincent et al., 2011b). Fourier analysis confirms that the expected order of accuracy is $p + 1$, but linear problems on uniform meshes with periodic boundary conditions can exhibit super-convergence.

Section 5 summarises the implementation aspects of FR schemes. The majority of operations can be cast in terms of matrix–matrix multiplications which can deferred to special kernels. The open source PyFR software developed at Imperial College London by Witherden et al. (2014) exploits this to enable it to target a range of hardware platforms. Results obtained with PyFR confirm that when run at orders of four or more the FR method is particularly well suited to modern computing platforms on which arithmetic operations are very cheap, and memory fetches are relatively expensive. PyFR has exhibited performance at around 50% of machine peak in benchmark calculations on massively parallel GPU-based clusters.

Some applications to the Navier–Stokes equations are presented in Section 6. These indicate that at moderate Reynolds numbers the FR method can be used to perform accurate large eddy simulations without the introduction of a sub-grid model. This provides a path towards practical simulations of a variety of important industrial applications such as the design of the low-pressure turbine stages of jet engines.

REFERENCES

Allaneau, Y., Jameson, A., 2011. Connections between the filtered discontinuous Galerkin method and the flux reconstruction approach to high order discretizations. Comput. Methods Appl. Mech. Eng. 200 (49), 3628–3636.

Arnold, D.N., 1982. An interior penalty finite element method with discontinuous elements. SIAM J. Numer. Anal. 19 (4), 742–760.

Asthana, K., Jameson, A., 2015. High-order flux reconstruction schemes with minimal dispersion and dissipation. J. Sci. Comput. 62 (3), 913–944.

Asthana, K., Watkins, J., Jameson, A., 2016. SIAM J. Numer. Anal. (accepted for publication).

Balan, A., May, G., Schöberl, J., 2011. A stable spectral difference method for triangles. In: 49th AIAA Aerospace Sciences Meeting including the New Horizons Forum and Aerospace Exposition, Aerospace Sciences Meetings.

Bassi, F., Rebay, S., 1997. A high-order accurate discontinuous finite element method for the numerical solution of the compressible Navier-Stokes equations. J. Comput. Phys. 131 (2), 267–279.

Bloor, M.S., 1964. The transition to turbulence in the wake of a circular cylinder. J. Fluid Mech. 19, 290–304.

Breuer, M., 1998. Large eddy simulation of the subcritical flow past a circular cylinder. Int. J. Numer. Methods Fluids 28 (9), 1281–1302.

Canuto, C., Hussaini, M.Y., Quarteroni, A., Zang, T.A., 2006. Spectral Methods: Fundamentals in Single Domains. Springer, Berlin.

Castonguay, P., Vincent, P.E., Jameson, A., 2011. Application of high-order energy stable flux reconstruction schemes to the Euler equations. In: AIAA paper, 686.

Castonguay, P., Vincent, P.E., Jameson, A., 2012. A new class of high-order energy stable flux reconstruction schemes for triangular elements. J. Sci. Comput. 51 (1), 224–256.

Castonguay, P., Williams, D.M., Vincent, P.E., Jameson, A., 2013. Energy stable flux reconstruction schemes for advection-diffusion problems. Comput. Methods Appl. Mech. Eng. 267, 400–417.

Cockburn, B., Shu, C.W., 1998. The local discontinuous Galerkin method for time-dependent convection-diffusion systems. SIAM J. Numer. Anal. 35 (6), 2440–2463.

De Grazia, D., Mengaldo, G., Moxey, D., Vincent, P.E., Sherwin, S.J., 2014. Connections between the discontinuous Galerkin method and high-order flux reconstruction schemes. Int. J. Numer. Methods Fluids 75 (12), 860–877. http://dx.doi.org/10.1002/fld.3915.

Gao, H., Wang, Z.J., 2009. A high-order lifting collocation penalty formulation for the Navier-Stokes equations on 2D mixed grids. In: 19th AIAA Computational Fluid Dynamics. San Antonio, TX.

Garai, A., Diosady, L.T., Murman, S.M., Madavan, N., 2016. DNS of flow in a low-pressure turbine cascade with elevated inflow turbulence using a discontinuous-Galerkin spectral-element method. In: Proceedings of ASME Turbo Expo 2016.

Gassner, G.J., 2013. A skew-symmetric discontinuous Galerkin spectral element discretization and its relation to SBP-SAT finite difference methods. SIAM J. Sci. Comput. 35 (3), A1233–A1253. http://dx.doi.org/10.1137/120890144.

Harten, A., Engquist, B., Osher, S., Chakravarthy, S.R., 1987. Uniformly high order accurate essentially non-oscillatory schemes, III. J. Comput. Phys. 71 (2), 231–303.

Hesthaven, J.S., Warburton, T., 2008. Nodal Discontinuous Galerkin Methods: Algorithms, Analysis, and Applications. vol. 54. Springer-Verlag, New York.

Hillewaert, K., de Wiart, C.C., Verheylewegen, G., Arts, T., 2014. Assessment of a high-order discontinuous Galerkin method for the direct numerical simulation of transition at low-Reynolds number in the T106C high-lift low pressure turbine cascade. In: ASME Turbo Expo 2014: Turbine Technical Conference and Exposition. American Society of Mechanical Engineers, pp.V02BT39A034–V02BT39A034.

Huynh, H.T., 2007. A flux reconstruction approach to high-order schemes including discontinuous Galerkin methods. In: 18th AIAA Computational Fluid Dynamics Conference, Fluid Dynamics and Co-located Conferences.

Jameson, A., 2010. A proof of the stability of the spectral difference method for all orders of accuracy. J. Sci. Comput. 45 (1–3), 348–358.

Jameson, A., Ou, K., 2011. 50 years of transonic aircraft design. Prog. Aerosp. Sci. 47 (5), 308–318.

Jameson, A., Vincent, P.E., Castonguay, P., 2011. On the non-linear stability of flux reconstruction schemes. J. Sci. Comput. 50 (2), 434–445.

Karniadakis, G., Sherwin, S.J., 2005. Spectral/hp Element Methods for Computational Fluid Dynamics. Oxford University Press, Oxford.

Kennedy, C.A., Carpenter, M.H., Lewis, R.M., 2000. Low-storage, explicit Runge-Kutta schemes for the compressible Navier-Stokes equations. Appl. Numer. Math. 35 (3), 177–219.

Kopriva, D.A., Kolias, J.H., 1996. A conservative staggered-grid Chebyshev multidomain method for compressible flows. J. Comput. Phys. 125 (1), 244–261.

Kravchenko, A.G., Moin, P., 2000. Numerical studies of flow over a circular cylinder at $re_d = 3900$. Phys. Fluids 12, 403–417.

Lehmkuhl, O., Rodriguez, I., Borrell, R., Oliva, A., 2013. Low-frequency unsteadiness in the vortex formation region of a circular cylinder. Phys. Fluids 25 (8), 3165–3168.

Li, Y., Wang, Z.J., 2013. Evaluation of optimized CPR schemes for computational aeroacoustics benchmark problems. In: AIAA Paper 2013-2689.

Liu, X.D., Osher, S., Chan, T., 1994. Weighted essentially non-oscillatory schemes. J. Comput. Phys. 115 (1), 200–212.

Ma, X., Karamanos, G.S., Karniadakis, G.E., 2000. Dynamics and low-dimensionality of a turbulent near wake. J. Fluid Mech. 310, 29–65.

Mengaldo, G., Grazia, D.D., Moxey, D., Vincent, P.E., Sherwin, S.J., 2015. Dealiasing techniques for high-order spectral element methods on regular and irregular grids. J. Comput. Phys. 299, 56–81.http:/dx.doi.org/10.1016/j.jcp.2015.06.032.

Mengaldo, G., De Grazia, D., Vincent, P.E., Sherwin, S.J., 2016. On the connections between discontinuous Galerkin and Flux Reconstruction schemes: extension to curvilinear meshes. J. Sci. Comput. 67 (3), 1272–1292.http:/dx.doi.org/10.1007/s10915-015-0119-z.

Michlek, J., Monaldi, M., Arts, T., 2012. Aerodynamic performance of a very high lift low pressure turbine airfoil (T106c) at low Reynolds and high mach number with effect of free stream turbulence intensity. J. Turbomach. 134 (6), 061009.

Norberg, C., 1998. LDV measurements in the near wake of a circular cylinder. Int. J. Numer. Methods Fluids 28 (9), 1281–1302.

Pacciani, R., Marconcini, M., Arnone, A., Bertini, F., 2011. An assessment of the laminar kinetic energy concept for the prediction of high-lift, low-Reynolds number cascade flows. Proc. Inst. Mech. Eng. Part A: J. Power Energy 225 (7), 995–1003.

Park, J.S., Kim, C., 2016. Hierarchical multi-dimensional limiting strategy for correction procedure via reconstruction. J. Comput. Phys. 308, 57–80. http://dx.doi.org/10.1016/j.jcp.2015.12.020.

Parnaudeau, P., Carlier, J., Heitz, D., Lamballais, E., 2008. Experimental and numerical studies of the flow over a circular cylinder at Reynolds number 3900. Phys. Fluids 20 (8), 085101.

Peraire, J., Persson, P.O., 2008. The compact discontinuous Galerkin (CDG) method for elliptic problems. SIAM J. Sci. Comput. 30 (4), 1806–1824.

Ranocha, H., Öffner, P., Sonar, T., 2016. Summation-by-parts operators for correction procedure via reconstruction. J. Comput. Phys. 311, 299–328. http://dx.doi.org/10.1016/j.jcp.2016.02.009.

Raviart, P.A., Thomas, J.M., 1977. A mixed finite element method for 2-nd order elliptic problems. In: Galligani, I., Magenes, E. (Eds.), Mathematical Aspects of Finite Element Methods. Springer, Berlin, pp. 292–315.

Reed, W.H., Hill, T.R., 1973. Triangular mesh methods for the neutron transport equation. Los Alamos Scientific Laboratory, Los Alamos. Technical Report LA-UR-73-479.

Roe, P.L., 1981. Approximate Riemann solvers, parameter vectors, and difference schemes. J. Comput. Phys. 43 (2), 357–372.

Roshko, A., 1953. On the development of turbulent wakes from vortex streets. California Institute of Technology. Technical Report NACA TR 1191.

Sheshadri, A., Jameson, A., 2016. On the stability of the flux reconstruction schemes on quadrilateral elements for the linear advection equation. J. Sci. Comput. 67 (2), 769–790.

Solin, P., Segeth, K., Dolezel, I., 2003. Higher-Order Finite Element Methods. CRC Press, Boca Raton, FL.

Sun, Y., Wang, Z.J., Liu, Y., 2007. High-order multidomain spectral difference method for the Navier-Stokes equations on unstructured hexahedral grids. Commun. Comput. Phys. 2 (2), 310–333.

Taylor, M.A., Wingate, B.A., Bos, L.P., 2005. Several new quadrature formulas for polynomial integration in the triangle. Mathematics. ArXiv:math/0501496.

Uranga, A., Persson, P.O., Drela, M., Peraire, J., 2011. Implicit large eddy simulation of transition to turbulence at low Reynolds numbers using a discontinuous Galerkin method. Int. J. Numer. Methods Eng. 87 (1–5), 232–261.

Vermeire, B.C., Nadarajah, S., 2013. Adaptive IMEX time-stepping for ILES using the correction procedure via reconstruction scheme. In: 21st AIAA Computational Fluid Dynamics Conference, Fluid Dynamics and Co-located Conferences.

Vermeire, B.C., Nadarajah, S., 2015. Adaptive IMEX schemes for high-order unstructured methods. J. Comput. Phys. 280, 261–286.

Vermeire, B.C., Cagnone, J.S., Nadarajah, S., 2013. ILES using the correction procedure via reconstruction scheme. In: 51st AIAA Aerospace Sciences Meeting including the New Horizons Forum and Aerospace Exposition, Aerospace Sciences Meetings.

Vermeire, B.C., Nadarajah, S., Tucker, P.G., 2014. Canonical test cases for high-order unstructured implicit large Eddy simulation. In: 52nd Aerospace Sciences Meeting. AIAA SciTech.

Vermeire, B.C., Nadarajah, S., Tucker, P.G., 2016. Implicit large eddy simulation using the high-order correction procedure via reconstruction scheme. Int. J. Numer. Methods Fluids1097-0363. http:/dx.doi.org/10.1002/fld.4214.

Vincent, P.E., Jameson, A., 2011. Facilitating the adoption of unstructured high-order methods amongst a wider community of fluid dynamicists. Math. Model. Nat. Phenom. 6 (3), 97–140.

Vincent, P.E., Castonguay, P., Jameson, A., 2011. Insights from von Neumann analysis of high-order flux reconstruction schemes. J. Comput. Phys. 230 (22), 8134–8154.

Vincent, P.E., Castonguay, P., Jameson, A., 2011. A new class of high-order energy stable flux reconstruction schemes. J. Sci. Comput. 47 (1), 50–72.

Vincent, P.E., Witherden, F.D., Farrington, A.M., Ntemos, G., Vermeire, B.C., Park, J.S., Iyer, A.S., 2015. PyFR: next-generation high-order computational fluid dynamics on many-core hardware. In: AIAA Paper 2015-3050.

Wang, Z.J., Gao, H., 2009. A unifying lifting collocation penalty formulation including the discontinuous Galerkin, spectral volume/difference methods for conservation laws on mixed grids. J. Comput. Phys. 228 (21), 8161–8186.

Williams, D.M., 2013. Energy stable high-order methods for simulating unsteady, viscous, compressible flows on unstructured grids (Ph.D. thesis). Stanford University.

Williams, D.M., Jameson, A., 2013. Energy stable flux reconstruction schemes for advection-diffusion problems on tetrahedra. J. Sci. Comput. 59 (3), 721–7591.

Williams, D.M., Castonguay, P., Vincent, P.E., Jameson, A., 2013. Energy stable flux reconstruction schemes for advection-diffusion problems on triangles. J. Comput. Phys. 250, 53–76.

Williamson, C.H.K., 1988. The existence of two stages in the transition to three dimensionality of a cylinder wake. Phys. Fluids 31, 3165–3168.

Williamson, C.H.K., 1996. Vortex dynamics in the cylinder wake. Annu. Rev. Fluid Mech. 28, 477–539.

Witherden, F.D., Vincent, P.E., 2014. An analysis of solution point coordinates for flux reconstruction schemes on triangular elements. J. Sci. Comput. 61 (2), 398–423.

Witherden, F.D., Farrington, A.M., Vincent, P.E., 2014. PyFR: an open source framework for solving advection-diffusion type problems on streaming architectures using the flux reconstruction approach. Comput. Phys. Commun. 185 (11), 3028–3040.

Witherden, F.D., Vermeire, B.C., Vincent, P.E., 2015. Heterogeneous computing on mixed unstructured grids with PyFR. Comput. Fluids 120, 173–186.

Witherden, F.D., Park, J.S., Vincent, P.E., 2016. An analysis of solution point coordinates for flux reconstruction schemes on tetrahedral elements. J. Sci. Comput. 69 (2), 905–920.

Wood, J.R., Strasizar, T., Hathaway, M., 1990. Test Case E/CA-6, Subsonic Turbine Cascade T106. Test Cases for Computation of Internal Flows, AGARD-AR-275, July.

Wozniak, B.D., Witherden, F.D., Russell, F.P., Vincent, P.E., Kelly, P.H.J., 2016. Gimmik generating bespoke matrix multiplication kernels for accelerators: application to high-order computational fluid dynamics. Comput. Phys. Commun. 202, 12–22.

Yu, M., Wang, Z.J., 2013. On the connection between the correction and weighting functions in the correction procedure via reconstruction method. J. Sci. Comput. 54 (1), 227–244.

Chapter 11

Linear Stabilization for First-Order PDEs

A. Ern* and J.-L. Guermond†

*Université Paris-Est, CERMICS (ENPC), Marne-la-Vallée cedex 2, France

†Texas A&M University, College Station, TX, United States

Chapter Outline

ABSTRACT

We analyze finite-element-based stabilization techniques for first-order PDEs within the framework of symmetric Friedrichs' systems, including residual-based methods like Galerkin/least-squares and fluctuation-based methods like continuous interior penalty, local projection stabilization and subgrid viscosity.

Keywords: First-order PDEs, Linear stabilization, Galerkin least-squares, Streamline diffusion, Subgrid viscosity

2010 MSC: 35F05, 35F15, 65N12, 65N30, 65J10

Handbook of Numerical Analysis, Vol. 17. http://dx.doi.org/10.1016/bs.hna.2016.09.017

1 FRIEDRICHS' SYSTEMS

The objective of this section is to present the theory of the symmetric positive systems of first-order linear PDEs. This theory has been developed by Friedrichs (1958) to study transonic flows. Friedrichs wanted to handle within a single functional framework PDEs that are partly elliptic and partly hyperbolic, and for this purpose he developed a formalism that goes beyond the traditional classification of PDEs into elliptic, parabolic and hyperbolic types. Friedrichs' formalism is very powerful and encompasses several model problems. Important examples are the advection–reaction equation, the div-grad problem related to Darcy's equations and the curl–curl problem related to Maxwell's equations. This theory is an important key to understand stabilization techniques for first-order PDEs. All the theoretical arguments are presented assuming that the functions are complex valued.

1.1 Basic Ideas and Model Problem

Let D be a strongly Lipschitz domain in $\mathbb{R}^d$. We consider functions defined over D with values in $\mathbb{C}^m$, $m \geq 1$. Let $\mathcal{B},\mathcal{C} \in \mathbb{C}^{m\times m}$ be two Hermitian matrices, i.e. $\mathcal{B} = \mathcal{B}^{\mathrm{H}}$, $\mathcal{C} = \mathcal{C}^{\mathrm{H}}$, where $\mathcal{Z}^{\mathrm{H}}$ is the Hermitian transpose of $\mathcal{Z}$; we say that $\mathcal{B} \geq \mathcal{C}$ if and only if $X^{\mathrm{H}}\mathcal{B}X \geq X^{\mathrm{H}}\mathcal{C}X$ for all $X \in \mathbb{C}^m$.

Let $\mathcal{K}$, $\{\mathcal{A}^k\}_{k\in\{1:\,d\}}$ be a family of $(d+1)$ fields on D with values in $\mathbb{C}^{m\times m}$. We assume that these fields satisfy the following key assumptions:

$$\text{Boundedness}: \ \mathcal{K}, \{\mathcal{A}^k\}_{k\in\{1:d\}}, \text{ and } \mathcal{X} \text{ are in } L^\infty(D; \mathbb{C}^{m\times m}), \tag{1a}$$

$$\text{Symmetry}: \ \mathcal{A}^k = (\mathcal{A}^k)^{\mathrm{H}} \text{ for all } k \in \{1:d\}, \text{ a.e. in } D, \tag{1b}$$

$$\text{Positivity}: \ \exists \mu_0 > 0 \text{ s.t. } \mathcal{K} + \mathcal{K}^{\mathrm{H}} - \mathcal{X} \geq 2\mu_0 \mathbb{I}_m \text{ a.e. in } D. \tag{1c}$$

In (1c), $\mathbb{I}_m$ denotes the identity matrix in $\mathbb{C}^{m\times m}$ and $\mathcal{X} := \sum_{k=1}^d \partial_k \mathcal{A}^k$ where $\partial_k := \frac{\partial}{\partial x_k}$. Note that $\mathcal{X} = \mathcal{X}^{\mathrm{H}}$ owing to (1b). We now define two differential operators A and A_1 such that

$$Av := \mathcal{K}v + A_1 v, \quad A_1 v := \sum_{k\in\{1:\,d\}} \mathcal{A}^k \partial_k v, \quad \forall v \in C^1(\overline{D};\mathbb{C}^m). \tag{2}$$

In what follows, we assume that the fields $\{\mathcal{A}^k\}_{k\in\{1:\,d\}}$ have a bounded trace at the boundary ∂D, and we introduce the boundary field $\mathcal{N} \in L^\infty(\partial D; \mathbb{C}^{m\times m})$ such that $\mathcal{N} := \sum_{k\in\{1:\,d\}} n_k \mathcal{A}^k_{|\partial D}$, where $(n_k)_{k\in\{1:\,d\}}$ are the Cartesian components of the outward unit normal $\boldsymbol{n}$. Note that $\mathcal{N} = \mathcal{N}^{\mathrm{H}}$ owing to (1b).

Let $L := L^2(D; \mathbb{C}^m)$ and let us denote $(f, g)_L := \int_D g^{\mathrm{H}} f \,\mathrm{d}x$ for any f, $g \in L$; note that $(f, g)_L = \overline{(g, f)_L}$. Similarly we introduce $L^\partial := L^2(\partial D; \mathbb{C}^m)$ with the inner product $(f, g)_{L^\partial} := \int_{\partial D} g^{\mathrm{H}} f \,\mathrm{d}s$. Integration by parts using the

(Hermitian) inner product in L is a key tool in the analysis of Friedrichs' systems. To formalize this idea we define the formal adjoint $\tilde{A}$ of A such that

$$\tilde{A}v := (\mathcal{K}^{\mathrm{H}} - \mathcal{X})v - A_1 v = (\mathcal{K} + \mathcal{K}^{\mathrm{H}} - \mathcal{X})v - Av, \quad \forall v \in C^1(\overline{D};\mathbb{C}^m). \tag{3}$$

Lemma 1 (Integration by parts). *The following holds for all* $v, w \in C^1(\bar{D};\mathbb{C}^m)$*:*

$$(Av, w)_L = (v, \tilde{A}w)_L + (\mathcal{N}v, w)_{L^\partial}, \tag{4}$$

$$\Re\big((Av, v)_L\big) \geq \mu_0 \| v \|_L^2 + \frac{1}{2}(\mathcal{N}v, v)_{L^\partial}. \tag{5}$$

The lower bound (5) says that the sesquilinear form $(Av, w)_L$ is L-coercive up to a boundary term. The key idea of Friedrichs is to enforce a suitable boundary condition to gain positivity on the boundary term. This is done by assuming that there exists another boundary field $\mathcal{M} \in L^\infty(\partial D; \mathbb{C}^{m\times m})$ satisfying the following two algebraic properties a.e. on ∂D:

$$\mathcal{M} \text{ is nonnegative} : \Re(\xi^{\mathrm{H}}\mathcal{M}\xi) \geq 0 \text{ for all } \xi \in \mathbb{C}^m, \tag{6a}$$

$$\ker(\mathcal{M} - \mathcal{N}) + \ker(\mathcal{M} + \mathcal{N}) = \mathbb{C}^m. \tag{6b}$$

Since any function v satisfying $(\mathcal{M} - \mathcal{N})v_{|\partial D} = 0$ also verifies $(\mathcal{M}v, v)_{L^\partial} \in \mathbb{R}$, we infer using (6a) in (5) that

$$\Re\big((Av, v)_L\big) \geq \mu_0 \| v \|_L^2 + \frac{1}{2}(\mathcal{M}v, v)_{L^\partial} \geq \mu_0 \| v \|_L^2 . \tag{7}$$

Given $f \in L$, our goal is to find a function $u : D \to \mathbb{C}^m$ such that

$$Au = f \text{ in } D, \quad (\mathcal{M} - \mathcal{N})u = 0 \text{ on } \partial D. \tag{8}$$

Under the assumptions (1) and (6), Friedrichs proved: (i) the uniqueness of the strong solution $u \in C^1(\bar{D};\mathbb{C}^m)$ satisfying $(Au, v)_L = (f, v)_L$ for all $v \in L$ and $(\mathcal{M} - \mathcal{N})u = 0$ on ∂D; (ii) the existence of a so-called ultraweak solution $u \in L$ such that $(u, \tilde{A}v)_L = (f, v)_L$ for all $v \in C^1(\bar{D};\mathbb{C}^m)$ such that $(\mathcal{M}^{\mathrm{H}} + \mathcal{N})v = 0$ on ∂D. In Section 2, we introduce a mathematical setting relying on boundary operators instead of boundary fields to define a notion of weak solution for (8), and we prove well-posedness of the said formulation by using the BNB Theorem.

1.2 Example 1: Advection–Reaction Equation

Let $\mu \in L^\infty(D;\mathbb{R})$ and let $\boldsymbol{\beta} \in \boldsymbol{L}^\infty(D;\mathbb{R}^d)$ be such that $\nabla \cdot \boldsymbol{\beta} \in L^\infty(D;\mathbb{R})$. Given $f \in L := L^2(D;\mathbb{R})$, we want to find $u : D \to \mathbb{R}$ such that

$$\mu u + \boldsymbol{\beta} \cdot \nabla u = f \text{ in } D. \tag{9}$$

This equation models the transport of a solute of concentration u by a flow field with velocity $\boldsymbol{\beta}$, linear reaction coefficient μ ($\mu \geq 0$ corresponds to depletion) and source term f. To recover Friedrichs' formalism, we set $m = 1$, $\mathcal{K} = \mu$ and $\mathcal{A}^k = \beta_k$ for all $k \in \{1{:}\, d\}$, where $(\beta_k)_{k\in\{1:d\}}$ denote the Cartesian components of $\boldsymbol{\beta}$. The assumption (1a) holds since $\mu \in L^\infty(D;\mathbb{R})$, $\beta_k \in L^\infty(D;\mathbb{R})$ for all $k \in \{1{:}\, d\}$, and $\mathcal{X} = \nabla\cdot\boldsymbol{\beta} \in L^\infty(D;\mathbb{R})$. The assumption (1b) is trivially satisfied since $m = 1$. Finally, the assumption (1c) is satisfied provided we assume that

$$\mu_0 := \operatorname*{ess\,inf}_{x\in D} \left(\mu - \frac{1}{2}\nabla\cdot\boldsymbol{\beta}\right)(\boldsymbol{x}) > 0. \tag{10}$$

The boundary field is $\mathcal{N} = \boldsymbol{\beta}\cdot\boldsymbol{n}$, and the integration by parts formula (4) is a reformulation of $\int_D ((\nabla\cdot\boldsymbol{\beta})vw + v(\boldsymbol{\beta}\cdot\nabla w) + w(\boldsymbol{\beta}\cdot\nabla v))\,\mathrm{d}x = \int_{\partial D}(\boldsymbol{\beta}\cdot\boldsymbol{n})vw\,\mathrm{d}s$.

To enforce a suitable boundary condition, we need to consider the sign of $(\boldsymbol{\beta}\cdot\boldsymbol{n})$ at the boundary. We define the inflow boundary $\partial D^- = \{\boldsymbol{x} \in \partial D \mid (\boldsymbol{\beta}\cdot\boldsymbol{n})(\boldsymbol{x}) < 0\}$, the outflow boundary $\partial D^+ = \{\boldsymbol{x} \in \partial D \mid (\boldsymbol{\beta}\cdot\boldsymbol{n})(\boldsymbol{x}) > 0\}$, and the characteristic boundary $\partial D^0 = \{\boldsymbol{x} \in \partial D \mid (\boldsymbol{\beta}\cdot\boldsymbol{n})(\boldsymbol{x}) = 0\}$. Then, the inflow boundary condition $u = 0$ on ∂D^- can be enforced by using the boundary field $\mathcal{M} = |\boldsymbol{\beta}\cdot\boldsymbol{n}|$ which satisfies (6). Finally, the L-coercivity property (7) becomes

$$(Av, v)_L \geq \mu_0 \| v \|_L^2 + \frac{1}{2}\int_{\partial D} |\boldsymbol{\beta}\cdot\boldsymbol{n}| v^2\,\mathrm{d}s.$$

1.3 Example 2: Maxwell's Equations

We consider the time-harmonic version of Maxwell's equations in the low-frequency regime where the displacement currents are negligible. Let σ be the electrical conductivity, μ the magnetic permeability, $\omega > 0$ the angular frequency and $i^2 = -1$. We assume that $\mu, \sigma \in L^\infty(D;\mathbb{R})$, and for simplicity, that both μ and σ are real valued. Given $\boldsymbol{j} \in \boldsymbol{L}^2(D) := L^2(D;\mathbb{C}^3)$ and setting $\widetilde{\mu} = \omega\mu$, we want to find functions $\boldsymbol{E}: D \to \mathbb{C}^3$ and $\boldsymbol{H}: D \to \mathbb{C}^3$ such that

$$\sigma\boldsymbol{E} - \nabla\times\boldsymbol{H} = \boldsymbol{j} \text{ in } D, \quad i\widetilde{\mu}\boldsymbol{H} + \nabla\times\boldsymbol{E} = \boldsymbol{0} \text{ in } D. \tag{11}$$

To recover Friedrichs' formalism, we set $m = 6$, $u := (\boldsymbol{E}, \boldsymbol{H})$, $\mathcal{K} = e^{i\theta}\begin{pmatrix} \sigma\mathbb{I}_3 & \mathbb{O}_3 \\ \mathbb{O}_3 & \widetilde{\mu}\,\mathbb{I}_3 \end{pmatrix}$ with $\theta = \frac{\pi}{4}$, and $\mathcal{A}^k = \begin{pmatrix} \mathbb{O}_3 & -e^{i\theta}\mathbb{J}^k \\ e^{-i\theta}\mathbb{J}^k & \mathbb{O}_3 \end{pmatrix}$, for all $k \in \{1{:}\, d\}$, where $\mathbb{I}_3$ and $\mathbb{O}_3$ are the identity and null matrix in $\mathbb{C}^{3\times3}$, respectively, and $\mathbb{J}^k_{ij} = \varepsilon_{ikj}$, for all $i, j, k \in \{1, 2, 3\}$, with ε_{ikj} the Levi-Civita symbol. The assumption (1a) holds since $\sigma, \mu \in L^\infty(D;\mathbb{R})$ and $\mathcal{X}$ is the null matrix in $\mathbb{C}^{6\times6}$. The assumption (1b) holds since, $\mathbb{J}^k$ being skew-symmetric, we have $(-e^{i\theta}\mathbb{J}^k)^{\mathrm{H}} = -e^{-i\theta}(\mathbb{J}^k)^{\mathrm{T}} = e^{-i\theta}\mathbb{J}^k$. Finally, the assumption (1c) is satisfied provided we assume that

$$\sigma_{\flat,D} := \operatorname*{ess\,inf}_{x\in D} \sigma(\boldsymbol{x}) > 0, \qquad \widetilde{\mu}_{\flat,D} := \operatorname*{ess\,inf}_{x\in D} \widetilde{\mu}(\boldsymbol{x}) > 0. \tag{12}$$

The boundary field is $\mathcal{N}=\begin{pmatrix} \mathbb{O}_3 & e^{i\theta}\mathbb{T} \\ -e^{-i\theta}\mathbb{T} & \mathbb{O}_3 \end{pmatrix}$, where $\mathbb{T}_{ij}=\sum_{k=1}^{3} n_k \varepsilon_{ijk}$, for all $i, j \in\{1, 2, 3\}$. Note that the definition of $\mathbb{T}$ implies that $\mathbb{T}\boldsymbol{\xi}=\boldsymbol{\xi}\times\boldsymbol{n}$ for all $\boldsymbol{\xi}\in\mathbb{C}^3$. The integration by parts formula (4) results from $\int_D(\boldsymbol{b}\cdot(\nabla\times\boldsymbol{E})-\boldsymbol{E}\cdot(\nabla\times\boldsymbol{b}))\,\mathrm{d}x=\int_{\partial D}\boldsymbol{b}\cdot(\boldsymbol{n}\times\boldsymbol{E})\,\mathrm{d}s$.

The boundary conditions $\boldsymbol{H}\times\boldsymbol{n}_{|\partial D}=\boldsymbol{0}$ and $\boldsymbol{E}\times\boldsymbol{n}_{|\partial D}=\boldsymbol{0}$ can be enforced, respectively, by using the boundary fields $\mathcal{M}_H=\begin{pmatrix} \mathbb{O}_3 & -e^{i\theta}\mathbb{T} \\ -e^{-i\theta}\mathbb{T} & \mathbb{O}_3 \end{pmatrix}$ and $\mathcal{M}_E=\begin{pmatrix} \mathbb{O}_3 & e^{i\theta}\mathbb{T} \\ e^{-i\theta}\mathbb{T} & \mathbb{O}_3 \end{pmatrix}$, which both satisfy (6), and the coercivity property (7) becomes

$$\Re(A(\boldsymbol{E},\boldsymbol{H}),(\boldsymbol{E},\boldsymbol{H}))_{L^2(D;\mathbb{C}^6)}\geq\frac{1}{\sqrt{2}}\left(\sigma_{\flat,D}\,\|\boldsymbol{E}\|^2_{\boldsymbol{L}^2(D)}+\tilde{\mu}_{\flat,D}\,\|\boldsymbol{H}\|^2_{\boldsymbol{L}^2(D)}\right).$$

2 WEAK FORMULATION AND WELL-POSEDNESS FOR FRIEDRICHS' SYSTEMS

The aim of this section is to devise a weak formulation of Friedrichs' systems for which well-posedness can be established by using the Banach–Nečas–Babuška (BNB) Theorem which provides necessary and sufficient conditions for well-posedness in the form of inf–sup conditions, see Ern and Guermond (2004, Thm. 2.6). The material is inspired from a series of papers by Ern and Guermond (2006a,b).

2.1 The Graph Space

We consider the space $S:=C_0^\infty(D;\mathbb{C}^m)$, composed of the smooth $\mathbb{C}^m$-valued fields compactly supported in D, and the Hilbert space $L:=L^2(D;\mathbb{C}^m)$, which we use as pivot space (i.e. $L\equiv L'$). While other functional settings could be considered, we will see in the forthcoming sections that L^2 plays a prominent role in a large class of stabilized finite element techniques.

The operators A and $\tilde{A}$ defined in (2) and (3), respectively, are each bounded in S with values in L and the following holds: There is c such that

$$(A\phi,\psi)_L=(\phi,\tilde{A}\psi)_L,\quad\forall\phi,\psi\in S,\tag{13a}$$

$$\|(A+\tilde{A})\phi\|_L\leq c\,\|\phi\|_L\quad\forall\phi\in S.\tag{13b}$$

The equality (13a) follows from Lemma 1, while (13b) follows from the definitions of A and $\tilde{A}$ and the boundedness property (1a). Let us define the inner product $(\cdot,\cdot)_V:=\mu_0(\cdot,\cdot)_L+\mu_0^{-1}(A_1(\cdot),A_1(\cdot))_L$ and let the induced norm be denoted by $\|\cdot\|_V$ with $\|\cdot\|_V^2 = \mu_0 \|\cdot\|_L^2 + \mu_0^{-1} \| A_1(\cdot) \|_L^2$ (the scaling factors μ_0 and μ_0^{-1} are introduced so that both terms have coherent units).

Let V_S be the completion of S with respect to the norm $\|\cdot\|_V$, i.e. $V_S = \bar{S}^V$. Using L as pivot space leads to $S \subset V_S \hookrightarrow L \equiv L' \hookrightarrow V_S' \subset S'$, where S' is the algebraic dual of S and L', V_S' are topological duals. By density, the operators A and $\tilde{A}$ can be extended to bounded linear operators from V_S to L; we say that V_S is the *minimal domain* of A and $\tilde{A}$. Owing to (13), we infer by density that $(A\phi, \psi)_L = (\phi, \tilde{A}\psi)_L$, for all $\phi, \psi \in V_S$. Let now $v \in L$; then, Av can be defined in V_S' by setting $\langle Av, \phi \rangle_{V_S', V_S} = (v, \tilde{A}\phi)_L$, for all $\phi \in V_S$. This definition allows us to extend A to a bounded linear operator from L to V_S'. Similarly we define $\langle \tilde{A}v, \phi \rangle_{V_S', V_S} = (v, A\phi)_L$, for all $v \in L$ and all $\phi \in V_S$. Since $L \subset V_S'$, it makes sense to define the *graph space* (or *maximal domain* of A and $\tilde{A}$) as

$$V := \{v \in L;\ A_1 v \in L\}. \tag{14}$$

By construction, $A \in \mathcal{L}(V; L)$, $\tilde{A} \in \mathcal{L}(V; L)$.

Proposition 1 (Hilbert space). *The graph space V is a Hilbert space when equipped with the inner product $(\cdot, \cdot)_V$. The norm $\|\cdot\|_V$ is called the graph norm.*

2.2 The Boundary Operators

Since A_1 is a first-order differential operator, defining the trace at the boundary of a function in the graph space V is not straightforward. The trace can be given a meaning in $H^{-\frac{1}{2}}(\partial D; \mathbb{C}^m)$, see Rauch (1994). However, this meaning is not suitable for the weak formulation we have in mind; this is why we now introduce two additional operators N and M to replace the boundary fields $\mathcal{N}$ and $\mathcal{M}$. We define the operator $N \in \mathcal{L}(V; V')$ by (compare with (4))

$$\langle Nv, w \rangle_{V',V} := (Av, w)_L - (v, \tilde{A}w)_L, \quad \forall v, w \in V. \tag{15}$$

This definition makes sense since both A and $\tilde{A}$ are in $\mathcal{L}(V;L)$. Moreover, the operator N is self-adjoint since (15) can be rewritten as

$$\langle Nv, w \rangle_{V',V} = (\mathcal{X}v, w)_L + (A_1 v, w)_L + (v, A_1 w)_L, \tag{16}$$

so that $\langle Nv, w \rangle_{V',V} = \overline{\langle Nw, v \rangle_{V',V}}$. Furthermore, we have $V_S \subset \ker(N)$ and $\operatorname{im}(N) \subset V_S^{\perp} = \{v' \in V' | \forall \phi \in V_S, \langle v', \phi \rangle_{V_S', V_S} = 0\}$. Actually, as proved in Ern et al. (2007), one has $\ker(N) = V_S$, $\operatorname{im}(N) = V_S^{\perp}$. The fact that $\ker(N) = V_S$ means that N is a *boundary operator*.

Boundary conditions in Friedrichs' systems can be formulated by assuming that there exists an operator $M \in \mathcal{L}(V;V')$ such that

$$M \text{ is monotone, i.e. } \Re(\langle Mv, v\rangle_{V',V}) \geq 0 \text{ for all } v \in V, \tag{17a}$$

$$\ker(N-M) + \ker(N+M) = V. \tag{17b}$$

Let $M^* \in \mathcal{L}(V; V')$ denote the adjoint operator of M, so that $\langle M^*w, v\rangle_{V',V} = \overline{\langle Mv, w\rangle_{V',V}}$. It is proved in Ern et al. (2007) that, under the assumptions (17),

$$\ker(N) = \ker(M) = \ker(M^*) \text{ and } \operatorname{im}(N) = \operatorname{im}(M) = \operatorname{im}(M^*).$$

In particular, M is a *boundary operator*, just like N.

2.3 Well-Posedness

Given $f \in L$, the problem we want to solve (compare with (8)) is to find

$$u \in V_0 := \ker(M-N) \text{ such that } Au = f \text{ in } L. \tag{18}$$

To recast this problem into a weak form, we introduce the sesquilinear form $a(v, w) := (Av, w)_L$, for all $(v, w) \in V \times L$. Letting $\ell(w) := (f, w)_L$, we consider the following weak problem:

$$\begin{cases} \text{Find } u \in V_0 \text{ such that} \\ a(u,w) = \ell(w), \quad \forall w \in L. \end{cases} \tag{19}$$

Theorem 1 (Well-posedness). *Let N be defined by (15) and assume (1), then $\Re(a(v,v)) \geq \mu_0 \| v \|_L^2 + \frac{1}{2}\langle Nv, v\rangle_{V',V}, \forall v \in V$. Moreover, let M satisfy (17), then $\Re(a(v,v)) \geq \mu_0 \| v \|_L^2 + \frac{1}{2}\Re(\langle Mv, v\rangle_{V',V}) \geq \mu_0 \| v \|_L^2, \forall v \in V_0$. If (1) and (17) hold, then the model problem (19) is well-posed, i.e. $A: V_0 \to L$ is an isomorphism.*

Remark 1 (Positivity assumption (1c)). This assumption can be relaxed if the missing control on $\|v\|_L$ can be recovered from an estimate on $\|A_1 v\|_L$. This is possible in the context of elliptic PDEs in mixed form by invoking a Poincaré-type inequality. Furthermore, everything that is said hereafter holds true by assuming that $A = K + A_1$ where K is a bounded operator on L satisfying the assumption $((K+K^*)v - \mathcal{X}v, v)_L \geq 2\mu_0 \| v \|_L^2$. The formal adjoint is then defined by $\tilde{A}v = K^*v - \mathcal{X}v + A_1 v$. For instance let $D = (0, a)\times(-1, 1)$, $a > 0$, and let $K: L \to L$, with $L = L^2(D;\mathbb{R})$, be such that $Kv(x,y) = v(x,y) - \frac{\sigma}{2}\int_{-1}^{+1} v(x,\xi)\,d\xi$ where $\sigma \in [0, 1)$. Then $((K+K^*)v, v)_L = 2(Kv, v)_L \geq 2 \| v \|_L^2 - 2\sigma \| v \|_L^2 = 2(1-\sigma) \| v \|_L^2$. This is the type of structure one encounters when solving the neutron transport equation.

Example 1 (Advection–reaction). The bilinear form a is

$$a(v,w) = \int_D (\mu v w + (\boldsymbol{\beta} \cdot \nabla v) w)\,\mathrm{d}x, \quad \forall v \in V, \quad \forall w \in L^2(D;\mathbb{R}),$$

with $V = \{v \in L^2(D;\mathbb{R}) \mid \boldsymbol{\beta} \cdot \nabla v \in L^2(D;\mathbb{R})\}$. Moreover,

$$\langle Nv, w\rangle_{V',V} = \int_D ((\nabla \cdot \boldsymbol{\beta}) v w + w(\boldsymbol{\beta} \cdot \nabla v) + v(\boldsymbol{\beta} \cdot \nabla w))\,\mathrm{d}x.$$

A result on traces of functions in V is needed to link N with $\mathcal{N} = \boldsymbol{\beta} \cdot \boldsymbol{n}$. Such a result is not straightforward, since the trace theorem for functions in $H^s(D;\mathbb{R})$, $s > \frac{1}{2}$, cannot be applied. It is shown in Ern and Guermond (2006a) that if the inflow and outflow boundaries are well separated, i.e. $\min_{(\boldsymbol{x},\boldsymbol{y}) \in \partial D^- \times \partial D^+} \|\boldsymbol{x} - \boldsymbol{y}\|_{\ell^2(\mathbb{R}^d)} > 0$, then the trace operator $\gamma : C^0(\bar{D}) \to C^0(\partial D)$ such that $\gamma(v) = v_{|\partial D}$ can be extended to a bounded linear operator from V to $L^2_{|\boldsymbol{\beta}\cdot\boldsymbol{n}|}(\partial D;\mathbb{R})$, where the subscript $|\boldsymbol{\beta}\cdot\boldsymbol{n}|$ means that the measure $\mathrm{d}s$ is replaced by $|\boldsymbol{\beta}\cdot\boldsymbol{n}|\,\mathrm{d}s$. This result implies that $\langle Nv, w\rangle_{V',V} = \int_{\partial D} \mathcal{N} v w\,\mathrm{d}s$ for all v, $w \in V$. Furthermore, the inflow boundary condition $u = 0$ on ∂D^- can be enforced by means of the boundary operator $M \in \mathcal{L}(V;V')$ defined by $\langle Mv, w\rangle_{V',V} = \int_{\partial D} |\boldsymbol{\beta} \cdot \boldsymbol{n}| v w\,\mathrm{d}s$, which satisfies (17). Note that the separation assumption cannot be circumvented if one wishes to work with traces in $L^2_{|\boldsymbol{\beta}\cdot\boldsymbol{n}|}(\partial D;\mathbb{R})$, regardless of the regularity of $\boldsymbol{\beta}$. For instance, let $D = \{(x_1,x_2) \in \mathbb{R}^2 \mid 0 < x_2 < 1 \text{ and } |x_1| < x_2\}$ with $\boldsymbol{\beta} = (1, 0)^{\mathrm{T}}$. One can verify that the function $u(x_1,x_2) = x_2^\alpha$ is in V for $\alpha > -1$, but $u_{|\partial D} \in L^2(|\boldsymbol{\beta}\cdot\boldsymbol{n}|;\, \partial D)$ only if $\alpha > -\frac{1}{2}$.

Example 2 (Maxwell). The sesquilinear form a is

$$a(v,w) = \int_D \left(e^{i\theta} \sigma \boldsymbol{E} \cdot \bar{\boldsymbol{e}} + i e^{-i\theta}\, \tilde{\mu} \boldsymbol{H} \cdot \bar{\boldsymbol{b}} - e^{i\theta} (\nabla \times \boldsymbol{H}) \cdot \bar{\boldsymbol{e}} + e^{-i\theta} (\nabla \times \boldsymbol{E}) \cdot \bar{\boldsymbol{b}}\right) \mathrm{d}x,$$

for all $v = (\boldsymbol{E}, \boldsymbol{H}) \in V$ and all $w = (\boldsymbol{e}, \boldsymbol{b}) \in L$ (note that we use the Euclidean dot product and write the complex conjugate explicitly), with $V = \boldsymbol{H}(\mathrm{curl};\, D) \times \boldsymbol{H}(\mathrm{curl};\, D)$, $\boldsymbol{H}(\mathrm{curl};D) = \{\boldsymbol{A} \in L^2(D;\mathbb{C}^3); \nabla \times \boldsymbol{A} \in L^2(D;\mathbb{C}^3)\}$, and $L = L^2(D;\mathbb{C}^6)$. Moreover,

$$\langle N(\boldsymbol{E},\boldsymbol{H}), (\boldsymbol{e},\boldsymbol{b})\rangle_{V',V} = e^{i\theta} t(\boldsymbol{H},\boldsymbol{e}) - e^{-i\theta} t(\boldsymbol{E},\boldsymbol{h}),$$

where $t(\boldsymbol{A},\boldsymbol{a}) = \int_D (\boldsymbol{A} \cdot (\nabla \times \bar{\boldsymbol{a}}) - (\nabla \times \boldsymbol{A}) \cdot \bar{\boldsymbol{a}})\,\mathrm{d}x$. Since $\boldsymbol{E} \times \boldsymbol{n}$ and $\boldsymbol{H} \times \boldsymbol{n}$ are in $\boldsymbol{H}^{-1/2}(\partial D)$, if $\boldsymbol{e}$ and $\boldsymbol{b}$ are in $\boldsymbol{H}^1(D)$, we have $\langle N(\boldsymbol{E},\boldsymbol{H}), (\boldsymbol{e},\boldsymbol{b})\rangle_{V',V} = e^{i\theta} \langle \boldsymbol{H} \times \boldsymbol{n}, \boldsymbol{e}\rangle_{\boldsymbol{H}^{-\frac{1}{2}},\boldsymbol{H}^{\frac{1}{2}}} - e^{-i\theta} \langle \boldsymbol{E} \times \boldsymbol{n}, \boldsymbol{b}\rangle_{\boldsymbol{H}^{-\frac{1}{2}},\boldsymbol{H}^{\frac{1}{2}}}$. The boundary condition $\boldsymbol{H} \times \boldsymbol{n} = \boldsymbol{0}$ can be enforced by means of the boundary operator $\langle M(\boldsymbol{E},\, \boldsymbol{H}),\, (\boldsymbol{e},\, \boldsymbol{b})\rangle_{V',V} = -e^{i\theta} t(\boldsymbol{H},\, \boldsymbol{e}) - e^{-i\theta} t(\boldsymbol{E},\, \boldsymbol{h})$, which satisfies (17).

3 RESIDUAL-BASED STABILIZATION

This section is concerned with the approximation of Friedrichs' systems using H^1-conforming finite elements in a standard Galerkin setting. The main issue one faces in this context is to achieve stability. At the continuous level, the proof of Theorem 1 shows that one needs to consider the first-order derivative $A_1 v$ as test function to control the graph norm of a function v. Unfortunately, this possibility is lost when working with H^1-conforming finite elements since the first-order derivative of v can no longer be represented by discrete test functions. As a result, one needs to devise suitable stabilization mechanisms. Those presented in this section are inspired by the least-squares (LS), or minimal residual, technique from linear algebra. The LS approximation gives optimal error estimates in the graph norm, but, unfortunately, gives suboptimal L^2-error estimates in most situations. The Galerkin/least-squares (GaLS) method improves the situation by combining the standard Galerkin approach with the LS technique and mesh-dependent weights. GaLS gives quasi-optimal L^2-error estimates and optimal graph-norm estimates. We further improve GaLS in the next section by introducing a boundary penalty technique that enforces boundary conditions weakly in the spirit of the theory of Friedrichs' systems.

3.1 Least-Squares Formulation

Given $f \in L$, let us consider the model problem (19). This problem is well-posed, see Theorem 1. The LS version of problem (19) is the following:

$$\begin{cases} \text{Find } u \in V_0 \text{ such that} \\ a^{\mathrm{LS}}(u, w) := (Au, Aw)_L = (f, Aw)_L, \quad \forall w \in V_0. \end{cases} \tag{20}$$

Observe that the test space is the same as the solution space in (20). Since $A : V_0 \to L$ is an isomorphism, requiring that $(Au, Aw)_L = (f, Aw)_L$ for all $w \in V_0$ is equivalent to ask that $(Au, w)_L = (f, w)_L$ for all $w \in L$. Hence, the problems (19) and (20) are equivalent. Actually, the well-posedness of (20) is a direct consequence of the Lax–Milgram Lemma, since there are real numbers $0 < \alpha \leq \varpi < \infty$ such that $\alpha \|v\|_V \leq \|Av\|_L \leq \varpi \|v\|_V$ for all $v \in V_0$.

Proposition 2 (V_0-coercivity). *a^{LS} is bounded and coercive on V_0.*

Remark 2 (Minimal residual). Consider the functional $\mathfrak{J} : V_0 \to \mathbb{R}$ defined by $\mathfrak{J}(v) := \frac{1}{2} \| Av - f \|_L^2$ for all $v \in V_0$. The Fréchet derivative of $\mathfrak{J}$ is such that $D\mathfrak{J}(v)(w) = \Re((Av - f, Aw)_L)$ for all $w \in V_0$, i.e. the problem (20) amounts to $D\mathfrak{J}(v) = 0$. Since the functional $\mathfrak{J}$ is strictly convex, the solution u of (20) is the global minimizer of $\mathfrak{J}$ over V_0. This LS technique is well known in the linear algebra context where it can be traced back to Gauss and Legendre. Starting from the linear system $\mathcal{A}U = B$ with $\mathcal{A}$ invertible and multiplying by $\mathcal{A}^{\mathrm{H}}$ leads to the so-called normal equations $(\mathcal{A}^{\mathrm{H}}\mathcal{A})U = \mathcal{A}^{\mathrm{H}}B$ where the matrix $\mathcal{A}^{\mathrm{H}}\mathcal{A}$ is Hermitian positive-definite.

3.2 Least-Squares Approximation Using Finite Elements

We assume that, for all $h > 0$, we have at hand a finite-dimensional space $V_{h0} \subset V_0$ built by using a shape-regular mesh sequence $(\mathcal{T}_h)_{h>0}$ and a finite element of degree $k \geq 1$. For simplicity, we consider the equal-order case for all the solution components. The space V_{h0} is H^1-conforming and composed of continuous, piecewise polynomial functions in $\bar{D}$. Let us assume now that we have at hand a quasi-interpolation operator $\mathcal{I}_{h0} : V_0 \rightarrow V_{h0}$ with optimal local approximation properties: There is a uniform constant c such that

$$\| v - \mathcal{I}_{h0}(v) \|_{L(K)} + h_K \| \nabla(v - \mathcal{I}_{h0}(v)) \|_{L(K)} \leq c\, h_K^{1+r} |v|_{H^{1+r}(D_K, \mathbb{C}^m)}, \qquad (21)$$

for all $r \in [0,\ k]$, all $v \in H^{1+r}(D, \mathbb{C}^m) \cap V_0$, and all $K \in \mathcal{T}_h$, with $L(K) := L^2(K; \mathbb{C}^m)$ and where D_K is the interior of the set composed of all the mesh cells having a nonempty intersection with K.

We construct a discrete counterpart of (20) as follows:

$$\begin{cases} \text{Find } u_h \in V_{h0} \text{ such that} \\ a^{\mathrm{LS}}(u_h, w_h) = (f, A w_h)_L, \quad \forall w_h \in V_{h0}. \end{cases} \qquad (22)$$

Theorem 2 (Well-posedness and error bound). *The problem (22) has a unique solution u_h, and the following error bound holds:*

$$\| u - u_h \|_V \leq \frac{\varpi}{\alpha} \inf_{v_h \in V_{h0}} \| u - v_h \|_V. \qquad (23)$$

Using (21), we infer the following approximation result in the graph norm: $\| u - \mathcal{I}_{h0}(u) \|_V \leq c\, \mu_0^{-\frac{1}{2}} \phi_D h^r |u|_{H^{1+r}(D;\, \mathbb{C}^m)}$, with $\phi_D := \max(\beta_D, \mu_0 h)$ and $\beta_D = \max_{k \in \{1:d\}} \| \mathcal{A}^k \|_{L^\infty(D; \mathbb{C}^{m \times m})}$. Assuming $u \in H^{1+r}(D; \mathbb{C}^m)$ and using the above approximation result, we infer that

$$\mu_0^{\frac{1}{2}} \| u - u_h \|_L + \mu_0^{-\frac{1}{2}} \| A_1(u - u_h) \|_L \leq c\, \mu_0^{-\frac{1}{2}} \phi_D h^r |u|_{H^{1+r}(D; \mathbb{C}^m)}. \qquad (24)$$

When $r = k$, the estimate on $\|A_1(u - u_h)\|_L$ is optimal, but the estimate on $\|u - u_h\|_L$ is *suboptimal* by one order. It is sometimes possible to improve the L-norm error estimate by means of the Aubin–Nitsche duality argument, but this is not systematic since, very often, first-order PDEs do not have a smoothing property. For instance, this improvement is possible for the one-dimensional transport equation and for Darcy's equation.

The LS technique has gained popularity in the numerical analysis community at the beginning of the 1970s following a series of papers by Bramble and

Schatz (1970, 1971), although it was already popular in the Russian literature (see Džiškariani, 1968; Lučka, 1969).

3.3 Galerkin/Least-Squares

In this section, we devise and analyze a GaLS approximation introduced in Hughes et al. (1989). A nonsymmetric variant known under the names Streamline Upwind Petrov–Galerkin (SUPG) or streamline diffusion method has been introduced in Brooks and Hughes (1982) and analyzed in Johnson et al. (1984), see Example 4.

We define the following local quantities:

$$\beta_K = \max_{k\in\{1:d\}} \| \mathcal{A}^k \|_{L^\infty(K;\mathbb{C}^{m\times m})}, \tag{25}$$

$$\tau_K = \left(\max\left(\beta_K h_K^{-1}, \mu_0\right)\right)^{-1} = \min\left(\beta_K^{-1} h_K, \mu_0^{-1}\right), \tag{26}$$

for all $K \in \mathcal{T}_h$, where μ_0 is defined in (1c) (the second equality is meaningful if β_K is nonzero; if $\beta_K = 0$, then $\tau_K = \mu_0^{-1}$). For instance, for the advection–reaction equation, μ_0 is the reciprocal of a time, β_K is a local velocity and τ_K is a local time scale. With a slight abuse of notation, we define the piecewise constant function $\tau : D \to \mathbb{R}$ such that $\tau_{|K} = \tau_K$ for all $K \in \mathcal{T}_h$. In what follows, we consider the Euclidean (or Hermitian) norm denoted $\|\cdot\|_{\ell^2}$ for $\mathbb{C}^{m\times m}$-valued fields, we set $\|\cdot\|_{L^\infty(D;\mathbb{C}^{m\times m})} = \| \, \|\cdot\|_{\ell^2} \, \|_{L^\infty(D;\mathbb{R})}$ and we assume for simplicity that

$$\max\left(\| \mathcal{K} \|_{L^\infty(D;\mathbb{C}^{m\times m})}, \| \mathcal{X} \|_{L^\infty(D;\mathbb{C}^{m\times m})}\right) \le c_{\mathcal{K},\mathcal{X}} \mu_0, \tag{27}$$

and we hide the factor $c_{\mathcal{K},\mathcal{X}}$ in the generic constants used in the error analysis.

We consider the finite element setting of Section 3.2. We define the following discrete sesquilinear forms on $V_{h0} \times V_{h0}$:

$$a_h(v_h, w_h) := (Av_h, w_h)_L + r_h(v_h, w_h), \quad r_h(v_h, w_h) := (Av_h, \tau A w_h)_L. \tag{28}$$

The sesquilinear form $(Av_h, w_h)_L$ is the Galerkin part of the formulation and the term $r_h(v_h, w_h)$ is the least-squares part. The role of r_h is to stabilize the formulation. We consider the following discrete problem:

$$\begin{cases} \text{Find } u_h \in V_{h0} \text{ such that} \\ a_h(u_h, w_h) = \ell_h(w_h) := (f, w_h + \tau A w_h)_L, \quad \forall w_h \in V_{h0}. \end{cases} \tag{29}$$

As usual the four steps of the analysis consist of (i) establishing stability, (ii) estimating the consistency error, (iii) proving a boundedness estimate and (iv) using the approximation properties of finite elements. We set $V_\flat = V_0 + V_{h0}$ and observe that $V_\flat = V_0$ since the approximation is V_0-conforming. Proceeding in the spirit of Strang's Second Lemma for the error analysis, we extend the sesquilinear form a_h to $V_0 \times V_{h0}$, and we equip the space V_0 with the norms:

$$\| v \|_{V_\flat}^2 := \mu_0 \| v \|_L^2 + \frac{1}{2}|v|_M^2 + \| \tau^{\frac{1}{2}} A v \|_L^2, \quad \| v \|_{V_{\flat\sharp}}^2 := \| v \|_{V_\flat}^2 + \| \tau^{-\frac{1}{2}} v \|_L^2, \tag{30}$$

with the boundary seminorm $|v|_M^2 := \Re(\langle Mv, v\rangle_{V',V})$.

Theorem 3 (Convergence). *(i) The discrete sesquilinear form a_h satisfies $\Re(a_h(v_h,v_h)) \geq \| v_h \|_{V_\flat}^2$, for all $v_h \in V_{h0}$. Consequently, the discrete problem (29) is well-posed. (ii) The discrete problem (29) is exactly consistent. (iii) There is c, uniform with respect to h, such that, $|a_h(v,w_h)| \leq c \, \| v\|_{V_{\flat\sharp}} \, \| w_h\|_{V_\flat}$ for all $(v, w_h) \in V_0 \times V_{h0}$. (iv) Let u be the unique solution to (19) and let u_h be the unique solution to (29). There is c, uniform with respect to h, such that*

$$\| u - u_h\|_{V_\flat} \leq c \inf_{v_h \in V_{h0}} \| u - v_h\|_{V_{\flat\sharp}}. \tag{31}$$

Moreover $\| u - u_h \|_{V_\flat}^2 \leq c \sum_{K\in\mathcal{T}_h} \max(\beta_K, \mu_0 h_K) h_K^{2r+1} |u|_{H^{1+r}(D_K;\mathbb{C}^m)}^2$ if $u \in H^{1+r}(D;\mathbb{C}^m)$, for all $r \in [0, k]$. Upon denoting $\phi_D := \max(\beta_D, \mu_0 h)$ and $\beta_D := \max_{K\in\mathcal{T}_h} \beta_K$, this implies in particular that $\| u - u_h\|_{V_\flat} \leq c \, \phi_D^{\frac{1}{2}} h^{r+\frac{1}{2}} |u|_{H^{1+r}(D;\mathbb{C}^m)}$.

Assuming $u \in H^{k+1}(D;\mathbb{C}^m)$, the above result implies that

$$\mu_0^{\frac{1}{2}} \| u - u_h\|_L + \| \tau^{\frac{1}{2}} A_1 (u - u_h)\|_L \leq c \, \phi_D^{\frac{1}{2}} h^{k+\frac{1}{2}} |u|_{H^{k+1}(D;\mathbb{C}^m)}.$$

Observe that the estimate on $\|u - u_h\|_L$ is improved by half a power in h when compared to that obtained with the LS technique, and the estimate on $\|A_1(u - u_h)\|_L$ is now a localized version of the LS estimate (24).

Example 3 (Advection–reaction). Consider the PDE $\mu u + \boldsymbol{\beta}\cdot\nabla u = f$ with the inflow boundary condition $u = 0$ on ∂D^-, see Section 1.2. Assume that all the mesh boundary faces are a subset of either ∂D^- or $\partial D \backslash \partial D^-$. Let $P_k^g(\mathcal{T}_h)$ be the H^1-conforming finite element space constructed on the mesh $\mathcal{T}_h$ using finite elements of degree $k \geq 1$ (Ern and Guermond, 2016). Set $V_{h0} := \{v_h \in P_k^g(\mathcal{T}_h) \mid v_{h|\partial D^-} = 0\}$. The GaLS discretization consists of seeking $u_h \in V_{h0}$ such that

$$\int_D (\mu u_h + \boldsymbol{\beta}\cdot\nabla u_h) w_h \, dx + \int_D \tau(\mu u_h + \boldsymbol{\beta}\cdot\nabla u_h)(\mu w_h + \boldsymbol{\beta}\cdot\nabla w_h) \, dx = \ell_h(w_h),$$

for all $w_h \in V_{h0}$, with $\tau_K = \min(\beta_K^{-1} h_K, \mu_0^{-1})$, $\beta_K = \| \boldsymbol{\beta}\|_{L^\infty(K)}$, and with right-hand side $\ell_h(w_h) = \int_D f w_h \, dx + \int_D \tau f(\mu w_h + \boldsymbol{\beta}\cdot\nabla w_h) \, dx$. Provided $u \in H^{1+r}(D)$, $r \in [0, k]$, and with $\phi_D := \max(\| \boldsymbol{\beta}\|_{L^\infty(D)}, \mu_0 h)$, Theorem 3 gives

$$\mu_0^{\frac{1}{2}} \| u - u_h\|_{L^2(D)} + \| \tau^{\frac{1}{2}} \boldsymbol{\beta}\cdot\nabla(u - u_h)\|_{L^2(D)} \leq c \, \phi_D^{\frac{1}{2}} h^{r+\frac{1}{2}} |u|_{H^{1+r}(D)}.$$

Example 4 (SUPG). Assume that $h_K \le \beta_K \mu_0^{-1} \min(1, \frac{1}{2}\frac{\mu_0^2}{\mu_\infty^2})$ with $\mu_\infty = \| \mathcal{K} \|_{L^\infty(D;\mathbb{C}^{m\times m})}$, for all $K \in \mathcal{T}_h$. The same error estimate as in the GaLS approximation is obtained by considering the following discrete problem: Find $u_h \in V_{h0}$ such that $a_h^{\mathrm{SUPG}}(u_h, w_h) = (f, w_h + \tau A_1 w_h)_L$ for all $w_h \in V_{h0}$ with the SUPG-stabilized sesquilinear form $a_h^{\mathrm{SUPG}}(v_h, w_h) = (Av_h, w_h)_L + (Av_h, \tau A_1 w_h)_L$.

Example 5 (Maxwell). Consider the PDEs $\sigma \boldsymbol{E} - \nabla \times \boldsymbol{H} = \boldsymbol{f}$ and $i\tilde{\mu}\boldsymbol{H} + \nabla \times \boldsymbol{E} = \boldsymbol{0}$ with the boundary condition $\boldsymbol{H} \times \boldsymbol{n} = \boldsymbol{0}$. Define the reference length scale $\ell_* = (\sigma_{\flat,D}\tilde{\mu}_{\flat,D})^{-\frac{1}{2}}$. Set $\boldsymbol{W}_h := \boldsymbol{P}_k^g(\mathcal{T}_h)$ and $\boldsymbol{W}_{h0} := \{\boldsymbol{b}_h \in \boldsymbol{W}_h \mid \boldsymbol{b}_h \times \boldsymbol{n}_{|\partial D} = \boldsymbol{0}\}$. The GaLS approximation amounts to finding $(\boldsymbol{E}_h, \boldsymbol{H}_h) \in V_{h0} := \boldsymbol{W}_h \times \boldsymbol{W}_{h0}$ such that

$$\int_D \Big((\sigma \boldsymbol{E}_h - \nabla \times \boldsymbol{H}_h) \cdot \bar{\boldsymbol{e}}_h + (i\tilde{\mu}\boldsymbol{H}_h + \nabla \times \boldsymbol{E}_h) \cdot \bar{\boldsymbol{b}}_h \Big) \,\mathrm{d}x$$

$$+ \int_D \tilde{\mu}_{\flat,D}^{-1} \tau (i\tilde{\mu}\boldsymbol{H}_h + \nabla \times \boldsymbol{E}_h) \cdot (-i\tilde{\mu}\bar{\boldsymbol{b}}_h + \nabla \times \bar{\boldsymbol{e}}_h) \,\mathrm{d}x$$

$$+ \int_D \sigma_{\flat,D}^{-1} \tau (\sigma \boldsymbol{E}_h - \nabla \times \boldsymbol{H}_h) \cdot (\sigma \bar{\boldsymbol{e}}_h - \nabla \times \bar{\boldsymbol{b}}_h) \,\mathrm{d}x = \ell_h(w_h),$$

for all $w_h = (\boldsymbol{e}_h, \boldsymbol{b}_h) \in V_{h0}$, with local weights $\tau_K = \min(\ell_*^{-1} h_K, 1)$, and right-hand side $\ell_h(w_h) = \int_D \boldsymbol{j} \cdot \bar{\boldsymbol{e}}_h \,\mathrm{d}x + \int_D \sigma_{\flat,D}^{-1} \tau \boldsymbol{j} \cdot (\sigma \bar{\boldsymbol{e}}_h - \nabla \times \bar{\boldsymbol{b}}_h) \,\mathrm{d}x$. Provided $(\boldsymbol{E}, \boldsymbol{H}) \in \boldsymbol{H}^{1+r}(D) \times \boldsymbol{H}^{1+r}(D)$, $r \in [0, k]$, Theorem 2, combined with the approximation properties of V_{h0}, yields

$$\begin{aligned}
&\sigma_{\flat,D}^{\frac{1}{2}} \| \boldsymbol{E} - \boldsymbol{E}_h \|_{\boldsymbol{L}^2(D)} + \tilde{\mu}_{\flat,D}^{\frac{1}{2}} \| \boldsymbol{H} - \boldsymbol{H}_h \|_{\boldsymbol{L}^2(D)} + \tilde{\mu}_{\flat,D}^{-\frac{1}{2}} \| \tau^{\frac{1}{2}} \nabla \times (\boldsymbol{E} - \boldsymbol{E}_h) \|_{\boldsymbol{L}^2(D)} \\
&\quad + \sigma_{\flat,D}^{-\frac{1}{2}} \| \tau^{\frac{1}{2}} \nabla \times (\boldsymbol{H} - \boldsymbol{H}_h) \|_{\boldsymbol{L}^2(D)} \le c\phi_D^{\frac{1}{2}} h^{r+\frac{1}{2}} \Big(|\boldsymbol{E}|_{\boldsymbol{H}^{1+r}(D)} + |\boldsymbol{H}|_{\boldsymbol{H}^{1+r}(D)} \Big),
\end{aligned}$$

with $\phi_D - \max(\ell_*, h)$.

4 BOUNDARY PENALTY FOR FRIEDRICHS' SYSTEMS

It is not always possible, or easy, to build V_0-conforming finite elements; think for instance of a boundary condition enforcing the value of the normal or tangential component of a vector field at the boundary of a domain that is not a rectangular parallelepiped. The goal of this section is twofold: First, to show how to enforce boundary conditions weakly in Friedrichs' systems; second, to combine this approach with the GaLS stabilization. The boundary penalty technique introduced herein will be used again in Section 5.

4.1 Model Problem

We now consider the sesquilinear form

$$\tilde{a}(v,w) := (Av, w)_L + \frac{1}{2}\langle (M-N)v, w\rangle_{V',V}, \quad \forall v,w \in V. \tag{32}$$

The last term on the right-hand side is used to enforce the boundary condition $u \in \ker(M-N)$ weakly. Owing to this additional term, the test functions are now restricted to be in the graph space V; i.e. taking test functions in L is no longer legitimate. The model problem that we consider is the following:

$$\begin{cases} \text{Find } u \in V \text{ such that} \\ \tilde{a}(u,w) = (f, w)_L, \quad \forall w \in V. \end{cases} \tag{33}$$

If u solves (33), taking w in $C_0^\infty(D;\mathbb{C}^m)$ implies that $Au = f$ in $L^2(D;\mathbb{C}^m)$; then, we have $\langle (M-N)u, w\rangle_{V',V} = 0$ for all $w \in V$, i.e. $u \in \ker(M-N)$.

Lemma 2 (L-coercivity and well-posedness). *The sesquilinear form $\tilde{a}$ defined by (32) is such that $\Re(\tilde{a}(v,v)) \geq \mu_0 \| v \|_L^2 + \frac{1}{2}|v|_M^2$, for all $v \in V$. Problem (33) is well-posed, and its unique solution is the unique solution to (19).*

4.2 Boundary Penalty Method

We are interested in a V-conforming approximation of the model problem (33). For this purpose, we assume that, for all $h > 0$, we have at hand an H^1-conforming finite-dimensional space $V_h \subset V$, built by using a shape-regular mesh sequence $(\mathcal{T}_h)_{h>0}$ and a finite element of degree $k \geq 1$, and a quasi-interpolation operator $\mathcal{I}_h : V \to V_h$ with optimal local approximation properties: There is a uniform constant c such that

$$\| v - \mathcal{I}_h(v) \|_{L(K)} + h_K \| \nabla(v - \mathcal{I}_h(v)) \|_{L(K)} \leq c\, h_K^{1+r} |v|_{H^{1+r}(D_K,\mathbb{C}^m)}, \tag{34}$$

for all $r \in [0, k]$, all $v \in H^{1+r}(D,\mathbb{C}^m)$, and all $K \in \mathcal{T}_h$.

Our starting point is the sesquilinear form $\tilde{a}$ defined in (32). At the discrete level, we would like to localize the term $\langle (M-N)v, w\rangle_{V',V}$ at the boundary faces $F \in \mathcal{F}_h^\partial$. Therefore, we assume that there are boundary fields $\mathcal{M}$ and $\mathcal{N}$ in $L^\infty(\partial D;\mathbb{C}^{m\times m})$ such that

$$\langle Mv, w\rangle_{V',V} = (\mathcal{M}v, w)_{L^\partial}, \quad \langle Nv, w\rangle_{V',V} = (\mathcal{N}v, w)_{L^\partial}, \tag{35}$$

for all $v,w \in V^s := H^s(D;\mathbb{C}^m)$ with $s > \frac{1}{2}$ and $L^\partial := L^2(\partial D;\mathbb{C}^m)$; whence,

$$\tilde{a}(v,w) = (Av, w)_L + \frac{1}{2}((\mathcal{M}-\mathcal{N})v, w)_{L^\partial}, \quad (v,w) \in V^s \times V^s. \tag{36}$$

The field $\mathcal{M}$ is such that $\Re((\mathcal{M}v, v)_{L^\partial}) \geq 0$, since the operator M is monotone. But it may occur that $\Re((\mathcal{M}v, v)_{L^\partial}) = 0$ (this happens for second-order

PDEs in mixed form). To gain some control on the boundary values, we introduce an additional boundary penalty field $\mathcal{S}^\partial \in L^\infty(\partial D; \mathbb{C}^{m\times m})$ and we define the following sesquilinear form on $V^s \times V^s$:

$$\begin{aligned}\check{a}(v,w) &:= \tilde{a}(v,w) + (\mathcal{S}^\partial v, w)_{L^\partial} \\ &= (Av, w)_L + \frac{1}{2}((\mathcal{M} - \mathcal{N})v, w)_{L^\partial} + (\mathcal{S}^\partial v, w)_{L^\partial}.\end{aligned} \tag{37}$$

In what follows, we use a subscript F to denote the restriction of a boundary field to $F \in \mathcal{F}_h^\partial$ and we set $L(F) := L^2(F; \mathbb{C}^m)$. We define the local boundary seminorm $|v|^2_{\mathcal{M}_F} := (\mathcal{M}_F v, v)_{L(F)}$ and we set $\rho_F := \| \mathcal{M}_F \|_{L^\infty(F;\mathbb{C}^{m\times m})}$. We assume for simplicity that

$$\rho_F \le c_{\mathcal{M}} \beta_{K_F}, \quad \forall F \in \mathcal{F}_h^\partial, \tag{38}$$

where $K_F \in \mathcal{T}_h$ is the mesh element such that $F = \partial K_F \cap \partial D$. The design conditions on $\mathcal{S}^\partial$ are as follows: There is c, uniform with respect to h, such that the following holds for all v, $w \in L(F)$ and all $F \in \mathcal{F}_h^\partial$.

$$\mathcal{S}_F^\partial \text{ is Hermitian and positive semidefinite,} \tag{39a}$$

$$\ker(\mathcal{M}_F - \mathcal{N}_F) \subset \ker(\mathcal{S}_F^\partial), \tag{39b}$$

$$|v|_{\mathcal{S}_F^\partial} \le c\, \rho_F^{\frac{1}{2}} \| v\|_{L(F)}, \tag{39c}$$

$$|((\mathcal{M}_F - \mathcal{N}_F)v, w)_{L(F)}| \le c(|v|_{\mathcal{M}_F} + |v|_{\mathcal{S}_F^\partial})\rho_F^{\frac{1}{2}} \| w\|_{L(F)}, \tag{39d}$$

$$|((\mathcal{M}_F + \mathcal{N}_F)v, w)_{L(F)}| \le c\, \rho_F^{\frac{1}{2}} \| v\|_{L(F)}(|w|_{\mathcal{M}_F} + |w|_{\mathcal{S}_F^\partial}). \tag{39e}$$

The assumption (39a) implies that the local boundary seminorm $|y|^2_{\mathcal{S}_F^\partial} := (\mathcal{S}_F^\partial v, v)_{L(F)}$ is well defined and that $(\mathcal{S}_F^\partial v, w)_{L(F)} \le |v|_{\mathcal{S}_F^\partial} |w|_{\mathcal{S}_F^\partial}$. The assumption (39b) is tailored to ensure exact consistency. The other assumptions (39c)–(39e) are stability properties. Note that (39d) and (39e) turn out to be equivalent; both properties are presented since they are useful in the analysis.

Example 6 (Advection–reaction). Since $\mathcal{M}_F = |\boldsymbol{\beta} \cdot \boldsymbol{n}_F|$ for all $F \in \mathcal{F}_h^\partial$, we can take $\mathcal{S}_F^\partial = 0$. The properties (39a)–(39c) are obvious and (39d) results from the Cauchy–Schwarz inequality since $\frac{1}{2}\int_F (|\boldsymbol{\beta}\cdot \boldsymbol{n}_F| - \boldsymbol{\beta}\cdot\boldsymbol{n}_F) vw\, ds \le \| |\boldsymbol{\beta}\cdot\boldsymbol{n}_F|^{\frac{1}{2}} v\|_{L^2(F)} \rho_F^{\frac{1}{2}} \| w\|_{L^2(F)}$.

Example 7 (Maxwell). Consider the boundary condition $\boldsymbol{H}\times\boldsymbol{n} = \boldsymbol{0}$ on ∂D. Recalling the matrix $\mathbb{T} \in \mathbb{R}^{3\times 3}$ from Section 1.3, the properties (39) are satisfied by taking $\mathcal{S}_F^\partial = \begin{pmatrix} \mathbb{O}_3 & \mathbb{O}_3 \\ \mathbb{O}_3 & \alpha \mathbb{T}^{\mathrm{T}}\mathbb{T} \end{pmatrix}$, for all $F \in \mathcal{F}_h^\partial$, with a parameter $\alpha > 0$. This means that the tangential component of $\boldsymbol{H}$ is penalized at the boundary.

4.3 Galerkin Least-Squares Stabilization with Boundary Penalty

We define the following discrete sesquilinear form on $V_h \times V_h$:

$$\check{a}_h(v_h,w_h) = \tilde{a}(v_h,w_h) + (Av_h, \tau Aw_h)_L, \tag{40}$$

that is to say $\check{a}_h(v_h,w_h) = (Av_h, w_h)_L + \frac{1}{2}((\mathcal{M} - \mathcal{N})v_h, w_h)_{L^\partial} + (\mathcal{S}^\partial v_h, w_h)_{L^\partial} + (Av_h, \tau Aw_h)_L$. We consider the following discrete problem:

$$\begin{cases} \text{Find } u_h \in V_h \text{ such that} \\ \check{a}_h(u_h,w_h) = (f, w_h + \tau Aw_h)_L, \quad \forall w_h \in V_h. \end{cases} \tag{41}$$

Let us set $V_\flat = V^s + V_h$. Notice that $V_\flat = V^s$ since the approximation is H^1-conforming. We extend the sesquilinear form $\check{a}_h$ to $V^s \times V_h$ and we equip the space V^s with the following norms:

$$\| v \|_{V_\flat}^2 := \mu_0 \| v \|_L^2 + \frac{1}{2}|v|_{\mathcal{M}}^2 + |v|_{\mathcal{S}^\partial}^2 + \| \tau^{\frac{1}{2}} Av \|_L^2, \tag{42a}$$

$$\| v \|_{V_{\flat\sharp}}^2 := \| v \|_{V_\flat}^2 + \| \tau^{-\frac{1}{2}} v \|_L^2 + \| \rho^{\frac{1}{2}} v \|_{L^\partial}^2, \tag{42b}$$

with boundary seminorms $|v|_{\mathcal{M}}^2 := \Re((\mathcal{M}v, v)_{L^\partial})$ and $|v|_{\mathcal{S}^\partial}^2 := \Re((\mathcal{S}^\partial v, v)_{L^\partial})$, and $\rho \in L^\infty(\partial D)$ is defined by $\rho_{|F} := \rho_F$ for all $F \in \mathcal{F}_h^\partial$.

Theorem 4 (Convergence). *(i) The discrete sesquilinear form $\check{a}_h$ satisfies $\Re(\check{a}_h(v_h,v_h)) \geq \| v_h \|_{V_\flat}^2$, for all $v_h \in V_h$. Consequently, the discrete problem (41) is well-posed. (ii) Assume that the exact solution u is in V^s. Then, the discrete problem (41) is exactly consistent. (iii) There is c, uniform with respect to h, such that $|\check{a}_h(v,w_h)| \leq |\check{a}(v,w_h)| + |(Av, \tau Aw_h)_L| \leq c \, \| v \|_{V_{\flat\sharp}} \| w_h \|_{V_\flat}$ for all $(v, w_h) \in V^s \times V_h$. (iv) Let u and u_h be the unique solutions to (19) and (41), respectively. Then, there is c, uniform with respect to h, such that*

$$\| u - u_h \|_{V_\flat} \leq c \inf_{v_h \in V_h} \| u - v_h \|_{V_{\flat\sharp}}. \tag{43}$$

Moreover, $\| u - u_h \|_{V_\flat}^2 \leq c \sum_{K \in \mathcal{T}_h} \max(\beta_K, \mu_0 h_K) h_K^{2r+1} |u|_{H^{1+r}(D_K;\mathbb{C}^m)}^2$ if $u \in H^{1+r}(D;\mathbb{C}^m)$, $r \in [0, k]$. This implies that $\| u - u_h \|_{V_\flat} \leq c \phi_D^{\frac{1}{2}} h^{r+\frac{1}{2}} |u|_{H^{1+r}(D;\mathbb{C}^m)}$.

5 FLUCTUATION-BASED STABILIZATION

This section presents a unified analysis of various techniques for the approximation of first-order PDEs using H^1-conforming finite elements. The gradient of a function in an H^1-conforming space generally exhibits jumps across the mesh interfaces. This means that only one part of the gradient can be controlled by test functions from this space; the remainder, which can be viewed as a fluctuation, needs to be controlled by some stabilization mechanism. Three stabilization techniques are considered herein: the continuous interior penalty (CIP), the local projection stabilization (LPS) and the subgrid viscosity (SGV). CIP penalizes the jump of the gradient across the mesh interfaces. LPS and SGV are both based on a two-scale decomposition of the discrete space consisting of a sum of resolved scales and fluctuations. LPS penalizes the fluctuations of the gradient, whereas SGV penalizes the gradient of the fluctuations. Throughout this section, the boundary conditions are enforced weakly by the boundary penalty technique introduced in Section 4.2.

5.1 Abstract Theory for Fluctuation-Based Stabilization

Let us consider the finite element setting introduced in Section 4.2. Let β_K and τ_K as defined in (25) and (26). Recall that β_K is a local velocity scale and τ_K is local time scale. We define the global quantity $\beta_D = \max_{K\in\mathcal{T}_h}\beta_K$, and we introduce a second local weighting parameter $\check{\tau}_K$ such that

$$\min(\beta_D^{-1}h_K, \mu_0^{-1}) \leq \check{\tau}_K \leq \tau_K, \quad \forall K \in \mathcal{T}_h. \tag{44}$$

We will take $\check{\tau}_K = \min(\beta_D^{-1}h_K, \mu_0^{-1})$ for the CIP stabilization and $\check{\tau}_K = \tau_K$ for the LPS and SGV stabilizations. With a slight abuse of notation, we define the piecewise constant function $\check{\tau}: D \to \mathbb{R}$ such that $\check{\tau}_{|K} = \check{\tau}_K$ for all $K \in \mathcal{T}_h$; the piecewise constant function $\tau: D \to \mathbb{R}$ is defined similarly.

We additionally assume that all the fields $\{\mathcal{A}^k\}_{k\in\{1:d\}}$ are piecewise Lipschitz on a partition of D and that the meshes are compatible with this partition, implying that the fields $\{\mathcal{A}^k_{|K}\}_{k\in\{1:d\}}$ are Lipschitz for all $K \in \mathcal{T}_h$. We denote by $L_{\mathcal{A}}$ the largest Lipschitz constant of these fields. To simplify the tracking of the model parameters in the analysis, we assume that

$$\max(\|\mathcal{K}\|_{L^\infty(D;\mathbb{C}^{m\times m})}, \|\mathcal{X}\|_{L^\infty(D;\mathbb{C}^{m\times m})}, L_{\mathcal{A}}) \leq c_{\mathcal{K},\mathcal{X},\mathcal{A}}\mu_0, \tag{45}$$

and we hide the nondimensional factor $c_{\mathcal{K},\mathcal{X},\mathcal{A}}$ in the generic constant c.

The boundary conditions are enforced by using the boundary penalty method from Section 4.2, i.e. we assume that there is $\mathcal{S}^\partial \in L^\infty(\partial D; \mathbb{C}^{m\times m})$ satisfying (39) for any boundary face $F \in \mathcal{F}_h^\partial$, with $\rho_F = \|\mathcal{M}_F\|_{L^\infty(F;\mathbb{C}^{m\times m})}$. We assume that there is a uniform constant $c_{\mathcal{M}}$ such that $\rho_F \leq c_{\mathcal{M}}\beta_{K_F}$ for all $F \in \mathcal{F}_h^\partial$ with $F = \partial K_F \cap \partial D$, see (38); we will hide the nondimensional factor $c_{\mathcal{M}}$ in the generic constant c. Our starting point is the following sesquilinear form, see (37):

$$\check{a}(v,w) = (Av, w)_L + \tfrac{1}{2}((\mathcal{M} - \mathcal{N})v, w)_{L^\partial} + (\mathcal{S}^\partial v, w)_{L^\partial}, \quad \forall (v,w) \in V^s \times V^s. \tag{46}$$

The main idea is to augment the sesquilinear form $\check{a}$ with a stabilization sesquilinear form s_h and to consider the following discrete problem:

$$\begin{cases} \text{Find } u_h \in V_h \text{ such that} \\ a_h(u_h, w_h) = (f, w_h)_L, \quad \forall w_h \in V_h, \end{cases} \tag{47}$$

with

$$a_h(v_h, w_h) := \check{a}(v_h, w_h) + s_h(v_h, w_h). \tag{48}$$

To stay somewhat general, we only require that s_h be defined on $V_h \times V_h$. Loosely speaking, the purpose of s_h is to control the difference between $A_1 v_h$ and a suitable representative of $A_1 v_h$ in V_h. We consider the following design requirements on the bilinear form s_h, where c_1, c_2, $c_3 > 0$ are uniform with respect to h:

(i) s_h is Hermitian positive semidefinite and satisfies $|v_h|_{\mathcal{S}} := s_h(v_h, v_h)^{\frac{1}{2}} \leq c_1 \| \check{\tau}^{-\frac{1}{2}} v_h \|_L$ for all $v_h \in V_h$.

(ii) There exists a linear map $\mathcal{J}_h : V_h \to V_h$ such that, for all $v_h \in V_h$,

$$c_2 \| \check{\tau}^{-\frac{1}{2}} \mathcal{J}_h(v_h) \|_L^2 \leq \| \check{\tau}^{\frac{1}{2}} A_1 v_h \|_L^2 + \mu_0 \| v_h \|_L^2 + |v_h|_{\mathcal{S}}^2, \tag{49a}$$

$$c_2 \| \check{\tau}^{\frac{1}{2}} A_1 v_h \|_L^2 \leq \Re((A_1 v_h, \mathcal{J}_h(v_h))_L) + \mu_0 \| v_h \|_L^2 + |v_h|_{\mathcal{S}}^2. \tag{49b}$$

(iii) $|\mathcal{I}_h(v)|_{\mathcal{S}} \leq c_3 \left(\sum_{K \in \mathcal{T}_h} (\check{\tau}_K^{-1} h_K) h_K^{2r+1} |v|^2_{H^{1+r}(D_K;\mathbb{C}^m)} \right)^{\frac{1}{2}}$ for all $r \in [0, k]$ and all $v \in H^{1+r}(D; \mathbb{C}^m)$ with $\mathcal{I}_h$ satisfying (34).

The error analysis is done in the spirit of Strang's First Lemma. This approach is the most general since it does not require that s_h be extended beyond $V_h \times V_h$. We consider the space $V_\flat = V^s + V_h$; note that $V_\flat = V^s$ since V_h is H^1-conforming. We define the following norms on V^s:

$$\| v \|_{V_\flat}^2 := \mu_0 \| v \|_L^2 + \frac{1}{2} |v|_{\mathcal{M}}^2 + |v|_{\mathcal{S}^\partial}^2 + \| \check{\tau}^{\frac{1}{2}} A_1 v \|_L^2, \tag{50a}$$

$$\| v \|_{V_{\flat\sharp}}^2 := \| v \|_{V_\flat}^2 + \| \check{\tau}^{-\frac{1}{2}} v \|_L^2 + \| \rho^{\frac{1}{2}} v \|_{L^\partial}^2. \tag{50b}$$

The first norm is used to establish the inf–sup stability of $\check{a}$ on $V_h \times V_h$ (and well-posedness) and the second one to prove the boundedness of $\check{a}$ on $V^s \times V_h$. Up to the change of τ by $\check{\tau}$, these norms are the same as those used in Section 4.3 for the GaLS stabilization with boundary penalty.

Theorem 5 (Convergence). *(i) Under the design conditions (i)–(ii)–(iii) for s_h, there is $\alpha > 0$, uniform with respect to h, such that the following holds:*

$$\alpha(\| v_h \|_{V_\flat} + |v_h|_{\mathcal{S}}) \leq \sup_{w_h \in V_h} \frac{\Re(a_h(v_h, w_h))}{\| w_h \|_{V_\flat} + |w_h|_{\mathcal{S}}}, \quad \forall v_h \in V_h. \tag{51}$$

Consequently, the discrete problem (47) is well-posed. (ii) There is c, uniform with respect to h, such that $|\check{a}(v,w_h)| \le c \, \| v\|_{V_{\flat\sharp}} \, \| w_h\|_{V_\flat}$ *holds for all* $(v, w_h) \in V^s \times V_h$. *(iii) Let u be the unique solution to (19) and let* u_h *be the unique solution to (47) with* s_h *satisfying the design conditions (i)–(ii)–(iii) above. There is c, uniform with respect to h, such that*

$$\| u - u_h\|_{V_\flat} \le c \inf_{v_h \in V_h} \left(\| u - v_h\|_{V_{\flat\sharp}} + |v_h|_S\right). \tag{52}$$

Moreover, $\| u - u_h \|_{V_\flat}^2 \le c\sum_{K\in\mathcal{T}_h} (\check{\tau}_K^{-1} h_K) h_K^{2r+1} |u|^2_{H^{1+r}(D_K;\mathbb{C}^m)}$ *if* $u \in H^{1+r}(D;\mathbb{C}^m)$, $r \in [0, k]$ *(note that* $\max(\beta_K, \mu_0 h_K) \le \check{\tau}_K^{-1} h_K \le \max(\beta_D, \mu_0 h_K)$*).*

In the next section we show how the above theory can be used to analyze the stability and convergence properties of the CIP, LPS and SGV methods.

5.2 Continuous Interior Penalty

The key idea in CIP stabilization (also termed edge stabilization in the literature) is to penalize the jump of $A_1 v_h$ across the mesh interfaces. This idea has been introduced in Burman (2005) and Burman and Hansbo (2004). We refer to Burman and Ern (2007a,b) for the *hp* analysis and extensions to Friedrichs' systems, and we refer to Ern and Guermond (2013) for extensions in the context of nonlinear conservation laws.

We set $\check{\tau}_K := \min(\beta_D^{-1} h_K, \mu_0^{-1})$ for all $K \in \mathcal{T}_h$. Let us take $V_h = P_k^g(\mathcal{T}_h;\mathbb{C}^m)$. Let $\mathcal{J}_h^{g,av}$ be the nodal averaging operator mapping onto $P_k^g(\mathcal{T}_h;\mathbb{C}^m)$ defined and analyzed in Ern and Guermond (2016), and let $\phi \in P_1^g(\mathcal{T}_h;\mathbb{R})$ be defined by $\phi(z) = \mathrm{card}(\mathcal{T}_z)^{-1} \sum_{K\in\mathcal{T}_z} \check{\tau}_K$ with $\mathcal{T}_z := \{K \in \mathcal{T}_h \mid z \in K\}$ for any mesh vertex z.

Lemma 3. *Define* $(\underline{A}_1 v_h)_{|K} := \sum_{k\in\{1:d\}} \underline{\mathcal{A}}_K^k \partial_k v_{h|K}$ *for all* $K \in \mathcal{T}_h$ *and all* $v_h \in V_h$, *where* $\underline{\mathcal{A}}_K^k := \dfrac{1}{|K|}\displaystyle\int_K \mathcal{A}^k \,\mathrm{d}x$. *Let* $\check{\tau}_F := \max(\check{\tau}_{K_l}, \check{\tau}_{K_r})$ *and* $\beta_F := \max(\beta_{K_l}, \beta_{K_r})$ *for all* $F = \partial K_l \cap \partial K_r \in \mathcal{F}_h^{\circ}$. *Then the sesquilinear forms*

$$s_h^{\mathrm{CIP}}(v_h, w_h) = \sum_{F\in\mathcal{F}_h^{\circ}} \check{\tau}_F h_F ([\![A_1 v_h]\!]_F, [\![A_1 w_h]\!]_F)_{L(F)}, \tag{53a}$$

$$s_h^{\mathrm{CIP}}(v_h, w_h) = \sum_{F\in\mathcal{F}_h^{\circ}} \check{\tau}_F h_F ([\![\underline{A}_1 v_h]\!]_F, [\![\underline{A}_1 w_h]\!]_F)_{L(F)}, \tag{53b}$$

$$s_h^{\mathrm{CIP}}(v_h, w_h) = \sum_{F\in\mathcal{F}_h^{\circ}} \beta_F h_F^2 ([\![\nabla v_h]\!]_F, [\![\nabla w_h]\!]_F)_{\mathbf{L}(F)}, \tag{53c}$$

all satisfy the conditions (i)–(ii) with $\mathcal{J}_h(v_h) = \mathcal{J}_h^{g,av}(\phi \underline{A}_1 v_h)$, *and the condition (iii) for* $r \ge 1$.

Remark 3 (Time-dependent case). The choice (53c) is interesting for time-dependent fields $\mathcal{A}^k$ since the matrix associated with (53c) can then be assembled only once, which is not the case for (53a) and (53b). Note that in (53c), only the normal component of the gradient can actually jump across F since functions in V_h are continuous.

5.3 Two-Scale Stabilization, Local Projection and Subgrid Viscosity

We present in this section two closely related stabilization techniques known in the literature as LPS and SGV. The SGV technique has been introduced in Guermond (1999, 2001a,b) for monotone operators and semigroups. The LPS technique has been introduced in Becker and Braack (2001) and Braack and Burman (2006) for Stokes and convection-diffusion equations; see also Matthies et al. (2007, 2008). LPS and SGV both rely on a two-scale decomposition of the discrete space V_h, leading to the notions of resolved and fluctuating (or subgrid) scales. Both stabilization techniques introduce a LS penalty: LPS penalizes the fluctuation of the gradient and SGV penalizes the gradient of the fluctuation. The notion of scale separation and subgrid scale dissipation is similar in spirit to the spectral viscosity technique introduced by Tadmor (1989) to approximate nonlinear conservation equations by means of spectral methods. This notion is also found in the Orthogonal Subscale Stabilization technique of Codina (2002).

5.3.1 *The Two-Scale Decomposition*

The starting point is a two-scale decomposition of V_h into the form

$$V_h = R_h + B_h, \tag{54}$$

where the sum is not necessarily direct. The discrete space R_h is viewed as the space of the resolved scales, and B_h is viewed as the space of the fluctuating (or subgrid) scales. It is important to realize that the degrees of freedom attached to B_h only serve to achieve stability, and that the approximation error is controlled by the best approximation in the space of the resolved scales R_h (and not in the full space V_h). We assume the following local approximation property in R_h: There is a quasi-interpolation operator $\mathcal{I}_h^R : V \to R_h$ and a constant c, uniform with respect to h, such that

$$\| v - \mathcal{I}_h^R(v)\|_{L(K)} + h \,\| \nabla(v - \mathcal{I}_h^R(v))\|_{L(K)} \le c\, h^{1+r} |v|_{H^{1+r}(D_K;\mathbb{C}^m)}, \tag{55}$$

for all $r \in [0, k]$, all $v \in H^{1+r}(D;\mathbb{C}^m)$, and all $K \in \mathcal{T}_h$.

Since functions in R_h are continuous, piecewise polynomials, the components of their gradients belong to a broken finite element space $G_h = \oplus_{K\in\mathcal{T}_h} G_K$, where functions in G_K are supported in K, i.e. $\partial_i r_h \in G_h$ for all $r_h \in R_h$ and all $i \in \{1{:}\,d\}$. We assume that the space of the fluctuating scales can also be localized in the form $B_h = \oplus_{K\in\mathcal{T}_h} B_K$, where the functions in B_K are supported in K (one may think of members of B_K as bubble-type functions, see the examples below). We define the local L-orthogonal projections $\pi_K^B : L(K) \to B_K$ and $\pi_K^G : L(K) \to G_K$ for all $K \in \mathcal{T}_h$ and the global counterparts $\pi_h^B : L \to B_h$ and $\pi_h^G : L \to G_h$ such that $\pi_{h|K}^B = \pi_K^B$ and $\pi_{h|K}^G = \pi_K^G$.

The key assumption linking the local gradient space G_K to the local fluctuation space B_K is the following inf–sup condition introduced in Guermond (1999, 2001b) (see also Matthies et al., 2007): There is $\gamma > 0$, uniform with respect to h, such that, for all $K \in \mathcal{T}_h$,

$$\inf_{g\in G_K} \sup_{b\in B_K} \frac{\Re\left(\int_K b^{\mathrm{H}} g \,\mathrm{d}x\right)}{\| g\|_{L(K)} \| b\|_{L(K)}} \geq \gamma, \tag{56}$$

or, equivalently, $\gamma \| g\|_{L(K)} \leq \| \pi_K^B g\|_{L(K)}$ for all $g \in G_K$. In what follows, we consider the local weighting parameter $\check{\tau}_K = \tau_K = \min(\beta_K h_K^{-1}, \mu_0^{-1})$ for all $K \in \mathcal{T}_h$.

We now describe three constructions of H^1-conforming finite element spaces of degree $k \geq 1$ which all satisfy the above assumptions. (1) In the first example, the space of the resolved scales is defined by $R_h = P_k^g(\mathcal{T}_h; \mathbb{C}^m)$, the H^1-conforming finite element space based on $\mathcal{T}_h$, so that $G_h = P_{k-1}^b(\mathcal{T}_h; \mathbb{C}^m)$ and G_K is composed of $\mathbb{C}^m$-valued polynomials of degree at most $(k - 1)$ on affine meshes. Following Guermond (1999) for $k \in \{1, 2\}$ and Matthies et al. (2007) for all $k \geq 1$, we take $B_K = b_K G_K$ where b_K is the $H_0^1(K)$-bubble function proportional to the product of the $(d + 1)$ barycentric coordinates over K; see the panels in the upper row in Fig. 1. (2) Instead of working with bubble functions, one can use hierarchical meshes (Guermond, 1999; Matthies et al., 2007). In this case, the construction starts from the mesh defining the space of the resolved scales, say $\check{\mathcal{T}}_h$. Assume for simplicity that $\check{\mathcal{T}}_h$ is composed of simplices, then the mesh $\mathcal{T}_h$ defining V_h is built by barycentric refinement, i.e. for any $K \in \check{\mathcal{T}}_h$, $(d + 1)$ new simplices are created by joining the barycentre of K to its $(d + 1)$ vertices. Then we take $V_h = P_k^g(\mathcal{T}_h; \mathbb{C}^m)$ and $R_h = P_k^g(\check{\mathcal{T}}_h; \mathbb{C}^m)$, so that $G_h = P_{k-1}^b(\check{\mathcal{T}}_h; \mathbb{C}^m)$, see the panels in the second row in Fig. 1. For any $K \in \check{\mathcal{T}}_h$, choose $g := \dim(G_K)$ shape functions of V_h with support in K, say $\varphi_1^K, \ldots, \varphi_g^K$ and set $B_K = \mathrm{span}\{\varphi_1^K, \ldots, \varphi_g^K\}$. The practical advantage of this construction is that V_h is a standard finite element space. (3) Finally, we mention the two-scale decomposition considered in Guermond (1999) for $k \in \{1, 2\}$ which also offers the advantage of V_h being a standard finite element space; a schematic representation of the scale decomposition is shown in the panels in the last row in Fig. 1. The analysis (not considered herein) is somewhat more involved since the fluctuating scales are represented by functions possibly supported on two adjacent mesh cells.

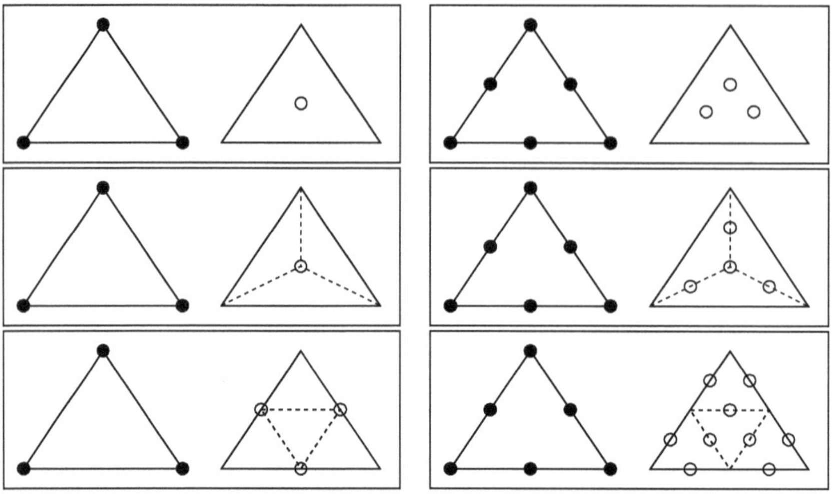

FIG. 1 Examples of two-scale finite elements. In each panel, the resolved scales are on the left and the fluctuating scales are on the right. The resolved scales are either $\mathbb{P}_1$ (*left column*) or $\mathbb{P}_2$ (*right column*) Lagrange elements. The upper panels illustrate the use of a standard bubble function to build the fluctuating scales; the central and lower panels illustrate the use of piecewise polynomial bubble functions on a submesh with the same size (central panel) or half the size (bottom panel) as that of the resolved scales space.

5.3.2 Local Projection Stabilization

Lemma 4. *Assume that (56) holds. Let $\underline{A}_1 v_h$ be defined as in Lemma 3 and $\beta : D \to \mathbb{R}$ be such that $\beta_{|K} := \beta_K$ for all $K \in \mathcal{T}_h$. Define the fluctuation operator $\kappa_h^G = I_L - \pi_h^G$, where I_L is the identity operator in L. Then, the sesquilinear forms*

$$s_h^{\mathrm{LPS}}(v_h, w_h) = (\check{\tau}\kappa_h^G(\underline{A}_1 v_h), \kappa_h^G(\underline{A}_1 w_h))_L, \tag{57a}$$

$$s_h^{\mathrm{LPS}}(v_h, w_h) = (\beta^2\check{\tau}\kappa_h^G(\nabla v_h), \kappa_h^G(\nabla w_h))_L, \tag{57b}$$

both satisfy the assumptions (i)–(ii)–(iii) with $\mathcal{J}_h(v_h) = \check{\tau}\pi_h^B\pi_h^G(A_1 v_h)$.

Remark 4 (Use of $\kappa_h^G(A_1 v_h)$). When the fields $\mathcal{A}^k$ are not piecewise constant, setting $s_h^{\mathrm{LPS}}(v_h, w_h) = (\check{\tau}\kappa_h^G(A_1 v_h), \kappa_h^G(A_1 w_h))_L$ is somewhat delicate since $|\mathcal{I}_h^R(u)|_S$ no longer vanishes. Bounding this quantity requires strong regularity assumptions on the fields $\mathcal{A}^k$.

5.3.3 Subgrid Viscosity

In the SGV method, the two-scale decomposition of V_h is assumed to be direct and L-stable, i.e. it is assumed that there is $\gamma_R > 0$, uniform with respect to h, such that

$$V_h = R_h \oplus B_h, \quad \gamma_R \|\pi_h^R v_h\|_L \leq \|v_h\|_L, \quad \forall v_h \in V_h. \tag{58}$$

Letting $\pi_h^R : V_h \to R_h$ be the oblique projector based on (58), we define the fluctuation operator $\kappa_h^R := I_{V_h} - \pi_h^R$, where I_{V_h} the identity in V_h. Just as for LPS stabilization, we can choose $R_h = P_k^g(\mathcal{T}_h)$. Then, G_h is the broken finite element space $P_{k-1}^b(\mathcal{T}_h)$, i.e. $G_K = \mathbb{P}_{k-1,d}$ on simplicial affine meshes (d-variate polynomials of order at most $k-1$). The simple choice $B_K = b_K G_K$ is only possible for $k \le d$, since otherwise the decomposition (58) is no longer direct. For $k \ge d+1$, a simple possibility to get around this technicality is to set $B_K = b_K^\alpha G_K$ with α equal to $\dfrac{k+1}{d+1}$ or to the smallest integer larger than $\dfrac{k}{d+1}$, see also Guermond (1999, prop. 4.1).

Lemma 5. *Assume that (56) holds. Let $\beta : D \to \mathbb{R}$ be such that $\beta_{|K} := \beta_K$ for all $K \in \mathcal{T}_h$. Then the sesquilinear forms*

$$s_h^{\mathrm{SGV}}(v_h, w_h) = (\check{\tau} A_1(\kappa_h^R v_h), A_1(\kappa_h^R w_h))_L, \tag{59a}$$

$$s_h^{\mathrm{SGV}}(v_h, w_h) = (\check{\tau} \underline{A}_1(\kappa_h^R v_h), \underline{A}_1(\kappa_h^R w_h))_L, \tag{59b}$$

$$s_h^{\mathrm{SGV}}(v_h, w_h) = (\beta^2 \check{\tau} \nabla(\kappa_h^R v_h), \nabla(\kappa_h^R w_h))_L, \tag{59c}$$

all satisfy the assumptions (i)–(ii)–(iii) with $\mathcal{J}_h(v_h) = \check{\tau} \pi_h^B \underline{A}_1(\pi_h^R(v_h))$.

REFERENCES

Becker, R., Braack, M., 2001. A finite element pressure gradient stabilization for the Stokes equations based on local projections. Calcolo 38 (4), 173–199.

Braack, M., Burman, E., 2006. Local projection stabilization for the Oseen problem and its interpretation as a variational multiscale method. SIAM J. Numer. Anal. 43 (6), 2544–2566.

Bramble, J.H., Schatz, A.H., 1970. Rayleigh-Ritz-Galerkin-methods for Dirichlet's problem using subspaces without boundary conditions. Comm. Pure Appl. Math. 23, 653–675.

Bramble, J.H., Schatz, A.H., 1971. Least squares for 2*m*th order elliptic boundary-value problems. Math. Comp. 25, 1–32.

Brooks, A.N., Hughes, T.J.R., 1982. Streamline Upwind/Petrov-Galerkin formulations for convective dominated flows with particular emphasis on the incompressible Navier-Stokes equations. Comput. Methods Appl. Mech. Engrg. 32, 199–259.

Burman, E., 2005. A unified analysis for conforming and nonconforming stabilized finite element methods using interior penalty. SIAM J. Numer. Anal. 43 (5), 2012–2033 (electronic).

Burman, E., Ern, A., 2007a. A continuous finite element method with face penalty to approximate Friedrichs' systems. M2AN Math. Model. Numer. Anal. 41 (1), 55–76.

Burman, E., Ern, A., 2007b. Continuous interior penalty *hp*-finite element methods for advection and advection-diffusion equations. Math. Comp. 76 (259), 1119–1140.

Burman, E., Hansbo, P., 2004. Edge stabilization for Galerkin approximations of convection-diffusion-reaction problems. Comput. Methods Appl. Mech. Engrg. 193 (15-16), 1437–1453.

Codina, R., 2002. Stabilized finite element approximation of transient incompressible flows using orthogonal subscales. Comput. Methods Appl. Mech. Engrg. 191 (39-40), 4295–4321.

Džiškariani, A.V., 1968. The least square and Bubnov-Galerkin methods. ž. Vyčisl. Mat. i Mat. Fiz. 8, 1110–1116.

Ern, A., Guermond, J.-L., 2004. Theory and Practice of Finite Elements. Applied Mathematical Sciences, vol. 159. Springer-Verlag, New York. xiv+524.

Ern, A., Guermond, J.-L., 2006a. Discontinuous Galerkin methods for Friedrichs' systems. I. General theory. SIAM J. Numer. Anal. 44 (2), 753–778.

Ern, A., Guermond, J.-L., 2006b. Discontinuous Galerkin methods for Friedrichs' systems. II. Second-order elliptic PDEs. SIAM J. Numer. Anal. 44 (6), 2363–2388.

Ern, A., Guermond, J.-L., 2013. Weighting the edge stabilization. SIAM J. Numer. Anal. 51 (3), 1655–1677.

Ern, A., Guermond, J.-L., 2016. Finite element quasi-interpolation and best approximation. ESAIM: Math. Model. Numer. Anal. preprint available at http://arxiv.org/abs/1505.06931.

Ern, A., Guermond, J.-L., Caplain, G., 2007. An intrinsic criterion for the bijectivity of Hilbert operators related to Friedrichs' systems. Comm. Partial Differ. Eq. 32, 317–341.

Friedrichs, K.O., 1958. Symmetric positive linear differential equations. Comm. Pure Appl. Math. 11, 333–418.

Guermond, J.-L., 1999. Stabilization of Galerkin approximations of transport equations by subgrid modeling. M2AN Math. Model. Numer. Anal. 33 (6), 1293–1316.

Guermond, J.-L., 2001a. Subgrid stabilization of Galerkin approximations of linear contraction semi-groups of class C^0 in Hilbert spaces. Numer. Methods Part. Diff. Eq. 17 (1), 1–25.

Guermond, J.-L., 2001b. Subgrid stabilization of Galerkin approximations of linear monotone operators. IMA J. Numer. Anal. 21, 165–197.

Hughes, T.J.R., Franca, L.P., Hulbert, G.M., 1989. A new finite element formulation for computational fluid dynamics: VIII. The Galerkin/Least-Squares method for advection-diffusive equations. Comput. Methods Appl. Mech. Engrg. 73, 173–189.

Johnson, C., Nävert, U., Pitkäranta, J., 1984. Finite element methods for linear hyperbolic equations. Comput. Methods Appl. Mech. Engrg. 45, 285–312.

Lučka, A.J., 1969. The rate of convergence to zero of the residual and the error for the Bubnov-Galerkin method and the method of least squares. In: Proc. Sem. Differential and Integral Equations, No. I (Russian), Akad. Nauk Ukrain. SSR Inst. Mat., Kiev, Ukraine, pp. 113–122.

Matthies, G., Skrzypacz, P., Tobiska, L., 2007. A unified convergence analysis for local projection stabilisations applied to the Oseen problem. M2AN Math. Model. Numer. Anal. 41 (4), 713–742.

Matthies, G., Skrzypacz, P., Tobiska, L., 2008. Stabilization of local projection type applied to convection-diffusion problems with mixed boundary conditions. Electron. Trans. Numer. Anal. 32, 90–105.

Rauch, J., 1994. Boundary value problems with nonuniformly characteristic boundary. J. Math. Pures Appl. 73 (4), 347–353.

Tadmor, E., 1989. Convergence of spectral methods for nonlinear conservation laws. SIAM J. Numer. Anal. 26 (1), 30–44.

Chapter 12

Least-Squares Methods for Hyperbolic Problems

P. Bochev* and M. Gunzburger†
*Center for Computing Research, Sandia National Laboratories, Albuquerque, NM, United States[1]
†Florida State University, Tallahassee, FL, United States

Chapter Outline

1. Sandia National Laboratories is a multiprogram laboratory operated by Sandia Corporation, a wholly owned subsidiary of Lockheed Martin Corporation, for the US Departmentof Energy's National Nuclear Security Administration under contract DE-AC04-94AL85000.

Handbook of Numerical Analysis, Vol. 17. http://dx.doi.org/10.1016/bs.hna.2016.07.002

ABSTRACT
Partial differential equations (PDE) problems are often intrinsically connected to the unconstrained minimization of a quadratic energy functional. The associated Rayleigh–Ritz variational principles provide an attractive setting for the development of finite element methods. Least-squares finite element methods (LSFEMs) aim to provide a Rayleigh–Ritz-like variational setting for any PDE, even if it is not associated with an unconstrained optimization principle. To this end, LSFEMs depend on an artificial, external energy-type principle (a least-squares principle) defined by summing up the equation residuals measured in suitable Hilbert space norms. In this chapter, we discuss the application of these ideas to the numerical solution of hyperbolic PDEs.

Keywords: Least-squares principles, Finite element methods, Hyperbolic problems, Conservation laws

AMS Classification Codes: 65N30, 65N12, 65N40, 65K10, 35L02

1 INTRODUCTION

Partial differential equations (PDEs) problems are often intrinsically connected to the *unconstrained minimization* of a quadratic energy functional. The associated Rayleigh–Ritz variational principles provide an attractive setting for the development of finite element methods because they lead to strongly coercive variational problems that are not subject to strict stability conditions such as the Ladyzhenskaya–Babuska–Brezzi (LBB) condition (Brezzi, 1974). The resulting discrete equations give rise to symmetric and positive definite linear systems that can be solved effectively by preconditioned conjugate gradient, multigrid, and other methods.

Least-squares finite element methods (LSFEMs) aim to provide a Rayleigh–Ritz-like variational setting for any PDE, even if it is not associated with an unconstrained optimization principle. To this end, LSFEMs depend on an artificial, external energy-type principle (*a least-squares principle*) defined by summing up the equation residuals measured in suitable Hilbert space norms. This approach is consistently capable of recovering most if not all of the advantages of the Rayleigh–Ritz setting over a wide range of problems.

In this section, we briefly review the basic tenets of least-squares methods for PDE problems and then, in Section 2 we specialize these ideas to hyperbolic problems. Let $\Omega \subset \mathbb{R}^d$, $d = 1, 2, 3$ be a bounded open region with Lipschitz continuous boundary $\partial\Omega$. For simplicity we consider linear boundary value problems

$$\mathcal{L}u = f \ \text{ in } \Omega \quad \text{and} \quad \mathcal{B}u = g \ \text{ in } \partial\Omega \tag{1}$$

specified by a linear differential operator $\mathcal{L}$ acting on functions u defined on Ω and a linear operator $\mathcal{B}$ acting on functions u defined on the boundary $\partial\Omega$. We assume that there exist Hilbert spaces $X = X(\Omega)$, $Y = Y(\Omega)$, and $B = B(\partial\Omega)$ such that $\mathcal{L}: X \mapsto Y$ and $\mathcal{B}: X \mapsto B$ satisfy the *stability bound*

$$C_1 \| u\|_X \leq \| \mathcal{L}u\|_Y + \| \mathcal{B}u\|_B \leq C_2 \| u\|_X \quad \forall\ u \in X \tag{2}$$

with constants C_1 and C_2 independent of u. The stability bound (2) is essential to the existence of a well-posed, externally defined energy-type principle for (1), which provides a Rayleigh–Ritz-like setting for the development of finite element methods. In other words, a bound such as (2) is a fundamental requirement for the formulation of a well-posed least-squares principle for the PDE (1) and for this reason we refer to it as the *energy balance* of this problem. To define such a principle we consider the residual energy functional (the least-squares functional (LSF))

$$J(u;f,g) = \| \mathcal{L}u - f \|_Y^2 + \| \mathcal{B}u - g \|_B^2 \tag{3}$$

and the unconstrained minimization problem (continuous least-squares principle (CLSP))

$$\text{minimize} \quad J(v;f,g) \ \text{ over } \ v \in X\,. \tag{4}$$

The energy balance (2) implies that the LSF is norm-equivalent, i.e., it satisfies

$$C_1 \| v\|_X \leq J(v;0,0) \leq C_2 \| v\|_X \quad \forall\ v \in X \tag{5}$$

for some positive constants C_1 and C_2 having values independent of v. This in turn implies that (4) is a well-posed minimization principle having a unique minimizer u_{LS}. Indeed, the first-order optimality condition for (4) is given by the variational equation

$$\text{seek } u \in X \text{ such that } \ Q(u;v) = F(v) \qquad \forall v \in X\,, \tag{6}$$

where $Q(u;v) = (\mathcal{L}u, \mathcal{L}w)_Y + (\mathcal{B}u, \mathcal{B}w)_B$ and $F(v) = (f, \mathcal{L}w)_Y + (g, \mathcal{B}w)_B$. The norm-equivalence (5) implies that the bilinear form $Q(u;v)$ is strongly coercive, in fact, since it is symmetric $Q(u;v)$ defines an equivalent inner product $((u,\ v))$ on $X \times X$. Existence and uniqueness of the minimizer u_{LS} then follows from the Lax–Milgram lemma.

Restriction of (3) to a conforming subspace $X^h \subset X$ yields a *discrete least-squares principle* (DLSP)

$$\text{minimize} \quad J(v^h;f,g) \ \text{ over } \ v^h \in X^h\,. \tag{7}$$

It is straightforward to see that the associated Euler–Lagrange equation given by

$$\text{seek } u^h \in X^h \text{ such that } \ Q(u^h;v^h) = F(v^h) \quad \forall v^h \in X^h \tag{8}$$

is a well-posed problem having a unique solution u^h_{LS}. Standard elliptic finite element arguments then yield the quasi optimal error estimate

$$\| u - u^h_{LS}\|_X \leq C \inf_{v^h \in X^h} \| u - v^h\|_X. \tag{9}$$

Finally, given a basis $\{\phi_i^h\}$ of X, one can show that (8) is equivalent to a linear algebraic system of equations with a symmetric and positive definite coefficient matrix.

Although this necessarily brief summary does not capture all of the intricacies involved in setting up a least-squares principle that is both well-posed and practical, it does convey the fundamental idea behind LSFEMs and the key properties that make them an attractive alternative for many PDE problems. We refer to Bochev and Gunzburger (2009) for further details.

2 LSFEM FOR HYPERBOLIC PROBLEMS

Although variational methods, including finite element methods, have been exceptionally successful for elliptic and parabolic problems, their application to hyperbolic PDEs has met with considerable difficulties. Some of these difficulties have been overcome by ideas such as Discontinuous Galerkin methods (Reed and Hill, 1973) and algebraic flux correction (Kuzmin et al., 2005) enabling finite element methods with properties approaching those of specialized finite volume and finite difference schemes.

The status of LSFEMs for hyperbolic problems largely mirrors this situation. Although the idea of replacing a hyperbolic PDE by an attractive Rayleigh–Ritz-like setting given by an LSF (3) and an associated CLSP (4) is very appealing and has attracted attention early on, see, e.g., Chen and Fix (1986a,b), Cox et al. (1983), and Wilders (1988), its straightforward application, without proper accounting for the distinctions between elliptic and hyperbolic PDEs, may lead to less than satisfactory methods. The root cause for these complications is the fact that for hyperbolic PDEs Hilbert spaces are not necessarily the only or the best choice for defining the energy balance (2) that is essential to LSFEMs. As a result, formulation of LSFEMs for this class of problems has followed two parallel pathways, one for which residuals are minimized in Banach space norms[a] and the other for which minimization is carried in Hilbert spaces.

A thorough discussion of minimization problems in Banach spaces is beyond the scope of this chapter. Nonetheless, since this setting shows great promise we provide some examples that interested readers can follow further on their own. Due to the limited space we also restrict the main focus of this chapter to conservation laws, which are ubiquitous in important science and engineering problems. Section 3 introduces the model equations, followed, in Section 4, by derivation of energy balances in Hilbert and Banach spaces. In Section 6 we present a collection of LSFEMs derived from "true" LSF, while Section 7 focusses on methods derived from residual minimization in $L^1(\Omega)$. Finally, Section 8 considers a class of LSFEMs that aim to recover the advantages of $L^1(\Omega)$ principles, while retaining attractive properties of Hilbert space minimization such as differentiability of the functionals. The chapter concludes with a brief discourse on practical issues and a summary of conclusions and recommendations.

[a]Of course in this case the associated minimization principles are not "least-squares" in the strict sense of this word but rather L^p minimization problems.

3 CONSERVATION LAWS

Conservation laws are PDEs that can be written in the following canonical abstract form:

$$\begin{cases} \dfrac{\partial u}{\partial t} + \nabla \cdot F(u) = f & \text{in } \Omega \times (0,T) \\ \mathcal{B}u = h & \text{on } \Gamma_- \times (0,T) \\ u = u_0 & \text{in } \Omega \text{ at } t = 0\,, \end{cases} \tag{10}$$

where F is a *flux* function, u is a dependent variable, and Γ_- is a subset of $\partial\Omega$ with a positive measure. The simplest example of a conservation law is the linear scalar advection equation

$$\frac{\partial \phi}{\partial t} + \nabla \cdot (\boldsymbol{b}\phi) = f \tag{11}$$

for which $u = \phi$ and $F(\phi) = \boldsymbol{b}\phi$ with $\boldsymbol{b}(\boldsymbol{x})$ denoting a given vector. This equation models the transport of a scalar quantity, such as the concentration, or a passive tracer by the given velocity field $\boldsymbol{b}$.

In some cases, it is convenient to write conservation laws in the *nonconservative*, first-order system form

$$\frac{\partial v}{\partial t} + \mathbb{A}_1 \frac{\partial v}{\partial x} + \mathbb{A}_2 \frac{\partial v}{\partial y} + \mathbb{A}_3 \frac{\partial v}{\partial z} + \vec{b} = 0\,, \tag{12}$$

where v denotes a vector of n primitive variables, $\mathbb{A}_i$, $i = 1, 2, 3$, denote square $n \times n$ matrices, and $\vec{b}$ denotes an n-vector. In general, both the $\mathbb{A}_i$s and $\vec{b}$ depend on $\boldsymbol{x}$ and u.

There have been many computational studies of LSFEMs for the model problems described earlier. Most of the theoretical results though have been confined to (11) and its steady state *advection–reaction* version

$$\begin{cases} \nabla \cdot (\boldsymbol{b}\phi) + c\phi = f & \text{in } \Omega \\ \phi = g & \text{on } \Gamma_-, \end{cases} \tag{13}$$

where f, c, and g are given functions and Γ_- is the inflow[b] part of $\partial\Omega$ given by

$$\Gamma_- = \{\boldsymbol{x} \in \partial\Omega \mid \boldsymbol{n}(\boldsymbol{x}) \cdot \boldsymbol{b}(\boldsymbol{x}) < 0\}. \tag{14}$$

The second term in the differential equation can be a "true" (linear) reaction term, or a time discretization term resulting from solving (11) by the method of lines. Throughout this chapter we assume that the advective field $\boldsymbol{b}(\boldsymbol{x})$ is of class $C^1(\overline{\Omega})$ and the reaction coefficient $c(\boldsymbol{x})$ is a bounded measurable function on Ω. We also use the notation $\nabla_{\boldsymbol{b}}\phi = \boldsymbol{b} \cdot \nabla\phi$ for the streamwise derivative of ϕ and the following inner product and norm, respectively,

---

[b]The outflow boundary is $\Gamma_+ = \partial\Omega \setminus \Gamma_-$.

$$(\phi, \psi)_{-b} = \int_{\Gamma_-} \phi\psi\, \boldsymbol{n} \cdot \boldsymbol{b}\, d\Gamma \quad \| \phi \|_{-b}^2 = \int_{\Gamma_-} \phi^2 \, |\boldsymbol{n} \cdot \boldsymbol{b}| \, d\Gamma$$

defined on Γ_-. Finally, we recall the following definition.

Definition 1. A point $\tilde{x}$ on the surface s defined by the implicit equation $s(\mathrm{x}) = 0$ is noncharacteristic for the first-order system (12) if

$$\det \left(\mathbb{A}_1 \frac{\partial s}{\partial x} + \mathbb{A}_2 \frac{\partial s}{\partial y} + \mathbb{A}_3 \frac{\partial s}{\partial z} \right) \neq 0 \quad \text{at}\, \boldsymbol{x} = \tilde{\boldsymbol{x}}\, .$$

A surface is noncharacteristic if it is noncharacteristic at every point $\boldsymbol{x}$. A surface s is noncharacteristic for (13) if $\boldsymbol{b} \cdot \nabla s \neq 0$. Geometrically, this means that, at any point of s, the advective vector $\boldsymbol{b}$ is not in the tangent plane to the surface at that point.

4 ENERGY BALANCES

The advection–reaction operator in (13) is a limit case, as $\epsilon \to 0$, of the advection–diffusion–reaction operator $-\epsilon\Delta\phi + \nabla \cdot (\boldsymbol{b}\phi) + c\phi$. Although solutions of this operator may have internal and/or boundary layers they remain in $H^1(\Omega)$ as long as $\epsilon > 0$. In contrast, solutions of (13) propagate any discontinuity in the boundary data along characteristics so that they are not, in general, in $H^1(\Omega)$. Thus, the absence of a diffusion term in (13) significantly changes the functional setting for this equation by making a Banach space such as L^1 a more natural choice than $H^1(\Omega)$.

In what follows we will consider both settings for the energy balance of (13) even though, as mentioned earlier, the Banach space setting does not lead to a bona fide least-squares principle but rather to an L^p minimization problem. The two settings are similar structurally in the sense that they both use the graph space of the advection–reaction operator $\mathcal{L}\phi \equiv \nabla \cdot (\boldsymbol{b}\phi) + c\phi$ as a solution space X. This means that regardless of the setting, solutions of (13) are not required to have weak first derivatives, except along the streamlines. The main difference is in the type of spaces where the solution and its streamline derivative are expected to belong, i.e., in the type of spaces used to construct X. In Section 4.1, we assume that $\mathcal{L}\phi$ is *square integrable*, which results in a Hilbertian graph space. In Section 4.2, this requirement is relaxed by assuming that $\mathcal{L}\phi$ is only *integrable*, leading to a Banach graph space.

4.1 Energy Balances in Hilbert Spaces

The Hilbertian setting for the energy balances in this section is similar to the settings employed in Bochev and Choi (2001b), Eriksson et al. (1996), Houston et al. (1999), Johnson (1992), Johnson et al. (1984), Lazarov et al. (1997), and Suli (1997), and we refer to these papers for technical details regarding the proofs of the results below. In the Hilbertian setting the advection–reaction operator is considered as a map $\mathcal{L}: X_H \mapsto L^2(\Omega)$, where

$$X_H = \{\phi \in L^2(\Omega) \mid \nabla\cdot(\boldsymbol{b}\phi) + c\phi \in L^2(\Omega)\} \tag{15}$$

is the graph space of $\mathcal{L}$. X_H is Hilbert space when equipped with the graph norm

$$\|\phi\|_H = \left(\|\phi\|_0^2 + \|\mathcal{L}\phi\|_0^2\right)^{1/2} = \left(\|\phi\|_0^2 + \|\nabla\cdot(\boldsymbol{b}\phi) + c\phi\|_0^2\right)^{1/2}. \tag{16}$$

One can show that this graph norm is equivalent to the "energy" norm

$$|||\phi||| = \left(\|\nabla\cdot(\boldsymbol{b}\phi) + c\phi\|_0^2 + \|\phi\|_{-\boldsymbol{b}}^2\right)^{1/2}. \tag{17}$$

This equivalence follows from a trace inequality and a Poincaré-type inequality for functions in X_H.

Lemma 1 (Trace inequality). *Assume that $\partial\Omega$ is of class C^1 and Γ_- is noncharacteristic. Then, there exists a continuous trace operator $\gamma : X_H \mapsto L^2(\Gamma_-)$ such that, for all $\phi \in X_H$,*

$$\|\phi\|_{0,\Gamma_-} \le C_T \|\phi\|_H, \tag{18}$$

where C_T is a positive constant. Furthermore, for any $\phi \in X_H$, we have the Green's formula

$$\int_\Omega \phi\nabla\cdot(\boldsymbol{b}\phi)\, d\Omega = \frac{1}{2}\left(\int_\Omega \phi^2\nabla\cdot\boldsymbol{b}dx + \int_{\partial\Omega}\phi^2\boldsymbol{n}\cdot\boldsymbol{b}\, d\partial\Omega\right). \tag{19}$$

Lemma 2 (Poincaré-type inequality). *Assume that $\partial\Omega$ is of class C^1, Γ_- is noncharacteristic, and that*[c]

$$c + \frac{1}{2}\nabla\cdot\boldsymbol{b} \ge \gamma_0 > 0. \tag{20}$$

Then, there exists a positive constant C_P such that, for all $\phi \in X_H$,

$$\|\phi\|_0 \le C_P|||\phi|||, \tag{21}$$

where $|||\cdot|||$ is the energy norm defined in (17).

The following theorem establishes the energy balance for (13) in the graph space X_H.

Theorem 1. *Assume that $\partial\Omega$ is of class C^1, Γ_- is noncharacteristic, and that (20) holds. Then, there exist positive constants C_1 and C_2 such that, for all $\phi \in X_H$,*

$$C_1\|\phi\|_H \le |||\phi||| \le C_2\|\phi\|_H. \tag{22}$$

Remark 1. Lemma 1 implies that functions belonging to X_H have well-defined traces in $L^2(\Gamma_-)$, provided the boundary is C^1 regular. This assumption can be

[c]In many applications $\nabla\cdot\boldsymbol{b} = 0$ in which case (20) reduces to the simple condition $c \ge \gamma_0 > 0$.

dropped by using a somewhat more complicated boundary norm (see De Sterck et al., 2005a,b) defined by

$$\| \phi \|_{-\ell}^2 = \int_{\Gamma_-} \phi^2 \, \ell(\boldsymbol{x}(s)) |\boldsymbol{b} \cdot \boldsymbol{n}| / \|\boldsymbol{b}\| \, d\Gamma \,, \tag{23}$$

where $\ell(\boldsymbol{x})$ denotes the length of the streamline defined by the vector field $\boldsymbol{b}$ connecting Γ_- to Γ_+. This boundary norm (23) satisfies a trace inequality similar to (18) and, if used in (17) (in lieu of $\|\cdot\|_{-\boldsymbol{b}}$), gives rise to an energy norm that satisfies a Poincaré inequality similar to (21). As a result, the energy balance in Theorem 1 also holds for this modified energy norm. □

4.2 Energy Balances in Banach Spaces

In the Banach space setting the advection–reaction operator is considered as a mapping $\mathcal{L}: X_B \mapsto L^1(\Omega)$, i.e., we have relaxed the assumption in Section 4.1 that $\mathcal{L}\phi$ is *square integrable* to an assumption that it is only *integrable*. The appropriate graph space for this setting is given by

$$X_B = \{\phi \in L^1(\Omega) \ | \ \nabla \cdot (\boldsymbol{b}\phi) + c\phi \in L^1(\Omega)\} \,. \tag{24}$$

When X_B is equipped with the graph norm

$$\| \phi \|_B = \| \phi \|_{L^1(\Omega)} + \| \mathcal{L}\phi \|_{L^1(\Omega)} \,, \tag{25}$$

it becomes a Banach space. It can be shown that the advection–reaction operator is an isomorphism $X_B \mapsto L^1(\Omega)$; see Guermond (2004) and the references cited therein. Using this fact, one can prove the following result.

Theorem 2. *There exists a positive constant* C *such that, for all* $\phi \in X_B$,

$$C \| \phi \|_B \leq \| \mathcal{L}\phi \|_{L^1} \,. \tag{26}$$

This result is the Banach space counterpart[d] of the Hilbertian energy balance (22). The Banach space setting is potentially attractive for problems that have discontinuous solutions. However, because residual minimization in $L^1(\Omega)$ leads to nondifferentiable functionals, direct use of this setting for the numerical solution of hyperbolic PDEs has been very limited; see Guermond (2004), Jiang (1993), and Lavery (1989). In Section 7, we discuss approaches that replace direct minimization in $L^1(\Omega)$ by a sequence of regularized $L^2(\Omega)$ minimization problems.

5 CONTINUOUS LEAST-SQUARES PRINCIPLES

The Hilbertian setting from Section 4.1 fits in the abstract least-squares framework outlined in Section 1, i.e., it provides a true Rayleigh–Ritz-like

[d]The upper bound in terms of $\|\phi\|_B$ is trivial.

foundation for the development of finite element methods for conservation laws. This is an important advantage of the Hilbertian setting, which allows us to develop well-posed CLSP for the advection–reaction problem by following the exact same procedure as described in Section 1. In particular, (22) implies that the CLSP

$$\text{minimize } J_H(\phi;f,g) = \| \nabla\cdot(\boldsymbol{b}\phi) + c\phi - f \|_0^2 + \| \phi - g \|_{-\boldsymbol{b}}^2 \quad \text{over } \phi \in X_H \tag{27}$$

has a unique minimizer $\phi \in X_H$ that satisfies the Euler–Lagrange equation: *seek* $\phi \in X_{\mathrm{H}}$ *such that*

$$Q(\phi;\psi) = F(\phi) \quad \forall \psi \in X_B . \tag{28}$$

In (28) the bilinear form Q and the functional F are given by

$$Q(\phi;\psi) = \int_\Omega (\nabla\cdot(\boldsymbol{b}\phi) + c\phi)(\nabla\cdot(\boldsymbol{b}\psi) + c\psi)\, d\Omega + \int_{\Gamma_-} \phi\psi \, |\boldsymbol{n}\cdot\boldsymbol{b}|\, d\Gamma, \text{ and}$$

$$F(\psi) = \int_\Omega f(\nabla\cdot(\boldsymbol{b}\psi) + c\psi)\, d\Omega + \int_{\Gamma_-} g\psi \, |\boldsymbol{n}\cdot\boldsymbol{b}|\, d\Gamma$$

respectively. Another important advantage of the Hilbertian setting for (13) is the practicality of (28). Because the LSF in (27) uses only standard $L^2(\Omega)$ norms and $\mathcal{L}$ contains only first-order derivatives[e] discretization of the weak problem requires at most C^0 continuous finite element spaces. On the other hand, as is shown in Section 9.2, the main drawback of Hilbert spaces is the strong diffusivity of the least-squares weak equation that has to be compensated by additional "adjustments" to (27).

According to Remark 1,

$$\text{minimize } J_H(\phi;f,g) = \| \nabla\cdot(\boldsymbol{b}\phi) + c\phi - f \|_0^2 + \| \phi - g \|_{-\ell}^2 \quad \text{over } \phi \in X_H, \tag{29}$$

where $\|\cdot\|_{-\ell}$ is the norm defined in (23), is another well-posed CLSP for the advection–reaction equation (13) whose norm equivalence does not require the C^1 assumption on $\partial\Omega$. However, because the characteristic length $\ell(\boldsymbol{x})$ required in the definition of $\|\cdot\|_{-\ell}$ may be difficult to compute for some problems, this CLSP is less practical than (27). Thus, in the following sections, we only consider LSFEMs based on (27).

[e]In Section 6.2 we will see an example of an *impractical* least-squares formulation, which involves higher-order derivatives and requires *impractical* C^1 finite element spaces for its conforming discretization. The tension between norm-equivalence and practicality of the finite element spaces necessary for the conforming discretization of the associated CLSP is a recurring theme in the formulation of LSFEMs (Bochev and Gunzburger, 2009, section 2.2.3). This tension is the strongest for PDEs involving second and higher order derivatives in which case formulation of practical LSFEM usually requires transitioning to an equivalent first-order system form of the PDE.

The most obvious disadvantage of the Banach space setting of Section 4.2 is that $L^1(\Omega)$ minimization is much more difficult to implement compared to $L^2(\Omega)$ minimization. Also, "least-squares" principles in the Banach space setting do not fit in the abstract framework of Section 1. Nevertheless, based on the result of Theorem 2, we can consider the following continuous "least-squares" principle for (13):

$$\text{minimize } J_B(\phi;f) = \| \nabla \cdot (\boldsymbol{b}\phi) + c\phi - f \|_{L^1(\Omega)} \text{ over } X_{B,g} = \{\phi \in X_B | \phi = g \text{ on } \Gamma_-\}. \tag{30}$$

Although the framework in Section 1 is not applicable to this minimization problem, it is equivalent to (13) in the sense that they have the same solution.

5.1 Extension to Time-Dependent Conservation Laws

Extension of the Continuous Least Squares Principles (CLSPs) (27) and (30) to time-dependent conservation laws can be accomplished by discretizing first in time and then minimizing the residual of the semi-discrete[f] equation in X_H or X_B, respectively. This approach has the advantage of reducing the solution of the time-dependent PDE to a solution of a sequence of unconstrained minimization problems. For practical reasons, most existing methods have only treated the Hilbertian setting. It is worth mentioning that the same strategy for the development of LSFEMs has been used in related application contexts such as time-dependent advection–diffusion–reaction problems (Toledo and Ruas, 2011) and the shallow water equations (Liang and Hsu, 2009).

One of the first examples of this approach is Jiang and Carey (1988a), which considers (11) in one space dimension; in this case, $\boldsymbol{b} = b$, a scalar function. The solution is advanced in time by solving a sequence of CLSPs

$$\text{minimize } J(\phi) = \| (\phi - \overline{\phi}) + \Delta t \left(\theta \frac{d}{dx}(b\phi) + (1-\theta)\frac{d}{dx}(b\overline{\phi}) \right) \|_0^2 \text{ over } \phi \in X_H \tag{31}$$

for the semi-discrete in time equations, obtained by an application of the generalized θ-method. In (31), $\overline{\phi}$ denotes the solution at time t and ϕ the unknown solution at time $t + \Delta t$. In Jiang and Carey (1988a), this approach is also extended to one-dimensional nonlinear conservation laws by using the same discretization in time and the following (nonquadratic) functional:

$$J(\phi) = \| (\phi - \overline{\phi}) + \Delta t \left(\theta \frac{d}{dx} F(\phi) + (1-\theta)\frac{d}{dx} F(\overline{\phi}) \right) \|_0^2 . \tag{32}$$

[f]For space-time least-squares formulations, see Bell and Surana (1994, 1996), Nguen and Reynen (1984), Perrochet and Azerad (1995), and De Sterck et al. (2005b).

It can be shown (see Jiang and Carey, 1988a) that the resulting LSFEMs are equivalent to a particular form of the Taylor–Galerkin approach (Donea, 1984; Lohner et al., 1984) and are unconditionally stable for linear problems.

A further example is the method of Jiang and Carey (1990) for the two-dimensional compressible Euler equations. This method uses the nonconservative form (12) of the Euler equations and the backward-Euler method in time, i.e., given a solution $\bar{v}$ at time t, the solution v at time $t + \Delta t$ is determined by solving the CLSP

$$\text{minimize } \; J(v) = \left\| (v - \bar{v}) + \Delta t\left(\mathbb{A}_1 \frac{\partial v}{\partial x} + \mathbb{A}_2 \frac{\partial v}{\partial y} \right) \right\|_0^2 \quad \text{over } v \in X_H . \tag{33}$$

Obviously, extensions of (30) to time-dependent problems can be obtained from that CLSP by measuring residuals in the $L^1(\Omega)$ norm and changing the minimization space to X_B.

6 LSFEM IN A HILBERT SPACE SETTING

In this section, we consider LSFEMs derived from the CLSP (27). The first group of methods is based on conforming DLSPs that use proper finite element subspaces of the graph space X_H. The second group includes several nonconforming methods.

6.1 Conforming LSFEMs

A conforming DLSP for the advection–reaction equation is given by

$$\text{minimize } \; J_H(\phi^h; f, g) = \| \nabla \cdot (\boldsymbol{b}\phi^h) + c\phi^h - f \|_0^2 + \| \phi^h - g \|_{-b}^2 \quad \text{over } \phi^h \in X_H^h, \tag{34}$$

where X_H^h is a finite element subspace of the graph space X_H. The standard C^0 nodal finite element space of polynomial order r, denoted here by $S_h^r(\Omega)$, is trivially a subspace of the graph space X_H, and so it can be used in (34). The resulting DLSP fits in the abstract framework of Section 1 and so, it has a unique minimizer ϕ^h. Furthermore, one can show the following result; see Bochev and Gunzburger (2009) for details.

Theorem 3. *Let $X_H^h = S_h^r(\Omega)$ for some integer* r $\geq$ *1. Then*

$$|||\phi - \phi^h||| \leq \inf_{\psi^h \in X_H^h} |||\phi - \psi^h|||. \tag{35}$$

If $\phi \in X_H \cap H^{r+1}(\Omega)$, then

$$|||\phi - \phi^h||| \leq Ch^r |\phi|_{r+1} . \quad \square \tag{36}$$

The semi-discrete in time CLSPs presented in Section 5.1 can also be discretized by standard C^0 finite element spaces. The original formulations of these methods in Jiang and Carey (1988a, 1990) correspond to the DLSPs

$$\text{minimize } J(\phi^h) = \left\| (\phi^h - \overline{\phi}^h) + \Delta t \left(\theta \frac{d}{dx}(b\phi^h) + (1-\theta)\frac{d}{dx}(b\overline{\phi}^h) \right) \right\|_0^2 \quad (37)$$
$$\text{over } \phi^h \in X^h = S_h^r(\Omega)$$

and

$$\text{minimize } J(v^h) = \left\| (v^h - \bar{v}^h) + \Delta t \left(\mathbb{A}_1 \frac{\partial v^h}{\partial x} + \mathbb{A}_2 \frac{\partial v^h}{\partial y} \right) \right\|_0^2 \text{ over } v^h \in X^h = [S_h^r(\Omega)]^4, \quad (38)$$

respectively.

6.2 Nonconforming Methods

If the exact solution of (13) is sufficiently smooth, Theorem 3 indicates that standard C^0 finite element spaces are completely satisfactory. However, if the boundary data have a discontinuity, it is shown in Lowrie and Roe (1994) that residual minimization in the norms of X_H or X_B over C^0 finite element spaces lead to solutions that necessarily spread the discontinuity over several elements and are not monotone; see Section 9.2 for a further discussion. One potential remedy is to replace the standard finite element space by a discontinuous space and a nonconforming version of (34).

Nonlinear conservation laws can develop discontinuous solutions even if the data are smooth. In such cases, the methods defined by (37) and (38) may break down. In Jiang and Carey (1988b), it was observed that for high-speed compressible flow problems, the presence of nonlinear instabilities as the developing shock steepens leads to a failure of (37). For the Euler equations, Jiang and Carey (1990) report that (38) is stable and gives nonoscillatory shock profiles, as long as the time step is sufficiently large to maintain the Courant number in the range 10–50. However, outside this range, this method was prone to instabilities.

The fix proposed in Jiang and Carey (1988b, 1990) was to augment the functionals in (37) and (33) by an $H^1(\Omega)$-seminorm of the nonlinear residual. However, the resulting $H^1(\Omega)$-seminorm least-squares method requires C^1 finite element spaces, which renders it *impractical* in more than one space dimension. Our second example of a nonconforming DLSP is a *practical* version of this method, which uses a "broken" H^1-seminorm term.

6.2.1 Discontinuous LSFEM

Let $\mathcal{T}_h$ denote a partition of $\Omega \subset \mathbb{R}^d$, $d = 2, 3$, into finite elements κ and let $\Sigma^h = \mathcal{C}_{d-1}(\mathcal{T}_h)$ denote the set of all oriented[g] inter-element interfaces having

[g]For $d = 2, 3$ an interface $\mathcal{C} \in \mathcal{C}_{d-1}(\mathcal{T}_h)$ is oriented by choosing a unit normal vector $\boldsymbol{n}_c$ at some interior point $\boldsymbol{x} \in \mathcal{C}$.

a unit normal $\boldsymbol{n}_c$. The first example of a nonconforming LSFEM is the discontinuous method proposed in De Sterck et al. (2005a) in which minimization is carried over the finite element space[h] $[S_h^r](\Omega) = \cup_\kappa S^r(\kappa)$. For $\phi^h \in [S_h^r](\Omega)$, we define the interface semi-norm

$$\| \phi^h \|_{\Sigma^h}^2 = \sum_{C \in \mathcal{C}_{d-1}(\mathcal{T}_h)} \int_C [\phi^h]^2 \, \omega_c^h \, |\boldsymbol{b} \cdot \boldsymbol{n}_c| \, d\Gamma, \tag{39}$$

where ω_c^h is a mesh-dependent weight and $[\,\cdot\,]$ is the jump function. The seminorm (39) vanishes if ϕ^h is a C^0 function. The discontinuous LSFEM of De Sterck et al. (2005a) is defined by the following mesh-dependent DLSP:

$$\text{minimize } J_H^h(\phi^h; f, g) = \sum_{\kappa \in \mathcal{T}_h} \| \nabla \cdot (\boldsymbol{b}\phi^h) + c\phi^h - f \|_{0,\kappa}^2 + \| \phi^h - g \|_{-\boldsymbol{b}}^2 + \| \phi^h \|_{\Sigma^h}^2$$
$$\text{over } \phi^h \in [S^r](\Omega). \tag{40}$$

The interface term $\| \phi^h \|_{\Sigma^h}$ promotes interelement continuity in a direction parallel to $\boldsymbol{b}$, while allowing for larger jumps in a direction perpendicular to $\boldsymbol{b}$. This mimics the continuity properties of solutions to (13) which may jump across the characteristics, but remain continuous along them.

6.2.2 $H^1(\Omega)$ LSFEM

The second example of a nonconforming LSFEM is the *practical* version of the $H^1(\Omega)$ method in Jiang and Carey (1988a), which was motivated by parallels between least-squares and multiobjective optimization problems. For a one-dimensional nonlinear conservation law, the original $H^1(\Omega)$ method replaces (32) by the following LSF:

$$J_1(\phi) = \| R(\phi) \|_0^2 + |\alpha R(\phi)|_1^2, \tag{41}$$

where $0 < \alpha \ll 1$ is a penalty parameter, $|\cdot|_1$ is the $H^1(\Omega)$ seminorm, and

$$R(\phi) = (\phi - \overline{\phi}) + \Delta t \left(\theta \frac{d}{dx} F(\phi) + (1-\theta) \frac{d}{dx} F(\overline{\phi}) \right)$$

is the residual of the semi-discrete equation. The penalty term involving the $H^1(\Omega)$ seminorm provides additional artificial viscosity proportional to Δt times the linearized flux function at $\overline{\phi}$. Conforming discretization of this functional requires C^1 finite element spaces and so, it is not practical[i] in more than one space dimension. This problem can be circumvented by considering instead the following nonconforming version of (41):

---

[h]Here, $\cup_\kappa S^r(\kappa)$ refers to the union of finite element spaces S^r defined over the individual elements κ.
[i]Construction of C^1 finite element spaces on general unstructured grids in two and three dimensions is nontrivial.

$$\hat{J}_1(\phi^h) = \| R(\phi^h) \|_{0,\Omega}^2 + \sum_{\kappa \in \mathcal{T}_h} |\alpha R(\phi^h)|_{1,\kappa}^2 .$$

This idea was used in Jiang and Carey (1990) to develop a practical extension of the $H^1(\Omega)$ method for the Euler equations.

7 RESIDUAL MINIMIZATION METHODS IN A BANACH SPACE SETTING

Methods for the approximate solution of PDEs based on residual minimization in Banach spaces are extremely rare. Arguably, the first example of this approach is the $L^1(\Omega)$-minimization method for one-dimensional conservation laws considered in Lavery (1988, 1989). Despite the initial promise shown by $L^1(\Omega)$ methods, their reliance on mathematical programming to compute the minimizers meant that they could not compete with more conventional approaches. As a result, $L^1(\Omega)$ methods remained dormant until, in Guermond (2004), a regularization procedure was formulated that allowed to approximate the minimizer by gradient methods. Although it is not clear if $L^1(\Omega)$ methods will ever become a practical alternative to other methods, they are an intriguing example of residual minimization that has influenced the development of LSFEMs for conservation laws. Thus, in this section, we offer a brief survey of $L^1(\Omega)$ methods and then, in Section 8, examine a class of LSFEMs motivated by them.

7.1 An $L^1(\Omega)$ Minimization Method

We describe the $L^1(\Omega)$ approach of Lavery (1988, 1989) using the following simple one dimensional problem:

$$\frac{d\phi}{dx} = 0 \ \text{ on } (0,1)\,, \ \ \phi(0) = g_0\,, \ \text{ and } \ \phi(1) = g_1\,. \tag{42}$$

For $g_0 \neq g_1$, the "physically" meaningful solution of (42) is a "compression shock" at $x = 1$ given by $\phi(x) = g_0$ for $0 \le x < 1$ and $\phi(1) = g_1$. This solution is the limit, as $\epsilon \to 0$, of the solution of the singularly perturbed elliptic equation $-\epsilon\phi_{xx} + \phi_x = 0$ along with the boundary conditions in (42).

The $L^1(\Omega)$ method (Lavery, 1988, 1989) starts with a conventional finite volume discretization of the following regularly perturbed version of (42):

$$\frac{d\phi}{dx} + 2\epsilon\phi = 0 \ \text{ on } (0,1), \ \ \phi(0) = g_0\,, \ \text{ and } \ \phi(1) = g_1\,. \tag{43}$$

Assuming that (0, 1) has been partitioned into N (not necessarily uniform) subintervals (x_{i-1}, x_i), $i = 1, 2, \ldots, N$, the discrete equations are given by

$$\phi_i - \phi_{i-1} + \epsilon h_i(\phi_i + \phi_{i-1}) = 0 \ \text{ for } i = 1,2,\ldots,N\,, \quad \phi_0 = g_0, \quad \phi_N = g_1\,, \tag{44}$$

where $h_i = x_i - x_{i-1}$ is the length of the ith interval. The next step is to seek solution of this over-determined $N \times (N-1)$ linear system for the unknown values ϕ_i, $i = 1, \ldots, N-1$, by minimizing the ℓ_1 norm of the residual in (43):

$$J_{\ell_1}(\vec{\phi}) = \sum_{i=1}^{N} |\phi_i - \phi_{i-1} + \epsilon h_i(\phi_i + \phi_{i-1})| \,. \tag{45}$$

It is easy to see that without the perturbation term, any monotone grid function $\{\psi_i\}_{i=0}^{N}$ that satisfies $\psi_0 = g_0$ and $\psi_N = g_1$ is a minimizer of this functional.

The last step in the $L^1(\Omega)$ approach is to recast (45) as a linear programming problem and solve it by a discrete optimization algorithm. We can write (45) in the form

$$\mathbb{A}\vec{\phi} = \vec{g} \,,$$

where

$$\vec{\phi} = (\phi_1, \ldots, \phi_{N-1})^{\mathsf{T}}, \qquad \vec{g} = ((1 - \epsilon h_1)g_0, 0, \ldots, 0, -(1 + \epsilon h_N)g_1)^{\mathsf{T}},$$

and $\mathbb{A}$ is an $N \times (N-1)$ matrix. Let $\vec{r} = \vec{g} - \mathbb{A}\vec{\phi}$ denote the residual of this linear system. Finding the minimizer of the ℓ_1 functional (45) is equivalent to finding the solution $\vec{\phi}$ of the following linear programming problem (see Lavery, 1989):

$$\text{minimize} \quad \|\vec{r}\|_{\ell_1} \quad \text{subject to} \quad \vec{z}^{\mathsf{T}}\vec{r} = \vec{z}^{\mathsf{T}}\vec{g} \,, \tag{46}$$

where $\vec{z}$ is a vector that spans the null space of $\mathbb{A}$ and

$$\|\vec{r}\|_{\ell^1} = \sum_{i=1}^{N} |r_i| \tag{47}$$

is the ℓ_1 vector norm on $\mathbb{R}^N$. The ℓ_1 algorithm consists of finding the null-space vector[j] $\vec{z}$ and solving (46). It turns out that this solution is given by

$$\phi_i = g_0 \prod_{k=1}^{i} \frac{1 - \epsilon h_k}{1 + \epsilon h_k}, \qquad i = 1, 2, \ldots, N-1 \,,$$

so that the discontinuity is confined to the last cell (x_{N-1}, x_N). In Lavery (1989), this method was also applied to a time-independent version of the inviscid Burger equation.

7.2 Regularized $L^1(\Omega)$ Minimization Method

In the approach of Lavery (1989), the $L^1(\Omega)$ minimization problem is solved by linear programming methods that do not require differentiability. To enable

[j]For the system (44), this vector can be found explicitly; this is not the case for a system obtained with the usual perturbation choice $-\epsilon\phi_{xx}$; see Lavery (1989).

application of more conventional (and potentially more efficient) gradient-based algorithms, a regularization of the $L^1(\Omega)$ functional that renders it Fréchet differentiable was proposed in Guermond (2004). The starting point is the discrete $L^1(\Omega)$ principle

$$\text{minimize } J_B(\phi^h;f) = \| \nabla \cdot (\boldsymbol{b}\phi^h) + c\phi^h - f \|_{L^1(\Omega)} \text{ over } \phi^h \in X^h_{B,g} = S^1_h \cap X_{B,g} \tag{48}$$

obtained by restriction of (30) to a standard C^0 linear, bilinear or trilinear finite element space. This problem is conforming because $X^h_{B,g} \subset X_{B,g}$. However, the functional $J_B(\phi^h;f)$ is not differentiable and so, the resulting "DLSP" does not fit in the abstract least-squares framework in Section 1. The following theorem is proved in Guermond (2004).

Theorem 4. *The following holds true for the conforming* $L^1(\Omega)$ *minimization principle (48).*

1. *The problem (48) has no local minimizers and at least one global minimizer.*
2. *All minimizers have the stability property* $\| \phi^h \|_B \leq C \| f \|_{L^1(\Omega)}$.
3. *All minimizers satisfy the a priori error bound* $\| \phi - \phi^h \|_B \leq \inf_{\psi^h \in X^h_{B,g}} \| \phi - \psi^h \|_B$.

The last assertion in Theorem 4 is similar to the best approximation property (9) of conforming DLSPs. However, as in Section 7.1, the $L^1(\Omega)$ principle (48) is ill-suited for numerical computation. It can have multiple minimizers[k] and the $L^1(\Omega)$ functional is not differentiable. In Section 7.1, the first issue was dealt with by using the nonsingularly perturbed equation (43) and the second issue was circumvented by using linear programming. The approach in Guermond (2004) allows the simultaneous resolution of both of these issues by using the *regularized* formulation

$$\text{minimize } J_{B,\epsilon}(\phi^h_\epsilon;f) = \int_\Omega \| \nabla \cdot (\boldsymbol{b}\phi^h_\epsilon) + c\phi^h_\epsilon - f \|_{\ell^1,\epsilon}\, d\Omega \text{ over } \phi^h \in X^h_{B,g}, \tag{49}$$

where, for $\epsilon > 0$ and $\vec{x} \in \mathbb{R}^n$,

$$\| \vec{x} \|_{\ell^1,\epsilon} = \sum_{i=1}^{n} \frac{|\vec{x}_i|^2}{|\vec{x}_i| + \epsilon}$$

is a regularization of the ℓ_1 norm defined in (47). The regularized norm is Fréchet differentiable so that (49) can be solved by gradient methods.

[k]Recall that the parent principle of (48), i.e., the $L^1(\Omega)$ minimization problem (30), does have a unique minimizer. The fact that its conforming discretization can have multiple minimizers is another key distinction between the Banach and Hilbertian settings for conservation laws.

The solution of the original problem is obtained as a limit of the sequence $\{\phi_\epsilon^h\}$ as $\epsilon \to 0$. The following results are proved in Guermond (2004).

Theorem 5. *The solutions of the regularized problem (49) possess the stability property*

$$\| \phi - \phi_\epsilon^h \|_B \leq C\left(\epsilon c(\| f \|_{L^1}) + \min_{\psi^h \in X_{B,g}^h} \| \phi - \psi^h \|_B \right),$$

where $c(\cdot)$ *is a continuous function. Moreover, up to a subsequence, every limit* ϕ_0^h *of* $\{\phi_\epsilon^h\}$ *is a solution of the nonregularized discrete* $L^1(\Omega)$ *principle (48).*

For further discussions of theoretical and computational properties of $L^1(\Omega)$-based finite element methods, including a definition of an iterative algorithm that determines minimizers of (48) by solving a sequence of regularized problems we refer to Guermond (2004) and Guermond and Popov (2004).

8 LSFEMs BASED ON ADAPTIVELY WEIGHTED $L^2(\Omega)$ NORMS

We now turn attention to a class of formulations that aim to combine the best properties of the Banach and Hilbertian space settings in Sections 4.1–4.2. The resulting LSFEMs do not give up the differentiability of the functionals found in bona fide least-squares principles, yet they are also capable of significantly reducing their inherent diffusivity and produce solutions that are qualitatively similar to those one can obtain through L^1 minimization. To accomplish this, these functionals use adaptively weighted $L^2(\Omega)$ norms to mimic the behaviour of L^1 minimization problems. The improved quality of the solutions does not come for free though. While a conventional conforming DLSP such as (34), or even a nonconforming but Hilbertian principle such as (40) require a solution of a single minimization problem, the methods in this section require the solution of a *sequence* of such problems, even if the PDE problem itself is linear.

8.1 An Iteratively Reweighted LSFEM

Our first example is the adaptively reweighted LSFEM considered in Jiang (1993). This method was motivated by the approach in Lavery (1988, 1989) and was purposely designed to mimic the properties of the L^1 minimization formulations in these papers. The key idea rests on the fact that although diffusive, a conventional DLSP such as (34) can serve as a reliable error indicator to detect solution discontinuities. Once the elements near the discontinuity are identified, the adaptively reweighted LSFEM reduces their contributions to the LSF by applying suitably defined weights to the element residuals.

By unweighting the "shocked" elements, the least-squares principle essentially disregards the equations associated with these elements. The unweighting process can be repeated in an iterative manner until some desired convergence tolerance is satisfied. The net result is a finite element solution that approximates well-smooth portions of the exact solution without having the typical over/undershoots near the discontinuities, i.e., it resembles a minimizer of an $L^1(\Omega)$ functional.

The iteratively reweighted least-squares method of Jiang (1993) consists of solving, for $k = 1, 2, \ldots$, the sequence of mesh-dependent DLSPs

$$\begin{aligned}&\text{minimize } J_{\omega_k}(\phi_k^h;f) = \| \sqrt{\omega_k}(\nabla\cdot(\boldsymbol{b}\phi_k^h) + c\phi_k^h - f) \|_0^2 + \| \sqrt{\omega_k}(\phi - g) \|_{0,\Gamma_-}^2 \\ &\qquad\qquad \text{over } \phi^h \in X_k^h = S_h^1\end{aligned} \tag{50}$$

until $\| \phi_k^h - \phi_{k-1}^h \|_0 \leq tol$. For $k = 1$, the weight ω_1 is set to 1, i.e., (50) is a standard, conforming $L^2(\Omega)$ least-squares principle. For subsequent steps,

$$\omega_k = \frac{1}{|R(\phi_{k-1}^h)|^6} \quad \text{for } k = 2, 3, \ldots,$$

where $R(\phi_{k-1}^h) = \nabla\cdot(\boldsymbol{b}\phi_{k-1}^h) + c\phi_{k-1}^h - f$ is the residual of the least-squares solution from the previous step. The result is an iterative solution procedure that resembles the solution algorithm of Guermond (2004) for (49). However, the reweighted method uses bona fide least-squares principles rather than a regularized ℓ_1 norm.

It is possible to implement the iteratively reweighted procedure using other error monitor functions; see Jiang (1993). There are no theoretical results regarding the convergence properties and accuracy of this method. Nevertheless, computational results given in Jiang (1993) show essentially monotone solutions and shock resolution within a narrow layer of elements.

8.2 A Feedback LSFEM

The feedback least-squares method of Choi (2000) for the model advection–reaction problem (13) is our second example of a least-squares formulation that aims to combine the best properties of Banach and Hilbert space settings. The method itself had been prompted by the adaptively reweighted LSFEM in the last section and so, the two methods are close relatives. The key differences between the methods include (i) the use of a statistical approach by the feedback method, adopted from Carmo and Galeao (1991), to locate the discontinuity, and (ii) a weight function based on the solution gradient rather than on its residual.

Assume that g has a jump discontinuity at $\boldsymbol{x} \in \Gamma_-$ that is propagated by the solution ϕ along the characteristic $\chi \subset \Omega$. Similarly to the iteratively reweighted method from Section 8.1, the feedback LSFEM relies on the residual of the finite element solution ϕ^h to locate χ. However, in the feedback method, the unweighting of the LSF is confined to a *discontinuity set* M_χ containing the elements that are near χ. To define this set, each element $\kappa \in \mathcal{T}_h$ is ranked using the following quantities.

1. The mean residual for the element κ:

$$\ll R \gg_\kappa = \int_\kappa |R_\kappa(\phi^h)|/\mu(\kappa)\, d\Omega \quad \forall \kappa \in \mathcal{T}_h,$$

where $\mu(\kappa)$ is the element measure and $R_\kappa(\phi) = \left(\nabla \cdot (\boldsymbol{b}\phi^h) + c\phi^h - f\right)|_\kappa$ is the element residual.

2. The mean residual for the finite element partition $\mathcal{T}_h$:

$$\ll R \gg_h = \frac{1}{N_e} \sum_{\kappa \in \mathcal{T}_h} \ll R \gg_\kappa,$$

where N_e is the number of elements in $\mathcal{T}_h$.

3. The mean deviation for $\mathcal{T}_h$:

$$\ll \delta \gg_h = \left(\frac{1}{N_e} \sum_{\kappa \in \mathcal{T}_h} (\ll R \gg_h - \ll R \gg_\kappa)^2 \right)^{1/2}.$$

An initial set M_0 is defined by including all elements κ whose mean element residual exceeds the mean residual for $\mathcal{T}_h$, plus a term proportional to the mean deviation:

$$M_0 = \{\kappa \in \mathcal{T}_h \mid \ll R \gg_\kappa \ \geq \ \ll R \gg_h + \epsilon \ll \delta \gg_h\}. \tag{51}$$

The positive parameter ϵ may be used to adjust the sensitivity of this detection criterion. For triangular partitions $\mathcal{T}_h$, the set M_χ is constructed from M_0 using the following recursive process (see Fig. 1).

1. Tag all elements in M_0 by 1 and all elements in $\mathcal{T}_h \backslash M_0$ by 0.

2. Update element tags according to the following rules: initialize $m_\Omega = 3$;
- if κ has tag 0 and m_Ω adjacent elements have tags 1, set the tag of κ to 1;
- if κ has tag 1 and all adjacent elements have tag 0, set the tag of κ to 0;
- set $m_\Omega = 2$ and repeat until no tags change.

3. M_χ is the set of all elements whose tag equals 1.

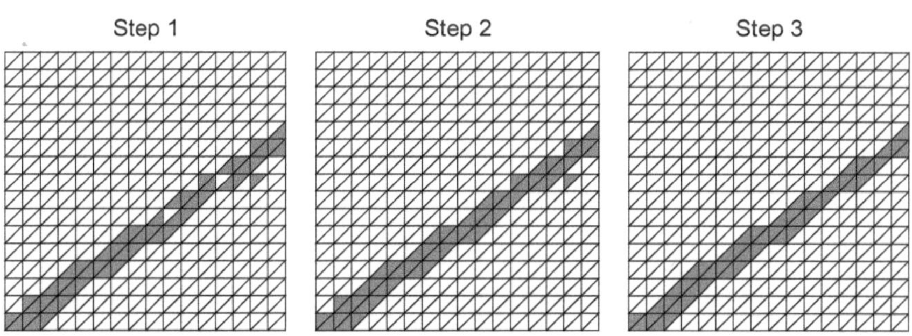

FIG. 1 The *shaded triangles* in the *leftmost* plot are in the set M_0. The *centre plot* shows an intermediate set obtained at step 2. The *rightmost plot* shows the set M_χ.

This procedure is used to define the feedback LSFEM as follows.

1. Set $\omega_\kappa = 1$ for all $\kappa \in \mathcal{T}_h$ and compute a minimizer ϕ_0^h of (50).
2. Use ϕ_0^h to construct the set M_χ.
3. Set

$$\omega_\kappa = \begin{cases} 1 & \text{for } \kappa \notin M_\chi \\ \dfrac{1}{1+\alpha \| \nabla \phi_0^h \|_\kappa} & \text{for } \kappa \in M_\chi, \end{cases}$$

where α is an amplifying constant (see Jiang, 1993) and solve (50).

If necessary, step 3 or steps 2–3 can be repeated until the relative error between two consecutive solutions becomes smaller than some prescribed tolerance.

Usually, the construction of M_χ from M_0 takes just few steps which are cheaper than solving the LSFEM problem itself. As a result, the feedback method tends to be more efficient than the iteratively reweighted approach for which the first couple of iterations are essentially used to determine the same information as provided by the discontinuity set M_χ.

9 EXAMPLES

In this section, we provide few examples that illustrate the use LSFEMs for hyperbolic PDEs.

9.1 Approximation of Smooth Solutions

Insofar as smooth solutions of hyperbolic PDEs are concerned, the conforming LSFEMs in Section 6.1 are more than adequate. Theorem 3 asserts that such LSFEMs for (13) are optimal in the energy norm (17). Moreover, computational evidence suggests that these methods are also optimally accurate in the L^2 norm; see Table 1 for an example. Considering that a least-

TABLE 1 Comparison of Convergence Rates for Streamline-Upwind Petrov–Galerkin (SUPG) and the Conforming LSFEM (34) for (13) Using the Manufactured Solution $\phi = \exp(x)\sin(x)\sin(y)$, Finite Element Partition of $\Omega = [0, 1]^2$ into Triangles, and $X^h = S_h^2$ (Standard C^0 Piecewise Quadratic Nodal Elements on Triangles)

	$L^2(\Omega)$ Error	$H^1(\Omega)$ Error	$\|\nabla_b\phi - \nabla_b\phi^h\|_0$
SUPG	3.0000	2.0024	1.9970
LSFEM	3.0055	2.0087	2.0377

squares principle for (13) embeds this hyperbolic equation into an elliptic problem, this should not come as a complete surprise: it is well known that energy norm estimates for elliptic PDEs can be supplemented by optimal L^2 bounds by using a duality argument. However, rigorously proving optimal $L^2(\Omega)$ estimates for (34) is a different matter because the least-squares dual problem: *seek* $\psi \in X_H$ *such that*

$$Q(\phi;\psi) = (f,\phi) \quad \forall \phi \in X_H\,, \tag{52}$$

where $Q(\cdot;\cdot)$ is the bilinear form in (28), is not sufficiently regular. As a result, a standard elliptic duality argument does not apply. To see this, assume that $\nabla\cdot\boldsymbol{b} = 0$, $c = 1$, $g = 0$, and X_H is constrained by the boundary condition on Γ_-. Define $\Gamma_{++} = \{\boldsymbol{x} \in \partial\Omega \mid \boldsymbol{n}(\boldsymbol{x}) \cdot \boldsymbol{b}(\boldsymbol{x}) > 0\}$. Formal integration by parts in (52) reveals that the strong form of this equation is given by the following boundary value problem

$$\begin{cases} -\nabla\cdot(\boldsymbol{b}(\nabla_{\boldsymbol{b}}\phi+\phi)) + (\nabla_{\boldsymbol{b}}\phi+\phi) = f & \text{in } \Omega \\ \phi = 0 & \text{on } \Gamma_- \\ \nabla_{\boldsymbol{b}}\phi + \phi = 0 & \text{on } \Gamma_{++}, \end{cases} \tag{53}$$

where $\nabla_{\boldsymbol{b}}\phi$ is the streamwise derivative defined in Section 3. The differential operator in (53) is degenerate elliptic because it only controls the second derivatives along the streamlines. As a result, it does not have the full elliptic regularity necessary for the standard duality argument. However, using a perturbed version of (53), one can establish *improved* $L^2(\Omega)$ estimates for the conforming LSFEM (34); see Bochev and Choi (2001b).

Theorem 6. *Assume that* Ω *is a rectangular domain,* $\mathbf{b} = (1, 0)^T$, *(13) has a solution* $\phi \in H^{r+1}(\Omega)$, *and* ϕ^h *is minimizer of the compliant DLSP (34). Then,*

$$\| \phi - \phi^h \|_0 \leq C h^{k+1/3} \| \phi \|_{k+1}\,. \tag{54}$$

This theorem can be extended to more general domains and advective vectors, but at present it does not appear that the exponent $k + 1/3$ can be improved to $k + 1$.

9.2 Approximation of Discontinuous Solutions

There are very few theoretical studies of LSFEMs for problems with less regular solutions. Not much is known about the error behaviour except when the flow is grid-aligned. In this case, error estimates were derived in De Sterck et al. (2005b).

The paper (Wilders, 1988) is perhaps the first systematic numerical investigation of what we refer to as "conforming LSFEM" in this chapter. The study in Wilders (1988) involves the Burgers equation in two dimensions and finds that "the accuracy of the numerical solution of a first-order conservation law by a least-squares method is disappointing," but can be remedied by using a special treatment of the nodes on the characteristic boundaries.

A different aspect of LSFEMs for hyperbolic problems was investigated in Lowrie and Roe (1994). This paper focussed on the ability of residual minimization principles to provide sharp resolution and correct placement of solution discontinuities. Among the findings of this paper is the result that a conforming Hilbertian least-squares principle with *fixed norms* cannot simultaneously resolve the discontinuity within a single mesh interval and compute a monotone solution. The conclusion drawn in Lowrie and Roe (1994) was that this may be possible for least-squares principles defined along the lines described in Section 8, i.e., by using adaptively weighted $L^2(\Omega)$ norms.

Computational studies of LSFEMs in Choi (2000) and Bochev and Choi (2001a) lend further credence to the observation that Hilbert space setting is perhaps less than satisfactory for the approximation of discontinuous solutions of conservation laws. In particular, these studies reveal a strong dependence of the dissipation in the LSFEM on the angle between the advective velocity and the grid lines with the worst case scenario occurring when this angle equals $\pi/4$. Fig. 2 shows typical results for a cone advected in two directions aligned with the grid lines and a third one in which the angle between $\boldsymbol{b}$ and the grid lines is exactly $\pi/4$. The latter exhibits an excessive amount of artificial diffusion which renders the LSFEM solution of little practical use. These and other similar results in the literature further reinforce the conclusion of Lowrie and Roe (1994) that straightforward LSFEMs defined in "static" Hilbert spaces, without suitable additional modifications, are not appropriate for discontinuous solutions.

On the other hand, LSFEM formulations in "dynamic" Hilbert spaces, such as the methods in Section 8, appear fully capable of providing sharp, nonoscillatory, essentially monotone resolution of solution discontinuities, even when using higher than linear polynomial finite element spaces. Figs. 3 and 4 show two such examples computed by the feedback LSFEM. The figures also provide comparisons of the feedback LSFEM solution with the solution of the classical SUPG (Hughes and Brooks, 1982) and the conforming LSFEM (34).

In these examples Ω is the unit square, $\mathcal{T}_h$ is a uniform partition of Ω into triangles,[1] and $X^h = S_h^2$ (piecewise quadratic finite elements). Fig. 3 shows results for a constant advection case where

$$\boldsymbol{b} = (1, \tan 35^\circ)^{\mathsf{T}} \quad \text{and} \quad g(x,y) = \begin{cases} 2 & \text{on}\,\Gamma_L \\ 1 & \text{on}\,\Gamma_B. \end{cases} \tag{55}$$

and Fig. 4 correspond to circular advection:

[1]$\mathcal{T}_h$ is defined by partitioning Ω into squares using $n \times n$ uniformly spaced grid lines in the x and y directions after which each square is divided into two triangles formed by the sides of the square and the diagonal from the bottom left to the top right vertex.

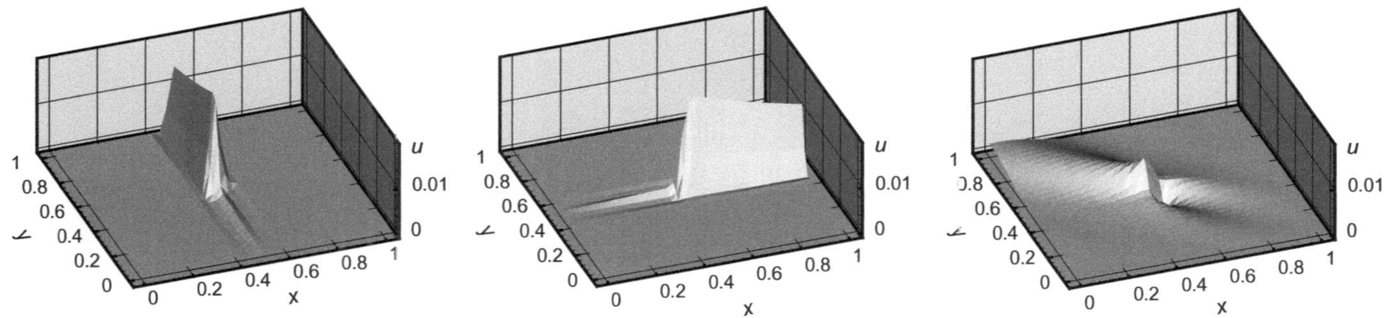

FIG. 2 Dependence on the amount of artificial diffusion in (34) on the *grid* direction. From *left* to *right* the plots correspond to $\boldsymbol{b} = (0, 1)^{\mathsf{T}}$, $\boldsymbol{b} = (1, 0)^{\mathsf{T}}$, and $\boldsymbol{b} = (-\sqrt{2}/2, \sqrt{2}/2)^{\mathsf{T}}$, respectively. In the first two cases $\boldsymbol{b}$ is aligned with the *grid lines*, whereas in the third case it forms a $\pi/4$ angle with these *lines*.

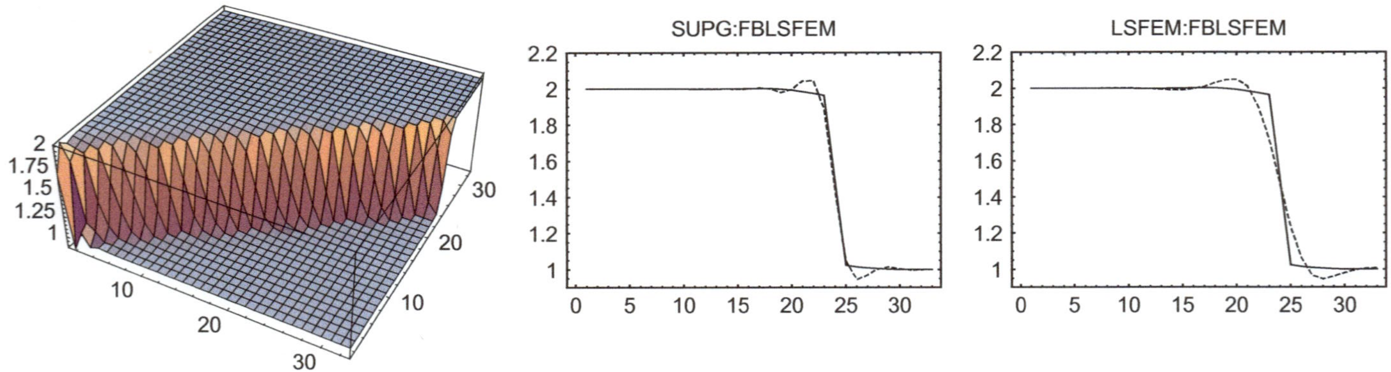

FIG. 3 *Left*: feedback LSFEM solution for constant advection (55) using quadratic C^0 finite elements on a uniform mesh $\mathcal{T}_h$ comprising 512 identical *triangular* elements. *Centre*: horizontal profiles at $y = 0.5$ of the feedback LSFEM solution (*solid line*) and SUPG (*dotted line*). *Right*: horizontal profiles at $y = 0.5$ of the feedback LSFEM solution (*solid line*) and conforming LSFEM (34) (*dotted line*).

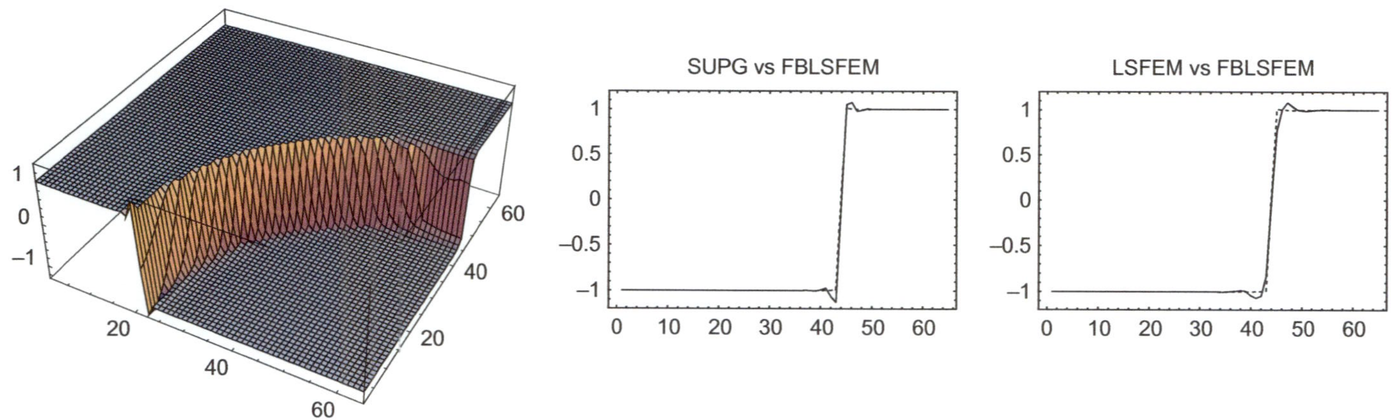

FIG. 4 *Left*: feedback LSFEM solution for circular advection (56) using quadratic C^0 finite elements on a uniform mesh $\mathcal{T}_h$ comprising 2048 identical triangular elements. *Centre*: outflow profiles of the feedback LSFEM solution (*dotted line*) and SUPG (*solid line*). *Right*: outflow profiles of the feedback LSFEM solution (*dotted line*) and conforming LSFEM (34) (*solid line*).

$$\boldsymbol{b}=(-y,x)^{\mathsf{T}} \quad \text{and} \quad g(x,y)=\begin{cases} -1 & \text{on } \Gamma_B \text{ and } x<43/64 \\ 1 & \text{on } \Gamma_B \text{ and } x\geq 43/64 \\ 1 & \text{on } \Gamma_R. \end{cases} \tag{56}$$

In (55) and (56),

$$\Gamma_L=\{(x,y)\,|\,x=0;\ 0\leq y\leq 1\}, \quad \Gamma_R=\{(x,y)\,|\,x=1;\ 0\leq y\leq 1\},$$

and

$$\Gamma_B=\{(x,y)\,|\,y=0;\ 0\leq x\leq 1\}.$$

The iteratively reweighted LSFEM in Section 8.1 yields similar solutions. These results demonstrate that LSFEM formulations in "dynamic" Hilbert spaces can indeed deliver nonoscillatory, essentially monotone solutions that provide highly accurate, sharp resolution of solution discontinuities.

10 A SUMMARY OF CONCLUSIONS AND RECOMMENDATIONS

The main message of this chapter is that solving hyperbolic PDEs by LSFEMs is an achievable, but by no means simple, task. Of course, this is also true for other methods such as SUPG, finite difference, and finite volume methods, and merely reflects the fact that hyperbolic PDEs are more difficult to solve than elliptic PDEs.

The theoretical results and computational examples presented in this chapter strongly suggest that, for hyperbolic PDEs, least-squares principles defined with respect to "static" Hilbert spaces are not the most appropriate choice without further modifications. This applies with equal force to both conforming and nonconforming LSFEMs because such methods, even when combined with grid refinement and higher-order elements (De Sterck et al., 2005b; Taghaddosi et al., 1999), are not capable of producing monotone solutions. Indeed, a comparison between the conforming (34) and discontinuous (40) LSFEMs in De Sterck et al. (2005b) reveals that both methods tend to smear discontinuities at about the same rate. In other words, merely switching to discontinuous elements is not enough to offset the natural dissipation present in least-squares formulations based on "static" Hilbert spaces. The same study indicates that discontinuity smear can be reduced by using higher-order elements but that such elements do not eliminate the over/undershoots in the "static" Hilbert space least-squares solution.

The most promising approaches appear to be either explicitly or implicitly tied to residual minimization problems in Banach spaces. Computationally, the $L^1(\Omega)$ method of Guermond (2004) recovers what is essentially a viscosity solution of the conservation law. The adaptively reweighted $L^2(\Omega)$ norm LSFEMs of Section 8 also perform very well by using "dynamic" Hilbert spaces, and are easier to implement and use than the $L^1(\Omega)$ method. However,

TABLE 2 Summary Properties of Select LSFEMs for the Advection–Reaction Problem (13)

Method→ Property↓	Conforming (34)	Reweighted LSFEM (Section 8.1)	Feedback LSFEM (Section 8.2)	Regularized $L^1(\Omega)$ (49)
Provably optimal	√	–	–	√
Monotone	–	√	√	√
Solution cost	Low	High	Medium	Medium
Coding effort	Simple	Simple	Simple	Not as simple

both $L^1(\Omega)$ and "dynamic" LSFEMs are not yet at a stage where they can truly compete with more established approaches that have a much longer history of use in practice.

Table 2 compares and contrasts properties of select methods from this chapter.

ACKNOWLEDGEMENTS

This material is based upon work supported by the US Department of Energy, Office of Science, Office of Advanced Scientific Computing Research.

REFERENCES

Bell, B.C., Surana, K.S., 1994. A space-time coupled *p*-version least-squares finite element formulation for unsteady fluid dynamics problems. Int. J. Numer. Meth. Eng. 37, 3545–3569.

Bell, B.C., Surana, K.S., 1996. A space-time coupled *p*-version least-squares finite element formulation for unsteady two-dimensional Navier-Stokes equations. Int. J. Numer. Meth. Eng. 39, 2593–2618.

Bochev, P., Choi, J., 2001a. A comparative study of least-squares, SUPG and Galerkin methods for convection problems. Int. J. Comput. Fluid Dyn. 15 (2), 127–146.

Bochev, P., Choi, J., 2001b. Improved least-squares error estimates for scalar hyperbolic problems. Comput. Meth. Appl. Math. 1 (2), 115–124.

Bochev, P., Gunzburger, M., 2009. Least-Squares Finite Element Methods, Applied Mathematical Sciences, vol. 166. Springer Verlag, New York.

Brezzi, F., 1974. On existence, uniqueness and approximation of saddle-point problems arising from Lagrange multipliers. Model. Math. Anal. Numer. 21, 129–151.

Carmo, E.G.D., Galeao, A., 1991. Feedback Petrov-Galerkin methods for convection-dominated problems. Comput. Meth. Appl. Mech. Eng. 88, 1–16.

Chen, T.F., Fix, G.J., 1986a. Least-squares finite element simulation of transonic flows. ICASE 86-27, NASA, Institute for Computer Applications in Science and Engineering (ICASE), NASA Langley Research Center, Hampton, VA.

Chen, T.F., Fix, G.J., 1986b. Least squares finite element simulation of transonic flows. Appl. Numer. Math. 2 (3), 399–408. ISSN 0168-9274. http://dx.doi.org/10.1016/0168-9274(86)90042-5. http://www.sciencedirect.com/science/article/pii/0168927486900425.

Choi, J., 2000. The Least-Squares Method for Hyperbolic Problems (Ph.D. thesis). Department of Mathematics, The University of Texas at Arlington.

Cox, C.L., Fix, G.J., Gunzburger, M.D., 1983. A least-squares finite element scheme for transonic flow around harmonically oscillating wings. J. Comp. Phys. 51 (3), 387–403.

De Sterck, H., Manteuffel, T., McCormick, S., Olson, L., 2005a. Least-squares finite element methods and algebraic multigrid solvers for linear hyperbolic PDEs. SIAM J. Sci. Comput. 26 (1), 31–54. ISSN 1064-8275. http://dx.doi.org/10.1137/S106482750240858X.

De Sterck, H., Manteuffel, T., McCormick, S., Olson, L., Philadelphia, P.A., 2005b. Numerical conservation properties of $H(div)$-conforming least-squares finite element methods for the Burgers equation. SIAM J. Sci. Comput. 26 (5), 1573–1597. ISSN 1064-8275. http://dx.doi.org/10.1137/S1064827503430758.

Donea, I., 1984. A Taylor-Galerkin method for convective transport problems. Int. J. Numer. Meth. Eng. 20, 101–119.

Eriksson, K., Estep, D., Hansbo, P., Johnson, C., 1996. Computational Differential Equations. Cambridge University Press, Cambridge, United Kingdom.

Guermond, J.-L., 2004. A finite element technique for solving first order PDE's in L^1. SIAM J. Numer. Anal. 42 (2), 714–737.

Guermond, J.-L., Popov, B., 2004. Linear advection with ill-posed boundary conditions via L^1-minimization. SIAM J. Numer. Anal. 42 (2), 714–737.

Houston, P., Mackenzie, J.A., Suli, E., Warnecke, G., 1999. A posteriori error analysis for numerical approximation of Friedrichs systems. Numer. Math. 82, 433–470.

Hughes, T.J.R., Brooks, A., 1982. Streamline upwind/Petrov-Galerkin formulation for convection dominated flows with particular emphasis on the incompressible Navier-Stokes equations. Comput. Meth. Appl. Mech. Eng. 32, 199–259.

Jiang, B.-N., 1993. Non-oscillatory and non-diffusive solution of convection problems by the iteratively reweighted least-squares finite element method. J. Comp. Phys. 105 (1), 108–121.

Jiang, B.-N., Carey, G.F., 1988a. Least-squares finite elements for first-order hyperbolic systems. Int. J. Num. Meth. Eng. 26, 81–93.

Jiang, B.-N., Carey, G.F., 1988b. A stable least-squares finite element method for nonlinear hyperbolic problems. Int. J. Num. Meth. Fluids 8, 933–942.

Jiang, B.-N., Carey, G.F., 1990. Least-squares finite element methods for compressible Euler equations. Int. J. Num. Meth. Fluids 10, 557–568.

Johnson, C., 1992. Numerical Solution of Partial Differential Equations by the Finite Element Method. Cambridge University Press.

Johnson, C., Navert, U., Pitkäranta, J., 1984. Finite element methods for linear hyperbolic problems. Comput. Meth. Appl. Mech. Eng. 45, 285–312.

Kuzmin, D., Löhner, R., Turek, S. (Eds.) 2005. Flux-Corrected Transport. Principles, Algorithms and Applications. Springer Verlag, Berlin, Heidelberg.

Lavery, J.E., 1988. Nonoscillatory solution of the steady-state inviscid Burgers equation by mathematical programming. J. Comp. Phys. 79, 436–448.

Lavery, J.E., 1989. Solution of steady-state one-dimensional conservation laws by mathematical programming. SIAM J. Numer. Anal. 26 (5), 1081–1089. http://dx.doi.org/10.1137/0726060. http://link.aip.org/link/?SNA/26/1081/1.

Lazarov, R., Tobiska, L., Vassilevski, P., 1997. Streamline-diffusion least-squares mixed finite element methods for convection-diffusion problems. East-West J. Numer. Math. 5 (4), 249–264.

Liang, S.-J., Hsu, T.-W., 2009. Least-squares finite-element method for shallow-water equations with source terms. Acta Mec. Sin. 25 (5), 597–610. ISSN 1614-3116. http://dx.doi.org/10.1007/s10409-009-0250-x.

Lohner, R., Morgan, K., Zienkiewicz, O.C., 1984. The solution of nonlinear hyperbolic equation systems by the finite element method. Int. J. Numer. Meth. Fluids 4, 1043–1063.

Lowrie, R.B., Roe, P.L., 1994. On the numerical solution of conservation laws by minimizing residuals. J. Comp. Phys. 113, 304–308.

Nguen, H., Reynen, J., 1984. A space-time least-squares finite element scheme for advection-diffusion equations. Comput. Meth. Appl. Mech. Eng. 42, 331–342.

Perrochet, P., Azerad, P., 1995. Space-time integrated least-squares: solving a pure advection equation with a pure diffusion operator. J. Comp. Phys. 117, 183–193.

Reed, W.H., Hill, T.R., 1973. Triangular mesh methods for the neutron transport equation. La-ur-73-479, Los Alamos Scientific Laboratory, Los Alamos, NM.

Suli, E., 1997. A posteriori error analysis and adaptivity for finite element approximations of hyperbolic problems. Technical Report 97/21. Oxford University Computing Laboratory, Oxford, England.

Taghaddosi, F., Habashi, W., Guevremont, G., Ait-Ali-Yahia, D., 1999. An adaptive least-squares method for the compressible Euler equations. Int. J. Numer. Meth. Fluids 31, 1121–1139.

Toledo, R.C.L., Ruas, V., 2011. Numerical analysis of a least-squares finite element method for the time-dependent advection-diffusion equation. J. Comput. Appl. Math. 235 (12), 3615–3631. ISSN 0377-0427. http://dx.doi.org/10.1016/j.cam.2011.02.022. http://www.sciencedirect.com/science/article/pii/S0377042711001105.

Wilders, P., 1988. On the accuracy of least-squares finite elements for a first-order conservation equation. Int. J. Numer. Methods Fluids 8 (8), 957–964. ISSN 1097-0363. http://dx.doi.org/10.1002/fld.1650080807.

Chapter 13

Staggered and Colocated Finite Volume Schemes for Lagrangian Hydrodynamics

R. Loubère*, P.-H. Maire† and B. Rebourcet‡

*Institut de Mathématiques de Toulouse and CNRS, Toulouse Cedex 9, France

†CEA/CESTA, Le Barp, France

‡CEA/DAM Ile de France, Arpajon Cedex, France

Chapter Outline

Handbook of Numerical Analysis, Vol. 17. http://dx.doi.org/10.1016/bs.hna.2016.07.003

ABSTRACT

We present the two main types of Finite Volume Lagrangian schemes named: staggered-grid hydrodynamics (SGH) and colocated Lagrangian hydrodynamics (CLH). Both are devoted to solve the hydrodynamic conservation laws and extended system in multidimension on general grid. They are funded on common paradigms, such as the need to solve the conservation laws consistently, the respect of thermodynamics and other closure relations with a moving and deforming geometry, and an accurate treatment of wave propagation and material behaviour. The scope of Lagrangian formalism covers fluid and solid mechanics, astrophysics and even cosmology and many communities shared those concerns. We propose a summary of key aspects of these numerical schemes attempting a unified framework. We refer to some works illustrating the subject, a more complete list could be found in Barlow, A.J., Maire, P.-H., Rider, W.J., Rieben, R.N., Shashkov, M.J., 2016. Arbitrary Lagrangian–Eulerian methods for modelling high-speed compressible multimaterial flows. J. Comput. Phys. 322, 603–665. http://dx.doi.org/10.1016/j.jcp.2016.07.001.

Keywords: Lagrangian hydrodynamics, Colocated scheme, Staggered scheme, Finite Volume, Riemann nodal solver, Artificial viscosity

AMS Classification Codes: 65M08, 65M22, 65M60, 65Z05, 76L05, 76N15

1 HISTORICAL BACKGROUND ON LAGRANGIAN COMPUTATIONAL FLUID DYNAMICS

The Lagrangian formulation of the equations of hydrodynamics has an old and venerable history. The very first numerical calculations that resemble modern computer simulations employed fluid equations in the Lagrangian frame of reference in 1D (Bethe et al., 1944; von Neumann, 1944) and were performed at the Los Alamos National Laboratory by von Neumann and Richtmyer. The Lagrangian formulation in this context focuses on solving the physical conservation laws with an accurate treatment of multifluid flows at high Mach number, consistently with a moving and deforming geometry. A Lagrangian formulation is well suited for tracking shock and compression waves and for exactly capturing the interfaces between different materials. Because the grid motion is that of the matter, admissible accuracy is reached maintaining a constant number of cells. The main design requirement of any Lagrangian numerical method consists in the accurate rendering of the thermodynamical evolution of each material, that is, the transformation of the kinetic energy into internal energy in a consistent way with the second law of thermodynamics. Moreover, constructing an accurate enough numerical scheme saving computer resources is still today of paramount importance.

The very first answer to the foregoing design principles was the one-dimensional Lagrangian VNR (von Neumann–Richtmyer) finite-difference scheme (Richtmyer and Morton, 1967; von Neumann and Richtmyer, 1950) in the 1940s. This scheme has settled important paradigms. The kinematic variables were located at the mesh vertex of a grid, while thermodynamic ones at cell centres, hence referred to as a staggered placement of variables, whence the term staggered-grid hydrodynamics (SGH). Accordingly, a staggered time discretization, where velocity is halftime step centred, was employed insuring nominally second-order accuracy in both time and space. While working with the partial differential equations (PDE) form of the Euler equations that only expresses the isotropic energy balance by way of internal energy equation, a so-called artificial viscosity term, say q, was conceived by von Neumann and effectively designed and tested by Richtmyer. This a posteriori added artificial viscosity had a dual purpose: damp the dispersive effect[a] of second-order discretization especially in zones with steep gradient and add a positive source term for internal energy mimicking local entropy deposition. For the sake of conservation which fits with both Rankine–Hugoniot jump relations and second principle, the q-term is written as a nonlinear pressure potential. The design principles of the VNR scheme revealed to be practically successful, and the whole approach was kept unmodified for several years.

Early attempts to extend the VNR scheme for two-dimensional problems appear in Livermore National Laboratory in 1953 (DeWitt, 1953), it gives birth to an amount of works accounted in the so-called green book (Alder et al., 1964) published in 1964. Here began the SGH saga for which Wilkins' contribution is of particular interest because it describes most of the concerns and advances made at this time: the extension to solid mechanics and to dynamic nonconformal mappings for sliding, an accurate and stable algorithm for cylindrical geometry now called area weighting.

In that context, the occurrence of nonphysical mesh motions, "hourglass modes," not completely damped by the numerical dissipation of classical artificial viscosity[b] motivated the development of specific cures: subcelling (Browne and Wallick, 1971), appropriate tensor q-term (refer to Schulz's contribution in Alder et al., 1964) and hourglass filtering (Chan, 1975).[c]

When most of numerical method developers were trained in physics, in the 1970–80s people from the applied mathematics community started to contribute to the subject. Their standard was Finite Element (FE) method (Flanagan and Belytschko, 1981; Goudreau and Hallquist, 1982; Lascaux, 1976). Starting from the usual VNR nonconservative PDE form with q-term, they introduced a $Q_1 - P_0$ mixed FE discretization and they interpreted the VNR legacy scheme as a lumped approximation. That approach allowed to define a functional framework of subcelling and hourglass filtering in terms of numerical quadrature. While Lagrangian numerical methods were still the main brick under study, their intrinsic unsuitability to deal with highly sheared

[a]The "modified equation" of the VNR scheme can be found in Yanenko et al. (1983) showing diffusive and dispersive terms induced by the discretization.

[b]See one of the first description of this phenomenon on page 203 of Alder et al. (1964).

[c]Notice also the saving procedures for tangled mesh described in Sofronov et al. (1984) and Pais and Caruso (1990).

flows led to the development of Arbitrary Lagrangian–Eulerian (ALE) numerical methods with the seminal paper of Hirt et al. in 1974. These methods were written in term of operators splitting: a Lagrangian phase, a rezone phase to enhance the geometrical quality of the computational grid followed by a remap phase (Lagrange-Remap splitting) to conservatively interpolate the physical variables from the old Lagrangian grid to the new rezoned one. Nonetheless intrinsic and fundamental defects of the Lagrangian numerical scheme were still present, mainly related to their original formulation in internal energy, leading, de facto, to violate conservation of total energy. All earlier mentioned numerical strategies are based on the same kind of variables distribution: staggered in time and staggered in space. Furthermore as momentum and mass conservation equations are solved explicitly, the nonlinear internal energy equation is implicit. This involves a consistent update of thermodynamic quantities: internal energy, pressure and temperature.

Soon, Lagrange, Lagrange-Remap ALE and Euler schemes were based on colocated in time and staggered in space discretization, refer to Trulio and Trigger (1961), Donea et al. (1982), DeBar (1974), Youngs (1982) and Holian et al. (1989). The pros of such variables mapping where all data are defined at the same time is the consistent writing of the remap phase and a straightforward definition of energy conservation. Hence, a new step forward emerged when a two-dimensional staggered-grid compatible discrete Lagrangian hydrodynamic numerical scheme enjoying total energy conservation was designed lying on new concepts. The first important one was to realize that the discrete equations in Lagrangian form can be generally derived from variational principles naturally leading to the concept of support operator, which states that the discrete equations must obey the global properties of the continuum ones to be considered as valid discretizations (Samarskii et al., 1981). As such they will mimic conservation properties in their discrete analogues. Consequently when one specifies in discrete form a vector differential operator, then, the conjugate operator can not but be necessarily derived from it. A second useful concept was enlightened by Burton in 1990 who has discretized the Lagrangian equation on a staggered grid utilizing subgrid quantities termed corner/subcell masses and forces, from which the cell and nodal masses, and the total force acting on a node, are constructed. This has completed the main concepts leading to a first version of the scheme nowadays known as the discrete, compatible formulation of Lagrangian hydrodynamics.[d] Sequels can be recently found in Llor et al. (2016).

Still coming from the Eulerian side, slope limiting procedures allowed to greatly improve wave propagation with a low rate of numerical dissipation (concept of TVD (total variation diminishing) artificial viscosity), refer to

[d]As quoted by Caramana et al. in Bauer et al. (2006): *The word "discrete" has been inserted to emphasize that these equations are essentially created in discrete form, as opposed to being the discretization of a system of PDE's…*

Christensen (1991) and Benson (1992). Moreover, FE, being still alive, gave birth to industrial codes coupling complex modellings for miscellaneous applications in any dimension, see for instance Goudreau and Hallquist (1982). FE methods also lead to greatly improved accuracy for instance refer to the ultimate works (Dobrev et al., 2010; Guermond et al., 2016). In particular, the increase of the order of accuracy in space and time allowed to consider curvilinear elements, improving dramatically the symmetry preservation and allowing coarser meshes.

Colocated Lagrangian schemes have been designed by S.K. Godunov in 1953 (Godunov, 1959; Godunov et al., 1979). Little noticed, the very first 1D Godunov scheme was Lagrangian. The basic difference with the SGH concept relies on the integral writing of the Euler equation under conservative form. This empowers the time and space colocalization of variables and forces to characterize variable jumps and fluxes at cell boundaries with the help of the Riemann problem instead of adding an extra artificial viscosity. This idea has opened the path to the nowadays extremely popular Finite Volume Godunov scheme in the Eulerian framework.

Based on fluxes computation across cell faces, the CAVEAT multidimensional Lagrangian extension of the Godunov scheme has been tested in the 1980s in Los Alamos (Adessio et al., 1986). While in 1D the sole defects appearing in the original scheme affect accuracy and entropy production in simple waves, it turns out that the volume fluxes computing in CAVEAT are not compatible with the Lagrangian motion of the grid; in other words the Geometrical Conservation Law (GCL) is not fulfilled. One must recall here that Godunov et al. (1979) assigned half of his book on the way to build a good mesh in order to fit a physical solution, suggesting the necessity of an ALE-like strategy to extend his algorithm in 2D. Although this incompatibility reveals itself problematic (Dukowicz and Meltz, 1992), the full conservation along with a cell-centred placement of variables were appreciated properties to design indirect or Lagrange-Remap ALE codes on fixed (Dukowicz and Baumgardner, 2000) or dynamic (Dukowicz et al., 1989) grid connectivities. With an SGH Lagrangian scheme the remap stage is more demanding if not intricate because the staggered placement of variables forces the remap to act not only on the primal cells but also on the dual ones which are in general nonconvex. The GCL incompatibility has been reinvestigated with more success starting in 2002 (Després and Mazeran, 2003). A breakthrough concerning the compatibility between flux discretization and vertex velocity computation has been obtained by Després and Mazeran in 2005 via a so-called nodal solver which allows the GCL compatibility along with the conservations of total energy and momentum. This original approach, based on one-point quadrature for the pressure gradient term, has, however, been improved by Maire et al. (Abgrall et al., 2004; Maire et al., 2007) to overcome defects on cells with large cell aspect ratio and in Després and Labourasse (2012) to eradicate hourglass instabilities. After 2005 this colocated Lagrangian hydrodynamic (CLH) scheme has been extended successively to multimaterial ALE (Galera et al., 2010), 2D r-z geometry (Cheng and Shu, 2010; Maire, 2009a), 3D geometry (Carré et al., 2009; Maire and Nkonga, 2009; Morgan et al., 2015), ALE with mesh adaptation (Delpino, 2011; Loubère et al., 2010), material strength modelling (Maire et al., 2013 and reference herein), sliding (Bertoluzza et al., 2016; Morgan et al., 2013) and discontinuous Galerkin framework (Vilar et al., 2014). Some alternative cell-centred Lagrangian formulations have been also proposed in Cheng and Shu (2007) and Barlow (2013). It is worth mentioning that from the theoretical point of view the weak consistency of the CLH formulation has been demonstrated in Després (2010).

Today, the compatible SGH and the CLH numerical schemes seem to be successful tools to be used in several Lagrangian or indirect ALE multidimensional simulations involving multimaterial hydrodynamic system of conservation laws. The readers interested in these topics might find a more detailed presentation in the recent review paper (Barlow et al., 2016). Thereby we present in the following the formal derivation of SGH and CLH numerical methods with an emphasize on their intrinsic characteristics. Our goal is not to have an exhaustive and detailed description but rather to enlighten the main concepts which link or separate those two schemes.

In Section 2, we introduce the governing conservation laws of the Lagrangian hydrodynamics. We focus on the gas dynamic system in planar geometry without ALE treatment. Section 3 present the compatible discretization of the GCL and the properties of the underlying discrete divergence and gradient operators. Section 4 is devoted to the description of the SGH scheme, whereas the final section presents the CLH scheme.

2 LAGRANGIAN HYDRODYNAMICS

2.1 Physical Conservation Laws Written Under Integral Form

Let $\omega(t)$ be a moving region of the d-dimensional Euclidean space filled by an inviscid, nonheat conducting compressible fluid characterized by ρ, $\mathbf{u}$, p and e, which are, respectively, its mass density, velocity, thermodynamic pressure and specific total energy, i.e., total energy per unit mass. The Lagrangian representation is characterized by a control volume $\omega(t)$ moving with the fluid velocity. In this framework the conservation laws of mass, momentum and total energy write (Gurtin et al., 2010)

$$\frac{d}{dt}\int_{\omega(t)} \rho \,\mathrm{d}v = 0, \tag{1a}$$

$$\frac{d}{dt}\int_{\omega(t)} \rho\mathbf{u} \,\mathrm{d}v + \int_{\partial\omega(t)} p\mathbf{n} \,\mathrm{d}s = \mathbf{0}, \tag{1b}$$

$$\frac{d}{dt}\int_{\omega(t)} \rho e \,\mathrm{d}v + \int_{\partial\omega(t)} p\mathbf{n}\cdot\mathbf{u} \,\mathrm{d}s = 0, \tag{1c}$$

where $\partial\omega(t)$ denotes the boundary of $\omega(t)$ and $\mathbf{n}$ its unit outward normal. In the momentum equation (1b), $p\mathbf{n}$ represents the force per unit area exerted on the surface element ds. In the total energy equation (1c), $p\mathbf{n}\cdot\mathbf{u}$ is the work rate of this surface force. For a point located on the boundary of $\omega(t)$, i.e., $\mathbf{x}\in\partial\omega(t)$, its path is determined by means of the trajectory equation

$$\frac{d\mathbf{x}}{dt}=\mathbf{u}(\mathbf{x},t),\quad \mathbf{x}(0)=\mathbf{X}, \tag{2}$$

where $\mathbf{X}$ is its position at $t = 0$. We point out that the mass conservation equation (1a) amounts to write that the mass of the convecting region $\omega(t)$ is constant and equal to the mass contained in $\omega(0)$. This is one of the main features of the Lagrangian representation.

The system of conservation laws, (1), has been written assuming a sufficient smoothness of the physical variables over the convecting region $\omega(t)$. In the presence of discontinuities such as shock waves, the conservation laws are supplemented by jump conditions that hold at the shock surface. The corresponding relations known as Rankine–Hugoniot conditions express the conservation of mass, momentum and total energy at the discontinuity surface (Gurtin et al., 2010).

2.2 Thermodynamic Closure

The thermodynamic closure of the foregoing system is achieved binding the thermodynamic variables $\tau=\frac{1}{\rho}$, p and $\varepsilon=e-\frac{1}{2}\mathbf{u}^2$, which is the specific internal energy, with the relation $\varepsilon=\varepsilon(\tau,\eta)$ expressing the specific internal energy in terms of the specific volume and the specific entropy η. The pressure $p=p(\tau,\eta)$ and the temperature $\theta=\theta(\tau,\eta)$ are given as first derivatives of the specific internal energy by means of the equation of state

$$p(\tau,\eta)=-\left(\frac{\partial\varepsilon}{\partial\tau}\right)_\eta \quad\text{and}\quad \theta(\tau,\eta)=\left(\frac{\partial\varepsilon}{\partial\eta}\right)_\tau, \tag{3}$$

in accordance with the fundamental Gibbs relation

$$\mathrm{d}\varepsilon=-p\mathrm{d}\tau+\theta\mathrm{d}\eta. \tag{4}$$

The temperature being nonnegative, one can express η in terms of ε by inverting $\varepsilon=\varepsilon(\tau,\eta)$. It is thus possible to write the equation of state under the incomplete form $p=p(\tau,\varepsilon)$. Assuming the convexity of ε with respect to (τ,η) implies that $\left(\frac{\partial p}{\partial\tau}\right)_\eta<0$ and enables us to define the isentropic sound speed

$$a^2=-\tau^2\left(\frac{\partial p}{\partial\tau}\right)_\eta. \tag{5}$$

This in turn ensures the hyperbolicity of the gas dynamics equations.

Finally, the selection of physically admissible discontinuous solutions relies on the second law of thermodynamics, which requires that entropy does not decrease. Thus, to ensure the thermodynamic consistency with the second law of thermodynamics, the foregoing system of conservation laws is supplemented by the entropy imbalance

$$\frac{d}{dt}\int_{\omega(t)} \rho\eta \, \mathrm{d}v \geq 0. \tag{6}$$

We observe that the above inequality boils down to an equality for smooth solutions.

2.3 Physical Conservation Laws Written Under Local Form

The Reynolds transport formula (RTF) is of paramount importance to derive the local form of the conservation laws, refer to Gurtin et al. (2010). It expresses the time rate of change of the integral over $\omega(t)$ of a physical quantity attached to the fluid as follows

$$\frac{d}{dt}\int_{\omega(t)} f(\mathbf{x},t) \, \mathrm{d}v = \int_{\omega(t)} \frac{Df}{Dt} + f\nabla \cdot \mathbf{u} \, \mathrm{d}v, \tag{7}$$

where $f = f(\mathbf{x}, t)$ is a variable (scalar, vector, or tensor) characterizing the fluid and $\frac{D}{Dt}() = \frac{\partial}{\partial t}() + \mathbf{u} \cdot \nabla()$ is the material derivative. In the present case the RTF has been written for a smooth function. It can be easily extended to take into account the presence of a surface discontinuity, refer to Gurtin et al. (2010). Applying RTF to $f = \rho$ and using mass conservation equation (1a) leads to the continuity equation

$$\rho \frac{D}{Dt}\left(\frac{1}{\rho}\right) - \nabla \cdot \mathbf{u} = 0, \tag{8}$$

which is nothing but the local form of the mass conservation equation. In addition, combining (8) and (7) yields

$$\frac{d}{dt}\int_{\omega(t)} \rho f \, \mathrm{d}v = \int_{\omega(t)} \rho \frac{Df}{Dt} \, \mathrm{d}v. \tag{9}$$

Applying the foregoing identity to $f = \mathbf{u}$ (resp. e) and using (1b) (resp. (1c)) we arrive at the local form of the momentum and total energy equations

$$\rho \frac{D\mathbf{u}}{Dt} + \nabla p = \mathbf{0}, \tag{10a}$$

$$\rho \frac{De}{Dt} + \nabla \cdot (p\mathbf{u}) = 0. \tag{10b}$$

Let us point out that these PDE have been obtained assuming a sufficient smoothness of the flow variables. Dot multiplying the momentum equation by the velocity field $\mathbf{u}$ yields the kinetic energy equation. Subtracting the kinetic energy equation to the total energy equation (10b) leads to the local form of the internal energy equation

$$\rho\frac{D\varepsilon}{Dt}+p\nabla\cdot\mathbf{u}=0. \tag{11}$$

Finally, combining the previous equation with the continuity equation (8) leads to the alternative form of the internal energy equation

$$\rho\frac{D\varepsilon}{Dt}+\rho p\frac{D}{Dt}\left(\frac{1}{\rho}\right)=0, \tag{12}$$

which is also the conservation of entropy, $\rho\theta\frac{D\eta}{Dt}=0$, thanks to the fundamental Gibbs relation (4). Let us point out that the foregoing conservation equation of entropy is valid uniquely for smooth flows and must be replaced by the entropy inequality (6) otherwise.

2.4 Geometrical Conservation Law

The direct consequence of the RTF for $f=1$ is the GCL that governs the time rate of change of the volume of $\omega(t)$

$$\frac{d}{dt}\int_{\omega(t)}\mathrm{d}v-\int_{\partial\omega(t)}\mathbf{u}\cdot\mathbf{n}\,\mathrm{d}s=0. \tag{13}$$

This conservation law is strongly linked to the trajectory equation (2). The quality of the spatial approximation of this equation is crucial for any numerical method discretizing Lagrangian hydrodynamics as we shall see it in the next section.

3 GCL AND RELATED DISCRETE OPERATORS

Starting from the GCL (13), we intend to construct not only its Finite Volume spatial discretization but also the discrete expressions of the divergence and gradient operators on general unstructured grids. These tools will be the cornerstone of the SGH and CLH schemes.

3.1 Grid Notation and Assumptions

Let the two-dimensional Euclidean space be equipped with the orthonormal basis $(\mathbf{e}_x, \mathbf{e}_y)$ supplemented by $\mathbf{e}_z = \mathbf{e}_x\times\mathbf{e}_y$. The convecting computational domain, $D(t)$, is paved using a collection of nonoverlapping conformal polygonal cells denoted by $\omega_c(t)$, i.e., $D(t)=\cup_c\omega_c(t)$. Each polygonal cell is entirely characterized by the set of its vertices $\mathcal{P}(c)$. A generic vertex is denoted by label p, thus, $\mathbf{x}_p = \mathbf{x}_p(t)$ is the vector position of vertex p at time $t > 0$.

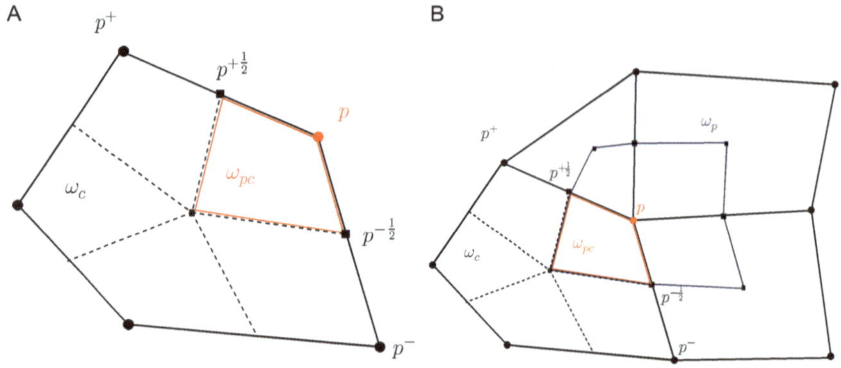

FIG. 1 Fragments of a polygonal grid and corresponding notations. (A) Primal polygonal cell and one of its subcell. (B) Dual polygonal cell.

In the counter-clockwise ordered list of vertices of cell ω_c, the vertex p^+ (resp. p^-) denotes the next (resp. previous) vertex with respect to p and $p^{\pm\frac{1}{2}}$ is the midpoint of $[p, p^{\pm}]$. The quadrangle obtained joining the cell centre c, the midpoint $p^{-\frac{1}{2}}$, the vertex p, the midpoint $p^{+\frac{1}{2}}$ and the cell centre defines the subcell ω_{pc}, refer to Fig. 1A. The set of subcells forms a partition of cell ω_c, that is $\omega_c = \bigcup_{p\in\mathcal{P}(c)}\omega_{pc}$. As such ω_{pc} is a quadrangular submesh for any kind of polygonal mesh. Collecting the subcells related to the generic vertex p allows us to define the dual cell

$$\omega_p = \bigcup_{c\in\mathcal{C}(p)} \omega_{pc},$$

where $\mathcal{C}(p)$ is the set of cells sharing vertex p. While polygonal primal cell, ω_c, is supposed to be convex, dual cells might be concave, refer to Fig. 1B. The motion of the polygonal primal grid is governed by the trajectory equation of each vertex p

$$\frac{d\mathbf{x}_p}{dt} = \mathbf{u}_p, \quad \mathbf{x}_p(0) = \mathbf{X}_p. \tag{14}$$

Here, $\mathbf{u}_p$ denotes the velocity of the vertex p.

3.2 Compatible Discretization of the GCL

According to (13) the time rate of change of the cell volume, $|\omega_c(t)|$, reads

$$\frac{d}{dt}|\omega_c(t)| = \int_{\partial\omega_c(t)} \mathbf{u}\cdot\mathbf{n}\,ds.$$

We shall construct the right-hand side of the above equation, by computing exactly the time rate of change of the polygonal cell volume in terms of the velocity of its vertices. Partitioning the polygonal primal cell into triangles, its volume expresses in terms of the vertex position as follows

$$|\omega_c(t)| = \sum_{p\in\mathcal{P}(c)} \frac{1}{2}(\mathbf{x}_p \times \mathbf{x}_{p^+}) \cdot \mathbf{e}_z. \tag{15}$$

Remarking that the cell volume is an explicit function of the vertex coordinates and applying the chain rule of composed derivatives leads to

$$\frac{d}{dt}|\omega_c(t)| = \sum_{p\in\mathcal{P}(c)} \frac{\partial|\omega_c(t)|}{\partial \mathbf{x}_p} \cdot \frac{d\mathbf{x}_p}{dt}.$$

Taking the gradient of (15) with respect to the vertex coordinate $\mathbf{x}_p$, we readily obtain

$$\frac{\partial|\omega_c(t)|}{\partial \mathbf{x}_p} = \frac{1}{2}[(\mathbf{x}_{p^+} - \mathbf{x}_p) \times \mathbf{e}_z + (\mathbf{x}_p - \mathbf{x}_{p^-}) \times \mathbf{e}_z].$$

We note that the right-hand side of the above equation is nothing but the half of sum of the two normals related to the edges impinging at vertex p, refer to Fig. 2

$$l_{pc}^- \mathbf{n}_{pc}^- = \frac{1}{2}(\mathbf{x}_p - \mathbf{x}_{p^-}) \times \mathbf{e}_z, \quad l_{pc}^+ \mathbf{n}_{pc}^+ = \frac{1}{2}(\mathbf{x}_{p^+} - \mathbf{x}_p) \times \mathbf{e}_z. \tag{16}$$

Therefore, the volume gradient is expressed as

$$\frac{\partial|\omega_c(t)|}{\partial \mathbf{x}_p} = l_{pc}^- \mathbf{n}_{pc}^- + l_{pc}^+ \mathbf{n}_{pc}^+ = l_{pc}\mathbf{n}_{pc}, \tag{17}$$

where $l_{pc}\mathbf{n}_{pc}$, the corner vector, is a fundamental geometrical object representing the gradient of the volume with respect to the vertex coordinate. By construction the corner vectors satisfy the fundamental geometrical identity

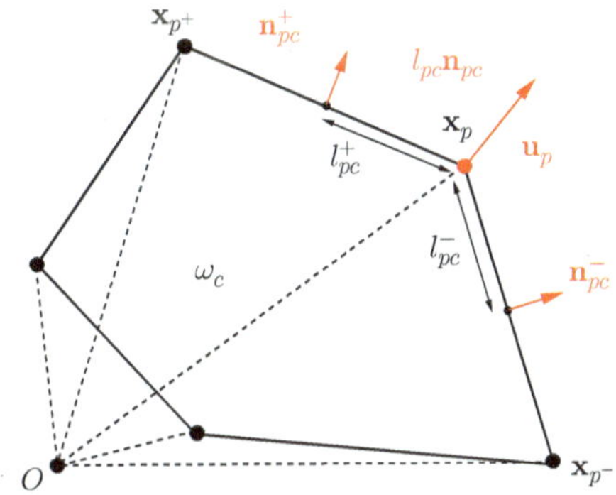

FIG. 2 Polygonal primal cell triangulation.

$$\sum_{p\in\mathcal{P}(c)} l_{pc}\mathbf{n}_{pc} = \mathbf{0}, \tag{18}$$

expressing that along time, $\partial\omega_c(t)$ remains a closed polygonal line. Gathering the foregoing results and recalling the trajectory equation (14), the time rate of change of cell volume is given by

$$\frac{d}{dt}|\omega_c(t)| = \sum_{p\in\mathcal{P}(c)} l_{pc}\mathbf{n}_{pc}\cdot\mathbf{u}_p. \tag{19}$$

Next, applying the mass conservation equation (1a) to the polygonal cell $\omega_c(t)$ implies that its mass $m_c = \int_{\omega_c(t)} \rho \, \mathrm{d}v$ remains constant along time. Denoting by ρ_c the volume average of the mass density over the polygonal cell c, the discrete mass conservation equation simply reads

$$m_c = \rho_c(t)|\omega_c(t)|. \tag{20}$$

Finally, substituting the above relation into (19) turns it into

$$m_c\frac{d}{dt}\left(\frac{1}{\rho_c}\right) - \sum_{p\in\mathcal{P}(c)} l_{pc}\mathbf{n}_{pc}\cdot\mathbf{u}_p = 0. \tag{21}$$

We have constructed a Finite Volume discretization of the GCL which is fully consistent with the volume variation resulting from the displacement of the vertex at velocity $\mathbf{u}_p$, i.e., with the trajectory equation $\frac{d\mathbf{x}_p}{dt} = \mathbf{u}_p$. Those formulas are the standard Finite Volume relations applied to $\nabla\cdot\mathbf{u}$ and define a consistent discretization of the volume flux $\int_{\partial\omega_c(t)} \mathbf{u}\cdot\mathbf{n}\,\mathrm{d}s$ that still holds for nonconvex cells. The semidiscrete formula (21) is then verified by any Lagrangian scheme characterized by a node-centred velocity.

3.3 Discrete Divergence and Gradient Operators

The compatible discretization of the GCL provides a natural definition of the discrete divergence operator defined on the primal polygonal grid ω_c

$$\mathcal{DIV}_c(\mathbf{u}) = \frac{1}{|\omega_c(t)|}\int_{\partial\omega_c(t)} \mathbf{u}\cdot\mathbf{n}\,\mathrm{d}s = \frac{1}{|\omega_c(t)|}\sum_{p\in\mathcal{P}(c)} l_{pc}\mathbf{n}_{pc}\cdot\mathbf{u}_p \tag{22}$$

This is the standard Finite Volume discretization for the divergence operator (Alder et al., 1964). It is nothing but the exact computation of the contour integral, $\int_{\partial\omega_c(t)} \mathbf{u}\cdot\mathbf{n}\,\mathrm{d}s$, for a linear velocity field. One can show (Maire, 2011) that it holds exactly for affine velocity fields. Following the same approach the discrete velocity gradient tensor is defined on the polygonal primal cell by

$$\mathcal{GRAD}_c(\mathbf{u}) = \frac{1}{|\omega_c(t)|}\sum_{p\in\mathcal{P}(c)} l_{pc}\mathbf{u}_p\otimes\mathbf{n}_{pc}, \tag{23}$$

where $\otimes$ denotes the tensor product. We easily check that the trace of the discrete velocity gradient tensor coincides with the discrete divergence of the velocity.

We conclude this section by deriving the expression of the discrete gradient operator over the dual cell ω_p. It is formally defined by

$$\mathcal{GRAD}_p(p) = \frac{1}{|\omega_p(t)|} \int_{\partial\omega_p(t)} p\mathbf{n}\,\mathrm{d}s. \tag{24}$$

Integrating the vectorial identity

$$p\nabla\cdot\mathbf{u} + \mathbf{u}\cdot\nabla p = \nabla\cdot(p\mathbf{u}), \tag{25}$$

over the domain $D(t)$ yields

$$\int_{D(t)} p\nabla\cdot\mathbf{u}\,\mathrm{d}v + \int_{D(t)} \mathbf{u}\cdot\nabla p\,\mathrm{d}v = \int_{\partial D(t)} p\mathbf{u}\cdot\mathbf{n}\,\mathrm{d}s.$$

Supposing for sake of simplicity that $\mathbf{u}\cdot\mathbf{n} = 0$ on $\partial D(t)$ leads to

$$\int_{D(t)} \mathbf{u}\cdot\nabla p\,\mathrm{d}v = -\int_{D(t)} p\nabla\cdot\mathbf{u}\,\mathrm{d}v.$$

The left-hand side of the above equation is discretized over the union of the dual cells, ω_p, whereas the right-hand side is discretized over the union of the primal cell and we get

$$\sum_p \int_{\omega_p} \mathbf{u}\cdot\nabla p\,\mathrm{d}v = -\sum_c \int_{\omega_c} p\nabla\cdot\mathbf{u}\,\mathrm{d}v.$$

Assuming that the velocity field is node centred, i.e., $\mathbf{u} = \mathbf{u}_p$ over ω_p, whereas the pressure field is cell centred, i.e., $p = p_c$ over ω_c, and substituting the definition of the discrete gradient and divergence operator the foregoing equation turns into

$$\sum_p |\omega_p|\mathbf{u}_p\cdot\mathcal{GRAD}_p(p) = -\sum_c |\omega_c|p_c\mathcal{DIV}_c(\mathbf{u}). \tag{26}$$

The above equation is the discrete counterpart of the continuous identity (25). It shows that the discrete gradient operator is the negative adjoint of the discrete divergence operator. Now, replacing the discrete divergence operator by its expression in terms of the corner normal yields

$$\sum_p |\omega_p|\mathcal{GRAD}_p(p)\cdot\mathbf{u}_p = -\sum_c p_c \sum_{p\in\mathcal{P}(c)} l_{pc}\mathbf{n}_{pc}\cdot\mathbf{u}_p.$$

Switching the summation in the right-hand side turns the above equation into

$$\sum_p |\omega_p|\mathcal{GRAD}_p(p)\cdot\mathbf{u}_p = -\sum_p \left(\sum_{c\in\mathcal{C}(p)} l_{pc}p_c\mathbf{n}_{pc}\right)\cdot\mathbf{u}_p.$$

The foregoing identity holds true for all $\mathbf{u}_p$ provided that the discrete gradient operator reads

$$\mathcal{GRAD}_p(p) = -\frac{1}{|\omega_p|}\sum_{c\in\mathcal{C}(p)} l_{pc}p_c\mathbf{n}_{pc}. \tag{27}$$

This construction of the discrete gradient operator has been obtained according to the methodology of mimetic finite difference discretization, refer to Lipnikov et al. (2014) for a complete presentation of this topic. It is worth mentioning that the construction of the SGH and CLH schemes relies on the use of the foregoing discrete divergence and gradient operators. However, such elementary discrete constructions of gradient and divergence involve severe defects for both of these schemes expressing: the rank deficiency of their composition operator, refer to Lipnikov et al. (2014) (this will not be discussed here) and the existence of so-called hourglass or checkerboard instabilities.

3.4 Hourglass Fixes

The discrete gradient (27) and the discrete divergence (22) are spoiled by nonphysical fields solution of:

$$\mathcal{GRAD}_p(p) = 0, \quad \mathcal{DIV}_c(\mathbf{u}) = \mathbf{0}. \tag{28}$$

For instance for a 2D n-faces polygon and for a scalar quantity, the number of those spurious solutions is $n - 3$, 3 being the number of point-values sufficient to define an affine field. The theoretical explanation relies in the fact that the *inf* $-$ *sup* or LBB condition (Girault and Raviart, 1986) is not satisfied. This translates the incompatibility between the discrete point definition of scalar and vector fields when subject to gradient and divergence operators. In terms of degrees of freedom, it expresses the difficulty to define a balance between pressure and velocity, depending on the problem dimension, the boundary conditions, the number and nature of the polygons. For Finite Volumes and $Q_1 - P_0$ Finite Elements on quadrilateral cells the pressure field is underintegrated with regard to the velocity field.[e] Practical recipes to eliminate such drawbacks lay on three principles:

- Decrease the number of degrees of freedom of the velocity;
- Increase those of the pressure;
- Adapt the diffusive properties of the numerical modelling.

We note in passing that the second and third items are common to SGH and CLH.

[e]At the contrary domain triangulation is associated to an over integrated pressure field and thus induces an unphysical stiff behaviour (mesh locking).

3.4.1 Hourglass Filtering

An hourglass filtering consists in smearing the velocity components by way of weighted averaging in order to eliminate the nonphysical high wave lengths. Being based on numerical considerations, the challenge is to match the physics in terms of energy conservation (from kinetic to internal), to respect wave spectrum and propagation. Linking all nodal velocities this procedure diminishes the accuracy of the discrete representation. We invite the reader to refer to the algorithm given in Flanagan and Belytschko (1981) for more details.

3.4.2 Subcelling

Early Browne and Wallick (1971), people notice compression of quadrilateral cells could be defined on the whole cell or locally on a cell partition, refer to Fig. 3. To get rid of that quadrilateral paradox, Browne and Wallick used a T-subcelling (triangular based) in order to define a counter local force to the local compression rate. Induced rigidity and low time step obliged later to introduce Q-subcelling (quadrangular based) allowing the occurrence of larger shear displacements (Burton, 1994).

3.4.3 Viscous Damping

The roots of the hourglass plague are mostly intrinsic to the elected Finite Volumes schemes (accuracy, consistency, dispersive properties), aspect ratio of the deformed Lagrangian mesh, the physical problems itself (boundary conditions, nontrivial terms in the modelling), but also due to computation

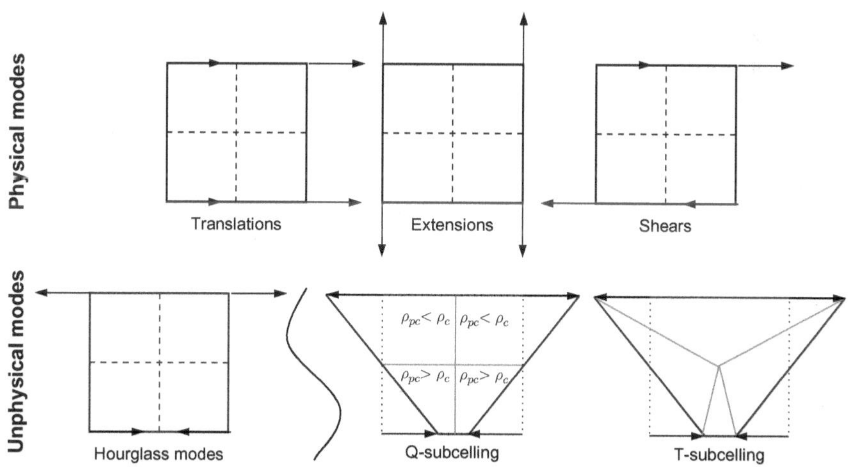

FIG. 3 Physical and unphysical modes of a quadrilateral cell. *Top*: translations, extensions and shears (considering symmetry). *Bottom*: unphysical hourglass modes. All but the two hourglass modes are physical but only for the hourglass modes do the subcell densities differ from the cell density. Two kinds of cell partitions: quadrilateral (Q) and triangular (T) subcellings.

choices (compiler, floating arithmetic). In that context, one tries to model a force that can alleviate the above flaws by modifying the acceleration term. From conservation constraints, that force derives from a potential and one then builds up a viscous pressure–stress term with the desired properties. In the following we will propose some ideas to design artificial viscosity formulations for SGH and shock pressure for CLH.

4 DISCRETE COMPATIBLE STAGGERED LAGRANGIAN HYDRODYNAMICS—SGH

In this section we have chosen to present one representative of the SGH schemes family: the discrete, compatible staggered-grid Lagrangian hydrodynamics scheme initially introduced in Caramana et al. (1998). This allows to summarize most of the characteristics of the 2D VNR scheme offsprings. Starting from the local form of the Lagrangian hydrodynamics equations (13), (10a) and (12), we construct a staggered discretization compatible with the GCL ensuring momentum and total energy conservation.

4.1 Notation and Assumptions

In what follows, we reuse the notations initially introduced in Section 3. The SGH scheme is characterized by a node-centred placement of the kinematic variables ($\mathbf{u}_p$ is the velocity of vertex p), whereas the thermodynamic variables such as the mass density, ρ_c, the pressure, p_c and the specific internal energy, ε_c are located at the cell centre. In this framework, the nodal velocity, $\mathbf{u}_p$, shall be computed solving a discrete momentum equation written over the dual cell ω_p associated to point p, refer to Fig. 1B.

Assuming a piecewise constant distribution of the mass density over the primal grids leads to initialize the mass density over subcell ω_{pc} as $\rho^0_{pc} = \rho^0_c$. In addition, supposing that the subcell is Lagrangian implies that their mass, $m_{pc} = \rho_{pc}(t)|\omega_{pc}(t)|$ are constant in time. The mass of the dual cell, ω_p, is naturally given by $m_p = \sum_{c\in\mathcal{C}(p)} m_{pc}$ and thus is also constant in time. Moreover, the total discrete mass is conserved, i.e., $\sum_c m_c = \sum_p m_p$. Finally the subcell and the dual cell mass density read

$$\rho_{pc}(t) = \frac{m_{pc}}{|\omega_{pc}(t)|}, \quad \text{and} \quad \rho_p(t) = \frac{m_p}{|\omega_p(t)|}. \tag{29}$$

4.2 Semidiscrete Compatible Discretization of the GCL

Utilizing the results obtained in Section 3 allows to write the semidiscrete GCL equation under the form

$$m_c \frac{d}{dt}\left(\frac{1}{\rho_c}\right) - |\omega_c(t)|\mathcal{DIV}_c(\mathbf{u}) = 0, \tag{30}$$

where the discrete cell-centred divergence operator is defined by (22). The semidiscrete GCL is compatible with the semidiscrete trajectory equation (14). Recalling that $m_c = \rho_c(t)|\omega_c(t)|$ leads to rewrite (30) under the alternative form

$$\frac{1}{|\omega_c(t)|}\frac{d}{dt}|\omega_c(t)| = \mathcal{DIV}_c(\mathbf{u}).$$

4.3 Semidiscrete Momentum Equation on the Dual Cell ω_p

The nodal velocity, $\mathbf{u}_p$, required for computing the displacement of vertex p is obtained by integrating the momentum equation over the dual cell ω_p

$$m_p \frac{d\mathbf{u}_p}{dt} + \int_{\partial\omega_p(t)} p\,\mathbf{n}\,\mathrm{d}s = \mathbf{0}.$$

By virtue of the definition of the discrete gradient operator (24) the above equation rewrites

$$m_p \frac{d\mathbf{u}_p}{dt} + |\omega_p| \mathcal{GRAD}_p(p) = \mathbf{0}. \tag{31}$$

This is nothing but the discrete analogue of the continuous PDE that governs momentum conservation (10a).

Next, substituting the expression of the discrete gradient operator (27) into the above equation leads to

$$m_p \frac{d\mathbf{u}_p}{dt} - \sum_{c \in \mathcal{C}(p)} l_{pc} p_c \mathbf{n}_{pc} = \mathbf{0}.$$

Introducing the subcell force

$$\mathbf{f}_{pc} = l_{pc} p_c \mathbf{n}_{pc}, \tag{32}$$

one obtains the following alternative form of the semidiscrete momentum equation

$$m_p \frac{d\mathbf{u}_p}{dt} - \sum_{c \in \mathcal{C}(p)} \mathbf{f}_{pc} = \mathbf{0}. \tag{33}$$

This corresponds to Wilkins' discretization (Alder et al., 1964) that applies to cylindrical geometry (area weighting) and has been extended in 3D by Pracht (1975).

We conclude this section exhibiting a sufficient condition on the subcell force to ensure the conservation of the global momentum $\mathcal{Q} = \sum_p m_p \mathbf{u}_p$. Thanks to the semidiscrete momentum equation and ignoring the boundary terms contribution, the time rate of change of the global momentum writes

$$\frac{d}{dt}\mathcal{Q} = \sum_p \sum_{c \in \mathcal{C}(p)} \mathbf{f}_{pc} = \sum_c \sum_{p \in \mathcal{P}(c)} \mathbf{f}_{pc}.$$

It is thus obvious that the global momentum is conserved provided that the subcell forces satisfies

$$\sum_{p\in\mathcal{P}(c)} \mathbf{f}_{pc} = \mathbf{0}. \tag{34}$$

Utilizing the definition of the pressure subcell force (32) and by virtue of the geometrical identity (18), it is clear that (34) is satisfied and thus the global momentum is conserved.

4.4 Semidiscrete Internal Energy Equation on the Primal Cell ω_c

The semidiscrete internal energy equation over the primal cell ω_c is constructed as a sufficient condition to enforce total energy conservation over the computational domain. For the sake of conciseness, the demonstration is performed ignoring the boundary terms contribution. Let φ denotes a generic thermodynamic variable, then its cell-centred value is its mass average over the primal cell ω_c, that is

$$\varphi_c = \frac{1}{m_c}\int_{\omega_c} \rho\varphi \, \mathrm{d}v. \tag{35}$$

Applying this definition to the specific internal energy, it is clear that the internal energy over the domain D is obtained by summing the contribution of the primal cells as follows

$$\mathcal{E} = \int_D \rho\varepsilon \, \mathrm{d}v = \sum_c m_c \varepsilon_c.$$

The velocity field being located at the nodes, the kinetic energy over the domain D is computed by collecting the contribution of the dual cells as follows

$$\mathcal{K} = \int_D \frac{1}{2}\rho\mathbf{u}^2 \, \mathrm{d}v = \sum_p \frac{1}{2} m_p \mathbf{u}_p^2.$$

If E denotes the total energy over the domain D, then the total energy conservation ignoring boundary contribution simply writes $E = \mathcal{K} + \mathcal{E} = \text{const.}$, which after time differentiation yields

$$\frac{d}{dt}(\mathcal{K} + \mathcal{E}) = 0. \tag{36}$$

The time rate of change of kinetic energy writes

$$\frac{d\mathcal{K}}{dt} = \sum_p \mathbf{u}_p \cdot m_p \frac{d\mathbf{u}_p}{dt}.$$

Substituting the semidiscrete momentum equation (33) in the right-hand side of the above equation yields

$$\frac{d\mathcal{K}}{dt}=\sum_{p}\sum_{c\in\mathcal{C}(p)}\mathbf{f}_{pc}\cdot\mathbf{u}_p=\sum_{c}\sum_{p\in\mathcal{P}(c)}\mathbf{f}_{pc}\cdot\mathbf{u}_p.$$

Combining this result with the total energy conservation (36) we arrive at

$$\sum_{c}\left[m_c\frac{d\varepsilon_c}{dt}+\sum_{p\in\mathcal{P}(c)}\mathbf{f}_{pc}\cdot\mathbf{u}_p\right]=0.$$

We conclude that the total energy over the domain D is conserved provided that the specific internal energy satisfies the semidiscrete equation

$$m_c\frac{d\varepsilon_c}{dt}+\sum_{p\in\mathcal{P}(c)}\mathbf{f}_{pc}\cdot\mathbf{u}_p=0. \tag{37}$$

We observe that the semidiscrete internal energy evolves thanks to the work made by the subcell forces acting on momentum. Momentum and internal energy balances are said to be compatible, refer to Caramana et al. (1998) and Shashkov (1996). We note in passing that the sufficient condition (34) not only ensures the conservation of momentum but also guarantees that (37) remains invariant under a Galilean transformation.

In addition, substituting the subcell force expression (32) in terms of the cell-centred pressure and the corner normal and recalling the definition of the divergence operator over the primal cell (22) leads to recast the semidiscrete internal energy equation into

$$m_c\frac{d\varepsilon_c}{dt}+|\omega_c(t)|p_c\mathcal{DIV}_c(\mathbf{u})=0. \tag{38}$$

Finally, discrete total energy conservation is ensured by the adjoint relationship, refer to (26), between the discrete divergence and gradient operators and the consistent force and work terms in the momentum and internal energy balances. Eq. (38) is the discrete analogue of the PDE (11) that governs internal energy at the continuous level.

4.5 Compatible Discretization of Additional Subcell Forces

The semidiscrete compatible staggered hydrodynamics discretization which satisfies the global conservation of total energy consists of the semidiscrete momentum equation (33) and the semidiscrete internal energy equation (37) along with the trajectory equation (14) and the mass conservation equation (20).

Three subcell forces are mandatory when building any staggered Lagrangian hydrodynamics scheme. The first subcell force corresponds to the pressure gradient and has been already defined in (32). It will be now denoted by $\mathbf{f}_{pc}^{\text{press}}$.

The foregoing compatible discretization has provided us the semidiscrete internal energy equation (38), which combined with the semidiscrete GCL (30) results in

$$\frac{d\varepsilon_c}{dt} + p_c \frac{d}{dt}\left(\frac{1}{\rho_c}\right) = 0.$$

This shows that the material within cell ω_c follows an isentropic path which is consistent with smooth flows. On the other hand, to capture properly discontinuous flows, such as shock waves, we must design a dissipative mechanism known as artificial viscosity to ensure that the kinetic energy is converted into internal energy consistently with the second law of thermodynamics. The corresponding viscous force will be denoted by $\mathbf{f}_{pc}^{\mathrm{q}}$.

Last, an antihourglass subcell force $\mathbf{f}_{pc}^{\boxtimes}$ can be designed to fight back parasitical grid motions, refer to Section 3.4.1 and Barlow et al. (2016) for more details about this subtle topic.

Therefore, the total subcell force $\mathbf{f}_{pc}$ that applies onto point p from cell c summons a true physical process bound to the pressure, a modelling of irreversible transformation with the help of the artificial viscosity and a pure numerical compound: the antihourglass force

$$\mathbf{f}_{pc} = \mathbf{f}_{pc}^{\mathrm{press}} + \mathbf{f}_{pc}^{\mathrm{q}} + \mathbf{f}_{pc}^{\boxtimes}. \tag{39}$$

Remark 1. In Section 4.3, we have demonstrated that a sufficient condition for global momentum conservation is $\sum_{p\in\mathcal{P}(c)} \mathbf{f}_{pc} = \mathbf{0}$ for all cell c. We note that some of subcelling modellings, especially for $\mathbf{f}_{pc}^{\mathrm{q}}$, might not fit this condition and thus might lead to the creation of an artificial inner cell force.

4.5.1 Artificial Viscosity Force

Because both $\mathcal{GRAD}_p(p)$ and $\mathcal{DIV}_c(\mathbf{u})$ operators are formally second-order centred on deformed grids, Gibbs phenomenon is present coupled to a centred discretization in time (see Section 4.6) and the VNR scheme applied to the full system of gas dynamics induces a high level of dispersion errors (Yanenko et al., 1983). A straightforward way to cure this flaw is to introduce a dissipative pressure term coming from a Riemann solver as it has been interpreted in Christensen (1991). Noticing that it can be expressed (Dukowicz, 1985) as an expansion in terms of velocity jump, the two first terms fitting the 1D legacy definition of von Neumann, Richtmyer and Landshoff (Landshoff, 1955; von Neumann and Richtmyer, 1950)

$$q_c = c_1 \rho_c a_c |\Delta\mathbf{u}| + c_2 \rho_c (\Delta\mathbf{u})^2, \tag{40}$$

where $\Delta\mathbf{u}$ is the velocity jump between the two nodes of an 1D cell c, a_c and ρ_c are, respectively, the isentropic sound speed and the density and c_1, c_2 two constants of the order of unity usually estimated by means of the Kurapatenko formula (Kurapatenko, 1967) deduced from the Rankine–Hugoniot conditions. The force ∇q_c is then the combination of two diffusion terms, respectively, linear and nonlinear, damping selectively short wave lengths.

In the 2D case, let q_c be a generic scalar artificial viscosity (see for instance Wilkins, 1999, section 4.2.8), the corresponding subcell force is $\mathbf{f}_{pc}^{\mathrm{q}} = l_{pc} q_c \mathbf{n}_{pc}$ and the semidiscrete internal energy equation (37) becomes

$$m_c \frac{d\varepsilon_c}{dt} + (\mathbf{f}_{pc}^{\text{pres}} + \mathbf{f}_{pc}^{\text{q}}) \cdot \mathbf{u}_p = 0.$$

Substituting the expression of the pressure subcell force (32) and utilizing the semidiscrete GCL (30) allows us to rewrite the above equation as

$$m_c \left[\frac{d\varepsilon_c}{dt} + p_c \frac{d}{dt}\left(\frac{1}{\rho_c}\right) \right] = -\sum_{p \in \mathcal{P}(c)} \mathbf{f}_{pc}^{\text{q}} \cdot \mathbf{u}_p.$$

Combining the above equation, the discrete analogue of the Gibbs relation (4) and the discrete divergence operator (22) we arrive at the following evolution equation for the specific entropy

$$m_c \theta_c \frac{d\eta_c}{dt} = -\sum_{p \in \mathcal{P}(c)} \mathbf{f}_{pc}^{\text{q}} \cdot \mathbf{u}_p = -|\omega_c| q_c \mathcal{DIV}_c(\mathbf{u}).$$

The consistency with the second law of thermodynamics implies to tune q_c in such a manner $-q_c \ |\omega_c| \mathcal{DIV}_c(\mathbf{u})$ does not become negative. More precisely, the artificial viscous subcell force is such that

$$\mathbf{f}_{pc}^{\text{q}} = \begin{cases} l_{pc} q_c \mathbf{n}_{pc} & \text{if } \mathcal{DIV}_c(\mathbf{u}) \leq 0, \\ 0 & \text{if } \mathcal{DIV}_c(\mathbf{u}) > 0. \end{cases} \tag{41}$$

We observe that the artificial viscous subcell force is active only for cells undergoing compression.

Remark 2. The artificial viscosity is a pressure potential mimicking the solution of a Riemann problem. Its expression induced a nonlinear diffusion of the velocity field erasing intrinsic perturbation effects due to the full-centred scheme. As a potential the work of the associated force $\mathbf{f}_{pc}^{\text{q}}$ insures conservation and positive entropy deposition.

To get a multidimensional extension, Alder et al. (1964) quoted several desired properties that a proper artificial viscosity should enjoy

- Dissipativity: artificial viscosity must always act to decrease kinetic energy.
- Galilean invariance: artificial viscosity should vanish uniformly (smoothly) as the velocity field tends to become constant.
- Self-similar motion invariance: artificial viscosity should vanish for a uniform contraction and/or a rigid rotation.
- Wave front invariance: artificial viscosity should have no effect along a wave front of constant phase.
- Viscous force continuity: artificial viscous force should go to zero continuously as compression vanishes and expansion develops, and remains zero for the latter.
- Mesh invariance: artificial viscosity is asymptotically independent of the mesh.

The quest of grabbing the universal 2D expression for the artificial viscosity mainly followed four paths.

4.5.2 Edge q

Introduced by Schulz (Alder et al., 1964) q-terms are computed face by face and the resulting subcell force coming from two faces contributions applied on each node. This popular q has been improved in term of dissipation by TVD limiting (Arminjon and Dervieux, 1993) and is closely related to T-subcelling.

4.5.3 Oriented q

Introduced in Wilkins (1999), an 1D-like expression of q in a computed shock direction $\boldsymbol{\nu}$ follows (40) with $\Delta\mathbf{u} = L_\nu\partial_\nu(\mathbf{u}\cdot\boldsymbol{\nu})$, where L_ν is a characteristic length related to the shock direction. Despite its fit to physical request concerning the anisotropy of shock propagation, its lack of robustness hinders application to sheared flows.

4.5.4 Tensor q

Many people expressed the artificial viscosity in terms of stress–strain relations trying to define a stress tensor selecting principal directions of the flows in order to induce the appropriate entropy deposition without creating hourglass tangential modes, refer to Tscharnuter and Winkler (1979), Campbell and Shaskhov (2001), Kolev and Rieben (2010) and Höller et al. (2014). For instance, let $\mathbb{Q}_c$ denotes the tensorial artificial viscosity within cell c. The corresponding artificial viscous force and its related entropy equation write

$$\mathbf{f}_{pc}^{\mathrm{q}} = \begin{cases} -l_{pc}\mathbb{Q}_c\mathbf{n}_{pc} & \text{if } \delta_c \le 0, \\ 0 & \text{if } \delta_c > 0. \end{cases} \quad m_c\theta_c\frac{d\eta_c}{dt} = \begin{cases} |\omega_c|[\mathbb{Q}_c : \mathcal{GRAD}_c(\mathbf{u})], & \text{if } \delta_c \le 0, \\ 0 & \text{if } \delta_c > 0. \end{cases}$$

Here, $\delta_c = \mathcal{DIV}_c(\mathbf{u})$ and : denotes the inner product of tensors defined by $\mathbb{T} : \mathbb{Q} = \mathrm{tr}(\mathbb{T}^t\mathbb{Q})$. The tensor $\mathbb{Q}_c$ might be written under the form

$$\mathbb{Q}_c = \rho_c(\tilde{c}_1 a_c L_c + \tilde{c}_2|\mathcal{DIV}_c(\mathbf{u})|L_c^2)\mathcal{GRAD}_c(\mathbf{u}). \tag{42}$$

One easily checks that the entropy deposition corresponding to the above formulation is always positive. However, one major drawback lies on the determination of the characteristic grid length, L_c, which is not easy to evaluate on distorted polygonal cell. This again comes from the PDE formulation of the SGH schemes.

4.5.5 Subcelled q

Since the 1970s, see for instance Browne and Wallick (1971), people try to define more selective nodal forces with regard to anisotropic cell compression to avoid hourglass instabilities appearing during shock propagation. The idea of subcell partitioning allows to compute several pressures or/and artificial viscosities in order to balance the ill distributed degrees of freedom between velocities and pressures, refer to Section 3.4. Finite Element method greatly helped to set the problem in term of level of quadrature (Cheng and Tian, 2015).

4.6 Time Discretization

Contrarily to the full staggered VNR scheme, a necessary condition to maintain compatibility between force and work is to integrate explicitly momentum and internal energy equation, this later being nonlinear in pressure and energy. In order to preserve a formal second order of accuracy in time, a predictor–corrector sequence is employed.

4.6.1 Predictor–Corrector Algorithm

Providing a small enough time step $\Delta t = t^{n+1} - t^n$ and reminding that all quantities are time centred, the predictor and corrector steps are

Predictor step

$$m_p \frac{\widehat{\mathbf{u}}_p - \mathbf{u}_p^n}{\Delta t} - \sum\nolimits_{c \in \mathcal{C}(p)} \mathbf{f}_{pc}^n = \mathbf{0},$$

$$m_c \frac{\widehat{\varepsilon}_c - \varepsilon_c^n}{\Delta t} + \sum\nolimits_{p \in \mathcal{P}(c)} \mathbf{f}_{pc}^n \cdot \mathbf{u}_p^n = 0,$$

$$\frac{\widehat{\mathbf{x}}_p - \mathbf{x}_p^n}{\Delta t} = \mathbf{u}_p^n.$$

Corrector step

$$m_p \frac{\mathbf{u}_p^{n+1} - \mathbf{u}_p^n}{\Delta t} - \sum\nolimits_{c \in \mathcal{C}(p)} \mathbf{f}_{pc}^{n+1/2} = \mathbf{0},$$

$$m_c \frac{\varepsilon_c^{n+1} - \varepsilon_c^n}{\Delta t} + \sum\nolimits_{p \in \mathcal{P}(c)} \mathbf{f}_{pc}^{n+1/2} \cdot \mathbf{u}_p^{n+1/2} = 0,$$

$$\frac{\mathbf{x}_p^{n+1} - \mathbf{x}_p^n}{\Delta t} = \mathbf{u}_p^{n+1/2}.$$

where for the corrector step halftime-centred variables are computed as $A^{n+1/2} = \frac{1}{2}(\widehat{A} + A^n)$ for $A = \mathbf{u}, \varepsilon, \mathbf{x}$. New cell and subcell volumes $|\omega_c^{n+1/2}|$, $|\omega_{cp}^{n+1/2}|$ as well as geometrical vectors $l_{pc}^{n+1/2}\mathbf{n}_{pc}^{n+1/2}$, are updated by means of the new vertex positions $\mathbf{x}^{n+1/2}$. Pressure at $t^{n+1/2}$ comes from the equation of state thanks to $\varepsilon_c^{n+1/2}$ and $\rho_c^{n+1/2} = m_c / |\omega_c^{n+1/2}|$. Consequently all subcell forces can be recomputed at $t^{n+1/2}$.

4.6.2 Time Step Monitoring

Lagrangian discrete schemes require a time step monitoring which rely on two criteria: the first one limits the cell volume variation and ensures the positivity of the cell volume under compression, whereas the second one is a kind of CFL (Courant–Friedrichs–Lewy) condition, which ensures the positivity of the updated specific internal energy. Practically the formula used to monitor the time step evolution is

$$\Delta t = \min\left(C_m \Delta t^n, C_v \Delta t_v, C_\varepsilon \Delta t_\varepsilon\right). \quad (43)$$

Here, Δt_v limits the cell volume variation and ensures the positivity of the cell volume under compression, whereas Δt_ε is computed to ensure the positivity of the updated specific internal energy, refer to Barlow et al. (2016). In addition, C_m, C_v and C_ε are user-defined parameters. C_m is a multiplicative coefficient usually set to $C_m = 1.05$, whereas C_ε is a safety coefficient fixed to $C_\varepsilon = 0.3$ and C_v might be set to $C_v = 0.1$. This time step monitoring ensures the positivity of the updated mass density and specific internal energy refer to Barlow et al. (2016).

5 DISCRETE COLOCATED LAGRANGIAN HYDRODYNAMICS—CLH

Starting from the updated Lagrangian integral formulation (1) of gas dynamics equations, we describe how to construct a cell-centred Finite Volume discretization (CLH) that conserves mass, momentum and total energy.

5.1 Notation and Assumptions

In CLH, we now apply the whole set of Eqs (1b) and (1c) and the GCL (13) to the polygonal cell ω_c. The semidiscrete system of conservation laws is

$$m_c \frac{d}{dt}\left(\frac{1}{\rho_c}\right) - \int_{\partial\omega_c(t)} \mathbf{u}\cdot\mathbf{n}\, \mathrm{d}s = 0, \tag{44a}$$

$$m_c \frac{d\mathbf{u}_c}{dt} + \int_{\partial\omega_c(t)} p\mathbf{n}\, \mathrm{d}s = \mathbf{0}, \tag{44b}$$

$$m_c \frac{de_c}{dt} + \int_{\partial\omega_c(t)} p\mathbf{n}\cdot\mathbf{u}\, \mathrm{d}s = 0. \tag{44c}$$

In the above system, the cell-centred values of the physical variables are defined according to (35). Here, one solves the equation for total energy e_c from which the specific internal energy of cell c is a back-product: $\varepsilon_c = e_c - \frac{1}{2}\mathbf{u}_c^2$. Because all physical variables are cell centred, this approach is referred as CLH.

The motion of the polygonal grid is governed by the trajectory equation of each vertex p, refer to (14). With CLH, the velocity field is represented on the computational domain by both cell velocities $\mathbf{u}_c$ inside $\omega_c(t)$ and node velocity $\mathbf{u}_p$ upon $\omega_p(t)$ for each p in $\mathcal{P}(c)$; these two fundamental quantities must consistently coexist.

To complete the discretization of system (44) we shall define firstly a relevant approximation of the numerical fluxes of the discrete conservation laws (44a), (44b) and (44c), and secondly the vertex velocity $\mathbf{u}_p$ to displace the computational grid. These two questions will be addressed according to the design principles:

- *GCL compatibility*. The volume flux approximation shall provide a time rate of change of the cell volume compatible with the geometric volume computed directly from the displacement of the vertices.

- *Entropy production.* The numerical fluxes are computed to ensure that a local entropy inequality be satisfied at the semidiscrete level.
- *Momentum and total energy conservation.* The numerical fluxes are computed to ensure that momentum and total energy are conserved over the whole computational domain up to the boundary conditions.

Recall that the semidiscrete form of the GCL has been already discussed in Section 3.2. We will detail in Section 5.6 the impact of time discretization.

5.2 Subcell Force-Based Discretization

Here, we reuse the notations introduced in Section 3.1. The CLH subcell force exerted on the outer boundary of the subcell ω_{pc}, refer to Fig. 1A, is defined by

$$\mathbf{F}_{pc} = -\int_{\partial\omega_{pc}\cap\partial\omega_c} p\mathbf{n}\,\mathrm{d}s. \tag{45}$$

This is the numerical flux related to the momentum. Employing this definition the momentum and total energy fluxes rewrite

$$\int_{\partial\omega_c(t)} p\mathbf{n}\,\mathrm{d}s = -\sum_{p\in\mathcal{P}(c)} \mathbf{F}_{pc}, \quad \text{and} \quad \int_{\partial\omega_c(t)} p\mathbf{n}\cdot\mathbf{u}\,\mathrm{d}s = -\sum_{p\in\mathcal{P}(c)} \mathbf{F}_{pc}\cdot\mathbf{u}_p.$$

Finally, we recast the system of conservation laws (44) under the semidiscrete form

$$m_c\frac{d}{dt}\left(\frac{1}{\rho_c}\right) - \sum_{p\in\mathcal{P}(c)} l_{pc}\mathbf{n}_{pc}\cdot\mathbf{u}_p = 0, \tag{46a}$$

$$m_c\frac{d\mathbf{u}_c}{dt} - \sum_{p\in\mathcal{P}(c)} \mathbf{F}_{pc} = \mathbf{0}, \tag{46b}$$

$$m_c\frac{de_c}{dt} - \sum_{p\in\mathcal{P}(c)} \mathbf{F}_{pc}\cdot\mathbf{u}_p = 0. \tag{46c}$$

To close the above semidiscrete system, it remains to determine both $\mathbf{F}_{pc}$ and $\mathbf{u}_p$ in terms of the cell-centred physical variables.

5.3 Local Entropy Inequality

From the Gibbs relation (4), the discrete analogue of the time rate of change of entropy within the polygonal cell ω_c reads

$$m_c\theta_c\frac{d\eta_c}{dt} = p_c m_c\frac{d}{dt}\left(\frac{1}{\rho_c}\right) - \mathbf{u}_c\cdot m_c\frac{d\mathbf{u}_c}{dt} + m_c\frac{de_c}{dt},$$

where $\theta_c > 0$ is the averaged temperature of the cell ω_c. Substituting the semi-discrete conservation laws (46a), (46b), (46c) and geometric identity (18) into the above equation yields

$$m_c\theta_c\frac{d\eta_c}{dt} = \sum_{p\in\mathcal{P}(c)} (\mathbf{F}_{pc} + l_{pc}p_c\mathbf{n}_{pc})\cdot(\mathbf{u}_p - \mathbf{u}_c). \tag{47}$$

To satisfy the entropy imbalance the right-hand side of the above equation has to be positive. Thus, the thermodynamic consistency of the CLH scheme with the second law of thermodynamic is ensured provided that there exists a positive definite 2×2 corner matrix, $\mathbb{M}_{pc}$, such that the subcell force writes

$$\mathbf{F}_{pc} = -l_{pc}p_c\mathbf{n}_{pc} + \mathbb{M}_{pc}(\mathbf{u}_p - \mathbf{u}_c). \tag{48}$$

The foregoing equation is a sufficient condition enforcing the positiveness of entropy production of the CLH scheme. Dimensional analysis shows that the dimension of the corner matrix $\mathbb{M}_{pc}$ should be homogeneous to a mass per unit time. Moreover, it should satisfy the principle of material frame indifference (Gurtin et al., 2010). Replacing (48) into (47) allows us to express the entropy production term as a quadratic function of the velocity jump $\mathbf{u}_p - \mathbf{u}_c$

$$m_c\theta_c\frac{d\eta_c}{dt} = \sum_{p\in\mathcal{P}(c)} \mathbb{M}_{pc}(\mathbf{u}_p - \mathbf{u}_c)\cdot(\mathbf{u}_p - \mathbf{u}_c) \geq 0.$$

and is directly governed by the corner matrix $\mathbb{M}_{pc}$ whose expression is investigated in Section 5.5.

5.4 Conservation of Total Energy and Momentum

The expression (48) of the subcell force defines a node-based Finite Volume discretization of numerical fluxes, through the used of $\mathbf{F}_{pc}$. Contrarily to the usual face-based FV discretization, the conservation of total energy and momentum is not trivial. Here, we shall derive a condition to ensure that total energy and momentum are globally conserved.

The conservation of total energy over the computational domain neglecting boundary conditions simply reads $\sum_c m_c\frac{de_c}{dt} = 0$. Replacing the semidiscrete equation of total energy (46c) in this equation and exchanging the order of summation from primal cells to dual cells yields

$$\sum_c \sum_{p\in\mathcal{P}(c)} \mathbf{F}_{pc}\cdot\mathbf{u}_p = \sum_p \sum_{c\in\mathcal{C}(p)} \mathbf{F}_{pc}\cdot\mathbf{u}_p = 0,$$

where $\mathcal{C}(p)$ is the set of cells surrounding node p. The foregoing condition shall be satisfied for all nodal velocity field $\mathbf{u}_p$ to ensure conservation of total energy over the computation grid at the semidiscrete level. Therefore, the semidiscrete CLH scheme conserves total energy if and only if the following condition is fulfilled

$$\sum_{c\in\mathcal{C}(p)} \mathbf{F}_{pc} = \mathbf{0}, \text{ for all point } p. \tag{49}$$

We point out that this condition also implies the conservation of momentum over the computational domain.

5.5 Nodal Solver

Combining the expression of the subcell force satisfying the entropy inequality (48) and condition (49) allows us to write the following vectorial equation that determines uniquely the nodal velocity

$$\mathbb{M}_p \mathbf{u}_p = \sum_{c\in\mathcal{C}(p)} (l_{pc} p_c \mathbf{n}_{pc} + \mathbb{M}_{pc} \mathbf{u}_c), \text{ for all point } p, \tag{50}$$

where $\mathbb{M}_p$ is the sum of the corner matrices surrounding node p, i.e., $\mathbb{M}_p = \sum_{c\in\mathcal{C}(p)} \mathbb{M}_{pc}$ which is also symmetric positive definite and thus invertible and admits a unique solution. Eq. (50) is a system of two equations with respect to components of $\mathbf{u}_p$. Indeed, being given some choice of $\mathbb{M}_{pc}$ matrix, all the terms in the right-hand side of (50) are known, because p_c and $\mathbf{u}_c$ are primary unknowns. Moreover, once (50) is solved, the nodal velocity is expressed in terms of the primary unknowns. Thus, replacing $\mathbf{u}_p$ in the subcell force expression (48), shows that $\mathbf{F}_{pc}$ is expressed in terms of p_c and $\mathbf{u}_c$ and the space discretization is achieved.

Recalling that $\mathbb{M}_{pc}$ has the dimension of a mass per unit time, it is worth mentioning that (50) can be interpreted as an instantaneous balance of momentum at node p. More precisely, introducing the SGH subcell force, $\mathbf{f}_{pc}$, this equation rewrites as

$$\mathbb{M}_p \mathbf{u}_p = \sum_{c\in\mathcal{C}(p)} \mathbf{f}_{pc} + \mathbf{f}_{pc}^{\text{vis}}, \tag{51}$$

where $\mathbf{f}_{pc} = l_{pc} p_c \mathbf{n}_{pc}$ is the pressure force, and $\mathbf{f}_{pc}^{\text{vis}} = \mathbb{M}_{pc} \mathbf{u}_c$ is the viscous subcell force induced by the nodal solver and characterized by the viscous matrix $\mathbb{M}_{pc}$. Here, the main difference with the SGH balance of momentum (33) is due to the presence of the nodal mass matrix, $\mathbb{M}_p$, instead of the scalar nodal mass, m_p.

The subcell force being not unique, several solvers have been proposed. Hereafter, we briefly list three of them:

- The GLACE (Godunov-type LAgrangian scheme Conservative for total Energy) solver (Després and Mazeran, 2005) is characterized by a one-point quadrature of the pressure gradient and relies on the computation of one nodal corner pressure writing an approximation of the Riemann invariant, $dp + zdu$, where $z = \rho a$ denotes the acoustic impedance, in the direction of the unit corner normal $\mathbf{n}_{pc}$. The corresponding viscous corner matrix reads

$$\mathbb{M}_{pc}^{\mathrm{GL}} = z_c l_{pc}(\mathbf{n}_{pc} \otimes \mathbf{n}_{pc}). \tag{52}$$

It is a symmetric positive rank one matrix which is only semidefinite.

- The EUCCLHYD (Explicit Unstructured Cell-Centred Lagrangian HYDrodynamics) solver (Maire et al., 2007) has been constructed to improve the robustness of the GLACE solver. It is based on a two-point quadrature of the pressure gradient which requires the computation of two nodal corner pressures obtained thanks to the approximation of the Riemann invariant, $dp + zdu$, in the direction of the two unit normals $\mathbf{n}_{pc}^{-}$ and $\mathbf{n}_{pc}^{+}$, refer to Fig. 2. The corresponding symmetric positive definite viscous matrix reads

$$\mathbb{M}_{pc}^{\mathrm{EU}} = z_c[l_{pc}^{-}(\mathbf{n}_{pc}^{-} \otimes \mathbf{n}_{pc}^{-}) + l_{pc}^{+}(\mathbf{n}_{pc}^{+} \otimes \mathbf{n}_{pc}^{+})]. \tag{53}$$

- The CCH (cell-centred hydrodynamics) solver (Burton et al., 2013) is also characterized by a two-point quadrature of the pressure gradient. The nodal corner pressures are evaluated writing a jump relation in the shock unit direction $\boldsymbol{\nu}_{pc}$ at the cell corner. The corresponding viscous corner matrix is spherical and reads

$$\mathbb{M}_{pc}^{\mathrm{CC}} = z_c(l_{pc}^{-}|\boldsymbol{\nu}_{pc}^{-} \cdot \mathbf{n}_{pc}^{-}| + l_{pc}^{+}|\boldsymbol{\nu}_{pc}^{+} \cdot \mathbf{n}_{pc}^{+}|)\mathbb{I}_{\mathrm{d}}, \tag{54}$$

where $\mathbb{I}_{\mathrm{d}}$ is the identity matrix and $\boldsymbol{\nu}_{pc} = \dfrac{\mathbf{u}_p - \mathbf{u}_c}{|\mathbf{u}_p - \mathbf{u}_c|}$.

Notice nonlinear variants rely on the use of a generalized impedance based on the Dukowicz approximate Riemann solver (Dukowicz, 1985; Maire, 2009b) instead of the acoustic impedance.

To fight a too great instable tendency, subzonal entropy applied when the local compression of a T-subcell reaches a high level was introduced in Després and Labourasse (2012). This addition of a source of local dissipation through a supplementary subcell force is unfortunately spoiled by the chosen triangle partitioning that induced some locking.

5.6 First-Order Time Discretization

The first-order explicit time integration of the system (46) of semidiscrete conservation laws over the time interval $[t^n, t^{n+1}]$ leads to

$$m_c\left[\left(\frac{1}{\rho_c}\right)^{n+1} - \left(\frac{1}{\rho_c}\right)^{n}\right] - \Delta t \sum_{p \in \mathcal{P}(c)} (l_{pc}\mathbf{n}_{pc})^{n+\frac{1}{2}} \cdot \mathbf{u}_p^n = 0, \tag{55a}$$

$$m_c(\mathbf{u}_c^{n+1} - \mathbf{u}_c^n) - \Delta t \sum_{p \in \mathcal{P}(c)} \mathbf{F}_{pc}^n = \mathbf{0}, \tag{55b}$$

$$m_c(e_c^{n+1} - e_c^n) - \Delta t \sum_{p \in \mathcal{P}(c)} \mathbf{F}_{pc}^n \cdot \mathbf{u}_p^n = 0. \tag{55c}$$

Here, the superscript n denotes the value of the physical variables at time t^n and $\Delta t = t^{n+1} - t^n$ is the time step. The superscript $n+\frac{1}{2}$ utilized in (55a) corresponds to a time-centred integration performed at $t^n + \frac{\Delta t}{2}$. This particular choice is motivated by the need to satisfy rigorously the GCL at the discrete level, refer to Maire (2011).

The nodal velocity, $\mathbf{u}_p^n$, and the subcell force, $\mathbf{F}_{pc}^n$, are computed by means of the nodal solver. Being given a particular choice of the viscous matrix, $\mathbb{M}_{pc}^n$, the nodal velocity is obtained solving the following system $\mathbb{M}_p^n \mathbf{u}_p^n = \sum_{c \in \mathcal{C}(p)} l_{pc}^n p_c^n \mathbf{n}_{pc}^n + \mathbb{M}_{pc}^n \mathbf{u}_c^n$, where the nodal mass matrix reads $\mathbb{M}_p^n = \sum_{c \in \mathcal{C}(p)} \mathbb{M}_{pc}^n$. Then, the subcell force is computed by means of $\mathbf{F}_{pc}^n = l_{pc}^n p_c^n \mathbf{n}_{pc}^n + \mathbb{M}_{pc}^n (\mathbf{u}_p^n - \mathbf{u}_c^n)$, using the expression of the nodal velocity coming from the previous linear system. Note that all numerical fluxes have been determined from primary unknowns and geometric quantities evaluated at time t^n.

Once the nodal velocity $\mathbf{u}_p^n$ is computed, the explicit time integration of the trajectory equation is available through

$$\mathbf{x}_p^{n+1} = \mathbf{x}_p^n + \Delta t \mathbf{u}_p^n. \tag{56}$$

The cell volume, $|\omega_c^{n+1}|$, is directly computed from the geometry at time t^{n+1} and the updated mass density might be computed either from mass conservation, $\rho_c^{n+1} = \frac{m_c}{|\omega_c^{n+1}|}$, or from the discrete GCL (55a). These two approaches are strictly equivalent provided that the volume flux is exactly integrated with respect to time, refer to Maire (2011).

The updated cell-centred, p_c^{n+1}, is obtained from the equation of state $p_c^{n+1} = p(\rho_c^{n+1}, \varepsilon_c^{n+1})$. The updated specific internal energy is given by subtracting the kinetic energy to the total energy: $\varepsilon_c^{n+1} = e_c^{n+1} - \frac{1}{2}(\mathbf{u}_c^{n+1})^2$.

5.7 Second-Order Extension

The simplest approach to construct a second-order time discretization relies on a two-stage time integrator such as a predictor–corrector procedure; however, a one-stage time integrator might also be employed to construct the second-order extension. In this alternative approach, the numerical fluxes are evaluated at time $t^{n+\frac{1}{2}} = t^n + \frac{\Delta t}{2}$, where Δt is the current time step, by means of a Taylor expansion. They are computed by means of the nodal solver using nodal extrapolated values of the cell-centred pressure and velocity. In addition, the time-centred numerical fluxes are computed employing the Generalized Riemann Problem methodology, introduced by Ben-Artzi and Falcovitz (2003). This approach has been successfully extended to two-dimensional Lagrangian hydrodynamics (Maire, 2009b).

The monotonicity of the piecewise linear reconstruction is ensured by means of a limiting procedure, which ensures that the nodal extrapolated values of the physical variables remain inside the bounds determined by their piecewise constant values taken in the local neighbourhood of the target cell. Note that the above limiting algorithm is well suited for scalar fields. It may be straightforwardly extend to vector and tensor fields applying it separately to each component. However, such a procedure is frame dependent and thus leads to rotational symmetry distortion. Namely, component-wise limiters do not preserve symmetry since a rotation of the coordinate axis produces different results. This flaw is particularly crucial in the framework of Lagrangian hydrodynamics since we are dealing with moving mesh discretizations that are particularly sensitive to symmetry breaking. Thus, for Lagrangian hydrodynamics the limiting procedure for vector fields must be frame indifferent.

A very promising approach to address this issue is the vector image polygon methodology introduced by Luttwak and Falcovitz in 2011. This method provides a rigorous mathematical framework to extend the notion of discrete maximum principle to vector fields. It consists in constructing the convex hull of the vector(tensor)-space points corresponding to the neighbouring cell-centred vectors (tensors). If the nodal extrapolated vector (tensor) lies inside the convex hull, then the piecewise linear representation is monotonicity preserving, otherwise, a limiting procedure is required. This limiting procedure consists in modifying the slope by projecting the nodal extrapolated value on the boundary of the convex hull. This methodology has been successfully applied in many works, refer for instance to Hoch and Labourasse (2014).

ACKNOWLEDGEMENTS

The authors wish to thank Rémi Abgrall, Stéphane Del Pino, Bruno Després, Emmanuel Labourasse and Misha Shashkov for their fruitful and relevant comments.

REFERENCES

Abgrall, R., Breil, J., Maire, P.-H., Ovadia, J., 2004. Un schéma centré pour l'hydrodynamique Lagrange bidimesionnelle. Technical Report, Laboratoire de Recherche Correspondant LRC CEAMAB, Université de Bordeaux. Available at https://hal.inria.fr/hal-01278264/document.

Adessio, F.L., Carroll, D.E., Dukowicz, K.K., Johnson, J.N., Kashiwa, B.A., Maltrud, M.E., Ruppel, H.M., 1986. CAVEAT: a computer code for fluid dynamics problems with large distortion and internal slip. Technical Report LA-10613-MS, Los Alamos National Laboratory.

Alder, B., Fernbach, S., Rotenberg, M. (Eds.), 1964. Methods in Computational Physics, Advances in Research and Applications, Fundamental Method in Hydrodynamics, vol. 3. Academic Press, New York, London.

Arminjon, P., Dervieux, A., 1993. Construction of TVD-like artificial viscosities on 2D arbitrary FEM grid. J. Comput. Phys. 106, 116–178.

Barlow, A.J., 2013. A high order cell centred dual grid Lagrangian Godunov scheme. Comput Fluids 83, 15–24, ISSN 0045-7930. http://dx.doi.org/10.1016/j.compfluid.2013.02.009. Numerical methods for highly compressible multi-material flow problems. http://www.sciencedirect.com/science/article/pii/S0045793013000649.

Barlow, A.J., Maire, P.-H., Rider, W.J., Rieben, R.N., Shashkov, M.J., 2016. Arbitrary Lagrangian-Eulerian methods for modeling high-speed compressible multimaterial flows. J. Comput. Phys. 322, 603–665. http://dx.doi.org/10.1016/j.jcp.2016.07.001.

Bauer, A.L., Burton, D.E., Caramana, E.J., Loubère, R., Shashkov, M.J., Whalen, P.P., 2006. The internal consistency, stability, and accuracy of the discrete, compatible formulation of Lagrangian hydrodynamics. J. Comput. Phys. 218 (2), 572–593.

Ben-Artzi, M., Falcovitz, J., 2003. Generalized Riemann problems in Computational Fluids Dynamics. Cambridge University Press, Cambridge, UK, New York.

Benson, D.J., 1992. Computational methods in Lagrangian and Eulerian hydrocodes. Comput. Meth. Appl. Mech. Eng. 99, 235–394.

Bertoluzza, S., Del Pino, S., Labourasse, E., 2016. A conservative slide line method for cell-centered semi-Lagrangian and ALE schemes in 2D. ESAIM Math. Model. Numer. Anal. 50 (1), 187–214.

Bethe, H.A., Christy, R., Davis, R.R., Feynman, R.P., Frankel, S., Konopinski, E.J., Livesay, N., Metropolis, N., Nelson, E., 1944. IBM calculations of implosion hydrodynamics. Technical Report, Los Alamos National Laboratory report, LA-94, unclassified, cover only.

Browne, P.L., Wallick, K.B., 1971. The reduction of mesh tangling in two-dimensional Lagrangian hydrodynamics codes by the use of viscosity, artificial viscosity, and T.T.S. (Temporary Triangular Subzoning for long thin zones). Los Alamos National Laboratory. Technical Report LA-4740-MS.

Burton, D.E., 1990. Exact conservation of energy and momentum in staggered-grid hydrodynamics with arbitrary connectivity. In: Advances in the Free Lagrange Method. Springer-Verlag, New York.

Burton, D.E., 1994. Multidimensional discretization of conservation laws for unstructured polyhedral grids. Technical Report UCRL-JC-118306, Lawrence Livermore National Laboratory.

Burton, D.E., Carney, T.C., Morgan, N.R., Sambasivan, S.K., Shashkov, M.J., 2013. A cell-centered Lagrangian Godunov-like method for solid dynamics. Comput. Fluids 83, 33–47.

Campbell, J.C., Shaskhov, M.J., 2001. A tensor artificial viscosity using a mimetic finite difference algorithm. J. Comput. Phys. 172 (4), 739–765.

Caramana, E.J., Burton, D.E., Shashkov, M.J., Whalen, P.P., 1998. The construction of compatible hydrodynamics algorithms utilizing conservation of total energy. J. Comput. Phys. 146 (1), 227–262.

Carré, G., Delpino, S., Després, B., Labourasse, E., 2009. A cell-centered Lagrangian hydrodynamics scheme in arbitrary dimension. J. Comput. Phys. 228 (14), 5160–5183.

Chan, R.K.C., 1975. A generalized arbitrary Lagrangian-Eulerian method for incompressible flows with sharp interfaces. J. Comput. Phys. 17, 311–331.

Cheng, J., Shu, C.-W., 2007. A high order ENO conservative Lagrangian type scheme for the compressible Euler equations. J. Comput. Phys. 227 (2), 1567–1596, ISSN 0021-9991.

http://dx.doi.org/10.1016/j.jcp.2007.09.017. http://www.sciencedirect.com/science/article/pii/S002199910700424X.

Cheng, J., Shu, C.-W., 2010. A cell-centered Lagrangian scheme with the preservation of symmetry and conservation properties for compressible fluid flows in two-dimensional cylindrical geometry. J. Comput. Phys. 229 (19), 7191–7206, ISSN 0021-9991. http://dx.doi.org/10.1016/j.jcp.2010.06.007. http://www.sciencedirect.com/science/article/pii/S0021999110003220.

Cheng, J., Tian, B., 2015. Elimination of hourglass distortions by means of the Lagrangian-Gauss-point-mass method for compressible fluid flows. Comput. Fluids 116, 148–157, ISSN 0045-7930. http://dx.doi.org/10.1016/j.compfluid.2015.04.013. http://www.sciencedirect.com/science/article/pii/S0045793015001280.

Christensen, R.B., 1991. Godunov methods on a staggered mesh—an improved artificial viscosity. Technical Report UCRL-JC-105269, Lawrence Livermore National Laboratory.

DeBar, R.B., 1974. Fundamentals of the KRAKEN code. Technical Report UCIR-76, Lawrence Livermore Laboratory.

Delpino, S., 2011. Metric-based mesh adaptation for 2D Lagrangian compressible flows. J. Comput. Phys. 1793–1821.

Després, B., 2010. Weak consistency of the cell-centered Lagrangian GLACE scheme on general meshes in any dimension. Comput. Methods Appl. Mech. Eng. 199 (41–44), 2669–2679, ISSN 0045-7825. http://dx.doi.org/10.1016/j.cma.2010.05.010. http://www.sciencedirect.com/science/article/pii/S0045782510001593.

Després, B., Labourasse, E., 2012. Stabilization of cell-centered compressible Lagrangian methods using subzonal entropy. J. Comput. Phys. 231 (20), 6559–6595, ISSN 0021-9991. http://dx.doi.org/10.1016/j.jcp.2012.04.018. http://www.sciencedirect.com/science/article/pii/S0021999112001994.

Després, B., Mazeran, C., 2003. Symmetrization of Lagrangian gas dynamics in dimension two and multidimensional solvers. Compt. Rend. Acad. Sci. Mécan. 331, 475–480.

Després, B., Mazeran, C., 2005. Lagrangian gas dynamics in two dimensions and Lagrangian systems. Arch. Ration. Mech. Anal. 178, 327–372.

DeWitt, B.S., 1953. A numerical method for two-dimensional Lagrangian hydrodynamics. Lawrence Livermore National Laboratory. UCRL-4250. http://www.computer-history.info/Page2.dir/pages/NumericalMethod.pdf.

Dobrev, V.A., Ellis, T.E., Kolev, T., Rieben, R.N., 2010. Curvilinear finite elements for Lagrangian hydrodynamics. Int. J. Numer. Methods Fluid 65, 1295–1310.

Donea, J., Giuliani, S., Halleux, J.-P., 1982. An arbitrary Lagrangian-Eulerian finite element method for transient dynamic fluid-structure interactions. Comput. Meth. Appl. Mech. Eng. 33, 689–723.

Dukowicz, J.K., 1985. A general, non-iterative Riemann solver for Godunov's method. J. Comput. Phys. 61, 119–137.

Dukowicz, J.K., Baumgardner, J., 2000. Incremental remapping as a transport/advection algorithm. J. Comput. Phys. 160, 318–335.

Dukowicz, J.K., Meltz, B., 1992. Vorticity errors in multidimensional Lagrangian codes. J. Comput. Phys. 99, 115–134.

Dukowicz, J.K., Cline, M.C., Addessio, F.S., 1989. A general topology Godunov method. J. Comput. Phys. 82, 29–63.

Flanagan, D.P., Belytschko, T., 1981. A uniform strain hexahedron and quadrilateral with orthogonal hourglass control. Int. J. Numer. Meth. Eng. 17, 679–706.

Galera, S., Maire, P.-H., Breil, J., 2010. A two-dimensional unstructured cell-centered multi-material ALE scheme using VOF interface reconstruction. J. Comput. Phys. 229, 5755–5787.

Girault, V., Raviart, P.A., 1986. Finite Element Methods for Navier-Stokes Equations. Springer-Verlag, Berin, Heidelberg, ISBN 3540157964.

Godunov, S.K., 1959. A difference scheme for numerical solution of discontinuous solution of hydrodynamic equations. Math. Sbornik 47, 271–306. (In Russian. Translated US Joint Publ. Res. Service, JPRS 7226, 1969).

Godunov, S.K., Zabrodine, A., Ivanov, M., Kraiko, A., Prokopov, G., 1979. Résolution numérique des problèmes multidimensionnels de la dynamique des gaz. MIR, Moscow.

Goudreau, G.L., Hallquist, J.O., 1982. Recent developments in large-scale finite element Lagrangian hydrocode technology. Comput. Meth. Appl. Mech. Eng. 33, 725–757.

Guermond, J.-L., Popov, B., Tomov, V., 2016. Entropy-viscosity method for the single material Euler equations in Lagrangien frame. Comput. Meth. Appl. Mech. Eng. 300, 402–426.

Gurtin, M.E., Fried, E., Anand, L., 2010. The Mechanics and Thermodynamics of Continua. Cambridge University Press, Cambridge, UK, New York.

Hirt, C.W., Amsden, A.A., Cook, J.L., 1974. An arbitrary Lagrangian-Eulerian computing method for all flow speeds. J. Comput. Phys. 14, 227–253. Reprinted in *J. Comput. Phys.*, 135:203-216, 1997.

Holian, K.S., Mandell, D.A., Adams, T.F., Addessio, F.L., Baumgardner, J.R., Mosso, S.J., 1989. MESA: a 3-D computer code for armor-anti-armor applications. In: Proc. of the Supercomputing World Conference, San Francisco, California.

Höller, H., Koskela, A., Dorfi, E., Benger, W., 2014. Artificial viscosity in comoving curvilinear coordinates: towards a differential geometrically consistent implicit advection scheme. Comput. Astrophys. Cosmol. 1, 2.

Hoch, P., Labourasse, E., 2014. A frame invariant and maximum principle enforcing second-order extension for cell-centered ALE schemes based on local convex hull preservation. Int. J. Numer. Methods Fluids 76 (12), 1043–1063, ISSN 1097-0363.

Kolev, T., Rieben, R.N., 2010. A tensor artificial viscosity using a finite element approach. J. Comput. Phys. 228 (22), 8336–8366.

Kurapatenko, V.F., 1967. On difference methods for the equations of hydrodynamics. In: Janenko, N.N. (Ed.), Difference Methods for Solutions of Problems of Mathematical Physics. I. American Mathematical Society, Providence, RI, p. 116.

Landshoff, R., 1955. A numerical method for treating fluid flow in the presence of shocks. Technical Report LA-1930, Los Alamos National Laboratory.

Lascaux, P., 1976. Lectures on numerical methods for time dependent equations. applications to fluid flow problems. Technical Report, Tata Institute of Fundamental Research. Available at www.math.tifr.res.in/publ/ln/tifr52.pdf.

Lipnikov, K., Manzini, G., Shashkov, M., 2014. Mimetic finite difference method. J. Comput. Phys. 257, 1163–1227.

Llor, A., Claisse, A., Fochesato, C., 2016. Energy preservation and entropy in Lagrangian space- and time-staggered hydrodynamic schemes. J. Comput. Phys. 309, 324–349.

Loubère, R., Maire, P.-H., Shashkov, M., Breil, J., Galera, S., 2010. ReALE: a reconnection-based arbitrary-Lagrangian-Eulerian method. J. Comput. Phys. 229, 4724–4761.

Luttwak, G., Falcovitz, J., 2011. Slope limiting for vectors: a novel limiting algorithm. Int. J. Numer. Methods Fluid 65, 1365–1375.

Maire, P.-H., 2009a. A high-order cell-centered Lagrangian scheme for compressible fluid flows in two-dimensional cylindrical geometry. J. Comput. Phys. 228, 6882–6915.

Maire, P.-H., 2009b. A high-order cell-centered Lagrangian scheme for two-dimensional compressible fluid flows on unstructured meshes. J. Comput. Phys. 228, 2391–2425.

Maire, P.-H., 2011. Contribution to the numerical modeling of Inertial Confinement Fusion. Habilitation à Diriger des Recherches, Université de Bordeaux. Available at https://tel.archives-ouvertes.fr/tel-00589758/.

Maire, P.-H., Nkonga, B., 2009. Multi-scale Godunov-type method for cell-centered discrete Lagrangian hydrodynamics. J. Comput. Phys. 228 (3), 799–821.

Maire, P.-H., Abgrall, R., Breil, J., Ovadia, J., 2007. A cell-centered Lagrangian scheme for two-dimensional compressible flow problems. SIAM J. Sci. Comput. 29 (4), 1781–1824.

Maire, P.-H., Abgrall, R., Breil, J., Loubère, R., Rebourcet, B., 2013. A nominally second-order cell-centered Lagrangian scheme for simulating elastic-plastic flows on two-dimensional unstructured grids. J. Comput. Phys. 235, 626–665.

Morgan, N.R., Kenamond, M.A., Burton, D.E., Carney, T.C., Ingraham, D.J., 2013. An approach for treating contact surfaces in Lagrangian cell-centered hydrodynamics. J. Comput. Phys. 250, 527–554, ISSN 0021-9991. http://dx.doi.org/10.1016/j.jcp.2013.05.015. http://www.sciencedirect.com/science/article/pii/S002199911300346X.

Morgan, N.R., Waltz, J.I., Burton., D.E., Charest, M.R., Canfield, T.R., Wohlbier, J.G., 2015. A point-centered arbitrary Lagrangian Eulerian hydrodynamic approach for tetrahedral meshes. J. Comput. Phys. 290, 239–273, ISSN 0021-9991. http://dx.doi.org/10.1016/j.jcp.2015.02.024. http://www.sciencedirect.com/science/article/pii/S0021999115000947.

Pais, V.A., Caruso, A., 1990. A collisional method to correct the mesh distortion in two-dimensional Lagrangian hydrocodes. Comput. Phys. Commun. 58, 55–62.

Pracht, W.E., 1975. Calculating three-dimensional fluid flows at all speeds with an Eulerian-Lagrangian computing mesh. J. Comput. Phys. 17, 132–159.

Richtmyer, R.D., Morton, K.W., 1967. Difference Methods for Initial Value Problems, second ed. Interscience Publishers, New York.

Samarskii, A.A., Tishkin, V.F., Favorskii, A.P., Shashkov, M.J., 1981. Operational finite difference schemes. Diff. Equ. 17, 854–862.

Shashkov, M., 1996. Conservative Finite-Difference Methods on General Grids. CRC Press, Boca Raton, FL.

Sofronov, I.D., Rasskazova, V.V., Nesterenko, L.V., 1984. The use of nonregular nets for solving two-dimensional non-stationary problems in gas dynamics. In: Yanenko, N.N., Shokin, Y.I. (Eds.), Numerical Methods in Fluid Dynamics. MIR, Moscow, pp. 82–121.

Trulio, J.G., Trigger, K.R., 1961. Numerical solution of the one-dimensional lagrangian hydrodynamics equations. Technical Report UCRL-6267, Lawrence Radiation Laboratory.

Tscharnuter, W.M., Winkler, K.H., 1979. A method for computing self-gravitating gas flows with radiation. Comput. Phys. Commun. 18, 171–199.

Vilar, F., Maire, P.-H., Abgrall, R., 2014. A discontinuous Galerkin discretization for solving the two-dimensional gas dynamics equations written under total Lagrangian formulation on general unstructured grids. J. Comput. Phys. 276, 188–234.

von Neumann, J., 1944. Proposal and analysis of a new numerical method for the treatment of hydrodynamical shock problems. Collected works, 6, 361–379.

von Neumann, J., Richtmyer, R.D., 1950. A method for the numerical calculations of hydrodynamical shocks. J. Appl. Phys. 21, 232–238.

Wilkins, M.L., 1999. Computer Simulation of Dynamic Phenomena. Springer-Verlag, Berlin, Heidelberg.

Yanenko, N.N., Fedotova, Z.I., Tusheva, L.A., Shokin, Y., 1983. Classification of difference schemes of gas dynamics by the method of differential approximation. Comput. Fluids 11 (3), 187–206.

Youngs, D.L., 1982. Time dependent multi-material flow with large fluid distortion. In: Morton, K.W., Baines, M.J. (Eds.), Numerical Methods for Fluid Dynamics. Academic Press, New York, pp. 274–285.

Chapter 14

High-Order Mass-Conservative Semi-Lagrangian Methods for Transport Problems

J.-M. Qiu[1]
[1]*University of Houston, Houston, United States*

Chapter Outline

ABSTRACT

Semi-Lagrangian methods have been under great development over the past few decades with many applications in fluid and kinetic simulations, weather forecasting, interface tracking, etc. They are designed based on a fixed set of computational mesh with information propagating along characteristic curves, taking advantages of high spatial resolution in an Eulerian approach and computational efficiency with large time-stepping in a Lagrangian method for transport problems.

In this chapter, we first give an overview of *high-order mass-conservative* semi-Lagrangian methods with various spatial discretization techniques. We present two formulations of semi-Lagrangian schemes: one is the semi-Lagrangian finite difference weighted essentially nonoscillatory scheme and the other is the semi-Lagrangian discontinuous Galerkin method. Both methods preserve the mass conservation and can be naturally designed to be of high-order spatial accuracy for high-dimensional problems via dimensional splitting. We briefly discuss limiting techniques for preserving positivity and maximum principle when such properties are desired. Extensions of high-order mass-conservative schemes to 2-D without dimensional splitting are still

Handbook of Numerical Analysis, Vol. 17. http://dx.doi.org/10.1016/bs.hna.2016.06.002

considered an open research topic. We discuss possible strategies and the associated difficulties in numerical stability or efficient implementation. Finally, we present several standard linear test examples and nonlinear applications (e.g. the Vlasov–Poisson system and incompressible Euler system) to showcase performance of semi-Lagrangian methods.

AMS Classification Code: 65

Keywords: Semi-Lagrangian, WENO, Discontinuous Galerkin, Mass conservation, Positivity preserving, Stability, Vlasov–Poisson, Navier–Stokes equations

1 INTRODUCTION

The semi-Lagrangian (SL) methodology has been under great development for more than three decades. It has been shown in many applications to be advantageous, compared with traditional Eulerian and Lagrangian approaches. In fact, it is designed to take advantage of both approaches. An Eulerian scheme, based on a fixed set of numerical mesh, is formulated by directly discretizing the PDE at a fixed spatial location. Such schemes, due to the Courant–Friedrichs–Lewy (CFL) condition (i.e. numerical domain of dependence should include the physical domain of dependence), often have quite restrictive time step constraint. Lagrangian scheme is designed by following macroparticle trajectories. The scheme has no CFL time step constraint. However, the initially regularly spaced macroparticles often got distorted after a few time steps, leading to either an overclustering or sparsely spaced grid that could no longer support important features of the solution. The SL method inherits the high spatial resolution from an Eulerian method based on a fixed set of computational grid and avoids the particle remeshing procedure in a Lagrangian method. It traces characteristics to propagate information as in a Lagrangian method in order to avoid the CFL time step constraint in an Eulerian method. The method has been very popular in various applications such as passive advection in weather forecasting (Staniforth and Côté, 1991; Lin and Rood, 1996; Yabe et al., 2001; Lauritzen et al., 2010; Guo et al., 2014), Navier–Stokes simulations of fluid dynamics (Pironneau, 1982; Süli, 1988; Xiu and Karniadakis, 2001), kinetic Vlasov–Poisson (Filbet et al., 2001; Filbet and Sonnendrücker, 2003; Besse and Sonnendrücker, 2003; Qiu and Christlieb, 2010; Crouseilles et al., 2010; Rossmanith and Seal, 2011; Guo and Qiu, 2013) and Boltzmann simulations (Russo et al., 2012), Hamilton–Jacobi equation (Falcone and Ferretti, 2013), interface front tracking (Enright et al., 2005; Strain, 1999), etc.

In this chapter, we focus on SL schemes for transport problems. A scheme is considered to be SL, if it has the following three components:

1. A solution space: the solution space can be point values, integrated mass (cell averages) or a piecewise continuous or discontinuous polynomial functions living on a set of computational grid, corresponding to an SL finite difference scheme (Sonnendrücker et al., 1999; Carrillo and Vecil, 2007;

Qiu and Shu, 2011a), an SL finite volume scheme (Lin and Rood, 1996; Lauritzen et al., 2010; Filbet et al., 2001; Crouseilles et al., 2010), a characteristic or Lagrangian–Galerkin method (Pironneau, 1982; Morton et al., 1988), SL spectral element method (Giraldo, 1998; Xiu and Karniadakis, 2001) and SL discontinuous Galerkin method (Rossmanith and Seal, 2011; Qiu and Shu, 2011c; Guo et al., 2014), respectively.

2. In each of the time step evolution from t^n to t^{n+1}, information is propagated along characteristics. Characteristics tracing is the main task in this step. For a passive advection with a prescribed velocity field, an ODE solver such as a Runge–Kutta (RK) method is usually employed for tracing characteristics curves with high-order temporal accuracy for long-time integration. For a nonlinear problem such as those in fluid and kinetic simulations, tracing of characteristics curves could be a more challenging task. It may take more computational effort to track characteristics with desired precision (Qiu and Russo, 2016), which is beyond the scope of this paper.
3. The evolved solution is projected back onto the solution space, updating the numerical solution at t^{n+1}. There are various ways of doing so depending on the solution space. For example, a finite difference scheme can be formulated based on a differential form of the equation, with backward characteristics tracing as well as high-order spatial interpolation techniques such as cubic spline (Sonnendrücker et al., 1999) and weighted essentially nonoscillatory (WENO) interpolation (Carrillo and Vecil, 2007; Qiu and Christlieb, 2010). A finite volume scheme can be formulated based on an integral form of the equation, with backward characteristics tracing and polynomial reconstruction procedures (Filbet et al., 2001; Crouseilles et al., 2010; Qiu and Shu, 2011b). For a discontinuous Galerkin method, the corresponding moments or nodal values of polynomials have to be updated via characteristics tracing (Rossmanith and Seal, 2011; Qiu and Shu, 2011c) based on a weak formulation of the problem.

For physical relevant long-time simulations, the following properties of SL methods are highly desired. It is indeed very challenging to design schemes with all properties mentioned below, hence the very active research in this direction:

- Consistency and high-order accuracy in space and in time. High-order accuracy in space is usually realized by using different interpolation and reconstruction techniques or by using piecewise polynomials as solution space. High-order temporal accuracy lies on high-order characteristics tracing, either by a direct application of high-order ODE integrators for a passive advection or for a careful design of characteristics tracking in the nonlinear setting (Qiu and Russo, 2016).
- Mass conservation. For certain applications, e.g., in weather forecasting and in kinetic simulations, mass conservation is an indispensable property especially when the solution becomes under-resolved by the existing mesh or for long-time simulations. For finite volume and finite element

methods, it is more natural to enforce the discrete mass conservation, by working with an integral form for finite volume (Lin and Rood, 1996) or by using the weak formulation with the constant being in the test function space for finite element methods (Guo et al., 2014). However, for a finite difference scheme, enforcing mass conservation is less natural and more challenging in the scheme design. In fact, many existing SL finite difference schemes rely on characteristics tracing and local interpolation; they do not necessarily have mass conservation.

- Stability with large time-stepping for efficiency. The main advantage of SL methods, compared with Eulerian ones, is the efficiency due to large time-stepping sizes. However, it was found in many different settings that, even though the numerical domain of dependence includes the physical domain of dependence, instabilities could happen due to inexact evaluation of spatial integral (Morton et al., 1988), or due to the large error occurs in evaluating the time integral of flux functions. In 1-D settings, such instability can be partially cured by taking exact integrations, e.g., see the work in Qiu and Shu (2011a,c). However, for high-dimensional problems, the stability issue is more pronounced and there is limited work to understand and resolve such stability issues when time-stepping size is extra large.
- Positivity or maximum principle preserving for robustness. It is of practical interest to preserve either maximum principle or positivity of the numerical solution at the discrete level, whenever such concepts are relevant. Recently, there have been different limiting techniques for such purpose (Filbet et al., 2001; Crouseilles et al., 2010; Qiu and Shu, 2011c; Xiong et al., 2014).
- Preservation of the uniform flow. For incompressible velocity field: the solution being constant is a steady state solution. However, not all the numerical schemes would numerically preserve such steady state. Note that if the scheme is maximum principle preserving, the uniform flow will be preserved numerically for incompressible flow field. For 2-D rotational problem, the preservation of uniform flow can be done by enforcing maximum principle. However, for the swirling deformation flow (LeVeque, 1996), the preservation is more computationally challenging especially when the mass conservation needs to be enforced.

There have been many review papers and books devoted to SL schemes (McDonald, 1991; Staniforth and Côté, 1991; Lin and Rood, 1996; Falcone and Ferretti, 2013). The chapter differs from other review papers, by emphasizing high-order and mass-conservative aspects of the algorithm, which are important when a numerical mesh cannot fully resolve solution structures and in long-time integration. The rest of the chapter is organized as follows. We present two versions of mass-conservative SL methods: one in the finite difference framework by approximating the differential form of a conservative equation and the other is in the finite element framework by working with a weak formulation. We comment on limiting procedures for positivity or maximum principle preserving properties, stability issues as well as extensions to high-dimensional problems. We present several standard test cases on passive advection and on nonlinear kinetic and fluid simulations.

2 MASS-CONSERVATIVE SL SCHEMES

In this chapter, we focus on high-order spatial discretization in the SL scheme formulation, as tracking characteristics accurately in time could be problem dependent and should be discussed case by case. For schemes with first- or second-order spatial accuracy, the SL finite volume scheme is a very popular choice and the mass-conservation property can be enforced naturally in the scheme design (see Lin and Rood, 1996; Lauritzen et al., 2010 for its successful application to global transport simulations and Filbet et al., 2001; Crouseilles et al., 2010 to Vlasov–Poisson simulations). When higher than second-order spatial accuracy is considered for high-dimensional problem, the evaluation of fluxes along cell boundaries with a high-order reconstruction procedure could be more involved in implementation and cost more computational time in a finite volume scheme. In contrast, the SL finite difference scheme offers some convenience for high-dimensional implementation by working with point values allowing dimension-by-dimension interpolations or dimensional splitting. Meanwhile, the mass conservation in a finite difference scheme can be enforced via a scheme in flux-difference form. In Section 2.1, we present the formulation of an SL finite difference WENO scheme. In Section 2.2, we present the SL discontinuous Galerkin method, which achieves high-order spatial accuracy with piecewise polynomial functions as its solution space. In our presentations, we start from a 1-D problem with a variable coefficient in the following conservative form,

$$\phi_t + (a(x,t)\phi)_x = 0. \tag{1}$$

We consider backward characteristics tracing, which is done by numerically integrating the following ODE

$$\frac{dX(t)}{dt} = a(x,t), \tag{2}$$

given arrival points from a set of numerical mesh. We note that when $a(x, t)$ varies, the solution *does not* stay constant along characteristics. In fact, many of the SL schemes in the literature are designed based on an advective form, for which the solution stays constant along characteristics. We recognize the importance of mass-conservation property, thus working with a conservative form (1). We also note the positivity-preserving property of the solution, if the initial condition is positive. In the special case when $a(x, t) \equiv a$ is a constant, then there is the maximum principle preserving property, i.e., the solution would stay within the bounds of initial conditions. We will discuss limiting techniques for preserving positivity or maximum principle of numerical solutions. The 1-D algorithms can be extended to high-D simulations by dimensional splitting. When applicable, we will also comment on the associated issues and challenges when generalized to high-D problems without splitting.

2.1 SL Finite Difference WENO Scheme

Usually the SL finite difference scheme consists of two main steps: characteristics tracing in time and high-order spatial interpolations, such as cubic spline (Sonnendrücker et al., 1999) and WENO (Carrillo and Vecil, 2007; Qiu and Russo, 2016). Many of these schemes do not have the mass-conservation property, which could be very important when the solutions become underresolved and for long-time simulations. Below we present a finite difference scheme with the mass-conservation property for a 1-D problem (Qiu and Shu, 2011a).

We perform a spatial discretization of the 1-D domain by a set of uniform grid points $\{x_i\}_{i=1}^N$. We formulate the SL finite difference scheme by integrating Eq. (1) over $[t^n, t^{n+1}]$,

$$\phi(x,t^{n+1}) = \phi(x,t^n) - \left(\int_{t^n}^{t^{n+1}} a(x,\tau)\phi(x,\tau)d\tau\right)_x.$$

Evaluating the above equation at the grid point x_i gives

$$\phi_i^{n+1} = \phi_i^n - \left(\int_{t^n}^{t^{n+1}} a(x,\tau)\phi(x_i,\tau)d\tau\right)_x = \phi_i^n - \mathcal{F}_x|_{x=x_i} \tag{3}$$

where we let

$$\mathcal{F}(x) = \int_{t^n}^{t^{n+1}} a(x,\tau)\phi(x,\tau)d\tau. \tag{4}$$

We introduce a function $\mathcal{H}(x)$, whose sliding average is $\mathcal{F}(x)$ (Shu, 2009), i.e.

$$\mathcal{F}(x) = \frac{1}{\Delta x}\int_{x-\frac{\Delta x}{2}}^{x+\frac{\Delta x}{2}} \mathcal{H}(\xi)d\xi. \tag{5}$$

Taking the x derivative of the above equation gives

$$\mathcal{F}_x = \frac{1}{\Delta x}\left(\mathcal{H}(x+\frac{\Delta x}{2}) - \mathcal{H}(x-\frac{\Delta x}{2})\right).$$

Therefore Eq. (3) can be written in a conservative form as

$$\phi_i^{n+1} = \phi_i^n - \frac{1}{\Delta x}\left(\mathcal{H}(x_{i+\frac{1}{2}}) - \mathcal{H}(x_{i-\frac{1}{2}})\right), \tag{6}$$

where $\mathcal{H}\left(x_{i+\frac{1}{2}}\right)$ is called the flux function. Notice that, from Eqs (1) to (6) there is no numerical discretization involved yet. Similar to the idea in a finite difference WENO scheme (Shu, 2009), $\mathcal{H}\left(x_{i+\frac{1}{2}}\right)$ can be reconstructed from its neighboring cell averages

$$\bar{\mathcal{H}}_j = \frac{1}{\Delta x}\int_{x_{j-\frac{1}{2}}}^{x_{j+\frac{1}{2}}} H(\xi)d\xi \overset{(5)}{=} \mathcal{F}(x_j), \quad j = i-p, \cdots, i+q. \tag{7}$$

Now the remaining question is how to obtain $\{\mathcal{F}(x_i)\}_{i=1}^N$ (or $\{\bar{\mathcal{H}}_i\}_{i=1}^N$) in an SL fashion by propagating information along characteristics. In Fig. 1, we show a backward characteristic curve emanating from a grid point at time level t^{n+1}, say (x_i, t^{n+1}) with the foot located at (x_i^*, t^n). Since we consider Eq. (1) in a conservative form, along characteristic curves, the solution is not necessary a constant. To resolve this, we consider a region Ω_i bounded by three points (x_i, t^{n+1}), (x_i, t^n) and (x_i^*, t^n) (see Fig. 1). We apply the integral form of Eq. (1) over the region Ω_i,

$$\int_{\Omega_i} \phi_t + (a(x,t)\phi)_x = 0. \tag{8}$$

By the Divergence Theorem, the left-hand side of the above equation can be written in the following explicit form:

$$\begin{aligned}\int_{\Omega_i} \phi_t + (a(x,t)\phi)_x &= \int_{\partial\Omega_i} \phi \cdot n_t + a(x,t)\phi \cdot n_x \\ &= -\int_{x_i^*}^{x_i} \phi(\xi, t^n)d\xi + \int_{t^n}^{t^{n+1}} a(x,\tau)\phi(x_i,\tau)d\tau.\end{aligned} \tag{9}$$

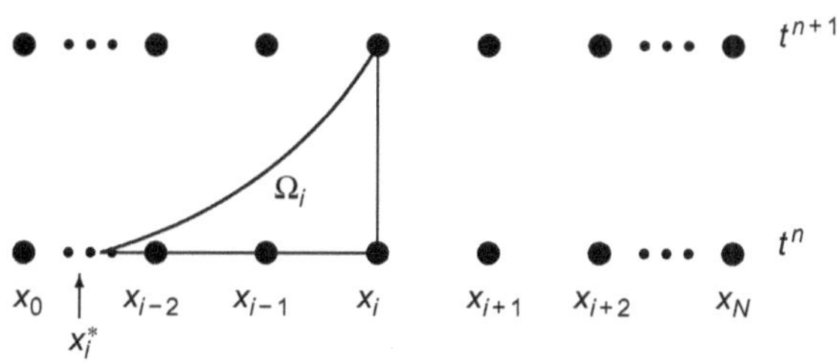

FIG. 1 SL finite difference scheme approximates Eq. (1).

From Eqs (8) and (9),

$$\bar{\mathcal{H}}_i = \mathcal{F}(x_i) = \int_{t^n}^{t^{n+1}} a(x,\tau)\phi(x_i,\tau)d\tau = \int_{x_i^*}^{x_i} \phi(\xi,t^n)d\xi, \qquad (10)$$

where $\int_{x_i^*}^{x_i} \phi(\xi,t^n)d\xi$ can be reconstructed from $\{\phi_i^n\}_{i=1}^N$. Here, the propagation of the information along characteristics is not as explicit as those in the traditional SL finite difference scheme. The benefit is the mass conservation by working with a conservative flux-difference form (6). By applying the Divergence Theorem on Eq. (8), the information on $[x_i^*, x_i] \times t^n$ is propagated over to the region of $\{x_i\} \times [t^n, t^{n+1}]$. Such propagation of information is in the spirit of SL. In summary, an SL finite difference scheme in evolving the solution from t^n to t^{n+1} to approximate Eq. (1) can be designed as follows:

Step 1: At each of the grid point (x_i, t^{n+1}), trace the characteristic backward to (x_i^*, t^n) by numerically integrating Eq. (2).
Step 2: Reconstruct $\{\bar{\mathcal{H}}_i\}_{i=1}^N$ with $\bar{\mathcal{H}}_i \overset{(10)}{=} \int_{x_i^*}^{x_i} \phi(\xi,t^n)d\xi$ from $\{\phi_i^n\}_{i=1}^N$. We use $\mathcal{R}_1$ to denote this reconstruction procedure

$$\bar{\mathcal{H}}_i = \mathcal{R}_1[x_i^*, x_i](\phi_{i-p_1}^n, \cdots, \phi_{i+q_1}^n), \qquad (11)$$

where $i - p_1, \ldots, i + q_1$ indicate the stencil used in the reconstruction, and $\mathcal{R}_1[a,b]$ indicates the reconstruction of $\int_a^b \phi(\xi,t)d\xi$.
Step 3: Reconstruct $\{\mathcal{H}(x_{i+\frac{1}{2}})\}_{i=0}^N$ from $\{\bar{\mathcal{H}}_i\}_{i=1}^N$ as in a standard WENO scheme (Shu, 2009). We use $\mathcal{R}_2$ to denote this reconstruction procedure

$$H(x_{i+\frac{1}{2}}) = \mathcal{R}_2(\bar{\mathcal{H}}_{i-p_2}, \ldots, \bar{\mathcal{H}}_{i+q_2}), \qquad (12)$$

where $i - p_2, \ldots, i + q_2$ indicate the stencil used in the reconstruction.
Step 4: Update the solution $\{\phi_i^{n+1}\}_{i=1}^N$ by Eq. (6) with $H(x_{i\pm\frac{1}{2}})$ computed in the previous step.

The scheme automatically conserves mass locally by working with the flux-difference form (6). We left the details of two WENO reconstruction procedures $\mathcal{R}_1$ and $\mathcal{R}_2$ to Qiu and Shu (2011a), but only highlight a few crucial ingredients in the scheme design and implementation:

- The WENO reconstruction realizes a robust and nonoscillatory capture of discontinuities or under-resolved solutions. The high-order WENO procedures could have been applied to $\mathcal{R}_1$ and $\mathcal{R}_2$ separately. However, the two individual reconstruction steps lead to not only very wide reconstruction stencils but also numerical instability, if the time step is larger than that restricted by the Eulerian CFL. A combination of the two reconstruction procedures with compact stencil is proposed in Qiu and Shu (2011a). It was proved that, with compact stencil, the updated solution ϕ_i^{n+1} depends only locally around the foot of characteristics ϕ_i^* (not the entire interval of

$[x_i^*, x_i]$) and numerical stability is numerically observed for arbitrary large time step evolution.

- In the case of a constant coefficient $a(x, t) \equiv a$, WENO procedures for the combined operator of $\mathcal{R}_1$ and $\mathcal{R}_2$ can be designed. For example, there are WENO3 and WENO5 schemes proposed in Qiu and Shu (2011a). However, when the coefficient $a(x, t)$ is not constant, positive linear weights for optimal WENO accuracy do not exist. Thus, a essentially nonoscillatory (ENO) procedure could be used, e.g., ENO2 and ENO3. One can also use the WENO procedure but with nonlinear weights computed base on some prescribed (but not optimal) linear weights, e.g., WENO2 and WENO3(2). For more details for such reconstruction procedures, we refer to Qiu and Shu (2011a).

2.1.1 Mass Conservation, Maximum Principle and Positivity-Preserving Numerical Stability

The WENO procedures are applied to the reconstruction of flux functions in a flux-difference formulation of the scheme; thus the local mass conservation is automatically preserved. When the advection speed is constant, there is the maximum principle. In Qiu and Shu (2011b), a high-order maximum principle preserving flux limiter is applied to the SL finite difference scheme and further applied to Vlasov applications. It can be shown that the solution to the 1-D problem (1) stays positive, if the initial condition does so. Although there is no theoretical stability result available, numerical stability is observed for arbitrary large time-stepping size. In fact, it was shown in Qiu and Shu (2011a) that the scheme with time step size larger than that of the CFL (CFL > 1) is equivalent to the scheme with time step smaller than CFL (CFL < 1) together with shifting the solution by whole grid points. When the maximum principle preserving limiters are applied, the L^∞ stability is guaranteed (Xiong et al., 2014).

2.1.2 Extension to 2-D Problems

This 1-D SL algorithm can be extended to 2-D problems by dimensional splitting. For incompressible flow fields $\mathbf{a} = (a_1, a_2)$, the conservative form

$$\phi_t + \nabla \cdot (\mathbf{a}\phi) = 0, \quad \text{(conservative form)} \tag{13}$$

is equivalent to the advective form

$$\phi_t + \mathbf{a} \cdot \nabla\phi = 0, \quad \text{(advective form)}. \tag{14}$$

The solution of these equations preserves the total mass and maximum principle. There are two ways of splitting the equation: conservative and advective splitting. For conservative splitting based on (13), we would solve the following two 1-D equations sequentially

$$\phi_t + (a_1\phi)_x = 0, \quad \phi_t + (a_2\phi)_y = 0. \tag{15}$$

Conservative splitting could locally preserve the mass but not the maximum principle, while the advective splitting could not preserve the total mass but some limiting procedures can be used to preserve the maximum principle. Different splittings could be used depends on practical needs. For example, the authors in Huot et al. (2003) showed numerical instabilities when the advective splitting is used without the discrete mass-conservation property, while the mass conservation may not be as essential in interface tracking. There are special cases, e.g., for a rotational problem and the Vlasov–Poisson system, where the conservative and advective splittings are equivalent. In this situation, both maximum principle preserving and local mass-conservation properties can be achieved (Xiong et al., 2014). The operator splitting is known to subject to some splitting errors. Strang splitting is a very popular choice for its efficiency and second-order accuracy. It is done by starting solving one equation with half a time step, followed by solving the other equation for a time step, and finally solving the first equation for another half a time step (Strang, 1968). Efforts have been made to improve the accuracy in operator splitting by using more intermediate stages (Yoshida, 1990; Rossmanith and Seal, 2011) or using the integral deferred correction framework (Christlieb et al., 2014).

Alternatively, one can solve the 2-D problem directly without dimensional splitting. The main difficulty associated with mass-conservative SL direct solvers is the stability. In particular, to preserve the mass conservation, one would work with a flux-difference form. In 1-D, the evaluation of the time integration of flux functions is performed almost exactly via the combination of two reconstruction procedures (11) and (12). In 2-D, the numerical evaluation of flux functions with time integration via quadrature rules would lead to stability issues when the time step size is larger than that from the CFL bound.

2.1.3 Comparison with a Mass-Conservative Finite Volume SL Scheme

To formulate an SL finite volume scheme, similar characteristic tracing and polynomial reconstruction procedures could be applied. For a 1-D problem (1), the finite volume procedure would be simpler, as only one reconstruction procedure, $\mathcal{R}_1$ in Eq. (11), would be involved. For 2-D problems, if the dimensional splitting procedure is applied, at best second-order spatial accuracy can be achieved, as the shearing velocity $\mathbf{a}$ varies within each individual mesh cell. A direct solver for 2-D problems without dimensional splitting, similar to that for the finite difference case, could suffer from stability issues due to inexact evaluation of flux functions. The conservative SL multi-tracer transport scheme proposed in Lauritzen et al. (2010) offers a nice way of resolving the stability issues by transforming the evaluation of area integrals into line integrals. Again, the high-order polynomial reconstruction procedure in 2-D could be computational involved. The DG scheme discussed in the following could be an effective and efficient alternative when higher than second-order spatial accuracy is desired.

2.2 Mass-Conservative SL DG Scheme

The SL DG schemes are designed in a very different fashion compared with finite difference or finite volume methods. In fact, there are different formulations of SL DG schemes based on moment projection, flux-difference form, etc. (Rossmanith and Seal, 2011; Qiu and Shu, 2011c; Restelli et al., 2006; Guo et al., 2014). Many of these methods are equivalent to each other for 1-D problems with constant coefficients. We choose to present the characteristic DG method in Guo et al. (2014) for the 1-D problem in a conservative form (1) based on introducing an adjoint problem for test functions. We assume a spatial discretization of the 1-D domain as nonoverlapping cells (elements) such that $\cup I_j = \cup[x_{j-\frac{1}{2}}, x_{j+\frac{1}{2}}]$ with the mid-point $x_j = \frac{1}{2}(x_{j-\frac{1}{2}} + x_{j+\frac{1}{2}})$ and the mesh size $h_j = \max(x_{j+\frac{1}{2}} - x_{j-\frac{1}{2}})$. A DG solution space is a vector space

$$V_h^k = \{v_h : v_h|_{I_j} \in P^k(I_j)\}, \quad P^k(I_j) \text{ is the set of polynomials with degree up to } k.$$

To update the solution at the time-level t^{n+1} over a cell I_j from the solution at t^n, we follow the weak formulation of characteristic Galerkin method of Childs and Morton (1990) and Russell and Celia (2002). Specifically, we let the test function $\psi(x, t)$ satisfy the adjoint problem with $\forall \Psi \in P^k(I_j)$,

$$\begin{cases} \psi_t + a(x,t)\psi_x = 0, \\ \psi(t = t^{n+1}) = \Psi(x). \end{cases} \tag{16}$$

For the above advective equation, the solution stays constant along a characteristic trajectory. It can be shown that

$$\frac{d}{dt}\int_{I_j(t)} \phi(x,t)\psi(x,t)dx = 0, \tag{17}$$

where $I_j(t)$ is a dynamic interval bounded by characteristics emanating from cell boundaries of I_j at $t = t^{n+1}$. Eq. (17) can be proved by the following:

$$\begin{aligned}
\frac{d}{dt}\int_{I_j(t)} \phi(x,t)\psi(x,t)dx - {} & \frac{dx}{dt}\Big|_{x_{j+\frac{1}{2}}} \phi(x_{j+\frac{1}{2}}(t),t)\psi(x_{j+\frac{1}{2}}(t),t) \\
& - \frac{dx}{dt}\Big|_{x_{j-\frac{1}{2}}} \phi(x_{j-\frac{1}{2}}(t),t)\psi(x_{j-\frac{1}{2}}(t),t) \\
& - \int_{I_j(t)} (a(x,y,t)\phi(x,t))_x \psi(x,t)dx + \int_{I_j(t)} \phi(x,t)\psi_t(x,t)dx \\
= {} & a\phi\psi|_{(x_{j+\frac{1}{2}}(t),y,t)} - a\phi\psi|_{(x_{j-\frac{1}{2}}(t),y,t)} \\
& - \left(a\phi\psi|_{(x_{j+\frac{1}{2}}(t),y,t)} - a\phi\psi|_{(x_{j-\frac{1}{2}}(t),y,t)} \right) \\
& + \int_{I_j(t)} a(x,y,t)\phi(x,t)\psi_x(x,t)dx \\
& - \int_{I_j(t)} \phi(x,t)a(x,y,t)\psi_x(x,t)dx \\
= {} & 0.
\end{aligned}$$

An SL time discretization of (17) leads to

$$\int_{I_j} \phi^{n+1} \Psi dx = \int_{I_j^*} \phi(x,t^n)\psi(x,t^n)dx, \tag{18}$$

where $I_j^* \doteq [x^*_{j-\frac{1}{2}}, x^*_{j+\frac{1}{2}}]$ with $x^*_{j\pm\frac{1}{2}}$ being the foot of trajectory emanating from $(x_{j\pm\frac{1}{2}}, t^{n+1})$ at time t^n. For implementation, one could use a standard monomial basis $\{1, \xi, \cdots, \xi^k\}$ with $\xi = \frac{x - x_j}{h_j}$ for the polynomial space $P^k(I_j)$ and express ϕ^{n+1} as $\sum_{l=0}^{k} c_l^{n+1} \xi^l$ with coefficients $c_l^{n+1}, l = 0, \cdots k$ to be updated. To update the numerical solution ϕ^{n+1} on a cell I_j, we have to evaluate the R.H.S. of Eq. (18) for all test functions Ψ in the monomial basis, by performing the following procedures. We note that one can choose to work with the orthogonal Legendre basis alternatively in implementation:

1. Locate the foot of trajectory $x^*_{j\pm\frac{1}{2}}$ (see Fig. 2 (left)). We numerically solve the trajectory Eq. (2) emanating from the cell boundaries at $(x_{j\pm 1/2}, t^{n+1})$ to the previous time step $(x^*_{j\pm 1/2}, t^n)$ by a high-order numerical integrator such as a classical fourth-order Runge–Kutta method.
2. Let $I_j^* = [x_{j-1/2}, x_{j+1/2}]$. Detect intervals/subintervals within $I_j^* = \cup_l I^*_{j,l}$, which are all the intersections between I_j^* and the grid elements (l is the index for subinterval). For example, in Fig. 2 (left), there are two subintervals: $I^*_{j,1} = [x^*_{j-1/2}, x_{j-1/2}]$ and $I^*_{j,2} = [x_{j-1/2}, x^*_{j+1/2}]$.
3. Locate the $(k + 1)$ local Gauss–Legendre–Lobatto (GLL) points over each $I^*_{j,l}$, which are mapped from the standard GLL points defined on the reference interval $[-1, 1]$ by an affine transformation. We denote them as $x^*_{j,l,ig}$ (ig is the index for GLL points). See the red (dark gray in the print version) circles as four GLL points per subinterval in Fig. 2 (right).
4. Trace trajectories forward in time from $(x^*_{j,l,ig}, t^n)$ to $(x_{j,l,ig}, t^{n+1})$. Especially, similar to the final-value problem above, we use a high-order time integrator to numerically solve an initial value problem (2) with the initial

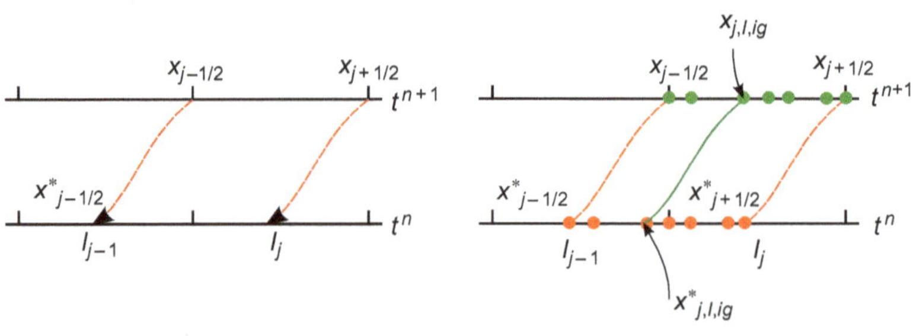

FIG. 2 Schematic showing the 1-D SL DG scheme, as described in the text. Steps i and ii (*left*); steps iii and iv (*right*). Four GLL points per cell are used as an example.

values $x(t^n) = x^*_{j,l,ig}$ (see the green (gray in the print version) curve and circles in Fig. 2 (right)). From the advective form of the adjoint problem (16), one has

$$\psi(x^*_{j,l,ig}, t^n) = \Psi(x_{j,l,ig}).$$

5. Use the GLL quadrature rule to evaluate

$$\int_{I_j^*} \phi(x,t^n)\psi(x,t^n)dx \approx \sum_l \left(\sum_{ig} w_{ig}\phi(x^*_{j,l,ig}, t^n)\Psi(x_{j,l,ig})\Gamma(I^*_{j,l}) \right), \tag{19}$$

with w_{ig} being the quadrature weights for a unit length interval and $\Gamma(I^*_{j,l})$ being the length of interval $I^*_{j,l}$.

6. Finally, find $\phi^{n+1} \in P^k(I_j)$, s.t. (18) is satisfied for all Ψ in the monomial basis of $P^k(I_j)$ with the R.H.S. term evaluated as described above.

2.2.1 Mass Conservation, Maximum Principle and Positivity-Preserving Stability and Error Estimate

In this formulation, the discrete mass-conservation property can be proved by letting the test function $\Psi = 1$ and using the fact that the numerical quadrature rule used in implementation is exact for polynomials up to degree k. Another desired property is the positivity preserving for the numerical solution, knowing that the exact solution stays positive if the initial condition is positive. It can be easily checked that the updated cell average at t^{n+1} (taking $\Psi = 1$ in Eq. 19) stays positive, if the numerical solution (piecewise polynomial function) at t^n is positive. To ensure the numerical solution as piecewise polynomial functions stays positive, we apply a bound preserving (BP) limiter (Zhang and Shu, 2010; Qiu and Shu, 2011c). The procedure of the BP filter can be viewed as ensuring the positivity of the numerical solution by a linear rescaling around the cell averages, with the assumption that the cell averages are positivity preserving. In particular, the numerical solution is modified from $\phi(x)$ to $\tilde{\phi}(x)$ such that it maintains the high-order accuracy of the original approximation, conserves the cell average (mass) and preserves positivity:

$$\tilde{\phi}(x) = \theta(\phi(x) - \overline{\phi}) + \overline{\phi}, \quad \theta = \min\left\{ \left| \frac{\overline{\phi}}{m' - \overline{\phi}} \right|, 1 \right\}, \tag{20}$$

where $\overline{\phi}$ is the cell average of the numerical solution and m' is the minimum of $\phi(x)$ over a given cell. A formal proof can be found in Zhang and Shu (2010) (Lemma 2.4). The proposed SL DG method with the BP filter enjoy the L^1 (mass) conservation, the proof of which can be found in Qiu and Shu (2011c). L^1L^2 stability analysis and error estimate of the SL DG method are provided in Qiu and Shu (2011c).

2.2.2 Extension to 2-D Problems

The above 1-D algorithm can be extended to 2-D via dimension splitting based on a Cartesian mesh. It is important to take the shearing of advection velocities over a cell into account in order to design a scheme that has high-order spatial accuracy. With such considerations, one could consider the DG solution space as $(k+1)^2$ point values at Gaussian nodes per cell. Take the rotational problem $\phi_t + y\phi_x - x\phi_y = 0$ as an example, the 2-D algorithm based on dimensional splitting is the following:

1. In each of the rectangular cell, locate $k + 1$ Gaussian quadrature nodes in both x- and y-directions as (x_{ig}, y_{jg}), $ig, jg = 1, \ldots, k + 1$. For example, see Fig. 3 for the case of $k = 3$.
2. Evolve $\phi_t + y\phi_x - x\phi_y = 0$ by dimensional splitting. For the split equation, e.g., $\phi_t + y\phi_x$, evolve the 1-D problems at different y_{jg}Õs with different advection velocities, see Fig. 3. For each y_{jg}, the $k + 1$ point values are mapped to a P^k polynomial per cell (the unique interpolation polynomial of degree up to k); then the 1-D problem is evolved by the 1-D SL DG scheme; finally the evolved P^k polynomial is mapped back to the $k + 1$ point values at Gaussian nodes to update the solution. Similar procedures can be applied to advection in the y-direction.

The algorithm can be generalized to second order by Strang splitting (Strang, 1968). Note that the BP limiter is applied separately in each direction and the resulting scheme can preserve positivity (see Qiu and Shu, 2011c). On the other hand, it is difficult to design a numerical scheme that preserves a constant field in the dimensional splitting framework, e.g., for the swirling deformation flow presented in the following section.

The 1-D formulation (18) can be directly generalized to 2-D simulations without dimensional splitting. The main difficulty is the implementation: thc upstream of a rectangular cell can be of any shape and evaluation of the area integral as in the R.H.S. of (18) over an irregular shape can be tricky in order

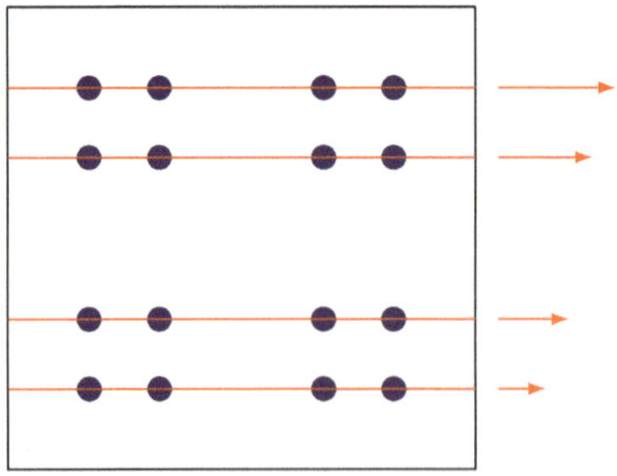

FIG. 3 The 2-D SL DG scheme with P^3 polynomial space via operator splitting.

to preserve mass conservation and high-order spatial accuracy. This is still an open area for research. The ideas in Lauritzen et al. (2010) in transforming the evaluation of area integrals into line integrals may be helpful for developing truly 2-D high-order SL DG algorithms.

2.2.3 Comparison with SL Finite Difference and Finite Volume Schemes

The SL DG scheme is based on a weak formulation. This is different from the differential form for a finite difference scheme and the integral form for a finite volume scheme. They all have mass-conservation properties. On one hand, the DG enjoys the compactness with ease in handling boundary conditions and the flexibility of using nonuniform meshes; on the other hand, the WENO reconstructions in finite difference or finite volume formulations are more robust for capturing discontinuities without oscillations.

3 STANDARD TEST SETS

In this section, we present a set of standard test problems for evaluating the performance of an SL scheme. It is considered a standard practice to compute the errors and orders of convergence for smooth problems and test the robustness in resolving complicated structures when the solution becomes under-resolved. To ensure numerical stability beyond the Eulerian CFL constraint, one wants to check the schemes' performance when the time step size is larger than the CFL restriction. We selectively present a few numerical results from either SL finite difference WENO or SL DG methods for illustration of schemes' performance.

3.1 1-D Problems

Example 1 (1-D linear transport with a constant advection speed).

$$\phi_t + \phi_x = 0. \tag{21}$$

This test problem with smooth initial data is usually used to check the order of convergence for schemes with CFL restricted time step and with extra large time step. In Table 1, we provide errors and orders of convergence for the SL finite difference WENO and SL DG schemes with time step size 2.2 times that from the CFL constraint. Expected orders of convergence are observed. One could also test this example with piecewise continuous initial data such as those in Shu and Osher (1988) to check the robustness of a proposed scheme in capturing discontinuities or under-resolved solutions. We skip to present the numerics here due to the space limit.

Example 2 (1-D transport with variable advection speed).

$$u_t + (\sin(x)u)_x = 0 \quad x \in [0, 2\pi]. \tag{22}$$

TABLE 1 Order of Accuracy for (21) with $\phi(x, t=0) = \sin(x)$ at $T = 20$

—	WENO3		WENO5		DG1		DG2		DG3	
Mesh	*Error*	*Order*	*Error*	*Order*	*Error*	*Order*	*Error*	*Order*	*Error*	*Order*
40	1.03E−2	—	1.18E−5	—	2.70E−3	—	2.60E−5	—	2.50E−7	—
80	2.66E−3	1.96	3.63E−7	5.03	6.48E−4	2.06	3.23E−6	3.00	1.49E−8	4.07
160	6.52E−4	2.02	1.12E−8	5.02	1.54E−4	2.07	4.04E−7	3.00	9.09E−10	4.04

CFL = 2.2. DG1, DG2 and DG3 are for SL DG schemes with piecewise P^1, P^2 and P^3 polynomial, respectively.

The initial condition is $u(x, 0) = 1$ and the boundary condition is periodic. The exact solution is

$$u(x,t) = \frac{\sin\left(2\tan^{-1}\left(e^{-T}\tan\left(\frac{x}{2}\right)\right)\right)}{\sin(x)}.$$

Table 2 presents the errors and orders of convergence for the SL finite difference scheme with various reconstruction procedures for Eq. (22). Table 3 presents the errors and orders of convergence for the SL DG scheme with a set of nonuniform mesh for this example. Expected orders of convergence are observed. For both schemes, we use time step sizes that are 2 to 3 times that of the CFL.

3.2 2-D Linear Passive Advection Problems

We present a collection of 2-D linear passive advection problems on Cartesian meshes. They can be further generalized to sphere geometry for global transport simulations (Lauritzen et al., 2010; Guo et al., 2014).

Example 3 (2-D linear transport).

$$u_t + u_x + u_y = 0. \tag{23}$$

For 2-D linear transport equations, an SL method is essentially a shifting procedure. Since the x-shifting and y-shifting operators commute, there is no dimensional splitting error in time and the spatial error is the dominant error. Similar performance as those for the 1-D example is expected. We skip presenting numerical results, but refer readers to our earlier papers (Qiu and Shu, 2011a,c).

Example 4 (Rigid body rotation).

$$u_t - (yu)_x + (xu)_y = 0, \quad x \in [-\pi, \pi], \quad y \in [-\pi, \pi]. \tag{24}$$

The equation rotates the initial condition with a constant angular velocity. We consider two sets of initial conditions.One is the smooth 'cos bell'

$$u(t=0,x,y) = \begin{cases} \cos^6(r) & \text{if } r < \pi/2 \\ 0 & \text{otherwise} \end{cases} \tag{25}$$

with $r = \sqrt{(x-1)^2 + (y+0.2)^2}$. Numerical solutions after a full revolution from SL DG schemes with piecewise P^3 polynomial are reported in Table 4. The second-order splitting error in time is observed when the spatial mesh is fine enough, i.e., when time error dominates. When the time step is small enough, spatial error dominates and high-order (fourth) spatial accuracy is observed. We note that for spatial mesh 96 × 96 with CFL = 0.25, we do not quite observe fourth-order spatial accuracy. This is because the error is

TABLE 2 Schemes with ENO2, WENO2, ENO3 and WENO3(2) Reconstructions for (22) with $u(x, t = 0) = 1$ at $T = 1$ with CFL = 3

	ENO2		WENO2		ENO3		WENO3(2)	
Mesh	*Error*	*Order*	*Error*	*Order*	*Error*	*Order*	*Error*	*Order*
40	9.83E−3	—	3.81E−3	—	2.26E−3	—	4.61E−4	—
80	3.99E−3	1.30	9.20E−4	2.05	4.95E−4	2.19	2.65E−5	4.12
160	1.17E−3	1.77	2.06E−4	2.16	9.28E−5	2.42	1.27E−6	4.37
320	5.38E−4	1.12	4.77E−5	2.11	2.50E−5	1.89	5.89E−8	4.44
640	1.60E−4	1.75	1.14E−5	2.06	4.25E−6	2.56	4.13E−9	3.83

TABLE 3 Positivity-Preserving SL DG Scheme with Different V_h^k for (22) with $u(x, t = 0) = 1$ at $T = 1$ with CFL = 2.2

L^2 error	$k = 0$		$k = 1$		$k = 2$		$k = 3$	
Mesh	*Error*	*Order*	*Error*	*Order*	*Error*	*Order*	*Error*	*Order*
40	1.77E−1	1.04	1.01E−2	2.38	7.26E−4	2.62	5.17E−5	4.66
80	9.50E−2	0.90	2.76E−3	1.87	9.71E−5	2.90	3.32E−6	3.96
160	4.77E−2	0.99	6.98E−4	1.98	1.18E−5	3.04	2.07E−7	4.00
320	2.44E−2	0.97	1.84E−4	1.92	1.52E−6	2.95	1.42E−8	3.87

The numerical meshes are nonuniform based on a 10% perturbation of uniform meshes.

TABLE 4 Second-Order Split: L^2 Error and Spatial/Temporal Order of Accuracy for Eq. (24) with Initial Data (25) with Different Spatial Meshes and CFLs

CFL	1		0.5		0.25	
Mesh	*Error*	*Spat./ temp. Order*	*Error*	*Spat./ temp. Order*	*Error*	*Spat./ temp. Order*
24 × 24	5.76E−04	—/—	2.41E−04	—/1.25	2.02E−04	—/0.25
48 × 48	1.45E−04	1.99/—	3.29E−05	2.87/2.08	1.17E−05	4.10/1.48
96 × 96	3.71E−05	1.96/—	8.90E−06	1.89/2.05	2.09E−06	2.48/2.08

SL DG method with piecewise P^3 solution space.

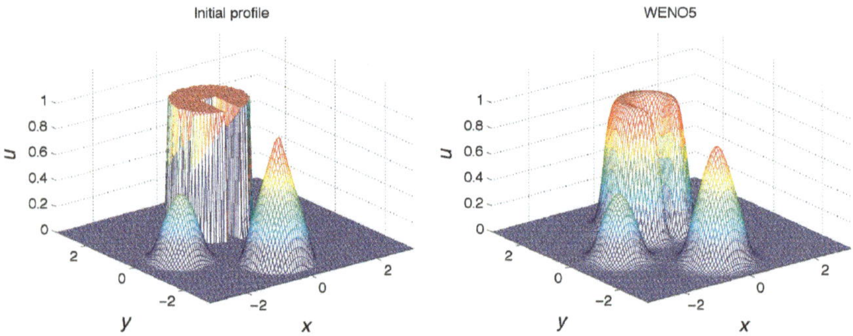

FIG. 4 *Left*: Plots of the initial profile. *Right*: Plots of the SL finite difference WENO5 solution for Eq. (24) with CFL = 2.2 at $T = 12\pi$. The numerical mesh is 100 × 100.

still contaminated by the temporal splitting error. When the CFL = 0.1, the error is reduced to $5.42E - 07$, which confirms fourth (≈ 4.42)-order spatial accuracy.

Another set of initial condition contains a slotted disk, a cone as well as a smooth hump, see the left plot in Fig. 4. It is considered a challenging test. We present the numerical solutions after six full revolutions from the fifth-order SL finite difference WENO schemes with six full revolution in the right plot of Fig. 4. SL schemes with WENO3, WENO5, ENO2, WENO2, ENO3 and WENO3(2) reconstruction operators are plotted in slides benchmarked with exact solution in Fig. 5. With all reconstructions, nonoscillatory capturing of discontinuities is observed. However, schemes with high-order reconstruction, such as WENO5 and WENO3(2), are observed to be less dissipative, therefore outperform schemes with lower order reconstructions. Due to the

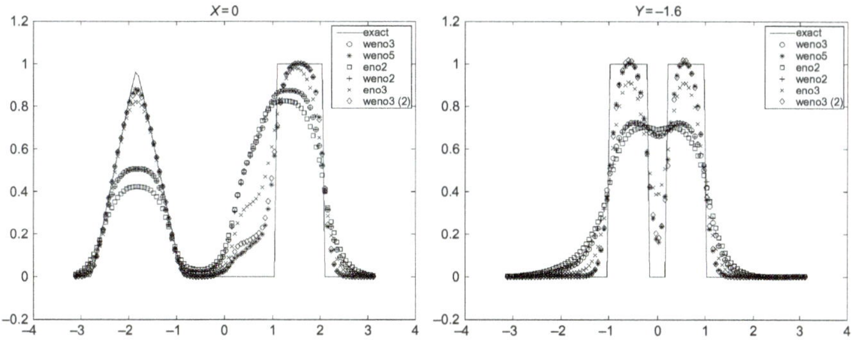

FIG. 5 Plots of slides of numerical solution for Eq. (24) at $X = 0$ (*left*), $Y = -1.6$ (*right*) with CFL $= 2.2$ at $T = 12\pi$. The numerical mesh is 100×100.

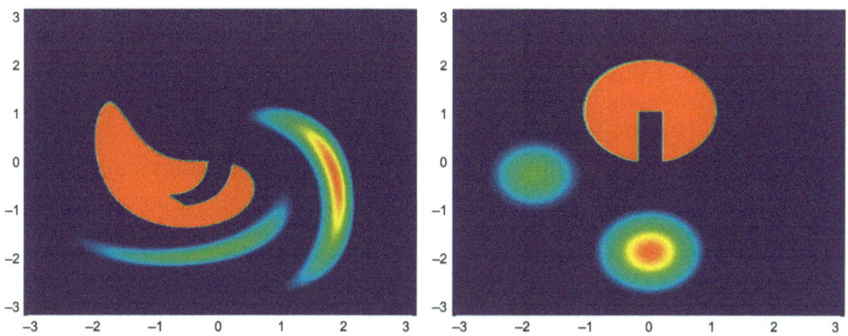

FIG. 6 Contour plots of the numerical solution of SL DG scheme 96×96 with P^3 solution space for Eq. (26). The time step size is 2.2 larger than that of the regular Eulerian CFL restriction. *Left*: Time is 0.75 when the initial profile is greatly deformed; *right*: time is 1.5 when the initial profile is recovered.

space limitation, we skip to present the numerical results from the SL DG scheme, but refer readers to our earlier work in Qiu and Shu (2011c).

Example 5 (Swirling deformation flow (LeVeque, 1996)).

$$
\begin{aligned}
&u_t - \left(\cos^2\left(\tfrac{x}{2}\right)\sin(y)g(t)u\right)_x + \left(\sin(x)\cos^2\left(\tfrac{y}{2}\right)g(t)u\right)_y = 0,\\
&x \in [-\pi, \pi], \quad y \in [-\pi, \pi],
\end{aligned}
\tag{26}
$$

where $g(t) = \cos(\pi t/T)\pi$ with $T = 1.5$. The initial condition is the same as that in Example 4. The initial profile is being deformed largely at time 0.75 and returns to its original shape at time 1.5. We showed in Fig. 6 the solution snapshots at these two times from a well-resolved SL DG simulation. We refer readers to Qiu and Shu (2011a,c) for more simulation results of SL finite difference WENO and SL DG methods.

4 NONLINEAR VLASOV-SL DG AND INCOMPRESSIBLE EULER SYSTEM

4.1 Vlasov–Poisson Simulations

The Vlasov–Poisson (VP) system, from the collisionless plasma application, models the dynamics of charged particles via its distribution function $f(t, \mathbf{x}, \mathbf{v})$. It reads as follows:

$$\frac{\partial f}{\partial t} + \mathbf{v} \cdot \nabla_{\mathbf{x}} f + \mathbf{E}(t,\mathbf{x}) \cdot \nabla_{\mathbf{v}} f = 0, \quad \text{(Vlasov)} \tag{27}$$

$$\mathbf{E}(t,\mathbf{x}) = -\nabla_{\mathbf{x}}\phi(t,\mathbf{x}), \qquad -\Delta_{\mathbf{x}}\phi(t,\mathbf{x}) = \rho(t,\mathbf{x}), \quad \text{(Poisson)} \tag{28}$$

where $\mathbf{E}$ is the electric field and ϕ is the self-consistent electrostatic potential. f describes the probability of finding a charged particle (e.g. an electron) with velocity $\mathbf{v} \in \mathbb{R}^3$ at position $\mathbf{x} \in \mathbb{R}^3$ at time t. It couples to the long range fields via the charge density, $\rho(t,\mathbf{x}) = \int_{\mathbb{R}^3} f(t,\mathbf{x},\mathbf{v})d\mathbf{v} - 1$, where we assume uniformly distributed infinitely massive ions in the background. The VP system describes the following physical process: electrons are moving with velocities $\mathbf{v}$ in a constant ion background; at the same time, their velocities $\mathbf{v}$ are being accelerated or decelerated by self-induced electric field $\mathbf{E}$ determined by the Poisson's equation. For the VP system, several norms are being analytically preserved in time. We hope that our numerical solutions can preserve these analytically conserved quantities as much as possible:

1. L^p norm, $\forall 1 \le p < \infty$: $\frac{d}{dt}\int_v \int_{\mathbf{x}} f(\mathbf{x},\mathbf{v},t)^p d\mathbf{x}d\mathbf{v} = 0$.
2. Entropy: $\frac{d}{dt}\int_v \int_{\mathbf{x}} f(\mathbf{x},\mathbf{v},t)\log(f(\mathbf{x},\mathbf{v},t))d\mathbf{x}d\mathbf{v} = 0$.
3. Energy: $\frac{d}{dt}\left(\int_{\mathbf{v}} \int_{\mathbf{x}} f(\mathbf{x},\mathbf{v},t)\mathbf{v}^2 d\mathbf{x}d\mathbf{v} + \int_x E^2(\mathbf{x},t)d\mathbf{x}\right) = 0$.

One may track relative deviations of these conserved quantities numerically, which is considered a good measurement of the quality of a numerical scheme.

The dimensional splitting for the VP system was originally proposed in Cheng and Knorr (1976). The splitting reduces the high-dimensional nonlinear Vlasov equation into lower-dimensional linear advection equations, allowing direct application of existing SL methods. The splitting decouples the simultaneous 'particle moving' and 'velocity accelerating/decelerating' processes as two separate processes: first particles are moving with constant speed $\mathbf{v}$, and then the particles do not move while their velocities are accelerated/decelerated by the self-induced electric field determined by Poisson's equation. The time splitting form of Eq. (27) is,

$$\frac{\partial f}{\partial t} + \mathbf{v} \cdot \nabla_{\mathbf{x}} f = 0, \tag{29}$$

$$\frac{\partial f}{\partial t} + \mathbf{E}(t,\mathbf{x}) \cdot \nabla_{\mathbf{v}} f = 0. \tag{30}$$

Standard test cases for the one-dimensional VP system include the Landau damping and two-stream instabilities. Boundary conditions are periodic in the x-direction and zero in the v-direction. We selectively present a few of our numerical results from positivity-preserving SL DG schemes. For results from many other schemes, we refer to Sonnendrücker et al. (1999), Filbet et al. (2001), Filbet and Sonnendrücker (2003), Crouseilles et al. (2010), Qiu and Christlieb (2010), Qiu and Shu (2011b), Rossmanith and Seal (2011), Qiu and Shu (2011c), Christlieb et al. (2014), Guo and Qiu (2013) and Xiong et al. (2014).

4.1.1 Landau Damping

$$f(x,v,t=0)=\frac{1}{\sqrt{2\pi}}(1+\alpha\cos(kx))\exp\left(-\frac{v^2}{2}\right), \tag{31}$$

with $k=0.5$ and $v_{max}=5$. For weak Landau damping $\alpha=0.01$ and for strong Landau damping $\alpha=0.5$. For the weak Landau damping, it is known from linearized analysis that the electric field will be damped exponentially fast in time. This is verified numerically (see the time evolution of the L^2 norm of the electric field as presented in the left plot of Fig. 7). The correct damping of the electric field is numerically observed benchmarked with the theoretical value $\gamma=0.1533$ (the solid line in the same plots) up to some time. The deviation of electric field damping at a later time is mainly due to insufficient numerical resolution (see discussions in Cheng et al., 2013). For the Strong Landau damping, the time evolution of the electric field is plotted in the right plot of Fig. 7. The numerical results, from SL DG schemes with different polynomial spaces, are observed to be consistent with each other. The initial damping rate of the electric field ($t<10$) is observed to be linear and is around $\gamma=0.0846$. For these examples, the time evolution of theoretically preserved L^P norm, energy and entropy could be plotted to evaluate the performance of different schemes; we refer readers to Qiu and Shu (2011c) for details.

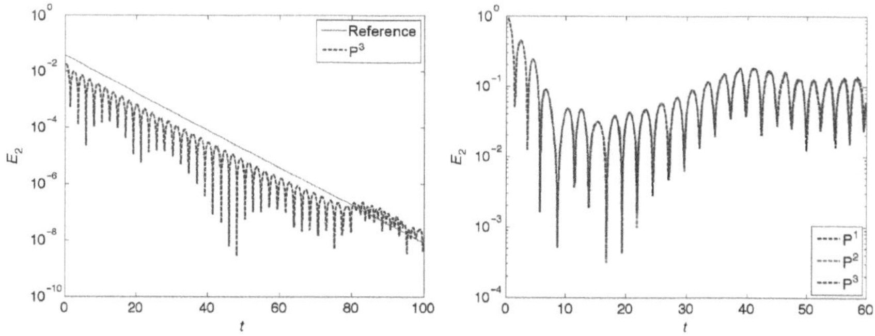

FIG. 7 Landau damping: time evolution of electric field in L^2 norm for weak Landau damping (*left*) and strong Landau damping (*right*).

4.1.2 *Two-Stream Instability (Filbet and Sonnendrücker, 2003)*

$$f(x,v,t=0)=\frac{2}{7\sqrt{2\pi}}(1+5v^2)\\(1+\alpha((\cos(2kx)+\cos(3kx))/1.2+\cos(kx)))\exp\left(-\frac{v^2}{2}\right), \tag{32}$$

where $\alpha = 0.01$, $k = 0.5$ and the length of the domain in the x direction is $L=\frac{2\pi}{k}$ and in the v direction $v_{max} = 5$. Fig. 8 shows numerical solutions of phase space profiles at $T = 53$ from the SL DG scheme with different solution spaces V_h^k. Consistent numerical results are observed. The higher degree polynomials in the solution space, the better resolution is shown. Numerical solution from SL DG scheme with P^1, but with refined mesh is also shown as the reference solution in Fig. 8, indicating the convergence of the numerical solution. Fig. 9 shows relative deviations of discrete L^1 norm, L^2 norm,

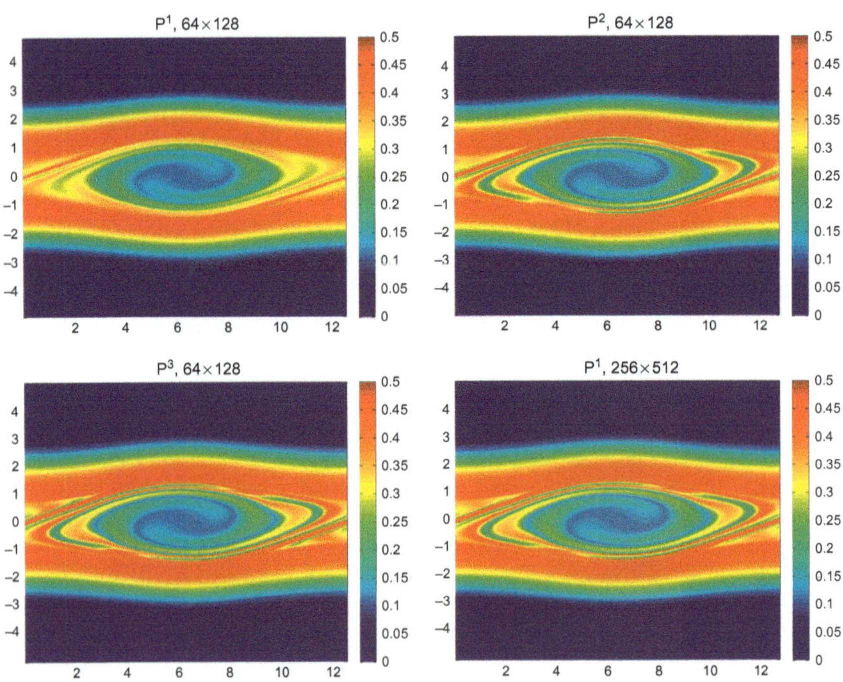

FIG. 8 Phase space plots of the two-stream instability at $T = 53$ using SL DG with P^1 (*upper left*), P^2 (*upper right*), P^3 (*lower left*) with mesh 64 × 128 and CFL = 3 for all test cases. The *lower right figure* is a reference solution produced by the scheme with P^1 solution space but with refined mesh 256 × 512.

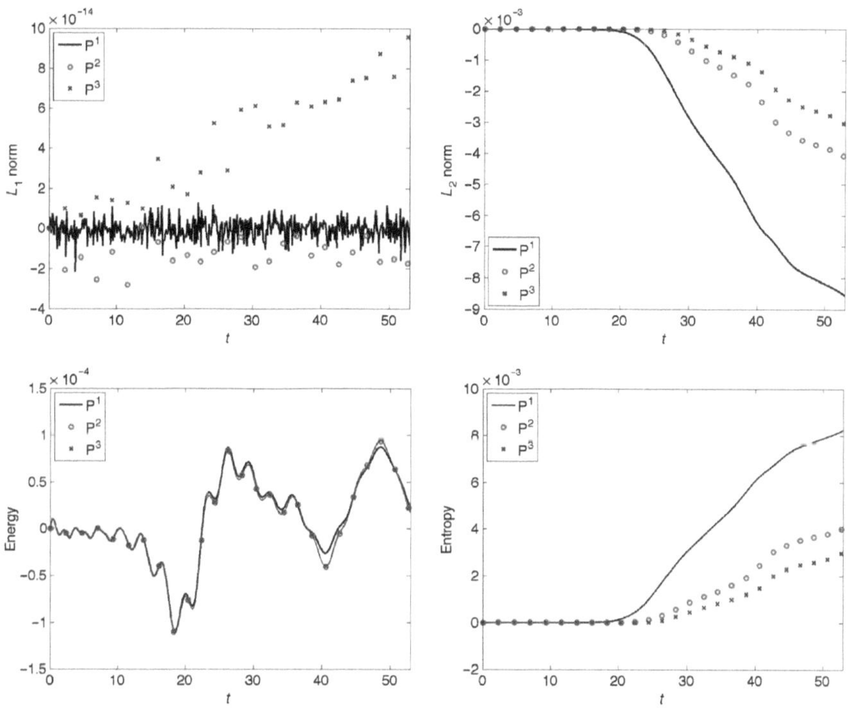

FIG. 9 Two-stream instability: the relative deviations of L^1 (*upper left*) and L^2 (*upper right*) norms as well as the discrete kinetic energy (*lower left*) and entropy (*lower right*) of the numerical solution from the corresponding initial values in time.

kinetic energy and entropy in time. It is observed that all the schemes are conservative in the L^1 norm due to the mass conservation and positivity-preserving properties. Schemes with higher oder in general perform better in preserving relevant physical norms. Note that relative deviation is defined to be the deviation away from the corresponding initial value divided by the magnitude of the initial value.

4.2 Guiding Center Model for a Kelvin–Helmholtz Instability

The guiding center model describes highly magnetized plasma in the transverse plane of a tokamak (Crouseilles et al., 2010). We consider the equation

$$\rho_t + \mathbf{E}^{\perp} \cdot \nabla \rho = 0, \tag{33}$$

where ρ is the particle density function, and $\mathbf{E}^{\perp} = (-E_y, E_x)$ is the electrostatic field satisfying the Poisson's equation

$$\Delta \Phi = \rho, \quad \mathbf{E} = -\nabla \Phi. \tag{34}$$

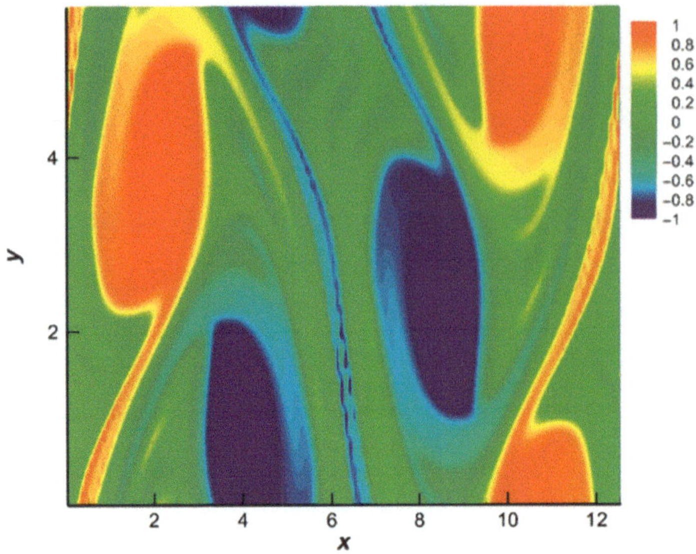

FIG. 10 Contour plots of the numerical solutions for the Kelvin–Helmholtz instability. $N_x \times N_y = 128 \times 128$.

We consider the initial condition $\rho(x,y,0) = \sin(y) + 0.015\cos(kx)$ and periodic boundary conditions. We let $k = 0.5$, thereby creating a Kelvin–Helmholtz instability. In the simulations, we set a mesh as $N_x \times N_y = 128 \times 128$. We use a third-order SL finite difference WENO scheme as a base scheme to achieve third-order spatial accuracy. The contour profile of the numerical solution at time $T = 40$ is shown in Fig. 10. Here we make the time step size small enough so that the spatial error is the dominant error.

4.3 2-D Incompressible Euler (Bell et al., 1989)

The incompressible Euler equation in vorticity stream-function formulation can be solved by dimensional split SL schemes. The system reads

$$\omega_t + \nabla \cdot (\mathbf{u}\omega) = 0, \quad x \in [0, 2\pi], \quad y \in [0, 2\pi], \tag{35}$$

where $\mathbf{u} = \nabla^{\perp}\Phi = (-\partial_y\Phi, \partial_x\Phi)$ with Φ satisfies the Poisson's equation $\Delta\Phi = \omega$. The boundary conditions are periodic. Two sets of initial condition are considered. One is

$$\omega(x,y,0) = \begin{cases} \delta\cos(x) - \dfrac{1}{\rho}\operatorname{sech}^2((y - \pi/2)/\rho), & y \le \pi, \\ \delta\cos(x) + \dfrac{1}{\rho}\operatorname{sech}^2((3\pi/2 - y)/\rho), & y > \pi, \end{cases} \tag{36}$$

where $\delta = 0.05$ and $\rho = \dfrac{\pi}{15}$, and the other is

$$\omega(x,y,0) = \begin{cases} -1, & x \in \left[\frac{\pi}{2}, \frac{3\pi}{2}\right],\ y \in \left[\frac{\pi}{4}, \frac{3\pi}{4}\right], \\ 1, & x \in \left[\frac{\pi}{2}, \frac{3\pi}{2}\right],\ y \in \left[\frac{5\pi}{4}, \frac{7\pi}{4}\right], \\ 0, & \text{otherwise.} \end{cases} \tag{37}$$

For the first set of initial condition (36), Fig. 11 gives the numerical solutions of SL finite difference scheme coupled with WENO2 (upper row) and WENO3 (lower row) reconstruction operators at $T = 6$ (left) and $T = 8$ (right) with numerical mesh 243×243. We let the time step size equals the Eulerian CFL for accuracy (not stability). For the second set of initial condition (37), Fig. 12 gives the numerical solution at $T = 5$ (left) and $T = 10$ (right) from the schemes with the same simulation parameters. The simulation results are consistent when benchmarked with the results in Weinan and Shu (1994), Liu and Shu (2000) and Zhang and Shu (2010). The solution structure is observed to be slightly better resolved with higher order reconstruction operators.

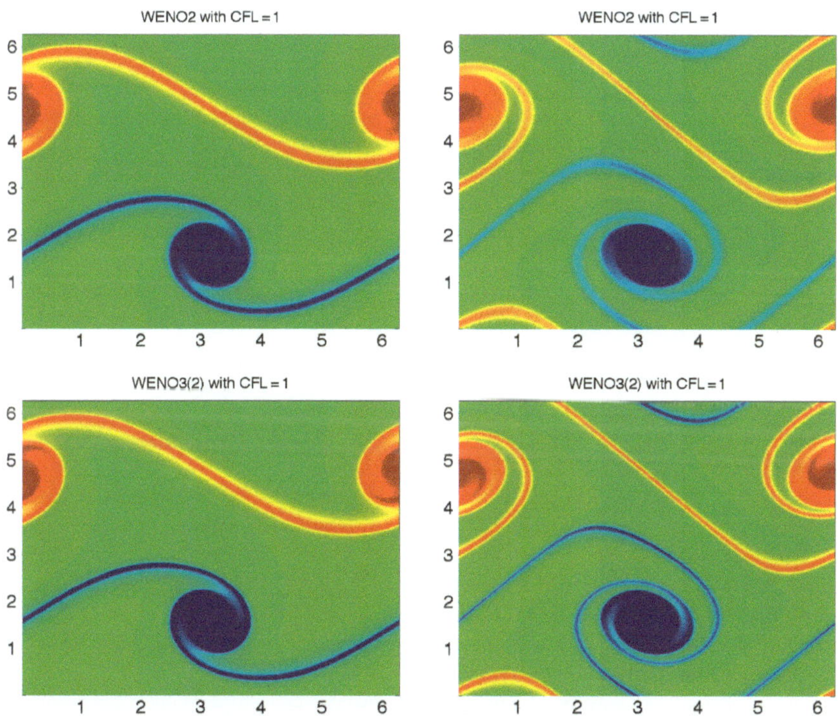

FIG. 11 Plots of the numerical solution of Eq. (35) with initial condition (36) at $T = 6$ (*left*) and $T = 8$ (*right*). The numerical mesh is 243×243 and CFL = 1. The SL scheme is coupled with WENO2 (*upper row*) and WENO3(2) (*lower row*) reconstruction operators.

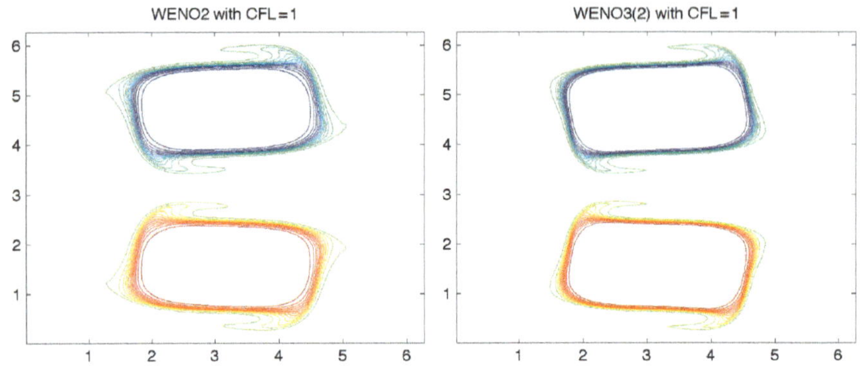

FIG. 12 Plots of the numerical solution of Eq. (35) with initial condition (37) at $T = 10$. The numerical mesh is 243×243 and CFL $= 1$. The SL scheme is coupled with WENO2 (*upper row*) and WENO3(2) (*lower row*) reconstruction operators. Being plotted are 30 equally spaced contours curves for $\omega \in [-1.1, 1.1]$.

ACKNOWLEDGEMENTS

Research supported by NSF DMS-1217008 and DMS-1522777.

REFERENCES

Bell, J.B., Colella, P., Glaz, H.M., 1989. A second-order projection method for the incompressible Navier-Stokes equations. J. Comput. Phys. 85 (2), 257–283.

Besse, N., Sonnendrücker, E., 2003. Semi-Lagrangian schemes for the Vlasov equation on an unstructured mesh of phase space. J. Comput. Phys. 191 (2), 341–376.

Carrillo, J.A., Vecil, F., 2007. Nonoscillatory interpolation methods applied to Vlasov-based models. SIAM J. Sci. Comput. 29 (3), 1179–1206.

Cheng, C.Z., Knorr, G., 1976. The integration of the Vlasov equation in configuration space. J. Comput. Phys. 22 (3), 330–351.

Cheng, Y., Gamba, I.M., Morrison, P.J., 2013. Study of conservation and recurrence of Runge-Kutta discontinuous Galerkin schemes for Vlasov-Poisson systems. J. Sci. Comput. 56 (2), 319–349.

Childs, P.N., Morton, K.W., 1990. Characteristic Galerkin methods for scalar conservation laws in one dimension. SIAM J. Numer. Anal. 27 (3), 553–594.

Christlieb, A., Guo, W., Morton, M., Qiu, J.M., 2014. A high order time splitting method based on integral deferred correction for semi-Lagrangian Vlasov simulations. J. Comput. Phys. 267, 7–27.

Crouseilles, N., Mehrenberger, M., Sonnendrücker, E., 2010. Conservative semi-Lagrangian schemes for Vlasov equations. J. Comput. Phys. 229 (6), 1927–1953.

Enright, D., Losasso, F., Fedkiw, R., 2005. A fast and accurate semi-Lagrangian particle level set method. Comput. Struct. 83 (6), 479–490.

Falcone, M., Ferretti, R., 2013. Semi-Lagrangian approximation schemes for linear and Hamilton-Jacobi equations. vol. 133. SIAM. ISBN: 978-1-61197-304-4. http://dx.doi.org/10.1137/1.9781611973051.

Filbet, F., Sonnendrücker, E., 2003. Comparison of Eulerian Vlasov solvers. Comput. Phys. Commun. 150 (3), 247–266.

Filbet, F., Sonnendrücker, E., Bertrand, P., 2001. Conservative numerical schemes for the Vlasov equation. J. Comput. Phys. 172 (1), 166–187.

Giraldo, F.X., 1998. The Lagrange-Galerkin spectral element method on unstructured quadrilateral grids. J. Comput. Phys. 147 (1), 114–146.

Guo, W., Qiu, J.M., 2013. Hybrid semi-Lagrangian finite element-finite difference methods for the Vlasov equation. J. Comput. Phys. 234, 108–132.

Guo, W., Nair, R.D., Qiu, J.M., 2014. A conservative semi-Lagrangian discontinuous Galerkin scheme on the cubed sphere. Mon. Weather Rev. 142 (1), 457–475.

Huot, F., Ghizzo, A., Bertrand, P., Sonnendrücker, E., Coulaud, O., 2003. Instability of the time splitting scheme for the one-dimensional and relativistic Vlasov-Maxwell system. J. Comput. Phys. 185 (2), 512–531.

Lauritzen, P.H., Nair, R.D., Ullrich, P.A., 2010. A conservative semi-Lagrangian multi-tracer transport scheme (CSLAM) on the cubed-sphere grid. J. Comput. Phys. 229 (5), 1401–1424.

LeVeque, R.J., 1996. High-resolution conservative algorithms for advection in incompressible flow. SIAM J. Numer. Anal. 33, 627–665.

Lin, S.J., Rood, R.B., 1996. Multidimensional flux-form semi-Lagrangian transport schemes. Mon. Weather Rev. 124 (9), 2046–2070.

Liu, J.G., Shu, C.W., 2000. A high-order discontinuous Galerkin method for 2D incompressible flows. J. Comput. Phys. 160, 577–596.

McDonald, A., 1991. Semi-Lagrangian methods. Meteorological Service. Technical note, https://books.google.com/books?id=I_yLNQAACAAJ.

Morton, K.W., Priestley, A., Suli, E., 1988. Stability of the Lagrange-Galerkin method with non-exact integration. RAIRO-Modél. Math. Anal. Numér. 22 (4), 625–653.

Pironneau, O., 1982. On the transport-diffusion algorithm and its applications to the Navier-Stokes equations. Numer. Math. 38 (3), 309–332.

Qiu, J.M., Christlieb, A., 2010. A conservative high order semi-Lagrangian WENO method for the Vlasov equation. J. Comput. Phys. 229 (4), 1130–1149.

Qiu, J.M., Russo, G., 2016. A high order multi-dimensional characteristic tracing strategy for the Vlasov-Poisson system. J. Sci. Comput. http://arxiv.org/abs/1602.08663. Submitted for publication.

Qiu, J.M., Shu, C.W., 2011a. Conservative high order semi-Lagrangian finite difference WENO methods for advection in incompressible flow. J. Comput. Phys. 230 (4), 863–889.

Qiu, J.M., Shu, C.W., 2011b. Conservative semi-Lagrangian finite difference WENO formulations with applications to the Vlasov equation. Commun. Comput. Phys. 10 (4), 979–1000.

Qiu, J.M., Shu, C.W., 2011c. Positivity preserving semi-Lagrangian discontinuous Galerkin formulation: theoretical analysis and application to the Vlasov-Poisson system. J. Comput. Phys. 230 (23), 8386–8409.

Restelli, M., Bonaventura, L., Sacco, R., 2006. A semi-Lagrangian discontinuous Galerkin method for scalar advection by incompressible flows. J. Comput. Phys. 216 (1), 195–215.

Rossmanith, J.A., Seal, D.C., 2011. A positivity-preserving high-order semi-Lagrangian discontinuous Galerkin scheme for the Vlasov-Poisson equations. J. Comput. Phys. 230 (16), 6203–6232.

Russell, T.F., Celia, M.A., 2002. An overview of research on Eulerian-Lagrangian localized adjoint methods (ELLAM). Adv. Water Resour. 25 (8), 1215–1231.

Russo, G., Santagati, P., Yun, S.B., 2012. Convergence of a semi-Lagrangian scheme for the BGK model of the Boltzmann equation. SIAM J. Numer. Anal. 50 (3), 1111–1135.

Shu, C.W., 2009. High order weighted essentially nonoscillatory schemes for convection dominated problems. SIAM Rev. 51 (1), 82–126.

Shu, C.-W., Osher, S., 1988. Efficient implementation of essentially non-oscillatory shock-capturing schemes. J. Comput. Phys. 77 (2), 439–471.

Sonnendrücker, E., Roche, J., Bertrand, P., Ghizzo, A., 1999. The semi-Lagrangian method for the numerical resolution of the Vlasov equation. J. Comput. Phys. 149 (2), 201–220.

Staniforth, A., Côté, J., 1991. Semi-Lagrangian integration schemes for atmospheric models—a review. Mon. Weather Rev. 119 (9), 2206–2223.

Strain, J., 1999. Semi-Lagrangian methods for level set equations. J. Comput. Phys. 151 (2), 498–533.

Strang, G., 1968. On the construction and comparison of difference schemes. SIAM J. Numer. Anal. 5 (3), 506–517.

Süli, E., 1988. Convergence and nonlinear stability of the Lagrange-Galerkin method for the Navier-Stokes equations. Numer. Math. 53 (4), 459–483.

Weinan, E., Shu, C.W., 1994. A numerical resolution study of high order essentially non-oscillatory schemes applied to incompressible flow. J. Comput. Phys. 110 (1), 39–46.

Xiong, T., Qiu, J.M., Xu, Z., Christlieb, A., 2014. High order maximum principle preserving semi-Lagrangian finite difference WENO schemes for the Vlasov equation. J. Comput. Phys. 273, 618–639.

Xiu, D., Karniadakis, G.E., 2001. A semi-Lagrangian high-order method for Navier-Stokes equations. J. Comput. Phys. 172 (2), 658–684.

Yabe, T., Tanaka, R., Nakamura, T., Xiao, F., 2001. An exactly conservative semi-Lagrangian scheme (CIP-CSL) in one dimension. Mon. Weather Rev. 129 (2), 332–344.

Yoshida, H., 1990. Construction of higher order symplectic integrators. Phys. Lett. A 150 (5), 262–268.

Zhang, X., Shu, C.W., 2010a. On maximum-principle-satisfying high order schemes for scalar conservation laws. J. Comput. Phys. 229, 3091–3120.

Chapter 15

Front-Tracking Methods

D. She, R. Kaufman, H. Lim, J. Melvin, A. Hsu and J. Glimm
Stony Brook University, Stony Brook, NY, United States

Chapter Outline

ABSTRACT

Front tracking is the use of surfaces or lower dimensional manifolds as computational degrees of freedom in a numerical algorithm. Its purpose is to improve the resolution of discontinuities or steep gradients in the solution variables or in the laws of physics which describe them. Thermal or concentration discontinuities, and thermodynamic phase discontinuities, often poorly handled by Eulerian advection schemes, may benefit from the use of front tracking. Other examples include discontinuities or strong gradients in opacity, conductivity, permeability and material strength.

We present the front-tracking algorithm in a formulation which includes two important recent developments, namely (cell-by-cell) conservation and an application programming interface for ease of insertion of tracking into client codes. Ongoing work to improve the late time robustness of the solution, still in progress, is outlined here.

We also present an overview of solved problems, based on the front-tracking algorithm. We discuss in general terms the problem classes for which the algorithm is beneficial as well as those for which it seems to offer little benefit. This important distinction is the topic of ongoing research.

Keywords: Front tracking, Conservation laws, Multiphase flow

AMS Classification Codes: 35L65, 74S10, 76T99

Handbook of Numerical Analysis, Vol. 17. http://dx.doi.org/10.1016/bs.hna.2016.07.004

1 INTRODUCTION

Front-tracking (FT) employs discrete surfaces as computational degrees of freedom within a numerical algorithm. It is a marriage of computational surface geometry with numerical difference methods. Numerical surface geometry is a well-established subject (Jiao and Zha, 2008; Wang et al., 2009; Zhou, 2014). Here, the surface is not an isolated object, and we focus on its coupling to a volume filling numerical algorithm.

FT has a wide range of applications. It is the unique method presently demonstrated to avoid systematic errors in an important class of problems revolving around turbulent mixing (Glimm et al., 2013; Lim et al., 2010a,b). The benefit to be derived from sharp resolution of interfaces and steep gradients runs through broad classes of problems, including cardiac electrophysiology (Xue et al., 2016), resin transfer molding (fibre-reinforced plastic) (Chui et al., 1995, 1997), primary breakup of a diesel fuel jet (Bo et al., 2011), deposition and etching in the manufacture of semiconductors (Glimm et al., 1999b), the tracking of cloud boundaries in meteorology (Huang, 2015), models of targets for high energy particle accelerators (Glimm et al., 2000) and mixing models for chemically reacting flows (Zhou et al., 2012), just to mention applications addressed in our own work.

FT is a natural method for the simulation of multiphase fluid flows, with bubbly flow (Unverdi and Tryggvason, 1992) as a prominent application. The FT algorithm (Tryggvason et al., 2001) was extended (Pivello et al., 2014) to AMR grids. After a nonconservative interpolation step, these authors apply a conservative correction, updating the interface position. In Nguyen et al. (2005), gas–gas and gas–water interface problems are simulated in 1D and 2D using the conservative FT approach of Glimm et al. (2001, 2003).

Mao and coworkers propose a conservative FT algorithm for 2D (Mao, 2000, 2007; Ullah et al., 2013). As with VOF, they use interface reconstruction, but in contrast to VOF, they connect the interface fragments. They simulate Haas–Sturtevant's bubble-shock interaction experiments (Ullah et al., 2011) and also compare to a 2D RM problem (Grove et al., 1993) which has experimental, theoretical and numerical data.

FT has been adapted to the discontinuous Galerkin method (Nguyen et al., 2010). The interface is represented by a collection of edges which are element boundaries and therefore is approximated by high-order polynomials. Their method leads to mass conservation error of the order 10^{-6} for the problem under study, a 2D two-phase simulations of droplet deformation under acoustic excitation. FT has been used to model parachute inflation and descent (Kim et al., 2013; Shi et al., 2015), with detailed comparison to experiment. The above is intended as a partial review of recent articles based on the front-tracking method; we do not claim that it is complete or fully representative.

For a review of VOF, discussing the numerical method, its implementation, and its advantages and disadvantages, refer to Hirt and Nichols (1981), Gueyffier et al. (1999) and Gopala and van Wachem (2008).

We present FT through an application programming interface (API), which we call FTI (Kaufman et al., 2016). This allows ease of insertion of FT into a client code. It is available for download from the url www.ams.sunysb.edu/fti at present for nonconservative tracking only.

FTI introduces a modification into client code to improve the resolution of solutions which are discontinuous at phase boundary surfaces or which have steep gradients or discontinuities at the Lagrangian advection of an initial contact (thermal or concentration) isosurface. We present here an FT algorithm which is cell-by-cell conservative and which is formally first order accurate for cells containing a solution discontinuity located at a tracked front. A straightforward extension yields an algorithm which is second order in time for these cells. To simplify the presentation of the ideas, we only present a first order in time version of FTI. Typical untracked codes have $\mathcal{O}(1)$ errors for such cells. We retain the higher order accuracy of the client code for cells removed from the front.

FT models the conceptually sharp boundary between distinct immiscible fluids or fluid phases. As a model of an subsurface concentration or temperature level, i.e., for miscible multiphase flow, we simplify the physics, and track the Lagrangian advection of an initialized isosurface. Our presentation is focused on the introduction of tracking into hyperbolic equations, but the case of elliptic and parabolic equations may also benefit from tracking. Our own preference is the immersed boundary method, which we have used in modelling of electrical signals in cardiac tissue (Xue, 2015).

FT had its origin with the book of Richtmyer and Morton (1967). Its early success introduced the use of ghost fluid cells (Glimm et al., 1980a,b), since widely used by others. With cell-by-cell conservative tracking, as explained here, the conceptual framework adheres to conventional ideas, while the nonstandard ghost cell construction is nearly eliminated from the algorithm.

The detailed development of the algorithm is explained in Section 2. Examples of FT and a discussion of its scope of utility are given in Section 3. Conclusions are given in Section 4.

2 FT AS A NUMERICAL ALGORITHM

2.1 FTI Overview

FTI uses a client/server model, in which the server provides geometric functionality (front advection, remeshing and topology changes) in the form of a library to be compiled with client code, while the client provides functions concerning physical variables such as states, fluxes and front points, as well as their dynamic evolution. The geometry algorithms will be discussed briefly, with details to be found in Glimm et al. (1999a), Zhou (2014), Bo et al. (2011) and references cited there. FTI provides reference implementations of the client algorithms. It hides the detail and complexity of the server functionality from the client, facilitating FT implementation in a variety of applications with reduced effort and without the need for "reinventing the wheel." Reference implementations provided include a complete FTI installation for the HEDP code FLASH (2D only).

Cell-by-cell conservation is applicable when the governing equations themselves are conservative, as with the conservation law

$$\frac{\partial U}{\partial t} + \nabla \cdot \mathbf{F}(U) = \nabla \epsilon \nabla U \tag{1}$$

for the quantity U. Conservation is achieved at a discrete level through flux balances, but now enforced at the level of space–time cut cells, that is within portions of a space–time cell cut by the moving interface and lying on one side of it. Denoting cut cell integrals of U at the old and new time levels of the space–time cut cell as Bottom and Top, we reformulate the standard conservative differencing as

$$\text{Top} = \text{Bottom} + \sum \text{Sides}, \tag{2}$$

where Side denotes the integral of flux $\mathbf{n} \cdot \mathbf{F}(U) + sU$ through a cut cell side moving with speed in normal direction s and spatial normal $\mathbf{n}$. Included in the RHS of (2) is a flux integral over the curved Lagrangian space–time surface defined by the moving front.

We encounter stability (Courant–Friedrichs–Lewy [CFL]) limits associated with possible small volume cut cell tops, resolved by merger of cut cells with adjacent cut cells on the same side of the front. The algorithm was discussed in Glimm et al. (2003), Liu et al. (2007) and Kaufman et al. (2016); here we present an improved plan.

The selection of (2) as the algorithm for evaluation of flux integrals and (cut) cell-averaged state values vs use of a client algorithm depends on the location of the cell boundary or of the cell to the front, and for this reason, we specify first the algorithm for front point propagation.

We assume that the client solver is directionally split into three 1D sweeps along the lines of (2), considered for full, not cut cells. We further assume that in each sweep, the calculation of each of the two individual fluxes can be isolated separately within the client code, each with its own stencil. We only discuss the hyperbolic algorithm $\epsilon = 0$, as in many cases, tracking is not needed or recommended for solution of the diffusion step in an operator split algorithm.

Denote the interior states at position $\mathbf{x}_i$ at time t_n as U_i^n. Further, denote the jth front point at time t_n as $\mathbf{p}_j^n$. We define $\mathbf{V}(U, \mathbf{x})$ to be the velocity field defined by the interior state interpolated to the point $\mathbf{x}$. The (two-sided) interpolation to a front point defines a pair of front states, one for each side of the front, which serve as input to a Riemann problem, solved in the direction of the front normal. The pair of inner solution states of the Riemann problem are called the front states. This choice is physics dependent, and corresponds to the tracking of a contact discontinuity. The resulting velocity $\mathbf{V}$ is double valued, but its normal component $V_* = \mathbf{n} \cdot \mathbf{V}$, with $\mathbf{n}$ the front normal vector, is single valued. We calculate the point position $\mathbf{p}^{n+1}$ at time t_{n+1} using the values at t_n,

$$\mathbf{p}_j^{n+1} = \mathbf{p}_j^n + \Delta t \mathbf{n} V_*(U^n, \mathbf{p}_j^n) \tag{3}$$

With the old and new time level front locations known, we can now define the logic which governs the choice of solver to yield the flux and state update algorithms.

We introduce three labels, FAR, NEAR and ON, which characterize the relation of a cell or cell face to the front in terms of components, an indicator of the side of a front, to be introduced as follows (Fig. 1):

FAR. All cell centres of the stencil have a common component value. The client solver is used without modification.

NEAR. The two cell centres immediately adjacent to the cell or cell face, and on either side of it, are on a common side of the front, but some of the other stencil points have the opposite component. The stencil is modified by a ghost cell extrapolation. This means that the state values in the stencil cells with the opposite component are replaced by front states of the same component, with the front states coming from the closest front crossing of the line through cell centres with the front. The modified stencil is unique (it is independent of which cell center adjacent to the flux boundary is being updated), with the differencing conservative (but modified from, and perhaps of lower order than, the client difference solver).

ON. Otherwise the cell or cell face is ON. It is also (reclassified as) ON if it is crossed by the front. The nonconservative algorithm uses ghost cell extrapolation to modify the stencils and then follows the client solver. The conservative algorithm uses (2) at the cut cell level.

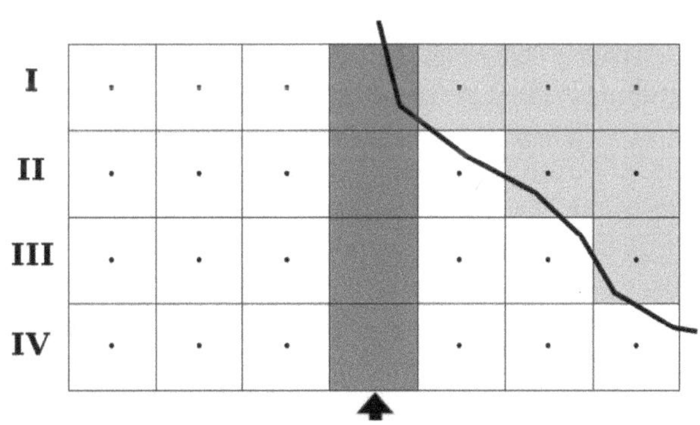

FIG. 1 Four x-direction stencils numbered I–IV are shown. The central location of the stencil is marked by an *arrow* at the bottom of the figure and is the point being updated. These center cells are labelled as: (I) ON, (II) & (III) NEAR and (IV) FAR. The *grey cells* whose centres lie across the front will be filled in with ghost states.

Passive tracking is the choice to override the FTI selection (NEAR or ON) and its flux, using instead the client algorithm. Passive tracking uses Sections 2.2.1 and 2.2.2.
Nonconservative tracking uses the ghost cell algorithm in ON cells, see Section 2.2.
Conservative tracking follows (2) for the ON space–time mesh cell solution update, after merging cells with small tops if necessary; see Section 2.3.

2.2 Application Specific (Client) Algorithms, Nonconservative Tracking

FTI offers reference implementations of the client algorithms for regular grids. For the FLASH code, we offer a full implementation, also as a reference implementation.

2.2.1 Components, the Front Defined Topology and One-Sided Interpolation

The front has two sides (components). These extend by continuity to the surrounding space. If the surface is closed (without boundary), the extension is globally unique, the case considered here. In this case, the components label or determine the topology coming from the partition of space which the front defines. If the surface is open with a boundary, the definition is unambiguous only near the front and not near the boundary. The components are used in the front aware interpolation algorithm and in the selection and modification of the client solver for the interior states.

The algorithm to determine components on a cell basis starts at the front, for which an FTI function is provided, and extends outward (with a depth specified according to the finite difference stencil size) by a marching cubes algorithm.

The construction starts by allocation of a place holder value set to 0 for each grid cell. An $\mathcal{O}(N)$ (N being the number of triangles [tris] in the front) algorithm determines which cells are crossed by triangles of the front. For these cells, the value 0 is modified to another value (ONFRONT). In a marching cubes manner, we move a specified distance (determined by the stencil size) outward from the ONFRONT cells. From an ONFRONT cell to a neighbour not on front, an FTI function searches for the nearest interface point and determines which side of the interface it belongs to. The component value (e.g., 1 or 2) is substituted for the place holder 0 for this cell. At the next level, moving from front adjacent cells to their neighbours, we use the local continuity of the component value to reset the place holder. This continues to any desired depth of search.

With the components now defined, we construct one-sided interpolation. This is constructed relative to the dual grid to the simulation grid; components and state values are defined on dual grid cell corners. To find the state at an arbitrary point, we identify the dual grid cell which contains the point. We then interpolate from as many of the corner states of this dual grid cell as have the correct component. We take the first to succeed of the four following constructions (with a detailed definition for regular grids) (Fig. 2):

8 points available: Trilinear interpolation on a cube of 8 points.

4–7 points available: We select 4 of the available points to construct a tetrahedron (tet) which contains the target point. We then interpolate linearly to the target.

4–7 points available: No such tet contains the target point, so 4 points are selected closest to the target and linear extrapolation is used.

Fewer points available: Constant extrapolation: the value of the nearest available point is used.

2.2.2 *Front States and Front Point Propagate*

A state is a set of the primitive solution variables. At each front point, we require two states, one for each side of the front. We first define "outer" front states, constructed by front aware interpolation of the interior states to the

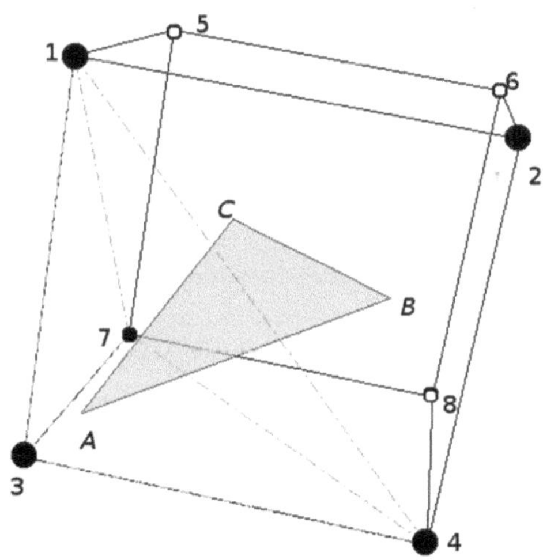

FIG. 2 Points on the dual grid cell corners labelled 1–8, with one component labelled with *solid points*, the other as *hollow points*. An interface *triangle ABC* is also shown. We discuss one-sided interpolation to the interface points *A–C*. The solid component for points *A* and *B* can be interpolated from tets. Point *C* is outside of any tet constructed from the 5 solid points, and must be extrapolated from the 4 nearest points. To interpolate in the hollow component, since only 3 points are available, each point will take it value from the closest hollow corner, i.e., *A* from (8), *B* from (6), and *C* from (5).

front point. The next step is physics dependent. Assuming that the front is the location of an isosurface in a contact discontinuity, we solve a Riemann problem with the two outer front states as data and select the two states adjacent to the contact discontinuity in the Riemann solution as the two-sided front states.

FTI supports a second order in time update, composed of predictor and corrector first-order Euler steps for propagation of front points, as in (3), with a client supplied velocity field.

2.2.3 Ghost States and Nonconservative Tracking

We assume the client difference solver is given by 1D spatial direction sweeps and that the differencing has the conservative form (2) considered at the level of full, not cut cells. We further assume that we can isolate a single face flux from the client solver to accomplish the update. Each time like (side) face of a space–time cell is labelled as FAR, NEAR or ON according to the previous rules. Ghost cell values are used in the NEAR case (and for nonconservative differencing, in the ON case). The solution updates are achieved by modification of stencil state values so that all values are obtained from a single side of the front, by extrapolation as needed from front states coming from the proper side of the front. Ghost states may reduce the order of accuracy of the client solver. The algorithm is conservative for NEAR cells but not conservative for ON cells.

2.3 Client Algorithms, Conservative Tracking

The key to conservative tracking is (2). In this formula, there are three types of integrals to be evaluated: cut cell volumes, cell side flux integrals and Lagrangian surface (2D space and time) flux integrals, as discussed in Section 2.3.5.

We summarize the conservative algorithm time step, specialized to the ON cells and faces:

C. Dynamic update. Client code.
- **C1.** Find front normal velocity, old time level, Section 2.2.3
- **C2.** Propagate front (3), Section 2.2.3
- **C4.** Merge small cell tops, Section 2.3.1
- **C3.** Compute cut cell fluxes using old time data, Section 2.3.5
- **C4.** Compute cut cell states at the new time level (2).

S. Server code
- **S1.** Process new front
- **S2.** Update components

2.3.1 Small Cell Merger

The cut cell volumes with small or no tops lead to a numerical instability or severe restriction on the stability limit. We resolve the problem by merging the cut cells with adjacent ones on the same side of the front until all merged

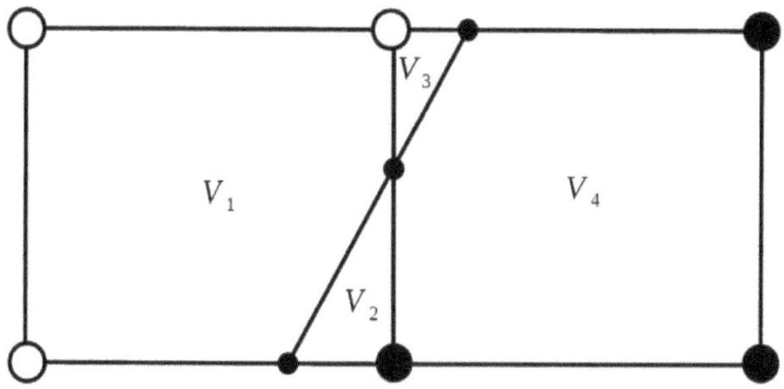

FIG. 3 2D space–time example to illustrate volume merging of cut cells. The cut cell volume V_3 with small top is merged with the adjacent cut cell volume V_1 and likewise the volume V_2 with no top is merged with V_4.

cut cells have a sufficiently large top. This algorithm of volume merging is illustrated in Fig. 3.

2.3.2 *Interior State Propagation, Conservative Tracking*

Cell faces are classified as FAR, NEAR or ON the front. Fluxes into the FAR and NEAR cells and fluxes through NEAR cell faces are evaluated as in Section 2.2.2, using the client solver and (for NEAR faces) a modified stencil. However, cells with one or more ON faces are be treated differently, with the conservative differencing (2) interpreted at the level of cut cells to prevent a loss of conservation.

2.3.3 *Cut Cell Polyhedral Volumes*

As preparation for the computation of cut cell flux integrals and cut cell tops, Section 2.3.5, we present formulas to calculate areas and volumes of cut cell polyhedra in 2D and 3D. The cut cell regions are disjoint unions of connected polyhedra. A closed planar polygon $\{(x_i, y_i)|i = 1, 2, \ldots, n\}$ in 2D has an area,

$$\text{area} = \frac{1}{2}\sum_{i=1}^{n}(x_i y_{i+1} - x_{i+1} y_i), \tag{4}$$

where $(x_{n+1}, y_{n+1}) = (x_1, y_1)$. A polyhedron in 3D with planar faces has a volume,

$$\text{vol} = \frac{1}{3}\sum_i S_i d_i, \tag{5}$$

where S_i is the surface area of face i of the polygon and $d_i = (x, y, z)\cdot\mathbf{n}_i$ with (x, y, z) on the face i and $\mathbf{n}_i$ its normal. The area is expressed as a sum in terms adjacent edges of the polyhedra, and the volume is expressed in terms of the areas of its bounding faces. Both formulas apply to unions of disconnected polyhedra, which is one of the cases in our calculation. The faces in our use of (5) result from tris of the interface, partially cut off by the space–time cell boundaries, and are planar for this reason.

2.3.4 Directional Tri Propagate

To aid in the computation of cut cell fluxes through sides of space–time cells, we propagate tris, not to the next time level, but to the next mesh space–time hyperplane in a given direction. We call this operation directional propagation, and we apply it to tris, to obtain a list of tris which meet any given space–time mesh hyperplane. For each point, we allocate and fill six coordinate points (space–time coordinates) and additional logical variables. Each of the six coordinates is the time and location that the propagating point (when extended for positive or negative times) reaches the next or previous mesh space–time hypersurface in a specified direction. The logical variables record whether this time of arrival occurs within the current time step, $[t_n, t_{n+1}]$. Each tri, through access to its point structure, then can determine which of its three points if any propagate (for example, in the $+x$ direction) to which space–time cells for this mesh hyperplane. We build a hash table to invert this relation and have a list $+x$ directionally propagating to a given space–time mesh cell face. This list of tris is the input to the side flux calculation.

2.3.5 Cut Cell Top, Side Flux and Lagrangian Surface Integrals

We compute three types of integrals: namely, state cut cell volumes for the top surface (the new time level), cut cell flux averages for the sides of the space–time mesh cell and flux through the Lagrangian moving front between the time levels.

All integrals are in 3D, either 2D spatially and time or 3D spatially. All integrands are continuous and in principal smooth, other than the presence of untracked discontinuities (not addressed by the front tracking). Specifically, the discontinuities or steep solution gradients at the tracked front do not enter into the integrands. Thus these are evaluated to second-order accuracy by the quadrature rule volume times the integrand evaluated at the centroid and to first order by evaluation of the integrand at any point in the mesh cell. For the top, we only compute volumes, and the integrand is 1. For the side faces, the integrand is $\mathbf{n} \cdot \mathbf{F}$, with the sign convention that $\mathbf{n}$ points into the cell, consistent with (2).

The needed formulas for cut cell volumes are given in Section 2.3.3. We treat the top and side integrals on a common basis, and within the point propagate step, we record which points move (totally or in part) to a given top or side cell. Thus each top or side cell has a list of tris. These tris are truncated if needed to lie entirely within the space–time mesh cell, and the formulas (4 and 5) are used to complete the calculation via a sum over tris.

For the Lagrangian moving 2D front, we first need to specify the integration measure, so that the flux through the surface is defined as a density relative to this measure. It is convenient to chose the surface measure $dtd\mathbf{x}^2$. We have two possible values for the flux density, depending on which cut cell volume and component is being integrated. The two flux densities are equal in magnitude but opposite in sign and have the form $\mathbf{n}\cdot\mathbf{F} + sU$, where $s = \mathbf{v}\cdot\mathbf{n}$. Here $\mathbf{n}$ is the front normal as a function of time, and pointing into the region being integrated; s is a front speed and is positive if the component (cut cell volume) being integrated is increasing with time at this location, again consistent with (2).

We compute the $dtd\mathbf{x}^2$ volume of the Lagrangian moving interface directly. with a sum over distinct tris. The 3D space–time volume is a union of the triangular prisms built over each tri of the interface. The total volume is a sum over all contributing triangular prisms. The volume of such a prism, if not cut by the faces of a space–time cell, has an elementary evaluation

$$\text{vol} = (\Delta t/2)(\text{area top} + \text{area bottom}). \qquad (6)$$

Each edge of the prism can cross at most two cell faces during a time step, and so there are at most six such prism edge crossings of space–time cell faces. Between such crossings, the formula (6) is still valid, with the truncated tri areas to be used, and the time interval between crossings replacing Δt. The triangles to which this formula applies come from the old time level, and the set of tris meeting the cell bottom face or propagating into one of its sides, both previously calculated.

2.4 Geometric (FTI) Algorithms

The geometric algorithms are not conservative and may introduce small conservation errors.

2.4.1 Interface Smoothing

We have two options for interface smoothing. One lists bad triangles in a queue (as too large, too small or with one or two vertex angles which are too small). These are processed, by splitting the large triangles, merging the small ones or those with poor aspect ratios in with neighbours. The other smoothing option moves front points to adjust the interface triangles without adding or removing points.

2.4.2 Elimination of Self-Intersections

We test for self-intersections of the front by testing all pairs of triangles for intersections. The algorithm is reduced from $\mathcal{O}(N^2)$ in the number N of triangles to $\mathcal{O}(N)$ by construction of lists of tris which cross each given mesh block. Intersections are then tested only among pairs of tris which cross a given mesh block.

Intersections, when detected, are placed within a bounding box. This interface in the bounding box is replaced with a simpler and untangled interface, in a construction called grid-based reconstruction (Glimm et al., 1999a). There remains a gap between the triangles removed and those restored inside the bounding box. This gap is filled in a sequential manner, in an algorithm described in Bo et al. (2011). The algorithm is not conservative and will eventually be replaced with a better one.

2.4.3 Robust Parallel Communication of Front Data

At late time the interface is often extremely complex at a grid level. A generic strategy to overcome this deficiency is mesh refinement, which is supported within FTI if supported by the client code. Parallel communication of interface patches to create updated ghost cell extension of the interface required match and identification of the communicated points. Floating point comparison of points is not robust, and for this reason, FTI supports unique logical variables associated with each point, allowing robust communication even at late times. Further solutions are physics dependent. For miscible displacement, the tracked isosurface will typically no longer present a steep gradient at late time and selective untracking (a future development) is proposed. For immiscible displacement, there is a surface-related disjuncture pressure as a correction to the Navier–Stokes equation to be added to the momentum equation. This term retards the close approximation and eventual intersection of adjacent portions of the interface. It thus retards merger of droplets and bubbles, and if given an artificially increased value (numerical disjuncture pressure) will retard the growth of subcell complexity.

3 SCIENTIFIC USES OF FT

3.1 Benchmark Problems

A detailed comparison of FT to the level set method was presented in Du et al. (2006), in which the high quality of FT could be seen, through a series of standard problems (rotating slotted disks, twisting and untwisting circles or spheres, etc.). Rather than repeating elements of this study, we analyse just what these comparisons test.

The tests come with an analytically prescribed velocity field. For FT, the tests are then solely dependent on three parameters: the order of time integration of an ODE derived from the velocity field, the interface processing frequency and its order of accuracy, with the temporal and spatial discretization linked by a CFL condition. Here we document this statement through consideration of a simple benchmark problem: deformation of a circle into a complicated shape and a subsequent reversal, back (hopefully) to the original circle. Errors can then be estimated as a norm difference between the original circle and the final shape. We document an error estimate of the form

$$\text{Error} = \mathcal{O}(h^n) + f^{-1}\mathcal{O}(h^m), \tag{7}$$

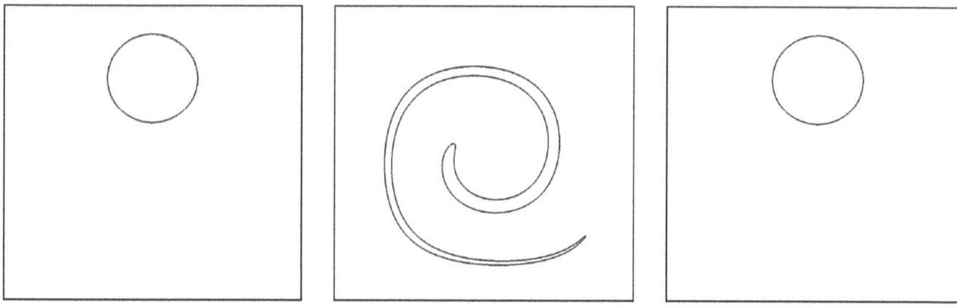

FIG. 4 *Left*: *original circle*; *mid*: highly deformed shape; *right:* final figure, for the choices $n = 4$, $m = 2, f = 50$ and $h = 0.002$ (in units of the circle diameter). The velocity field is from Du et al. (2006).

TABLE 1 Convergence Rates for Solutions to the Deformed Circle Problem via the FT Algorithm, Showing Common Rather than Asymptotic Levels of Mesh Refinement

n	f	m	Observed Convergence Order	Predicted Convergence (7)
$n = 2$	$f = 20$	$m = 1$	1	$h^2 + 0.05 \times h$
$n = 2$	$f = 50$	$m = 1$	1	$h^2 + 0.02 \times h$
$n = 4$	$f = 50$	$m = 2$	2	$h^4 + 0.02 \times h^2$

where h is the spatial resolution of the front, n is the order of accuracy of the time integration, and m is the order of geometrical interpolation for insertion of new front points, which occurs at a remeshing frequency f. The results are illustrated in Fig. 4.

The convergence can be accessed from Table 1, a limited, nonasymptotic exploration of the n, f, m parameter space.

3.2 Verification and Validation Examples

We summarize results concerning simulations of acceleration driven fluid instabilities. These are known as Rayleigh–Taylor instabilities in the case of constant acceleration and Richtmyer–Meshkov instabilities in the case of impulsive (shock wave) acceleration. Tables 2–4 are taken from Glimm et al. (2016).[a]

[a]Reprinted from Glimm, J., Plohr, B., Lim, H., Hu, W., Sharp, D.H., 2016. Large eddy simulation, turbulent transport and the renormalization group. Ann. Math. Sci. Appl. 1, 149–180, with permission from International Press.

TABLE 2 Variabilities in α from a Variety of Experimental and Numerical Sources

Experimental Variabilities	
Due to initial conditions	5–30%
Due to experimental fluid transport properties	20%
Numerical issues: ILES	
ILES to experiment discrepancy (Dimonte et al., 2004)	100%
ILES to ILES simulation discrepancy (Dimonte et al., 2004)	50%
Numerical issues: LES/SGS/FT	
Numerical variation from transport coefficients (Lim et al., 2010a,b)	5%
FT/LES/SGS to experiment discrepancy (Lim et al., 2010b)	5%

3.3 A Complex Physics Example

We study turbulent mixing in Inertial Confinement Fusion (ICF) simulations, comparing three simulations, the Lagrangian code HYDRA from LLNL and the Eulerian code FLASH, with and without front tracking. ICF is based on implosion of a small capsule, driven by a strong radiation field, to cause ablation of the outer surface layer of the capsule. The capsule is constructed with an outer layer of plastic (CH), a middle layer of deuterium–tritium (DT) ice, and an inner region of DT gas.

The code comparison study will be reported fully elsewhere. ICF has three classical instabilities, an ablation driven modified RT instability, followed by a shock-driven RM instability, and at stagnation, with strong deceleration, an RT instability. We find that the RM phase is RT stable, meaning that the RM instabilities do not grow and might even decrease in time. Thus we analyze the RM phase of the implosion in terms of a diffusion equation, i.e., using fluid transport theory.

On this basis, the strong difference between the tracked and untracked Eulerian calculations is ascribed to numerical concentration diffusion and is a mesh resolution effect. The tracked calculation mimics the behaviour of the Lagrangian calculation, preventing numerical diffusion of the concentration at the tracked front. All three simulations omit physical concentration diffusion. The differences start with untracked Eulerian diffusion of concentration, mixing the cold outer shell DT (fuel) of the ICF capsule with the CH ablator. As this interface is close to an RT instability threshold (a time dependent radius), the further the CH penetrates into the cold outer shell,

TABLE 3 Comparison of FT Simulation to Experiment

Ref.	Exp.	Sim. Ref.	α_{exp}	α_{sim}	Discrepancy
Smeeton and Youngs (1987)	#112	Lim et al. (2010a)	0.052	0.055	6%
Smeeton and Youngs (1987)	#105	Glimm et al. (2013)	0.072	0.076 ± 0.004	0%
Smeeton and Youngs (1987) and Read (1984)	10 exp.	George et al. (2006)	0.055–0.077	0.066	0%
Ramaprabhu and Andrews (2004)	Air–He	Liu et al. (2006)	0.065–0.07	0.069	0%
Mueschke (2008)	Hot–cold	Lim et al. (2010a) and Glimm et al. (2013)	0.070 ± 0.011	0.075	0%
Mueschke (2008)	Salt–fresh	Glimm et al. (2013)	0.085 ± 0.005	0.084	0%

Discrepancy refers to the comparison of results outside of uncertainty intervals, if any, as reported.

TABLE 4 The Role of Front Tracking (FT) and LES with Subgrid Models (LES/SGS) for the Modelling Solution Details (Mixing Zone Edges, Cut Cell Thermodynamics and Second Moments of the Solution State Variables) for Several Cases of RT and RT Instabilities

	MZ Edge	Cut Cell Thermodynamics	2nd Moments
RT (fluids)			
High *Re*	FT/LES/SGS (Dimonte et al., 2004; Glimm et al., 2013; Lim et al., 2010a)	U	U
Low *Re*	N (Glimm et al., 2016; Mueschke and Schilling, 2009)	U	FT (Glimm et al., 2016)
RM (fluids)	FT (Section 3.3); N (Masser, 2007)	FT (Masser, 2007)	FT (Section 3.3)

The entry FT/LES/SGS or FT indicates the code feature of importance for the indicated problem. The entry N indicates neither FT nor SGS/LES appear to be important, while U indicates an unknown case.

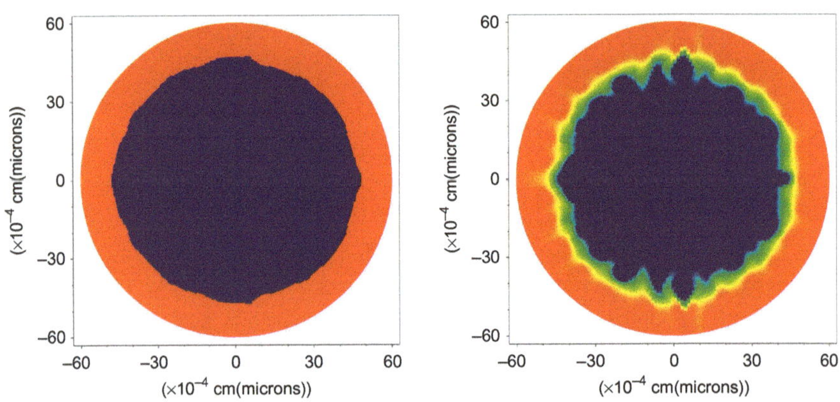

FIG. 5 CH concentration in an ICF implosion near the time of the stagnation instability for a coarse grid simulation. *Left*: tracked Eulerian. *Right*: untracked Eulerian simulation. The *left figure* shows no numerical diffusion of concentration, consistent with the reference simulation (HYDRA), as a by product of its Lagrangian formulation. All three simulations omit effects of physical concentration diffusion, presumed to lead to results intermediate between those shown here.

the more likely it is to cross the threshold and mix extensively via the strong RT instability in the cold shell DT. This effect is illustrated in Fig. 5, comparing the CH concentration levels in the tracked and untracked simulations. The resolution is ≈ 0.5 μ per cell.

The result is a systematic difference in the ICF thermodynamic state of the cold shell and the prediction of mixing at late time. Thus, in conclusion, tracking of the Eulerian LES version of this problem prevents excessive concentration (CH vs DT) diffusion in the cold shell at all mesh levels. While numerical diffusion can be improved by mesh refinement, the required grids can quickly become computationally infeasible, especially in three dimensions.

4 CONCLUSIONS

We have presented the FT algorithm. In its nonconservative form, the algorithm is supported by an API (FPI), for ease of insertion into a client code. The API has been used to insert FT into the plasma physics code FLASH. Both the nonconservative and the conservative forms of the algorithm are outlined here. We have noted the wide range of problems with naturally occurring discontinuities or steep gradients, for which the method is ideal. References to numerous solved problems are included. The exact scope of the method is an ongoing topic of research, but basically it is of value when the solution gradients or discontinuities are large enough to cause serious errors when using standard Eulerian methods. Future progress with the late time complexity of the tracked front is ongoing. For miscible fluids, the discontinuities often become less severe at late time, so that selective untracking of portions of the front appears to be a desirable option.

ACKNOWLEDGEMENTS

Among the scores of students and faculty collaborators, we have had the pleasure of working with on the development of front tracking, it would be remiss not to mention especially the roles of Oliver McBryan, Dan Marchesin, John Grove and David Sharp. This work is supported in part by Leland Stanford Junior University (subaward with DOE as prime sponsor), the U.S. Department of Energy via Los Alamos National Laboratory contract number 228022, and the Army Research Organization grant number W911NF1310249.

REFERENCES

Bo, W., Liu, X., Glimm, J., Li, X., 2011. A robust front tracking method: verification and application to simulation of the primary breakup of a liquid jet. SIAM J. Sci. Comput. 33, 1505–1524.

Chui, W.K., Glimm, J., Tangerman, F.M., Jardine, A.P., Madsen, J.S., 1995. Modeling of resin transfer molding. In: Proceedings of the First Regional Symposium on Manufacturing Science and Technology. Stony Brook, New York.

Chui, W.K., Glimm, J., Tangerman, F.M., Jardine, A.P., Madsen, J.S., Donnellan, T.M., Leek, R., 1997. Process modeling in resin transfer molding as a method to enhance product quality. SIAM Rev. 39 (4), 714–727.

Dimonte, G., Youngs, D.L., Dimits, A., Weber, S., Marinak, M., Wunsch, S., Garsi, C., Robinson, A., Andrews, M., Ramaprabhu, P., Calder, A.C., Fryxell, B., Bielle, J., Dursi, L., MacNiece, P., Olson, K., Ricker, P., Rosner, R., Timmes, F., Tubo, H., Young, Y.-N.,

Zingale, M., 2004. A comparative study of the turbulent Rayleigh-Taylor instability using high-resolution three-dimensional numerical simulations: the Alpha-Group collaboration. Phys. Fluids 16, 1668–1693.

Du, J., Fix, B., Glimm, J., Jia, X., Li, X., Li, Y., Wu, L., 2006. A simple package for front tracking. J. Comput. Phys. 213, 613–628.

George, E., Glimm, J., Li, X.-L., Li, Y.-H., Liu, X.-F., 2006. Influence of scale-breaking phenomena on turbulent mixing rates. Phys. Rev. E 73, 016304.

Glimm, J., Marchesin, D., McBryan, O., 1980a. Subgrid resolution of fluid discontinuities II. J. Comput. Phys. 37, 336–354.

Glimm, J., Marchesin, D., McBryan, O., 1980b. Statistical fluid dynamics: unstable fingers. Comm. Math. Phys. 74, 1–13.

Glimm, J., Grove, J.W., Li, X.-L., Zhao, N., 1999a. Simple front tracking. In: Chen, G.-Q., DiBenedetto, E. (Eds.), Contemporary Mathematics, vol. 238. American Mathematical Society, Providence, RI, pp. 133–149.

Glimm, J., Simanca, S.R., Tan, D.C., Tangerman, F.M., VanDerWoude, G., 1999b. Front tracking simulations of ion deposition and resputtering. SIAM J. Sci. Comput. 20, 1905–1920.

Glimm, J., Kirk, H., Li, X.L., Pinezich, J., Samulyak, R., Simos, N., 2000. Simulation of 3D fluid jets with application to the Muon Collider target design. In: Rahman, M., Brebbia, C.A. (Eds.), Advances in Fluid Mechanics III, vol. 26. WIT Press, Southampton, Boston, pp. 191–200.

Glimm, J., Li, X.-L., Liu, Y.-J., Zhao, N., 2001. Conservative front tracking and level set algorithms. Proc. Natl. Acad. Sci. 98, 14198–14201.

Glimm, J., Li, X.-L., Liu, Y.-J., Xu, Z.L., Zhao, N., 2003. Conservative front tracking with improved accuracy. SIAM J. Numer. Anal. 41, 1926–1947.

Glimm, J., Sharp, D.H., Kaman, T., Lim, H., 2013. New directions for Rayleigh-Taylor mixing. Phil. Trans. R. Soc. A 371, 20120183. Los Alamos National Laboratory Preprint LA-UR 11-00423 and Stony Brook University Preprint SUNYSB-AMS-11-01.

Glimm, J., Plohr, B., Lim, H., Hu, W., Sharp, D.H., 2016. Large eddy simulation, turbulent transport and the renormalization group. Ann. Math. Sci. Appl. 1, 149–180. Los Alamos Preprint LA-UR-12-26149. Stony Brook University Preprint Number SUNYSB-AMS-15-05.

Gopala, V.R., van Wachem, B.G.M., 2008. Volume of fluid methods for immiscible-fluid and free-surface flows. Chem. Eng. J. 141 (13), 204–221.

Grove, J.W., Holmes, R., Sharp, D.H., Yang, Y., Zhang, Q., 1993. Quantitative theory of Richtmyer-Meshkov instability. Phys. Rev. Lett. 71 (21), 3473–3476.

Gueyffier, D., Li, J., Nadim, A., Scardovelli, R., Zaleski, S., 1999. Volume-of-fluid interface tracking with smoothed surface stress methods for three-dimensional flows. J. Comput. Phys. 152 (2), 423–456.

Hirt, C., Nichols, B., 1981. Volume of fluid (VOF) method for the dynamics of free boundaries. J. Comput. Phys. 39, 201–225.

Huang, Y.-T., 2015. A Stochastic Formulation of Short Term Cloud Cover Forecasts (Ph.D. thesis). State University of New York at Stony Brook.

Jiao, X., Zha, H., 2008. Consistent computation of first and second-order differential quantities for surface meshes. In: ACM Solid and Physical Modeling Symposium, 159–170.

Kaufman, R., Lim, H., Glimm, J., 2016. Conservative front tracking: the algorithm, the rationale and the API. In: Bulletin of the Institute of Mathematics, Academia Sinica New Series, vol. 11, pp. 115–130. Stony Brook University Preprint SUNYSB-AMS-15-01.

Kim, J.-D., Li, Y., Li, X., 2013. Simulation of parachute FSI using the front tracking method. J. Fluids Struct. 37, 100–119.

Lim, H., Iwerks, J., Glimm, J., Sharp, D.H., 2010a. Nonideal Rayleigh-Taylor mixing. Proc. Natl. Acad. Sci. 107 (29), 12786–12792. Stony Brook University Preprint SUNYSB-AMS-09-05 and Los Alamos National Laboratory Preprint LA-UR 09-06333.

Lim, H., Iwerks, J., Yu, Y., Glimm, J., Sharp, D.H., 2010b. Verification and validation of a method for the simulation of turbulent mixing. Phys. Scr. T142, 014014. Stony Brook University Preprint SUNYSB-AMS-09-07 and Los Alamos National Laboratory Preprint LA-UR 09-07240.

Liu, X.-F., George, E., Bo, W., Glimm, J., 2006. Turbulent mixing with physical mass diffusion. Phys. Rev. E 73, 056301.

Liu, J., Lim, H.-K., Glimm, J., Li, X., 2007. A conservative front tracking method in N-dimensions. J. Sci. Comp. 31, 213–236. Stony Brook University preprint number SUNYSB-AMS-06-04.

Mao, D., 2000. Toward front-tracking based on conservation in two space dimensions. SIAM J. Sci. Comput. 22, 113–151.

Mao, D., 2007. Towards front-tracking based on conservation in two space dimensions II, tracking discontinuities in capturing fashion. J. Comput. Phys. 226, 1550–1588.

Masser, T.O., 2007. The Effects of Temperature Equilibrium in Mixed Cell Hydrodynamics (Ph.D. thesis). State University of New York at Stony Brook.

Mueschke, N.J., 2008. Experimental and Numerical Study of Molecular Mixing Dynamics in Rayleigh-Taylor Unstable Flows (Ph.D. thesis). Texas A and M University.

Mueschke, N., Schilling, O., 2009. Investigation of Rayleigh-Taylor turbulence and mixing using direct numerical simulation with experimentally measured initial conditions. I. Comparison to experimental data. Phys. Fluids 21, 1–19. 014106.

Nguyen, V.T., Cheong, K.B., Peraire, J., 2005. A Conservative Front Tracking Algorithm. https://dspace.mit.edu/bitstream/handle/1721.1/7376/HPCES002.pdf?sequence=1.

Nguyen, V.T., Peraire, J., Khoo, B., Persson, P., 2010. A discontinuous Galerkin front tracking method for two phase flows with surface tension. Comput. Fluids 39, 1–14.

Pivello, M., Villar, M., Serfaty, E., Roma, A., Silveira-Neto, A., 2014. A fully adaptive front tracking method for the simulation of two phase flows. Int. J. Multiphase Flow 58, 72–82.

Ramaprabhu, P., Andrews, M., 2004. Experimental investigation of Rayleigh-Taylor mixing at small Atwood numbers. J. Fluid Mech. 502, 233–271.

Read, K.I., 1984. Experimental investigation of turbulent mixing by Rayleigh-Taylor instability. Physica D 12, 45–58.

Richtmyer, R., Morton, K., 1967. Difference Methods for Initial Value Problems, second ed. Interscience, New York.

Shi, Q., Reasor, D., Gao, Z., Li, X., Charles, R.D., 2015. On the verification and validation of a spring fabric for modeling parachute inflation. J. Fluids Struct. 58, 20–39.

Smeeton, V.S., Youngs, D.L., 1987. Experimental investigation of turbulent mixing by Rayleigh-Taylor instability (part 3). AWE Report Number 0 35/87.

Tryggvason, G., Bunner, B., Esmaeeli, A., Juric, D., Al-Rawahi, N., Tauber, W., Han, J., Nas, S., Jan, Y.J., 2001. A front-tracking method for the computations of multiphase flow. J. Comput. Phys. 169, 708–759.

Ullah, M.A., Gao, W., Mao, D., 2011. Richtmyer-Meshkov instabilities using conservative front-tracking method. Appl. Math. Mech. 32, 119–132.

Ullah, M.A., Gao, W., Mao, D., 2013. Towards front-tracking based on conservation in two space dimensions III, tracking interfaces. J. Comput. Phys. 242, 268–303.

Unverdi, S.O., Tryggvason, G., 1992. A front-tracking method for viscous, incompressible, multi-fluid flows. J. Comput. Phys. 100 (1), 25–37.

Wang, D., Clark, B.L., Jiao, X., 2009. An analysis and comparison of parameterization based computation differential quantities for discrete surfaces. Comput. Aided Geom. Des. 26, 510–527.

Xue, S., 2015. A Sharp Boundary Model for Electrocardiac Simulations (Ph.D. thesis). State Univ. of New York at Stony Brook.

Xue, S., Lim, H., Glimm, J., Fenton, F.H., Cherry, E.M., 2016. Sharp boundary electrocardiac simulations. SISC 38, B100–B117. Stony Brook University Preprint SUNYSB-AMS-15-03.

Zhou, Y., 2014. Front Tracking Method with High-Order Enhancement and Its Application in Two-Phase Micromixing of Incompressible Viscous Fluids (Ph.D. thesis). State Univ. of New York at Stony Brook.

Zhou, Y., Ray, N., Lim, H., Wang, S., de Almeida, V.F., Glimm, J., Li, X.-L., Jiao, X., 2012. Development of a front tracking method for two-phase micromixing of incompressible viscous fluids with interfacial tension in solvent extraction. Technical Report ORNL/TM-2012/28. Oak Ridge National Laboratory.

Chapter 16

Moretti's Shock-Fitting Methods on Structured and Unstructured Meshes

A. Bonfiglioli*, R. Paciorri†, F. Nasuti† and M. Onofri†
*Scuola di Ingegneria, Università degli Studi della Basilicata, Potenza, Italy
†Dip. di Ingeneria Meccanica e Aerospaziale, Università degli studi di Roma "La Sapienza", Rome, Italy

Chapter Outline

ABSTRACT

Over the last 25 years shock fitting methods have been almost completely abandoned and replaced by the prevailing shock-capturing methods. Only a few research groups have continued to use and to develop these techniques. A few months after the death of Gino Moretti, the man who contributed to the development of shock-fitting methods from the 1960s until the 1980s, it is useful to describe the state-of-the-art of these methods, from the techniques developed for structured meshes up to recent developments for unstructured grids. The chapter aims to spread knowledge of these little known techniques, especially among the latest generations of CFDers that grew up in a context exclusively dominated by shock-capturing methods.

Keywords: Compressible flows, Shock-fitting, Shock-capturing, Structured meshes, Unstructured meshes

2010 MSC: 00-01, 99-00

Handbook of Numerical Analysis, Vol. 17. http://dx.doi.org/10.1016/bs.hna.2016.09.011

1 INTRODUCTION

Gino Moretti passed away peacefully on 15 March 2015 at the age of 98. He was one of the most active pioneers of modern computational fluid dynamics (CFD) and his contributions set some of the milestones in the evolution of the simulation of compressible high-speed flows, in particular by providing original contributions on the importance to take into account wave propagation direction in the discretisation of the equations, and on the treatment of shock waves or, more generally, of the discontinuities that may occur in a flow field.

This chapter aims at providing a concise memory of his main contributions on the development of the "Shock-Fitting" technique, and at describing some of its recent evolutions, which can be considered the direct and modernised heritage of the original technique.

2 SHOCK-FITTING, UPWINDING AND MODERN SHOCK-CAPTURING SCHEMES

Shock waves occur very frequently in nature and in technological applications. Their presence characterises compressible flows not only in aeronautics and aerospace but also in other areas of theoretical and applied physics and engineering. In all flows where shock waves occur, they play an important role that affects the overall flow behaviour.

At the dawn of CFD, two alternative approaches were introduced to simulate flows with shocks: shock-fitting and shock-capturing. The former was able to provide very accurate solutions and to be very efficient in terms of computational cost. Unfortunately it also had clear limitations in simulating three-dimensional flow fields and complex shock patterns.

Shock-capturing discretisations lay their foundations in the mathematical theory of weak solutions, which allows the computation of all type of flows, including those with shocks, using the same discretisation of the conservation-law-form of the governing equations at all grid cells. This yields obvious consequences in terms of coding simplicity, since a single numerical scheme is used and the same set of operations is repeated within all control volumes of the mesh, no matter how complicated the flow might be.

As a consequence, nowadays shock-capturing schemes are the most widespread choice for practical fluid-dynamics simulations involving shock waves.

Coding simplicity does not come for free, however, and shock-capturing solutions of flows featuring strong shock waves are often characterised by the appearance of bizarre anomalies and sometimes by large numerical errors. The deficiencies exhibited by state-of-the-art shock capturing discretisations are in some cases sufficient to lead the author of a recent review paper (Pirozzoli, 2011) on numerical methods for high-speed flows to conclude that "these limitations, related to the misrepresentation of discontinuities on a mesh with finite spacing, can only be overcome by some form of shock-fitting".

Shock-fitting consists first in locating and then tracking the discontinuities in the flow field. These are treated as boundaries between regions where a smooth solution to the governing partial differential equations (PDEs) exists. The flow variables on the two sides of the discontinuities are evaluated analytically by enforcing the Rankine–Hugoniot (R-H) jump relations, which are simple algebraic equations connecting the states on both sides of the discontinuity and its local speed. Then, this solution is used to compute the space–time evolution of the discontinuity, that is, to track its motion.

Shock-fitting methods enjoyed a remarkable popularity in the early CFD era thanks to the efforts of Gino Moretti and his collaborators. To understand the reasons behind Moretti's interest in the development of the shock-fitting technique and also to take into due consideration the full extent of Moretti's contribution to the development of CFD, it would be useful to reread some of his papers (Moretti, 1987b, 1988). The reader would discover that Moretti's attention was not only focused on the shock-fitting technique but also on the choice of the most suitable form of the governing PDEs to be discretised when dealing with shock waves and other discontinuities. Indeed, Moretti (2002) was convinced that the "codes derived from the conservation equations are clumsier, slower and less efficient than codes based on Riemann's characteristics equations". Since compressible flows are dominated by wave propagation phenomena and the characteristic formulation of the governing PDEs is the one that best reveals the contribution of the various waves that travel within the flow field, this particular form of the governing equations is the most natural and "clever" choice to build numerical schemes for compressible flows, because of its capability to follow the main physical characteristics of the flow behaviour. In particular, the choice to respect carefully the physical nature of the phenomenon inspired Moretti to introduce the upwind differencing for his λ (Moretti, 1979) and FAST (Moretti, 1985) schemes.

It is important to underline that these schemes were proposed and successfully used by Moretti in many applications in the 1970s, at a time when most numerical methods were based on central differencing. Modern shock-capturing methods largely in use today do indeed take into account wave propagation phenomena, but were developed some years later, probably in response to the pressure of Moretti's criticisms. Therefore, Moretti anticipated the massive use of upwind methods in CFD, while remaining faithful to the use of the characteristic equations. As pointed out by Roe in 2011, he "waged a long rearguard action at the Polytechnic Institute of Brooklyn in support of fitting methods and he was involved with ICASE during the years when many of current capturing methods were being forged". Therefore, even though Moretti continued to dislike modern shock-capturing methods, he contributed to the development of these methods by showing the importance of accounting for wave propagation phenomena when developing numerical schemes for compressible flows.

What Roe defines as "a long rearguard action" was actually a coherent path followed without taking shortcuts or avoiding difficulties. The coherence in the use of numerical schemes based on the characteristic equations led Moretti to a continued development of the shock-fitting technique. Since the characteristic equations are not valid across the shocks, shocks cannot be treated with the same numerical scheme used in smooth flow regions. For this reason, the development of a general-purpose shock-fitting technique became the element characterising Moretti's overall scientific production.

Moretti's studies were only carried out on structured grids. In fact he was active over a time frame, ranging from the mid-1960s to the late-1980s, when almost exclusively structured-grid discretisations were used. When he retired, by the end of the 1980s, unstructured-grid schemes started to be applied to CFD problems, unveiling new unpredictable opportunities for building more robust and versatile shock-fitting schemes. Within the structured-grid framework, shock-fitting methods took two different approaches: boundary shock-fitting, see Section 3, and floating shock-fitting, see Section 4. After retirement Moretti's genius was still active and he envisaged a future development of shock fitting based on different tracking techniques coming from the development of new approaches in different fields like, for instance, those of video games or that of fuzzy logic. He was not wrong as it took one more decade to see some of his second generation disciples developing a new technique based on the exploitation of unstructured grid properties, see Section 5. The present paper only focusses its attention towards the application of shock-fitting methods to two- and three-dimensional flows. A recent review of shock-fitting methods for one-dimensional flows, along with some interesting applications, can be found in chapters 2 and 4 of the book by Salas (2009).

3 BOUNDARY SHOCK-FITTING

By introducing in 1966 the "boundary" shock-fitting approach to simulate blunt body flows, Moretti and Abbett (1966) turned a complex problem that used to be solved with expensive and approximate computational techniques into a problem that could be solved quickly and exactly.

This effective solution strategy for solving blunt body flows relies on two key ingredients: (i) the shock-fitting technique is used to compute the bow shock and (ii) the steady flow field within the shock-layer is asymptotically approached by time-marching the unsteady Euler equations, rather than trying to solve directly the steady Euler equations.

The first ingredient is the topic of this section, where it will be described at some depth; the second ingredient is nowadays routinely used to solve steady compressible flow problems because it transforms a mixed hyperbolic–elliptic system (the steady Euler equations) into a purely hyperbolic one (the unsteady Euler equations), with the advantage that the same numerical scheme can be used in both the subsonic and supersonic regions of the shock layer.

As shown in Fig. 1A, the basic idea of the "boundary" shock-fitting approach applied to blunt body problems is to map the shock-layer onto a

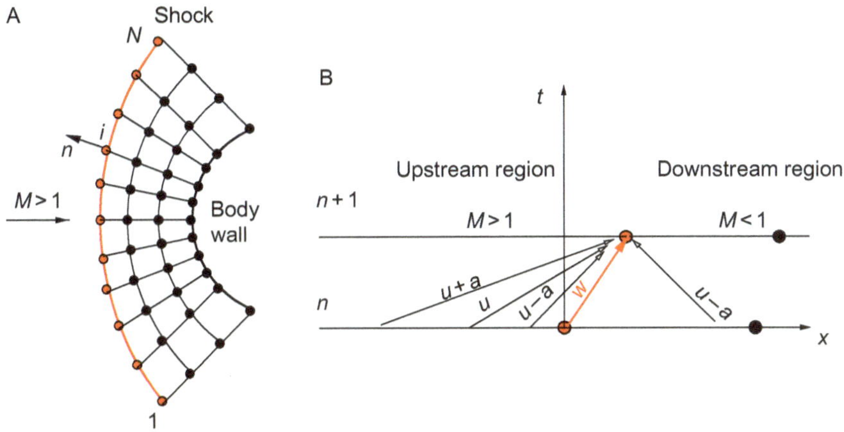

FIG. 1 Boundary shock fitting applied to the blunt body problem. (A) Body-fitted computational domain. (B) One-dimensional problem around the shock-point.

single block of a structured mesh in which the body wall is a boundary and the bow shock is the opposite boundary.

At each time step, the shock position and its upstream and downstream states need to be updated from time level t to $t + \Delta t$, along with the solution inside the entire shock-layer. A gasdynamic solver is used to update all the interior grid-points and, using the appropriate boundary conditions, also the boundary points, except those that are located along the bow shock, shown using red circles in Fig. 1, which will be hereafter referred to as "shock-points". The computation of the shock-points requires an ad hoc procedure, called "boundary" shock-fitting, which can be summarised in the following three steps.

1. *Computation of the unit vector normal to the shock.* The unit vector normal to the shock in shock-point i is numerically computed from its geometrical location and those of its neighbouring shock-points, while taking into account the domain of dependence of shock-point i. In other words, if the shock-downstream state is subsonic, the normal unit vector is computed using centred finite differences, but if the shock-downstream state is supersonic, upwind finite differences are used instead.
2. *Computation of the shock-points.* By projecting the solution along the normal to each shock-point, the problem of the computation of a two-dimensional shock-line made of N shock-points is transformed into N one-dimensional problems, as sketched in Fig. 1B. The shock-point that at time t is at the origin of the reference frame, see Fig. 1B, moves with a yet unknown speed w over the time interval $[t, t+\Delta t]$. The flow state at time $t + \Delta t$ ahead of the shock can be correctly updated by the gasdynamic solver, without applying any boundary condition, because all characteristic

lines point downstream, see Fig. 1B, and, therefore, take information into the shock. Moreover, in the blunt body problem under scrutiny, the upstream state does not change in time and it is uniform upstream of the bow shock. By contrast, the downstream state at time $t + \Delta t$ is yet unknown, because the gasdynamic solver is not able to compute the correct shock-downstream state, unless further boundary conditions, corresponding to the characteristic lines that point away from the shock, are supplied. However, within the shock-downstream region there is an acoustic wave that points upstream along the characteristic line of slope $u - a$, so that the gasdynamic solver is capable of correctly computing the Riemann variable:

$$a - \delta u = R, \quad \delta = \frac{\gamma - 1}{2} \tag{1}$$

conveyed along this characteristic line, without applying any boundary condition. The downstream state (three unknowns in the shock-normal reference frame) and the shock speed, w, can now be computed by solving a nonlinear system of algebraic equations made of Eq. (1) and the R-H jump relations, which provide three further equations.

3. *Shock displacement*. The shock position is updated to time $t + \Delta t$ using the shock speed, w, computed in the previous step. Of course, the displacement of the shock-points causes the deformation of the entire block-grid, which can, however, be easily regenerated by modifying the nodal distribution along the wall-normal coordinate line, moving from the shock towards the wall.

It is evident that the boundary shock-fitting technique is particularly simple to implement in existing gasdynamic codes, since it only requires the coding of a boundary condition that accounts for the presence of the fitted shock; no other change is necessary inside the computational kernel of the solver. For this reason, the application of the boundary shock-fitting technique to the blunt-body problem has found its way well outside the circle of Moretti's disciples, see, e.g., Weilmuenster and Hamilton (1979), Menne (1995), Zhong (1998), Kopriva (1999) and Najafi et al. (2014), for a more recent example.

Although it is less known, the boundary shock-fitting technique can be applied to flows different from the blunt body problem, such as the DNS of compressible turbulence (Ma and Zhong, 2003a,b, 2005; Sesterhenn et al., 2005), possibly characterised by a nonuniform shock-upstream state (Rawat and Zhong, 2010, 2011) and flows characterised by multiple shocks and even by shock-shock interactions. When multiple shocks are present, the embedded shocks are treated as interior boundaries that bound different blocks within a multiblock grid setting. Fig. 2 shows a nontrivial application of the boundary shock-fitting method to the computation of the hypersonic flow past a double ellipse in which both the bow and the embedded shocks are fitted. Other relevant applications of the boundary shock-fitting approach can be found in

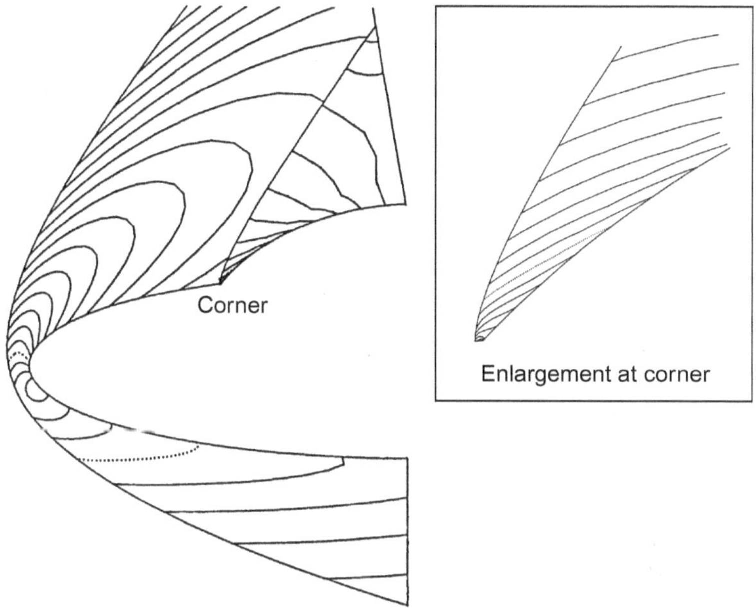

FIG. 2 Hypersonic flow past a double ellipse computed by means of a multiblock, boundary shock-fitting method.

Marconi and Salas (1973), Yamamoto and Karashima (1982) and Morton and Paisley (1989).

Despite these significant results, the boundary shock-fitting approach does not allow to treat arbitrarily complex flow problems, because of the topological limitations inherent in the use of structured multiblock grids. Moreover, since the shocks move, the handling of the motion and deformation of the various blocks may quickly became a "topological nightmare" (Moretti, 1988); this observation led Moretti to introduce the so-called "floating" shock-fitting approach.

4 FLOATING SHOCK-FITTING

A step forward towards a technique capable of more easily handling complex flow configurations, including shock interactions, was therefore undertaken by Moretti with the development of the floating shock-fitting technique, which led to the outstanding results presented in Moretti (1988) and relevant to typical shock interaction test cases. Some details of the technique have only been published in Moretti (1987c). In floating shock-fitting, the discontinuities are allowed to move (float) freely over a fixed background structured grid. A shock front is described by its intersections with grid lines, which give rise to shock-points, as shown in Fig. 3.

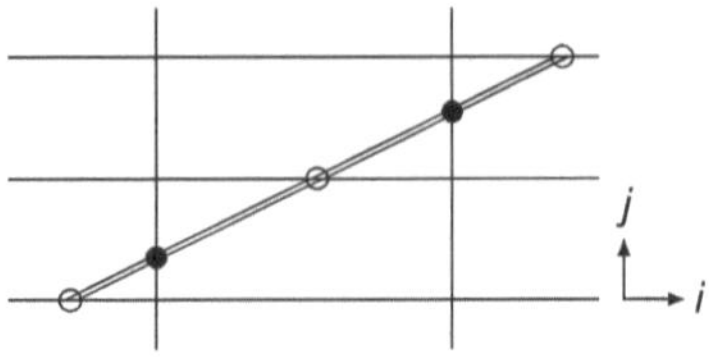

FIG. 3 Shock-points: ○ = x-shock, ● = y-shock.

Moretti's claim was that shock-fitting codes were simple and provided fast and accurate solutions if coupled with a suitable solver of the Euler equations, as the λ-scheme proposed by himself. He started from the assumption that shock-fitting simplicity was obvious in 1D problems, where the shock depends on its environment through a Riemann variable which can be correctly computed on both sides of the shock. In a one-dimensional problem, the values of the variables in the mesh node on the low pressure side of the shock $()_A$ and the value of one Riemann variable (see Eq. 1) in the mesh node on the high-pressure side are correctly computed by λ-scheme (Moretti, 1987a). On the other hand, the values on the high-pressure side of the shock $()_B$ will be updated by means of the R-H relations once the shock strength M_s, defined as:

$$M_s = \frac{|u_A - w|}{a_A} \tag{2}$$

has been computed. To evaluate M_s, a new variable Σ is introduced (Moretti and Di Piano, 1983), computed from the variables on the low pressure side and the Riemann variable on the high-pressure side of the shock as:

$$\Sigma = (a_B + \delta|u_A - u_B|)/a_A \tag{3}$$

Since Σ can also be expressed, using the R-H jump relations, as a monotonic increasing function of M_s:

$$\Sigma = \frac{\left[\sqrt{(\gamma M_s^2 - \delta)(1 + \delta M_s^2)} + \delta(M_s^2 - 1)\right]}{(1+\delta)M_s} \tag{4}$$

the shock strength M_s is obtained by inverting (4).

According to Moretti, simplicity is not lost in two dimensions, where the equations are only slightly altered by a slope factor and one more R-H condition, as successfully demonstrated by the boundary shock-fitting technique described in Section 3. The topological problem of evaluation of shock slope for a shock placed anywhere on a two-dimensional computational field was solved by suitably analysing the shock-point neighbourhood. When a shock occurs in a two-dimensional field it can cross one or both families of coordinate lines. Therefore we can identify x and y shocks accordingly. Moreover, for each of them one can identify the high-pressure side of the shock.

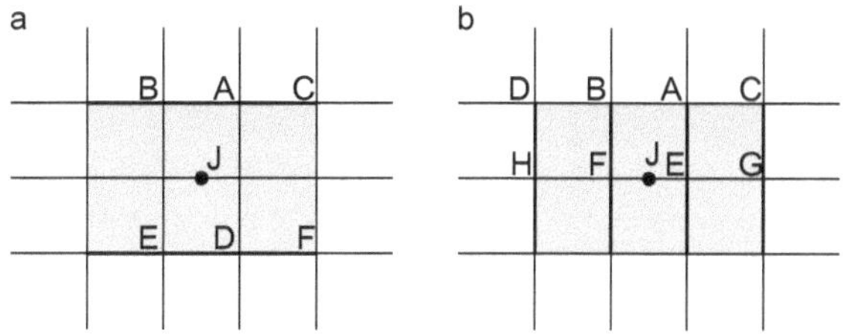

FIG. 4 Shock-point and its neighbourhood. (a) Scan for neighbouring x-shocks. (b) Scan for neighbouring y-shocks.

The original technique proposed by Moretti considered that some portions of the shock are better defined by their intersection with one family of coordinate lines, other portions by their intersection with the opposite family. In an attempt of further generalisation of the technique, Moretti's disciples tracked all intersections of shocks with coordinate lines (Nasuti and Onofri, 1996) eventually reaching an equivalent procedure and similar results. The solution of the topological problem of shock slope evaluation was found by storing information relevant to shock-points in single arrays without ordering them in any particular way. Shock slope was therefore obtained looking to the immediate neighbourhood of the shock-point. Let us consider for instance the x shock-point J sketched in Fig. 4. Its neighbourhood is made by the six cells surrounding it. The task of identifying which is a neighbouring shock-point, if any exist, of a given x shock-point as J in Fig. 4, is carried out as follows. A double array JSX(N, M) is set equal to zero at all grid nodes but the node at the right-hand side of a shock-point where it assumes the value J. There will be therefore a correspondence between the shock-point J and the grid nodes bracketing it (N, M) and (N + 1, M). With this data structure it is easy to find neighbours. More precisely, at first the neighbourhood of J is scanned for x-shocks (looking to left of upper grid nodes, A, B, C and of lower grid nodes D, E, F of Fig. 4a) and then, if upper nodes search fails, nodes A, B, C and D of Fig. 4b are scanned in search of y-shocks, whereas if lower nodes search fails, nodes E, F, G and H of Fig. 4b are scanned in search of y-shocks. At the end of the process an upper and a lower shock-point are identified that will be used to evaluate shock slope in J. Another way of coding the neighbourhood shock is reported in Nasuti and Onofri (1996). Being all intersections of shock and grid lines available with this latter approach the neighbourhood to be analysed is made of only two cells (see Fig. 5). For the sake of brevity we will only refer to the latter approach to describe the procedure to evaluate the shock slope.

In case of ordinary shock-points, which are defined as shock-points J with no more than two shock-points in their neighbourhood, one (J_i, $i = 1, 3, 5$) in the mesh above and/or one (J_j, $j = 2, 4, 6$) in the mesh below, the connection of the shock-points around J used to compute θ_s (the shock angle with respect

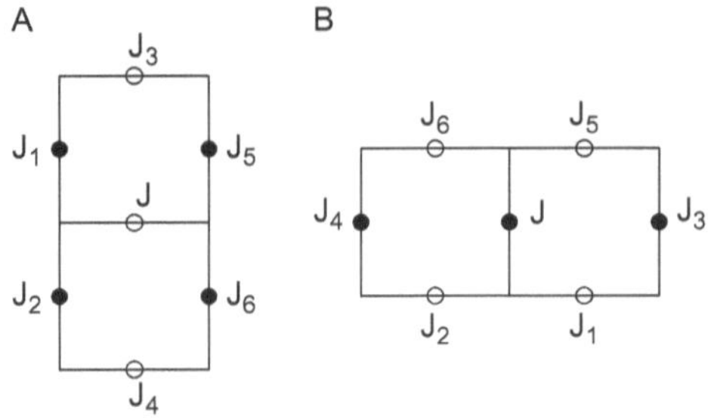

FIG. 5 Possible shock-points around: (A) an x-shock; (B) a y-shock.

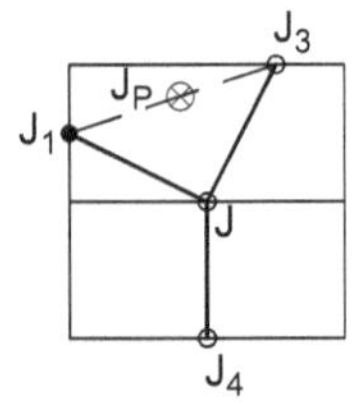

FIG. 6 Triple shock-point.

to coordinate line directions) should be carried out following a physical criterion based on the domain of dependence, as also addressed in Section 3. From this point of view, two cases are possible:

1. *Supshock* when the velocity component along the shock direction ($\tilde{v}$) is such that $|\tilde{v}_A| > a_A$ and $|\tilde{v}_B| > a_B$ (note that $\tilde{v}_A = \tilde{v}_B$). In this case no signal can propagate upstream in the direction tangential to the shock, because the flow is supersonic in that direction at both sides of the shock. As a consequence, no J_i which is downstream of J along the shock can affect the computation of θ_s in J. Therefore, only the shock-point located upstream of J is used to compute θ_s, and the direction J–J_i is taken as the shock direction.
2. *Subshock* when $|\tilde{v}_A| < a_A$ or $|\tilde{v}_B| < a_B$ ($\tilde{v}_A = \tilde{v}_B$). In this case θ_s is computed taking the direction J_i–J_j.

The technique is also able to deal with shock interaction so also triple or multiple shock-points may occur. An x-shock is defined as a triple shock-point if either two of J_1, J_3 and J_5 or two of J_2, J_4 and J_6 exist (Fig. 5A). Let us consider the example of Fig. 6. If J, J_4 and J_3 have the high-pressure side on the right, and J_1 on the upper side, then J is actually a triple point and requires a specific treatment, whereas J_1 and J_3 are ordinary shock-points. θ_s is computed taking the direction J_P–J_4 as the shock direction, where J_P is the middle point of the line J_1–J_3. During the computation, particularly in the case of transient flows, it may happen that many shock-points are close to each other. If there are both two or three compatible shock-points above J and two or three compatible shock-points below J (Fig. 5A), θ_s is obtained with a procedure similar to that used for triple points, considering a suitable averaging of compatible shock-point positions. The floating shock-fitting technique is completed by specific procedures able to move shock-points along the grid according to computed shock velocity, detect the formation of new shocks (Moretti and Valorani, 1988; Nasuti and Onofri, 1996) and the disappearance of shocks as they weaken. Moreover, a special cure was needed when coding to avoid the calculation of differences across the discontinuities that could affect the solution during transients.

Summarising, each time iteration of the integration process can be organised in six main steps:

1. Integration of the equations over all the mesh nodes, according to the λ-scheme.
2. Detection of the formation of new shocks.
3. Calculation of the local slope of the shocks.
4. Enforcement of the R-H jump conditions between the two sides of each shock-point.
5. Displacement of the shock-points, according to the velocity of propagation of the shock front.
6. Check of crossing of shock and grid lines introducing markers (Moretti, 1988) or connecting shock-points (Nasuti and Onofri, 1996), to avoid differencing through discontinuities.

The procedure can also include an analogous contact-discontinuities fitting, if it is of interest for a specific problem. It is not much different than for shocks and is omitted here. Details can be found in Nasuti and Onofri (1996).

The solution of more complicated, two-dimensional flow fields can hardly be obtained relying on a single block. The task becomes even harder if one considers a code based on orthogonal grids. A needed step towards the extension of the floating shock-fitting technique was therefore to include an overlapping multiblock technique. In the approach presented in Nasuti (2003), integration at the boundary embedded in the analysed flow field is carried out as in all other nodes thanks to the introduction of suitable ghost nodes and exploitation of bilinear interpolation, and of a specific interpolation procedure for floating shock-points. More specifically, two rings of ghost nodes are defined as shown in Fig. 7, whose position is obtained by smoothly prolonging the inner node distribution outside of the block, so that the solution at the ghost nodes can be interpolated from that of the neighbouring block. Such nodes are required to lie inside the neighbouring block, or in other words, there can be no mesh-size holes between neighbouring blocks. The generic ghost point A lies inside a real cell BCED of the neighbouring block (block 2) and the generic variable at the point A is obtained by the bilinear interpolation of its value in the nodes BCED. Then, at the next iteration, the solution at the boundary nodes can be updated as done at the inner nodes, because the missing derivatives are

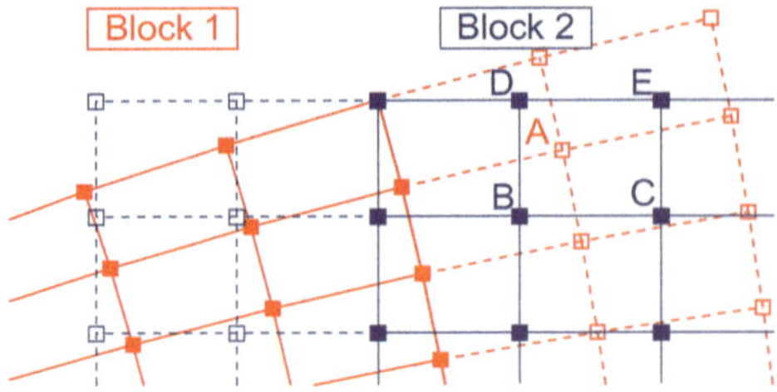

FIG. 7 Ghost nodes (*hollow squares*) introduced to ensure solution continuity through a block boundary internal to the analysed flow field.

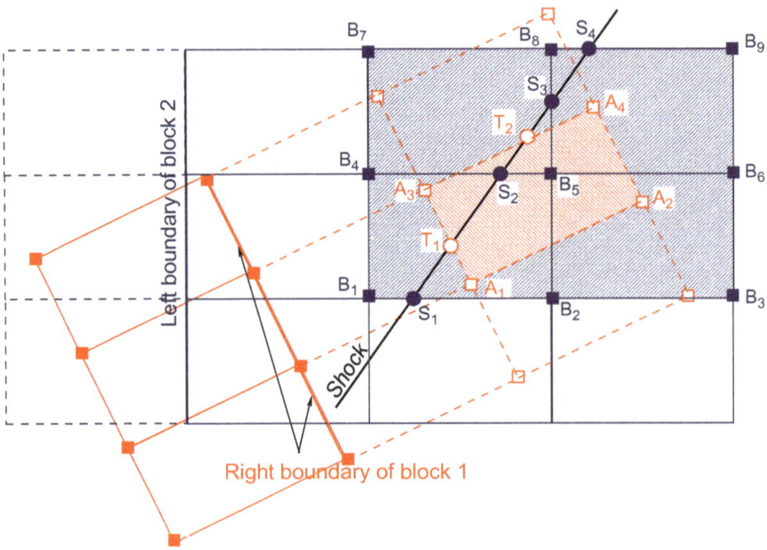

FIG. 8 Introduction of ghost shock-points.

obtained from the solution in the inner and ghost nodes, the latter bringing information from the neighbouring blocks.

The use of ghost nodes allows introducing also ghost shock-points needed to follow the general shock-fitting procedure to transfer information between blocks. In particular, the introduction of ghost shock-points is necessary to compute the local shock slope of real shock-points at the block boundary and to avoid interpolating across shocks. The procedure used to find ghost shock-points is illustrated by the example shown in Fig. 8, where the problem is to find the ghost shock-points in the prolongation of block 1, when the shock position is known by the real shock-points in block 2. A possible approach is to consider for each ghost cell $A_1A_2A_4A_3$ a quadrilateral $B_1B_3B_9B_7$, that is the smallest group of cells large enough to include the whole ghost cell. Once the quadrilateral is defined, it is possible to look for shock-points along its boundaries, and, in case some are found, connect them by straight lines and find the intersection of these lines with the boundaries of the ghost cell. In the example of Fig. 8 the procedure finds the shock-points S_1 and S_4, and the intersection of S_1S_4 with A_1A_3 and A_3A_4 provides the ghost shock-points T_1 and T_2, respectively.

4.1 Floating Shock-Fitting Results

The classical validation test cases of complex shock patterns obtained by floating shock-fitting techniques are reported in Moretti (1988), Moretti (1987c) and Nasuti and Onofri (1996). They address Euler test cases, like transonic profiles, unsteady Mach reflections, regular reflections, supersonic air intake and shock diffraction.

As a first example, consider the case of steady state regular reflection obtained with the coding of Nasuti and Onofri (1996). A supersonic flow with a Mach number $M = 3$ enters a 2D plane duct, whose geometry displays two regions with constant area sections joined through a ramp of 10 degree slope, and having a contraction ratio of 0.65. The Mach number computed with a 70×20 grid is shown in Fig. 9.

A second example is the case of Mach reflection over a concave corner of 10 degree, as obtained with the coding of Nasuti and Onofri (1996). A plane shock ($M_s = 6.69$) followed by an inviscid supersonic flow ($M = 1.75$) moves from the left in quiescent air. As it enters the converging ramp a simple Mach reflection occurs. Before the reflected shock reaches the lower wall, the flow behaves as in an unbounded field, and the solution is pseudostationary, i.e., similar to itself in time, with the triple point moving along a straight line. The shock evolution in time is shown in Fig. 10. The straight line overlapping

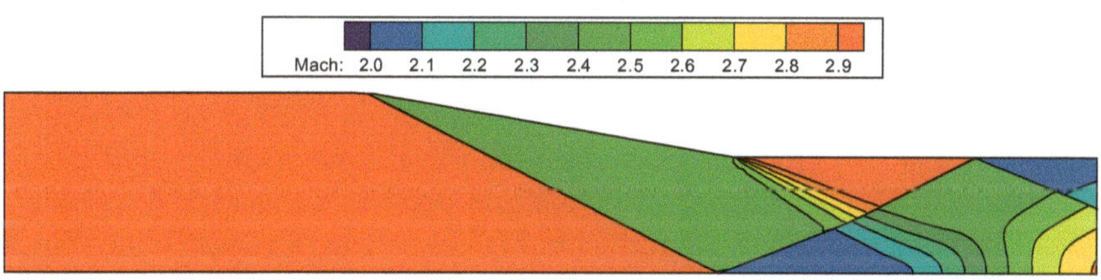

FIG. 9 Steady inviscid regular reflection of an oblique shock computed by floating shock fitting.

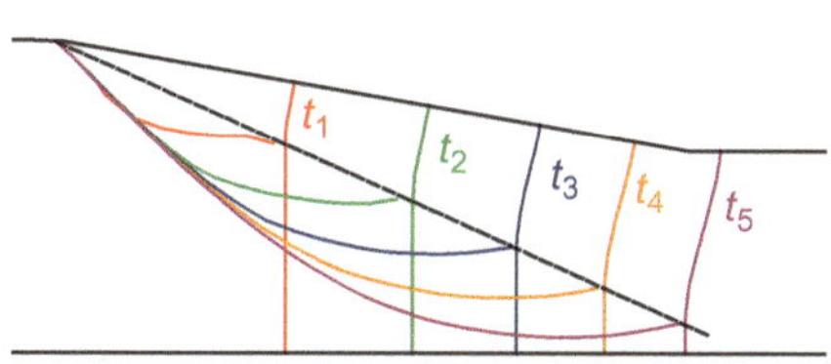

FIG. 10 Unsteady inviscid Mach reflection: comparison of experimental data fit line for triple point evolution and computed shock at different times (shock lines are obtained connecting shock-points).

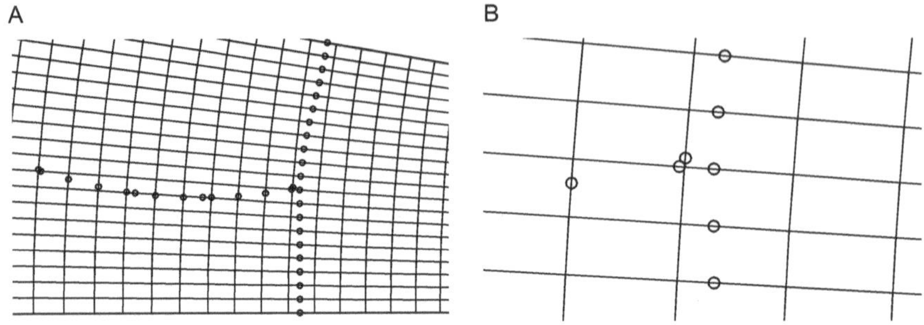

FIG. 11 Unsteady inviscid Mach reflection: computed shock-points and the underlying grid lines at time t_3 (see Fig. 10). (A) Mach reflection shock points. (B) Detail of shock points at the triple point.

the computed triple point solution is drawn according to the experimental measure fit reported in Ben-Dor (1992). Details of the position of computed shock-points at a given time are shown in Fig. 11.

Another test case concerns the inviscid solution of the flow field relevant to a cylindrical converging shock wave. This test allows the assessment of the capability of the fitting technique to predict the movement of flow discontinuity fronts forming any angle with the coordinate lines. The flow starts by removing a cylindrical diaphragm separating two regions of quiescent air at the same temperature but with a pressure in the external side of the cylindrical diaphragm four times greater than in the inner side. A cylindrical converging shock wave is generated, followed by a converging contact discontinuity, together with diverging expansion waves. Both shock and contact discontinuity have been fitted. The calculation has been performed on a Cartesian grid, with the origin of the frame of reference placed on the axis of the cylindrical diaphragm. The position of the discontinuities computed at three different times is shown in Fig. 12. The circular shape of discontinuities is preserved all along the computation (see Fig. 13) up to the time when the shock wave includes only a very small number of cells inside. The comparison between the analytical and computed shock intensity is also reported in Fig. 14, where the dashed line shows the reference pressure ratio through the shock (Chisnell, 1957) and the solid lines the computed pressure evolution in radial direction at different times.

4.2 Viscous Flows

Floating shock-fitting has also been used to simulate viscous flows including shock-boundary layer interactions. An example of a shock impinging on a viscous wall and thus reflected after shock-boundary layer interaction is presented and discussed in Moretti et al. (1993). Here the separated flow in an over-expanded rocket nozzle is shown as an example of viscous flow computation with floating shock-fitting. When a shock wave interacts with the

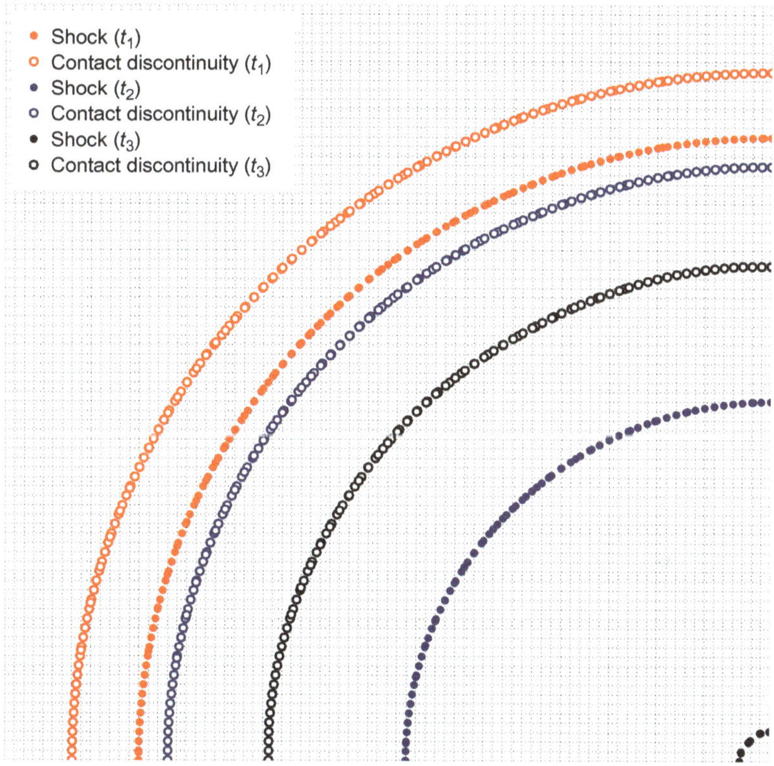

FIG. 12 Inviscid simulation of the propagation of a converging shock wave: computed shock and contact discontinuity points at three different times and underlying grid lines.

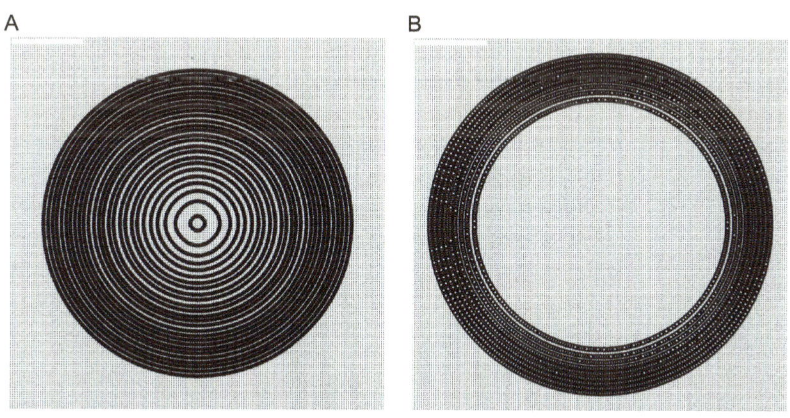

FIG. 13 Inviscid simulation of the propagation of a converging shock wave: computed shock-points (*left*) and contact discontinuity points (*right*).

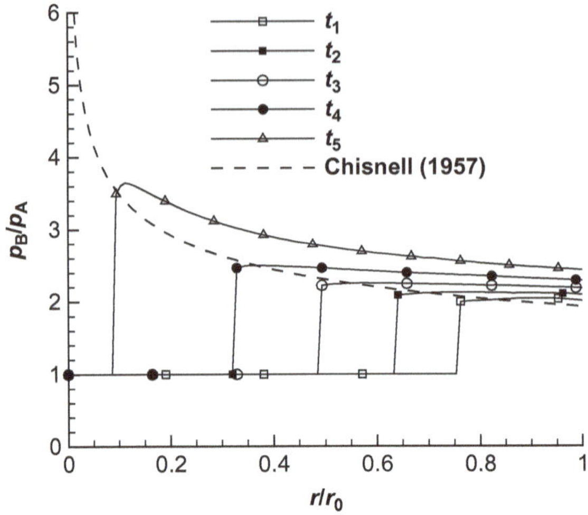

FIG. 14 Evolution of flow pressure in radial direction for a converging shock wave and comparison with the reference solution of converging shock pressure ratio evaluated in Chisnell (1957).

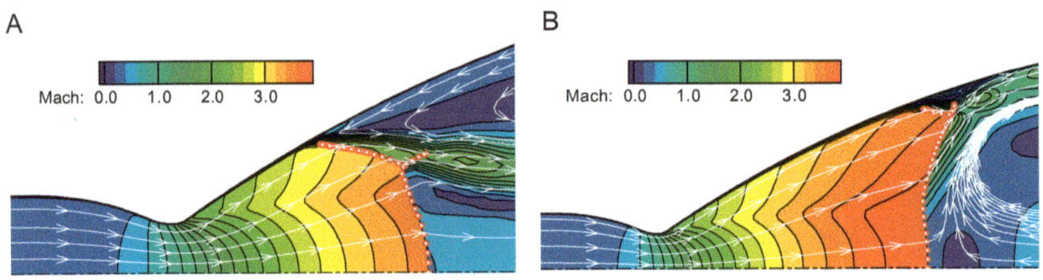

FIG. 15 Flow separation in an over-expanded nozzle with two different flow structures depending on the shock evolution in the core field and to its interaction with the boundary layer: Mach number contour lines, streamlines and shock-points. (A) Conventional flow separation structure. (B) Peculiar flow separation structure with flow reattachment.

boundary layer, floating shock-fitting is able to handle interaction exploiting a suitable check on shock intensity and so eliminating shock-points of vanishing intensity that fall in the depth of the boundary layer. Results as those reported in Fig. 15 have been obtained. On the left-hand side of the figure the flow field highlights the conventional flow separation structure. Because of the high over-expansion the flow separates at the wall and as a consequence of the change of direction of the streamline an oblique shock is generated that allows also the supersonic jet to recover the ambient pressure. The oblique shock generates as the result of converging characteristic lines and is well identified by the shock-points shown in the figure. The oblique shock eventually is reflected at the nozzle axis, and a Mach reflection occurs, where the Mach disk is not flat because of the nonuniform incoming flow. The right-hand side of Fig. 15 shows a different configuration which has been shown to occur for a family of bell-shaped nozzles by different CFD codes and by experiments. Due to upstream nonuniformity, the flow can separate from the axis and it is pushed towards the wall. In this case the separation bubble is restricted in a small zone and bounded by a supersonic jet. The computed shock-points highlight the shock curvature, the small dimension of the separation shock and the interaction between the reflected shock and the separated region.

4.3 Complex Flows

To demonstrate multiblock floating shock-fitting, we consider the computation of an aerospike nozzle flow field. This axisymmetric rocket nozzle is of interest in aerospace propulsion since the external expansion adapts itself to the ambient pressure with a consequent performance benefit (Nasuti and Onofri, 1998, 2001). The study has been carried out by partitioning the whole computational domain in 22 blocks and considering a coarse grid made of 28,828 cells and two further grids obtained by multiplying the number of cells in each direction by 1.5 and 2.0, respectively (Fig. 16). The resulting flow field obtained simulating in-flight conditions (free-stream Mach number, $M_\infty = 2$) is particularly interesting because of the presence of two supersonic afterbody regions, one behind the nozzle external shroud, the other behind the truncated spike, characterised by turbulent vortices, mixing layers and shocks. Fig. 17 shows that the three main fitted shocks smoothly cross the boundaries of the blocks (these latter are not shown in Fig. 17).

Another example of application is that of a dual-bell rocket nozzle operating in over-expanded condition that yields a large separated region (Fig. 18). The computed solution is clearly unaffected by the decomposition in blocks (whose boundaries are highlighted by black lines in Fig. 18) as shown by the supersonic jet region, the subsonic recirculating region and the fitted shock. The flow evolution during transition between operating modes with separated and attached

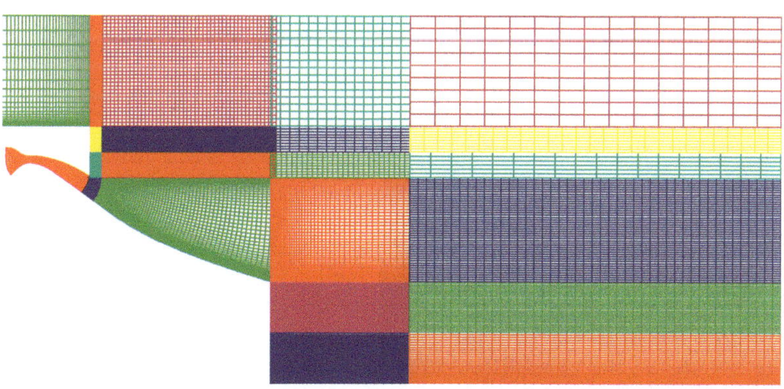

FIG. 16 Multiblock discretisation for the Reynolds averaged Navier–Stokes simulation of an axisymmetric aerospike nozzle flow.

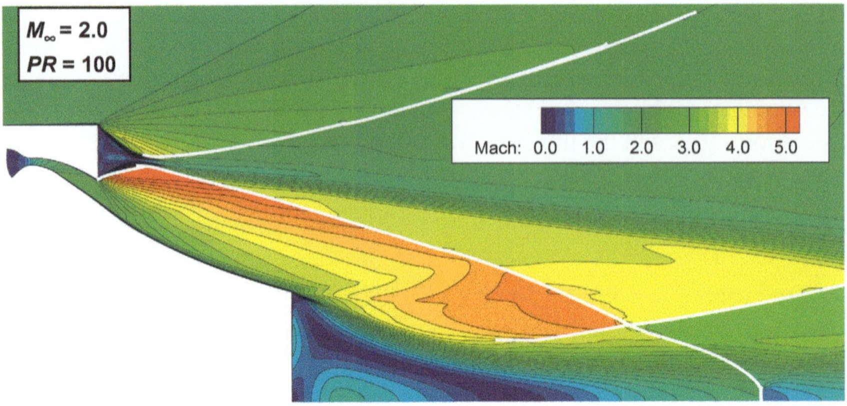

FIG. 17 Mach number isolines, and shocks computed for an axisymmetric aerospike nozzle flow.

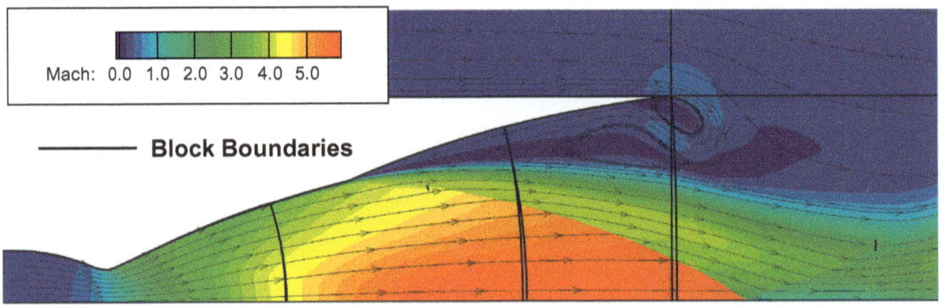

FIG. 18 Mach number isolines and streamlines computed for an over-expanded dual-bell rocket nozzle.

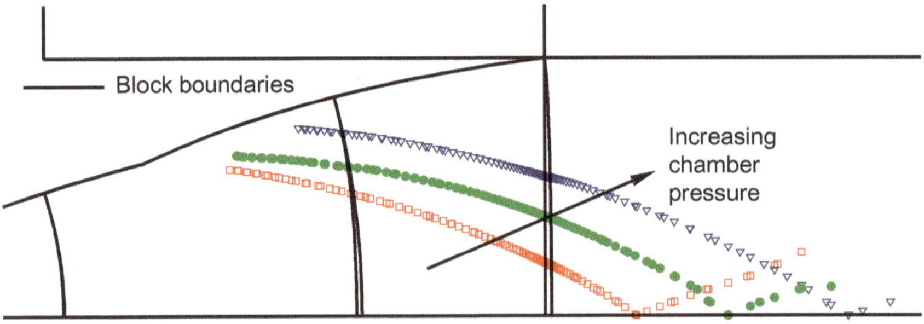

FIG. 19 Evolution of shock-points for increasing chamber pressure in a dual-bell nozzle.

flow is also of great practical interest for dual bell nozzles. To this goal an unsteady simulation obtained by doubling with a linear ramping chamber pressure in time has been carried out.

Fig. 19 shows the capability of the method of studying such flows by the expected smooth evolution of the shock towards the nozzle exit as chamber pressure increases in time.

These results indicate that although more cumbersome to develop than common CFD methods widespread in commercial and scientific communities, shock-fitting approaches retain their own merits that, if fully exploited, can yield the best result for some class of problems of interest, like flow transients with shocks moving through the flow field. The use of multiblock techniques allows extending the number of applications to more and more complex problems.

5 SHOCK-FITTING FOR UNSTRUCTURED GRIDS

The continuous advances in computer power and computational science that have taken place over the last decades have made the use of unstructured grids computationally affordable and algorithmically simpler. Taking advantage of this opportunity, a new shock-fitting technique for unstructured meshes has been developed in Italy by second generation of Moretti's disciples (Bonfiglioli et al., 2013; Paciorri and Bonfiglioli, 2009, 2011).

This novel, unstructured shock-fitting technique combines the coding simplicity of the "boundary" shock-fitting described in Section 3 with the capability of dealing with complex flows of the "floating" technique described in Section 4. Moreover, the unstructured version of the technique does not suffer from the strong topological limitations that plague boundary shock-fitting when implemented on structured grids and the coupling between the unstructured shock-fitting algorithm and existing gasdynamic solvers is algorithmically simple, as in the boundary shock-fitting technique for structured-grids.

5.1 Unstructured Shock-Fitting: Algorithmic Features

The approach is inherently time-dependent: both the solution and the grid change with time, due to the displacement of the fitted discontinuities. When a steady solution exists, the shock speed will asymptotically vanish and the tessellation of the flow domain will not any longer change.

At time t the set of dependent variables and grid velocity are available within all grid-points of a tessellation (made of triangles in 2D and tetrahedra in 3D) that covers the entire computational domain; this is what we call the "background" mesh. In addition to the background mesh, the fitted discontinuities (either shocks or contact discontinuities) are discretised using a collection of points which are mutually joined to form a connected series of line segments, as shown in Fig. 20A for the 2D case, or a triangulated surface in 3D, as shown in Fig. 20B. For example, a thick solid (yellow) line is used in Fig. 20A to mark the various fitted discontinuities that arise due to the interaction between two shocks of the same family: the two incident shocks, the resulting shock, a weak compression (or expansion) wave and the contact-discontinuity located between the former two. Fig. 20B, which refers to the three-dimensional, supersonic flow past a blunt-nosed object, shows the

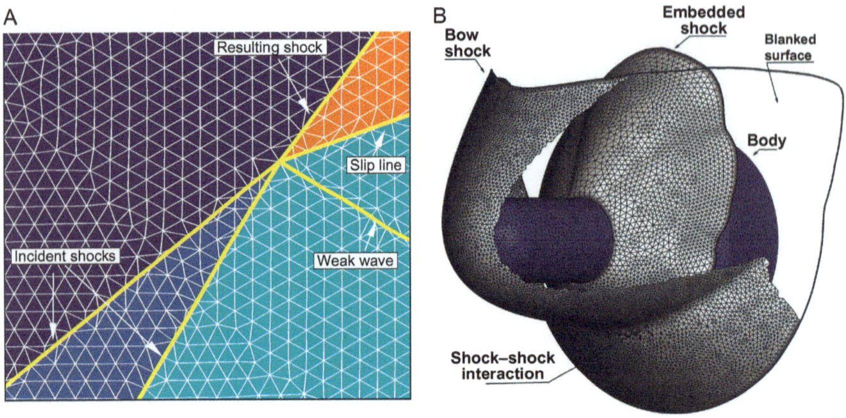

FIG. 20 Examples of fitted discontinuities on unstructured meshes. (A) Interaction of shocks belonging to the same family. (B) Supersonic flow over a blunt-nosed body.

triangulated surfaces used to fit the bow shock and the embedded shock that arises at the cylinder-flare junction. Although it may not evident from Fig. 20, each fitted discontinuity is a double-sided internal *boundary* of zero thickness; this is sketched in Fig. 21D, which refers to the case of a shock wave. Because the width of the discontinuity is negligible (it has been amplified in Fig. 21D to improve visibility), its two sides are discretised using the same polygonal curve or triangulated surface; each pair of nodes that face each other on the two sides of the discontinuity share the same geometrical location, but store different values of the dependent variables, one corresponding to the upstream state and the other to the downstream one. Moreover, a velocity vector normal to the discontinuity is assigned to each pair of grid-points on the fitted discontinuity: it represents the displacement velocity of the discontinuity. As shown in Fig. 21A, the spatial location of the fitted discontinuities is independent of the location of the grid-points that make up the background grid.

The process that leads from the available mesh and solution at time t to an updated mesh and solution at time $t + \Delta t$ can be split into seven steps that will be described in the following sections.

5.1.1 Cell Removal Around the Shock Front

In this first step, the fitted discontinuities are laid on top of the background mesh, as shown in Fig. 21A. All those cells that are crossed by the fitted discontinuities and those mesh points that are located too close to it are temporarily removed from the background mesh, as shown in Fig. 21B. We call "phantom" those grid-points of the background mesh (shown using dashed circles in Fig. 21B) that have been temporarily removed. All cells having at least one phantom node among their vertices are also removed from the

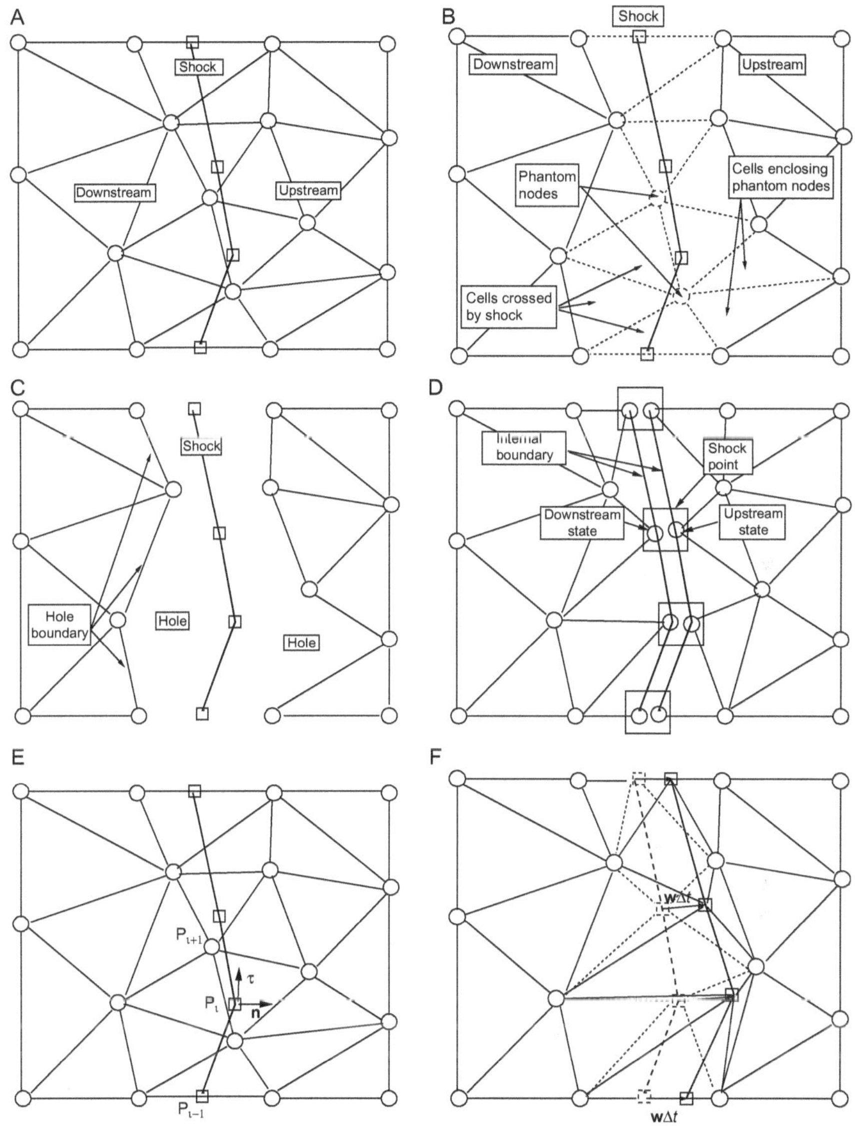

FIG. 21 Unstructured shock-fitting. (A) Shock front moving over the background triangular mesh at time *t*. (B) *Dashed lines mark* the cells to be removed; *dashed circles* denote the phantom nodes. (C) The background mesh is split into disjoint subdomains by a hole which encloses the shock. (D) The triangulation around the shock has been rebuilt. (E) Calculation of the shock-tangent and shock-normal unit vectors. (F) The shock displacement induces mesh deformation.

background triangulation; these are the cells shown using dashed edges in Fig. 21B. Further details concerning the criteria used to identify and remove the phantom nodes can be found in Paciorri and Bonfiglioli (2009) and Bonfiglioli et al. (2013).

5.1.2 Local Remeshing Around the Shock Front

Following the cell removal step, the background triangulation has been split into two or more disjoint subdomains, as shown in Fig. 21C. The hole dug by the fitted front is then remeshed using a constrained Delaunay tessellation (CDT): the edges/triangles that make up the fitted discontinuity and the boundary of the hole are both constrained to be part of the final tessellation; this is illustrated in Fig. 21D. Public domain software is used to construct the CDT (Shewchuk, 1996; Si, 2015) and, in 3D, to triangulate the fitted shock surfaces; observe that remeshing is localised around the discontinuities and, therefore, does not overload the algorithm in terms of CPU cost.

Upon completion of this stage, the computational domain is discretised using what we call the "shock-fitting" mesh, which differs from the background mesh only in the neighbourhood of the fitted discontinuities.

5.1.3 Computation of the Tangent and Normal Unit Vectors

In order to apply the jump relations, normal ($\mathbf{n}$) and tangent ($\boldsymbol{\tau}$) unit vectors are needed within each pair of grid-points located along the discontinuities, as also described for the boundary and floating shock-fitting, see Fig. 21E. These unit vectors are computed using finite-difference (FD) formulae which involve the coordinates of the shock-point itself and those of its neighbouring shock-points. Depending on the local flow regime, it may be necessary to use upwind-biased formulae to avoid geometrical instabilities along the fitted discontinuity, see also Section 4. Full details describing how to compute the normals to the discontinuity can be found in Paciorri and Bonfiglioli (2009) and Salas (2009) for the 2D case and in Bonfiglioli et al. (2013) for the 3D case.

5.1.4 Solution Update Using the Shock-Capturing Code

Using the shock-fitting mesh as input, a single time step calculation is performed using an unstructured, vertex-centred shock-capturing solver which returns updated nodal values at time $t + \Delta t$. Since the discontinuities are seen by the shock-capturing code as internal boundaries (of zero thickness) that move with the velocity of the discontinuity, there is no need to modify the spatial discretisation scheme already implemented in the PDEs solver to account for the presence of the fitted discontinuities. In practice, the shock-capturing solver is used as a black-box: it receives as input the shock-fitting grid, the nodal values of the solution and grid velocity at time t and returns the updated solution at time $t + \Delta t$.

The solution returned by the shock-capturing solver at time $t + \Delta t$ is however missing some boundary conditions on one or both sides of each discontinuity, depending on whether it is a shock or a contact. These missing pieces of information will be determined in the next step.

5.1.5 Enforcement of the Jump Relations

The missing pieces of information that are needed to correctly update the solution within all pairs of grid-points located on the discontinuities are obtained by enforcing the R-H jump relations; this also provides the local velocity of the discontinuity along its normal.

The R-H jump relations are a set of nonlinear algebraic equations that can be solved within each pair of grid-points located along the discontinuities by means of Newton–Raphson's algorithm. In order to match the number of unknowns with the available equations, one or more additional pieces of information are required within both or either of the two sides of the fitted discontinuity, depending on whether this is a shock or a contact discontinuity. In very much the same way described in Sections 3 and 4, these additional pieces of information are obtained from the characteristic formulation of the Euler equations and correspond to those characteristic quantities that are convected towards the discontinuity from the subdomain that is attached to that side of the discontinuity. Using an upwind-biased discretisation within the shock-capturing solver, as we do, one can reasonably assume that the spatial and temporal evolution of these characteristic quantities has been correctly computed. Full algorithmic details concerning the practical implementation of the jump relations for shocks and contact discontinuities is reported elsewhere and will not be repeated here: see Ivanov et al. (2010) and Paciorri and Bonfiglioli (2009, 2011) for the 2D case and Bonfiglioli et al. (2013) for the 3D case.

An ad hoc treatment is moreover required within those special points where different discontinuities interact; this is the case of triple and quadruple points, points where an impinging shock is reflected off a solid surface, etc. The algorithmic details are described in Ivanov et al. (2010) and Paciorri and Bonfiglioli (2011) for the 2D case, whereas the interaction among fitted discontinuities has not yet been dealt with in 3D.

5.1.6 Shock Displacement

The enforcement of the jump relations described in Section 5.1.5 provides the speed at which each pair of grid-points located on the discontinuity move along its local normal vector. The position of the discontinuity at time $t + \Delta t$ is computed in a Lagrangian manner by displacing all its grid-points, as shown in Fig. 21F where the dashed and solid lines represents the discontinuity at time t, resp. $t + \Delta t$. When simulating steady flows, this can be accomplished using the following first-order-accurate (in time) integration formula:

$$\mathbf{P}_i^{t+\Delta t} = \mathbf{P}_i^t + \mathbf{w}_i^t \Delta t \tag{5}$$

which returns the spatial coordinates of the i-th grid-point at time $t + \Delta t$. The low temporal accuracy of Eq. (5) does not affect the spatial accuracy of the steady state solution which only depends on the spatial accuracy of the gasdynamics solver and that of the tangent and normal unit vectors. On the contrary, when dealing with unsteady flows, the temporal accuracy of the shock motion has to be the same as that of the shock-capturing solver, i.e., second-order-accurate in our case. This can be accomplished using a predictor–corrector type temporal integration scheme, as described in Bonfiglioli et al. (2016), or a Runge–Kutta, multistep scheme.

Fig. 21F shows that even when the background mesh is fixed in space, the triangular cells that abut on the discontinuity have one of their edges that moves with the discontinuity, thus deforming the cell. This implies that the shock capturing solver that is used in Step 5.1.4 must be capable of handling moving meshes, i.e. it must be capable of solving the governing PDEs written using an Arbitrary Eulerian Lagrangian (ALE) formulation.

Finally, the time step Δt to be used in Eq. (5) to move the shock is chosen in such a way that during the time interval $[t, t+\Delta t]$ the shock will remain within the hole that it has dug in the background mesh. By doing so, none of the grid-points of the shock-fitting mesh will be overcome by the moving discontinuity, as shown in Fig. 21F.

5.1.7 Interpolation of the Phantom Nodes

Upon completion of the previous steps, all nodes of the shock-fitting mesh have been updated at time $t + \Delta t$. The shock-fitting mesh is made up of all the grid-points belonging to the fitted discontinuities and all nodes of the background mesh, except those that have been declared "phantom". Therefore, the nodal values within the phantom nodes have not been updated to time $t + \Delta t$. However, during the current time step, the discontinuity might have moved sufficiently far away from its previous position, that some of the phantom nodes may reappear in the shock-fitting mesh at the next time step. It follows that also the nodal values within the phantom nodes need to be updated to time $t + \Delta t$. This is easily accomplished by transferring the available solution at time $t + \Delta t$ from the current shock-fitting mesh to the grid-points of the background one, using linear interpolation. Once the phantom nodes have been updated, the shock-fitting mesh used in the current time interval has completed its task and can be removed.

At this stage the numerical solution has correctly been updated at time $t + \Delta t$ within all grid-points of the background tessellation and within all pairs of grid-points that belong to the fitted discontinuities. The next time interval can be computed restarting from the first step in Section 5.1.1 of the algorithm.

5.2 Unstructured Shock-Fitting: Applications

The unstructured shock-fitting technique described in Section 5 will be hereafter used to simulate steady and unsteady flows, both in two and three spatial dimensions. For some test-cases, a comparison will also be made with a shock-capturing calculation performed on the same background tessellation also used in the shock-fitting calculation. For both sets of simulations, the in-house, vertex-centred, shock-capturing solver that has been used (Bonfiglioli, 2000; Bonfiglioli and Paciorri, 2013) relies on Residual Distribution, or Fluctuation Splitting schemes (Abgrall, 2006; Deconinck et al., 1993).

5.2.1 *Transonic* ($M_\infty = 0.8$) Flow Past the NACA 0012 Airfoil

The superior accuracy that fitted shock-waves deliver over captured ones is illustrated by reference to a well documented external flow test-case, namely the two-dimensional, inviscid, transonic flow past the NACA 0012 airfoil at $\alpha_\infty = 0$ degree angle of incidence and free-stream Mach number equal to $M_\infty = 0.80$. A symmetric flow field has been chosen since it allows to compare the two approaches using a single numerical simulation: as shown in Fig. 22, the shock on the upper side of the profile has been fitted, whereas the one on the lower side has been captured. The numerical solution has been obtained starting from a background grid featuring 9912 triangles and 5024 mesh-points, 78 of which are placed along the airfoil's profile.

Fig. 22A shows the shock-fitting mesh at steady state: the mesh above the profile has been modified by the shock-fitting algorithm to accommodate the fitted shock and it only differs from the background triangulation in the immediate neighbourhood of the discontinuity; the mesh below the profile coincides with the background triangulation. Fig. 22B shows the computed Mach isocontour lines: it is clear that shock-fitting provides a much more realistic shock thickness than shock-capturing, using grids of nearly identical spatial resolution. For a more detailed comparison between the two different shock-modelling approaches it is worth having a closer look at the foot and the tip of the shock wave: the former is the point where the shock impinges on the profile and the latter is the region away from the profile where the shock disappears.

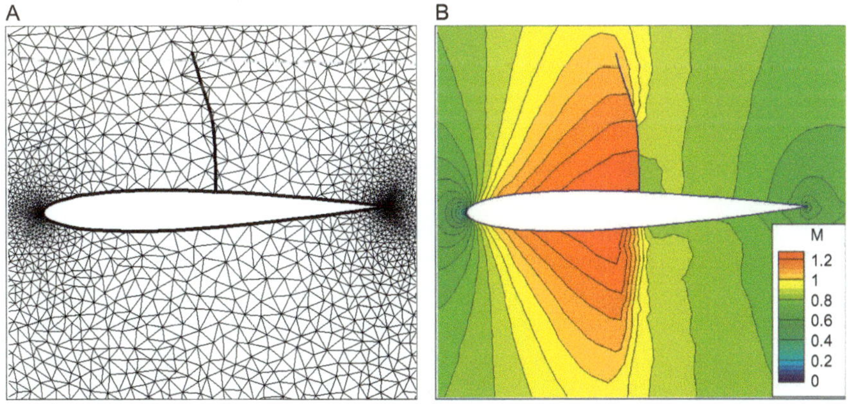

FIG. 22 NACA 0012 airfoil, $\alpha_\infty = 0$ degree, $M_\infty = 0.80$. (A) "Shock-fitting" mesh at steady-state. (B) Mach number field.

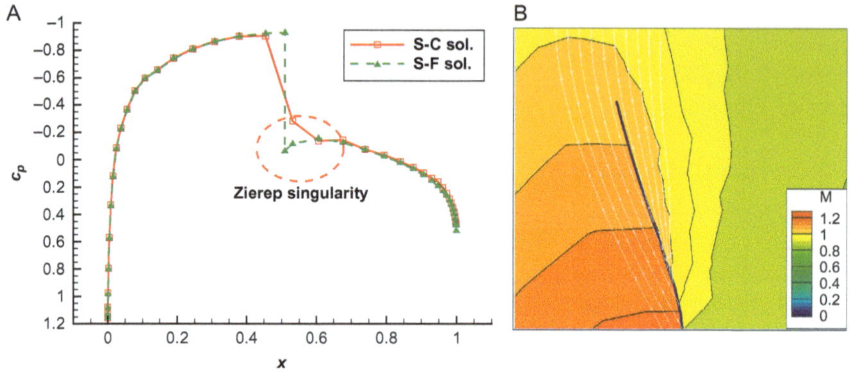

FIG. 23 NACA 0012 airfoil, $\alpha_\infty = 0$ degree, $M_\infty = 0.80$. (A) Pressure coefficient distribution. (B) Shock coalescence.

Fig. 23A compares the pressure coefficient C_p distributions computed along the profile using the two different modelling practices: not only the shock-capturing calculation predicts an unphysically large shock-thickness, it also completely misses the so-called Zierep (2003) singularity that occurs on the downstream foot of the shock which, by contrast, is picked up when the shock is fitted. This very same observation had been made more than 25 years ago by Morton and Paisley who performed a very similar calculation using a structured quadrilateral mesh on top of which a (small) mesh patch was adapted around a fitted shock "defining a line of adjustable mesh points treated as an internal boundary" (Morton and Paisley, 1989).

When looking at the tip of the shock line, see Fig. 23B, the shock-fitting solution clearly reveals that the shock is formed through the coalescence of those characteristic curves of the steady Euler equations that form an angle $-\mu$ with respect to the streamline, where μ is the Mach angle. These same features could hardly be seen in a shock-capturing solution, unless a much finer, or feature-adapted, mesh were used.

5.2.2 *Planar, Transonic Compressible Point-Source Flow*

Not only does shock-fitting allow us to compute accurate solutions on coarse meshes, even more importantly, it avoids most of the drawbacks that plague shock-capturing discretisations. These include: the reduction of the global order of accuracy within the regions of the flow field that are located downstream of a discontinuity (Carpenter and Casper, 1999; Casper and Carpenter, 1998; Engquist and Sjögreen, 1998) and the appearance of spurious numerical oscillations behind steady (Lee and Zhong, 1999) and moving (Johnsen et al., 2010) shock waves. We have investigated (Bonfiglioli and Paciorri, 2014) these issues using a planar, two-dimensional compressible source flow featuring an analytical solution. The computational domain consists in an annulus, see Fig. 24A, and the inflow/outflow conditions are

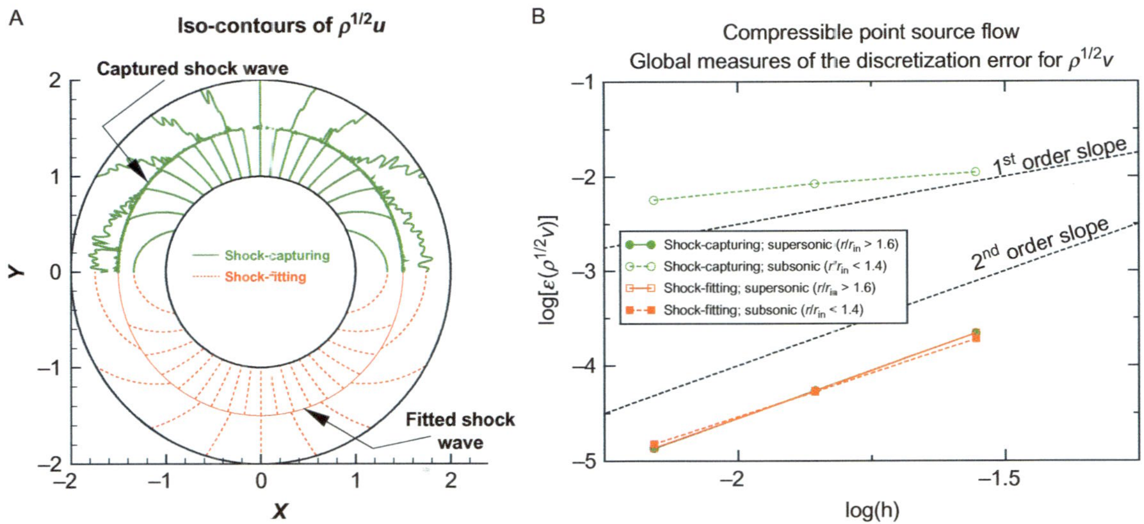

FIG. 24 Planar, compressible point-source flow. (A) Isocontours of $\sqrt{\rho}u$. (B) Order of convergence of $\sqrt{\rho}v$.

chosen in such a way that a shock is located half way between the supersonic inflow and the subsonic outflow boundaries. Three levels of nested triangular grids have been generated in order to perform a grid-convergence study.

A qualitative comparison between shock-capturing and shock-fitting is reported in Fig. 24A, which shows the isocontours of $\sqrt{\rho}u$ computed by means of shock-capturing in the upper half of the frame and by means of shock-fitting in the lower half. It is evident that the two different modelling practices provide identical solutions upstream of the shock-wave, whereas severe oscillations are evident in the solution computed by means of shock-capturing within the entire shock-downstream region. A more quantitative comparison between the two different approaches is reported in Fig. 24B, where the L_1-norm of the discretisation error ϵ for the variable $\sqrt{\rho}v$ is plotted against the mesh spacing h. The grid convergence analysis has been performed separately within two disjoint subdomains that do not encompass the shock-wave: one lies entirely within the supersonic region and the other within the supersonic one. It can be seen that both the shock-capturing and shock-fitting discretisations recover second-order upstream of the shock wave, whereas downstream of it the shock-fitting solution converges at design order, but the shock-capturing calculation falls even below first order.

5.2.3 *Type IV Shock–Shock Interaction*

The test cases presented so far are characterised by the presence of a single shock wave. The unstructured shock-fitting technique is, however, capable of simulating steady flows featuring different kinds of shock–shock and shock–wall interactions, both in two (Ivanov et al., 2010; Paciorri and Bonfiglioli, 2011) and three (Bonfiglioli et al., 2012, 2013) spatial dimensions.

A type IV (Edney, 1968) interaction between an incident shock and a bow shock has been simulated using both shock-capturing and shock-fitting. The shock-fitting result is shown in Fig. 25A: observe that the various shocks, triple points and slip-lines that characterise this complex flow have all been fitted. Fig. 25 allows to compare the shock-fitting result with a shock-capturing calculation obtained on an anisotropically adapted (Dolejší, 1998) mesh. The two calculations appear to be of comparable quality: both feature smooth contours within the shock layer, even though small wavelength disturbances are visible in the proximity of the bow shock in the shock-capturing solution. It is, however, worth underlining that the shock-fitting technique requires roughly half the number of grid-points and triangles than that required by shock-capturing computed on the anisotropically adapted mesh. A more thorough comparison between shock-fitting and shock-capturing on feature-adapted meshes can be found in Ivanov et al. (2010).

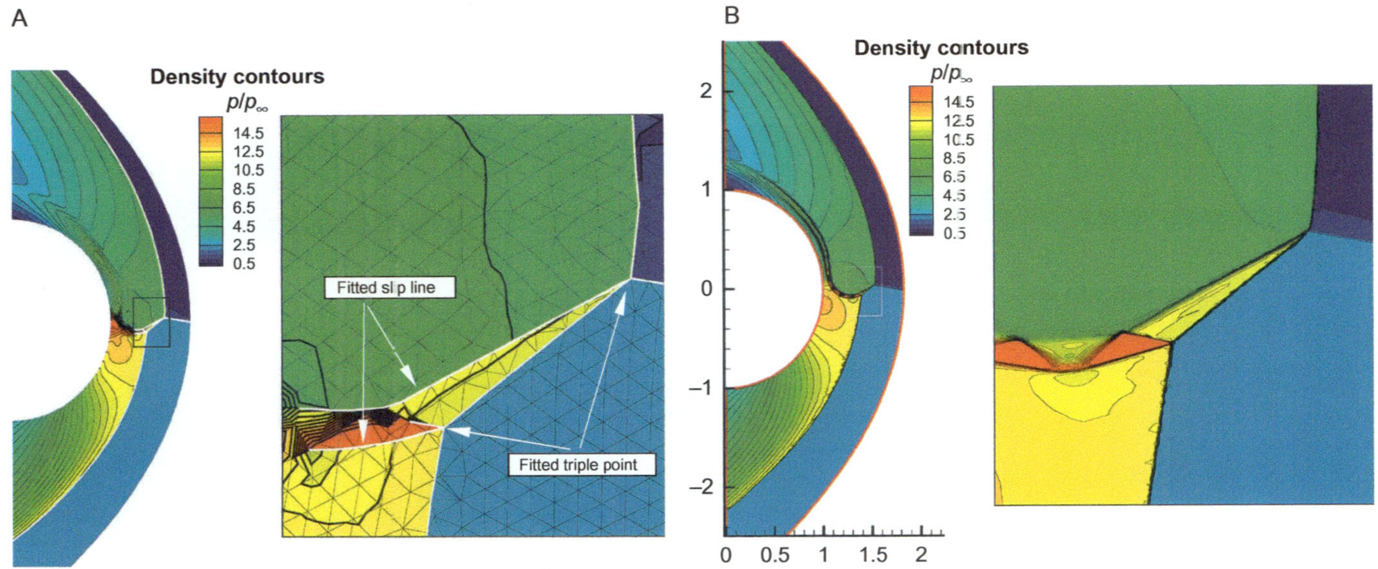

FIG. 25 Type IV shock–shock interaction: density isocontour lines. (A) Shock-fitting solution. (B) Shock-capturing solution on an anisotropically-adapted mesh.

5.2.4 *Hypersonic ($M_\infty = 24$) and Supersonic ($M_\infty = 4.04$) Three-Dimensional Blunt-Body Flows*

The shock-fitting technique for unstructured grids has been successfully extended to the three-dimensional case (Bonfiglioli et al., 2013). Steady simulations of supersonic and hypersonic flows past three-dimensional bodies are shown in Fig. 26. Fig. 26A shows pressure isocontours within three cross-flow planes and over the body of the European Agency IXV vehicle, flying at $M_\infty = 24$ and 45 degree angle of attack. In this simulation only the bow shock has been fitted, whereas it has been left to the shock-capturing solver to capture the embedded shocks that arise at the flap-body junction and their interaction with the fitted bow shock.

Fig. 26B shows pressure isocontours within the symmetry plane and on the surface of a cylinder with an hemispherical nose and a conical flare. Free-stream conditions are: $M_\infty = 4.04$ and 20 degree angle of attack. In this calculation not only the bow shock, but also the embedded shock that arises at the cylinder-cone junction have been fitted. In contrast to the two-dimensional case, whereby we are capable of fitting various kinds of shock–shock interactions, this capability is currently unavailable in three dimensions, primarily because of the limited capabilities of the software currently used for surface remeshing. Therefore, in the simulation shown in Fig. 26B the interaction between the embedded and the bow shocks has been captured, rather than fitted.

5.2.5 *Un-Steady, Two-Dimensional Flows*

By increasing to second order in time the accuracy of the Lagrangian motion of the discontinuities, as described in Section 5.1.6, the unstructured shock-fitting technique can also be used to simulate unsteady flows in a time-accurate

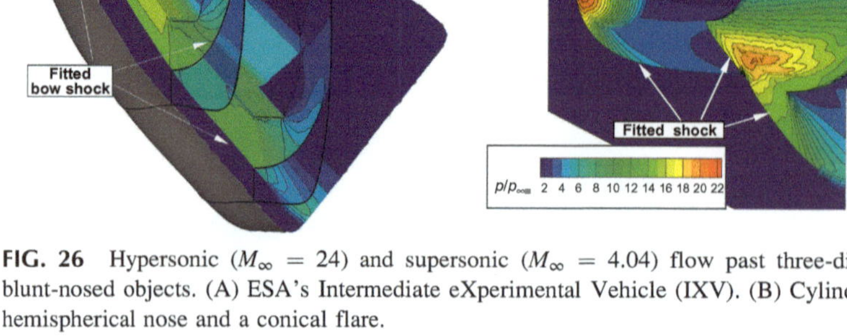

FIG. 26 Hypersonic ($M_\infty = 24$) and supersonic ($M_\infty = 4.04$) flow past three-dimensional, blunt-nosed objects. (A) ESA's Intermediate eXperimental Vehicle (IXV). (B) Cylinder with an hemispherical nose and a conical flare.

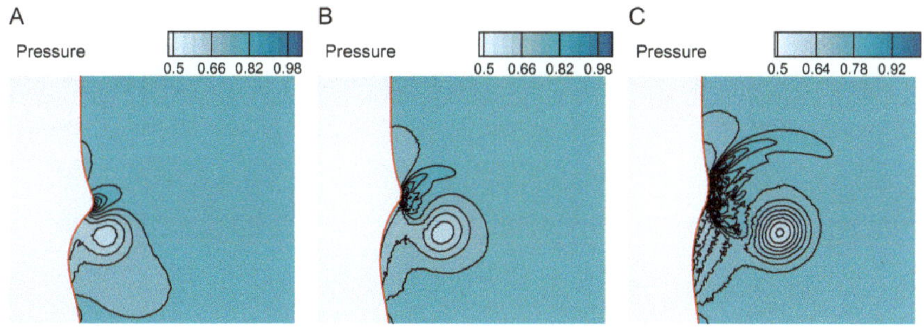

FIG. 27 Shock-vortex interaction: pressure isocontour lines. (A) $t = 0.3$. (B) $t = 0.4$. (C) $t = 0.5$.

manner. At present, however, the algorithm does not have the capability to recognise and deal with, for example, the formation of a new shock or a new interaction point; in other words, the technique can only simulate transient flows that do not undergo topological changes of the fitted-shock pattern.

We illustrate the present capabilities of the unstructured shock-fitting technique by reference to the interaction between a stationary shock and a moving vortical structure. A uniform, supersonic stream carries a vortex towards a stationary normal shock; their interaction gives rise to a Mach reflection with the formation of two triple points and two secondary shocks. Fig. 27, which shows the temporal evolution of the pressure field, clearly reveals how severely the fitted shock is deformed by the passage of the vortex.

Because of the aforementioned algorithmic limitations, only the normal shock has been fitted in the simulation, whereas the two triple points and the two secondary shocks that are visible in Fig. 27B and C have been captured. Further details concerning this specific test-case, including comparisons with shock-capturing calculations, are reported in Bonfiglioli et al. (2016).

Clearly, the capability to recognise and treat the topological changes that often occur in unsteady flows is a fundamental ingredient of the unstructured shock-fitting which we are currently developing. The task is certainly complex, but the availability of a substantial body of literature (see, to cite just a few Moretti and Valorani, 1988; Nasuti and Onofri, 1996), published by the "floating" shock-fitting community on the subject of automatic shock detection certainly provides a solid starting point.

Preliminary efforts in this direction are currently under way and will be hereafter documented using the last test-case to be presented, which consists in a planar, moving shock that impinges on a 58 degree wedge, as shown in Fig. 28. Once the normal shock has overtaken the corner of the wedge, a new curved shock and a regular reflection appear, giving rise to a different shock topology.

In the numerical simulation, the topological changes have been introduced shortly after the planar shock has overtaken the corner. This is achieved by adding a shock which is initially made up of only three shock-points, as shown in

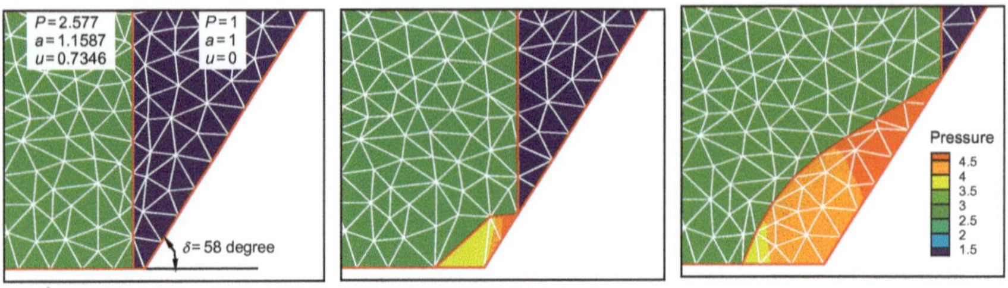

FIG. 28 Regular reflection: frames corresponding to increasing time.

the central frame of Fig. 28: two are located along the boundaries and the third one inside the flow field; the reflected and the incident shocks are joined in a point that moves over the wedge. The shape of the curved shock and the states of these new shock-points are computed using simple analytical relations obtained from the reflection of isentropic waves. Starting from this approximate solution, the shock-fitting algorithm is capable of correctly evolving the solution based on this new topology: this is clearly shown in the right frame of Fig. 28.

Building upon our seminal work on unstructured shock-fitting, and owing to its modularity, which greatly simplifies the coupling with existing CFD solvers, other groups have supplemented their own CFD codes with unstructured shock-fitting capabilities. A "boundary" version of our unstructured algorithm, whereby only the bow shock is fitted and the mesh connectivity is preserved while the bow shock moves to reach its steady state, has been documented in Azevedo et al. (2011) and Gnoffo et al. (2013). Azevedo et al. (2011) use shock-fitting to support their conjecture that carbuncles are created due to the nonphysical structure of numerical shock waves. Gnoffo et al. (2013) have implemented bow-shock-fitting capabilities within NASA's FUN3D code to study stagnation heating anomalies in hypersonic flows. Unstructured-grid implementations of the shock-fitting technique have also been reported by two different teams in China, although their algorithmic details can hardly be inferred from Wu et al. (2010) and Liu et al. (2015), since only the abstracts are available in English.

6 CONCLUSIONS

Shock-capturing and shock-fitting methods have been used to simulate flows characterised by the presence of shock waves and other gasdynamic discontinuities since the dawn of CFD. Their merits and weaknesses became clear soon after their first appearance. Shock-capturing methods are relatively simple to code, which allows to build general-purpose computer codes, able to simulate any type of flow field, regardless of the presence of discontinuities. The poor quality of the solutions obtained using these methods, when discontinuities are present, is their main weakness.

On the contrary, shock-fitting methods are able to provide high-quality solutions, even on rather coarse meshes, but their implementation in a general-purpose code capable of simulating complex flows can be very difficult from an algorithmic point of view. In Moretti's words (Moretti, 2002): "In conclusion, the only difficulty in shock-fitting stems not from fluid mechanics but from topology (this happens, however, only when multiple discontinuities are present) and, consequently, from computational logic".

Indeed, over the last 30 years, shock-fitting methods have experienced an increasingly widespread ostracism, mainly caused by the fear of the difficulties posed by the implementation of the shock-fitting logic. The efforts of many CFD developers have rather been focussed on counteracting the flaws and shortcomings of the shock-capturing technique; in doing so, however, shock-capturing algorithms have lost some, if not most, of their algorithmic simplicity. Moreover, despite these great efforts that involved a large number of scientists, the shock-capturing drawbacks have not been completely overcome.

Numerous practical applications of different shock-fitting methods developed by the authors over the years, using both structured- and unstructured-grid gasdynamics solvers, have been presented in this chapter, demonstrating that these methods can be effectively used to simulate complex two- and three-dimensional flow fields. Obtaining these results required a considerable effort in the implementation of the "computational logic" that lays behind the shock-fitting idea; however, the quality of the obtained results fully repaid these efforts. A beautiful analogy that well summarises the view-point taken by the shock-fitting developers can again be found in Moretti's words (Moretti, 2002): "To a baby, learning to walk takes a year and a half; the baby, however, recognises the advantages of walking over crawling, works hard and finally is rewarded for the rest of its life".

As a conclusion, we believe it may be time to refocus the efforts of a larger number of CFD developers on a technique that is able to ensure results that are qualitatively superior to those that shock-capturing methods can deliver. As summarised in Roe's words Roe (2011): "a reappraisal of fitting methods is timely", since"a revival of shock-fitting is one of the few remaining possibilities for revolutionary change [in high-resolution methods]".

REFERENCES

Abgrall, R., 2006. Residual distribution schemes: current status and future trends. Comput. Fluids 35 (7), 641–669. ISSN 0045-7930. http://dx.doi.org/10.1016/j.compfluid.2005.01.007. http://www.sciencedirect.com/science/article/pii/S0045793005001738 (Special issue dedicated to professor Stanley G. Rubin on the occasion of his 65th birthday).

Azevedo, J.L., Ramalho, M.V., Azevedo, J.H., 2011. Further investigation into the origin of the carbuncle phenomenon in aerodynamic simulations. In: Aerospace Sciences Meetings. American Institute of Aeronautics and Astronautics, Reston, VA, pp. 1–33. http://dx.doi.org/10.2514/6.2011-1184.

Ben-Dor, G., 1992. Shock Wave Reflection Phenomena. Springer-Verlag, New York, 75 pp.

Bonfiglioli, A., 2000. Fluctuation splitting schemes for the compressible and incompressible Euler and Navier-Stokes equations. Int. J. Comput. Fluid D. 14 (1), 21–39. http://dx.doi.org/10.1080/10618560008940713.

Bonfiglioli, A., Paciorri, R., 2013. A mass-matrix formulation of unsteady fluctuation splitting schemes consistent with Roe's parameter vector. Int. J. Comput. Fluid D. 27 (4–5), 210–227. http://dx.doi.org/10.1080/10618562.2013.813491.

Bonfiglioli, A., Paciorri, R., 2014. Convergence analysis of shock-capturing and shock-fitting solutions on unstructured grids. AIAA J. 52 (7), 1404–1416. ISSN 0001-1452. http://dx.doi.org/10.2514/1.J052567.

Bonfiglioli, A., Paciorri, R., Di Mascio, A., 2012. The role of mesh generation, adaptation and refinement on the computation of flows featuring strong shocks. Model. Simul. Mater. Sci. Eng. 2012 Article ID 631276, 15 pp. http://dx.doi.org/10.1155/2012/631276. http://downloads.hindawi.com/journals/mse/2012/631276.pdf.

Bonfiglioli, A., Grottadaurea, M., Paciorri, R., Sabetta, F., 2013. An unstructured, three-dimensional, shock-fitting solver for hypersonic flows. Comput. Fluids 73, 162–174. ISSN 0045-7930. http://dx.doi.org/10.1016/j.compfluid.2012.12.022. http://www.sciencedirect.com/science/article/pii/S0045793013000121.

Bonfiglioli, A., Paciorri, R., Campoli, L., 2016. Unsteady shock-fitting for unstructured grids. Int. J. Numer. Meth. Fl. 81 (4), 245–261. ISSN 1097-0363. http://dx.doi.org/10.1002/fld.4183.

Carpenter, M.H., Casper, J.H., 1999. Accuracy of shock capturing in two spatial dimensions. AIAA J. 37 (9), 1072–1079.

Casper, J., Carpenter, M., 1998. Computational considerations for the simulation of shock-induced sound. SIAM J. Sci. Comput. 19 (3), 813–828. http://dx.doi.org/10.1137/S1064827595294101.

Chisnell, R.F., 1957. The motion of a shock wave in a channel, with application to cylindrical and spherical shock waves. J. Fluid Mech. 2, 286–298.

Deconinck, H., Paillère, H., Struijs, R., Roe, P.L., 1993. Multidimensional upwind schemes based on fluctuation-splitting for systems of conservation laws. Comput. Mech. 11, 323–340. ISSN 0178-7675. http://dx.doi.org/10.1007/BF00350091.

Dolejší, V., 1998. Anisotropic mesh adaptation for finite volume and finite element methods on triangular meshes. Comput. Vis. Sci. 1 (3), 165–178. ISSN 1432-9360. http://dx.doi.org/10.1007/s007910050015.

Edney, B., 1968. Anomalous heat transfer and pressure distributions on blunt bodies at hypersonic speeds in the presence of an impinging shock, TR 115. Flygtekniska Forsoksanstalten (The Aeronautical Research Institute of Sweden), Stockholm.

Engquist, B., Sjögreen, B., 1998. The convergence rate of finite difference schemes in the presence of shocks. SIAM J. Numer. Anal. 35 (6), 2464–2485. http://dx.doi.org/10.1137/S0036142997317584.

Gnoffo, P.A., Wood, W.A., Kleb, B., Alter, S.J., Padilla, J., White, J.A., 2013. Functional equivalence acceptance testing of FUN3D for entry, descent, and landing applications. In: Fluid Dynamics and Co-located Conferences. American Institute of Aeronautics and Astronautics, http://dx.doi.org/10.2514/6.2013-2558.

Ivanov, M.S., Bonfiglioli, A., Paciorri, R., Sabetta, F., 2010. Computation of weak steady shock reflections by means of an unstructured shock-fitting solver. Shock Waves 20 (4), 271–284. ISSN 0938-1287. http://dx.doi.org/10.1007/s00193-010-0266-y.

Johnsen, E., Larsson, J., Bhagatwala, A.V., Cabot, W.H., Moin, P., Olson, B.J., Rawat, P.S., Shankar, S.K., Sjögreen, B., Yee, H., Zhong, X., Lele, S.K., 2010. Assessment of high-resolution methods for numerical simulations of compressible turbulence with shock waves. J. Comput.

Phys. 229 (4), 1213–1237. ISSN 0021-9991. http://dx.doi.org/10.1016/j.jcp.2009.10.028. http://www.sciencedirect.com/science/article/B6WHY-4XH5MS8-5/2/78a62583706d9d63ab6a070b68e362ac.

Kopriva, D.A., 1999. Shock-fitted multidomain solution of supersonic flows. Comput. Method Appl. M. 175 (3–4), 383–394. ISSN 0045-7825. http://dx.doi.org/10.1016/S0045-7825(98)00362-4. http://www.sciencedirect.com/science/article/B6V29-40B2XGX-9/2/b6765e2254ee3f82e2ddfab1daa07890.

Lee, T.K., Zhong, X., 1999. Spurious numerical oscillations in simulation of supersonic flows using shock-capturing schemes. AIAA J. 37 (3), 313–319.

Liu, J., Zou, D., Xu, C., 2015. An unsteady shock-fitting technique based on unstructured moving grids. Kongqi Donglixue Xuebao 33 (1), 10–16. http://dx.doi.org/10.7638/kqdlxxb-2014.0109.

Ma, Y., Zhong, X., 2003a. Receptivity of a supersonic boundary layer over a flat plate. Part 1. Wave structures and interactions. J. Fluid Mech. 488, 31–78. http://dx.doi.org/10.1017/S0022112003004786. http://journals.cambridge.org/action/displayAbstract?fromPage=online&aid=164075&fulltextType=RA&fileId=S0022112003004786.

Ma, Y., Zhong, X., 2003b. Receptivity of a supersonic boundary layer over a flat plate. Part 2. Receptivity to free-stream sound. J. Fluid Mech. 488, 79–121. http://dx.doi.org/10.1017/S0022112003004798. http://journals.cambridge.org/action/displayAbstract?fromPage=online&aid=164077&fulltextType=RA&fileId=S0022112003004798.

Ma, Y., Zhong, X., 2005. Receptivity of a supersonic boundary layer over a flat plate. Part 3. Effects of different types of free-stream disturbances. J. Fluid Mech. 532, 63–109. ISSN 1469-7645. http://dx.doi.org/10.1017/S0022112005003836. http://journals.cambridge.org/article_S0022112005003836.

Marconi, F., Salas, M., 1973. Computation of three dimensional flows about aircraft configurations. Comput. Fluids 1, 185–195.

Menne, S., 1995. Computation of non-winged vehicle aerodynamics in the low supersonic range. In: Hunt, J.J., (Ed.), Aerothermodynamics for Space Vehicles, vol. 367, ESA Special Publication, Noordwijk, The Netherlands, pp. 73–78.

Moretti, G., 1979. The λ-scheme. Comput. Fluids 7, 191–205.

Moretti, G., 1985. Fast Euler solver for steady, one-dimensional flows. Comput. Fluids 13 (1), 61–81. http://dx.doi.org/10.1016/0045-7930(85)90033-7.

Moretti, G., 1987a. A technique for integrating two-dimensional Euler equations. Comput. Fluids 15 (1), 59–75.

Moretti, G., 1987b. Computation of flows with shocks. Annu. Rev. Fluid Mech. 19, 313–317.

Moretti, G., 1987c. Efficient calculation of two-dimensional, unsteady, compressible flows. In: Vichnevetsky, R., Stepleman, R.S. (Eds.), Advanced in Computer Methods for Partial Differential Equations—VI. IMACS, Liége, Belgium.

Moretti, G., 1988. Efficient Euler solver with many applications. AIAA J. 26 (6), 655–660. ISSN 0001-1452. http://dx.doi.org/10.2514/3.9950.

Moretti, G., 2002. Thirty-six years of shock fitting. Comput. Fluids 31 (4–7), 719–723.

Moretti, G., Abbett, M., 1966. A time-dependent computational method for blunt body flows. AIAA J. 4 (12), 2136–2141.

Moretti, G., Di Piano, M.T., 1983. An improved λ-scheme for one-dimensional flows. NASA-CR-3712.

Moretti, G., Valorani, M., 1988. Detection and fitting of two-dimensional shocks. Notes Numer. Fluid Mech. 20, 239–246.

Moretti, G., Marconi, F., Onofri, M., 1993. Shock boundary layer interactions computed by a shock fitting technique. In: Napolitano, M., Sabetta, F. (Eds.), Lecture Notes in Physics, vol. 414. Springer-Verlag, Berlin, pp. 345–350.

Morton, K., Paisley, M., 1989. A finite volume scheme with shock fitting for the steady Euler equations. J. Comput. Phys. 80 (1), 168–203. ISSN 0021-9991. http://dx.doi.org/10.1016/0021-9991(89)90094-6. http://www.sciencedirect.com/science/article/pii/0021999189900946.

Najafi, M., Hejranfar, K., Esfahanian, V., 2014. Application of a shock-fitted spectral collocation method for computing transient high-speed inviscid flows over a blunt nose. J. Comput. Phys. 257, Part A, 954–980. ISSN 0021-9991. http://dx.doi.org/10.1016/j.jcp.2013.09.037. http://www.sciencedirect.com/science/article/pii/S0021999113006517.

Nasuti, F., 2003. A multi-block shock-fitting technique to solve steady and unsteady compressible flows. In: Armfield, S., Morgan, P., Srinivas, K. (Eds.), Computational Fluid Dynamics 2002. Springer-Verlag, Berlin, pp. 217–222.

Nasuti, F., Onofri, M., 1996. Analysis of unsteady supersonic viscous flows by a shock-fitting technique. AIAA J. 34, 1428–1434.

Nasuti, F., Onofri, M., 1998. Methodology to solve flowfields of plug nozzles for future launchers. J. Propul. Power 14, 318–326.

Nasuti, F., Onofri, M., 2001. Analysis of in-flight behavior of truncated plug nozzles. J. Propul. Power 17 (4), 809–817.

Paciorri, R., Bonfiglioli, A., 2009. A shock-fitting technique for 2D unstructured grids. Comput. Fluids 38 (3), 715–726. ISSN 0045-7930. http://dx.doi.org/10.1016/j.compfluid.2008.07.007. http://www.sciencedirect.com/science/article/pii/S0045793008001503.

Paciorri, R., Bonfiglioli, A., 2011. Shock interaction computations on unstructured, two-dimensional grids using a shock-fitting technique. J. Comput. Phys. 230 (8), 3155–3177. ISSN 0021-9991. http://dx.doi.org/10.1016/j.jcp.2011.01.018. http://www.sciencedirect.com/science/article/pii/S0021999111000362.

Pirozzoli, S., 2011. Numerical methods for high-speed flows. Ann. Rev. Fluid Mech. 43 (1), 163–194. http://dx.doi.org/10.1146/annurev-fluid-122109-160718.

Rawat, P., Zhong, X., 2010. Numerical simulation of shock-turbulence interactions using high-order shock-fitting algorithms. In: Aerospace Sciences Meetings. American Institute of Aeronautics and Astronautics, http://dx.doi.org/10.2514/6.2010-114.

Rawat, P., Zhong, X., 2011. Direct numerical simulations of turbulent flow interactions with strong shocks using shock-fitting method. In: Aerospace Sciences Meetings. American Institute of Aeronautics and Astronautics, Reston, VA. 50 pp. http://dx.doi.org/10.2514/6.2011-649.

Roe, P., 2011. Review of the book "A Shock-Fitting Primer" by Manuel D. Salas. SIAM Rev. 53, 207–210.

Salas, M., 2009. A Shock-Fitting Primer, first ed. CRC Applied Mathematics & Nonlinear Science, Chapman & Hall, Boca Raton, FL.

Sesterhenn, J., Dohogne, J.F., Friedrich, R., 2005. Direct numerical simulation of the interaction of isotropic turbulence with a shock wave using shock-fitting. C. R. Mecanique 333 (1), 87–94. ISSN 1631-0721. http://dx.doi.org/10.1016/j.crme.2004.09.017. http://www.sciencedirect.com/science/article/pii/S1631072104002347 (High-order methods for the numerical simulation of vortical and turbulent flows).

Shewchuk, J.R., 1996. Triangle: engineering a 2D quality mesh generator and Delaunay triangulator. In: Lin, M.C., Manocha, D. (Eds.), Applied Computational Geometry: Towards Geometric Engineering. Lecture Notes in Computer Science, vol. 1148. Springer-Verlag, Berlin, pp. 203–222.

Si, H., 2015. TetGen, a Delaunay-based quality tetrahedral mesh generator. ACM Trans. Math. Softw. 41 (2), 11:1–11:36. ISSN 0098-3500. http://dx.doi.org/10.1145/2629697.

Weilmuenster, K.J., Hamilton, H.H., 1979. A hybridized method for computing high-Reynolds-number hypersonic flow about blunt bodies. NASA-TP-1497, L-12946. NASA Langley Research Center, Hampton, VA. http://hdl.handle.net/2060/19790025270.

Wu, Q., Zhong, Y.C., Hu, J., Yu, S.Z., 2010. Numerical simulation based on DDB fitting for 2D/axisymmetric triple shocks. Kongqi Donglixue Xuebao 28, 683–688.

Yamamoto, Y., Karashima, K., 1982. Floating shock fitting for three-dimensional inviscid supersonic flows. AIAA J. 20 (1), 9–17. ISSN 0001-1452. http://dx.doi.org/10.2514/3.7897.

Zhong, X., 1998. High-order finite-difference schemes for numerical simulation of hypersonic boundary-layer transition. J. Comput. Phys. 144 (2), 662–709. ISSN 0021-9991. http://dx.doi.org/10.1006/jcph.1998.6010. http://www.sciencedirect.com/science/article/B6WHY-45J58Y6-45/2/bc35eb5ba7f2d31873a4abab10755bc3.

Zierep, J., 2003. New results for the normal shock in inviscid flow at a curved surface. ZAMM - J. Appl. Math. Mech./Z. Angew. Math. Mech. 83 (9), 603–610. ISSN 1521-4001. http://dx.doi.org/10.1002/zamm.200310051.

Chapter 17

Spectral Methods for Hyperbolic Problems[1]

J.S. Hesthaven
EPFL-SB-MATHICSE-MCSS, Ecole Polytechnique Fédérale de Lausanne, Lausanne, Switzerland

Chapter Outline

ABSTRACT

We review spectral methods for the solution of hyperbolic problems. To keep the discussion concise, we focus on Fourier spectral methods and address key issues of accuracy, stability and convergence of the numerical approximations. Polynomial methods are discussed when these lead to qualitatively different schemes as, for instance, when boundary conditions are required. The discussion includes nonlinear stability and the use of filters and postprocessing techniques to minimize or overcome the Gibbs phenomenon.

1. This revised and updated chapter is based partly on work from the author's original article first published in the Journal of Computational and Applied Mathematics, Volume 128, Gottlieb and Hesthaven, Elsevier, 2001.

Handbook of Numerical Analysis, Vol. 17. http://dx.doi.org/10.1016/bs.hna.2016.09.007

Keywords: Spectral, Pseudospectral, Galerkin, Collocation, Penalty methods, Discontinuous solutions, Gibbs phenomenon, Stability, Filtering, Vanishing viscosity, Skew-Symmetric form

AMS Classification Codes: 65M70, 65M12, 35L65

1 INTRODUCTION

The theory and application of spectral methods for the solution of partial differential equations has traditionally focused on problems with a certain amount of inherent regularity of the solutions, e.g. elliptic/parabolic problems. The application that is perhaps most responsible for the widespread use of spectral methods is the incompressible Navier–Stokes equations (Canuto et al., 1988; Deville et al., 2002; Peyret, 2013).

At the heart of a spectral method lies the assumption that the solution, $u(x, t)$, to a partial differential equation can be expressed by a series of smooth basis functions as

$$u(x,t) \simeq u_h(x,t) = \sum_{n=0}^{N} \hat{u}_n(t)\phi_n(x). \tag{1}$$

The choice of the basis $\phi_n(x)$ and the way in which the expansion coefficients $\hat{u}_n(t)$ are computed results in different methods. Let us first assume that $\phi_n(x) : [a,b] \to \mathsf{R}$ is orthogonal in L^2_w such that

$$\hat{u}_n(t) = \frac{1}{\gamma_n}(u, \phi_n)_w, \tag{2}$$

where

$$(f,g)_w = \int_a^b f(x)g(x)w(x)\,dx, \quad \gamma_n = (\phi_n, \phi_n)_w, \tag{3}$$

and $w(x)$ is an L^1-integrable weight function. This defines the truncated projection of the function $u(x, t)$ as

$$\mathcal{P}_N u(x,t) = \sum_{n=0}^{N} \hat{u}_n(t)\phi_n(x). \tag{4}$$

To understand the accuracy of this truncated expansion, consider

$$\| u(x,t) - \mathcal{P}_N u(x,t) \|_w^2 = \sum_{n=N+1}^{\infty} \gamma_n \hat{u}_n^2,$$

i.e. the accuracy depends solely on the decay of the expansion coefficients. To understand their behaviour, assume that the basis satisfies

$$\left[\frac{d}{dx}q(x)\frac{d}{dx} + \lambda_n w(x)\right]\phi_n(x) = [\mathcal{L} + \lambda_n w(x)]\phi_n(x) = 0, \quad x \in [a,b]. \tag{5}$$

In the simplest case of $q(x) = w(x) = 1$, $x \in [0, 2\pi]$, the trigonometric functions, $\phi_n(x) = \exp(inx)$ satisfy this with $\lambda_n = n^2$. For the more general case of $p(x)$ and $w(x)$ with $x \in [-1, 1]$, we recover all the classic orthogonal polynomials provided $q(\pm 1) = 0$ (Szego, 1939). In this case, (5) is the singular Sturm–Liouville problem with $\lambda_n \propto n^2$. Prominent examples of these polynomial basis functions include Legendre and Chebyshev polynomials.

Under the assumption of (5), integration by parts of (2) yields

$$\hat{u}_n = \frac{1}{\gamma_n}(u, \phi_n)_w = \frac{-1}{\gamma_n \lambda_n}\left(\left[u q \phi_n' - u' q \phi_n \right]_a^b + \left(\frac{\mathcal{L}}{w} u, \phi_n \right)_w \right).$$

If we now further assume that the solution u and the basis ϕ_n is periodic in $[a, b]$, as for the trigonometric basis, or $q(a) = q(b) = 0$ as for the polynomial basis, we recover

$$\hat{u}_n = \frac{-1}{\gamma_n \lambda_n}\left(\frac{\mathcal{L}}{w} u, \phi_n \right)_w .$$

Under the assumption that u is sufficiently smooth and periodic, repeating this p times yields

$$\hat{u}_n = \frac{1}{\gamma_n}\left(\frac{-1}{\lambda_n} \right)^p \left(\left(\frac{\mathcal{L}}{w} \right)^p u, \phi_n \right)_w .$$

This we may now bound as

$$|\hat{u}_n| \leq \frac{1}{\gamma_n \lambda_n^p} \| u^{(2p)} \|_w \leq \frac{C}{\gamma_n n^{2p}} \| u^{(2p)} \|_w .$$

This highlights the direct connection between the regularity of the solution u and the decay of the expansion coefficients. This yields the estimate

$$\| u(x,t) - \mathcal{P}_N u(x,t) \|_w^2 \leq C N^{-p} \| u^{(p)} \|_w . \tag{6}$$

In the event that $u(x, t)$ is analytic we recover the remarkable property (Tadmor, 1986)

$$\| u - \mathcal{P}_N u \|_w \leq C N^{-p} \| u^{(p)} \|_w \sim C \frac{p!}{N^p} \| u \|_w \sim C e^{-cN} \| u \|_w ,$$

known as spectral accuracy or spectral convergence. This is the property that gives name to spectral methods.

The use of spectral methods for the solution of hyperbolic problems has traditionally been viewed as problematic and only more recently have such methods seen a wider use. The reasons for the perceived difficulty are several. Contrary to parabolic and elliptic problems, there is no physical dissipation inherent in the hyperbolic problem. This implies that even minor errors and under resolved phenomena can cause the scheme to become unstable.

Perhaps the most important reason, however, for the slow acceptance of spectral methods for solving hyperbolic conservation laws is the appearance of the Gibbs phenomenon as finite time discontinuities develop in the solution. Left alone, the nonlinear mixing of the Gibbs oscillations with the approximate solution will eventually cause the scheme to become unstable. Moreover, even if stability is maintained for sufficiently long time, the computed solution appears to be only first-order accurate in which case the use of a high-order method is questionable. More fundamental issues of conservation and the ability of the scheme to compute the correct entropy solution to conservation laws have also caused considerable concern among practitioners and theoreticians alike.

While many of these issues are genuine and require careful attention, they do not cause the spectral methods to fail if applied correctly. This was indicated already to in early work around 1980 (Gottlieb et al., 1981a; Kreiss and Oliger, 1979; Majda et al., 1978) where the first numerical solution of problems with discontinuous solutions and general nonlinear conservation laws were presented.

To understand the potential of spectral methods for solving conservation laws problems, we need to dig deeper into the development and analysis of these methods. To keep the discussion brief we focus on Fourier spectral methods and discuss key developments in this context. However, when appropriate, we revisit qualitative differences induced by the use of a polynomial basis. For further details, in particular for polynomials methods and more complex applications, we refer available texts and reviews (e.g. Ben-Yu, 1998; Bernardi and Maday, 1997; Boyd, 2001; Funaro, 2008; Gottlieb and Hesthaven, 2001; Karniadakis and Sherwin, 2013; Shen et al., 2011; Trefethen, 2000).

The remainder of this chapter is organized as follows. Next, we revisit the spectral expansion and different ways to express this. We also outline the key approximation results of the continuous and discrete expansions for smooth and nonsmooth functions. In Section 3 we introduce Fourier spectral methods and discuss their stability for linear problems. We shall also discuss polynomial methods and techniques for the imposing general boundary conditions, leading to additional complications. In Section 4 we return to the Fourier spectral methods, now with a focus on nonlinear problems and discuss stability and convergence for such problems. Section 5 discusses ways to overcome the impact of the Gibbs phenomenon on the global accuracy. Throughout the discussion we strive to include sufficient references to allow the reader to pursue more advanced topics.

2 THE SPECTRAL EXPANSION

We focus on spectral methods based on the Fourier expansion

$$\mathcal{P}_N u(x) = \sum_{n=-N}^{N} \hat{u}_n \exp(inx). \tag{7}$$

Here and in the following we suppress the explicit time-dependency of $u(x, t)$ for simplicity.

The expansion coefficients are obtained directly as

$$\hat{u}_n = \frac{1}{2\pi}(u, \exp(inx)) = \frac{1}{2\pi}\int_0^{2\pi} u(x)\exp(-inx)\,dx, \tag{8}$$

through the orthogonality of the basis in the inner product

$$(f, g) = \int_0^{2\pi} f\bar{g}\,dx, \quad \|f\|^2 = \int_0^{2\pi} |f|^2\,dx,$$

with the associated norm, $\|\cdot\|$.

Once the expansion is known, we can evaluare spatial derivatives of the function as

$$\frac{d^p u(x)}{dx^p} \simeq \frac{d^p \mathcal{P}_N u(x)}{dx^p} = \sum_{n=-N}^{N} (in)^p \hat{u}_n \phi_n(x) = \sum_{n=-N}^{N} \hat{u}_n^{(p)} \phi_n(x),$$

i.e. $\hat{u}_n^{(p)} = (in)^p \hat{u}_n$, for the approximation of an arbitrary derivative of a function, given by its Fourier coefficients.

2.1 Smooth Problems

We have already discussed the close connection between regularity of the function and accuracy of the truncated Fourier expansion. While the algebra involved is quantitatively different when a different basis and norm is used, the results for a basis comprising classic orthogonal polynomials is qualitative the same as in (6), i.e. there is a direct relationship between the accuracy of the spectral expansion and the regularity of the function being approximated.

Results similar to (6) are also available in higher norms. For the Fourier series we have (Hesthaven et al., 2007)

$$\| u - \mathcal{P}_N u\|_{W^p} \leq C(p,q)N^{p-q} \| u\|_{W^q}, \tag{9}$$

provided only that $0 \leq p \leq q$. Here, we have the Sobolev norm

$$\| u \|_{W^q}^2 = \sum_{s=0}^{q} \| u^{(s)} \|^2,$$

to measure the error on the derivatives. The results for the classic polynomials are qualitatively similar (Bernardi and Maday, 1997; Canuto and Quarteroni, 1982a).

Results for pointwise accuracy are harder to obtain. For the truncated Fourier series one recovers (Canuto et al., 1988)

$$\| u - \mathcal{P}_N u\|_{L^\infty} \leq C(q)(1 + \log N)N^{-q} \| u^{(q)} \|_{L^\infty},$$

where the L^∞-norm measures the maximum pointwise error. This indicates that we expect a poor pointwise accuracy for problems with low regularity. This happens both locally, where convergence is lost at discontinuous point, and in the entire domain containing the discontinuity due to the Gibbs phenomenon as discussed in more detail in Section 2.2.

The computation of the Fourier coefficients, $\hat{u}_n$, poses a problem as one cannot in general evaluate the inner product. The natural solution is to introduce a quadrature approximation to (8) on the form

$$\tilde{u}_n = \frac{1}{2N+1} \sum_{j=0}^{2N} u(x_j) \exp(-inx_j). \tag{10}$$

We recognize this as the trapezoidal rule with the equidistant grid

$$x_j = \frac{2\pi}{2N+1} j, \quad j = 0, \ldots, 2N. \tag{11}$$

This is known as the odd method, due to the odd number of grid points. Historically, an even number of points have been preferred, leading to minor quantitative differences but no qualitative differences. We refer to Hesthaven et al. (2007) for through discussion of this.

As N in (10) increases one hopes that $\tilde{u}_n$ is a good approximation to $\hat{u}_n$. To quantify this, we can express $\tilde{u}_n$ using $\hat{u}_n$ as

$$\tilde{u}_n = \hat{u}_n + \sum_{\substack{m=-\infty \\ m \neq 0}}^{m=\infty} \hat{u}_{n+2Nm},$$

where the second term is termed the aliasing error. The aliasing error reflects that certain basis components cannot be distinguished on a finite grid, causing highly oscillatory components to be misinterpreted as slowly varying basis components. While the analysis is more complex in the polynomial case, the introduction of the aliasing error by the grid remains qualitatively the same (Bernardi and Maday, 1997; Canuto and Quarteroni, 1982a).

Understanding the accuracy of the discrete expansion thus reduces to an analysis of the error caused by the aliasing error. For the Fourier basis, the analysis in Kreiss and Oliger (1979) shows that the aliasing error and the truncation error is of the same order, i.e. the result in (6) carries over to the case of interpolation in the Fourier case. This is likewise the case for the general result (9).

Since the use of the modal expansions requires the introduction of a finite grid one could question the need to consider special basis functions at all. Indeed, given a specific nodal set, x_j, we can construct a global interpolation

$$\mathcal{I}_N u(x) = \sum_{j=0}^{2N} u(x_j) l_j(x),$$

where the Lagrange interpolating polynomials, $l_j(x)$, takes the form

$$l_j(x) = \frac{q(x)}{(x - x_j)q'(x_j)}, \quad q(x) = \prod_{j=0}^{2N} (x - x_j).$$

Clearly, if the x_j's are distinct, $l_j(x)$ is uniquely determined as the polynomial of order $2N$, specified at $2N + 1$ points. We can directly explore this to approximate derivatives of $u(x)$. In particular, if we restrict attention to the approximation of the derivative of $u(x)$ at the grid points, x_j, we have

$$\left.\frac{du}{dx}\right|_{x_i} \simeq \left.\frac{d\mathcal{I}_N u}{dx}\right|_{x_i} = \sum_{j=0}^{2N} u(x_j) \left.\frac{dl_j}{dx}\right|_{x_i} = \sum_{j=0}^{2N} u(x_j) \mathrm{D}_{ij},$$

where D_{ij} is a differentiation matrix with the entries

$$\mathrm{D}_{ij} = \begin{cases} \dfrac{(-1)^{i+j}}{2} \left[\sin\left(\dfrac{\pi}{2N+1}(i-j) \right) \right]^{-1} & i \neq j. \\ 0 & i = j \end{cases} \tag{12}$$

The global nature of the interpolation implies that the differentiation matrix is full. We also note that D is skew-symmetric, a property that does not carry over to polynomial methods (Hesthaven et al., 2007).

2.2 Nonsmooth Problems

If the solution possesses significant regularity we can expect the spectral expansion to be highly efficient as a representation of the solution and its spatial derivatives.

However, for problems with only limited regularity the picture is more complex and the above results do not inform us much about the accuracy of the approximation of such solutions. In particular, if the solution is only piecewise smooth only convergence in mean is ensured while the question of pointwise convergence remains open.

It is by now a classical result that the Fourier series, Eq. (7), in the neighbourhood of a point of discontinuity, x_0, behaves as Gottlieb and Orszag (1977)

$$\mathcal{P}_N u\left(x_0 + \frac{2z}{2N+1} \right) \sim \frac{1}{2}\left[u(x_0^+) + u(x_0^-) \right] + \frac{1}{\pi}\left[u(x_0^+) - u(x_0^-) \right] \mathrm{Si}(z),$$

where z is a constant and $\mathrm{Si}(z)$ signifies the Sine-integral. Away from the point of discontinuity, x_0, we recover linear pointwise convergence as $\mathrm{Si}(z) \simeq \frac{\pi}{2}$ for z large. Close to the point of discontinuity, however, we observe that for any fixed value of z, pointwise convergence is lost regardless of the value of N. This nonuniform convergence and loss of pointwise convergence are the celebrated Gibbs phenomenon, and the oscillatory behaviour of the Sine-integral is the familiar Gibbs oscillations.

As we shall discuss in more detail in Section 5, recent results allow us to dramatically improve on this situation and even completely overcome the Gibbs oscillations to recover an exponentially accurate approximation to a piecewise analytic function, represented by its global expansion.

2.3 The Duality Between Modes and Nodes

While there is flexibility in the choice of the quadrature rules, used to compute the discrete expansion coefficients in the modal expansions, and similar freedom in choosing a nodal set on which to base the Lagrange interpolation polynomials, particular choices are awarded by insight.

Consider, as an example, the modal expansion, (7), with the expansion coefficients approximated as in (10). Inserting the latter directly into the former yields

$$
\begin{aligned}
\mathcal{I}_N u(x) &= \sum_{n=-N}^{N} \left[\frac{1}{2N+1} \sum_{j=0}^{2N} u(x_j) \exp(-inx_j) \right] \exp(inx) \\
&= \sum_{j=0}^{2N} u(x_j) \left[\frac{1}{2N+1} \sum_{n=-N}^{N} \exp\left(in(x-x_j)\right) \right] \\
&= \sum_{j=0}^{2N} u(x_j) \frac{1}{2N+1} \frac{\sin\left(\frac{1}{2}(2N+1)(x-x_j)\right)}{\sin\left(\frac{1}{2}(x-x_j)\right)} = \sum_{j=0}^{2N} u(x_j) h_j(x).
\end{aligned}
$$

Hence, provided the expansion coefficients are approximated by the trapezoidal rule, (10), we recover the interpolation. This particular combination of grid points and quadrature rules results in two mathematically equivalent, but computationally very different, ways of expressing the interpolation and hence the computation of spatial derivatives.

A similar result can be recovered for the orthogonal polynomials provided Gauss quadrature nodes are used (Canuto et al., 1988; Hesthaven et al., 2007).

3 SPECTRAL METHODS

Let us now turn the attention towards the solution of hyperbolic problems using spectral methods. Prominent examples of problems include Maxwells equations from electromagnetics, the Euler equations of gas-dynamics and the equations of elasticity. For the sake of simplicity we concentrate on methods for the scalar conservation law

$$
\frac{\partial u}{\partial t} + \frac{\partial f(u)}{\partial x} = 0,
$$

subject to appropriate boundary and initial conditions. For this initial discussion, we focus on problems for which the solution remains smooth and return to the nonsmooth case in Section 4.

We assume that the solution is given as

$$u(x,t) \simeq u_h(x,t) = \sum_{n=-N}^{N} \tilde{u}_n(t) \exp(inx),$$

where $\tilde{u}_n(t)$ represent the continuous or the discrete expansion coefficients. We can now define the residual

$$R_h(x,t) = \frac{\partial u_h}{\partial t} + \frac{\partial f(u_h)}{\partial x}.$$

Specifying exactly how this vanishes, hence stating in which sense u_h satisfies the conservation laws, gives rise to different families of methods with subtle differences.

3.1 Galerkin Methods

In the Galerkin approach, we require that the residual is orthogonal to the space spanned by the basis functions. For the Fourier case, this results in the scheme

$$\frac{d\hat{u}_n}{dt} = -\frac{1}{2\pi} \int_0^{2\pi} R_h(x,t) \exp(-inx)\, dx.$$

This we can also express as

$$\frac{\partial u_h}{\partial t} + \mathcal{P}_N \left(\frac{\partial f(u_h)}{\partial x} \right) = 0,$$

subject to the initial conditions

$$u_h(x,0) = \mathcal{P}_N u(x,0).$$

One observes that boundary conditions must be reflected in the approximation itself, i.e. in the Galerkin method, each of the basis functions in (1) must satisfy the boundary conditions. For periodic problems, the Fourier series automatically enforces this. However, for nontrivial boundary conditions, this may present a challenge albeit successful schemes have be formulated (Gottlieb and Orszag, 1977; Shen et al., 2011).

The stability of Galerkin schemes is closely related to the well-posedness of the conservation laws in the norm $\|\cdot\|$ (Gottlieb and Orszag, 1977). The practical difficulty with the Galerkin scheme is the need to evaluate the projection of the general flux. While this may be possible for certain simple fluxes, e.g. linear or polynomial fluxes, it is clearly not possible for more general cases. In such a case, one can no longer express the scheme without the use of quadratures. However, this introduces a grid, induces aliasing and suggests that we consider collocation methods.

3.2 Collocation Methods

To overcome the difficulties associated with exact evaluation of the inner products in the Galerkin method, we can change the statement on the residual. Let us define $2N + 1$ distinct collocation points, y_j, and require that the residual vanishes in a pointwise sense

$$R_h(y_j,t) = \left(\frac{\partial u_h}{\partial t} + \frac{\partial f(u_h)}{\partial x} \right) \bigg|_{y_j} = 0.$$

This results in $2N + 1$ equations for the $2N + 1$ unknowns. In principle, there are no restrictions on how y_j is chosen although the stability of the scheme is, to some extend, impacted by this Hesthaven et al. (2007). If we make the most natural choice that the interpolation points x_j and the collocation points y_j are the same we recover the classic collocation scheme

$$\mathcal{I}_N \frac{\partial u_h}{\partial t} + \mathcal{I}_N \frac{\partial f(u_h)}{\partial x} = 0,$$

which can also conveniently be expressed as

$$\frac{d}{dt}\boldsymbol{u} + \mathrm{D}\boldsymbol{f} = 0,$$

where $\boldsymbol{u}$ and $\boldsymbol{f}$ represents vectors of the solution and the flux, respectively, evaluated at the grid points.

To understand the stability of collocation schemes for hyperbolic problems, let us consider the linear problem

$$\frac{\partial u}{\partial t} + a(x)\frac{\partial u}{\partial x} = 0, \tag{13}$$

where $a > 0$ implies a rightward propagating wave and $a < 0$ corresponds to a leftward propagating wave. The Fourier collocation approximation becomes

$$\frac{d}{dt}\boldsymbol{u} + \mathrm{A}\mathrm{D}\boldsymbol{u} = 0, \tag{14}$$

where $\mathrm{A}_{ii} = a(x_i)$ is diagonal.

Define the discrete inner product and L^2-equivalent norm as

$$[f,g]_N = \frac{2\pi}{2N+1}\sum_{i=0}^{2N} f(x_i)g(x_i), \quad \|f\|_N^2 = [f,f]_N.$$

If we initially assume that $|a(x)| > 0$ (Fornberg, 1975; Gottlieb and Orszag, 1977; Kreiss and Oliger, 1972; Orszag, 1972; Pasciak, 1980), it is easy to see that for $\boldsymbol{v} = \mathrm{A}^{-1/2}\boldsymbol{u}$, we recover

$$\frac{d}{dt}\boldsymbol{v} + \mathrm{A}^{1/2}\mathrm{D}\mathrm{A}^{1/2}\boldsymbol{v} = 0,$$

such that

$$\frac{1}{2}\frac{d}{dt}\| v_h \|_N^2 = \frac{1}{2}\frac{d}{dt}\boldsymbol{u}^T \mathrm{A}^{-1}\boldsymbol{u} = 0,$$

since $\mathrm{A}^{1/2}\mathrm{D}\mathrm{A}^{1/2}$ is antisymmetric.

For the general case where $a(x)$ changes sign within the computational domain, the situation is more complex. The straightforward way to guarantee stability is to consider the skew-symmetric form (Kreiss and Oliger, 1972; Tadmor, 1987)

$$\frac{\partial u}{\partial t} + \frac{1}{2}\frac{\partial a(x)u}{\partial x} + \frac{1}{2}a(x)\frac{\partial u}{\partial x} - \frac{1}{2}a_x(x)u(x) = 0, \tag{15}$$

with the discrete form

$$\left.\frac{\partial u_h}{\partial t}\right|_{x_j} + \frac{1}{2}\left.\frac{\partial \mathcal{I}_N a(x)u_h}{\partial x}\right|_{x_j} + \frac{1}{2}a(x_j)\left.\frac{\partial u_h}{\partial x}\right|_{x_j} - \frac{1}{2}a_x(x_j)u_h(x_j) = 0.$$

Stability follows since

$$\frac{1}{2}\frac{d}{dt}\| u_h \|_N^2 \leq \frac{1}{2}\max_{x\in[0,2\pi]} |a_x(x)| \; \| u_h \|_N^2 .$$

The disadvantage of the skew-symmetric formulation is a doubling of the computational work.

The question of stability of the simple formulation, (14), for general $a(x)$ remained an open question for a long time, although partial results were known (Gottlieb et al., 1981b). The difficulty in resolving this issue is associated with the development of very steep spatial gradients which, for a fixed resolution, eventually introduce significant aliasing that affect the stability. By carefully examining the interplay between aliasing, resolution and stability, it was shown in Goodman et al. (1994) that the Fourier approximation is only algebraically stable (Gottlieb and Orszag, 1977), i.e.

$$\| u_h(t) \|_N \leq C(t)N \, \| u_h(0) \|_N, \tag{16}$$

or weakly unstable. The weak aliasing-driven instability spreads from the high modes through the aliasing and results in at most an $\mathcal{O}(N)$ amplification of the Fourier components of the solution. In other words, for well-resolved computations where these aliasing components are very small the computation will appear stable for all practical purposes. Furthermore, in Gottlieb and Hesthaven (2001) it is shown that a weak amount of filtering suffices to control the instability. We return to this in more detail in Section 4.

3.3 Interlude on Polynomial Methods and Boundary Conditions

Let us now briefly consider the more general initial boundary value problem

$$\begin{aligned} &\frac{\partial u}{\partial t}+\frac{\partial f(u)}{\partial x}=0, \\ &u(x,0)=g(t), \end{aligned} \tag{17}$$

posed on a finite domain which we take to be $[-1, 1]$ without loss of generality. For the problem to be well posed, we must specify boundary conditions on the form

$$\alpha u(-1,t)=f^{-}(t),\quad \beta u(1,t)=f^{+}(t).$$

Specification of α and β is related to the fluxfunction, e.g. if

$$x\frac{\partial f}{\partial u}<0,$$

at the boundary, information is incoming and a boundary condition must be given. For a system of equations, the equivalent condition is posed through the characteristic variables, i.e. characteristic waves entering the computational domain must be specified and, hence, require a boundary condition to ensure well-posedness of the problem (see Gustafsson et al. (1995) and Hesthaven et al. (2007) for further details).

What separates the polynomial approximation from the trigonometric schemes discussed so far is the need to impose boundary conditions to restrict the numerical solutions, $u_h(x, t)$, to satisfy the boundary conditions.

3.3.1 *Strongly Imposed Boundary Conditions*

In the classic approach one requires that the boundary conditions are imposed strongly, i.e. exactly. Hence, we shall seek a polynomial, $u_h(x, t)$, that satisfies (17) in a collocation sense at all internal grid points, x_j, as

$$\left.\frac{\partial u_h}{\partial t}\right|_{x_j}+\left.\frac{\partial \mathcal{I}_N f(u_h)}{\partial x}\right|_{x_j}=0,$$

while the boundary conditions are imposed exactly

$$\alpha u_h(-1,t)=f^{-}(t),\quad \beta u_h(1,t)=f^{+}(t).$$

If we again consider the wave equation, (13), the collocation scheme becomes

$$\left.\frac{\partial u_h}{\partial t}\right|_{x_j}+a(x_j)\left.\frac{\partial u_h}{\partial x}\right|_{x_j}=0,$$

at all interior grid points, i.e. for $a > 0$, $j \in [1, N]$, while $u_h(x_0, t) = f^{-}(t)$.

Establishing stability of the collocation scheme is considerably more challenging than for the Fourier collocation method. To expose the source of this difficulty, consider the simple wave equation, (13), with $a(x) = 1$ and subject to the conditions

$$u(x,0) = g(x), \quad u(-1,t) = 0.$$

A collocation scheme based on the Gauss–Lobatto nodes yields

$$\frac{d}{dt}\boldsymbol{u} = -\widetilde{\mathrm{D}}\boldsymbol{u}. \tag{18}$$

Here the matrix $\widetilde{\mathrm{D}}$ represents the polynomial differentiation matrix (Funaro, 2008; Gottlieb and Orszag, 1977; Hesthaven et al., 2007) modified to enforce the boundary condition strongly, i.e. by introducing zeros in the first row and column.

The strongly enforced boundary condition introduces the first main obstacle as any structure of the differentiation matrix is destroyed. This leaves us with the quadrature formula to establish stability. The straightforward quadrature formula, however, is closely related to the weighted inner product, $(f,g)_{L^2_w}$, in which the polynomials are orthogonal. With the exception of the Legendre polynomials, the norm associated with the inner product is not uniformly equivalent to the usual L^2-norm (Gottlieb and Orszag, 1977; Hesthaven et al., 2007). This loss of equivalence eliminates the straightforward use of the quadrature rules to establish stability as the corresponding norm is too weak. Thus, the two central techniques utilized for the Fourier methods are not directly applicable to the case of the polynomial collocation methods.

One approach is to construct a new inner product and associated norm, uniformly equivalent to L^2, and subsequently establish stability in this norm. This is the approach taken in Gottlieb et al. (1981b, 1987). The more general variable coefficient problem, (13), with $a(x)$ being smooth can be addressed using a similar approach. In particular, if $a(x)$ is smooth and uniformly bounded away from zero stability is established in the elliptic norm (Gottlieb et al., 1981b)

$$\frac{1}{2}\frac{d}{dt}\sum_{j=0}^{N-1} v_N^2(x_j)\frac{\widetilde{w}_j}{a(x_j)} \leq 0.$$

For the more general case of $a(x)$ changing sign the only known results are based on the skew-symmetric form (Canuto and Quarteroni, 1982b; Hesthaven et al., 2007), (15), although numerical experiments suggest that the straightforward Chebyshev collocation approximation of the wave equation with a variable coefficient behaves much as the Fourier approximation discussed above, i.e. if the solution is well resolved, the approximation is stable (Gottlieb and Orszag, 1977; Gottlieb et al., 1981b).

3.3.2 Weakly Imposed Boundary Conditions

The conceptual leap that leads one to consider other ways of imposing boundary conditions is the observation that it suffices to impose the boundary conditions to the order of the scheme, i.e. weakly.

This idea, put forward in the context of spectral methods in Canuto and Quarteroni (1982b) for the weak formulation and in Funaro and Gottlieb (1988) and Funaro and Gottlieb (1991) for the strong formulation considered here, has recently been developed further into a flexible technique to impose boundary conditions in pseudospectral approximations to a variety of problems (Carpenter and Gottlieb, 1996; Don and Gottlieb, 1994; Hesthaven, 1997, 1998, 2000; Hesthaven and Gottlieb, 1996, 1999).

In this setting, one seeks a polynomial solution, $u_h(x, t)$, to (17) that satisfy

$$\begin{aligned}\frac{\partial u_h}{\partial t}+\mathcal{I}_N\frac{\partial \mathcal{I}_N f(u_h)}{\partial x} &= -\tau^-\alpha Q^-(x)\left[u_h(-1,t)-\tilde{f}^-(t)\right]\\ &\quad -\tau^+\beta Q^+(x)\left[u_h(1,t)-\tilde{f}^+(t)\right],\end{aligned} \tag{19}$$

where we have introduced the polynomials, $Q^{\pm}(x)$, and the scalars, $\tau^{\pm}$.

To complete the scheme we must specify $Q^{\pm}(x)$ and define an approach by which to specify the scalar parameters, $\tau^{\pm}$. While the latter choice is dictated by requiring semidiscrete stability, the former choice of $Q^{\pm}(x)$ is associated with a great deal of freedom.

As an example, consider the approximation to the constant coefficient wave equation (13)

$$\frac{\partial u_h}{\partial t}+a\frac{\partial u_h}{\partial x}=-\tau^- aQ^-(x)[u_h(-1,t)-f(t)],$$

where $u_h(x, t)$ is based on the Legendre–Gauss–Lobatto points. A viable choice of $Q^-(x)$ is

$$Q^-(x)=\frac{(1-x)P'_N(x)}{2P'_N(-1)}=\begin{cases}1 & x=-1\\ 0 & x=x_j\neq -1\end{cases},$$

where x_j refers to the Legendre–Gauss–Lobatto points and $P_N(x)$ is the Legendre polynomial of order N. By requesting that the equation be satisfied in a collocation sense and the scheme be stable, we recover the scheme

$$\left.\frac{\partial u_h}{\partial t}\right|_{x_j}+a\left.\frac{\partial u_h}{\partial x}\right|_{x_j}=-a\frac{N(N+1)}{4}\frac{(1-x_j)P'_N(x_j)}{2P'_N(-1)}[u_h(-1,t)-f(t)].$$

Using the accuracy of the quadrature, one easily shows asymptotic stability. Although the boundary condition is imposed only weakly, the approximation is clearly consistent, i.e. if $u_h(x, t) = u(x, t)$ the penalty term vanishes identically. A key difference between the schemes with strongly and weakly imposed boundary conditions is that in the former case, stability is established after construction of the scheme whereas in the latter case, stability is guaranteed as a result of the construction of the scheme.

4 STABILITY AND CONVERGENCE OF NONLINEAR PROBLEMS

Turning to the development of spectral methods for nonlinear problems introduces a number of challenges. First of all, the use of standard energy methods to establish stability is no longer possible except in certain special cases. As a result of this, the question of convergence remains open and must be addressed in a different way.

4.1 Skew-Symmetric Form

If we consider the Fourier collocation scheme for Burgers equation

$$\frac{\partial u}{\partial t}+\frac{1}{2}\frac{\partial u^2}{\partial x}=0,$$

we seeking the approximate solution, $u_h(x, t)$, such that

$$\left.\frac{\partial u_h}{\partial t}\right|_{x_j}+\left.\frac{1}{2}\frac{\partial}{\partial x}\mathcal{I}_N u_h^2\right|_{x_j}=0. \tag{20}$$

Note that while the partial differential equation has the equivalent formulation

$$\frac{\partial u}{\partial t}+u\frac{\partial u}{\partial x}=0,$$

for smooth solutions, the corresponding nonconservative Fourier approximation

$$\left.\frac{\partial u_h}{\partial t}\right|_{x_j}+u_h(x_j)\left.\frac{\partial u_h}{\partial x}\right|_{x_j}=0,$$

is not equivalent to (20) and may behave differently due to the aliasing.

We cannot establish stability of these scheme using standard means. However, by writing it on skew-symmetric form

$$\frac{\partial u}{\partial t}+\frac{1}{3}\frac{\partial u^2}{\partial x}+\frac{1}{3}u\frac{\partial u}{\partial x}=0,$$

stability of the collocation approximation

$$\left.\frac{\partial u_h}{\partial t}\right|_{x_j}+\left.\frac{1}{3}\frac{\partial}{\partial x}\mathcal{I}_N u_h^2\right|_{x_j}+\frac{1}{3}u_h(x_j)\left.\frac{\partial u_h}{\partial x}\right|_{x_j}=0,$$

follows directly from the accuracy of the quadrature.

If we consider a general hyperbolic conservation law

$$\frac{\partial u}{\partial t} + \frac{\partial f(u)}{\partial x} = 0,$$

one can prove under light conditions on $f(u)$ that it may always be expressed on skew-symmetric form. This extends to many systems. In Tadmor (1984) it is shown that the existence of a skew-symmetric form is guaranteed for any system that has a convex entropy or is symmetrizable. This includes all major systems of conservation laws, e.g. the Euler equations.

This results suggests that one could simply express the conservation law on skew-symmetric form to ensure stability of the scheme. For problems with smooth solutions, this is indeed a powerful technique, although it doubles the computational cost. However, if we recall that for the scalar problem, the quadratic functional u^2 plays the role of both energy and entropy, we realize that the skew-symmetric form conserves entropy. For problems with shocks this is in violation of basic properties of the hyperbolic conservation laws. Hence, the skew-symmetric form is suitable only for problems with smooth solutions or in combination with additional dissipation.

4.2 Filtering for Stability

Maintaining stability of the numerical approximation becomes increasingly hard as the discontinuity evolves and generates energy with higher and higher frequency content. This process, amplified by the nonlinear mixing of the Gibbs oscillations and the numerical solution, eventually renders the scheme unstable or, if the scheme is expressed on skew-symmetric form, the solution wildly inaccurate.

Understanding the source of the stability problem, i.e. accumulation of high-frequency energy, suggests a possible solution is the introduction of a dissipative mechanism to remove the high-frequency components.

A classical way to accomplish this is to modify the original problem by adding artificial dissipation as

$$\frac{\partial u}{\partial t} + \frac{\partial f(u)}{\partial x} = \varepsilon(-1)^{p+1}\frac{\partial^{2p} u}{\partial x^{2p}}.$$

A direct implementation of this, however, may be costly and introduces additional stiffness which limits the stable time step (Gottlieb and Hesthaven, 2001; Hesthaven et al., 2007).

To seek a different path, let us modify the numerical solution, $u_h(x, t)$, by the use of a spectral filter as

$$\mathcal{F}_N u_h(x,t) = \sum_{n=-N}^{N} \sigma\left(\frac{n}{N}\right) \tilde{u}_n(t) \exp(inx). \tag{21}$$

To understand the impact of using the filter at regular intervals as a stabilizing mechanism, a procedure first proposed in Majda et al. (1978) and Kreiss and Oliger (1979), consider an exponential filter

$$\sigma(\eta) = \exp\left(-\alpha\eta^{2p}\right).$$

As discussed in Section 5.1 this filter allows for a dramatic improvement in the accuracy of the approximation away from points of discontinuity.

To appreciate its impact on stability, consider the generic initial value problem

$$\frac{\partial u}{\partial t} = \mathcal{L}u,$$

and the Fourier scheme

$$\frac{d}{dt}\boldsymbol{u} = \mathcal{L}_N\boldsymbol{u}.$$

Advancing the solution from $t = 0$ to $t = \Delta t$, followed by filtering, is expressed as

$$\boldsymbol{u}(\Delta t) = \mathcal{F}_N \exp\left(\mathcal{L}_N \Delta t\right)\boldsymbol{u}(0).$$

If we first assume that $\mathcal{L}_N$ represents the constant coefficient hyperbolic problem, i.e. $\mathcal{L} = a\frac{\partial}{\partial x}$, we recover that

$$\tilde{u}_n(\Delta t) = \exp\left(-\alpha\eta^{2p} + a(ik)\Delta t\right)\tilde{u}_n(0), \tag{22}$$

i.e. we are in fact computing the solution to the modified problem

$$\frac{\partial u}{\partial t} = a\frac{\partial u}{\partial x} - \alpha\frac{(-1)^p}{\Delta t N^{2p}}\frac{\partial^{2p} u}{\partial x^{2p}}.$$

The effect of the filter is thus equivalent to that of adding a small dissipative term to the original equation. However, the process of adding the dissipation through the filter is very simple.

For a general $\mathcal{L}$, e.g. with a variable coefficient or a nonlinear flux in which case $\mathcal{F}_N$ and $\mathcal{L}_N$ no longer commute, the modified equation being solved takes the form

$$\frac{\partial u}{\partial t} = \mathcal{L}u - \alpha\frac{(-1)^p}{\Delta t N^{2p}}\frac{\partial^{2p} u}{\partial x^{2p}} + \mathcal{O}(\Delta t^2),$$

by viewing the application of the filter as an operator splitting problem (Boyd, 1998; Don and Gottlieb, 1998).

It is clear that the filter has a stabilizing effect, established more rigorously for problem with smooth and nonsmooth initial data in Majda et al. (1978), Kreiss and Oliger (1979) and Goodman et al. (1994) for the Fourier approximation to the general variable coefficient problem, (13). In Gottlieb and

Hesthaven (2001) and Hesthaven et al. (2007) it is furthermore established that light filtering suffices to stability an aliasing-driven instability.

4.3 Vanishing Viscosity Techniques

The foundation of a convergence theory for the spectral approximations to hyperbolic conservation laws has been laid in Tadmor (1989), Maday and Tadmor (1989) and Chen et al. (1993) for the periodic case and extended in Maday et al. (1993) to the Legendre approximation and to the Chebyshev–Legendre scheme in Ma (1998b) and Ma (1998a).

To outline the basic elements of this convergence theory let us restrict ourselves to the periodic case. For the discrete approximation we must add a dissipative term that is strong enough to stabilize the approximation, yet small enough so as to not ruin the spectral accuracy of the scheme. In Tadmor (1989) and Maday and Tadmor (1989) the following spectral viscosity method was considered

$$\frac{\partial u_h}{\partial t}+\frac{\partial}{\partial x}\mathcal{P}_N(f(u_h))=\varepsilon_h(-1)^{p+1}\frac{\partial^p}{\partial x^p}\left[Q_m(x,t)*\frac{\partial^p u_h}{\partial x^p}\right], \tag{23}$$

where

$$\frac{\partial^p}{\partial x^p}\left[Q_m(x,t)*\frac{\partial^p u_h}{\partial x^p}\right]=\sum_{m<|n|\le N}(ik)^{2p}\hat{Q}_n\hat{u}_n\exp(inx).$$

To ensure stability m should not be taken too big. On the other hand, taking m too small will impact the accuracy in a negative way. An acceptable compromise is

$$m\sim N^\theta,\quad \theta<\frac{2p-1}{2p}.$$

Moreover, the smoothing factors, $\hat{Q}_n$, should only be activated for high modes as

$$\hat{Q}_n=1-\left(\frac{m}{|n|}\right)^{\frac{2p-1}{\theta}},$$

for $|n|>m$ and $\hat{Q}_n=1$ otherwise. Finally, we assume that the amplitude of the viscosity is small as

$$\varepsilon_h\sim\frac{C}{N^{2p-1}}.$$

Under these assumptions, one can prove for $p=1$ that the solution is bounded in $L^\infty[0,2\pi]$ and obtain the estimate (Tadmor, 1989)

$$\|u_h\|_{L^2[0,2\pi]}+\sqrt{\varepsilon_h}\left\|\frac{\partial u_h}{\partial x}\right\|_{L^2_{loc}}\le C.$$

Convergence to the correct entropy solution then follows from compensated compactness arguments (Maday and Tadmor, 1989; Tadmor, 1989).

To realize the connection between the spectral viscosity method and the use of filters discussed above, consider the simple case where $f(u) = au$. In this case, the solution to (23) is

$$\hat{u}_n(t) = \exp\left(inat - \varepsilon_h n^2 \hat{Q}_n\right)\hat{u}_n(0), \quad |n| > m,$$

which is equivalent to the effect of the filtering, albeit with a particular filter function.

For $p \neq 1$ a bound on the $L^\infty[0,2\pi]$ is no longer known. However, experience suggests that it is better to filter from the first mode but to employ a slower decay of the expansion coefficients, corresponding to taking $p > 1$. This yields the superviscosity (Tadmor, 1990) method in which one solves

$$\frac{\partial u_h}{\partial t} + \frac{\partial}{\partial x}\mathcal{P}_N f(u_h) = \varepsilon_h (-1)^{p+1}\frac{\partial^{2p} u_h}{\partial x^{2p}},$$

which is equivalent to the use of a high-order exponential filter.

5 POSTPROCESSING TECHNIQUES

A manifestation of the slow and nonuniform convergence of $\mathcal{I}_N u$ for a piecewise smooth functions is the linear decay of the global expansion coefficients, $\tilde{u}_n$. This observation also suggests that one could attempt to modify the global expansion coefficients to enhance the convergence rate of the spectral approximation. The key question to consider is exactly how one should modify the expansion to ensure enhanced convergence to the correct solution.

However, before doing so, it is worth understanding if the emergence of a shock and the Gibbs phenomenon effectively eliminates any hope of maintaining high-order accuracy.

Consider again

$$\frac{\partial u}{\partial t} + a(x)\frac{\partial u}{\partial x} = \frac{\partial u}{\partial t} + \mathcal{L}u = 0.$$

Both $a(x)$ and $u(x, t)$ are considered periodic and $a(x)$ is smooth. We have already established stability of this scheme, possibly by using filtering or through the skew-symmetric form.

We assume that the initial condition, $u(x, 0)$, is nonsmooth, resulting in the introduction of the Gibbs phenomenon.

Let us also introduce the adjoint problem

$$\frac{\partial v}{\partial t} - \mathcal{L}^* v = 0,$$

where $(\mathcal{L}u,v)=(u,\mathcal{L}^{*}v)$. We assume smooth initial conditions for the adjoint problem. A seminal result (Abarbanel et al., 1985) can be obtained as

$$(u_h(t),v(t))=(u(t),v(t))+\varepsilon, \tag{24}$$

where ε is very small and depends only on the smoothness of $v(x, t)$. Since the adjoint problem is smooth, this can be made arbitrarily small. This highlights, at least for the case of a variable coefficient problem with nonsmooth initial conditions, the possibility of recovering a high-order accurate solution, $u_h(t)$. However, this accuracy is not found directly in the solution $u_h(x, t)$, but, rather, in the moments of the solution.

While it is a surprising result, it is also an encouraging result. It clarifies that the Gibbs oscillations may look bad, but they do not destroy the attractive basic properties of the schemes—in particular, the properties related to the highly accurate propagation. This result can be extended to nonsmooth solutions and sources (Zudrop and Hesthaven, 2015) and suggests that we consider ways to recover a pointwise spectrally accurate solution from the oscillatory solution which is only pointwise first-order accurate. For Burgers equation, extensive computational results in Shu and Wong (1995) suggest that high-order accuracy is also retained in this case.

5.1 Filtering for Accuracy

We consider the filtered approximation, $\mathcal{F}_N u_h(x)$, of the form

$$\mathcal{F}_N u_h(x)=\sum_{n=-N}^{N}\sigma\left(\frac{n}{N}\right)\tilde{u}_n\exp(inx), \tag{25}$$

where $\sigma(\eta)$ is a real filter function with the following properties (Vandeven, 1991)

$$\sigma(\eta)=\begin{cases}\sigma(-\eta) & \\ \sigma(0)=1 & \\ \sigma^{(q)}(0)=0 & 1\le q\le 2p-1 \\ \sigma(\eta)=0 & |\eta|\ge 1\end{cases}. \tag{26}$$

If $\sigma(\eta)$ has at least $2p-1$ continuous derivatives, $\sigma(\eta)$ is termed a filter of order $2p$.

As the filter is nothing more than a low-pass filter, it is not surprising that the filtered function converges faster than the unfiltered filtered original expansion. To understand exactly how much the filter modifies the convergence rate, assume that $u(x)$ is piecewise C^{2p} with one discontinuity located at $x = \xi$. Let us furthermore assume that the filter is of order $2p$. Then the pointwise error of the filtered approximation is given as Vandeven (1991), Gottlieb and Shu (1997) and Hesthaven et al. (2007)

$$|u(x)-\mathcal{F}_N u_N(x)|\le C\frac{1}{N^{2p-1}d(x,\xi)^{2p-1}}K(u)+C\frac{\sqrt{N}}{N^{2p}}\|u^{(2p)}\|_{L_B^2},$$

where $d(x, \xi)$ measures the distance from x to the point of discontinuity, ξ, $K(u)$ is uniformly bounded away from the discontinuity and a function of $u(x)$ only. Also $\| \cdot \|_{L^2_B}$ signifies the broken L^2-norm.

While the details of the proof of this result are technical and can be found in Vandeven (1991), Gottlieb and Shu (1997) and Hesthaven et al. (2007), the interpretation of the result is simple, and perhaps somewhat surprising. It states that the convergence rate of the filtered approximation is determined solely by the order $2p$ of the filter $\sigma(\eta)$ and the regularity of the function, $u(x)$, away from the point of discontinuity. In particular, if the function $u(x)$ is piecewise analytic and the order of the filter increases with N, one recovers an exponentially accurate approximation to the unfiltered function everywhere except very close to the discontinuity (Gottlieb and Shu, 1997; Vandeven, 1991). Partial results for polynomial expansions suggest similar behaviour (Hesthaven and Kirby, 2008).

Spectral filtering of the expansion coefficients remains the most popular way of enhancing the convergence rate. An alternative is to improve the approximation by localizing the approximation close to the point of the discontinuity. This approach, known as physical space filtering, operates directly on the interpolating polynomials rather than the expansion coefficients. This is developed and applied with success in Gottlieb and Tadmor (1985), Gottlieb and Shu (1997), Tadmor and Tanner (2002) and Tadmor and Tanner (2005).

An alternative approach explores the superior properties of Padé-approximation to address the Gibbs phenomenon by a reprojection. This approach, developed in Driscoll and Fornberg (2001), Emmel et al. (2003), Kaber and Maday (2005) and Hesthaven et al. (2006), often yields excellent results, even at the point of discontinuity. However, the nonlinear nature of the Padé approximant makes its application complex.

5.2 Gegenbauer Reconstruction

Let us finally outline the key elements of a general theory that establishes the possibility of recovering a piecewise exponentially convergent approximation to a piecewise analytic function, having knowledge of the global expansion coefficients and the position of the discontinuities only.

The basic element of this approach is the identification of a new basis with very special properties and, subsequently, the expansion of the slowly convergent truncated global expansion in this new basis. Provided this new basis satisfies certain conditions, the new expansion has the remarkable property that it is exponentially convergent to the original piecewise analytic function even though its evaluation uses information from the slowly convergent global expansion.

We assume that there exists an interval $[a,b] \subset [0,2\pi]$ in which $u(x)$ is analytic and, furthermore, that the original truncated expansion is pointwise convergent in all of $[0, 2\pi]$ with the exception of a finite number of points. We introduce the scaled variable

$$\xi(x) = -1 + 2\frac{x-a}{b-a}.$$

Clearly, $\xi : [a,b] \to [-1,1]$.

Now define a new basis, $\psi_n^\lambda(\xi)$, which is orthogonal in the weighted inner product, $(\cdot,\cdot)_w^\lambda$ where λ signifies that the weight, $w(x)$, may depend on λ, i.e.

$$\left(\psi_k^\lambda, \psi_n^\lambda\right)_w^\lambda = \| \psi_n^\lambda \|_{L_w^2}^2 \, \delta_{kn} = \gamma_n^\lambda \delta_{kn}.$$

Furthermore, we require that if $v(\xi)$ is analytic then

$$\mathcal{P}_\lambda v(\xi) = \sum_{n=0}^{\lambda} \frac{1}{\gamma_n^\lambda} \left(v, \psi_n^\lambda\right)_w^\lambda \psi_n^\lambda(\xi),$$

is pointwise exponentially convergent as λ increases, i.e.

$$\| v - \mathcal{P}_\lambda v \|_{L^\infty} \le C e^{-c\lambda},$$

with $c > 0$. This is simply the statement of exponential convergence for a polynomial expansion of a analytic function.

A final condition sets this basis apart and is central in order to overcome the Gibbs phenomenon. We require that there exists a number $\beta < 1$, such that for $\lambda = \beta N$ we have

$$\left| \frac{1}{\gamma_n^\lambda} \left(\phi_k(x(\xi)), \psi_n^\lambda(\xi) \right)_w^\lambda \right| \, \| \psi_n^\lambda \|_{L^\infty} \le \left(\frac{\alpha N}{k} \right)^\lambda, \tag{27}$$

for $k > N$, $n \le \lambda$ and $\alpha < 1$. The interpretation of this condition is that the projection of the high modes of ϕ_k onto the basis, ψ_n^λ, is exponentially small in the interval, $\xi \in [-1, 1]$. In other words, by reexpanding the slowly decaying ϕ_n-based global expansion in the local ψ_n^λ-basis, an exponentially accurate local approximation is recovered. Moreover, this can be achieved everywhere in the domain where $u(x)$ is analytic. This latter condition on ψ_n^λ is termed the Gibbs condition to emphasize its close connection to the resolution of the Gibbs phenomenon (Gottlieb and Shu, 1997, 1998).

Provided only that the ψ_n^λ-basis, termed the Gibbs complementary basis, is complete we recover the key result

$$\left\| u(x) - \sum_{n=0}^{\lambda} \frac{1}{\gamma_n^\lambda} \left(\mathcal{P}_N u, \psi_n^\lambda \right)_w^\lambda \psi_n^\lambda(\xi(x)) \right\|_{L^\infty} \le C \exp(-cN),$$

where $\lambda = \beta N$ and $u(x)$ is analytic in the interval $[a, b]$.

In other words, if a Gibbs complementary basis exists it is possible to reconstruct a piecewise exponentially convergent approximation to a piecewise analytic function from the information contained in the original slowly converging global approximation. The only additional piece of information needed is the location of the points of discontinuity. The Gibbs phenomenon can be overcome.

A constructive approach to the identification of the complementary basis is currently unknown. The existence of such a basis, however, has been established by carefully examining the properties of the basis

$$\psi_n^{\lambda}(\xi) = C_n^{(\lambda)}(\xi),$$

where $C_n^{\lambda}(\xi)$ represent the Gegenbauer polynomials, also known as the symmetric Jacobi polynomials or the ultraspherical polynomials (Szego, 1939).

Using the Fourier basis, it must be established that

$$\left| \frac{1}{\gamma_n^{\lambda}} (\phi_k, \psi_n^{\lambda})_w^{\lambda} \right| \leq \left(\frac{\alpha N}{k} \right)^{\lambda},$$

for $k > N$, $0 < \alpha < 1$, and $n \leq \beta N = \lambda$. However, for this basis the inner product allows an exact evaluation

$$\frac{1}{\gamma_n^{\lambda}} (\phi_k, \psi_n^{\lambda}) = i^n \Gamma(\lambda) \left(\frac{2}{\pi k \varepsilon} \right)^{\lambda} (n+\lambda) J_{n+\lambda}(\pi \varepsilon k),$$

with $J_\nu(x)$ being the Bessel function and $\varepsilon = b - a$ measures the width of the interval. Using the properties of the Bessel function and the Stirling formula for the asymptotic of the Γ-function, the Gibbs condition is satisfied if (Gottlieb and Shu, 1997)

$$\beta = \frac{2\pi\varepsilon}{27}.$$

This establishes the existence of a Gibbs complementary basis to the Fourier basis (Gottlieb and Shu, 1997, 1998).

The extension to the polynomial case follows a similar approach and the Gegenbauer polynomials again play the role as the complementary basis (Gottlieb and Shu, 1997, 1998).

The reconstruction of piecewise smooth solutions to conservation laws as a post processing technique has been exploited in Don (1994), Don and Gottlieb (1998), Gelb and Gottlieb (1997) and Gelb and Tadmor (2000).

REFERENCES

Abarbanel, S., Gottlieb, D., Tadmor, E., 1985. Spectral Methods for Discontinuous Problems, vol. 177974. Langley Research Center, NASA, Hampton.

Ben-Yu, G., 1998. Spectral Methods and Their Applications. World Scientific, Singapore.

Bernardi, C., Maday, Y., 1997. Spectral methods. Handb. Numer. Anal. 5, 209–485.

Boyd, J.P., 1998. Two comments on filtering (artificial viscosity) for Chebyshev and Legendre spectral and spectral element methods: preserving boundary conditions and interpretation of the filter as a diffusion. J. Comput. Phys. 143 (1), 283–288.

Boyd, J.P., 2001. Chebyshev and Fourier Spectral Methods. Dover Publications, Inc., New York

Canuto, C., Quarteroni, A., 1982a. Approximation results for orthogonal polynomials in sobolev spaces. Math. Comput. 38 (157), 67–86.

Canuto, C., Quarteroni, A., 1982b. Error estimates for spectral and pseudospectral approximations of hyperbolic equations. SIAM J. Numer. Anal. 19 (3), 629–642.

Canuto, C., Hussaini, M.Y., Quarteroni, A., Zang, T.A., 1988. Spectral Methods in Fluid Dynamics. Springer-Verlag, New York.

Carpenter, M.H., Gottlieb, D., 1996. Spectral methods on arbitrary grids. J. Comput. Phys. 129, 74–86.

Chen, G.Q., Du, Q., Tadmor, E., 1993. Spectral viscosity approximations to multidimensional scalar conservation laws. Math. Comput. 61 (204), 629–643.

Deville, M.O., Fischer, P.F., Mund, E.H., 2002. High-Order Methods for Incompressible Fluid Flow, vol. 9. Cambridge University Press, Cambridge.

Don, W.S., 1994. Numerical study of pseudospectral methods in shock wave applications. J. Comput. Phys. 110 (1), 103–111.

Don, W.S., Gottlieb, D., 1994. The Chebyshev-Legendre method: implementing Legendre methods on Chebyshev points. SIAM J. Numer. Anal. 31 (6), 1519–1534.

Don, W.S., Gottlieb, D., 1998. Spectral simulation of supersonic reactive flows. SIAM J. Numer. Anal. 35 (6), 2370–2384.

Driscoll, T.A., Fornberg, B., 2001. A Padé-based algorithm for overcoming the Gibbs phenomenon. Numer. Algorithms 26 (1), 77–92.

Emmel, L., Kaber, S.M., Maday, Y., 2003. Padé-Jacobi filtering for spectral approximations of discontinuous solutions. Numer. Algorithms 33 (1–4), 251–264.

Fornberg, B., 1975. On a Fourier method for the integration of hyperbolic equations. SIAM J. Numer. Anal. 12 (4), 509–528.

Funaro, D., 2008. Polynomial Approximation Of Differential Equations, vol. 8. Springer Science & Business Media, Berlin.

Funaro, D., Gottlieb, D., 1988. A new method of imposing boundary conditions in pseudospectral approximations of hyperbolic equations. Math. Comput. 51 (184), 599–613.

Funaro, D., Gottlieb, D., 1991. Convergence results for pseudospectral approximations of hyperbolic systems by a penalty-type boundary treatment. Math. Comput. 57 (196), 585–596.

Gelb, A., Gottlieb, D., 1997. The resolution of the Gibbs phenomenon for spliced functions in one and two dimensions. Comput. Math. Appl. 33 (11), 35–58.

Gelb, A., Tadmor, E., 2000. Enhanced spectral viscosity approximations for conservation laws. Appl. Numer. Math. 33 (1), 3–21.

Goodman, J., Hou, T., Tadmor, E., 1994. On the stability of the unsmoothed Fourier method for hyperbolic equations. Numer. Math. 67 (1), 93–129.

Gottlieb, D., Hesthaven, J.S., 2001. Spectral methods for hyperbolic problems. J. Comput. Appl. Math. 128 (1), 83–131.

Gottlieb, D., Orszag, S.A., 1977. Numerical Analysis of Spectral Methods: Theory and Applications, vol. 26. Siam, Philadelphia.

Gottlieb, D., Shu, C.W., 1997. On the Gibbs phenomenon and its resolution. SIAM Rev. 39 (4), 644–668.

Gottlieb, D., Shu, C.W., 1998. A general theory for the resolution of the Gibbs phenomenon. Atti. Convegni. Lincei. 147, 39–48.

Gottlieb, D., Tadmor, E., 1985. Recovering pointwise values of discontinuous data within spectral accuracy. In: Progress and Supercomputing in Computational Fluid Dynamics. Springer, Boston, pp. 357–375.

Gottlieb, D., Lustman, L., Orszag, S.A., 1981a. Spectral calculations of one-dimensional inviscid compressible flows. SIAM J. Sci. Stat. Comp. 2 (3), 296–310.

Gottlieb, D., Orszag, S.A., Turkel, E., 1981b. Stability of pseudospectral and finite-difference methods for variable coefficient problems. Math. Comput. 37 (156), 293–305.

Gottlieb, D., Lustman, L., Tadmor, E., 1987. Stability analysis of spectral methods for hyperbolic initial-boundary value systems. SIAM J. Numer. Anal. 24 (2), 241–256.

Gustafsson, B., Kreiss, H.O., Oliger, J., 1995. Time Dependent Problems and Difference Methods, vol. 24. John Wiley & Sons, New Jersey.

Hesthaven, J.S., 1997. A stable penalty method for the compressible Navier-Stokes equations: II. One-dimensional domain decomposition schemes. SIAM J. Sci. Comput. 18 (3), 658–685.

Hesthaven, J.S., 1998. A stable penalty method for the compressible Navier-Stokes equations: III. Multidimensional domain decomposition schemes. SIAM J. Sci. Comput. 20 (1), 62–93.

Hesthaven, J.S., 2000. Spectral penalty methods. Appl. Numer. Math. 33 (1), 23–41.

Hesthaven, J.S., Gottlieb, D., 1996. A stable penalty method for the compressible Navier-Stokes equations: I. Open boundary conditions. SIAM J. Sci. Comput. 17 (3), 579–612.

Hesthaven, J.S., Gottlieb, D., 1999. Stable spectral methods for conservation laws on triangles with unstructured grids. Comput. Method Appl. M. 175 (3), 361–381.

Hesthaven, J.S., Kirby, R., 2008. Filtering in Legendre spectral methods. Math. Comput. 77 (263), 1425–1452.

Hesthaven, J.S., Kaber, S.M., Lurati, L., 2006. Padé-Legendre interpolants for Gibbs reconstruction. J. Sci. Comput. 28 (2–3), 337–359.

Hesthaven, J.S., Gottlieb, S., Gottlieb, D., 2007. Spectral Methods for Time-Dependent Problems, vol. 21. Cambridge University Press, Cambridge.

Kaber, S.M., Maday, Y., 2005. Analysis of some Padé-Chebyshev approximants. SIAM J. Numer. Anal. 43 (1), 437–454.

Karniadakis, G., Sherwin, S., 2013. Spectral/hp Element Methods for Computational Fluid Dynamics. Oxford University Press, Oxford, UK.

Kreiss, H.O., Oliger, J., 1972. Comparison of accurate methods for the integration of hyperbolic equations. Tellus 24 (3), 199–215.

Kreiss, H.O., Oliger, J., 1979. Stability of the Fourier method. SIAM J. Numer. Anal. 16 (3), 421–433.

Ma, H., 1998a. Chebyshev-Legendre spectral viscosity method for nonlinear conservation laws. SIAM J. Numer. Anal. 35 (3), 869–892.

Ma, H., 1998b. Chebyshev-Legendre super spectral viscosity method for nonlinear conservation laws. SIAM J. Numer. Anal. 35 (3), 893–908.

Maday, Y., Tadmor, E., 1989. Analysis of the spectral vanishing viscosity method for periodic conservation laws. SIAM J. Numer. Anal. 26 (4), 854–870.

Maday, Y., Kaber, S.M.O., Tadmor, E., 1993. Legendre pseudospectral viscosity method for non-linear conservation laws. SIAM J. Numer. Anal. 30 (2), 321–342.

Majda, A., McDonough, J., Osher, S., 1978. The Fourier method for nonsmooth initial data. Math. Comput. 32 (144), 1041–1081.

Orszag, S.A., 1972. Comparison of pseudospectral and spectral approximation. Stud. Appl. Math. 51 (3), 253–259.

Pasciak, J.E., 1980. Spectral and pseudospectral methods for advection equations. Math. Comput. 35 (152), 1081–1092.

Peyret, R., 2013. Spectral Methods for Incompressible Viscous Flow, vol. 148. Springer Science & Business Media, Berlin.

Shen, J., Tang, T., Wang, L.L., 2011. Spectral Methods: Algorithms, Analysis and Applications, vol. 41. Springer Science & Business Media, Berlin.

Shu, C.W., Wong, P.S., 1995. A note on the accuracy of spectral method applied to nonlinear conservation laws. J. Sci. Comput. 10 (3), 357–369.

Szego, G., 1939. Orthogonal Polynomials. vol. 23. American Mathematical Society, Providence, RI.

Tadmor, E., 1984. Skew-selfadjoint form for systems of conservation laws. J Math. Anal. Appl. 103 (2), 428–442.

Tadmor, E., 1986. The exponential accuracy of Fourier and Chebyshev differencing methods. SIAM J. Numer. Anal. 23 (1), 1–10.

Tadmor, E., 1987. Stability analysis of finite difference, pseudospectral and Fourier-Galerkin approximations for time-dependent problems. SIAM Rev. 29 (4), 525–555.

Tadmor, E., 1989. Convergence of spectral methods for nonlinear conservation laws. SIAM J. Numer. Anal. 26 (1), 30–44.

Tadmor, E., 1990. Shock capturing by the spectral viscosity method. Comput. Method Appl. Mech. Eng. 80 (1–3), 197–208.

Tadmor, E., Tanner, J., 2002. Adaptive mollifiers for high resolution recovery of piecewise smooth data from its spectral information. Found. Comput. Math. 2 (2), 155–189.

Tadmor, E., Tanner, J., 2005. Adaptive filters for piecewise smooth spectral data. IMA J. Numer. Anal. 25 (4), 635–647.

Trefethen, L.N., 2000. Spectral Methods in MATLAB, vol. 10. Siam, Philadelphia.

Vandeven, H., 1991. Family of spectral filters for discontinuous problems. J. Sci. Comput. 6 (2), 159–192.

Zudrop, J., Hesthaven, J.S., 2015. Accuracy of high order and spectral methods for hyperbolic conservation laws with discontinuous solutions. SIAM J. Numer. Anal. 53, 1857–1875.

Chapter 18

Entropy Stable Schemes

E. Tadmor
Center of Scientific Computation and Mathematical Modeling (CSCAMM), Department of Mathematics, Institute for Physical Science and Technology, University of Maryland, College Park, MD, United States

Chapter Outline

ABSTRACT

We review the topic of entropy stability of discrete schemes, finite-difference and finite-volume schemes, for the approximate solution of nonlinear systems of conservation laws. The question of entropy stability plays an important role in both, the theory and computation of such systems, which is reflected by the extensive literature on this topic. Here we focus on a several key ingredients in the study of entropy stable schemes. Our main theme is the investigation of entropy stability using a *comparison principle*. Thus for example, the entropy stability of scalar *monotone* schemes follows from a comparison with constant solutions, and the more general E-schemes are stable

Handbook of Numerical Analysis, Vol. 17. http://dx.doi.org/10.1016/bs.hna.2016.09.006

by a comparison with Godunov solvers. For system of conservation laws, entropy stability is investigated by comparing the amount of their *numerical viscosity* with that of certain *entropy conservative* fluxes. These ingredients are explored in the context of first- and second-order schemes and lend themselves to *higher-order methods* and multidimensional schemes on *unstructured grids.*

AMS Classification Codes: 65M12, 35L65, 65M06, 35R06

Keywords: Entropy inequality, Entropy stability, Monotone schemes, E-schemes, Entropy conservative schemes, Numerical viscosity, Euler and Navier–Stokes equations, High-order methods, Unstructured grids

1 ENTROPIC SYSTEMS OF CONSERVATION LAWS

We are concerned with discrete approximation to hyperbolic systems of conservation laws. These laws take the form,

$$\frac{\partial}{\partial t}\mathbf{u}+\sum_{j=1}^{d}\frac{\partial}{\partial x_j}\mathbf{f}^{(j)}(\mathbf{u})=0,\quad \mathbf{x}=(x_1,\ldots,x_d)\in\Omega,\qquad t\in\mathbb{R}_+, \tag{1}$$

governing the flow of n conservative variables,[a] $\mathbf{u}(x, t) = (u_1(x, t), \ldots, u_n(x, t))^\top$, by their fluxes, $\mathbf{f}(\mathbf{u})=\left(\mathbf{f}^{(1)},\ldots,\mathbf{f}^{(d)}\right)$, where $\mathbf{f}^{(j)}(\mathbf{u}):\mathbb{R}^n\mapsto\mathbb{R}^n$. We consider the Cauchy problem, where a solution of (1) is sought subject to prescribed initial data $\mathbf{u}_0(x)$, in either the whole space, $\Omega=\mathbb{R}^d$, or the periodic torus, $\Omega=\mathbb{T}^d$; in either case, there are no contributions from the boundaries. The system is hyperbolic if the eigenvalues of the $n\times n$ symbol, $\sum_j\xi_jA_j(\mathbf{u}),\ A_j(\mathbf{u}):=\partial_{\mathbf{u}}\mathbf{f}^{(j)}(\mathbf{u})$, are real for all real $\xi = (\xi_1, \ldots, \xi_d)$.

The progress in the development of mathematical theory for nonlinear systems of conservation laws was summarized over the years in a series of monographs and books, starting with the classical work by Courant and Hilbert (1962) and followed by Lax (1973), Smoller (1983), Whitham (1999), Serre (2000), Bressan (2000), Dafermos (2016), as well as a series of biannual conferences devoted to the theory, numerics and applications of hyperbolic problems (Hyp. series, 1984–2016).

The study of such systems was motivated, to a large extent, by the canonical example of the compressible Euler equations.

Example 1 (Euler equations). The compressible Euler equations given by

$$\frac{\partial}{\partial t}\begin{bmatrix}\rho\\ \mathbf{m}\\ E\end{bmatrix}+\sum_{j=1}^{d}\frac{\partial}{\partial x_j}\begin{bmatrix}m_j\\ v_j\mathbf{m}+p\\ v_j(E+p)\end{bmatrix}=0, \tag{2}$$

express the conservative flow of density ρ, the d-dimensional momentum $\mathbf{m} := \rho\mathbf{v}$, and (total) energy E, in terms of the fluxes $\mathbf{f}^{(j)}(\mathbf{u}) = (\rho v_j, \rho v_j\mathbf{v}{+}p,$

[a]Here and later, scalars are distinguished from vectors which are denoted by **bold** letters.

$v_j(E+p))^\top$, where the closure for the pressure is determined by the γ-law, $p := (\gamma - 1)(E - \rho|\mathbf{v}|^2/2)$.

Euler equations (2) admit yet another conservation law which is expressed in terms of the *specific entropy* $S := \ln(p\rho^{-\gamma})$,

$$\frac{\partial}{\partial t}(-\rho S) + \sum_{j=1}^{d} \frac{\partial}{\partial x_j}(-\rho v_j S) = 0. \tag{3}$$

The last equality, which follows by *formal* manipulations of (2), asserts the conservation of the entropy $\eta(\mathbf{u}) = -\rho S$ in terms of the entropy flux $F(\mathbf{u}) = -\rho\mathbf{v}S$. This motivates the notion of *entropy pairs* for general system of conservation laws.

1.1 Entropy Pairs

An *entropy pair* associated with (1) consists of a convex entropy $\eta : \mathbb{R}^n \mapsto \mathbb{R}$ and the corresponding entropy flux $\mathbf{F} = \left(F^{(1)}, \ldots, F^{(d)}\right) : \mathbb{R}^n \mapsto \mathbb{R}^d$, such that the following compatibility relations hold[b]

$$\eta'(\mathbf{u}) A_j(\mathbf{u}) = F^{(j)\prime}(\mathbf{u}), \quad A_j(\mathbf{u}) = \frac{\partial}{\partial \mathbf{u}} \mathbf{f}^{(j)}(\mathbf{u}), \ j = 1, \ldots, d. \tag{4}$$

The existence of such compatible entropy pair allows us to proceed with the following *formal* manipulation

$$\begin{aligned} \eta(\mathbf{u})_t + \nabla_{\mathbf{x}} \cdot \mathbf{F}(\mathbf{u}) &= \langle \eta'(\mathbf{u}), \mathbf{u}_t \rangle + \sum_{j=1}^{d} \left\langle F^{(j)\prime}(\mathbf{u}), \mathbf{u}_{x_j} \right\rangle \\ &= \left\langle \eta'(\mathbf{u}), \mathbf{u}_t + \sum_{j=1}^{d} A_j(\mathbf{u}) \mathbf{u}_{x_j} \right\rangle = \langle \eta'(\mathbf{u}), \mathbf{u}_t + \nabla_{\mathbf{x}} \cdot \mathbf{f}(\mathbf{u}) \rangle = 0. \end{aligned} \tag{5}$$

Thus, if $\mathbf{u}$ is a classical C^1-solution of the conservation law (1), then the pair $(\eta(\mathbf{u}), F(\mathbf{u}))$ forms a conservative extension of it, in complete analogy to the conservation of physical entropy in Euler equations (3). The convexity of $\eta = \eta(\cdot)$ signifies that $\eta(\mathbf{u})$ is a nontrivial extension, beyond the obvious conserved linear combinations $\mathbf{c} \cdot \mathbf{u}$. Thus for example, the judicious minus sign in (3) is chosen to make the corresponding Euler's entropy, $\eta(\mathbf{u}) = -\rho S$, a convex entropy function of the conservative variables ρ, $\mathbf{m}$ and E.

Nonlinear conservation laws may admit one or more entropy pairs, or none at all. This depends on whether there exists a Hessian $\eta''(\mathbf{u})$ which symmetrizes the Jacobians $A_j(\mathbf{u}) = \partial_{\mathbf{u}} \mathbf{f}^{(j)}(\mathbf{u})$. Observe that systems which do admit an entropic extension are necessarily (symmetric) hyperbolic: since terms on the right of the identity (which follows by differentiation of (4)),

[b] We let prime denotes the gradient w.r.t. to specified variable, $X'(\mathbf{u}) := (X_{u_1}, \ldots, X_{u_n})$.

$\eta'' A_j \equiv F_j'' - \eta' A_j'$, are symmetric, the A_j's are *symmetrizable* and hence have a complete set of real eigen-system. Scalar equations have all convex functions as admissible entropy functions. One-dimensional system in n unknowns admit entropy functions for $n = 2$, but the overdetermined symmerazibility condition, $\eta'' A_j = A_j^\top \eta''$, may fail for $n \times n$ systems with $n > 2$ unknowns. We focus our discussion on *entropic systems of conservation laws*—those that are endowed with at least one entropy pair. Most "physically relevant" systems, Euler equations, the shallow-water equations, MHD equations, etc., are entropic.

1.2 Entropy Inequality

The generic phenomena associated with these nonlinear equations are the finite-time breakdown of differentiability of their solutions. Thereafter, one must admit *weak solutions*, where (1) are interpreted in distribution sense (Dafermos, 2016). Among the possibly many weak solutions, physically relevant solutions are postulated as those realized by *vanishing viscosity* limits, $\mathbf{u} = \lim_{\epsilon \downarrow 0} \mathbf{u}^\epsilon$, where

$$\frac{\partial}{\partial t}\mathbf{u}^\epsilon + \sum_{j=1}^{d} \frac{\partial}{\partial x_j}\mathbf{f}^{(j)}(\mathbf{u}^\epsilon) = \epsilon \Delta \mathbf{u}^\epsilon. \tag{6}$$

In this context of weak solutions, one cannot proceed with the formal manipulations (5) which led to the entropy equality $\eta(\mathbf{u})_t + \nabla_\mathbf{x} \cdot \mathbf{F}(\mathbf{u}) = 0$. Instead, arguing along the lines of (5) while using the convexity of $\eta(\cdot)$, we end up with

$$\begin{aligned}\eta(\mathbf{u}^\epsilon)_t + \nabla_\mathbf{x} \cdot \mathbf{F}(\mathbf{u}^\epsilon) &= \langle \eta'(\mathbf{u}^\epsilon), \mathbf{u}^\epsilon_t + \nabla_\mathbf{x} \cdot \mathbf{f}(\mathbf{u}^\epsilon) \rangle \\ &= -\epsilon \langle \eta'(\mathbf{u}^\epsilon), \Delta \mathbf{u}^\epsilon \rangle \equiv \epsilon \Delta \eta(\mathbf{u}^\epsilon) - \epsilon \langle \nabla_\mathbf{x} \mathbf{u}^\epsilon, \eta''(\mathbf{u}^\epsilon) \nabla_\mathbf{x} \mathbf{u}^\epsilon \rangle \le \epsilon \Delta \eta(\mathbf{u}^\epsilon).\end{aligned}$$

It follows that boundedly a.e. limits of vanishing viscosity solutions satisfy the *entropy inequality*,

$$\eta(\mathbf{u})_t + \nabla_\mathbf{x} \cdot \mathbf{F}(\mathbf{u}) \le 0. \tag{7}$$

A weak solution of (1) is *entropic* if it satisfies the *entropy inequality* (7) *for all* admissible entropy pairs (η, $\mathbf{F}$) associated with (1). This notion of entropy solution is the cornerstone for the theory of hyperbolic systems of nonlinear conservation laws. We mention here the pioneering contributions (Godunov, 1961; Kruzkhov, 1970, §7; Friedrichs and Lax, 1971; Lax, 1957, 1971).

1.3 The One-Dimensional Setup

The entropy inequality involves an entropy flux whose components, $\mathbf{F} = (F^{(1)}, \ldots, F^{(d)})^\top$, are aligned with the Cartesian coordinates and sought to satisfy the compatibility requirement (4), one component at the time. We can therefore reduce the question of entropic solution to the one-dimensional case, where the conservative variables $\mathbf{u} = (u_1, \ldots, u_n)^\top$ are balanced by the flux $\mathbf{f}(\mathbf{u}) = (f_1(\mathbf{u}), \ldots, f_n(\mathbf{u}))^\top$,

$$\frac{\partial}{\partial t}\mathbf{u}(x,t) + \frac{\partial}{\partial x}\mathbf{f}(\mathbf{u}(x,t)) = 0, \quad (x,t) \in \Omega \times \mathbb{R}_+. \tag{8}$$

The system is augmented with (one or more) entropy inequalities

$$\frac{\partial}{\partial t}\eta(\mathbf{u}(x,t)) + \frac{\partial}{\partial x}F(\mathbf{u}(x,t)) \le 0, \tag{9}$$

which should hold for all admissible entropy pairs (η, F), satisfying the compatibility condition (4), $\eta' A = F'$, realizing the boundedly a.e. limits of vanishing viscosity limit (Bianchini and Bressan, 2005). Again, the prototype example is the one-dimensional Euler equations, where the density, momentum, $m := \rho v$, and energy, $\mathbf{u} = (\rho, m, E)^\top$ are balanced by the flux $\mathbf{f}(\mathbf{u}) = (\rho v, \rho v^2 + p, v(E+p))^\top$. One seeks entropic solutions which satisfy, in addition, the entropy inequality $(-\rho S)_t + (-\rho v S)_x \le 0$.

2 DISCRETE APPROXIMATIONS AND ENTROPY STABILITY

Weak solutions of (8) can be observed in terms of their (sliding) averages, $\bar{\mathbf{u}}(x,t) := \frac{1}{\Delta x}\int_{I_x} \mathbf{u}(y,t)\mathrm{d}y$ across the cell $I_x := \left[x - \frac{\Delta x}{2}, x + \frac{\Delta x}{2}\right]$. Integrating (8) over the *control volume* $\Omega_\Delta := I_x \times [t, t + \Delta t]$ we find

$$\frac{\bar{\mathbf{u}}(x,t+\Delta t) - \bar{\mathbf{u}}(x,t)}{\Delta t} = -\frac{1}{\Delta x}\left[\tilde{\mathbf{f}}\left(x + \frac{\Delta x}{2}\right) - \tilde{\mathbf{f}}\left(x - \frac{\Delta x}{2}\right)\right]. \tag{10}$$

This reflects the balance between the difference of spatial averages on the left and the *temporal* averages of fluxes on the right, $\tilde{\mathbf{f}}\left(x + \frac{\Delta x}{2}\right) := \int_{\tau=t}^{t+\Delta t} \mathbf{f}\left(\mathbf{u}\left(x + \frac{\Delta x}{2}, \tau\right)\right)\mathrm{d}\tau$.

We are interested in computation of approximate entropy solutions of (8) and (9). To this end we now *fix* the mesh ratio between a small time-step and the size of spatial cells, $\lambda := \frac{\Delta t}{\Delta x}$, and we consider the corresponding class of *conservative schemes* of the form

$$\mathbf{u}_\nu(t+\Delta t) = \mathbf{u}_\nu(t) - \frac{\Delta t}{\Delta x}\left(\mathbf{f}_{\nu+\frac{1}{2}} - \mathbf{f}_{\nu-\frac{1}{2}}\right). \tag{11a}$$

Here, $\mathbf{u}_\nu(t)$ denotes the discrete solution, viewed as an approximate *cell average* $\mathbf{u}_\nu(t) \approx \bar{\mathbf{u}}(\mathbf{x}_\nu, t)$ centred at (x_ν, t). At the heart of matter are the *numerical fluxes* depending on $2p$ neighbouring gridvalues,[c] $\mathbf{f}_{\nu+\frac{1}{2}} := \mathbf{f}(\mathbf{u}_{\nu-p+1}(t), \ldots, \mathbf{u}_{\nu+p}(t))$, which approximate the differential flux, $\mathbf{f}_{\nu+\frac{1}{2}} \approx \overline{\mathbf{f}}(x_{\nu+\frac{1}{2}}, t)$, and in particular, are consistent with the differential flux in the sense that

$$\mathbf{f}(\mathbf{u},\mathbf{u},\ldots,\mathbf{u}) \equiv \mathbf{f}(\mathbf{u}). \tag{11b}$$

The framework of conservative difference schemes (11) was initiated in the seminal paper of Lax and Wendroff (1960).

2.1 Examples

We mention four canonical examples.

Use forward differencing in time and centred differencing in space to discretize (8). The resulting so-called forward Euler scheme reads, $\mathbf{u}_\nu(t+\Delta t) = \mathbf{u}_\nu(t) - \dfrac{\Delta t}{2\Delta x}(\mathbf{f}(\mathbf{u}_{\nu+1}(t)) - \mathbf{f}(\mathbf{u}_{\nu-1}(t)))$, which is associated with the numerical flux $\mathbf{f}^{\mathrm{FE}}_{\nu+\frac{1}{2}} = \dfrac{1}{2}(\mathbf{f}(\mathbf{u}_\nu) + \mathbf{f}(\mathbf{u}_{\nu+1}))$. It is a prototype example for an *unstable* scheme due to amplification of high-modes and lack of numerical dissipation to tame their unbounded growth.

The *Lax–Friedrichs scheme* is the canonical example for a robust numerical solver

$$\mathbf{u}_\nu(t+\Delta t) = \frac{\mathbf{u}_{\nu+1}(t) + \mathbf{u}_{\nu-1}(t)}{2} - \frac{\Delta t}{2\Delta x}(\mathbf{f}(\mathbf{u}_{\nu+1}(t)) - \mathbf{f}(\mathbf{u}_{\nu-1}(t))). \tag{12}$$

It is associated with numerical flux[d] $\mathbf{f}^{\mathrm{LxF}}_{\nu+\frac{1}{2}} = \dfrac{1}{2}(\mathbf{f}(\mathbf{u}_\nu) + \mathbf{f}(\mathbf{u}_{\nu+1})) - \dfrac{1}{2\lambda}\Delta\mathbf{u}_{\nu+\frac{1}{2}}$.

The *Lax–Wendroff scheme* is the prototypical example of a second-order accurate finite-difference scheme

$$\begin{aligned}\mathbf{u}_\nu(t+\Delta t) = {} & \mathbf{u}_\nu(t) - \frac{\Delta t}{2\Delta x}(\mathbf{f}(\mathbf{u}_{\nu+1}(t)) - \mathbf{f}(\mathbf{u}_{\nu-1}(t))) \\ & + \frac{(\Delta t)^2}{2(\Delta x)^2}\left(A_{\nu+\frac{1}{2}}\Delta\mathbf{f}_{\nu+\frac{1}{2}} - A_{\nu-\frac{1}{2}}\Delta\mathbf{f}_{\nu-\frac{1}{2}}\right).\end{aligned} \tag{13}$$

Here, $A_{\nu+\frac{1}{2}}$ is the mid-value of the Jacobian such that $\Delta\mathbf{f}_{\nu+\frac{1}{2}} = A_{\nu+\frac{1}{2}}\Delta\mathbf{u}_{\nu+\frac{1}{2}}$. The corresponding flux is found to be $\mathbf{f}^{\mathrm{LxW}}_{\nu+\frac{1}{2}} = \dfrac{1}{2}(\mathbf{f}(\mathbf{u}_\nu) + \mathbf{f}(\mathbf{u}_{\nu+1})) - \dfrac{\lambda}{2}A^2_{\nu+\frac{1}{2}}\Delta\mathbf{u}_{\nu+\frac{1}{2}}$.

Godunov's scheme (Godunov, 1959) is the forerunner for the class of finite-volume schemes. It evolves a piecewise-constant approximate solution

---

[c]Remark that the numerical flux involves a stencil of $2p$ neighbouring grid values centred at half-indexed gridpoints, $\mathbf{f}(\cdot,\cdot,\ldots,\cdot) \rightsquigarrow \mathbf{f}_{\nu\pm\frac{1}{2}}$, and as such, could be clearly distinguished from the (same notation of) the differential flux tagged at integer indexed gridpoints, $\mathbf{f}(\cdot) \rightsquigarrow \mathbf{f}_\nu$.

[d]For a given gridfunction $\{X_\nu\}$ we let $\Delta X_{\nu+\frac{1}{2}}$ denote the forward difference, $\Delta X_{\nu+\frac{1}{2}} := X_{\nu+1} - X_\nu$, centred at $x_{\nu+\frac{1}{2}}$. Thus for example, $\Delta\mathbf{f}_{\nu+\frac{1}{2}} = \mathbf{f}(\mathbf{u}_{\nu+1}(t)) - \mathbf{f}(\mathbf{u}_\nu(t))$ and $\Delta\mathbf{u}_{\nu+\frac{1}{2}} = \mathbf{u}_{\nu+1}(t) - \mathbf{u}_\nu(t)$.

$\mathbf{u}_\Delta(x,t) := \sum_\nu \mathbf{u}_\nu(t) \mathbf{1}_{I_{x\nu}}(x)$, using the *exact* entropic solution operator. The "pushed-forward" solution, $\{\mathbf{u}_\Delta(x, \tau), \tau > t\}$ is then realized at $\tau = t + \Delta t$ in terms of its cell-averages, $\mathbf{u}_\nu(t+\Delta t) = \frac{1}{\Delta x}\int_{I_{x\nu}} \mathbf{u}_\Delta(y, t+\Delta t)\mathrm{d}y$: appealing to (10) we find that these cell averages, $\mathbf{u}_\nu(t + \Delta t)$, satisfy

$$\mathbf{u}_\nu(t+\Delta t) = \mathbf{u}_\nu(t) - \frac{\Delta t}{\Delta x}\left(\mathbf{f}\left(\mathbf{u}_R\left(x_{\nu+\frac{1}{2}}, t+\frac{\Delta t}{2}\right)\right) - \mathbf{f}\left(\mathbf{u}_R\left(x_{\nu-\frac{1}{2}}, t+\frac{\Delta t}{2}\right)\right)\right). \tag{14}$$

Here, $\mathbf{u}_R\left(x_{\nu+\frac{1}{2}}, t+\frac{\Delta t}{2}\right)$ is the centred value of the entropic solution for the Riemann fan which resolves the discontinuous jump from $\mathbf{u}_\ell = \mathbf{u}_{\nu-\frac{1}{2}}$ to $\mathbf{u}_r = \mathbf{u}_{\nu+\frac{1}{2}}$ at $(x_{\nu+\frac{1}{2}}, t)$.

2.2 Entropy Stability

Let (η, F) be an entropy pair associated with the system (8). The scheme (11) is *entropy stable* w.r.t. such a pair, if it satisfies a discrete entropy inequality analogous to the entropy inequality $\eta(\mathbf{u})_t + F(\mathbf{u})_x \leq 0$, namely, if

$$\eta(\mathbf{u}_\nu(t+\Delta t)) \leq \eta(\mathbf{u}_\nu(t)) - \frac{\Delta t}{\Delta x}\left(F_{\nu+\frac{1}{2}} - F_{\nu-\frac{1}{2}}\right). \tag{15}$$

Here, $F_{\nu+\frac{1}{2}} := F\left(\mathbf{u}_{\nu-p+1}(t), \ldots, \mathbf{u}_{\nu+p}(t)\right)$ is a *numerical entropy flux* which is consistent with the differential one, $F(\mathbf{u}, \mathbf{u}, \ldots, \mathbf{u}) = F(\mathbf{u})$.

The development of numerical methods for approximate solution of nonlinear conservation laws paralleled the development of the analytical theory. It was driven, to a large extent, by the need for scientific computation of stable, high-resolution simulations which in many cases, superseded the analytical theories at the time. We mention the pioneering work of von Neumann (Lax, 2014; von Neumann and Richtmyer, 1950). The progress in the development of numerical methods for nonlinear systems of conservation laws was summarized over the years in a series of monographs and books, starting with the classical work by Richtmyer and Morton (1967) and followed by LeVeque (1992), Godlewski and Raviart (1996), Cockburn et al. (1997), LeVeque (2002) and Gustafsson et al. (2013).

3 ENTROPY STABLE SCHEMES FOR SCALAR CONSERVATION LAWS

3.1 Monotone Schemes

A main feature of *scalar* conservation laws is *monotonicity*. Let $u_1(\cdot, t)$ and $u_2(\cdot, t)$ be two entropy solutions of the scalar law $u_t + f(u)_x = 0$ subject to two different initial data, u_{10} and u_{20}, and assume that u_{20} *dominates* u_{10}, denoted $u_{20} \succeq u_{10}$, in the sense that $u_{20}(x) \geq u_{10}(x) \quad \forall x$. Then $u_2(\cdot, t) \succeq u_1(\cdot, t)$. Namely, a relative ordering among entropy solutions propagates in time. This follows at once from the corresponding ordering of the viscosity solutions, $u_{20} \succeq u_{10} \quad \rightsquigarrow \quad u_2^\epsilon(\cdot, t) \succeq u_1^\epsilon(\cdot, t)$. We turn to the discrete case,

$$u_\nu(t+\Delta t) = u_\nu(t) - \frac{\Delta t}{\Delta x}\left(f_{\nu+\frac{1}{2}} - f_{\nu-\frac{1}{2}}\right), \quad f_{\nu+\frac{1}{2}} = f(u_{\nu-p+1}(t), \ldots, u_{\nu+p}(t)). \tag{16}$$

The scheme is *monotone* if $u_\nu(t + \Delta t)$ is an increasing function of its $2p + 1$ arguments, $(u_{\nu-p}(t), \ldots, u_{\nu+p}(t))$ on the right of (16). Let $\mathbf{u}(t) := (\ldots, u_{\nu-1}(t), u_\nu(t), u_{\nu+1}(t), \ldots)$ and $\boldsymbol{v}(t)$ be two different discrete states at time level t. Monotone schemes propagate their order, namely, if $\mathbf{u}(t) \succeq \boldsymbol{v}(t)$ in the sense that $u_\nu(t) \geq v_\nu(t) \ \forall \nu$, then $\mathbf{u}(t + \Delta t) \succeq \boldsymbol{v}(t + \Delta t)$.

The entropy stability of monotone schemes, originally due to Harten et al. (1976) and Sanders (1983), follows from a comparison with the constant solution,[e] $\mathbf{c} := (\ldots, c, c, c, \ldots)$. We recall here the elegant argument of Crandall and Majda (1980). Consider the discrete grid function $(\mathbf{u}(t) \vee \mathbf{c})_\nu := \max\{u_\nu(t), c\}$: since $\mathbf{u}(t) \vee \mathbf{c}$ dominates both — $\mathbf{u}(t)$ and $\mathbf{c}$, monotonicity implies

$$(u_\nu(t+\Delta t) \vee c) \leq (u_\nu(t) \vee c) - \frac{\Delta t}{\Delta x}\left(f_{\nu+\frac{1}{2}}(\mathbf{u}(t) \vee \mathbf{c}) - f_{\nu-\frac{1}{2}}(\mathbf{u}(t) \vee \mathbf{c})\right), l$$

where we abbreviate $f_{\nu+\frac{1}{2}}(\mathbf{u} \vee \mathbf{c}) := f(u_{\nu-p+1} \vee c, \ldots, u_{\nu+p} \vee c)$. Similarly, since $(\mathbf{u} \wedge \mathbf{c})_\nu = \min\{u_\nu, c\}$ is dominated by both $\mathbf{u}(t)$ and $\mathbf{c}$, it follows that

$$(u_\nu(t+\Delta t) \wedge c) \geq (u_\nu(t) \wedge c) - \frac{\Delta t}{\Delta x}\left(f_{\nu+\frac{1}{2}}(\mathbf{u}(t) \wedge \mathbf{c}) - f_{\nu-\frac{1}{2}}(\mathbf{u}(t) \wedge \mathbf{c})\right),$$

Taking the difference of the last two inequalities yields

$$|u_\nu(t+\Delta t) - c| \leq |u_\nu(t) - c| - \frac{\Delta t}{\Delta x}\left(F_{\nu+\frac{1}{2}} - F_{\nu-\frac{1}{2}}\right),$$

where $F_{\nu+\frac{1}{2}} := f_{\nu+\frac{1}{2}}(\mathbf{u}(t) \vee \mathbf{c}) - f_{\nu+\frac{1}{2}}(\mathbf{u}(t) \wedge \mathbf{c})$ is a numerical flux consistent with the entropy flux $F(u;c) = f(u \vee c) - f(u \wedge c)$. We conclude that monotone schemes are entropy stable with respect to the class of Kruzkov entropy pairs (Kruzkhov, 1970) $\eta(u;c) = |u - c|, \ F(u;c)) = \text{sign}(u - c)(f(u) - f(c))$.

Example 2 (3-point schemes). Consider the class of scalar schemes based on *3-point stencils*,

$$u_\nu(t+\Delta t) = u_\nu(t) - \frac{\Delta t}{\Delta x}(f(u_\nu(t), u_{\nu+1}(t)) - f(u_{\nu-1}(t), u_\nu(t))).$$

The scheme has monotone dependence on $u_{\nu\pm1}(t)$ if and only if its two-point flux, $f(u_\ell, u_r)$, is increasing in u_ℓ and respectively decreasing in u_r, abbreviated as $f(\uparrow, \downarrow)$. The monotone dependence on $u_\nu(t)$ follows from a CFL condition

[e] Observe that $\mathbf{c}$ is a steady solution of (16) for an arbitrary c.

$$\lambda\left(\partial_{u_\ell} f_{\nu+\frac{1}{2}}(u_\nu,\cdot) - \partial_{u_r} f_{\nu-\frac{1}{2}}(\cdot,u_\nu)\right) \le 1.$$

Thus, the LxF scheme is monotone but LxW scheme is not. The scalar Godunov scheme is monotone because it combines the exact entropic evolution operator together with projection to cell averages, $u(\cdot,t+\Delta t) \mapsto \sum_\nu \bar{u}_\nu(t) \mathbf{1}_{I_{x_\nu}}(x)$, which are both monotone. Indeed, the numerical flux of scalar Godunov schemes is given by Osher (1984, lemma 1.1)

$$f^{\mathrm{G}}(u_\nu, u_{\nu+1}) = \begin{cases} \min\limits_{u_\nu \le u \le u_{\nu+1}} f(u), & \text{if} \quad u_\nu < u_{\nu+1}, \\ \max\limits_{u_{\nu+1} \le u \le u_\nu} f(u), & \text{if} \quad u_\nu > u_{\nu+1}. \end{cases}$$

which is readily verified to be of monotone type $f^{\mathrm{G}}(\uparrow, \downarrow)$.

3.2 E-Schemes

In the particular case of scalar conservation laws, all convex η's are admissible entropy functions: the compatibility condition (4), $\eta' f' = F'$, merely recovers the corresponding entropy flux as $F(u) = \int^u \eta'(v) f'(v) \mathrm{d}v$. A discrete scheme is an *E-scheme* (Osher, 1984; Tadmor, 1984; Osher, 1985) if it is entropy stable w.r.t. *all convex* entropies. Godunov and Lax–Friedrichs schemes are primary examples: they are entropy stable w.r.t. all entropy pairs associated with an underlying conservation law, and in the particular case of scalar laws—w.r.t. all convex entropies. Furthermore, in Example 7 we show that Godunov scheme has the distinct feature of having the *least* numerical viscosity among those scalar schemes which are entropy stability w.r.t. *all* convex entropies. The characterization of the scalar E-class is accomplished by a comparison with Godunov scheme.

3.3 Numerical Viscosity I

To carry out this comparison (Tadmor, 1984), consider the class of discrete schemes written in the *viscosity form*

$$\begin{aligned} u_\nu(t+\Delta t) = \; & u_\nu(t) - \frac{\Delta t}{2\Delta x}(f(u_{\nu+1}(t) - f(u_{\nu-1}(t)) \\ & + \frac{\Delta t}{2\Delta x}\left(q_{\nu+\frac{1}{2}} \Delta u_{\nu+\frac{1}{2}} - q_{\nu-\frac{1}{2}} \Delta u_{\nu-\frac{1}{2}}\right). \end{aligned} \tag{17}$$

The role of $\{q_{\nu+\frac{1}{2}}\}$ as the *numerical viscosity coefficients* is revealed once we view (17) as an approximation to the *modified equation*, $u_t + f(u)_x = \dfrac{\Delta x}{2}(q u_x)_x$, with vanishing viscosity amplitude of order $\sim \dfrac{\Delta x}{2} q(\cdot)$.

When compared with (16), we observe that these are conservative schemes with numerical flux[f]

$$f_{\nu+\frac{1}{2}} = \frac{1}{2}(f(u_{\nu+1} + f(u_\nu)) - \frac{1}{2}q_{\nu+\frac{1}{2}}\Delta u_{\nu+\frac{1}{2}}. \tag{18}$$

Conversely, every 3-point scheme admits a viscosity form (17) with a numerical viscosity coefficient dictated by (18).[g] Thus, for example, $q^{\mathrm{LxF}}_{\nu+\frac{1}{2}} \equiv \frac{1}{\lambda}$ and

$$\begin{aligned} q^{\mathrm{G}}_{\nu+\frac{1}{2}} &= \max_{u \in C_{\nu+\frac{1}{2}}} \frac{f(u_{\nu+1}) + f(u_\nu) - 2f(u)}{u_{\nu+1} - u_\nu}, \\ C_{\nu+\frac{1}{2}} &:= [\min\{u_\nu, u_{\nu+1}\}, \max\{u_\nu, u_{\nu+1}\}]. \end{aligned} \tag{19}$$

The class of E-schemes consists of those schemes which contain *more* numerical viscosity than Godunov's, so that their numerical viscosity coefficient, $q^{\mathrm{E}}_{\nu+\frac{1}{2}}$, satisfies, (Tadmor, 1984; Makridakis and Perthame, 2003)

$$\lambda q^{\mathrm{G}}_{\nu+\frac{1}{2}} \leq \lambda q^{\mathrm{E}}_{\nu+\frac{1}{2}} \leq 1.$$

Indeed, such a scheme satisfies the discrete entropy inequality

$$\eta(u_\nu(t+\Delta t)) \leq \eta(u_\nu(t)) - \frac{\Delta t}{\Delta x}\left(F^{\mathrm{E}}_{\nu+\frac{1}{2}} - F^{\mathrm{E}}_{\nu-\frac{1}{2}}\right),$$

for an arbitrary convex entropy $\eta(\cdot)$. The numerical entropy flux F^{E} is given by a convex combination of the corresponding fluxes, $F^{\mathrm{G}}_{\nu+\frac{1}{2}}$ and $F^{\mathrm{LxF}}_{\nu+\frac{1}{2}}$.

Example 3. (Engquist–Osher scheme (Engquist and Osher, 1980)). The Engquist–Osher scheme is an example for an E-scheme. Its numerical viscosity coefficient, $q^{\mathrm{EO}} := \frac{1}{\Delta u_{\nu+\frac{1}{2}}}\int_{u_\nu}^{u_{\nu+1}} |f'(u)|\mathrm{d}u$, satisfies $q^{\mathrm{EO}}_{\nu+\frac{1}{2}} \geq q^{\mathrm{G}}_{\nu+\frac{1}{2}}$, under the CFL condition $\lambda q^{\mathrm{EO}}_{\nu+\frac{1}{2}} \leq 1$.

[f]In certain references, the numerical viscosity coefficient is rescaled *with* the mesh-ration λ, so that (17) reads $u_\nu(t+\Delta t) = u_\nu(t) - \frac{\Delta t}{2\Delta x}(f(u_{\nu+1}(t) - f(u_{\nu-1}(t)) + \frac{1}{2}\left(q_{\nu+\frac{1}{2}}\Delta u_{\nu+\frac{1}{2}} - q_{\nu-\frac{1}{2}}\Delta u_{\nu-\frac{1}{2}}\right)$, with a numerical flux $f_{\nu+\frac{1}{2}} = \frac{1}{2}(f(u_{\nu+1} + f(u_\nu)) - \frac{1}{2\lambda}q_{\nu+\frac{1}{2}}\Delta u_{\nu+\frac{1}{2}}$.

[g]Indeed, every *essentially* 3-point scheme in the sense that its flux satisfies $f(u_{\nu-p+1}, \ldots, u, u, \ldots, u_{\nu+p}) = f(u)$, admits the viscosity form (17).

4 SEMIDISCRETE SCHEMES FOR SYSTEMS OF CONSERVATION LAWS

We focus our attention on the *semidiscrete* limit, $\Delta t \downarrow 0$, where (11) recasts into the form (so-called method of lines)

$$\frac{\mathrm{d}}{\mathrm{d}t}\mathbf{u}_\nu(t) = -\frac{1}{\Delta x}\left(\mathbf{f}_{\nu+\frac{1}{2}} - \mathbf{f}_{\nu-\frac{1}{2}}\right), \quad \mathbf{f}_{\nu+\frac{1}{2}} = \mathbf{f}\left(\mathbf{u}_{\nu-p+1}(t), \ldots \mathbf{u}_{\nu+p}(t)\right). \tag{20}$$

We now fix an entropy pair, (η, F), and seek the corresponding entropy stability, where

$$\frac{\mathrm{d}}{\mathrm{d}t}\eta(\mathbf{u}_\nu(t)) \leq -\frac{1}{\Delta x}\left(F_{\nu+\frac{1}{2}} - F_{\nu-\frac{1}{2}}\right) \tag{21}$$

holds for a consistent numerical entropy flux $F_{\nu+\frac{1}{2}}$. In the particular case that entropy *equality* holds in (21), we say that the scheme (20) is *entropy conservative*.

To address the question on entropy stability w.r.t. to this pair, we seek special schemes which do not dissipate this entropy. These are the *entropy conservative* schemes constructed in Tadmor (1987). The study of entropy stability then proceeds using two main ingredients: (i) the use of the *entropy variables* which enables us to compare numerical viscosity *matrix* coefficients by the natural ordering of symmetric matrices; and (ii) comparison with the appropriate *entropy conservative* schemes. We discuss these two ingredients.

4.1 Entropy Variables (Godunov, 1961; Mock, 1980; see also Godunov and Peshkov, 2008)

Define the *entropy variables* $\boldsymbol{v} \equiv \boldsymbol{v}(\mathbf{u}) := \eta'(\mathbf{u})$. Thanks to the convexity of $\eta(\mathbf{u})$, the mapping $\mathbf{u} \rightarrow \boldsymbol{v}$ is one-to-one and hence we can make the (*local*) change of variables $\mathbf{u}_\nu = \mathbf{u}(\boldsymbol{v}_\nu)$. The scheme (11) then recasts into an equivalent form expressed in terms of the discrete entropy variables $\boldsymbol{v}_\nu = \boldsymbol{v}_\nu(t)$,

$$\frac{\mathrm{d}}{\mathrm{d}t}\mathbf{u}(\boldsymbol{v}_\nu(t)) = -\frac{1}{\Delta x}\left(\mathbf{f}_{\nu+\frac{1}{2}} - \mathbf{f}_{\nu-\frac{1}{2}}\right), \tag{22}$$

with a numerical flux[h] $\mathbf{f}_{\nu+\frac{1}{2}} = \mathbf{f}(\boldsymbol{v}_{\nu-p+1}, \ldots, \boldsymbol{v}_{\nu+p}) := \mathbf{f}(\mathbf{u}(\boldsymbol{v}_{\nu-p+1}), \ldots, \mathbf{u}(\boldsymbol{v}_{\nu+p}))$, consistent with the differential flux, $\mathbf{f}(\boldsymbol{v}, \boldsymbol{v}, \ldots, \boldsymbol{v}) = \mathbf{f}(\mathbf{u}(\boldsymbol{v}))$.

---

[h]We shall often abuse the notation using the same $\mathbf{f}(\cdot)$ as a vector function of the conservative variables $\mathbf{f}(\mathbf{u})$ and of the entropy variables, $\mathbf{f}(\mathbf{u}(\boldsymbol{v})) \rightsquigarrow \mathbf{f}(\boldsymbol{v})$, whenever their dependence is clear from context and there is no ambiguity.

4.2 Entropy Conservative Fluxes

We seek entropy conservative fluxes, denoted $\mathbf{f}^*_{\nu+\frac{1}{2}}$, such that

$$\frac{d}{dt}\mathbf{u}_\nu(t)+\frac{1}{\Delta x}\left(\mathbf{f}^*_{\nu+\frac{1}{2}}-\mathbf{f}^*_{\nu-\frac{1}{2}}\right)=0 \overset{?}{\rightsquigarrow} \frac{d}{dt}\eta(\mathbf{u}_\nu(t))+\frac{1}{\Delta x}\left(F_{\nu+\frac{1}{2}}-F_{\nu-\frac{1}{2}}\right)=0. \quad (23)$$

Premultiply both sides by $\eta'(\mathbf{u})$: we conclude that $\mathbf{f}^*_{\nu-\frac{1}{2}}$ is an *entropy conservative numerical flux* if preserves the structure of 'perfect differences' in the sense that $\left\langle \eta'(\mathbf{u}_\nu),\ \mathbf{f}^*_{\nu+\frac{1}{2}}-\mathbf{f}^*_{\nu-\frac{1}{2}}\right\rangle = F_{\nu+\frac{1}{2}}-F_{\nu-\frac{1}{2}}$. Expressed in terms of the entropy variables, $\boldsymbol{v}_\nu=\eta'(\mathbf{u}_\nu)$, the requirement that $\left\langle \boldsymbol{v}_\nu,\mathbf{f}^*_{\nu+\frac{1}{2}}-\mathbf{f}^*_{\nu-\frac{1}{2}}\right\rangle$ is a 'perfect difference' holds iff $\left\langle \boldsymbol{v}_{\nu+1}-\boldsymbol{v}_\nu,\mathbf{f}^*_{\nu+\frac{1}{2}}\right\rangle$ is a perfect difference. Specifically, the following identity holds,

$$\begin{aligned}&\frac{d}{dt}\eta(\mathbf{u}_\nu(t))+\frac{1}{\Delta x}\left(F_{\nu+\frac{1}{2}}-F_{\nu-\frac{1}{2}}\right)\\&\qquad\equiv\frac{1}{2\Delta x}\left[\left\langle\Delta\boldsymbol{v}_{\nu+\frac{1}{2}},\mathbf{f}_{\nu+\frac{1}{2}}\right\rangle-\Delta\psi_{\nu+\frac{1}{2}}\right]+\frac{1}{2\Delta x}\left[\left\langle\Delta\boldsymbol{v}_{\nu-\frac{1}{2}},\mathbf{f}_{\nu-\frac{1}{2}}\right\rangle-\Delta\psi_{\nu-\frac{1}{2}}\right].\end{aligned} \quad (24)$$

Here $F_{\nu+\frac{1}{2}}$ is a numerical entropy flux expressed in terms of the corresponding *entropy flux potential*,

$$\psi(\boldsymbol{v}):=\langle\boldsymbol{v},\mathbf{f}(\boldsymbol{v})\rangle-F(\mathbf{u}(\boldsymbol{v})). \quad (25)$$

This brings us to the following (Tadmor, 1987, §3).

(i) *[Entropy conservative scheme]*. The difference scheme (22) is entropy conservative so that (23) holds, if its numerical flux, denoted $\mathbf{f}_{\nu+\frac{1}{2}}=\mathbf{f}^*_{\nu+\frac{1}{2}}$, satisfies

$$\left\langle\boldsymbol{v}_{\nu+1}-\boldsymbol{v}_\nu,\mathbf{f}^*_{\nu+\frac{1}{2}}\right\rangle=\psi_{\nu+1}-\psi_\nu,\quad \psi_\nu=\langle\boldsymbol{v}_\nu,\mathbf{f}(\boldsymbol{v}_\nu)\rangle-F(\mathbf{u}(\boldsymbol{v}_\nu)) \quad (26)$$

(ii) *[Entropy stable schemes]*. Consider a numeral flux $\mathbf{f}_{\nu+\frac{1}{2}}$ of the form

$$\mathbf{f}_{\nu+\frac{1}{2}}=\mathbf{f}^*_{\nu+\frac{1}{2}}-\frac{1}{2}D_{\nu+\frac{1}{2}}(\boldsymbol{v}_{\nu+1}-\boldsymbol{v}_\nu),\quad D_{\nu+\frac{1}{2}}\geq 0; \quad (27)$$

Here, $\mathbf{f}^*_{\nu+\frac{1}{2}}$ is any entropy conservative flux satisfying (26) and $D_{\nu+\frac{1}{2}}$ is *any* positive definite symmetric matrix. Then the resulting scheme (20) is entropy stable,

$$\begin{aligned}&\frac{d}{dt}\eta(\mathbf{u}_\nu(t))+\frac{1}{\Delta x}\left(F^*_{\nu+\frac{1}{2}}-F^*_{\nu-\frac{1}{2}}\right)\\&\qquad=-\frac{1}{4\Delta x}\left\langle\Delta\boldsymbol{v}_{\nu-\frac{1}{2}},D_{\nu-\frac{1}{2}}\Delta\boldsymbol{v}_{\nu-\frac{1}{2}}\right\rangle-\frac{1}{4\Delta x}\left\langle\Delta\boldsymbol{v}_{\nu+\frac{1}{2}},D_{\nu+\frac{1}{2}}\Delta\boldsymbol{v}_{\nu+\frac{1}{2}}\right\rangle\leq 0.\end{aligned} \quad (28)$$

Remark that a general framework for *explicit* construction of entropy conservative fluxes (26) is outlined in Section 4.5. Together with (27), they provide an explicit recipe for constructing entropy stable schemes. In particular, the fluxes (27) satisfy the entropy stability condition in Osher (1984, lemma 3.1).

4.3 How Much Numerical Viscosity

The entropy inequality $\eta(\mathbf{u})_t + F(\mathbf{u})_x \leq 0$ is imposed as a stability condition which excludes nonphysically relevant shock discontinuities. In particular, the entropy decay follows $\int \eta(\mathbf{u}(x,t_2))\mathrm{d}\mathbf{x} \leq \int \eta(\mathbf{u}(x,t_1))\mathrm{d}\mathbf{x},\ \ t_2 > t_1$. The question is to quantify the inequality, namely—how much entropy decay will suffice? "physically relevant" entropy decay could be dictated by various mechanisms. We mention the most important two:

(i) *Physical diffusion.* The canonical example of the conservative Euler equations vs. the entropy decay dictated by Navier–Stokes equations. However, in practical simulations one does not often fully resolve the small scales governed by physical diffusion, and an "artificial" numerical viscosity is being used.

(ii) *Numerical viscosity.* According to (28), one can add *any* amount of numerical viscosity to enforce entropy stability. The goal is to add a judicious amount of vanishing viscosity so that in the resulting scheme admits additional desirable and often competing properties of high-resolution and nonoscillatory behaviour. A recent discussion along these lines with the arbitrarily high-order nonoscillatory ENO schemes can be found in Fjordholm et al. (2012, 2015).

4.4 Scalar Entropy Stability Revisited

We discuss the question of entropy stability for semidiscrete scalar schemes $\frac{\mathrm{d}}{\mathrm{d}t}u_\nu(t) = -\frac{1}{\Delta x}\left(f_{\nu+\frac{1}{2}} - f_{\nu-\frac{1}{2}}\right)$, which is expressed in its equivalent viscosity form (17),

$$\frac{\mathrm{d}}{\mathrm{d}t}u_\nu(t) = -\frac{1}{2\Delta x}(f(u_{\nu+1}) - f(u_{\nu-1})) + \frac{1}{2\Delta x}\left(q_{\nu+\frac{1}{2}}\Delta u_{\nu+\frac{1}{2}} - q_{\nu-\frac{1}{2}}\Delta u_{\nu-\frac{1}{2}}\right), \quad (29)$$

To simplify matters we now *fix the quadratic entropy* $\eta(u) = \frac{1}{2}u^2$, where the entropy variables coincide with the conservative variables, $v = u$. The corresponding entropy conservative flux is now uniquely determined as $f^*_{\nu+\frac{1}{2}} = \frac{\psi(u_{\nu+1}) - \psi(u_\nu)}{u_{\nu+1} - u_\nu}$, which can be expressed as

$$\begin{aligned} f^*_{\nu+\frac{1}{2}} &:= \frac{\psi(u_{\nu+1}) - \psi(u_\nu)}{u_{\nu+1} - u_\nu} \equiv \int_{\xi=-\frac{1}{2}}^{\frac{1}{2}} \psi'\left(u_{\nu+\frac{1}{2}}(\xi)\right)\mathrm{d}\xi \\ &= \frac{1}{2}(f(u_{\nu+1}) + f(u_\nu)) - \frac{1}{2}q^*_{\nu+\frac{1}{2}}\Delta u_{\nu+\frac{1}{2}}. \end{aligned}$$

Recall (18): we recognize $q^*_{\nu+\frac{1}{2}}$ as the entropy conservative numerical viscosity coefficient, which is given by

$$q^*_{\nu+\frac{1}{2}} := \int_{\xi=-\frac{1}{2}}^{\frac{1}{2}} 2\xi f'\left(u_{\nu+\frac{1}{2}}(\xi)\right)\mathrm{d}\xi, \quad u_{\nu+\frac{1}{2}}(\xi) := \frac{1}{2}(u_\nu + u_{\nu+1}) + \xi\Delta u_{\nu+\frac{1}{2}}.$$

The resulting entropy conservative scheme then takes the *viscosity form*

$$\frac{\mathrm{d}}{\mathrm{d}t}u_\nu(t) = -\frac{1}{2\Delta x}(f(u_{\nu+1}) - f(u_{\nu-1})) + \frac{1}{2\Delta x}\left(q^*_{\nu+\frac{1}{2}}\,\Delta u_{\nu+\frac{1}{2}} - q^*_{\nu-\frac{1}{2}}\,\Delta u_{\nu-\frac{1}{2}}\right). \quad (30)$$

The statement of entropy stability, (27) and (28), can be rephrased by stating that the conservative scheme (17) is entropy stable if it contains *more* viscosity than the entropy conservative scheme (30), in the sense that $q_{\nu+\frac{1}{2}} \geq q^*_{\nu+\frac{1}{2}}$. Indeed, the numerical flux associated with (29) can be expressed as

$$f_{\nu+\frac{1}{2}} = \frac{1}{2}(f(u_{\nu+1}) + f(u_\nu)) - \frac{1}{2}q_{\nu+\frac{1}{2}}\Delta u_{\nu+\frac{1}{2}} \equiv f^*_{\nu+\frac{1}{2}} + \frac{1}{2}\left(q_{\nu+\frac{1}{2}} - q^*_{\nu+\frac{1}{2}}\right)\Delta u_{\nu+\frac{1}{2}},$$

and entropy stability follows from (27) with $D_{\nu+\frac{1}{2}} = q_{\nu+\frac{1}{2}} - q^*_{\nu+\frac{1}{2}} \geq 0$.

The corollary above enables to verify the entropy stability of first- and second-order accurate schemes. A host of examples can be found in Tadmor (2003). We mention a couple of them.

Example 4 (Burgers' equation). Consider the inviscid Burgers' equation, $u_t + \left(\frac{1}{2}u^2\right)_x = 0$, augmented with the quadratic entropy inequality, $\left(\frac{1}{2}u^2\right)_t + \left(\frac{1}{3}u^3\right)_x = 0$. The entropy variable $\boldsymbol{v}(u) = u$ and entropy potential $\psi(v) := vf - F = \frac{1}{6}u^3$ yield the entropy conservative flux which is the "$\frac{1}{3}$"-rule

$$\frac{\mathrm{d}}{\mathrm{d}t}u_\nu(t) = -\frac{2}{3}\left(\frac{u^2_{\nu+1} - u^2_{\nu-1}}{4\Delta x}\right) - \frac{1}{3}\left(u_\nu\frac{u_{\nu+1} - u_{\nu-1}}{2\Delta x}\right) \leadsto \sum u^2_\nu(t)\Delta x = \sum u^2_\nu(0)\Delta x.$$

Example 5 (Lax–Wendroff viscosity (Lax and Wendroff, 1960)). Consider the case of a convex flux $f(u)$ and fix the quadratic entropy $\eta(u) = \frac{1}{2}u^2$. To see how much viscosity is required to guarantee the quadratic entropy stability,[i] we use the fact that the f' is increasing, leading to the upper bound

[i]We note in passing that quadratic entropy stability is sufficient to single out the unique physically relevant solution in the case of convex flux, e.g., Chen (2000).

$$
\begin{aligned}
q^*_{\nu+\frac{1}{2}} &= \int_{\xi=-\frac{1}{2}}^{\frac{1}{2}} 2\xi f'\left(u_{\nu+\frac{1}{2}}(\xi)\right)\mathrm{d}\xi \leq \frac{1}{4}\int_{\xi=-\frac{1}{2}}^{\frac{1}{2}} f''\left(u_{\nu+\frac{1}{2}}(\xi)\right)\mathrm{d}\xi \\
&= \frac{1}{4}(f'(u_{\nu+1}) - f'(u_\nu))_+.
\end{aligned}
$$

The resulting viscosity coefficient on the right is the second-order Lax–Wendroff viscosity proposed in Lax and Wendroff (1960) with numerical viscosity coefficient $q^{LxW}_{\nu+\frac{1}{2}} = \frac{1}{4}(f'(u_{\nu+1}) - f'(u_\nu))_+$. It follows that this version of LxW scheme is entropy stable, $\frac{1}{2}\frac{d}{dt}u_\nu^2(t) + \frac{1}{\Delta x}\left(F_{\nu+\frac{1}{2}} - F_{\nu-\frac{1}{2}}\right) \leq 0.$

4.5 Numerical Viscosity II

We extend the previous discussion to an arbitrary convex entropy. Let $= \eta'(u)$ denote the corresponding entropy variables. The starting point is the corresponding conservative entropy flux (26)

$$
f^*_{\nu+\frac{1}{2}} := \frac{\psi(u_{\nu+1}) - \psi(u_\nu)}{v_{\nu+1} - v_\nu} \equiv \int_{\xi=-\frac{1}{2}}^{\frac{1}{2}} \psi'\left(u(v_{\nu+\frac{1}{2}}(\xi))\right)\mathrm{d}\xi = \frac{1}{2}(f(u_{\nu+1}) + f(u_\nu)) - \frac{1}{2}p^*_{\nu+\frac{1}{2}}\Delta v_{\nu+\frac{1}{2}},
$$

which yields the entropy conservative schemes in its viscosity form

$$
\frac{\mathrm{d}}{\mathrm{d}t}u_\nu(t) = -\frac{1}{2\Delta x}(f(u_{\nu+1}) - f(u_{\nu-1})) + \frac{1}{2\Delta x}(p^*_{\nu+\frac{1}{2}}\Delta_{v\nu+\frac{1}{2}} - p^*_{\nu-\frac{1}{2}}\Delta_{v\nu-\frac{1}{2}}).
$$

Observe that the viscosity term on the right is expressed in terms of the jump in entropy variables, $\Delta_{v\nu+\frac{1}{2}}\}$. (Of course, in the case of quadratic entropy $v = u$ hence $p_{v+\frac{1}{2}} = q_{v+\frac{1}{2}}$ and $p^*_{v+\frac{1}{2}} = q^*_{v+\frac{1}{2}}$ recovering (29)). This motivates the general viscosity form

$$
\frac{\mathrm{d}}{\mathrm{d}t}u_\nu(t) = -\frac{1}{2\Delta x}(f(u_{\nu+1}) - f(u_{\nu-1})) + \frac{1}{2\Delta x}\left(p_{\nu+\frac{1}{2}}\Delta v_{\nu+\frac{1}{2}} - p_{\nu-\frac{1}{2}}\Delta v_{\nu-\frac{1}{2}}\right). \quad (31)
$$

corresponding to the vanishing viscosity $u_t + f(u)_x = \frac{\Delta x}{2}(pv_x)_x$. The η-entropy stability follows if and only if $p_{\nu+\frac{1}{2}} \geq p^*_{\nu+\frac{1}{2}}$. We conclude with a couple of examples.

Example 6 (Entropy conservative Toda flow). Consider the equation $u_t + (e^u)_x = 0$ augmented with exponential entropy pair, $(e^u)_t + (e^{2u})_x = 0$. The entropy variable associated with $\eta(u) = e^u$ are $\boldsymbol{v}(u) = e^u$, the entropy potential is $\psi(v) := vf - F = \frac{1}{2}v^2$, and we end up with the entropy conservative flux:

$$
f^*_{\nu+\frac{1}{2}} = \frac{\psi(v_{\nu+1}) - \psi(v_\nu)}{v_{\nu+1} - v_\nu} = \frac{\frac{1}{2}v^2_{\nu+1} - \frac{1}{2}v^2_\nu}{v_{\nu+1} - v_\nu} = \frac{1}{2}(v_\nu + v_{\nu+1}) = \frac{1}{2}(e^{u_\nu} + e^{u_{\nu+1}}).
$$

This leads to the *dispersive* centred scheme, interesting for its own sake, e.g., Lax (1986) and Deift and McLaughlin (1998)

$$\frac{\mathrm{d}}{\mathrm{d}t}u_\nu(t) + \frac{e^{u_{\nu+1}(t)} - e^{u_{\nu-1}(t)}}{2\Delta x} = 0,$$

which conserve the exponential entropy $\eta(u_\nu(t)) = e^{u_\nu(t)}$,

$$\begin{aligned}\frac{\mathrm{d}}{\mathrm{d}t}\sum_\nu e^{u_\nu(t)}\Delta x &= -\sum_\nu \frac{e^{u_\nu+u_{\nu+1}} - e^{u_\nu+u_{\nu-1}}}{2\Delta x}\Delta x \\ &= 0 \rightsquigarrow \sum_\nu \eta(u_\nu(t))\Delta x = \sum_\nu \eta(u_\nu(0))\Delta x.\end{aligned}$$

Example 7 (On the optimality of Godunov flux (Tadmor, 2003, Example 4.4)). We normalize the viscous term on the right of (31) in terms of the conservative variables

$$\frac{1}{2\Delta x}\left(\left(p_{\nu+\frac{1}{2}}\frac{\Delta v_{\nu+\frac{1}{2}}}{\Delta u_{\nu+\frac{1}{2}}}\right)\Delta u_{\nu+\frac{1}{2}} - \left(p_{\nu-\frac{1}{2}}\frac{\Delta v_{\nu-\frac{1}{2}}}{\Delta u_{\nu-\frac{1}{2}}}\right)\Delta u_{\nu-\frac{1}{2}}\right),$$

It follows that in order to maintain entropy stability w.r.t. all η's, we need to *maximize* the corresponding entropy viscous factors $p^*_{\nu+\frac{1}{2}}\left(\Delta v_{\nu+\frac{1}{2}}/\Delta u_{\nu+\frac{1}{2}}\right)$,

$$\sup_v \frac{f(u_\nu) + f(u_{\nu+1}) - 2f^*_{\nu+\frac{1}{2}}}{\Delta u_{\nu+\frac{1}{2}}}, \quad f^*_{\nu+\frac{1}{2}} = \int_{\xi=-\frac{1}{2}}^{\frac{1}{2}} f\left(u\left(v_{\nu+\frac{1}{2}}(\xi)\right)\right)\mathrm{d}\xi,$$

where the supremum is taken over *all increasing* $v = v(u)$. This is precisely the Godunov's viscosity coefficient (19). Thus, the scalar schemes which are entropy stable with respect to all convex entropies are precisely those that contain at least as much numerical viscosity as the Godunov scheme does.

4.6 Entropy Conservative Fluxes—Systems of Conservation Laws

Unlike the scalar case, there is more than one way to meet the requirement of entropy conservative flux, (26), for systems of conservation laws. In particular, one can set (Tadmor, 1987),

$$\mathbf{f}^*_{\nu+\frac{1}{2}} = \int_{\xi=-\frac{1}{2}}^{\frac{1}{2}} \mathbf{f}\left(\mathbf{u}(\boldsymbol{v}_{\nu+\frac{1}{2}}(\xi))\right)\mathrm{d}\xi, \tag{32}$$

integrated along the *straight-path* $\boldsymbol{v}_{\nu+\frac{1}{2}}(\xi) := \frac{1}{2}(\boldsymbol{v}_\nu + \boldsymbol{v}_{\nu+1}) + \xi\Delta\boldsymbol{v}_{\nu+\frac{1}{2}}$. Observe that when viewed as a function of the entropy variables, the $\boldsymbol{v}$-dependent flux $\mathbf{f}(\boldsymbol{v}) \equiv \mathbf{f}(\mathbf{u}(\boldsymbol{v}))$ becomes a gradient, $\mathbf{f}(\boldsymbol{v}) = \nabla_{\boldsymbol{v}}\psi(\boldsymbol{v})$ of the entropy potential, $\psi(\boldsymbol{v}) := \langle \boldsymbol{v}, \mathbf{f}(\boldsymbol{v})\rangle - F(\mathbf{u}(\boldsymbol{v}))$. Hence, the value of $\mathbf{f}^*$ is in fact, independent of the path of integration. In particular, a more accessible recipe, amenable for explicit evaluation of such fluxes is given by integration along a *piecewise-path* in phase-space, connecting the two neighbouring values $\boldsymbol{v}_\nu$ and $\boldsymbol{v}_{\nu+1}$. To this end, we begin at $\boldsymbol{v}^1 := \boldsymbol{v}_\nu$, and follow the intermediate steps $\boldsymbol{v}^{j+1} = \boldsymbol{v}^j + \left\langle \boldsymbol{\ell}^j_{\nu+\frac{1}{2}}, \Delta\boldsymbol{v}_{\nu+\frac{1}{2}} \right\rangle \mathbf{r}^j_{\nu+\frac{1}{2}}$ for $j = 1, 2, \ldots,$ ending at $\boldsymbol{v}^{n+1} = \boldsymbol{v}_{\nu+1}$. Here, $\{\mathbf{r}^j\}_{j=1}^n$ be an *arbitrary* set of n linearly independent n-directions, and let $\{\boldsymbol{\ell}^j\}_{j=1}^n$ denote the corresponding orthogonal set, $\langle \boldsymbol{\ell}^j, \mathbf{r}^k \rangle = \delta_{jk}$. (since the mapping $\mathbf{u} \mapsto \boldsymbol{v}$ is one-to-one, the path is mirrored in the usual phase space of conservative variables, starting with $\mathbf{u}^1_{\nu+\frac{1}{2}} = \mathbf{u}_\nu$ and ending with $\mathbf{u}^{n+1}_{\nu+\frac{1}{2}} = \mathbf{u}_{\nu+1}$). The entropy conservative flux $\mathbf{f}^*_{\nu+\frac{1}{2}}$ is then given by the *explicit* formula (Tadmor, 2003, theorem 6.1)

$$\frac{\mathrm{d}}{\mathrm{d}t}\mathbf{u}_\nu(t) = -\frac{1}{\Delta x}\left(\mathbf{f}^*_{\nu+\frac{1}{2}} - \mathbf{f}^*_{\nu-\frac{1}{2}}\right), \quad \mathbf{f}^*_{\nu+\frac{1}{2}} := \sum_{j=1}^{n} \frac{\psi(\mathbf{v}^{j+1}) - \psi(\boldsymbol{v}^j)}{\left\langle \boldsymbol{\ell}^j, \Delta\boldsymbol{v}_{\nu+\frac{1}{2}} \right\rangle}\boldsymbol{\ell}^j, \tag{33}$$

We demonstrate the above approach in the context of entropic Euler equations, with entropy pair $(\eta, F) = (-\rho S, -\rho v S)$.

Example 8 (Entropy conservative Euler flux (Tadmor and Zhong, 2006)). The entropy function $\eta(\mathbf{u}) = -\rho S$ induces the entropy variables, $\boldsymbol{v} = \eta'(\mathbf{u}) = \left(-{}^{E}/_{e} - S + \gamma + 1, {}^{q}/_{\theta}, -{}^{1}/_{\theta}\right)^\top$ expressed in terms of the internal energy $e := E - \frac{1}{2}\rho v^2 = C_V \rho\theta$. The corresponding entropy flux potential amounts to $\psi(\boldsymbol{v}) = \langle \boldsymbol{v}, \mathbf{f} \rangle - F(\mathbf{u}) = (\gamma - 1)m$ and the entropy conservative Euler flux is then given by $\mathbf{f}^*_{\nu+\frac{1}{2}} = (\gamma - 1)\sum_{j=1}^{3} \frac{m^{j+1} - m^j}{\langle \boldsymbol{\ell}^j, \Delta\boldsymbol{v}_{\nu+\frac{1}{2}} \rangle}\boldsymbol{\ell}^j$.

Example 9 (An affordable recipe for entropy conservative flux). An "affordable" entropy conservative flux for Euler equations was derived in by Ismail and Roe in 2009 by clever manipulation of the algebraic relations (26). Expressed in terms of the normalized vector $\mathbf{z} := \sqrt{\frac{\rho}{p}}(1, v, p)^\top$, the entropy conservative flux, $\mathbf{f}^*_{\nu+\frac{1}{2}} := (f^1, f^2, f^3)^\top$, is given by the explicit recipe, in terms of the averages $\bar{z}_{\nu+\frac{1}{2}} := \frac{1}{2}(z_\nu + z_{\nu+1})$ and $z^{ln}_{\nu+\frac{1}{2}} := \frac{\Delta z_{\nu+\frac{1}{2}}}{\Delta \log(z)_{\nu+\frac{1}{2}}}$,

$$f^1_{\nu+\frac{1}{2}} = (\bar{z}_2)_{\nu+\frac{1}{2}}(z_3)^{ln}_{\nu+\frac{1}{2}}, \quad f^2_{\nu+\frac{1}{2}} = \frac{(\bar{z}_3)_{\nu+\frac{1}{2}}}{(\bar{z}_1)_{\nu+\frac{1}{2}}} + \frac{(\bar{z}_2)_{\nu+\frac{1}{2}}}{(\bar{z}_1)_{\nu+\frac{1}{2}}} f^1_{\nu+\frac{1}{2}},$$

and

$$f^3_{\nu+\frac{1}{2}} = \frac{1}{2(\bar{z}_1)_{\nu+\frac{1}{2}}}\left(\frac{\gamma+1}{\gamma-1}\frac{1}{(z_1)^{ln}_{\nu+\frac{1}{2}}}f^1_{\nu+\frac{1}{2}} + (\bar{z}_2)_{\nu+\frac{1}{2}}f^2_{\nu+\frac{1}{2}}\right).$$

5 FULLY DISCRETE SCHEMES FOR SYSTEMS OF CONSERVATION LAWS

Godunov scheme (14) is based on "pushing-forward" an exact entropic solution, $\mathbf{u}_\Delta(\cdot,\tau)$ for $\tau > t$, subject to piecewise-constant data, $\mathbf{u}_\Delta(x,t) = \sum_\nu \mathbf{u}_\nu(t)\mathbf{1}_{I_{x_\nu}}(x)$. As such, $\mathbf{u}_\Delta(x,\cdot)$ satisfies the entropy inequality $\partial_\tau\eta(\mathbf{u}_\Delta) + \partial_x F(\mathbf{u}_\Delta) \le 0$. Integrated across the control volume $I_{x_\nu} \times [t, t+\Delta t]$ we find the balance between spatial and temporal averages,

$$\overline{\eta(\mathbf{u}_\Delta)}(x_\nu, t+\Delta t) \le \overline{\eta(\mathbf{u}_\Delta)}(x_\nu, t) - \frac{\Delta t}{\Delta x}\left(\widetilde{F}(\mathbf{u}_R)\left(x_{\nu+\frac{1}{2}}\right) - \widetilde{F}(\mathbf{u}_R)\left(x_{\nu-\frac{1}{2}}\right)\right).$$

But $\overline{\eta(\mathbf{u}_\Delta)}(x_\nu, t) = \eta(\mathbf{u}_\nu(t))$, and by Jensen's inequality,

$$\eta(\mathbf{u}_\nu(t+\Delta t)) = \eta(\overline{\mathbf{u}_\Delta})(x_{\nu+\frac{1}{2}}, t+\Delta t) \le \overline{\eta(\mathbf{u}_\Delta)}(x_\nu, t+\Delta t),$$

and we conclude that Godunov scheme is entropic for all admissible pairs,

$$\eta(\mathbf{u}_\nu(t+\Delta t)) \le \eta(\mathbf{u}_\nu(t)) - \frac{\Delta t}{\Delta x}\left(F_{\nu+\frac{1}{2}} - F_{\nu-\frac{1}{2}}\right),$$

$$F_{\nu\pm\frac{1}{2}} = \frac{1}{\Delta t}\int_{\tau=t}^{t+\Delta t} F\left(\mathbf{u}_R(x_{\nu\pm\frac{1}{2}}, \tau)\right)\mathrm{d}\tau.$$

Lax–Friedrich scheme, (12), can be interpreted as a Godunov scheme, where a piecewise-constant solution, $\mathbf{u}_\Delta(t) = \sum_\nu\left(\mathbf{u}_{\nu-1}(t)\mathbf{1}_{I_{x_{\nu-\frac{1}{2}}}}(x) + \mathbf{u}_{\nu+1}(t)\mathbf{1}_{I_{x_{\nu+\frac{1}{2}}}}(x)\right)$ is being "pushed-forward by the exact entropy solution operator, and then realized at $t+\Delta t$ by its averages over the *staggered grid*, $\mathbf{u}_\nu(t+\Delta t) = \frac{1}{\Delta x}\int_{I_{x_\nu}} \mathbf{u}_\Delta(x, t+\Delta t)\mathrm{d}x$. Arguing along the above lines for Godunov scheme, we find that LxF is an E-scheme (it is entropy stable w.r.t. all admissible entropy pairs η, F),

$$\eta(\mathbf{u}_\nu(t+\Delta t)) \le \eta(\mathbf{u}_\nu(t)) - \frac{\Delta t}{2\Delta x}(F(\mathbf{u}_{\nu+1}(t)) - F(\mathbf{u}_{\nu-1}(t))), \tag{34}$$

under the CFL condition[j] $\lambda\rho(A(\mathbf{u})) \le \frac{1}{2}$.

[j]We let $\rho(M) := \max_k|\lambda_k(M)|$ denote the *spectral radius* of a matrix M.

5.1 Numerical Viscosity III

Godunov and LxF are the prototype for the class of (essentially) 3-point schemes, which take the viscosity form

$$\mathbf{u}_\nu(t+\Delta t)=\mathbf{u}_\nu(t)-\frac{\Delta t}{2\Delta x}(\mathbf{f}(\mathbf{u}_{\nu+1}(t)-\mathbf{f}(\mathbf{u}_{\nu-1}(t)))+\frac{\Delta t}{2\Delta x}\left(P_{\nu+\frac{1}{2}}\Delta\boldsymbol{v}_{\nu+\frac{1}{2}}-P_{\nu-\frac{1}{2}}\Delta\boldsymbol{v}_{\nu-\frac{1}{2}}\right). \tag{35}$$

In the case of systems of conservation law, $P_{\nu+\frac{1}{2}}$ are $n\times n$ *matrix* numerical viscosity coefficients. The viscosity form for the entropy conservative schemes (32) is given in Tadmor (2003, §5) in terms of the *symmetric* Jacobian $\boldsymbol{B}(\boldsymbol{v})=\partial_{\boldsymbol{v}}\mathbf{f}(\mathbf{u}(\boldsymbol{v}))$,

$$\mathbf{f}^*_{\nu+\frac{1}{2}}=\frac{1}{2}(\mathbf{f}(\mathbf{u}_{\nu+1})-\mathbf{f}(\mathbf{u}_\nu))-\frac{1}{2}P^*_{\nu+\frac{1}{2}}\Delta\boldsymbol{v}_{\nu+\frac{1}{2}},\quad P^*_{\nu+\frac{1}{2}}:=\int_{\xi=-\frac{1}{2}}^{\frac{1}{2}}2\xi B\left(\mathbf{u}(\boldsymbol{v}_{\nu+\frac{1}{2}}(\xi))\right)d\xi.$$

The key point is the expression of the viscosity on the right of (35) in terms of the entropy variables which yields symmetric matrices, and which turn are amenable to a *comparison*: (35) is entropy stable if $P_{\nu+\frac{1}{2}}\geq P^*_{\nu+\frac{1}{2}}$ in the usual sense of ordering among symmetric matrices. For example, Lax–Friedrichs viscosity in (12), expressed in terms of the usual conservative variables is given by $Q^{\mathrm{LxF}}_{\nu+\frac{1}{2}}=\frac{1}{\lambda}I_{n\times n}$. Translated into the entropy variables, $\Delta\boldsymbol{v}_{\nu+\frac{1}{2}}=H_{\nu+\frac{1}{2}}\Delta\mathbf{u}_{\nu+\frac{1}{2}}$, we find,[k]

$$P^{LxF}_{\nu+\frac{1}{2}}=\frac{1}{\lambda}\left[\frac{\Delta\mathbf{u}_{\nu+\frac{1}{2}}}{\Delta\boldsymbol{v}_{\nu+\frac{1}{2}}}\right]:=\frac{1}{\lambda}H^{-1}_{\nu+\frac{1}{2}},\qquad H_{\nu+\frac{1}{2}}:=\int_{\xi=-\frac{1}{2}}^{\frac{1}{2}}\eta''(\mathbf{u}(\boldsymbol{v}_{\nu+\frac{1}{2}}(\xi))d\xi.$$

It dominates $P^*_{\nu+\frac{1}{2}}$ and hence LxF scheme entropy stable w.r.t. all admissible entropy function associated with (8).

We demonstrate the derivation of entropy stability for the more general class of schemes (35) by a comparison with the entropy conservative flux. We have

$$\eta(\mathbf{u}_\nu(t+\Delta t))=\eta(\mathbf{u}_\nu(t))-\frac{\Delta t}{\Delta x}\left(F^*_{\nu+\frac{1}{2}}(t)-F^*_{\nu-\frac{1}{2}}(t)\right)-\mathcal{E}^{(x)}_\nu+\mathcal{E}^{(t)}_\nu.$$

here, $F^*_{\nu+\frac{1}{2}}$ is the entropy conservative flux, $\mathcal{E}^{(x)}_\nu$ is the amount of spatial entropy dissipation quoted in (28),

---

[k]We use abbreviated notation for $\left[\frac{\Delta\boldsymbol{v}_{\nu+\frac{1}{2}}}{\Delta\mathbf{u}_{\nu+\frac{1}{2}}}\right]$ for any matrix such that $\Delta\boldsymbol{v}_{\nu+\frac{1}{2}}=\left[\frac{\Delta\boldsymbol{v}_{\nu+\frac{1}{2}}}{\Delta\mathbf{u}_{\nu+\frac{1}{2}}}\right]\Delta\mathbf{u}_{\nu+\frac{1}{2}}$. $H_{\nu+\frac{1}{2}}$ is such a matrix realized by integration along the usual straight path in phase space.

$$-\mathcal{E}_\nu^{(x)} = -\frac{\lambda}{4}\left\langle \Delta \boldsymbol{v}_{\nu-\frac{1}{2}}, D_{\nu-\frac{1}{2}} \Delta \boldsymbol{v}_{\nu-\frac{1}{2}} \right\rangle - \frac{\lambda}{4}\left\langle \Delta \boldsymbol{v}_{\nu+\frac{1}{2}}, D_{\nu+\frac{1}{2}} \Delta \boldsymbol{v}_{\nu+\frac{1}{2}} \right\rangle, \quad D_{\nu+\frac{1}{2}} := P_{\nu+\frac{1}{2}} - P^*_{\nu+\frac{1}{2}},$$

and $\mathcal{E}_\nu^{(t)} = \frac{1}{2}|\mathbf{u}_\nu(t+\Delta t) - \mathbf{u}_\nu(t)|^2$ is the entropy *production* due to the forward time-differencing. Thus, entropy stability is guaranteed if the former dominates the latter, and to this end, one needs to employ large enough numerical viscosity, $D_{\nu+\frac{1}{2}}$. How much is "enough"? observe that all the matrices involved are symmetric, and one is led to the matrix inequality

$$Q_{\nu+\frac{1}{2}} \ge |A_{\nu+\frac{1}{2}}| + 2|Q^*_{\nu+\frac{1}{2}}|, \qquad Q^*_{\nu+\frac{1}{2}} := P^*_{\nu+\frac{1}{2}} \left[\frac{\Delta \boldsymbol{v}_{\nu+\frac{1}{2}}}{\Delta \mathbf{u}_{\nu+\frac{1}{2}}}\right] = P^*_{\nu+\frac{1}{2}} H_{\nu+\frac{1}{2}}, \tag{36}$$

under CFL condition $\lambda|Q_{\nu+\frac{1}{2}}| \le \frac{1}{4}$.

An alternative approach of securing entropy stability is achieved by adding a minimal amount of numerical viscosity *correction* of Khalfallah and Lerat (1989): starting with a given viscosity matrix $Q_{\nu+\frac{1}{2}}$, we use the *scalar* correction wherever $P_{\nu+\frac{1}{2}}$ has a smaller entropy dissipation then required by $P^*_{\nu+\frac{1}{2}}$, quantified by how negative $\left(D_{\nu+\frac{1}{2}}\right)_-$ is,

$$Q^c_{\nu+\frac{1}{2}} := Q_{\nu+\frac{1}{2}} + \beta^c_{\mu+\frac{1}{2}} I_{n\times n}, \qquad \beta^c := \frac{\left|\left\langle \Delta \boldsymbol{v}_{\nu+\frac{1}{2}}, \left(D_{\nu+\frac{1}{2}}\right)_- \Delta \mathbf{u}_{\nu+\frac{1}{2}} \right\rangle\right|}{\left\langle \Delta \boldsymbol{v}_{\nu+\frac{1}{2}}, \Delta \mathbf{u}_{\nu+\frac{1}{2}} \right\rangle}.$$

One can readily verify that $\langle \Delta \boldsymbol{v}, Q^c_{\nu+\frac{1}{2}} \Delta \mathbf{u}\rangle \ge \langle \Delta \boldsymbol{v}, P^*_{\nu+\frac{1}{2}} \Delta \boldsymbol{v}\rangle$.

Roe scheme (Roe, 1981) is the canonical example for an "upwind scheme": one sets[1] $Q^{\text{Roe}}_{\nu+\frac{1}{2}} = |A_{\nu+\frac{1}{2}}|$ where $A_{\nu+\frac{1}{2}}$ is an averaged Jacobian such that $\Delta \mathbf{f}_{\nu+\frac{1}{2}} = A_{\nu+\frac{1}{2}} \Delta \mathbf{u}_{\nu+\frac{1}{2}}$. It has the attractive feature of keeping sharp resolution of shock discontinuities, whether they are physical or not, and it therefore fails to be entropy stable across steady rarefactions. The entropy stability condition (36) shows that one needs to add a minimal amount of numerical viscosity of order $|Q^*_{\nu+\frac{1}{2}}| \sim |\Delta \mathbf{u}_{\nu+\frac{1}{2}}|$ to enforce entropy stability.

[1]The function value of a diagonalizable matrix $M = T\Lambda T^{-1}$, is set as $h(M) = T\begin{pmatrix} h(\lambda_1) & \cdots & \\ & \ddots & \\ & & h(\lambda_n)\end{pmatrix}T^{-1}$. In particular, a mid-value Jacobian of entropic system $A_{\nu+\frac{1}{2}}$ is symmetrizable, hence diagonalizable (Barth, 1999) and $|A_{\nu+\frac{1}{2}}| := T\begin{pmatrix} |\lambda_1| & \cdots & \\ & \ddots & \\ & & |\lambda_n|\end{pmatrix}T^{-1}$.

5.2 A Homotopy Method

The entropy stability of LxF scheme was derived in Lax (1971) using a homotopy method, *independent* of the existence of entropic solution for Riemann problem. The general case (Tadmor, 2003, §8) implies entropy stability of (35) provided $Q_{\nu+\frac{1}{2}}$ is "large enough", $Q_{\nu+\frac{1}{2}} \geq \max_\xi |A\left(\mathbf{u}(\boldsymbol{v}_{\nu+\frac{1}{2}}(\xi))\right)|$, under the CFL condition $\lambda Q_{\nu+\frac{1}{2}} \leq \frac{1}{2}$. Observe that in this version of the so-called local Lax–Friedrichs scheme (Rusanov, 1961), the viscosity coefficient matrix domains *all* intermediate states rather than the one state offered by the Roe matrix $\boldsymbol{A}_{\nu+\frac{1}{2}} = \boldsymbol{B}_{\nu+\frac{1}{2}} \boldsymbol{H}_{\nu+\frac{1}{2}}$, $B_{\nu+\frac{1}{2}} = \int_{\xi=-\frac{1}{2}}^{\frac{1}{2}} B(\mathbf{u}(\boldsymbol{v}_{\nu+\frac{1}{2}}(\xi)))\mathrm{d}\xi$.

6 HIGHER-ORDER METHODS

The entropy conservative fluxes (32) and (33) are second-order accurate, leading to second-order entropy stable semidiscrete schemes. Extension to arbitrarily high-order entropy stable schemes was introduced in Fjordholm et al. (2012). The question of entropy stability for *fully-discrete* schemes is more intricate. Observe that the results in Section 5 compares with the first-order Roe numerical viscosity. A rigorous entropy stability analysis for fully-discrete *second-order* schemes can be found in Majda and Osher (1978, 1979) for modified Lax–Wendroff scheme, in Nessyahu and Tadmor (1990), Popov and Trifonov (2006) and Kurganov (2016) for Nessyahu–Tadmor scheme, in Osher and Tadmor (1988), Bouchut et al. (1996) and Coquel and LeFloch (1995) for the MUSCL scheme, in LeFloch and Rohde (2000), Chalons and LeFloch (2001), LeFloch et al. (2002) for high-order extensions based on caparison with entropy conservative fluxes, in Fjordholm et al. (2012) for the class of ENO-based schemes developed in Harten et al. (1987), Shu and Osher (1989), and in Jiang and Shu (1994), Qiu and Zhang (2016) and the references therein for DG method.

In practical applications, one proceeds by discretization of the entropy stable semidiscrete schemes using Runge–Kutta (RK) time integrators (Gottlieb and Ketcheso, 2016). The first- and second-order RK solvers are responsible for entropy *production*, and require entropy dissipation to compete with entropy production, making the overall fully-discrete scheme entropy stable. In contrast, the generic cases of third- and higher-order RK time integrators *retain* the entropy stability of the underlying semidiscrete scheme. The *linear* stability of third- and higher-order RK methods is well known for diagonalizable systems and was shown for general linear operators (Tadmor, 2002). This question of *nonlinear* entropy stability was demonstrated in Fjordholm et al. (2009, §4.2.5), but the rigorous entropy stability analysis for high-order RK solvers is, to our knowledge, completely open.

7 MULTIDIMENSIONAL SYSTEMS OF CONSERVATION LAWS

7.1 Cartesian Grids

When multidimensional conservation laws are discretized over grids which are *aligned with the Cartesian coordinates*, the question of entropy stability can be addressed along these coordinates, one dimension at the time. Thus, our one-dimensional setup applies.

Example 10 (Well balanced shallow-water equations). We consider the 2D shallow water equations, e.g., Xing (2017)

$$\frac{\partial}{\partial t}\mathbf{u}+\frac{\partial}{\partial x_1}\mathbf{f}^{(1)}(\mathbf{u})+\frac{\partial}{\partial x_2}\mathbf{f}^{(2)}(\mathbf{u})=-gh\nabla b(\mathbf{x}),\ \mathbf{u}:=[h,h\mathbf{v}]^\top,\ \mathbf{x}=(x_1,x_2)\in\Omega\subset\mathbb{R}^2,$$

which govern the motion of shallow-water with height h above bottom topography $b(\mathbf{x})$, and velocity field, $\mathbf{v}=(v_1,v_2)^\top$, driven by the convective fluxes, $\mathbf{f}^{(j)}=\left(hv_j,\ hv_1v_j+\frac{1}{2}gh^2\delta_{1j},\ hv_2v_j+\frac{1}{2}gh^2\delta_{2j}\right)^\top$,

$$h_t+(hv_1)_{x_1}+(hv_2)_{x_2}=0$$

$$(hv_1)_t+\left(hv_1^2+\frac{1}{2}gh^2\right)_{x_1}+(hv_1v_2)_{x_2}=-ghb_{x_1}$$

$$(hv_2)_t+(hv_2v_1)_{x_1}+\left(hv_2^2+\frac{1}{2}gh^2\right)_{x_2}=-ghb_{x_2}.$$

The entropy function is the total energy, $E(\mathbf{u})=\frac{1}{2}(gh(h+b)+h|\mathbf{v}|^2)$ with energy variables, $\boldsymbol{v}=(gh-\frac{1}{2}|\mathbf{v}|^2,v_1,v_2)^\top$. Observe that the shallow-water fluxes are quadratic in $z:=(h,\sqrt{h}v_1,\sqrt{h}v_2)^\top$. This enables a straightforward "affordable" algebraic approach for satisfying the energy conservative compatibility relation (26), $\left\langle \boldsymbol{v}_{\nu+1,\mu}-\boldsymbol{v}_{\nu,\mu}\,,\,\mathbf{f}^{(1)*}_{\nu+\frac{1}{2},\mu}\right\rangle=\psi(\boldsymbol{v}_{\nu+1,\mu})-\psi(\boldsymbol{v}_{\nu,\mu})$. Here we use the usual indexing of two-dimensional grid-functions attached to grid points $\mathbf{x}_{\nu,\mu}:=(x_{1\nu},x_{2\mu})$. Using the average values, $\bar{z}_{\nu+\frac{1}{2}}:={}^1/_2(z_\nu+z_{\nu+1})$, one finds the x_1-entropy conservative flux (Fjordholm et al., 2011)

$$\mathbf{f}^{(1)*}_{\nu+\frac{1}{2},\mu}=\begin{bmatrix}\bar{h}_{\nu+\frac{1}{2},\mu}(\overline{v_1})_{\nu+\frac{1}{2},\mu}\\ \bar{h}_{\nu+\frac{1}{2},\mu}(\overline{v_1})^2_{\nu+\frac{1}{2},\mu}+\frac{g}{2}\left(\overline{h^2}\right)_{\nu+\frac{1}{2},\mu}+g\overline{h(b_{x_1})}\bar{h}_{\nu+\frac{1}{2},\mu}(\overline{v_1})_{\nu+\frac{1}{2},\mu}(\overline{v_2})_{\nu+\frac{1}{2},\mu}\end{bmatrix}. \tag{37a}$$

Similar expression applies for the conservative flux $\mathbf{f}^{(2)*}_{\nu,\mu+\frac{1}{2}}$ in the x_2-direction. We end up with the energy conservative scheme

$$\frac{\mathrm{d}}{\mathrm{d}t}\mathbf{u}_{\nu,\mu}(t) = -\frac{1}{\Delta x_1}\left(\mathbf{f}^{(1)*}_{\nu+\frac{1}{2},\mu} - \mathbf{f}^{(1)*}_{\nu-\frac{1}{2},\mu}\right) - \frac{1}{\Delta x_2}\left(\mathbf{f}^{(2)*}_{\nu,\mu+\frac{1}{2}} - \mathbf{f}^{(2)*}_{\nu,\mu-\frac{1}{2}}\right). \tag{37b}$$

These schemes recover the precise energy balance, $E(\mathbf{u})_t + F^{(1)}(\mathbf{u})_{x_1} + F^{(2)}(\mathbf{u})_{x_2} = 0$, in terms of the energy fluxes $F^{(j)}(\mathbf{u}) = \frac{1}{2}\left(hv_j|\mathbf{v}|^2 + gh(h+b)\right)$.

7.2 Unstructured Grids

We consider a computational domain which is *partitioned* to a set of nonoverlaping cells, $\Omega_h = \bigcup_i C_i$. Let $\mathbf{n}_{ij} = \int_{\partial C_i \cap \partial C_j} \mathbf{n}\, \mathrm{d}\sigma$ be the unit normal on the non-empty interface pointing out of the control volume C_i. Note that when we sum the normals over all neighbouring cells $\mathcal{N}_i$ then $\sum_{j\in\mathcal{N}_i} \mathbf{n}_{ij} = 0$. The semidiscrete finite volume scheme of (1) reads,

$$\frac{\mathrm{d}}{\mathrm{d}t}\mathbf{u}_\nu = -\frac{1}{|C_\nu|}\sum_{\mu\in\mathcal{N}_\nu} \mathbf{f}_{\nu\mu},$$

where the 2-point numerical flux, $\mathbf{f}_{\nu\mu} = \mathbf{f}\left(\mathbf{u}_\nu(t), \mathbf{u}_\mu(t), \mathbf{n}_{\nu\mu}\right)$, is assumed to consistent, $\mathbf{f}(\mathbf{u}, \mathbf{u}, \mathbf{n}) = \mathbf{f}(\mathbf{u}) \cdot \mathbf{n}$. The scheme is conservative in the sense that $\sum_\nu |C_\nu|\, \mathbf{u}_\nu(t)$ is conserved in time, since $\mathbf{f}_{\nu\mu}(\mathbf{n}) = -\mathbf{f}_{\mu\nu}(-\mathbf{n})$.

The corresponding question of entropy stability for such schemes, satisfying the cell entropy inequality, $\frac{\mathrm{d}}{\mathrm{d}t}\eta(\mathbf{u}_\nu(t)) \leq -\frac{1}{|C_\nu|}\sum_{\mu\in\mathcal{N}_\nu} \mathbf{F}_{\nu\mu}$, was investigated by extension of the tools outlined above. In particular, the question of *scalar* entropy stability was studied in a long series of papers and we mention here (Barth, 1999; Eymard et al., 2000; Kroner et al., 1995; Sonar, 2016) and the references therein.

To design or investigate entropy stable fluxes for systems of conservation laws, one may proceed by comparing their numerical viscosities with entropy conservative schemes. A numerical flux $\mathbf{f}^*_{\nu\mu} = \mathbf{f}^*(\mathbf{u}_\nu, \mathbf{u}_\mu, \mathbf{n}_{\nu\mu})$ is *entropy conservative* if its components, projected along the normal directions, $\mathbf{f}_{\nu\mu} = \mathbf{f}^{(1)}_{\nu\mu}\mathbf{n}^{(1)}_{\nu\mu} + \mathbf{f}^{(2)}_{\nu\mu}\mathbf{n}^{(2)}_{ij}$, satisfy the compatibility relations,

$$\left\langle \boldsymbol{v}_\mu - \boldsymbol{v}_\nu, \mathbf{f}^{*(j)}_{\nu\mu} \right\rangle = \psi^{(j)}_\nu - \psi^{(j)}_\mu,$$

where $\psi^{(j)}(\mathbf{u})$ is the usual entropy potential $\psi^{(j)} := \left\langle \boldsymbol{v}, \mathbf{f}^{(j)} \right\rangle - F^{(j)},\ j = 1,2$.

Entropy stable fluxes then take the form $\mathbf{f}_{\nu\mu} = \mathbf{f}^*_{\nu\mu} - \frac{1}{2}D_{\nu\mu}(\boldsymbol{v}_\nu - \boldsymbol{v}_\mu)$ for positive-definite entropy dissipation matrices D's. The study of entropy stable schemes on two-dimensional unstructured grids by comparing numerical viscosities along these lines was carried out in Ray et al. (2016).

ACKNOWLEDGEMENTS

This review was written while visiting the ETH Institute for Theoretical Studies (ITS) and it is a pleasure to thank their hospitality. Research was supported in part by NSF grants DMS16-13911, RNMS11-07444 (Ki-Net) and ONR grant 00014-1512094.

REFERENCES

Barth, T., 1999. Numerical methods for gas-dynamics systems on unstructured meshes. In: Kroner, D., Ohlberger, M., Rohde, C. (Eds.), An Introduction to Recent Developments in Theory and Numerics of Conservation Laws, Lecture Notes in Computational Science and Engineering, vol. 5. Springer, New York City, pp. 195–285.

Bianchini, S., Bressan, A., 2005. Vanishing viscosity solutions of nonlinear hyperbolic systems. Ann. Math. 161, 223–342. http://dx.doi.org/10.4007/annals.2005.161.223.

Bouchut, F., Bourdarias, C., Perthame, B., 1996. A MUSCL method satisfying all the numerical entropy inequalities. Math. Comp. 65, 1439–1461. http://dx.doi.org/10.1090/S0025-5718-96-00752-1.

Bressan, A., 2000. Hyperbolic Systems of Conservation Laws. The One Dimensional Cauchy Problem. Oxford University Press, Oxford.

Chalons, C., LeFloch, P., 2001. A fully discrete scheme for diffusive-dispersive conservation law. Numer. Math. 89, 493–509. http://dx.doi.org/10.1007/PL00005476.

Chen, G.-Q., 2000. Compactness methods and nonlinear hyperbolic conservation laws: some current topics on nonlinear conservation laws. In: AMS/IP Stud. Adv. Math., vol. 15. American Mathematical Society, Providence, RI, pp. 33–75.

Cockburn, B., Johnson, C., Shu, C.-W., Tadmor, E., 1997. Advanced numerical approximation of nonlinear hyperbolic equations. In: Quarteroni, A. (Ed.), Lectures Given at the 2nd Session of C.I.M.E. Held in Cetraro, Italy, June 23–28. Lecture Notes in Mathematics, vol. 1697. Springer, Berlin. http://dx.doi.org/10.1007/BFb0096351.

Coquel, F., LeFloch, P., 1995. An entropy satisfying MUSCL scheme for systems of conservation laws. CR Acad. Sci. Paris Sér. I 320, 1263–1268.

Courant, R., Hilbert, D., 1962. Methods of Mathematical Physics. vol. II. John Wiley & Sons-Interscience, New York.

Crandall, M.G., Majda, A., 1980. Monotone difference approximations for scalar conservation laws. Math. Comp. 34, 1–21. http://dx.doi.org/10.1090/S0025-5718-1980-0551288-3.

Dafermos, C., 2016. Hyperbolic Conservation Laws in Continuum Physics. vol. 325. Springer, Berlin. http://dx.doi.org/10.1007/978-3-662-49451-6.

Deift, P., McLaughlin, K.T.R., 1998. A Continuum Limit of the Toda Lattice. vol. 131. Memoirs of the American Mathematical Society. x+216 pp. http://dx.doi.org/10.1090/memo/0624.

Engquist, B., Osher, S., 1980. Stable and entropy condition satisfying approximations for transonic flow calculations. Math. Comp. 34, 44–75. http://dx.doi.org/10.1090/S0025-5718-1980-0551290-1.

Eymard, R., Gallouet, T., Herbin, R., 2000. Finite volume methods. In: Ciarlet, P., Lions, J. (Eds.), Handbook of Numerical Analysis. vol. VII. North-Holland, Amsterdam, pp. 713–1020.

Fjordholm, U., Mishra, S., Tadmor, E., 2009. Energy preserving and energy stable schemes for the shallow water equations, Foundations of Computational Mathematics. In: Cucker, F., Pinkus, A., Todd, M. (Eds.), Proceedings of FoCM held in Hong Kong 2008, London Math. Soc. Lecture Notes Ser. 36393–139.

Fjordholm, U., Mishra, S., Tadmor, E., 2011. Well-balanced and energy stable schemes for the shallow water equations with discontinuous topography. J. Comput. Phys. 230, 5587–5609. http://dx.doi.org/10.1016/j.jcp.2011.03.042.

Fjordholm, U., Mishra, S., Tadmor, E., 2012. Arbitrarily high order accurate entropy stable essentially non-oscillatory schemes for systems of conservation laws. SIAM J. Numer. Anal. 50, 544–573. http://dx.doi.org/10.1137/110836961.

Fjordholm, U., Kappeli, R., Mishra, S., Tadmor, E., 2015. Construction of approximate entropy measure valued solutions for hyperbolic systems of conservation laws. Found. Comp. Math. 2015, 1–65. http://dx.doi.org/10.1007/s10208-015-9299-z.

Friedrichs, K.O., Lax, P.D., 1971. Systems of conservation laws with a convex extension. Proc. Nat. Acad. Sci. USA 68, 1686–1688.

Godlewski, E., Raviart, P.-A., 1996. Numerical Approximation of Hyperbolic Systems of Conservation Laws. Springer, New York.

Godunov, S.K., 1959. A difference scheme for numerical computation of discontinuous solutions of fluid dynamics. Mat. Sb. 47, 271–306.

Godunov, S.K., 1961. An interesting class of quasilinear systems. Dokl. Acad. Nauk. SSSR 139 (3), 521–523.

Godunov, S.K., Peshkov, I.M., 2008. Symmetrization of the nonlinear system of gas dynamics equations. Siberian Math. J. 49 (5), 829–834. http://dx.doi.org/10.1007/s11202-008-0081-1.

Gottlieb, S., Ketcheso, D.I., 2016. Time discretization techniques. Handbook of Numerical Analysis, vol. 17. Elsevier, Amsterdam, pp. 549–583.

Gustafsson, B., Kreiss, H.-O., Oliger, J., 2013. Time Dependent Problems and Difference Methods, second ed. Wiley, New Jersey.

Harten, A., Hyman, J.M., Lax, P.D., 1976. On finite difference approximations and entropy conditions for shocks. Comm. Pure Appl. Math. 29, 297–322. http://dx.doi.org/10.1002/cpa.3160290305.

Harten, A., Engquist, B., Osher, S., Chakravarty, S.R., 1987. Uniformly high order accurate essentially non-oscillatory schemes. J. Comput. Phys. 71, 231–303. http://dx.doi.org/10.1016/0021-9991(87)90031-3.

Hyp., series, 1984–2016. International conference of hyperbolic problems: theory, numerics and applications. http://www.cscamm.umd.edu/hyp2008/#history.

Ismail, F., Roe, P.L., 2009. Affordable, entropy-consistent Euler flux functions II: entropy production at shocks. J. Comput. Phys. 228, 5410–5436. http://dx.doi.org/10.1016/j.jcp.2009.04.021.

Jiang, G.-S., Shu, C.-W., 1994. On a cell entropy inequality for discontinuous Galerkin method. Math. Comp. 62, 531–538. http://dx.doi.org/10.1090/S0025-5718-1994-1223232-7.

Khalfallah, K., Lerat, A., 1989. Correction d'entropie pour des schémas numeriques approchant un système hyperbolique. CR Acad. Sci. Paris Sér. II 308, 815–820.

Kroner, D., Noelle, S., Rokyta, M., 1995. Convergence of higher order upwind finite volume schemes on unstructured grids for scalar conservation laws in several space dimensions. Numer. Math. 71 (4), 527–560.

Kruzkhov, S.N., 1970. First order quasilinear equations in several independent variables. USSR Math. Sbornik. 10 (2), 217–243. http://dx.doi.org/10.1070/SM1970v010n02ABEH002156.

Kurganov, A., 2016. Central schemes: A powerful black-box solver for nonlinear hyperbolic PDEs. Handbook of Numerical Analysis, vol. 17. Elsevier, Amsterdam, pp. 525–548.

Lax, P.D., 1957. Hyperbolic systems of conservation laws II. Comm. Pure Appl. Math. 10, 537–566. http://dx.doi.org/10.1002/cpa.3160100406.

Lax, P.D., 1971. Shock waves and entropy. In: Zarantonello, E. (Ed.), Contributions to Nonlinear Functional Analysis, Academic Press, New York, pp. 603–634.

Lax, P.D., 1973. Hyperbolic Systems of Conservation Laws and the Mathematical Theory of Shock Waves. vol. 11. SIAM Regional Conference Lectures in Applied Mathematics.

Lax, P.D., 1986. On dispersive difference schemes. Physica D. 18, 250–254. http://dx.doi.org/10.1016/0167-2789(86)90185-5.

Lax, P.D., 2014. John von Neumann: the early years, the years at Los Alamos and the road to computing. In: Modern Perspectives in Applied Mathematics: Theory and Numerics of PDEs. www.ki-net.umd.edu/tn60/2014_04_30_Lax_Banquet_talk.pdf.

Lax, P.D., Wendroff, B., 1960. Systems of conservation laws. Comm. Pure Appl. Math. 13, 217–237. http://dx.doi.org/10.1002/cpa.3160130205.

LeFloch, P., Rohde, C., 2000. High-order schemes, entropy inequalities and non-classical shocks. SIAM J. Numer. Anal. 37, 2023–2060. http://dx.doi.org/10.1137/S0036142998345256.

LeFloch, P., Mercier, J.M., Rohde, C., 2002. Fully discrete, entropy conservative schemes of arbitrary order. SIAM J. Numer. Anal. 40, 1968–1992. http://dx.doi.org/10.1137/S003614290240069X.

LeVeque, R., 1992. Numerical Methods for Conservation Laws. Lectures in Mathematics. Birkhäuser, Basel.

LeVeque, R., 2002. Finite Volume Methods for Hyperbolic Problems. Texts in Applied Mathematics. Cambridge University Press, Cambridge.

Majda, A., Osher, S., 1978. A systematic approach for correcting nonlinear instabilities: the Lax-Wendroff scheme for scalar conservation laws. Numer. Math. 30, 429–452. http://dx.doi.org/10.1007/BF01398510.

Majda, A., Osher, S., 1979. Numerical viscosity and the entropy condition. Comm. Pure Appl. Math. 32, 797–838. http://dx.doi.org/10.1002/cpa.3160320605.

Makridakis, C., Perthame, B., 2003. Sharp CFL, discrete kinetic formulation and entropy schemes for scalar conservation laws. SIAM J. Numer. Anal. 41 (3), 1032–1051. http://dx.doi.org/10.1137/S0036142902402997.

Mock, M.S., 1980. Systems of conservation of mixed type. J. Diff. Eqns 37, 70–88. http://dx.doi.org/10.1016/0022-0396(80)90089-3.

Nessyahu, H., Tadmor, E., 1990. Non-oscillatory central differencing for hyperbolic conservation laws. J. Comput. Phys. 87, 408–463. http://dx.doi.org/10.1016/0021-9991(90)90260-8.

Osher, S., 1984. Riemann solvers, the entropy condition, and difference approximations. SIAM J. Numer. Anal. 21, 217–235. http://dx.doi.org/10.1137/0721016.

Osher, S., 1985. Convergence of generalized MUSCL schemes. SIAM J Numer. Anal. 22, 947–961. http://dx.doi.org/10.1137/0722057.

Osher, S., Tadmor, E., 1988. On the convergence of difference approximations to scalar conservation laws. Math. Comp. 50, 19–51. http://dx.doi.org/10.1090/S0025-5718-1988-0917817-X.

Popov, B., Trifonov, O., 2006. One sided stability and convergence of the Nessyahu-Tadmor scheme. Numer. Math. 104, 539–559. http://dx.doi.org/10.1007/s00211-006-0015-4.

Qiu, J., Zhang, Q., 2016. Stability, error estimate and limiters of discontinuous Galerkin methods. Handbook of Numerical Analysis, vol. 17. Elsevier, Amsterdam, pp. 147–171.

Ray, D., Chandrashekara, P., Fjordholm, U.S., Mishra, S., 2016. Entropy stable scheme on two-dimensional unstructured grids for Euler equations. Comm. Comput. Phys. 19 (5), 1111–1140. http://dx.doi.org/10.4208/cicp.scpde14.43s.

Richtmyer, R., Morton, B., 1967. Difference Methods for Initial-Value Problems, second ed. Wiley-Interscience, New York.

Roe, P.L., 1981. Approximate Riemann solvers, parameter vectors and difference schemes. J. Comput. Phys. 43, 357–372. http://dx.doi.org/10.1016/0021-9991(81)90128-5.

Rusanov, V.V., 1961. Calculation of interaction of non-steady shock-waves with obstacles. J. Comput. Math. Phys USSR 1 (2), 304–320. http://dx.doi.org/10.1016/0041-5553(62)90062-9.

Sanders, R., 1983. On convergence of monotone finite difference schemes with variable spatial differencing. Math. Comp. 40, 91–106. http://dx.doi.org/10.1090/S0025-5718-1983-0679435-6.

Serre, D., 2000. Hyperbolic Conservation Laws. Vol I. Geometric Structures, Oscillations, and Initial-Boundary Value Problem; vol. II Hyperbolicity, Entropies, Shock Waves. Cambridge University Press, Cambridge.

Shu, C.W., Osher, S., 1989. Efficient implementation of essentially non-oscillatory schemes—II. J. Comput. Phys. 83, 32–78. http://dx.doi.org/10.1016/0021-9991(89)90222-2.

Smoller, J., 1983. Shock Waves and Reaction Diffusion Equations. Springer, Berlin.

Sonar, T., 2016. Classical finite volume methods. Handbook of Numerical Analysis, vol. 17. Elsevier, Amsterdam, pp. 55–76.

Tadmor, E., 1984. Numerical viscosity and the entropy condition for conservative difference schemes. Math. Comp. 43, 369–381. http://dx.doi.org/10.1090/S0025-5718-1984-0758189-X.

Tadmor, E., 1987. The numerical viscosity of entropy stable schemes for systems of conservation laws, I. Math. Comp. 49, 91–103. http://dx.doi.org/10.1090/S0025-5718-1987-0890255-3.

Tadmor, E., 2002. From semi-discrete to fully discrete: stability of Runge–Kutta schemes by the energy method. II. In: Estep, D., Tavener, S. (Eds.), Collected Lectures on the Preservation of Stability under Discretization, Lecture Notes from Colorado State University Conference, Fort Collins, CO, 2001, Proc. in Applied Math. 109, SIAM, 25–49.

Tadmor, E., 2003. Entropy stability theory for difference approximations of nonlinear conservation laws and related time dependent problems. Acta Numer. 42, 451–512. http://dx.doi.org/10.1017/S0962492902000156.

Tadmor, E., Zhong, W., 2006. Entropy stable approximations of Navier-Stokes equations with no artificial numerical viscosity. J. Hyperbolic DEs. 3, 529–559. http://dx.doi.org/10.1142/S0219891606000896.

von Neumann, J., Richtmyer, R.D., 1950. A method for the numerical calculation of hydrodynamic shocks. J. Appl. Phys. 21, 232–237. http://dx.doi.org/10.1063/1.1699639.

Whitham, G.B., 1999. Linear and Nonlinear Waves. Wiley-Interscience, New York.

Xing, Y., 2017. Numerical methods for the nonlinear shallow water equations. Handbook of Numerical Analysis, vol. 18. Elsevier, Amsterdam, article in press.

Chapter 19

Entropy Stable Summation-by-Parts Formulations for Compressible Computational Fluid Dynamics

M.H. Carpenter*, T.C. Fisher†, E.J. Nielsen*, M. Parsani‡, M. Svärd§ and N. Yamaleev¶

*NASA Langley Research Center, Hampton, VA, United States

†Sandia National Laboratories, Albuquerque, NM, United States

‡King Abdullah University of Science and Technology (KAUST), Extreme Computing Research Center (ECRC), Thuwal, Saudi Arabia

§University of Bergen, Bergen, Norway

¶Old Dominion University, Norfolk, VA, United States

Chapter Outline

Handbook of Numerical Analysis, Vol. 17. http://dx.doi.org/10.1016/bs.hna.2016.09.014

ABSTRACT
A systematic approach based on a diagonal-norm summation-by-parts (SBP) framework is presented for implementing entropy stable (SS) formulations of any order for the compressible Navier–Stokes equations (NSE). These SS formulations discretely conserve mass, momentum, energy and satisfy a mathematical entropy equality for smooth problems. They are also valid for discontinuous flows provided sufficient dissipation is added at shocks and discontinuities to satisfy an entropy inequality. Admissible SBP operators include all centred diagonal-norm finite-difference (FD) operators and Legendre spectral collocation-finite element methods (LSC-FEM). Entropy stable multiblock FD and FEM operators follows immediately via nonlinear coupling operators that ensure conservation, accuracy and preserve the interior entropy estimates. Nonlinearly stable solid wall boundary conditions are also available. Existing SBP operators that lack a stability proof (e.g. weighted essentially nonoscillatory) may be combined with an entropy stable operator using a comparison technique to guarantee nonlinear stability of the pair. All capabilities extend naturally to a curvilinear form of the NSE provided that the coordinate mappings satisfy a geometric conservation law constraint. Examples are presented that demonstrate the robustness of current state-of-the-art entropy stable SBP formulations.

Keywords: Nonlinear stability, Entropy analysis, Compressible Navier–Stokes, High-order summation-by-parts, Simultaneous-approximation-term, WENO, Contravariant stability

AMS Classification Codes: 35 (pdes), 65 (num analysis), 76 (fluids)

1 INTRODUCTION

High-order discrete operators lose robustness for solutions containing discontinuities or even under-resolved physical features (e.g. under-resolved turbulent flows). Although a variety of mathematically rigorous stabilization techniques have been developed for second-order methods (e.g. total variation diminishing limiters (LeVeque, 1992) and entropy stability (Tadmor, 1987)), extending these techniques to high-order formulations has been problematic.

The "tools of the trade" used to stabilize high-order formulations are (1) hyper-viscosity dissipation, (2) dealiasing, (3) over-integration, (4/5) filtering of the fluxes/solution and (6) limiters or some combination of the above; yet it is still difficult to add sufficient stabilization while retaining the design order of accuracy. high-order essentially nonoscillatory (ENO) (Harten et al., 1987; Shu and Osher, 1988) and weighted ENO (WENO) (Jiang and Shu, 1996; Liu et al., 1994) techniques are effective on structured meshes and achieve design-order accuracy away from captured discontinuities, with "nearly monotone" captured shocks. Nevertheless, they too suffer instabilities in less than ideal circumstances (i.e. curvilinear mapped grids or expansion of flows into vacuum).

A common impediment of the aforementioned stabilization techniques is their reliance on heuristics and experimentation rather than mathematics.

The ultimate goal of entropy stable summation-by-parts (SBP) research is to develop fully discrete operators with nonlinear stability estimates that are equivalent to those existing at the continuous level. Stronger discrete nonlinear estimates invariably yield enhanced robustness! Entropy stability (continuous/discrete) guarantees that the thermodynamic entropy has an upper bound provided that density and temperature remain positive and boundary data are well-posed and preserve the entropy estimate (and an entropy stable discretization is used in time).

Research over the years by many groups has contributed to a general procedure for implementing entropy conservative and entropy stable schemes of any order for the compressible Navier–Stokes equations (NSE). Nearly three decades ago, entropy conservative/stable schemes that discretely satisfy an entropy conservation/stability property are constructed by Tadmor (1987, 2003) for second-order finite-volume (FV) methods. These schemes are extended to high-order periodic domains by LeFloch et al. (2002). Ismail and Roe (2009) find a computationally efficient discrete entropy flux for the Euler equations, thereby removing a major computational obstacle. Fisher et al. (Carpenter et al., 2014, 2015; Fisher and Carpenter, 2013a) present multidomain/Legendre spectral collocation-finite element methods (LSC-FEM) proofs of entropy conservation and stability based on diagonal norm SBP operators of arbitrary order. Generalization to arbitrary Cartesian domains follows immediately using simultaneous-approximation-term (SAT) penalty-type interface conditions between adjoining domains (Carpenter et al., 1999, 2014, 2015). Nonlinear solid wall boundary conditions are introduced by Parsani et al. in 2015b. Entropy stable WENO schemes are presented in Fisher and Carpenter (2013a) as well as a comparison approach technique for combining SBP operators. Extension to generalized curvilinear coordinates appears in Fisher (2012).

The depth and breadth of recent SS-SBP contributions are realized by comparison with similar developments in the FEM literature. Entropy analysis of the *nonconservative* form of the compressible NSE first appears in the early work of Hughes et al. (1986); continuous entropy stability follows immediately from the symmetric form of the equations. Discretization using FEM approaches (e.g. Galerkin or Petrov–Galerkin) preserves the estimates assuming exact element integrations. Finite element entropy stability proofs for scalar nonlinear equations written in conservation form appear in many texts (e.g. see Hesthaven and Warburton, 2008) and are valid for elements of arbitrary order. Extension of weak-form FEMs to the compressible NSE in conservation form has been difficult to achieve. Indeed, a fundamental obstacle in the compressible Euler proofs is the requirement for integral exactness, a property that is all but impossible to achieve for high polynomial orders (e.g. $p \geq 3$). Dissipation could be used to suppress the instabilities arising from integration errors, but the size of the dissipation coefficient is not known a priori thereby introducing the heuristics one is trying to avoid.

Entropy stability proofs that are based on diagonal norm, SBP-SAT schemes (Carpenter et al., 2014, 2015; Fisher and Carpenter, 2013a), do not suffer this limitation and are remarkably *general*, *sharp* and *simple*. Indeed, the proofs are valid for all nondissipative, diagonal norm discretizations of the conservation form of the compressible NSE. The discrete operators are design-order accurate for the conservative variables (mass, momentum, energy) and also for the entropy; the entropy solution is design order close to the nonlinear isentropic manifold of the continuous Euler equations. Despite the generality and accuracy, all entropy stability proofs require only elementary theorems in linear algebra. No vector space abstractions (Sobolev, Hilbert or others) are used or required. Thus, construction of entropy conservative and stable formulations of arbitrary order for many popular discrete operators are possible; diagonal-norm, centred finite-difference (FD) and LSC-FEM being important examples.

The focus herein is a "high-level" overview of the mathematical concepts of entropy stability for diagonal norm, SBP-SAT operators, including illustrative examples that support the theory. The pivotal works (theorems) that facilitated recent contributions are presented, supplemented with the citations necessary to aid further investigation by interested readers. Only discretizations consistent with the Lax–Wendroff theorem are considered. Section 2 introduces the compressible NSE and the concept of convex entropy functions used to prove nonlinear stability. A comprehensive introduction to diagonal-norm SBP-SAT theory and operators is included in Section 3 and concludes with a discussion of the two diagonal norm, SBP operators used herein. Section 4 introduces the concepts of entropy consistency at the semidiscrete and fully discrete levels. Section 5 introduces and develops entropy stable interior interface couplings used to couple adjoining block (FD) or elements (LSC-FEM). Section 6 summarizes solid wall boundary conditions that preserve the nonlinear entropy estimate of the interior operator. Section 7 extends a conventional FD operator into a form amenable to the stencil biasing mechanics of WENO, including a description of a comparison technique used to ensure entropy stability of the resulting formulation (e.g. SSWENO). Section 8 extends the entropy analysis developed in Section 4 to the curvilinear form of the compressible NSE. Results from several high-order discrete entropy stable operators are presented in Section 9 followed by conclusions in Section 10.

2 THE COMPRESSIBLE NSE

2.1 Governing Equations

Consider the calorically perfect compressible NSE expressed in the form

$$\begin{aligned} q_t + (f_k)_{x_k} &= (f_k^{(v)})_{x_k}, \quad x \in \Omega, \quad t \in [0, \infty), \\ Bq &= g_b, \quad x \in \partial\Omega, \quad t \in [0, \infty), \\ q(x,0) &= g_0(x), \quad x \in \Omega, \end{aligned} \tag{1}$$

where the Cartesian coordinates, $x=(x_1,x_2,x_3)^\top$, and time, t, are independent variables, and index sums are implied. The equations govern the conservation of mass, k-momentum and energy in the domain. The vector of conserved variables is $q=[\rho,\ \rho u_k,\ \rho E]^\top$, while the vectors f_k and $f_k^{(v)}$ are the conserved inviscid and viscous fluxes, respectively. The boundary vector g_b is assumed to contain well-posed Dirichlet/Neumann data. A detailed description of the three-dimensional compressible NSE may be found elsewhere (Fisher, 2012).

2.2 Continuous Entropy Analysis

A convex extension of the compressible NSE is sought, with a volume integral that depends only on boundary data and dissipative terms. An extension of this form provides a mechanism for proving stability of the nonlinear system (Harten, 1983).

Define a convex scalar entropy function $S=S(q)$ for the system of Eq. (1) that when differentiated, simultaneously contracts all spatial fluxes as follows

$$S_q(f_k)_{x_k}=S_q(f_k)_q q_{x_k}=(F_k)_q q_{x_k}=(F_k)_{x_k};\quad k=1,2,3 \tag{2}$$

for each spatial coordinate. The components of the contracting vector, S_q, are the entropy variables denoted as $w^\top=S_q$, while $F_k(q)$ are the entropy fluxes in the k-direction. The convexity of the entropy guarantees that the Hessian $S_{qq}=w_q$ is symmetric positive definite and yields a one-to-one mapping from conservation variables, q, to entropy variables, $w^\top=S_q$.

A differential statement of entropy stability for the compressible NSE follows immediately by combining Eqs. (1) and (2). Contracting the system of Eq. (1) with the entropy variables S_q forms the differential form of the scalar entropy equation

$$\begin{aligned} S_q q_t+S_q(f_k)_{x_k}&=S_t+(F_k)_{x_k}=S_q(f_k)_{x_k}^{(v)}\\ &=\left(w^\top f_k^{(v)}\right)_{x_k}-w_{x_k}^\top f_k^{(v)}=\left(w^\top f_k^{(v)}\right)_{x_k}-w_{x_k}^\top \bar{c}_{kj}w_{x_j}. \end{aligned} \tag{3}$$

Integrating Eq. (3) over the domain Ω yields a global conservation statement for the entropy in the domain

$$\frac{d}{dt}\int_\Omega S dx_k=\left[w^\top f_k^{(v)}-F_k\right]_{\partial\Omega}-\int_\Omega w_{x_j}^\top \bar{c}_{kj}w_{x_k}dx_k. \tag{4}$$

The second integral term in the entropy equation (4) accounts for viscous dissipation and is always negative (Fisher and Carpenter, 2013a,b). Thus, the entropy equation is the convex extension of the original compressible NSE and serves as an integral measure of stability of the system.

A family of convex entropy functions S exists for the Euler equations with associated entropy variables w. All symmetrize the Jacobian of the inviscid flux: $f_w = (f_w)^\top$ (e.g. see Harten, 1983). Hughes et al. (1986) show, however, that only the entropy function $S = -\rho s$ with s defined by the thermodynamic relation

$$s = \frac{R}{\gamma - 1}\log\left(\frac{T}{T_0}\right) - R\log\left(\frac{\rho}{\rho_0}\right),$$

simultaneously symmetrizes all the inviscid and viscous coefficient matrices in Eq. (1), where T_0 and ρ_0 are the reference temperature and density, respectively. The entropy variables $w^\top = S_q$ for the compressible NSE are given by

$$w = \left[\frac{h}{T} - s - \frac{u_1^2 + u_2^2 + u_3^2}{2T}, \frac{u_1}{T}, \frac{u_2}{T}, \frac{u_3}{T}, -\frac{1}{T}\right]^\top, \tag{5}$$

where T, h and u, v, w are the temperature, specific enthalpy and velocity components, respectively.

The entropy variables, w given by Eq. (5), symmetrize Eq. (1) as can be demonstrated by allowing the entropy variables w to assume the role of a new dependent variable (i.e. $q = q(w)$). Expressing Eq. (1) in terms of w yields the system of equations

$$q_t + (f_k)_{x_k} - (f_k^{(v)})_{x_k} = q_w w_t + (f_k)_w w_{x_k} - (\bar{c}_{kj} w_{x_j})_{x_k} = 0; \quad k = 1,2,3 \tag{6}$$

with the symmetry conditions: $q_w = (q_w)^\top$, $(f_k)_w = (f_k)_w^\top$, $(\bar{c}_{kj}) = (\bar{c}_{jk})^\top$.

The symmetry of the matrices q_w and $(f_k)_w$, imply that the conservation variables, q, and fluxes, f_k, are Jacobians of scalar functions with respect to the entropy variables,

$$q^\top = \varphi_w, \quad (f_k)^\top = (\psi_k)_w, \tag{7}$$

where the nonlinear function, φ, is called the potential and ψ_k are called the potential fluxes (Tadmor, 2003). Just as the entropy function is convex with respect to the conservative variables (S_{qq} is SPD), the potential function is convex with respect to the entropy variables.

Godunov (1961) and Mock (1980) show that these nonlinear functions satisfy the relations

$$\varphi = w^\top q - S, \quad \psi_k = w^\top f_k - F_k. \tag{8}$$

See reference Harten (1983) for a detailed summary of both proofs.

3 SBP OPERATORS

3.1 Mimetic Operators

First derivative operators that satisfy the one-dimensional, SBP convention, discretely mimic the integration-by-parts property

$$\int_{x_L}^{x_R} \phi q_x dx = \phi q|_{x_L}^{x_R} - \int_{x_L}^{x_R} \phi_x q dx, \tag{9}$$

with ϕ an arbitrary test function. This mimetic property is achieved by constructing the first derivative approximation, $\mathcal{D}\phi$, with an operator in the form

$$\begin{aligned} &\mathcal{D} = \mathcal{P}^{-1}\mathcal{Q}, \quad \mathcal{P} = \mathcal{P}^{\top}, \ \zeta^{\top}\mathcal{P}\zeta > 0, \ \zeta \neq \mathbf{0}, \\ &\mathcal{Q}^{\top} = \mathcal{B} - \mathcal{Q}, \quad \mathcal{Q}\mathbf{1} = \mathbf{0}, \ \mathcal{B} = \mathrm{diag}(-1,0,\ldots,0,1), \end{aligned} \tag{10}$$

with ζ an arbitrary vector. Diagonal-norm SBP operators are considered exclusively herein; only they can be manipulated into the entropy stable SBP-SAT (SS-SBP-SAT) form.

Define on the interval $-1 \leq x \leq 1$, the vectors of discrete *solution points*

$$\mathbf{x} = [x_1, x_2, \ldots, x_{N-1}, x_N]^{\top}; \quad -1 \leq x_1, x_2, \ldots, x_{N-1}, x_N \leq 1 \tag{11}$$

and ϕ and $\mathbf{q}$ as the projections of continuous variables onto a grid $x_1, \ldots, x_N$

$$\phi^{\top} = (\phi(x_1), \phi(x_2), \ldots, \phi(x_N))^{\top}, \quad \mathbf{q} = (q(x_1), q(x_2), \ldots, q(x_N))^{\top}. \tag{12}$$

Discretizing the left-hand side of Eq. (10) by using (12), the mimetic SBP property is demonstrated,

$$\phi^{\top}\mathcal{P}\mathcal{P}^{-1}\mathcal{Q}\mathbf{q} = \phi^{\top}(\mathcal{B} - \mathcal{Q}^{\top})\mathbf{q} = \phi_N q_N - \phi_1 q_1 - \phi^{\top}\mathcal{D}^{\top}\mathcal{P}\mathbf{q}. \tag{13}$$

3.2 Complementary Grid and Telescopic Flux Form

Define a set of (N + 1) intermediate *flux points* prescribing bounding control volumes about each solution point as

$$\begin{aligned} &\bar{\mathrm{x}} = (\bar{x}_0, \bar{x}_1, \ldots \bar{x}_N)^{\top}, \ \bar{x}_0 = x_1, \ \bar{x}_N = x_N, \\ &\Delta\bar{\mathrm{x}} = \mathcal{P}\mathbf{1}; \ \mathbf{1} = (1, 1, \ldots, 1)^{\top}, \end{aligned} \tag{14}$$

with the $N \times (N + 1)$ matrix Δ is defined as

$$\Delta = \begin{pmatrix} -1 & 1 & 0 & 0 & 0 & 0 \\ 0 & -1 & 1 & 0 & 0 & 0 \\ 0 & 0 & \ddots & \ddots & 0 & 0 \\ 0 & 0 & 0 & -1 & 1 & 0 \\ 0 & 0 & 0 & 0 & -1 & 1 \end{pmatrix}. \tag{15}$$

The spacing between the flux points is implicitly defined by the norm $\mathcal{P}$; the diagonal contributions of $\mathcal{P}$ are equal to the spacing between flux points. Note that the solution and flux points coincide on the boundaries of the interval in Eq. (14). Thus, the flux at each boundary can be attributed to either the solution or flux point

$$\bar{\mathbf{f}}_0 = f(q_1), \quad \bar{\mathbf{f}}_N = f(q_N), \tag{16}$$

a necessary boundary duality needed when proving entropy stability.

All SBP derivative operators $\mathcal{D}$ may be manipulated into the telescopic flux form,

$$f_x(\mathbf{q}) = \mathcal{P}^{-1}\mathcal{Q}\mathbf{f} + \mathcal{T}_p = \mathcal{P}^{-1}\Delta\bar{\mathbf{f}} + \mathcal{T}_p, \tag{17}$$

where p is the order of the truncation error. The $N \times (N+1)$ matrix Δ provides the undivided difference of the two adjacent fluxes. (The original proof appears elsewhere (Fisher et al., 2011a).)

3.3 Extension to Multiple Dimensions

One-dimensional SBP operators naturally extend via tensor product arithmetic to multiple spatial dimensions. The multidimensional tensor product nomenclature used herein is defined as

$$\begin{aligned} \mathcal{D}_{x_1} &= \mathcal{D}_{N_1} \otimes \mathcal{I}_{N_2} \otimes \mathcal{I}_{N_3} \otimes \mathcal{I}_5, & \Delta_{x_1} &= \Delta_{N_1} \otimes \mathcal{I}_{N_2} \otimes \mathcal{I}_{N_3} \otimes \mathcal{I}_5 \\ \mathcal{P}_{x_1} &= \mathcal{P}_{N_1} \otimes \mathcal{I}_{N_2} \otimes \mathcal{I}_{N_3} \otimes \mathcal{I}_5, & \mathcal{P}_{\perp,1} &= \mathcal{I}_{N_1} \otimes \mathcal{P}_{N_2} \otimes \mathcal{P}_{N_3} \otimes \mathcal{I}_5 \\ \mathcal{P} &= \mathcal{P}_{N_1} \otimes \mathcal{P}_{N_2} \otimes \mathcal{P}_{N_3} \otimes \mathcal{I}_5, & \widehat{\mathcal{P}} &= \mathcal{P}_{N_1} \otimes \mathcal{P}_{N_2} \otimes \mathcal{P}_{N_3}, \end{aligned} \tag{18}$$

with similar definitions for $\mathcal{B}_{x_1}$ and $\mathcal{Q}_{x_1}$, as well as directions x_2 and x_3. Surface integration of scalar and vector fields are accomplished using the integration rules $\widehat{\mathcal{P}}_{\perp,k}$ and $\mathcal{P}_{\perp,k}$, respectively (i.e. $\mathcal{P}_{\perp,k} = \widehat{\mathcal{P}}_{\perp,k} \otimes I_5$).

3.4 Diagonal-Norm SBP Operators

Two popular SBP operators are used in the demonstrations reported herein: centred FD and LSC-FEM. Centred FD operators use the same (skew-symmetric) stencil throughout the interior of the domain. At boundaries, the centred stencil must be inwardly biased because data are generally not available outside the domain. Constructing design-order consistent, stable and conservative boundary closures was the first widely accepted application of the SBP mechanics (Strand, 1994). Either block- or diagonal-norm boundary closures may be constructed for centred FD operators. Although block-norm operators have superior accuracy properties: ($2p - 1$ vs p for $(2p)^{th}$-order interior stencils, $p \geq 1$), they do not satisfy the matrix commuting properties needed to prove entropy stability. Thus, diagonal norm centred FD operators are the exclusive focus herein.

Spectral collocation operators may be expressed in SBP form (Carpenter and Gottlieb, 1996; Carpenter et al., 2009; Hesthaven and Gottlieb, 1996). Not all spectral matrices, however, can be expressed as diagonal-norm SBP operators. For example, the mass matrix of a Legendre or a Chebyshev operator is full. Legendre collocation schemes, however, may be expressed as diagonal-norm SBP operators if the mass matrix is lumped onto the diagonal.

Thus, Legendre collocation schemes satisfies all the sufficient requirements for SBP operators admissible for an entropy stable form, making it now possible to construct entropy stable, spectrally accurate operators of arbitrary order on unstructured grids!

3.5 The Semidiscrete Operators With Boundary and Interface Conditions

Based on the previous discussion of SBP operators and their equivalent telescoping form, the semidiscrete form of the system of Eq. (1) becomes

$$\begin{aligned}\mathbf{q}_t &= -\mathcal{D}_{x_k}\mathbf{f}_k(\mathbf{q}) + \mathcal{D}_{x_k}[\bar{c}_{kj}]\mathcal{D}_{x_j}\mathbf{w} + \mathcal{P}^{-1}(\mathbf{g}_k^{(BC)} + \mathbf{g}_k^{(Int)}), \quad k=1,2,3 \\ &= \mathcal{P}^{-1}\Delta_k\left(-\bar{\mathbf{f}}_k + \bar{\mathbf{f}}_k^{(v)}\right) + \mathcal{P}^{-1}(\mathbf{g}_k^{(BC)} + \mathbf{g}_k^{(Int)}), \quad \mathbf{q}(x,0) = g_0(x), \quad x \in \Omega,\end{aligned} \tag{19}$$

with $\mathbf{g}^{(BC)}$ and $\mathbf{g}^{(Int)}$ enforcing boundary and interface conditions, respectively. (The bracket nomenclature, e.g., $[\bar{c}_{kj}]$, denotes a block-diagonal matrix with blocks that are 5×5.) Full implementation details of the symmetrized viscous Jacobian $\bar{c}_{kj}$ tensor are available elsewhere (Fisher, 2012; Fisher and Carpenter, 2013a). Implementation details for SAT penalty terms represented by $\mathbf{g}^{(BC)}$ and $\mathbf{g}^{(Int)}$ are available elsewhere (Berg and Nordström, 2011; Carpenter et al., 1994, 2014; Nordström et al., 2009; Parsani et al., 2015a,b; Svärd and Nordström, 2006; Svärd et al., 2007).

4 SEMIDISCRETE AND FULLY DISCRETE ENTROPY ANALYSIS

The semidiscrete, nonlinear entropy estimate is achieved by mimicking term by term the continuous estimate given in Eqs. (3) and (4). The analysis begins by contracting the discrete entropy variables: $\mathbf{w}^\top$, with the semidiscrete system of Eq. (19). The resulting global equation that governs the semidiscrete decay of entropy is given by Carpenter et al. (2014) and Parsani et al. (2015b)

$$\mathbf{w}^\top\mathcal{P}\frac{\partial \mathbf{q}}{\partial t} + \mathbf{w}^\top\mathcal{P}_{\perp,k}\Delta_{x_k}\bar{\mathbf{f}}_k = \mathbf{w}^\top\mathcal{P}_{\perp,k}\Delta_{x_k}\bar{\mathbf{f}}_k^{(V)} + \mathbf{w}^\top\mathcal{P}_{\perp,k}\mathbf{g}_k^{(BC)} + \mathbf{w}^\top\mathcal{P}_{\perp,k}\mathbf{g}_k^{(Int)}, \quad k=1,2,3, \tag{20}$$

where

$$\mathbf{w} = \left(w(q_{(1),(1),(1)})^\top, w(q_{(2),(1),(1)})^\top, \ldots, w(q_{(N),(N),(N)})^\top\right)^\top.$$

The entropy variables, $\mathbf{w}$, are defined at the solution points whereas the quantities with an over-bar, i.e., $\bar{\mathbf{f}}$ and $\bar{\mathbf{f}}_i^{(V)}$ are defined at the flux points.

Simplifying Eq. (20) into a semidiscrete convection–diffusion equation for the entropy, proceeds as follows. The semidiscrete time term is manipulated for diagonal-norm SBP operators as $\mathbf{w}^\top\mathcal{P}\partial\mathbf{q}/\partial t = \widehat{\mathbf{1}}^\top\mathcal{P}\partial\mathbf{S}/\partial t$ (or the local statement: $w_i^\top \mathcal{P}_i \partial q_i/\partial t = \mathcal{P}_i \partial S_i/\partial t, \forall i$). The viscous terms are first symmetrized as in (6) and then manipulated as

$$\mathbf{w}^\top \mathcal{P}_{\perp,k} \Delta_{x_k} \bar{\mathbf{f}}_k^{(V)} = \mathbf{w}^\top \mathcal{B}_k \mathcal{P}_{\perp,k} [\bar{c}_{kj}] \mathcal{D}_{x_j} \mathbf{w} - (\mathcal{D}_{x_k} \mathbf{w})^\top \mathcal{P}[\bar{c}_{kj}](\mathcal{D}_{x_j} \mathbf{w}), \quad k = 1,2,3. \tag{21}$$

The terms $\mathbf{w}^\top \mathcal{B}_k \mathcal{P}_{\perp,k}[\bar{c}_{kj}]\mathcal{D}_{x_j}\mathbf{w}$ account for diffusive fluxes on boundaries, while the terms $-(\mathcal{D}_{x_k}\mathbf{w})^\top \mathcal{P}[\bar{c}_{kj}](\mathcal{D}_{x_j}\mathbf{w})$ account for viscous dissipation. The matrix $\mathcal{P}_{\perp,k}$ provides quadrature weights in the plane orthogonal to the direction k. See reference Carpenter et al. (2014) for a complete derivation of the temporal and viscous manipulations.

The inviscid terms of Eq. (20) are globally entropy conservative in each coordinate direction k, if they satisfy

$$\mathbf{w}^\top \mathcal{P}_{\perp,k} \Delta \bar{\mathbf{f}}_k = \mathcal{P}_{\perp,k}(\bar{F}_k(q_N) - \bar{F}_k(q_1)) = \mathcal{P}_{\perp,k}(F_k(q_N) - F_k(q_1)) = \widehat{\mathbf{1}}^\top \mathcal{P}_{\perp,k} \Delta_{x_k} \overline{\mathbf{F}}_k, \quad k = 1,2,3. \tag{22}$$

A design-order nonentropy conservative discrete flux $\bar{\mathbf{f}}_k$ will in general only satisfy Eq. (22) to design order, but could lead to numerical instability if it is not entropy stable. A general procedure for constructing design order, entropy conservative fluxes, $\bar{\mathbf{f}}^{(S)}$, appears in references Fisher (2012) and Fisher and Carpenter (2013a). The design order local entropy conservative flux $\bar{f}_i^{(S)}$ is constructed in each coordinate direction k, by using a linear combination of q_{rs}-weighted, two-point entropy conservative fluxes $\bar{f}_S = \bar{f}_S(q_r, q_s)$. (The discrete entries in the r-row and s-column of the $\mathcal{Q}$ matrix are denoted q_{rs}.)

The following theorem establishes that the linear combinations of two-point entropy conservative fluxes preserve entropy stability for arbitrary one-dimensional, diagonal-norm SBP matrix $\mathcal{Q}$.

Theorem 1. *A high-order entropy conservative flux $\bar{\mathbf{f}}_i^{(S)}$ may be constructed as follows*

$$\bar{f}_i^{(S)} = \sum_{s=i+1}^{N} \sum_{r=1}^{i} 2 q_{(r,s)} \bar{f}_S(\boldsymbol{q}_r, \boldsymbol{q}_s), \quad 1 \le i \le N-1, \tag{23}$$

where $\bar{f}_S(\mathbf{q}_r, \mathbf{q}_s)$ is any two-point nondissipative function that satisfies the entropy conservation shuffle condition

$$(w_r - w_s)^\top \bar{f}_S(\boldsymbol{q}_r, \boldsymbol{q}_s) = \psi_r - \psi_s. \tag{24}$$

The high-order entropy conservative flux satisfies an additional local entropy flux conservation property,

$$\mathbf{w}^\top \mathcal{P}^{-1} \Delta \bar{f}^{(S)} = \mathcal{P}^{-1} \Delta \overline{F} = F_x(\boldsymbol{q}) + \mathcal{T}_d, \tag{25}$$

where $\mathcal{T}_d$ is a design-order truncation term, and the local entropy flux $\bar{F}_i$ is given by the expression

$$\overline{F}_i = \sum_{s=i+1}^{N} \sum_{r=1}^{i} 2q_{(r,s)} \left[\frac{(w_r + w_s)^{\top}}{2} \bar{f}_S(q_r, q_s) - \frac{(\psi_r + \psi_s)}{2} \right], \quad 1 \le i \le N-1. \quad (26)$$

Extension to multiple dimension follows immediately via tensor product arithmetic.
Proof. The proof is available elsewhere (Fisher and Carpenter, 2013b). □

The two-point entropy conservative flux $\bar{f}_S$ of Ismail and Roe (2009) is used in (23).

Combining all terms yields the semidiscrete entropy transport equation

$$\begin{aligned} \frac{d}{dt}\widehat{\mathbf{1}}^{\top}\widehat{\mathcal{P}}\mathbf{S} + \left\| \sqrt{[\bar{c}_{kj}]}\mathcal{D}_{x_j}\mathbf{w} \right\|_{\mathcal{P}}^2 = & \left[\mathbf{w}^{\top}\mathcal{B}_k\mathcal{P}_{\perp,k}[\bar{c}_{kj}]\mathcal{D}_{x_j}\mathbf{w} - \widehat{\mathbf{1}}^{\top}\widehat{\mathcal{P}}_{\perp,k}\Delta_{x_k}\overline{\mathbf{F}} \right]_k \\ & + \mathbf{w}_k^{\top}\mathcal{P}_{\perp,k}(\mathbf{g}_k^{(BC)} + \mathbf{g}_k^{(Int)}), \quad k = 1,2,3, \end{aligned} \quad (27)$$

which is the semidiscrete equivalent of the continuous entropy transport given in Eq. (4).

4.1 Fully Discrete Operators

The semidiscrete entropy stability does not necessarily lead to fully discrete stability as is usually the case for linear partial differential equations (PDEs). However, as noted by Tadmor (2003), entropy stability is enhanced by fully implicit time discretization. For example, the fully implicit backward Euler time discretization is unconditionally entropy stable and is responsible for additional entropy dissipation.

In contrast, explicit time discretization leads to entropy production. Thus, the entropy stability of explicit schemes hinges on a delicate balance between temporal entropy production and spatial entropy dissipation. The fully explicit Euler time discretization does not conserve entropy except in the case of linear fluxes (LeFloch and Rohde, 2000). Consequently, both the fully explicit and fully implicit Euler differencing do not respect (nonlinear) entropy conservation, independent of the spatial discretization. Fully discrete entropy conservation is offered, however, by Crank–Nicolson time differencing (Tadmor, 2003).

5 ENTROPY STABLE INTERIOR INTERFACE COUPLING

Discontinuous formulations do not enforce solution continuity across adjoining element interfaces. Thus, the interface penalty terms $\mathbf{g}_i^{(Int)}$ in (20) account for interface connections at adjoining elements and are formulated using an SAT approach (Parsani et al., 2015a). Inviscid coupling terms are first constructed that ensure entropy conservation across the interface. Then a characteristic-based dissipation term is added, that results in an upwinding of the interface fluxes. The viscous coupling terms are patterned after a local discontinuous Galerkin (LDG) approach (Cockburn and Shu, 1998) with internal penalty (IP) terms (Wheeler, 1978) added to damp neutral eigenmodes. An extensive analysis of all semidiscrete terms can be found in Fisher and Carpenter (2013a), Carpenter et al. (2014), Carpenter and Fisher (2013) and Parsani et al. (2014, 2015a,b).

An extension of Eq. (19) that includes interface coupling terms necessary to achieve design-order entropy stability is

$$\begin{aligned}\frac{d\mathbf{q}_L}{dt} = {} & \mathcal{P}^{-1}_{x_k,L}\left[-\Delta_{x_k,L}\bar{\mathbf{f}}_{k,L} + \mathcal{P}_{x_k,L}\mathcal{D}_{x_k,L}\mathbf{f}^{(V)}_{k,L}(\mathbf{w}_L, \boldsymbol{\Theta}_{j,L})\right] \\ & +\mathcal{P}^{-1}_{x_k,L}\left[\bar{\mathbf{f}}^{(-)}_k(\mathbf{w}^{(-)}_k) - f^{(SS)}_k\left(\mathbf{q}^{(-)},\mathbf{q}^{(+)}\right)\right]\mathbf{e}^{(-)}_k \\ & +\mathcal{P}^{-1}_{x_k,L}\left[+\Lambda^{(IP)}_k\left(\mathbf{w}^{(-)}_k - \mathbf{w}^{(+)}_k\right) - \frac{1}{2}(1+\alpha)\left(\mathbf{f}^{(V-)}_k - \mathbf{f}^{(V+)}_k\right)\right]\mathbf{e}^{(-)}_k\end{aligned} \tag{28a}$$

$$\boldsymbol{\Theta}_{k,L} = \mathcal{P}^{-1}_{x_k,L}\left[\mathcal{P}_{x_k,L}\mathcal{D}_{x_k}\mathbf{w}_L - \frac{1}{2}(1-\alpha)\left(\mathbf{w}^{(-)}_k - \mathbf{w}^{(+)}_k\right)\mathbf{e}^{(-)}_k\right] \tag{28b}$$

$$\begin{aligned}\frac{d\mathbf{q}_R}{dt} = {} & \mathcal{P}^{-1}_{x_k,R}\left[-\Delta_{x_k,R}\bar{\mathbf{f}}_{k,R} + \mathcal{P}_{x_k,R}\mathcal{D}_{x_k,R}\mathbf{f}^{(V)}_{k,R}(\mathbf{w}_R, \boldsymbol{\Theta}_{j,R})\right] \\ & +\mathcal{P}^{-1}_{x_k,R}\left[-\bar{\mathbf{f}}^{(-)}_k(\mathbf{w}^{(+)}_k) + f^{(SS)}_k\left(\mathbf{q}^{(-)},\mathbf{q}^{(+)}\right)\right]\mathbf{e}^{(+)}_k \\ & +\mathcal{P}^{-1}_{x_k,R}\left[+\Lambda^{(IP)}_k\left(\mathbf{w}^{(+)}_k - \mathbf{w}^{(-)}_k\right) + \frac{1}{2}(1-\alpha)\left(\mathbf{f}^{(V+)}_k - \mathbf{f}^{(V-)}_k\right)\right]\mathbf{e}^{(+)}_k\end{aligned} \tag{28c}$$

$$\boldsymbol{\Theta}_{k,R} = \mathcal{P}^{-1}_{x_k,R}\left[\mathcal{P}_{x_k,R}\mathcal{D}_{x_k}\mathbf{w}_R + \frac{1}{2}(1+\alpha)\left(\mathbf{w}^{(+)}_k - \mathbf{w}^{(-)}_k\right)\mathbf{e}^{(+)}_k\right], \tag{28d}$$

where $\boldsymbol{\Theta}_{k,\,L}$ and $\boldsymbol{\Theta}_{k,\,R}$ are the vectors of the gradient of the entropy variables on the left and right elements in the k direction. Note that they explicitly contain contributions from the adjoining element.

New viscous fluxes that include coupling terms, are defined as

$$\mathbf{f}^{(V)}_{k,L}(\mathbf{w}_L, \boldsymbol{\Theta}_{j,L}) = [\bar{c}_{kjL}]\boldsymbol{\Theta}_{j,L};\ \ \mathbf{f}^{(V)}_{k,R}(\mathbf{w}_R, \boldsymbol{\Theta}_{j,R}) = [\bar{c}_{kjR}]\boldsymbol{\Theta}_{j,R}. \tag{29}$$

Interface penalty vectors couple the solution and fluxes at adjoining interfaces. The vectors that isolate interface data on either side of an adjoining interface are defined as

$$\begin{aligned}\mathbf{e}^{(-)}_k &= \mathbf{e}_{N x_k}\mathbf{1};\ \ \mathbf{w}^{(-)}_k = \mathbf{e}_{N x_k}\mathbf{w}_{k,L};\ \ \mathbf{f}^{(V-)}_k = \mathbf{e}_{N x_k}\mathbf{f}^{(V)}_{k,L}; \\ \mathbf{e}^{(+)}_k &= \mathbf{e}_{1 x_k}\mathbf{1};\ \ \mathbf{w}^{(+)}_k = \mathbf{e}_{1 x_k}\mathbf{w}_{k,R};\ \ \mathbf{f}^{(V+)}_k = \mathbf{e}_{1 x_k}\mathbf{f}^{(V)}_{k,R}.\end{aligned}$$

An inviscid interface flux that preserves the entropy consistency of the interior high-order accurate spatial operators (Carpenter et al., 2014) on either side of the interface $f^{(SS)}_k\left(\mathbf{q}^{(-)},\mathbf{q}^{(+)}\right)$ is constructed as

$$f^{(SS)}_k\left(\mathbf{q}^{(-)},\mathbf{q}^{(+)}\right) = f^{(S)}_k\left(\mathbf{q}^{(-)},\mathbf{q}^{(+)}\right) + \Lambda^{(I)}_k\left(\mathbf{w}^{(+)}_k - \mathbf{w}^{(-)}_k\right), \tag{30}$$

where $f^{(S)}_k\left(\mathbf{q}^{(-)},\mathbf{q}^{(+)}\right)$ is the design-order entropy *conservative* inviscid interface flux given in Eq. (23), and $\Lambda^{(I)}_k$ is a negative semidefinite interface matrix with zero or negative eigenvalues.

Entropy stability analysis of Eq. (28) (see references Fisher and Carpenter (2013a), Carpenter et al. (2014), Parsani et al. (2015a) and Parsani et al.(2015b) for details) begins by contracting Eqs. (28a)–(28d) with $\mathbf{w}_L^\top \mathcal{P}_L$, $(\mathbf{f}_{k,L}^{(V)})^\top \mathcal{P}_L$, $n_R^\top \mathcal{P}_R$ and $(\mathbf{f}_{k,R}^{(V)})^\top \mathcal{P}_R$, respectively. Summing their contributions yields an expression for the time derivative of the entropy function S on the union of the two elements.

The semidiscrete analysis is simplified by assuming that entropy stable operators are used on all exterior interfaces of the adjoining elements; thus, their contributions can be neglected without loss of generality. Eq. (28) reduces to

$$\frac{d}{dt}\widehat{\mathbf{1}}^\top \widehat{\mathcal{P}}_L \mathbf{S}_L + \frac{d}{dt}\widehat{\mathbf{1}}^\top \widehat{\mathcal{P}}_R \mathbf{S}_R + 2\left[\left\|\sqrt{[\bar{c}_{kj_L}]}\boldsymbol{\Theta}_{j,L}\right\|_{\mathcal{P}_L}^2 + \left\|\sqrt{[\bar{c}_{kj_R}]}\boldsymbol{\Theta}_{j,R}\right\|_{\mathcal{P}_R}^2\right] = \boldsymbol{\Upsilon}_k, \tag{31a}$$

where

$$\Upsilon_k = \Upsilon_k^I + \Upsilon_k^V = (\mathbf{w}_k^{(-)} - \mathbf{w}_k^{(+)})^\top \mathcal{P}_{\perp,k}(\Lambda_k^{(I)} + \Lambda_k^{(IP)})(\mathbf{w}_k^{(-)} - \mathbf{w}_k^{(+)}). \tag{31b}$$

The viscous dissipation terms $\left\|\sqrt{[\bar{c}_{kj_L}]}\boldsymbol{\Theta}_{j,L}\right\|_{\mathcal{P}_L}^2$ and $\left\|\sqrt{[\bar{c}_{kj_R}]}\boldsymbol{\Theta}_{j,R}\right\|_{\mathcal{P}_R}^2$ are uniformly dissipative. Thus, entropy stability of Eq. (31) follows immediately if the term Υ_k is dissipative. Note that Υ_k is composed of both inviscid and viscous terms; i.e. $\boldsymbol{\Upsilon}_k = \boldsymbol{\Upsilon}_k^I + \boldsymbol{\Upsilon}_k^V$. The inviscid and viscous terms are bounded individually to guarantee that the inviscid terms are stable in the limit of $Re \to \infty$ (Parsani et al., 2015a,b).

6 ENTROPY STABLE SOLID WALL BOUNDARY CONDITIONS

Solid wall boundary conditions are now constructed for the compressible NSE, which preserve the nonlinear entropy stability of the interior operator. The boundary conditions are imposed weakly via an (SAT) penalty approach, motivated by the general interface coupling conditions (see Section 5), used to couple elements in the interior of the domain. The interior conditions combine an entropy stable characteristic-based coupling condition for the inviscid terms, with an LDG approach and IP procedure for the viscous terms.

Consider a hexahedral element and assume that the solid wall boundary condition is imposed on the face plane $k = 1$: $(0, x_2, x_3)$ of the hexahedron. With these assumptions, Eq. (27) reduces to

$$\frac{d}{dt}\widehat{\mathbf{1}}^\top \widehat{\mathcal{P}}\mathbf{S} - \widehat{\mathbf{1}}^\top \widehat{\mathcal{P}}_{\perp,1}\bar{\mathbf{F}}_1 + \mathbf{w}^\top \mathcal{P}_{\perp,1}\bar{\mathbf{f}}_1^{(V)} + \mathbf{DT} = \mathbf{w}^\top \mathcal{P}_{\perp,1}\mathbf{g}_1^{(BC)}, \tag{32}$$

with n accounting for all neglected dissipative terms. A penalty-type source term $n_1^{(BC)}$ is now constructed that weakly enforces the nonslip and thermal heat-flux wall boundary conditions. It is nonlinearly (entropy) stable for the compressible NSE in three spatial dimensions.

The penalty source term $\mathbf{g}_1^{(BC)}$ is constructed from three design-order terms that weakly enforce the boundary conditions:

$$\mathbf{g}_1^{(BC)} = -\left[\bar{\mathbf{f}}_1^{(I)}(\mathbf{q}) - \bar{\mathbf{f}}_1^{(S)}\left(\mathbf{q}, \mathbf{g}^{(E)}\right)\right] + \left[\bar{\mathbf{f}}_1^{(V)} - \bar{\mathbf{f}}_1^{(V,B)}\right] + \mathcal{M}\left[\mathbf{w} - \mathbf{g}^{(NS),Vel}\right]. \quad (33)$$

Each of the three-bracketed contributions is composed of the difference between a numerical state (the first component) and a physical state (the second component). The numerical state is strictly "numerical," while the physical state combines four independent physical boundary contributions and supplements the fifth component with numerical data.

The first bracket enforces the no-penetration Euler wall condition through the inviscid flux of the compressible Euler equations. The pointwise boundary state is formed by constructing an entropy conservative flux based on the numerical state $q_{(1)}$ and a manufactured boundary state given by the vector $g^{(E)}$:

$$g^{(E)} = \begin{pmatrix} 1 & 0 & 0 & 0 & 0 \\ 0 & -1 & 0 & 0 & 0 \\ 0 & 0 & 1 & 0 & 0 \\ 0 & 0 & 0 & 1 & 0 \\ 0 & 0 & 0 & 0 & 1 \end{pmatrix} q_{(1)} = \left(\rho_{(1)}, -(\rho u_1)_{(1)}, (\rho u_2)_{(1)}, (\rho u_3)_{(1)}, (\rho E)_{(1)}\right)^{\top}, \quad (34)$$

where $q_{(1)}$ is the vector of the conservative variables constructed from the numerical state at the boundary, and the numerical state at the boundary point on plane $k = 1$ is identified with the subscript $(\cdot)_{(1)}$. A two-point entropy conservative flux function is used to connect the two states (e.g. Ismail and Roe, 2009).

Theorem 2. *The inviscid flux penalty in Eq. (33)*

$$\mathbf{g}_1^{(BC)Inv} = -\left[\bar{\mathbf{f}}_1^{(I)}(\mathbf{q}) - \bar{\mathbf{f}}_1^{(S)}\left(\mathbf{q}, \mathbf{g}^{(E)}\right)\right]$$

is entropy conservative if the vector $\mathrm{g}^{(\mathrm{E})}$ *is defined as in (34).*

Proof. The inviscid penalty is equivalent to the one developed by Svärd and Özcan (2014) in the context of high-order FD operators. The proof is reported elsewhere (Parsani et al., 2015a; Svärd and Özcan, 2014). □

The second bracket enforces a boundary condition on the surface temperature and/or heat flux, facilitated by manufacturing a boundary viscous flux $\bar{f}_1^{(V,B)}$. Define the gradient of the entropy variables in the numerical state as Θ_{x_1}, Θ_{x_2} and Θ_{x_3}. Next, specify the thermal condition

$$\mathrm{g}(t) = \frac{\partial T}{\partial x_n}\frac{1}{T}, \quad (35)$$

where $\mathrm{g}(t)$ is a bounded function independent of the numerical solution, and x_n is the direction normal to the boundary. Define $\widetilde{\Theta}_{x_1}$ as

$$\widetilde{\Theta}_{x_1} = [\Theta_{x_1}(1), \Theta_{x_1}(2), \Theta_{x_1}(3), \Theta_{x_1}(4), \widetilde{\Theta}_{x_1}(5)]^{\top},$$

where $\widetilde{\Theta}_{x_1}(5)$ is computed as

$$\widetilde{\Theta}_{x_1}(5) = -g(t)w_{(1)}(5) = \frac{g(t)}{T_{(1)}} \tag{36}$$

to enforce a Neumann boundary condition, or

$$\widetilde{\Theta}_{x_1}(5) = \left(\frac{\partial T}{\partial x_1}\right)_{(1)} \frac{1}{g(t)} \tag{37}$$

to enforce a Dirichlet boundary condition (i.e. an isothermal wall). With these definitions, the manufactured viscous flux $\bar{f}_1^{(V,B)}$ is constructed as

$$\bar{f}_1^{(V,B)} = c_{11}\widetilde{\Theta}_{x_1} + c_{12}\Theta_{x_2} + \bar{c}_{13}\Theta_{x_3}, \tag{38}$$

where $\bar{c}_{1j}\, j = 1, 2, 3$ is constructed from the numerical state. Note that the adiabatic wall condition: $g(t) = 0$, yields $\widetilde{\Theta}_{x_1}(5) = 0$ in Eq. (36). As expected, the first two components of the viscous flux penalty in (33) are entropy conservative.

The third bracket enforces the no-slip wall Dirichlet boundary conditions ($u_1 = u_2 = u_3 = 0$) through a standard SAT approach. The manufactured boundary state $g^{(NS),\, Vel}$ is defined in terms of entropy variables as

$$g^{(NS),Vel} = \left(w_{(1)}(1), 0, 0, 0, w_{(1)}(5)\right)^{\top}, \tag{39}$$

where $w_{(1)}(1)$ and $w_{(1)}(5)$ are the first and the fifth components of the entropy vector constructed from the numerical state. Three boundary conditions are imposed in Eq. (39); all velocity components are set to zero at the wall. This is immediately clear by recalling that the entropy variables for the compressible NSE are defined as

$$\mathbf{w} = \left[\frac{h}{T} - s - \frac{u^2 + v^2 + w^2}{2T}, \frac{u}{T}, \frac{v}{T}, \frac{w}{T}, -\frac{1}{T}\right]^{\top},$$

The matrix $\mathcal{M}$ is defined as $\mathcal{M} = (I_N \otimes I_N \otimes I_N \otimes M)$ where M is a five-by-five matrix and is defined as

$$M = -\frac{\alpha^{(BC)}}{(\mathcal{P}_{\perp,1})} H\widetilde{c}_{11}H, \quad H = \mathrm{diag}(1,1,1,1,0), \tag{40}$$

where $\widetilde{c}_{11}$ has the functional form of the usual $\bar{c}_{11}$ matrix. The matrix $\widetilde{c}_{11}$ is constructed using the reference primitive variables used to nondimensionalize the compressible NSE. The coefficient $\alpha^{(BC)}$ is a positive value used to modify the strength of the SAT penalty term, and can be specified by the user. The factor $\left(\widehat{\mathcal{P}}_{\perp,1}\right)$ in the denominator is introduced to achieve the correct asymptotic order of accuracy; it allows an increase in the strength of M with increased resolution. As Reynolds number Re approaches ∞, the term smoothly recovers the no-penetration boundary condition that characterizes the Euler equations.

Theorem 3. *The viscous penalty terms in (33)*

$$\mathbf{g}_1^{(BC)Vis} = \left(\bar{\mathbf{f}}_1^{(V)} - \bar{\mathbf{f}}_1^{(V,B)}\right) + \mathcal{M}\left(\mathbf{w} - \mathbf{g}^{(NS),Vel}\right)$$

are entropy stable for any value of $g(t)$ *and any matrix* M *as defined in (40).*

Proof. The proof is reported elsewhere (Parsani et al., 2015a). □

7 ENTROPY STABLE WENO FORMULATIONS

Nondissipative numerical methods cannot effectively capture shocks. Shocks dissipate entropy and nondissipative numerical methods cannot mimic this dissipation. Herein, SBP-weighted essentially nonoscillatory (SBP-WENO) FD methods are used to introduce dissipation at shocks and other discontinuities. Stencil biasing mechanics follow references Yamaleev and Carpenter (2009b) and Yamaleev and Carpenter (2009a), while full implementation details are available in references Fisher et al. (2011b) and Carpenter et al. (2013).

An SBP-WENO operator is formulated based on an existing SBP difference operator in flux form,

$$\mathcal{Q}\mathbf{f} = \Delta\bar{\mathbf{f}}.$$

These fluxes that approximate the nondissipative first derivative operator are denoted *target fluxes*. The target fluxes are broken into a sum of fluxes on smaller stencils of width, p_w, denoted *candidate stencils*,

$$\bar{\mathbf{f}}_i = \sum_{k=1}^{n_s} \bar{d}_i^k \bar{\mathbf{f}}_i^{\mathcal{I}_k}, \quad i = 1, 2, \ldots, N-1, \tag{41}$$

where n_s is the number of candidate stencils needed to describe the target flux, $\mathbf{f}_{\mathrm{i}}^{\mathcal{I}_k}$, are the *candidate fluxes*, and $\bar{d}_i^k$ are the *target weights* that recover the target flux. The candidate stencil width is held constant for all fluxes in the domain. The number of candidate stencils needed to describe the fluxes, $\bar{\mathbf{f}}_i$, can vary, as the target fluxes do not all have the same stencil size when approaching the boundary.

WENO attempts to prevent the interpolated fluxes: n_i, from using data across discontinuities. This is achieved by replacing the target weights: $\bar{d}_i^k$, with the following nonlinear weights

$$\bar{\omega}_i^k = \frac{\bar{\alpha}_i^k}{\sum_{\ell} \bar{\alpha}_i^{\ell}}, \quad \bar{\alpha}_i^k = \bar{d}_i^k \left(1 + \frac{\bar{\tau}_i}{\bar{\beta}_i^k + \bar{\epsilon}_i}\right), \quad k = 1, \ldots, n_s. \tag{42}$$

The functional form of the nonlinear weights relies on the scaling parameter, $\bar{\epsilon}$, and dual stencil biasing parameters, $\bar{\tau}$ and $\bar{\beta}$. The parameter $\bar{\epsilon}$ is chosen to satisfy sufficient conditions for accuracy (Yamaleev and Carpenter, 2009a,b). The parameter $\bar{\tau}$ is a measure of the smoothness over the full stencil,

$$\bar{\tau}_i = \sum_{k=1}^{n_\tau} \left(\frac{\partial^{2p_w-1} u(\bar{x}_k)}{\partial x^{2p_w-1}} (\delta x)^{2p_w-1} \right)^2, \quad n_\tau = n_s - p_w, \tag{43}$$

where u is the unique polynomial fit of the solution over the global stencil. The parameter $\bar{\beta}$ is a measure of the smoothness over each individual candidate stencil,

$$\bar{\beta}_i^k = \sum_{\ell=1}^{p_w-1} (\delta x)^{2\ell} \left(\frac{\partial^\ell \varphi_i^k(\bar{x}_k)}{\partial x^\ell} \right)^2, \tag{44}$$

where $\varphi_i^k(x)$ is the unique order $(p_w - 1)$ polynomial fit of the solution over the candidate stencil, $\mathcal{I}_k$. The final WENO flux is given by the formula.

$$\bar{\mathbf{f}}_i^{(W)} = \sum_{k=1}^{n_s} \bar{\omega}_i^k \bar{\mathbf{f}}_i^{\mathcal{I}_k}, \quad i = 1,2,\ldots,N-1. \tag{45}$$

Although WENO operators smoothly bias the stencil away from discontinuities and shocks, they do not guarantee that the resulting scheme is entropy conservative/stable. The WENO operator may be combined with the entropy conservative target operator via a comparison technique (description to follow). The operator pair retains the dissipation of WENO and is provably entropy stable.

7.1 An Entropy Comparison Approach

A comparison approach compares the entropy dissipation of the WENO scheme and an entropy conservative/stable scheme constructed from the target operator. If the dissipation is less than the entropy conservative datum, then more dissipation is necessary. Conditions that guarantee entropy stability are now established.

A one-dimensional condition analogous to Eq. (22) that guarantees entropy stability is

$$\mathbf{w}^\top \mathcal{P} \mathbf{q}_t + F(q_N) - F(q_1) \leq \mathbf{w}^\top \mathbf{g}_b, \tag{46}$$

which is satisfied if the "baseline" entropy stable inviscid fluxes satisfy the comparison condition

$$\mathbf{w}^\top \Delta \bar{\mathbf{f}} \geq \widehat{\mathbf{1}}^\top \Delta \bar{\mathbf{F}} = \mathbf{w}^\top \Delta \bar{\mathbf{f}}^{(S)}. \tag{47}$$

The vector inequality given in Eq. (47) is valid provided that the following local conditions are satisfied

$$(w_{i+1} - w_i)^\top (\bar{\mathbf{f}}_i - \bar{\mathbf{f}}_i^{(S)}) \leq 0, \quad i = 1,2,\ldots,N-1. \tag{48}$$

The proof appears in reference Fisher and Carpenter (2013a). Extension to multiple dimensions follows immediately via tensor product arithmetic.

The constraint given in Eq. (48) enforces an entropy stability condition between two SBP operators with consistently located fluxes. Although WENO is ideally suited for capturing discontinuities, it is not necessary to use WENO as the companion operator; any consistent SBP operator is sufficient. (SBP-WENO operators exist for high-order FDs and P3 Legendre–Gauss–Lobatto spectral collocation. It is not known whether WENO operators exist for all SBP operators.) Although (48) is derived from the global property given in Eq. (47), it is not sufficiently sharp to guarantee a nonoscillatory solution. The companion operator must supply sufficient dissipation to achieve this goal.

8 CONSERVATION OF ENTROPY IN CURVILINEAR COORDINATES

8.1 Coordinate Transformations and Geometric Conservation Laws

Simulations on complex geometries are performed on the union of piecewise smooth subdomains/elements. Each subdomain $\Omega(x_j)$ is smoothly (one-to-one) mapped into a unit cube $\Omega(\xi_k)$ with $-1 \leq \xi_1, \xi_2, \xi_3 \leq 1$ (i.e. $x = x(\xi)$). The derivatives in the physical coordinates are then transformed into the computational coordinates using the chain rule and the transformation Jacobian J to yield the expressions

$$J\frac{\partial}{\partial x_\ell} = J\frac{\partial \xi_k}{\partial x_\ell}\frac{\partial}{\partial x_k} = J(\xi_k)_{x_\ell}\frac{\partial}{\partial x_k}; \quad J = Det\left(\frac{\partial \mathbf{x}}{\partial \boldsymbol{\xi}}\right),$$

with index sums implied on k. The transformed compressible NSE (1) in computational coordinates becomes

$$\begin{aligned}(Jq)_t + [J(\xi_k)_{x_\ell}(f_\ell - f_\ell^{(v)})]_{\xi_k} = (f_\ell - f_\ell^{(v)})[J(\xi_k)_{x_\ell}]_{\xi_k} = 0,\\ Bq = g_b, \quad \xi \in \partial\Omega(\xi), \quad t \in [0,\infty),\\ q(x,0) = g_0(x), \quad \xi \in \Omega(\xi),\end{aligned} \tag{49}$$

where the Cartesian coordinates: $\xi = (\xi_1, \xi_2, \xi_3)^{\top}$, and time: t, are independent variables, and index sums are implied. The geometric conservation law (GCL) terms $[J(\xi_k)_{x_\ell}]_{\xi_k}, \ell = 1,2,3$ on the right-hand side of Eq. (49) are identically zero at the continuous level.

The discrete version of the GCL terms are not in general zero. Elsewhere (Thomas and Lombard, 1979; Visbal and Gaitonde, 2002) discrete metric terms are constructed as

$$[J(\xi_k)_{x_\ell}]_{\xi_k} = +\mathcal{D}_{\xi_{\ell+2}}[\mathcal{D}_{\xi_{\ell+1}}(x_{k+1})][x_{k+2}]\mathbf{1} - \mathcal{D}_{\xi_{\ell+1}}[\mathcal{D}_{\xi_{\ell+2}}(x_{k+1})][x_{k+2}]\mathbf{1}, \quad (50)$$

and discretely satisfy the GCL expression

$$\mathcal{D}_{\xi_k}[J(\xi_k)_{x_\ell}]\mathbf{1} = \mathcal{D}_{\xi_k}[(\hat{a}_l^k)]\mathbf{1} = 0, \quad i = 1,2,3. \quad (51)$$

(Note the abbreviated metric definition $[J(\xi_k)_{x_\ell} = (\hat{a}_l^k)]$ introduced in Eq. (51).) The nomenclature of Eq. (50) assumes three-cyclic indices; e.g. $k + 2 = \text{mod}\,(k + 2, 3)$.

The semidiscretized, contravariant compressible NSE are

$$\hat{q}_t + \sum_{k=1}^{3} \mathcal{P}_{\xi_k}^{-1} \Delta_{\xi_k} [\hat{\bar{\mathbf{f}}}_k - \hat{\bar{\mathbf{f}}}_k^{(v)}] = \sum_{k=1}^{3} \mathcal{P}_{\xi_k}^{-1} \hat{\mathbf{g}}_k^{(BC)} + \mathcal{P}_{\xi_k}^{-1} \hat{\mathbf{g}}_k^{(Int)}, \quad x \in \Omega, \quad t \in [0, \infty),$$
$$Bq = g_b, \quad x \in \partial\Omega, \quad t \in [0, \infty),$$
$$q(x,0) = g_0(x), \quad x \in \Omega,$$

with (52)

$$\hat{\mathbf{q}} = [J]\mathbf{q}; \quad \widehat{\mathcal{D}}_{\xi_1} = \mathcal{P}^{-1}{}_{\xi_1} \mathcal{Q}_{\xi_1} = \mathcal{P}^{-1}{}_{\xi_1} \Delta_{\xi_1} \mathcal{I}_{\xi_1} = \mathcal{P}^{-1}{}_{\xi_1} \Delta_{\xi_1} \mathcal{I}_{\xi_1} \otimes \mathcal{I}_{\xi_2} \otimes \mathcal{I}_{\xi_3} \otimes \mathcal{I}_5,$$

and similar tensor product expressions for the other coordinate directions.

8.2 Curvilinear Conservation and Stability

Contracting the discrete entropy variables; $\mathbf{w}^\top$, with the semidiscrete equation (52) yields a global equation for the semidiscrete decay of entropy in the contravariant frame. The resulting expression is

$$\mathbf{w}^\top \mathcal{P} \frac{\partial \widehat{\mathbf{q}}}{\partial t} + \sum_{k=1}^{3} \mathbf{w}^\top \mathcal{P}_{\perp,k} \Delta_{\xi_k} [\widehat{\bar{\mathbf{f}}}_k - \widehat{\bar{\mathbf{f}}}_k^{(v)}] = \sum_{k=1}^{3} \mathbf{w}^\top \mathcal{P}_{\perp,k} \widehat{\mathbf{g}}_k^{(BC)} + \mathbf{w}^\top \mathcal{P}_{\perp,k} \widehat{\mathbf{g}}_k^{(Int)}. \quad (53)$$

Simplifying Eq. (53) into a semidiscrete convection–diffusion equation is analogous to the Cartesian case. The semidiscrete time term is manipulated for diagonal-norm SBP operators as $\mathbf{w}^\top \mathcal{P} \partial \widehat{\mathbf{q}} / \partial t = \widehat{\mathbf{1}}^\top \mathcal{P} \partial \widehat{\mathbf{S}} / \partial t$ with $\widehat{\mathbf{S}} = [J]\mathbf{S}$. The viscous terms are first symmetrized as in (6) and then manipulated as

$$\mathbf{w}^\top \mathcal{P}_{\perp,k} \Delta_{\xi_k} \widehat{\bar{\mathbf{f}}}_k^{(v)} = \mathbf{w}^\top \mathcal{P}_{\perp,k} \mathcal{B}_k [\widehat{c}_{kj}] \mathcal{D}_{\xi_j} \mathbf{w} - (\mathcal{D}_{\xi_k} \mathbf{w})^\top \mathcal{P} [\widehat{c}_{kj}] (\mathcal{D}_{\xi_j} \mathbf{w}). \quad (54)$$

The terms $\mathbf{w}^\top \mathcal{P}_{\perp,k} \mathcal{B}_k [\widehat{c}_{kj}] \mathcal{D}_{\xi_j} \mathbf{w}$ account for diffusive fluxes on contravariant boundaries, while the terms $-(\mathcal{D}_{\xi_k} \mathbf{w})^\top \mathcal{P} [\widehat{c}_{kj}] (\mathcal{D}_{\xi_j} \mathbf{w})$ account for viscous dissipation in computational space.

The contravariant inviscid terms of Eq. (53) are entropy conservative in each coordinate direction k, if they satisfy the constraints given in the following theorem:

Theorem 4. *A high-order, contravariant, entropy conservative flux* $\widehat{\bar{f}}_k^{(S)}, k = 1,2,3$, *may be constructed at point* i *using the expression*

$$\widehat{\bar{f}}_k^{(S)}\Big|_i = \sum_{s=i+1}^{N}\sum_{r=1}^{i} 2q_{(r,s)}\widehat{\bar{f}}_{(S)k}(\boldsymbol{q}_r,\boldsymbol{q}_s;(\hat{a}_l^k)_r,(\hat{a}_l^k)_s), \quad 1 \le i \le N-1,$$
$$= \sum_{s=i+1}^{N}\sum_{r=1}^{i} 2q_{(r,s)}\sum_{l=1}^{3}\bar{f}_S^l(\boldsymbol{q}_r,\boldsymbol{q}_s)\frac{(\hat{a}_l^k)_r+(\hat{a}_l^k)_s}{2} \tag{55}$$

where $\bar{f}_S^l(\mathbf{q}_\ell,\mathbf{q}_k)$ *is any two-point nondissipative function that satisfies the entropy consistency condition*

$$(w_r^\top - w_s^\top)\bar{f}_S^l(\boldsymbol{q}_r,\boldsymbol{q}_s) = \psi_r^l - \psi_s^l. \tag{56}$$

The high-order, contravariant, entropy conservative flux satisfies an additional local entropy consistency property provided the entropy consistency condition given in Eq. (56), which is derived using Tadmor's integration through phase space (Tadmor, 2003),

$$W\mathcal{P}_k^{-1}\Delta_{\xi_k}\widehat{\bar{\mathbf{f}}}_k^{(S)} = \mathcal{P}_k^{-1}\Delta_{\xi_k}\widehat{\bar{\mathbf{F}}}_k^{(S)} = \frac{\partial F(\mathbf{q})}{\partial \xi_k} + \mathcal{T}_p, \tag{57}$$

where p *is the order of the truncation error* $\mathcal{T}$. *The local entropy flux* $\widehat{\bar{F}}^{(S)}{}_k$ *is*

$$\widehat{\bar{F}}_k^{(S)}\Big|_i = \sum_{s=i+1}^{N}\sum_{r=1}^{i} 2q_{(r,s)}\widehat{\bar{F}}_{(S)k}(\boldsymbol{q}_r,\boldsymbol{q}_s;(\hat{a}_l^k)_r,(\hat{a}_l^k)_s), \quad 1 \le i \le N-1,$$
$$= \sum_{s=i+1}^{N}\sum_{r=1}^{i} 2q_{(r,s)}\sum_{l=1}^{3}\left[\frac{(w_r^\top + w_s^\top)}{2}\bar{f}_S^l(\boldsymbol{q}_r,\boldsymbol{q}_s) - \frac{(\psi_r^l+\psi_s^l)}{2}\right]\frac{(\hat{a}_l^k)_r+(\hat{a}_l^k)_s}{2}. \tag{58}$$

Proof. A brief sketch of the proof is presented. Details are available in reference Fisher (2012). The local flux differences $\Delta_{\xi_k}\widehat{\bar{\mathbf{f}}}_k^{(S)}$ at point i, in each coordinate direction k, follow immediately from Eq. (55)

$$\widehat{\bar{f}}_k^{(S)}\Big|_i - \widehat{\bar{f}}_k^{(S)}\Big|_{i-1} = \sum_{s=i+1}^{N}\sum_{r=1}^{i} 2q_{(r,s)}\widehat{\bar{f}}_{(S)k}(\mathbf{q}_r,\mathbf{q}_s) - \sum_{s=i}^{N}\sum_{r=1}^{i-1} 2q_{(r,s)}\widehat{\bar{f}}_{(S)k}(\mathbf{q}_r,\mathbf{q}_s),$$
$$= \sum_{j=1}^{N} 2q_{(i,j)}\widehat{\bar{f}}_{(S)k}(\mathbf{q}_r,\mathbf{q}_s;(\hat{a}_l^k)_r,(\hat{a}_l^k)_s), \quad 1 \le i \le N. \tag{59}$$

Using Eq. (59) to simplify Eq. (57) yields

$$\sum_{k=1}^{3} W\mathcal{P}_k^{-1}\Delta_{\xi_k}\widehat{\bar{\mathbf{f}}}_k^{(S)} = \sum_{k=1}^{3}\mathcal{P}_k^{-1}\sum_{j=1}^{N}\sum_{l=1}^{3} 2q_{ij}^k w_i^\top \bar{\mathbf{f}}_k^l \frac{(\hat{a}_l^k)_i + (\hat{a}_l^k)_j}{2}. \tag{60}$$

Adding and subtracting w_j in Eq. (60), and simplifying with Eq. (56) yields the expression

$$\sum_{k=1}^{3} W\mathcal{P}_k^{-1}\Delta_{\xi_k}\widehat{\mathbf{f}}_k^{(S)} = \sum_{k=1}^{3}\mathcal{P}_k^{-1}\sum_{j=1}^{N}\sum_{l=1}^{3} 2q_{ij}^k\left[\frac{(w_i^\top + w_j^\top)}{2} + \frac{(w_i^\top - w_j^\top)}{2}\right]\overline{\mathbf{f}}_k^l \frac{(\hat{a}_l^k)_i + (\hat{a}_l^k)_j}{2}$$
$$= \sum_{k=1}^{3}\mathcal{P}_k^{-1}\sum_{j=1}^{N}\sum_{l=1}^{3} 2q_{ij}^k\left[\frac{(w_i^\top + w_j^\top)}{2}\overline{\mathbf{f}}_k^l + \frac{(\psi_i^l - \psi_j^l)}{2}\right]\frac{(\hat{a}_l^k)_i + (\hat{a}_l^k)_j}{2}. \tag{61}$$

Consistency and the GCL terms are used to subtract zero

$$0 = \sum_{k=1}^{3}\sum_{j=1}^{N}\sum_{l=1}^{3} 2q_{ij}^k \psi_i^l \frac{(\hat{a}_l^k)_i + (\hat{a}_l^k)_j}{2}$$
$$= \sum_{l=1}^{3} 2\psi_i^l \sum_{k=1}^{3}\frac{(\hat{a}_l^k)_i}{2}\underbrace{\sum_{j=1}^{N} q_{ij}^k}_{\text{Consistency}=0} + \underbrace{\sum_{l=1}^{3} 2\psi_i^l \sum_{k=1}^{3}\sum_{j=1}^{N} q_{ij}^k \frac{(\hat{a}_l^k)_j}{2}}_{\text{GCL}=0} \tag{62}$$

from each row i of Eq. (61), to yield an expression of the desired form.

$$\sum_{k=1}^{3} W\mathcal{P}_k^{-1}\Delta_{\xi_k}\widehat{\mathbf{f}}_k^{(S)} = \sum_{k=1}^{3}\mathcal{P}_k^{-1}\sum_{j=1}^{N}\sum_{l=1}^{3} 2q_{ij}^k\left[\frac{(w_i^\top + w_j^\top)}{2}\overline{\mathbf{f}}_k^l - \frac{(\psi_i^l + \psi_j^l)}{2}\right]\frac{(\hat{a}_l^k)_i + (\hat{a}_l^k)_j}{2}$$
$$= \sum_{k=1}^{3}\mathcal{P}_k^{-1}\Delta_{\xi_k}\widehat{\overline{\mathbf{F}}}_k^{(S)}. \tag{63}$$

The entropy fluxes $\widehat{\overline{\mathbf{F}}}^{(S)}{}_k$ are design-order accurate if the two-point nondissipative fluxes $\overline{f}_S^l(\mathbf{q}_r,\mathbf{q}_s)$ satisfy

$$\overline{\mathbf{f}}_S(\mathbf{q}_r,\mathbf{q}_s) = \int_0^1 \mathbf{g}(\mathbf{w}(\mathbf{q}_r) + \xi(\mathbf{w}(\mathbf{q}_s) - \mathbf{w}(\mathbf{q}_r)))d\xi, \quad \mathbf{g}(\mathbf{w}(\mathbf{q})) = \mathbf{f}(\mathbf{q}). \tag{64}$$

A Cartesian coordinate proof is presented elsewhere (Fisher, 2012; Fisher and Carpenter, 2013a). Extension to curvilinear coordinates follows immediately (Fisher, unpublished).

The two-point entropy flux of Ismail and Roe (2009) has not been shown to be the result of an integration through phase space. A posteriori accuracy tests demonstrate design-order convergence for smooth problems.

9 RESULTS: ACCURACY AND ROBUSTNESS

Design-order accuracy of all discrete operators have been established elsewhere (Carpenter et al., 2014, 2015; Fisher and Carpenter, 2013a) on smooth test cases such as the isentropic Euler vortex and the viscous shock. Thus, the primary focus of this section is to demonstrate the robustness of the entropy stable formulations, by using one smooth and two discontinuous problems. The Taylor–Green vortex is used to test the behaviour of the algorithms on smooth under-resolved turbulent simulations. Next, a supersonic bluff-body flow is simulated using the entropy stable spectral collocation algorithm, run without the addition of any additional dissipation. The final supersonic simulation is run using the SSWENO FD algorithm. This problem demonstrates the SSWENO capability while testing the efficacy of the entropy stable correction terms.

9.1 Taylor–Green Vortex

Nonlinearly stable spectral collocation operators do not require ad hoc stabilization techniques (e.g. dealiasing, filtering, limiting, over-integration procedures) for simulations of under-resolved turbulent flows. To demonstrate this, the Taylor–Green vortex test case is used as a benchmark model problem and is solved on the periodic cube $[-\pi L \leq x, y, z \leq +\pi L]$. The initial condition is given by the following analytical expressions

$$\begin{aligned}
u_1 &= V_0 \sin\left(\frac{x_1}{L}\right) \cos\left(\frac{x_2}{L}\right) \cos\left(\frac{x_3}{L}\right), \\
u_2 &= -V_0 \cos\left(\frac{x_1}{L}\right) \sin\left(\frac{x_2}{L}\right) \cos\left(\frac{x_3}{L}\right), \\
u_3 &= 0, \\
p &= p_0 + \frac{\rho_0 V_0^2}{16}\left[\cos\left(\frac{2x_1}{L}\right) + \cos\left(\frac{2x_2}{L}\right)\right]\left[\cos\left(\frac{2x_3}{L}\right) + 2\right],
\end{aligned} \tag{65}$$

where u_1, u_2 and u_3 are the components of the velocity in the x_1-, x_2- and x_3-directions, p is the pressure and ρ is the density. The flow is initialized to be isothermal, i.e., $p/\rho = p_0/\rho_0 = RT_0$. The Reynolds number for this flow is defined as $Re = (\rho_0 V_0 L)/\mu$, where μ is the dynamic viscosity. Starting from the initial condition, the nonlinear interactions of different flow scales yield vortex breakdowns.

The compressible NSE in curvilinear coordinates are used in the simulation, with a Mach number of $M = 0.08$. The low Mach number allows for a reasonable comparison with the incompressible simulations reported in literature. The Reynolds number and Prandtl number are set to $Re = 1600$, $Pr = 0.71$, respectively. A simulation is run using a fully unstructured grid containing 42 hexahedron. The distribution of the element in the 2D plane is shown in Fig. 1.

Fig. 2 shows the kinetic energy dissipation rate of our computations with $p_{LGL} = 16$, and $p_{LGL} = 22$ and the reference data of de Wiart et al. (2014). (The subscript "*LGL*" denotes an SBP operator based on the Legendre–Gauss–Lobatto points.) The DOFs for the computation with $p_{LGL} = 16$ are too few to accurately resolve the flow field, however, the computation with a formally 17-order accurate scheme is stable through all the simulation. This is a feat unattainable with alternative approaches based on high-order accurate

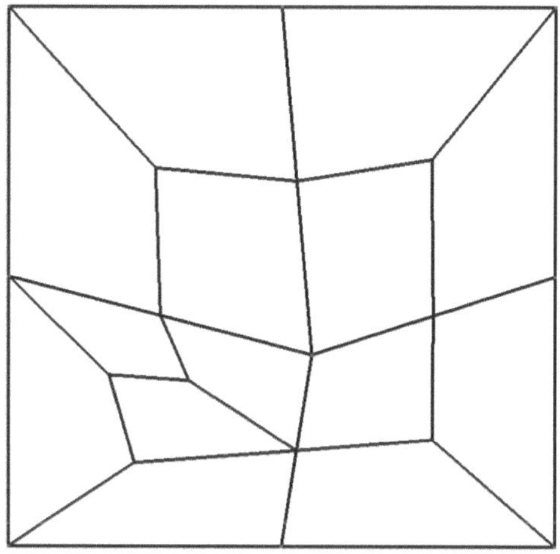

FIG. 1 Plane distribution of the elements used for the Taylor–Green vortex at $Re = 1600, M = 0.08$.

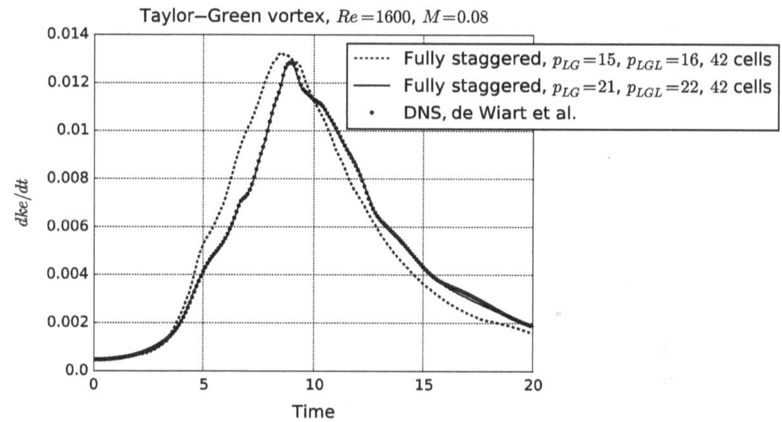

FIG. 2 Evolution of the time derivative of the kinetic energy for the Taylor–Green vortex at $Re = 1600$, $M = 0.08$; entropy stable LSC-FEM algorithm.

linear stable schemes without filtering, dealiasing, limiters, etc. (Bull and Jameson, 2014; Gassner and Beck, 2012). The robustness and accuracy of the entropy stable algorithm is attained also for higher-order discretization as indicated in Fig. 2 for the solution computed with $p_{LGL} = 22$. In this case, the computation with a formally 23-order accurate scheme compares very well with the DNS results.

9.2 Computation of a Square Cylinder in Supersonic Free Stream

A simulation of the flow past a 3D square cylinder at $Re_\infty = 10^4$ and $M_\infty = 1.5$ is presented, to provide numerical evidence of the robustness of entropy stable high-order spatial discretizations. The supersonic flow is characterized by a very large range of length scales, strong shocks and expansion regions that interact with each other, leading to complex flow patterns. During the past three decades, this fluid flow problem has been thoroughly investigated by several researchers for aerodynamic applications (see for instance, references Nakagawa (1987) and Birch et al. (2003)).

The domain of interest spans one square cylinder edge in the x_3 direction, and at the two planes perpendicular to this coordinate direction, periodic boundary conditions are used. The flow is computed on an unstructured grid with 43,936 hexahedron, using a fourth-order accurate entropy stable spectral collocation formulation. Ad hoc stabilization techniques (e.g. dealiasing, filtering, limiting, over-integration procedures) are not used. The only sources of dissipation in the simulation are (1) physical viscosity, (2) boundary conditions and (3) characteristic upwinding on element interfaces. The comparison approach is not used, and neither is any form of shock capturing technique nor slope-limiters. The body surface is considered adiabatic and the wall boundary conditions are imposed using the entropy stable approach presented in Section 6 and reference Parsani et al. (2015b).

The solution is initialized using a uniform flow at $M_\infty = 1.5$ with zero angle of attack. A strong shock is formed in front of the bluff body in the beginning of the simulation. The discontinuity moves upstream until it reaches a "stationary" position about 2.15 square cylinder edges from the frontal surface of the body. Additional weaker rarefaction waves originating from the four sharp corners of the body, interact with the subsonic regions formed near the walls. This complicated flow pattern yields the formation of shocklets in the wake of the square cylinder.

A global view of the Mach number contours at $t = 100$ is shown in Fig. 3. The shock is stationary by $t = 100$, and the flow past the square cylinder is completely unsteady, characterized by subsonic and supersonic regions. The formation of shocklets in the near wake region are clearly visible.

An extensive parametric study of Mach numbers ($1.1 < M < 1.8$) and grids (10–40 K elements) was performed using the supersonic square cylinder. The failure mode is characterized by a negative density that develops in the vicinity of the strong bow strong shock, a condition that is outside the scope of the bounded entropy estimate.

9.3 Supersonic Cylinder

A problem adapted from Chaudhuri et al. (2011) is used to simulate shock-vortex interactions in viscous supersonic flow around a cylinder in a duct. The geometry is simple by multiblock standards, but the bow shock can create

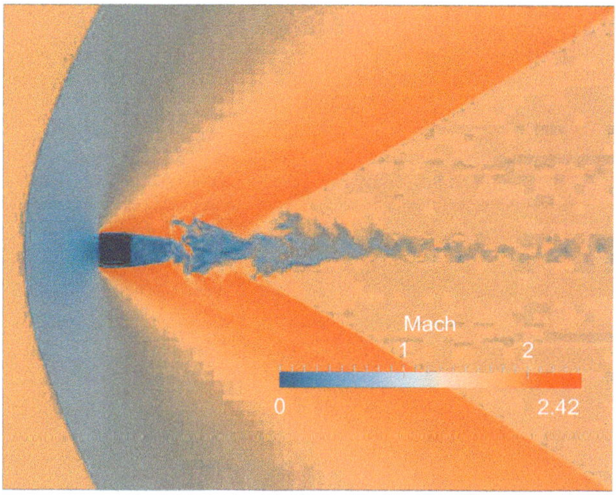

FIG. 3 Unsteady flow past a 3D square cylinder at $Re_\infty = 10^4$ and $M_\infty = 1.5$; fourth-order accurate spectral collocation is used with only interface dissipation; $t = 100$.

a stability problem for multiblock interfaces if the treatment lacks sufficient robustness.

The cylinder is located at $(x, y)/D = (0, 0)$. The duct inlet is located at $x/D = 3$ and the duct outlet is at $x/D = 21$. The top and bottom duct walls are located along $y/D = 3$, respectively. An O-type multiblock configuration is used to discretize the domain surrounding the cylinder. Five topological blocks are needed to fully discretize the domain. The cylinder walls are treated as adiabatic no slip walls. The duct walls are treated as slip walls and the grid spacing is approximately isotropic. The inflow is a uniform freestream condition at $M = 3.5$. The Prandtl number used is $Pr = 0.7$ and the Reynolds number based on diameter is 10^4. The fourth-order SSWENO FD algorithm is used in the study. A shock sensor is used to increase efficiency, by deactivating the stencil biasing mechanics in the WENO algorithm in regions where the solution is smooth.

Iso-contours of the density, Mach number, entropy and the shock sensor are shown in Fig. 4. Note that shock–vortex interactions pervade the entire wake region, and the shocks also propagate and interact throughout the domain. Many modes of large scale unsteadiness are observed where the reflected shocks move back and forth downstream of the cylinder and the vortex structures propagated through the domain exhibit different pairings for different times. It is recognized that the true physical problem would be a three-dimensional turbulent flow. However, this simulation suffices for the purpose of demonstrating the high Mach number capability of the multiblock SSWENO formulation.

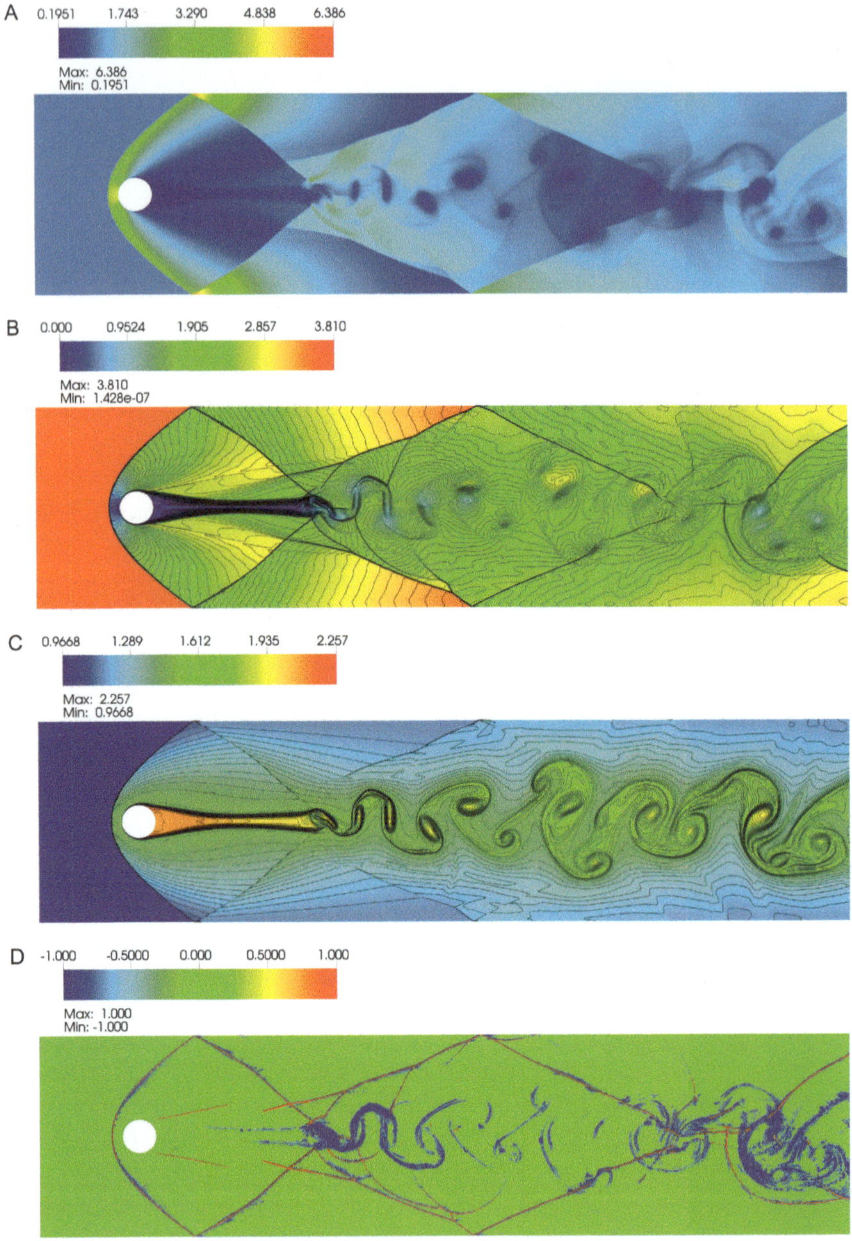

FIG. 4 The density, Mach number, entropy and shock sensor of a shock-cylinder interaction, demonstrating the shock capturing capabilities of the entropy stable WENO formulation.

10 CONCLUSIONS

A "high-level" overview is presented of the mathematical concepts of semidiscrete entropy stability for the compressible NSE. Recent contributions to the field prove that all diagonal norm, SBP-SAT operators may be used to implement an entropy conservative (Euler) or entropy stable (compressible Navier–Stokes) semidiscrete operator. Thus, entropy stable operators of arbitrary order may be constructed for the compressible NSE, which guarantee a bound on the thermodynamic entropy function $-\rho s$, given an a priori assumption on the pointwise positivity of density and temperature, and well-posed boundary data. Popular one-dimensional diagonal-norm SBP discrete operators include all centred FD and LSC-FEM operators. Extension to three-dimensional geometries via a curvilinear mapping of the domain follows immediately, provided that the discrete GCL terms are satisfied exactly.

LSC-FEM operators as well as multiblock FD operators need conservative, accurate and entropy stable interface operators to couple adjoining elements/blocks. An SAT framework is adopted to construct entropy stable interface coupling operators for the compressible NSE in three dimensions. Entropy conservative inviscid interface operators are first developed and then combined with sufficient dissipation to achieve local upwinding of the characteristic variables. Design-order accuracy is achieved for the viscous terms by using a Local DG—internal penalty approach.

New entropy stable solid wall boundary conditions are developed that discretely mimic the continuous analysis. Hydrodynamic and thermodynamic analyses motivate a no-slip condition combined with the heat entropy flow condition: $(\partial Ln[T]/\partial n)_{wall}$, to achieve stability for the continuous compressible NSE. Entropy stable numerical boundary conditions are developed that weakly enforce these solid wall boundary conditions. The newly proposed nonlinear boundary conditions are an important step towards completing the nonlinear stability analysis for the compressible NSE.

A multiblock SBP-WENO operator is ideal for simulations of somewhat complex geometries with strong shocks. A comparative approach is used to combine an entropy conservative/stable operator with the SBP-WENO operator, thereby guaranteeing an entropy bound for the SSWENO combination. Important implementation details of the combined algorithm are included.

Test cases demonstrate the robustness of the entropy stable SBP operators. An entropy stable spectral collocation approach with high polynomial orders (e.g. $7 \leq p \leq 27$) is used on a excessively coarse curvilinear mesh to perform under-resolved simulations of the Taylor–Green vortex test case. This scenario is known to be problematic for conventional operators unless additional stabilization techniques are included (e.g. filtering, over-integration, dealiasing). A fourth-order spectral collocation formulation is then used for simulations of $M = 1.5$ blunt-body flows, again with no ad hoc dissipation. Finally, a multiblock, entropy stable WENO FD algorithm is used to simulate a cylinder in a Mach 3.5 cross-flow. The simulation demonstrates the shock capturing capabilities of the WENO algorithm, the geometric flexibility of the multiblock FD approach and the efficacy of the comparison approach to combine conventional SBP operators with an entropy stable framework.

REFERENCES

Berg, J., Nordström, J., 2011. Stable Robin solid wall boundary conditions for the Navier-Stokes equations. J. Comput. Phys. 230, 7519–7532.

Birch, T., Prince, S.A., Simpson, G.M., 2003. An experimental and computational study of the aerodynamics of a square cross-section body at supersonic speeds. Tech. Rep. RTO-MP-069(I), DERA.

Bull, J.R., Jameson, A., 2014. Simulation of the compressible Taylor-Green vortex using high-order flux reconstruction schemes. In: 7th AIAA Theoretical Fluid Mechanics Conference, AIAA 2014-3210.

Carpenter, M.H., Fisher, T.C., 2013. Entropy stable spectral collocation schemes for the Navier-Stokes equations: discontinuous interfaces. Tech. Rep. NASA TM 218039.

Carpenter, M.H., Gottlieb, D., 1996. Spectral methods on arbitrary grids. J. Comput. Phys. 129, 74–86.

Carpenter, M.H., Gottlieb, D., Abarbanel, S., 1994. Time-stable boundary conditions for finite-difference schemes solving hyperbolic systems: methodology and application to high-order compact schemes. J. Comput. Phys. 111, 220–236.

Carpenter, M.H., Nordström, J., Gottlieb, D., 1999. A stable and conservative interface treatment of arbitrary spatial accuracy. J. Comput. Phys. 148, 341–365.

Carpenter, M.H., Nordström, J., Gottlieb, D., 2009. Revisiting and extending interface penalties for multi-domain summation-by-parts operators. J. Sci. Comput. 45, 118–150.

Carpenter, M.H., Fisher, T.C., Yamaleev, N.K., 2013. Boundary closures for sixth-order energy-stable weighted essentially non-oscillatory finite-difference schemes. In: Melnik, R., Kotsireas, I.S. (Eds.), Advances in Applied Mathematics, Modeling, and Computational Science. Springer, New York, pp. 117–160.

Carpenter, M.H., Fisher, T.C., Nielsen, E.J., Frankel, S.H., 2014. Entropy stable spectral collocation schemes for the Navier-Stokes equations: discontinuous interfaces. J. Sci. Comput. 36 (5), B835–B867.

Carpenter, M.H., Parsani, M., Fisher, T.C., Nielsen, E.J., 2015. Entropy stable staggered grid spectral collocation for the Burgers' and compressible Navier-Stokes equations. Tech. Rep. NASA TM 218990.

Chaudhuri, A., Hadjadj, A., Chinnayya, A., 2011. On the use of immersed boundary methods for shock/obstacle interactions. J. Comput. Phys. 230, 1731–1748.

Cockburn, B., Shu, C.W., 1998. The local discontinuous Galerkin method for time-dependent convection-diffusion systems. SIAM J. Numer. Anal. 35, 2440–2463.

de Wiart, C., Hillewaert, K., Duponcheel, M., Winckelmans, G., 2014. Assessment of a discontinuous Galerkin method for the simulation of vortical flows at high Reynolds number. Int. J. Numer. Methods Fluids 74 (7), 469–493.

Fisher, T.C., 2012. High-order L2 stable multi-domain finite difference method for compressible flows (Ph.D. thesis). Purdue University.

Fisher, T.C., Carpenter, M.H., 2013. High-order entropy stable finite difference schemes for nonlinear conservation laws: finite domains. J. Comput. Phys. 252, 518–557.

Fisher, T.C., Carpenter, M.H., 2013. High-order entropy stable finite difference schemes for nonlinear conservation laws: finite domains. Tech. Rep. NASA TM-217971.

Fisher, T.C., Carpenter, M.H., Nordström, J., Yamaleev, N.K., Swanson, R.C., 2011. Discretely conservative finite-difference formulations for nonlinear conservation laws in split form: theory and boundary conditions. Tech. Rep. NASA TM 2011-217307.

Fisher, T.C., Carpenter, M.H., Yamaleev, N.K., Frankel, S.H., 2011. Boundary closures for fourth-order energy stable weighted essentially non-oscillatory finite-difference schemes. J. Comput. Phys. 230, 3727–3752.

Gassner, G.J., Beck, A.D., 2012. On the accuracy of high-order discretizations for underresolved turbulence simulations. Theor. Comput. Fluid Dyn. 27 (3), 221–237.

Godunov, S.K., 1961. An interesting class of quasilinear systems. Dokl. Akad. Nauk SSSR 139 (3), 521–523.

Harten, A., 1983. On the symmetric form of systems of conservation laws with entropy. J. Comput. Phys. 49, 151–164.

Harten, A., Engquist, B., Osher, S., Chakravarthy, S.R., 1987. Uniformly high order accurate essentially non-oscillatory schemes, III. J. Comput. Phys. 0021-999.171 (2), 231–303. http://dx.doi.org/10.1016/0021-9991(87)90031-3. http://www.sciencedirect.com/science/article/pii/0021999187900313.

Hesthaven, J.S., Gottlieb, D., 1996. A stable penalty method for the compressible Navier-Stokes equations: I: open boundary conditions. SIAM J. Sci. Comput. 17, 579–612.

Hesthaven, J.S., Warburton, T., 2008. Nodal Discontinuous Galerkin Methods: Algorithms, Analysis, and Applications. Texts in Applied Mathematics, Springer, New York. ISBN 9780387720654.

Hughes, T.J.R., Franca, L.P., Mallet, M., 1986. A new finite element formulation for computational fluid dynamics: I. Symmetric forms of the compressible Euler and Navier-Stokes equations and the second law of thermodynamics. Comput. Methods Appl. Mech. Eng. 54, 223–234.

Ismail, F., Roe, P.L., 2009. Affordable, entropy-consistent Euler flux functions II: entropy production at shocks. J. Comput. Phys. 228, 5410–5436.

Jiang, G., Shu, C.W., 1996. Efficient implementation of weighted ENO schemes. J. Comput. Phys. 126, 202–228.

LeFloch, P.G., Rohde, C., 2000. High-order schemes, entropy inequalities, and nonclassical shocks. SIAM J. Numer. Anal. 37, 2023–2060.

LeFloch, P.G., Mercier, J.M., Rohde, C., 2002. Fully discrete, entropy conservative schemes of arbitrary order. SIAM J. Numer. Anal. 40, 1968–1992.

LeVeque, R., 1992. Numerical Methods for Conservation Laws. Birkhauser, Basel.

Liu, X.D., Osher, S., Chan, T., 1994. Weighted essentially non-oscillatory schemes. J. Comput. Phys. 0021-9991.115 (1), 200–212. http://dx.doi.org/10.1006/jcph.1994.1187. http://www.sciencedirect.com/science/article/pii/S0021999184711879.

Mock, M.S., 1980. Systems of conservation laws of mixed type. J. Differ. Equ. 37 (1), 70–88.

Nakagawa, T., 1987. Vortex shedding behind a square cylinder in transonic flows. J. Fluid Mech. 178, 303–323.

Nordström, J., Gong, J., van der Weide, E., Svärd, M., 2009. A stable and conservative high order multi-block method for the compressible Navier-Stokes equations. J. Comput. Phys. 228, 9020–9035.

Parsani, M., Carpenter, M.H., Nielsen, E.J., 2014. Entropy stable wall boundary conditions for the compressible Navier-Stokes equations. Tech. Rep. NASA TM 218282.

Parsani, M., Carpenter, M.H., Nielsen, E.J., 2015. Entropy stable discontinuous interfaces coupling for the three-dimensional compressible Navier-Stokes equations. J. Comput. Phys. 290, 132–138.

Parsani, M., Carpenter, M.H., Nielsen, E.J., 2015. Entropy stable wall boundary conditions for the three-dimensional compressible Navier-Stokes equations. J. Comput. Phys. 292, 88–113.

Shu, C.W., Osher, S., 1988. Efficient implementation of essentially non-oscillatory shock-capturing schemes. J. Comput. Phys. 0021-9991.77 (2), 439–471. http://dx.doi.org/10.1016/0021-9991(88)90177-5. http://www.sciencedirect.com/science/article/pii/0021999188901775.

Strand, B., 1994. Summation by parts for finite difference approximations for d/dx. J. Comput. Phys. 110, 47–67.

Svärd, M., Nordström, J., 2006. On the order of accuracy for difference approximations of initial-boundary value problems. J. Comput. Phys. 218, 333–352.

Svärd, M., Özcan, H., 2014. Entropy-stable schemes for the Euler equations with far-field and wall boundary conditions. J. Sci. Comput. 58 (1), 61–89.

Svärd, M., Carpenter, M.H., Nordström, J., 2007. A stable high-order finite difference scheme for the compressible Navier-Stokes equations, far-field boundary conditions. J. Comput. Phys. 225, 1020–1038.

Tadmor, E., 1987. The numerical viscosity of entropy stable schemes for systems of conservation laws. I. Math. Comput. 49, 91–103.

Tadmor, E., 2003. Entropy stability theory for difference approximations of nonlinear conservation laws and related time-dependent problems. Acta Numer 12, 451–512.

Thomas, P.D., Lombard, C.K., 1979. Geometric conservation law and its application to flow computations on moving grids. AIAA J. 17, 1030–1037.

Visbal, M.R., Gaitonde, D.V., 2002. On the use of higher-order finite-difference schemes on curvilinear and deforming meshes. J. Comput. Phys. 181, 155–185.

Wheeler, M., 1978. An elliptic collocation-finite element method with interior penalties. SIAM J. Numer. Anal. 15, 152–161.

Yamaleev, N.K., Carpenter, M.H., 2009. A systematic methodology for constructing high-order energy stable WENO schemes. J. Comput. Phys. 228, 4248–4272.

Yamaleev, N.K., Carpenter, M.H., 2009. Third-order energy stable WENO scheme. J. Comput. Phys. 228, 3025–3047.

Chapter 20

Central Schemes: A Powerful Black-Box Solver for Nonlinear Hyperbolic PDEs

A. Kurganov
Tulane University, New Orleans, LA, United States

Chapter Outline

ABSTRACT

We review a class of Godunov-type finite-volume methods for hyperbolic systems of conservation and balance laws—nonoscillatory central schemes. These schemes date back to 1950s, when the first-order Lax–Friedrichs scheme was introduced. The central Lax–Friedrichs scheme can be viewed as a simple alternative to the upwind Godunov scheme, which was also introduced in the 1950s. The main idea in the construction of both central and upwind first-order schemes is the same: use a global piecewise constant approximation of the solution at a certain time level and evolve it in time to the next time level exactly. The exact evolution is performed using the integral form of the studied system of PDEs. The difference is in one small detail—which is in fact not small at all—how to select the space–time control volume for the time evolution. The key idea in the construction of central schemes is to choose these control volume in such a way that no (localized) Riemann problems need to be solved at the evolution step. This makes central schemes particularly simple and universal numerical tool for general hyperbolic systems. On the other hand, central schemes are based on averaging the nonlinear waves rather than resolving them and thus they have larger numerical

Handbook of Numerical Analysis, Vol. 17. http://dx.doi.org/10.1016/bs.hna.2016.09.008

dissipation than their upwind counterparts. In order to increase the resolution achieved by central schemes, one has to increase their order. We describe how to design high-order nonoscillatory central schemes and also discuss how to further decrease their numerical dissipation without risking oscillations. The latter is achieved by utilizing some upwinding information (local speeds of propagation) within the framework of the Riemann-problem-solver-free central schemes and modifying the set of control volumes used for the time evolution. This leads to another type of central schemes—central-upwind schemes, whose derivation is reviewed in this work.

AMS Classification Codes: 65M08, 76M12, 35L65

Keywords: Hyperbolic systems of conservation and balance laws, Godunov-type finite-volume methods, Staggered central schemes, Central-upwind schemes, Piecewise polynomial reconstruction

1 A VERY BRIEF THEORETICAL BACKGROUND

We consider one-dimensional (1D) hyperbolic systems of conservation laws,

$$\boldsymbol{q}_t + \boldsymbol{f}(\boldsymbol{q})_x = \boldsymbol{0}, \tag{1}$$

subject to the prescribed initial data

$$\boldsymbol{q}(x,0) = \boldsymbol{q}_0(x). \tag{2}$$

Here, x is a space variable, t is time, $\boldsymbol{q} = \boldsymbol{q}(x, t)$ is a vector of unknown quantities in $\mathbb{R}^N$, $\boldsymbol{f}(\boldsymbol{q})$ is the flux vector.

It is well known that solutions of the initial value problems (IVPs) (1), (2) do not necessarily preserve their initial smoothness. Moreover, even when the initial data are infinitely smooth, solutions of these IVPs may break down and develop such nonsmooth structures as shock waves, contact discontinuities, rarefaction waves and even singular δ-shocks. In such a generic case, the nonsmooth solutions are nonclassical (weak) and they are to be understood in the sense of distributions. Namely, we say that $\boldsymbol{q}$ is a weak solution of the IVP (1), (2), if it satisfies the weak formulation of (1), (2),

$$\int_0^\infty \int_{\mathbb{R}} [\boldsymbol{q}(x,t)\varphi_x(x,t) + \boldsymbol{f}(\boldsymbol{q}(x,t))\varphi_t(x,t)]dx\,dt + \int_{\mathbb{R}} \boldsymbol{q}(x,0)\varphi(x,0)\,dx = 0,$$

for any smooth and compactly supported test function φ with supp $\varphi \subset \mathbb{R} \times [0, \infty)$. Weak solutions, however, are not unique and to single out the unique physically relevant solution, one needs to require the weak solution either to be the limit of the vanishing viscosity approximation of the studied hyperbolic system or satisfy a certain additional criterion such as an entropy condition. For this and other analytical results on nonlinear hyperbolic partial differential equations

(PDEs), we refer the reader, e.g., to Dafermos (2010), LeFloch (2002), Li et al. (1998), Serre (1999), Smoller (1994) and Zheng (2001).

2 FINITE-VOLUME FRAMEWORK

Finite-volume methods are based on the integral form of the system (1), which we integrate over a space–time control volume $[a, b] \times [c, d]$ to obtain the following system of integral equations:

$$\int_a^b \boldsymbol{q}(x,d)\,dx = \int_a^b \boldsymbol{q}(x,c)\,dx - \int_c^d [\boldsymbol{f}(\boldsymbol{q}(b,t)) - \boldsymbol{f}(\boldsymbol{q}(a,t))]dt. \tag{3}$$

If these equations are satisfied for any $a < b$ and $0 \le c \le d$, then the systems (1) and (3) are equivalent for piecewise smooth weak solutions. Therefore, Eq. (3) can be considered as a definition of a weak piecewise smooth solution, and we would like to point out that only piecewise smooth solutions—not general weak solutions—can be computed numerically.

In order to design a numerical method based on the integral equations (3), we will first introduce small scales in both space (Δx) and time (Δt) and take the space–time control volume to be $\left[x - \frac{\Delta x}{2}, x + \frac{\Delta x}{2}\right] \times [t, t + \Delta t]$, for which equation (3) after the division by Δx reads as

$$\bar{\boldsymbol{q}}(x, t + \Delta t) = \bar{\boldsymbol{q}}(x,t) - \lambda\left[\hat{\boldsymbol{f}}\left(x + \frac{\Delta x}{2}, t\right) - \hat{\boldsymbol{f}}\left(x - \frac{\Delta x}{2}, t\right)\right], \tag{4}$$

where $\lambda := \frac{\Delta t}{\Delta x}$,

$$\bar{\boldsymbol{q}}(x,t) := \frac{1}{\Delta x}\int_{x-\frac{\Delta x}{2}}^{x+\frac{\Delta x}{2}} \boldsymbol{q}(\xi,t)\,d\xi \tag{5}$$

is a so-called sliding average of $\boldsymbol{q}$, and

$$\hat{\boldsymbol{f}}(x,t) := \frac{1}{\Delta t}\int_t^{t+\Delta t} \boldsymbol{f}(\boldsymbol{q}(x,\tau))\,d\tau \tag{6}$$

is the averaged flux across x during the time interval $[t, t + \Delta t]$.

Formula (4) serves as a starting point of finite-volume evolution, which is performed in the following three steps. First, the computational domain is split into nonoverlapping intervals (in fact, one may also consider overlapping intervals as it was done in, e.g., Liu (2005), Liu et al. (2007), Liu et al. (2009)

and Yang et al. (2015)). Then, if a (global in space) approximate solution is available at time level t,. one may use the definition of a sliding average (5) to compute $\bar{q}(x,t)$ at the centers of the selected intervals. Finally, one should evaluate the averaged fluxes (6) to obtain the set of the sliding averages $\bar{q}(x,t+\Delta t)$ at the same points in space. Finite-volume evolution via an arbitrary set of space–time control volumes is schematically shown in the (x, t)-plane in Fig. 1.

Since the data obtained at the end of each finite-volume evolution step will consist of the sliding averages computed at a certain set of points, construction of a finite-volume method should begin with selection of grid intervals, which we denote by $C_j := (x_{j-\frac{1}{2}}, x_{j+\frac{1}{2}})$. For the sake of simplicity, let us assume that the mesh is uniform, that is, $x_{j+\frac{1}{2}} - x_{j-\frac{1}{2}} = \Delta x$ for all j and the cell centers are $x_j = x_{j-\frac{1}{2}} + \Delta x/2$. We now assume that the sliding averages over the grid cells are available at time level t^n. These sliding averages are called the cell averages and denoted by

$$\bar{q}_j^n : = \frac{1}{\Delta x} \int_{C_j} q(x,t^n)\, dx.$$

Equipped with the given set of cell averages, one can easily obtain a global piecewise constant approximation of the solution:

$$\tilde{q}^n(x) : = \sum_j \bar{q}_j^n \chi_{C_j}, \tag{7}$$

where χ_{C_j} is the characteristic function of the interval C_j. We now can complete the finite-volume evolution step, but the way the averaged fluxes,

$$\hat{f}(x,t^n) : = \frac{1}{\Delta t^n} \int_{t^n}^{t^{n+1}} f(q(x,t))\, dt, \quad t^{n+1} : = t^n + \Delta t^n, \tag{8}$$

are computed depends on the way the space–time control volumes are selected. This is a crucial point in the construction of finite-volume methods.

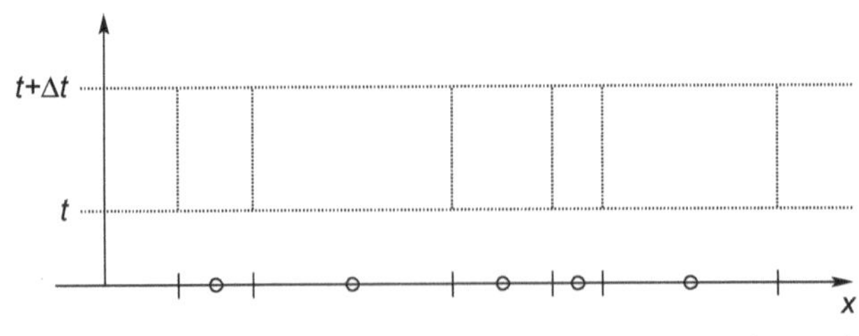

FIG. 1 Space–time control volumes. *Circles* "o" represent the centers of space intervals.

There are two major approaches, which lead to two different classes of schemes—upwind and central ones.

Remark 1. We note that the piecewise constant approximation (7) is only first-order accurate. In order to increase the order of accuracy, one needs to use higher-order interpolants instead of (7), as discussed in Section 5.

3 FIRST-ORDER UPWIND SCHEMES

The first finite-volume upwind scheme is the Godunov scheme proposed in Godunov (1959). The importance of this work is recognized in the fact that finite-volume methods are often called Godunov-type schemes.

In the upwind setting, the finite-volume evolution is carried out using the space–time control volumes $\left[x_{j-\frac{1}{2}}, x_{j+\frac{1}{2}}\right] \times [t^n, t^{n+1}]$ outlined in Fig. 2 (left). This means that, in fact, one simply needs to substitute the values $x = x_j$ and $t = t^n$ into (4) to obtain

$$\bar{\boldsymbol{q}}_j^{n+1} = \bar{\boldsymbol{q}}_j^n - \lambda\left[\hat{\boldsymbol{f}}\left(x_{j+\frac{1}{2}}, t^n\right) - \hat{\boldsymbol{f}}\left(x_{j-\frac{1}{2}}, t^n\right)\right], \tag{9}$$

and thus, in order to complete the construction of the scheme one has to evaluate the averaged fluxes

$$\hat{\boldsymbol{f}}(x_{j+\frac{1}{2}}, t^n) = \frac{1}{\Delta t^n}\int_{t^n}^{t^{n+1}} \boldsymbol{f}\left(\boldsymbol{q}\left(x_{j+\frac{1}{2}}, t\right)\right)\, dt. \tag{10}$$

After the integrals in (10) are (approximately) evaluated, one obtains the corresponding numerical fluxes denoted by $\boldsymbol{H}^n_{j+\frac{1}{2}} \approx \hat{\boldsymbol{f}}\left(x_{j+\frac{1}{2}}, t^n\right)$, and the resulting upwind scheme can be written as

$$\bar{\boldsymbol{q}}_j^{n+1} = \bar{\boldsymbol{q}}_j^n - \lambda^n\left[\boldsymbol{H}^n_{j+\frac{1}{2}} - \boldsymbol{H}^n_{j-\frac{1}{2}}\right], \quad \lambda^n := \frac{\Delta t^n}{\Delta x}. \tag{11}$$

In the classical Godunov scheme, the integrals in (10) are evaluated exactly. This is possible thanks to the one of the most important properties

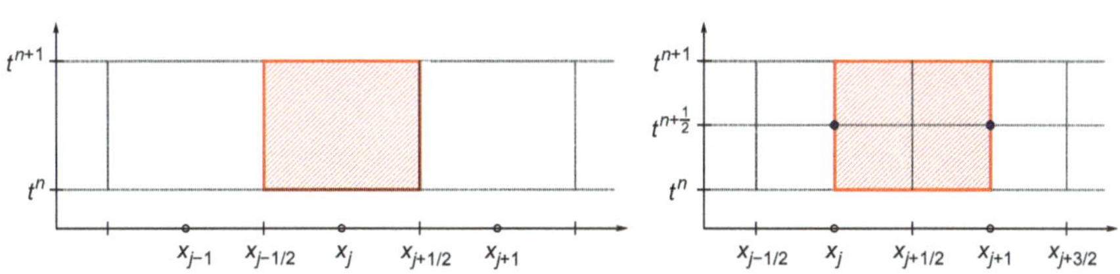

FIG. 2 Space–time control volumes: upwind (*left*) and central (*right*) settings.

of hyperbolic systems—finite speed of propagation, which is determined by the largest spectral radius of the Jacobian $A(\boldsymbol{q}) := \partial \boldsymbol{f}/\partial \boldsymbol{q}$ calculated over the entire computational domain at time $t = t^n$. Namely, the waves generated at the cell interfaces $x = x_{j+\frac{1}{2}}$ at time $t = t^n$, at which the approximate solution is a piecewise constant function $\boldsymbol{q}(x, t^n) = \hat{\boldsymbol{q}}^{\,n}(x)$, will not propagate faster than with the speed

$$a^n := \max_j \left\{\rho\left(A(\overline{\boldsymbol{q}}_j^n)\right)\right\}, \tag{12}$$

where $\rho(A)$ is a spectral radius of the matrix A. Therefore, if the time step is restricted by

$$\Delta t^n \le \frac{\Delta x}{a^n}, \tag{13}$$

the solution $\boldsymbol{q}\left(x_{j+\frac{1}{2}}, t\right)$ at the time interval $[t^n, t^{n+1}]$ needed to evaluate the time integral in (10) will not be affected by the (nonlinear) waves generated at other cell interfaces. Thus, in order to compute the required values of $\boldsymbol{q}\left(x_{j+\frac{1}{2}}, t\right)$, one has to solve the following Riemann problem: the system (1) subject to the initial data

$$\boldsymbol{q}(x, t^n) = \begin{cases} \overline{\boldsymbol{q}}_j^{\,n}, & \text{if} \quad x < x_{j+\frac{1}{2}}, \\ \overline{\boldsymbol{q}}_{j+1}^{\,n}, & \text{if} \quad x > x_{j+\frac{1}{2}}, \end{cases} . \tag{14}$$

prescribed at time $t = t^n$. It is well known (see, e.g., Dafermos, 2010; LeFloch, 2002; Li et al., 1998; Serre, 1999; Smoller, 1994; Zheng, 2001) that solutions of Riemann problems for hyperbolic systems of conservation laws are self-similar and therefore the corresponding wave propagation can be schematically shown using the straight lines originating at the points $\left(x_{j+\frac{1}{2}}, t^n\right)$ in the space–time plane, see Fig. 3 (left).

Remark 2. Note that in the literature, Riemann initial data are usually prescribed at time $t = 0$ and the breaking point of the initial data is typically $x = 0$. However,

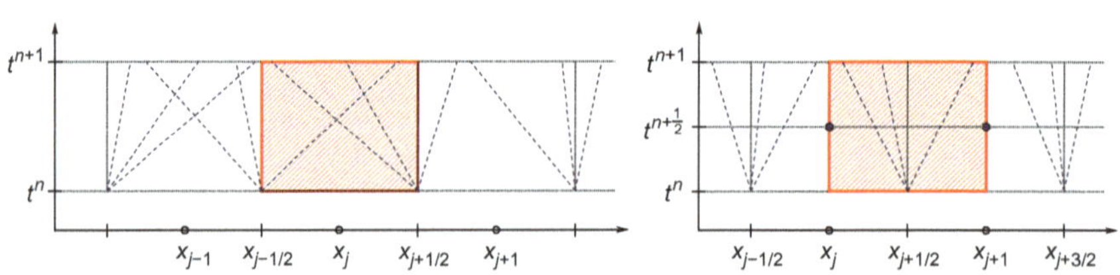

FIG. 3 Wave propagation in upwind (*left*) and central (*right*) settings. *Dashed lines* represent Riemann fans generated at each cell interface at time $t = t^n$.

due to the translation invariance of the hyperbolic systems (1), one can shift the data by $x_{j+\frac{1}{2}}$ and t^n without affecting any solution properties.

Since the self-similar solution of the Riemann problem (1), (14) is $\boldsymbol{q}(x,t)=\mathcal{R}^n_{j+\frac{1}{2}}(\xi)$, where the self-similar coordinate is $\xi=\left(x-x_{j+\frac{1}{2}}\right)/(t-t^n)$ and $t>t^n$, the solution $\boldsymbol{q}\left(x_{j+\frac{1}{2}},t\right)=\mathcal{R}^n_{j+\frac{1}{2}}(0)$ for all $t>t^n$ and therefore the integrals in (10) are immediately obtained. This results in the numerical flux $\boldsymbol{H}^n_{j+\frac{1}{2}}=\boldsymbol{f}\left(\mathcal{R}^n_{j+\frac{1}{2}}(0)\right)$ and the classical Godunov scheme can be written as

$$\bar{\boldsymbol{q}}_j^{n+1}=\bar{\boldsymbol{q}}_j^{n}-\lambda^n\left[\boldsymbol{f}\left(\mathcal{R}^n_{j+\frac{1}{2}}(0)\right)-\boldsymbol{f}\left(\mathcal{R}^n_{j-\frac{1}{2}}(0)\right)\right]. \tag{15}$$

Unfortunately, the Godunov scheme (15) has a serious drawback: it relies on the exact solution of the Riemann problem, which may be computationally expensive and very hard (or even impossible) to obtain analytically. The major difficulty in solving Riemann problems exactly is a very complicated structure of their solutions. Indeed, for the $N\times N$ system (1) the solution of the Riemann problem, in general, consists of N waves, which are not easy to compute analytically.

Since 1960s many approximate Riemann problem solvers have been designed as alternatives to the exact one used in (15). One of the possible strategies in constructing approximate solvers is to replace the exact wave structure with the approximate one. For instance, taking into account the fastest waves only, one obtains the Rusanov scheme (see Rusanov, 1961), which can be written in the form (11) with the following numerical flux:

$$\boldsymbol{H}^n_{j+\frac{1}{2}}=\frac{1}{2}\left[\boldsymbol{f}(\bar{\boldsymbol{q}}^n_{j+1})+\boldsymbol{f}(\bar{\boldsymbol{q}}_j^{n})\right]-\frac{a^n_{j+\frac{1}{2}}}{2}\left[\bar{\boldsymbol{q}}^n_{j+1}-\bar{\boldsymbol{q}}_j^{n}\right], \tag{16}$$

where $a^n_{j+\frac{1}{2}}$ are the local propagation speeds, computed using the spectral radius of the Jacobian as follows:

$$a^n_{j+\frac{1}{2}}=\max\left\{\rho\left(A(\bar{\boldsymbol{q}}^n_{j+1})\right),\rho\left(A(\bar{\boldsymbol{q}}_j^{n})\right)\right\}. \tag{17}$$

Considering both fastest left- and right-going waves results in a more accurate approximate Riemann problem solver, which leads to the Harten–Lax–van Leer (HLL) scheme (see Harten et al., 1983), which can also be put into the same form (11), but with a different numerical flux:

$$\boldsymbol{H}^n_{j+\frac{1}{2}}=\frac{a^{n+}_{j+\frac{1}{2}}\boldsymbol{f}(\bar{\boldsymbol{q}}_j^{n})-a^{n-}_{j+\frac{1}{2}}\boldsymbol{f}(\bar{\boldsymbol{q}}^n_{j+1})}{a^{n+}_{j+\frac{1}{2}}-a^{n-}_{j+\frac{1}{2}}}+\frac{a^{n+}_{j+\frac{1}{2}}a^{n-}_{j+\frac{1}{2}}}{a^{n+}_{j+\frac{1}{2}}-a^{n-}_{j+\frac{1}{2}}}\left[\bar{\boldsymbol{q}}^n_{j+1}-\bar{\boldsymbol{q}}_j^{n}\right], \tag{18}$$

where $a^{n\pm}_{j+\frac{1}{2}}$ are the one-sided local propagation speeds, computed using the largest and smallest eigenvalues of the Jacobian as follows:

$$a^{n+}_{j+\frac{1}{2}}=\max\left\{\lambda_N\left(A(\bar{\boldsymbol{q}}^n_{j+1})\right),\lambda_N\left(A(\bar{\boldsymbol{q}}^n_j)\right),0\right\},\quad a^{n-}_{j+\frac{1}{2}}=\min\left\{\lambda_1\left(A(\bar{\boldsymbol{q}}^n_{j+1})\right),\lambda_1\left(A(\bar{\boldsymbol{q}}^n_j)\right),0\right\}. \tag{19}$$

Here, $\lambda_1(A)\leq\cdots\leq\lambda_N(A)$ denote the ordered set of the eigenvalues of the matrix A. The HLL scheme was further improved by taking into account of the presence of slower (linear) contact waves in Einfeld (1988) and Toro et al. (1994), where modified HLL schemes were developed.

Remark 3. We note that both Rusanov and HLL schemes are, in fact, central schemes; see Remark 13.

An alternative approach for designing approximate Riemann problem solvers was proposed by Roe in 1981. The idea is to locally replace a complicated nonlinear hyperbolic system with a linear one, namely, to evaluate the numerical flux at $x=x_{j+\frac{1}{2}}$ by solving a linearized system with constant coefficients

$$\boldsymbol{q}_t+A^n_{j+\frac{1}{2}}\boldsymbol{q}_x=\boldsymbol{0}, \tag{20}$$

subject to the same initial data (14). Here, $A^n_{j+\frac{1}{2}}$ is a proper linearization of the Jacobian $A(\boldsymbol{q})$. The classical Roe scheme is obtained by constructing the Roe matrix $\hat{\boldsymbol{A}}^n_{j+\frac{1}{2}}$ satisfying the condition $\boldsymbol{f}(\bar{\boldsymbol{q}}^n_{j+1})-\boldsymbol{f}(\bar{\boldsymbol{q}}^n_j)=\hat{\boldsymbol{A}}^n_{j+\frac{1}{2}}(\bar{\boldsymbol{q}}^n_{j+1}-\bar{\boldsymbol{q}}^n_j)$, which guarantees that if the states $\bar{\boldsymbol{q}}^n_j$ and $\bar{\boldsymbol{q}}^n_{j+1}$ correspond to an isolated shock wave, its speed will coincide with the speed of the linearized wave, and taking $A^n_{j+\frac{1}{2}}=\hat{\boldsymbol{A}}^n_{j+\frac{1}{2}}$ in (20). For some systems, however, it is not easy to construct the Roe matrix. A much simpler approach was advocated in Buffard et al. (2000), Gallouët et al. (2002) and Masella et al. (1999), where the so-called VFRoe scheme was developed by simply taking $A^n_{j+\frac{1}{2}}=A(\boldsymbol{q}_{j+\frac{1}{2}})$, where $\boldsymbol{q}_{j+\frac{1}{2}}:=\frac{1}{2}(\bar{\boldsymbol{q}}^n_{j+1}+\bar{\boldsymbol{q}}^n_j)$.

For both the Roe and VFRoe schemes, the solution of the IVP (20) and (14) is self-similar and easy to obtain by diagonalizing the system (20) using the characteristic variables. We denote this solution by $\boldsymbol{q}(x,t)=\mathcal{L}^n_{j+\frac{1}{2}}(\xi)$ so that the numerical flux in (11) reduces to $\boldsymbol{H}^n_{j+\frac{1}{2}}=\boldsymbol{f}\left(\mathcal{L}^n_{j+\frac{1}{2}}(0)\right)$, and the resulting scheme reads as

$$\bar{\boldsymbol{q}}^{n+1}_j=\bar{\boldsymbol{q}}^n_j-\lambda^n\left[\boldsymbol{f}\left(\mathcal{L}^n_{j+\frac{1}{2}}(0)\right)-\boldsymbol{f}\left(\mathcal{L}^n_{j-\frac{1}{2}}(0)\right)\right]. \tag{21}$$

For a detailed description and derivation of the aforementioned and other popular approximate Riemann problem solvers as well as for their comparative study, we refer the reader to Toro (2009); see also Godlewski and Raviart (1996), Kröner (1997) and LeVeque (2002).

4 FIRST-ORDER CENTRAL SCHEMES

Central schemes can be considered as a simple Riemann-problem-solver-free alternative to upwind schemes. Unfortunately, a straightforward central-difference approximation of the space derivatives in (1) leads to the finite-difference scheme

$$\boldsymbol{q}_j^{n+1} = \boldsymbol{q}_j^n - \frac{\lambda^n}{2}\left[\boldsymbol{f}(\boldsymbol{q}_{j+1}^n) - \boldsymbol{f}(\boldsymbol{q}_{j-1}^n)\right], \tag{22}$$

which is known to be unconditionally unstable and thus prone to uncontrolled oscillations. In 1954, Lax (1954) and Friedrichs (1954) proposed a stabilized version of the scheme (22)—the celebrated Lax–Friedrichs scheme:

$$\boldsymbol{q}_j^{n+1} = \frac{\boldsymbol{q}_{j+1}^n + \boldsymbol{q}_{j-1}^n}{2} - \frac{\lambda^n}{2}\left[\boldsymbol{f}(\boldsymbol{q}_{j+1}^n) - \boldsymbol{f}(\boldsymbol{q}_{j-1}^n)\right]. \tag{23}$$

Note that in both (22) and (23), the evolved quantities are the point values of $\boldsymbol{q}$ at the grid nodes rather than the cell averages. Even though these schemes can be artificially put into the finite-volume form (11) with appropriate fluxes, they are not Godunov-type schemes in the sense that they cannot be rigorously derived using the finite-volume framework described in Section 2.

The first-order Godunov-type central scheme is obtained using exactly the same finite-volume evolution equations (4)–(6), which were used to design upwind schemes in Section 3, but sampled at a different set of points: $\left(x_{j+\frac{1}{2}}, t^n\right)$ instead of (x_j, t^n), as illustrated in Fig. 2. Compared to the upwind setting, the space–time control volumes $[x_j, x_{j+1}] \times [t^n, t^{n+1}]$ used in the construction of central schemes, is shifted by $\Delta x/2$, see Fig. 2 (right).

We stress that while the data at time level $t = t^n$ is given over the original grid $\left[x_{j-\frac{1}{2}}, x_{j+\frac{1}{2}}\right]$, the new computed solution will be realized over the staggered grid. Indeed, Eqs. (4)–(6) will now lead to

$$\bar{\boldsymbol{q}}_{j+\frac{1}{2}}^{n+1} = \frac{1}{\Delta x}\int\limits_{x_j}^{x_{j+1}} \widetilde{\boldsymbol{q}}^n(x)\,dx - \frac{\lambda^n}{\Delta t^n}\int\limits_{t^n}^{t^{n+1}} \left[\boldsymbol{f}(\boldsymbol{q}(x_{j+1},t)) - \boldsymbol{f}(\boldsymbol{q}(x_j,t))\right]dt. \tag{24}$$

The space integral on the right-hand side (RHS) of (24) is just an integral of a piecewise constant function (7) and thus can be exactly evaluated in a straightforward manner. The time integrals on the RHS of (24) are not easy to compute unless the timestep restriction is tightened and Δt^n is taken to be twice smaller compared to (13), namely,

$$\Delta t^n \leq \frac{\Delta x}{2a^n}. \tag{25}$$

As one can see in Fig. 3 (right), no waves generated at the cell interfaces can reach the vertical segments at $x = x_j$ for $t \in [t^n, t^{n+1}]$, and thus the solution of the system (1) subject to the piecewise constant initial data $\boldsymbol{q}(x,t^n)=\widetilde{\boldsymbol{q}}^{\,n}(x)$ remains constant there. Therefore, after evaluating all of the integrals in (24), we obtain the first-order staggered central scheme—staggered Lax–Friedrichs scheme (see Nessyahu and Tadmor, 1990):

$$\bar{\boldsymbol{q}}^{\,n+1}_{j+\frac{1}{2}}=\frac{\bar{\boldsymbol{q}}^{\,n}_{j+1}+\bar{\boldsymbol{q}}^{\,n}_{j}}{2}-\lambda^n\left[\boldsymbol{f}(\bar{\boldsymbol{q}}^{\,n}_{j+1})-\boldsymbol{f}(\bar{\boldsymbol{q}}^{\,n}_{j})\right]. \tag{26}$$

Central schemes (both Lax–Friedrichs and staggered Las-Friedrichs ones) are extremely simple and universal tool for solving hyperbolic systems of conservation laws. They, however, have a substantial disadvantage compared to the upwind schemes—their numerical viscosity is much larger (see, e.g., Tadmor, 1984a,b), which leads to excessive smearing of discontinuous and other "rough" parts of the computed solution. In fact, first-order upwind schemes are also quite diffusive and cannot provide high resolution of non-smooth parts of the solution (especially of linear contact waves) unless very small Δx and Δt^n are used and the latter may be computationally unaffordable. The way to enhance the resolution is to increase the order of the scheme.

5 HIGH-ORDER FINITE-VOLUME METHODS

We first note that both the Godunov (15) and staggered Lax–Friedrichs (26) schemes are based on the exact finite-volume evolution of the computed solution. Therefore, the loss of accuracy occurs at the approximation step, when the global first-order accurate piecewise constant interpolant (7) is reconstructed from the set of computed cell averages. In order to increase the (formal) order of accuracy of the approximation and thus of the entire scheme, one should replace the piecewise constant approximation with a higher-order one.

5.1 Second-Order Upwind Schemes

The first second-order Godunov-type scheme was introduced by van Leer in 1979, where the so-called MUSCL approach was proposed. It is based on a piecewise linear reconstruction

$$\widetilde{\boldsymbol{q}}^{\,n}(x):=\sum_j\left[\bar{\boldsymbol{q}}^{\,n}_j+(\boldsymbol{q}_x)^n_j(x-x_j)\right]\chi_{C_j}, \tag{27}$$

which will be second-order accurate provided the slopes $(\boldsymbol{q}_x)^n_j\approx\boldsymbol{q}_x(x_j,t^n)$ within at least first order of accuracy. To keep the resulting scheme from being too oscillatory, one has to use a nonlinear limiter in the computation of $(\boldsymbol{q}_x)^n_j$ to ensure that no large over- or undershoots of size $\mathcal{O}(1)$ are created at the cell interfaces. A library of such limiters is available; we refer the reader, e.g., to Godlewski and Raviart (1996), Kröner (1997), LeVeque

(2002), Lie and Noelle (2003), Nessyahu and Tadmor (1990), Sweby (1984) and van Leer (1979). They may be applied to the vector quantity $\boldsymbol{q}$ either in a componentwise manner, that is, to each component of $\boldsymbol{q}$ directly, or using the local characteristic decompositions as it was done, for instance, in Qiu and Shu (2002). The latter approach results in the schemes that are in general less oscillatory, but more computationally expensive.

When the piecewise constant interpolant is replaced with a piecewise linear one, the finite-volume evolution procedure must be modified. It is not so easy to do in the framework of upwind schemes, since now instead of solving the Riemann problems at each cell interface, one has to solve the generalized Riemann problem: the system (1) subject to the initial data

$$\boldsymbol{q}(x,t^n)=\begin{cases}\bar{\boldsymbol{q}}_j^n+(\boldsymbol{q}_x)_j^n(x-x_j), & \text{if} \quad x<x_{j+\frac{1}{2}},\\ \bar{\boldsymbol{q}}_{j+1}^n+(\boldsymbol{q}_x)_{j+1}^n(x-x_{j+1}), & \text{if} \quad x>x_{j+\frac{1}{2}},\end{cases} \tag{28}$$

prescribed at time $t=t^n$. For certain hyperbolic systems of conservation laws the exact solution of the generalized Riemann problem can be constructed; see Ben-Artzi and Falcovitz (2003). However, these solutions are very complicated and quite computationally expensive. As an alternative, one can design an approximate generalized Riemann problem solver, see, e.g., Godlewski and Raviart (1996), Kröner (1997), LeVeque (2002) and Toro (2009).

5.2 Second-Order Nessyahu–Tadmor Scheme

Another alternative is to switch to the central framework, in which the fact that the data are now piecewise linear does not lead to any substantial increase in the level of complexity since the solution still remains smooth at the cell centers for $t\in[t^n,t^{n+1}]$ provided the timestep restriction (25) with

$$a^n:=\max_x\left\{\rho\left(A(\widetilde{\boldsymbol{q}}^{\,n}(x))\right)\right\}$$

is satisfied. After noticing this, the second-order staggered central scheme—the Nessyahu–Tadmor scheme (Nessyahu and Tadmor, 1990)—is now designed as follows.

First, we evaluate the space integral on the RHS of (24) exactly, which is straightforward since we simply need to integrate two linear pieces:

$$\begin{aligned}\frac{1}{\Delta x}\int\limits_{x_j}^{x_{j+1}}\widetilde{\boldsymbol{q}}^{\,n}(x)\,dx&=\frac{1}{\Delta x}\left\{\int\limits_{x_j}^{x_{j+\frac{1}{2}}}\left[\bar{\boldsymbol{q}}_j^n+(\boldsymbol{q}_x)_j^n(x-x_j)\right]dx+\int\limits_{x_{j+\frac{1}{2}}}^{x+1}\left[\bar{\boldsymbol{q}}_{j+1}^n+(\boldsymbol{q}_x)_{j+1}^n(x-x_{j+1})\right]dx\right\}\\&=\frac{\bar{\boldsymbol{q}}_{j+1}^n+\bar{\boldsymbol{q}}_j^n}{2}+\frac{\Delta x}{8}\left[(\boldsymbol{q}_x)_j^n-(\boldsymbol{q}_x)_{j+1}^n\right].\end{aligned} \tag{29}$$

We then approximate the time integrals of the smooth functions of t on the RHS of (24) using the midpoint rule so that

$$\frac{1}{\Delta t^n}\int_{t^n}^{t^{n+1}} \boldsymbol{f}(\boldsymbol{q}(x_j,t))\,dt \approx \boldsymbol{f}\left(\boldsymbol{q}_j^{n+\frac{1}{2}}\right), \tag{30}$$

where the values of $\boldsymbol{q}$ at the points $(x_j,t^{n+\frac{1}{2}})$ marked by the filled circles in Figs. 2 and 3 on the right can be obtained using the Taylor expansion in t (which is valid since the solution is smooth there) in the following way:

$$\boldsymbol{q}_j^{n+\frac{1}{2}} = \tilde{\boldsymbol{q}}^n(x_j) + \frac{\Delta t^n}{2}\boldsymbol{q}_t(x_j,t^n) = \bar{\boldsymbol{q}}_j^n - \frac{\Delta t^n}{2}\boldsymbol{f}(\bar{\boldsymbol{q}}_j^n)_x. \tag{31}$$

Here, the numerical derivatives $\boldsymbol{f}(\bar{\boldsymbol{q}}_j^n)_x$ can be computed either using $\boldsymbol{f}(\bar{\boldsymbol{q}}_j^n)_x = A(\bar{\boldsymbol{q}}_j^n)(\boldsymbol{q}_x)_j^n$, or by applying the same limiter used in computing the slopes in (27) to the set of the flux values $\boldsymbol{f}(\boldsymbol{q}_j^n)$. Finally, we substitute (29)–(31) into (24) and obtain the Nessyahu–Tadmor scheme:

$$\bar{\boldsymbol{q}}_{j+\frac{1}{2}}^{n+1} = \frac{\bar{\boldsymbol{q}}_{j+1}^n + \bar{\boldsymbol{q}}_j^n}{2} + \frac{\Delta x}{8}\left[(\boldsymbol{q}_x)_j^n - (\boldsymbol{q}_x)_{j+1}^n\right] - \lambda^n\left[\boldsymbol{f}\left(\boldsymbol{q}_{j+1}^{n+\frac{1}{2}}\right) - \boldsymbol{f}\left(\boldsymbol{q}_j^{n+\frac{1}{2}}\right)\right]. \tag{32}$$

Remark 4. One can view the Nessyahu–Tadmor scheme as a predictor–corrector method, in which (31) is a first-order predictor and (32) is the second-order corrector.

Remark 5. Staggered central schemes have been extended to the case of multiple space dimensions both on Cartesian (Jiang and Tadmor, 1998) and unstructured (Arminjon et al., 1997) grids.

5.3 High-Order Schemes

In order to further increase the order of the finite-volume methods, one needs to further increase the accuracy of the piecewise polynomial reconstruction. Third-order schemes can be constructed using piecewise parabolic interpolants $\tilde{\boldsymbol{q}}^n(x)$. It should be pointed out though that it is much harder to ensure nonoscillatory properties of higher than second order piecewise polynomials. For instance, several third-order piecewise parabolic reconstructions satisfying the number of extrema nonincreasing property were introduced in Kurganov and Petrova (2001), Liu and Osher (1996) and Liu and Tadmor (1998). An alternative approach of constructing high-order essentially nonoscillatory interpolant is based on the idea of differentiating in the direction of smoothness, which was realized in so-called ENO reconstructions; see, e.g., Abgrall (1994), Cockburn et al. (1998) and Harten et al. (1987). Other popular high-order reconstructions are based on the idea of taking a linear combinations of several polynomial pieces (each of which is obtained by differentiating in different directions) with the weights inversely proportional to the their smoothness measured in Sobolev spaces. This leads to a class of weighted ENO (WENO) schemes; see, e.g., Cockburn et al. (1998), Jiang and Shu (1996), Shi et al. (2002), Shu (2003) and Shu (2009). Unlike their counterparts, WENO reconstructions employ polynomials of lower degree and thus they are not uniformly high-order—only the data needed for evolving solutions in time (point values of the solution at the cell interfaces) are computed within the desired high accuracy. Therefore, when applied in the staggered central framework, WENO reconstructions have to be modified to accurately approximate the integrals $\int_{x_j}^{x_{j+\frac{1}{2}}} \widetilde{\boldsymbol{q}}^n(x)\,dx$ and $\int_{x_{j+\frac{1}{2}}}^{x_{j+1}} \widetilde{\boldsymbol{q}}^n(x)\,dx$. This was achieved in Bianco et al. (1999), Levy et al. (1999), Levy et al. (2000) and Levy et al. (2002), where a class of central WENO (CWENO) schemes was introduced.

Remark 6. For finite-volume methods of higher than second order, reconstruction procedures based on nonlinear limiters are typically computationally expensive. One can alternatively enforce stability by adding an artificial viscosity and not using any limiters. This idea was first proposed in 1950 in von Neumann and Richtmyer (1950) and since then it was notably adopted in many works including Caramana et al. (1998), Shchepetkin and McWilliams (1998), Wilkins (1980) and among others. One, however, has to be careful since adding artificial viscosity terms may cause either the discontinuities to be oversmeared or the oscillations not to be sufficiently damped. Highly accurate and robust artificial viscosity methods with the viscosity coefficients being proportional to either the weak local residual (Kurganov and Liu, 2012) or entropy production (Guermond and Pasquetti, 2008; Guermond et al., 2011) have been recently proposed.

6 CENTRAL-UPWIND SCHEMES

Even though the use of higher-order reconstructions significantly improves the resolution achieved by both upwind and staggered central schemes, central schemes may suffer from excessive numerical viscosity, which is of order $\mathcal{O}((\Delta x)^{2r}/\Delta t^n)$, where r is the formal order of the scheme. In order to illustrate this point, we rewrite the simplest staggered central scheme—the first-order Lax–Friedrichs scheme—in the following equivalent form:

$$\frac{\boldsymbol{q}^{n+1}_{j+\frac{1}{2}} - \boldsymbol{q}^n_{j+\frac{1}{2}}}{\Delta t^n} + \frac{\boldsymbol{f}(\boldsymbol{q}^n_{j+1}) - \boldsymbol{f}(\boldsymbol{q}^n_j)}{\Delta x} = \frac{(\Delta x)^2}{8\Delta t^n} \cdot \frac{\boldsymbol{q}^n_{j+1} - 2\boldsymbol{q}^n_{j+\frac{1}{2}} + \boldsymbol{q}^n_j}{(\Delta x/2)^2}. \tag{33}$$

Note that here we have replaced the cell averages of $\boldsymbol{q}$ in (26) with the corresponding point values (this can be done since for both first- and second-order schemes, these quantities are equal). The terms on the left-hand side

(LHS) of (33) clearly approximate the corresponding terms on the LHS of (1) and the term on the RHS of (33) represents the numerical viscosity with the viscosity coefficient being $\mathcal{O}((\Delta x)^2/\Delta t^n)$.

Therefore, numerical viscosity present in staggered central schemes is particularly large when sufficiently small timesteps are enforced, for instance, due to the presence of (degenerate) diffusion and/or source terms, or if the final computational time is very large as it may be the case when steady-state solutions are to be captured. One can reduce the numerical dissipation by modifying the central finite-volume evolution procedure. This leads to a new class of Godunov-type Riemann-problem-solver-free central schemes—central-upwind schemes. In the remaining part of this section, we show the derivation of the second-order central-upwind scheme along the lines of Kurganov and Lin (2007).

The key idea is to select space–time control volumes in the finite-volume evolution procedure (3) adaptively depending on the size of Riemann fans generated at each cell interface. More precisely, we assume that, as before, the computed solution at time $t=t^n$ is available and realized in terms of the cell averages $\left\{\bar{\boldsymbol{q}}_j^{\,n}\right\}$ over the grid $\left[x_{j-\frac{1}{2}},x_{j+\frac{1}{2}}\right]$. We first introduce the following notations:

$$\begin{aligned}
\boldsymbol{q}^{n-}_{j+\frac{1}{2}} &:= \lim_{x\to x_{j+\frac{1}{2}}-} \widetilde{\boldsymbol{q}}^{\,n}(x) = \bar{\boldsymbol{q}}_j^{\,n} + \frac{\Delta x}{2}(\boldsymbol{q}_x)_j^n,\\
\boldsymbol{q}^{n+}_{j+\frac{1}{2}} &:= \lim_{x\to x_{j+\frac{1}{2}}+} \widetilde{\boldsymbol{q}}^{\,n}(x) = \bar{\boldsymbol{q}}_{j+1}^{\,n} - \frac{\Delta x}{2}(\boldsymbol{q}_x)_{j+1}^n
\end{aligned} \tag{34}$$

for the reconstructed one-sided point values of $\boldsymbol{q}$ at the points $(x_{j+\frac{1}{2}},t^n)$, and

$$\begin{aligned}
a^{n+}_{j+\frac{1}{2}} &= \max\left\{\lambda_N\left(A\left(\boldsymbol{q}^{n+}_{j+\frac{1}{2}}\right)\right),\lambda_N\left(A\left(\boldsymbol{q}^{n-}_{j+\frac{1}{2}}\right)\right),0\right\},\\
a^{n-}_{j+\frac{1}{2}} &= \min\left\{\lambda_1\left(A\left(\boldsymbol{q}^{n+}_{j+\frac{1}{2}}\right)\right),\lambda_1\left(A\left(\boldsymbol{q}^{n-}_{j+\frac{1}{2}}\right)\right),0\right\}
\end{aligned} \tag{35}$$

for the one-sided local speeds of propagation, which reduces to (19) in the case of the first-order piecewise constant reconstruction.

Remark 7. In fact, the estimate of one-sided local speed (35) is only true in the case of a convex flux $\boldsymbol{f}(\boldsymbol{q})$. In the nonconvex case, a more careful estimate is required; see Kurganov et al. (2007).

Remark 8. It might be impossible to exactly evaluate the largest and smallest eigenvalues of the Jacobian required in (35). In this case, one may use an upper bound on λ_N and a lower bound on λ_1 as it was done in, e.g., Kurganov and Petrova (2009) and Liu et al. (2015).

We then take the control volumes $\left[x^n_{j+\frac{1}{2},\ell},x^n_{j+\frac{1}{2},r}\right]\times[t^n,t^{n+1}]$ that contain all of the waves generated at time $t=t^n$ at the corresponding cell interfaces

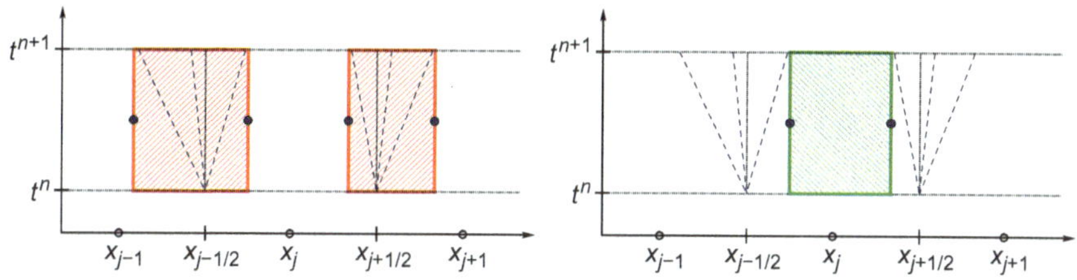

FIG. 4 Central-upwind control volumes over nonsmooth (*left*) and smooth (*right*) parts of the solution. *Dashed lines* represent Riemann fans generated at each cell interface at time $t=t^n$.

$x=x_{j+\frac{1}{2}}$; see Fig. 4 (left). The left and right boundaries of these control volumes, $x^n_{j+\frac{1}{2},\ell}:=x_j+a^{n-}_{j+\frac{1}{2}}\Delta t^n$ and $x^n_{j+\frac{1}{2},r}:=x_j+a^{n+}_{j+\frac{1}{2}}\Delta t^n$, respectively, are determined by the one-sided local speeds of propagation and therefore, the solution of the IVP (1) and (28) remains smooth at $x=x^n_{j+\frac{1}{2},\ell}$ and $x=x^n_{j+\frac{1}{2},r}$ for $t\in[t^n,t^{n+1}]$. Hence, the solution may be evolved to the time level $t=t^{n+1}$ exactly in the same manner as in the Nessyahu–Tadmor scheme, namely, we obtain the new cell averages, denoted by $\bar{\boldsymbol{q}}^{\,\text{int}}_{j+\frac{1}{2}}$, as follows. First, we use (3) to obtain

$$\bar{\boldsymbol{q}}^{\,\text{int}}_{j+\frac{1}{2}}=\frac{1}{x^n_{j+\frac{1}{2},r}-x^n_{j+\frac{1}{2},\ell}}\left\{\int\limits_{x^n_{j+\frac{1}{2},\ell}}^{x^n_{j+\frac{1}{2},r}}\widetilde{\boldsymbol{q}}^{\,n}(x)\,dx-\int\limits_{t^n}^{t^{n+1}}\left[\boldsymbol{f}\left(\boldsymbol{q}(x^n_{j+\frac{1}{2},r},t)\right)-\boldsymbol{f}\left(\boldsymbol{q}(x^n_{j+\frac{1}{2},\ell},t)\right)\right]dt\right\}. \tag{36}$$

Then, evaluating the first integral on the RHS of (36) exactly and using the midpoint rule for the flux integrals in (36), we arrive at

$$\begin{aligned}\bar{\boldsymbol{q}}^{\,\text{int}}_{j+\frac{1}{2}}=\frac{1}{a^{n+}_{j+\frac{1}{2}}-a^{n-}_{j+\frac{1}{2}}}&\left\{a^{n+}_{j+\frac{1}{2}}\boldsymbol{q}^n_{j+\frac{1}{2},r}-a^{n-}_{j+\frac{1}{2}}\boldsymbol{q}^n_{j+\frac{1}{2},\ell}+\frac{\Delta t^n}{2}\left[\left(a^{n-}_{j+\frac{1}{2}}\right)^2(\boldsymbol{q}_x)^n_j-\left(a^{n+}_{j+\frac{1}{2}}\right)^2(\boldsymbol{q}_x)^n_{j+1}\right]\right.\\&\left.-\left[\boldsymbol{f}\left(\boldsymbol{q}^{n+\frac{1}{2}}_{j+\frac{1}{2},r}\right)-\boldsymbol{f}\left(\boldsymbol{q}^{n+\frac{1}{2}}_{j+\frac{1}{2},\ell}\right)\right]\right\},\end{aligned} \tag{37}$$

where the point values of $\boldsymbol{q}$ at $\left(x^n_{j+\frac{1}{2},\ell},t^n\right)$ and $\left(x^n_{j+\frac{1}{2},r},t^n\right)$ are obtained from $\widetilde{\boldsymbol{q}}^{\,n}(x)$:

$$\boldsymbol{q}^n_{j+\frac{1}{2},\ell}=\bar{\boldsymbol{q}}^n_j+\left(\frac{\Delta x}{2}+a^{n-}_{j+\frac{1}{2}}\Delta t^n\right)(\boldsymbol{q}_x)^n_j,\quad \boldsymbol{q}^n_{j+\frac{1}{2},r}=\bar{\boldsymbol{q}}^n_{j+1}-\left(\frac{\Delta x}{2}-a^{n+}_{j+\frac{1}{2}}\Delta t^n\right)(\boldsymbol{q}_x)^n_{j+1} \tag{38}$$

and the point values of $\boldsymbol{q}$ at $\left(x^n_{j+\frac{1}{2},\ell},t^{n+\frac{1}{2}}\right)$ and $\left(x^n_{j+\frac{1}{2},r},t^{n+\frac{1}{2}}\right)$ are predicted using the corresponding Taylor expansions:

$$\boldsymbol{q}^{n+\frac{1}{2}}_{j+\frac{1}{2},\ell}=\boldsymbol{q}^n_{j+\frac{1}{2},\ell}-\frac{\Delta t^n}{2}\boldsymbol{f}(\bar{\boldsymbol{q}}^n_j)_x,\quad \boldsymbol{q}^{n+\frac{1}{2}}_{j+\frac{1}{2},r}=\boldsymbol{q}^n_{j+\frac{1}{2},r}-\frac{\Delta t^n}{2}\boldsymbol{f}(\bar{\boldsymbol{q}}^n_{j+1})_x. \tag{39}$$

As one can see in Fig. 4 (left), even though the control volumes $\left[x^n_{j+\frac{1}{2},\ell}, x^n_{j+\frac{1}{2},r}\right] \times [t^n, t^{n+1}]$ contain all of the Riemann fans, they do not cover the entire strip $\mathbb{R} \times [t^n, t^{n+1}]$ since there are gaps between these control volumes, $\left[x^n_{j-\frac{1}{2},r}, x^n_{j+\frac{1}{2},\ell}\right] \times [t^n, t^{n+1}]$, shown in Fig. 4 (right), where the solution is smooth. The solution there is evolved using the same integral form (3), which results in

$$\bar{\boldsymbol{q}}_j^{\,\mathrm{int}} = \bar{\boldsymbol{q}}_j^{\,n} + \frac{\Delta t^n}{2}\left(a^{n+}_{j+\frac{1}{2}} + a^{n-}_{j+\frac{1}{2}}\right)(\boldsymbol{q}_x)_j^n - \frac{\Delta t^n}{\Delta x - \left(a^{n+}_{j+\frac{1}{2}} - a^{n-}_{j+\frac{1}{2}}\right)\Delta t^n}\left[\boldsymbol{f}\left(\boldsymbol{q}^{n+\frac{1}{2}}_{j+\frac{1}{2},\ell}\right) - \boldsymbol{f}\left(\boldsymbol{q}^{n+\frac{1}{2}}_{j-\frac{1}{2},r}\right)\right], \tag{40}$$

where the predicted values of $\boldsymbol{q}$ are, as before, given by (39).

At this stage, the approximate solution at time level $t = t^{n+1}$ is realized in terms of its cell averages $\left\{\bar{\boldsymbol{q}}^{\,\mathrm{int}}_{j+\frac{1}{2}}\right\}$ and $\left\{\bar{\boldsymbol{q}}_j^{\,\mathrm{int}}\right\}$, distributed over the nonuniform mesh $\bigcup_j\left\{\left[x^n_{j-\frac{1}{2},r}, x^n_{j+\frac{1}{2},\ell}\right] \cup \left[x^n_{j+\frac{1}{2},\ell}, x^n_{j+\frac{1}{2},r}\right]\right\}$. This solution is quite accurate, but impractical since the number of cells is doubled in just one time step, and this is the reason why it was denoted by $\boldsymbol{q}^{\mathrm{int}}$, where "int" stands for the intermediate. In order to complete construction of the central-upwind scheme, we need to project these intermediate data back onto the original grid $\bigcup_j C_j$. To this end, we use the intermediate data to reconstruct a conservative, non-oscillatory, second-order piecewise linear interpolant

$$\begin{aligned}\widetilde{\boldsymbol{q}}^{\,\mathrm{int}}(x) := \sum_j \Bigg\{ &\left[\bar{\boldsymbol{q}}^{\,\mathrm{int}}_{j+\frac{1}{2}} + (\boldsymbol{q}_x)^{\mathrm{int}}_{j+\frac{1}{2}}\left(x - \frac{x^n_{j+\frac{1}{2},\ell} + x^n_{j+\frac{1}{2},r}}{2}\right)\right]\chi_{\left[x^n_{j+\frac{1}{2},\ell},\, x^n_{j+\frac{1}{2},r}\right]} \\ &+ \left[\bar{\boldsymbol{q}}_j^{\,\mathrm{int}} + (\boldsymbol{q}_x)_j^{\mathrm{int}}\left(x - \frac{x^n_{j-\frac{1}{2},r} + x^n_{j+\frac{1}{2},\ell}}{2}\right)\right]\chi_{\left[x^n_{j-\frac{1}{2},r},\, x^n_{j+\frac{1}{2},\ell}\right]}\Bigg\},\end{aligned} \tag{41}$$

and average it over the cells C_j to end up with

$$\begin{aligned}\bar{\boldsymbol{q}}_j^{\,n+1} = {} & \lambda^n a^{n+}_{j-\frac{1}{2}}\bar{\boldsymbol{q}}^{\,\mathrm{int}}_{j-\frac{1}{2}} + \left[1 + \lambda^n\left(a^{n-}_{j-\frac{1}{2}} - a^{n+}_{j+\frac{1}{2}}\right)\right]\bar{\boldsymbol{q}}_j^{\,\mathrm{int}} \\ & - \lambda^n a^{n-}_{j+\frac{1}{2}}\bar{\boldsymbol{q}}^{\,\mathrm{int}}_{j+\frac{1}{2}} + \frac{\lambda^n\Delta t^n}{2}\left[a^{n+}_{j+\frac{1}{2}}a^{n-}_{j+\frac{1}{2}}(\boldsymbol{q}_x)^{\mathrm{int}}_{j+\frac{1}{2}} - a^{n+}_{j-\frac{1}{2}}a^{n-}_{j-\frac{1}{2}}(\boldsymbol{q}_x)^{\mathrm{int}}_{j-\frac{1}{2}}\right].\end{aligned} \tag{42}$$

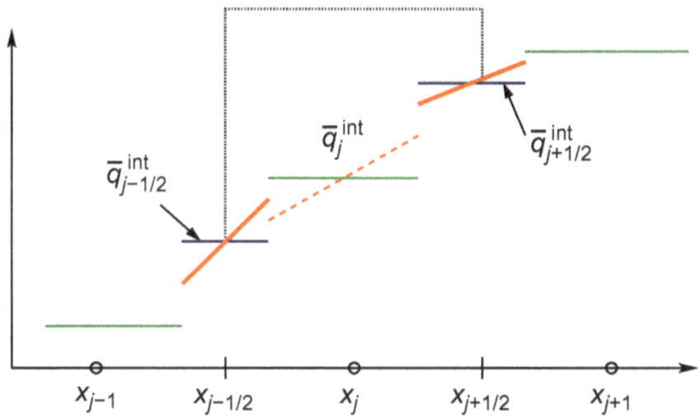

FIG. 5 Projection of the intermediate data onto the original grid.

In fact, we do not need to compute the slopes $(\boldsymbol{q}_x)_j^{\text{int}}$ in (41) since they are "averaged out" at the averaging step (42); see Fig. 5. The slopes $(\boldsymbol{q}_x)_{j+\frac{1}{2}}^{\text{int}}$ are to be computed using a nonlinear limiter. This may be done in several different ways. One of the sharpest possible approached, proposed in Kurganov and Lin (2007), uses the cell averages $\bar{\boldsymbol{q}}_{j+\frac{1}{2}}^{\text{int}}$ and the predicted point values of $\boldsymbol{q}$ at time level $t = t^{n+1}$:

$$\boldsymbol{q}_{j+\frac{1}{2},\ell}^{n+1} = \boldsymbol{q}_{j+\frac{1}{2},\ell}^{n} - \Delta t^n \boldsymbol{f}(\bar{\boldsymbol{q}}_j^n)_x, \quad \boldsymbol{q}_{j+\frac{1}{2},r}^{n+1} = \boldsymbol{q}_{j+\frac{1}{2},r}^{n} - \Delta t^n \boldsymbol{f}(\bar{\boldsymbol{q}}_{j+1}^n)_x \tag{43}$$

with $\boldsymbol{q}_{j+\frac{1}{2},\ell}^{n}$ and $\boldsymbol{q}_{j+\frac{1}{2},r}^{n}$ given by (38). The slopes are then obtained using the minmod limiter (see, e.g., Lie and Noelle, 2003; Nessyahu and Tadmor, 1990; Sweby, 1984; van Leer, 1979):

$$(\boldsymbol{q}_x)_{j+\frac{1}{2}}^{\text{int}} = \text{minmod}\left(\frac{\bar{\boldsymbol{q}}_{j+\frac{1}{2}}^{\text{int}} - \boldsymbol{q}_{j+\frac{1}{2},\ell}^{n+1}}{\delta}, \frac{\boldsymbol{q}_{j+\frac{1}{2},r}^{n+1} - \bar{\boldsymbol{q}}_{j+\frac{1}{2}}^{\text{int}}}{\delta}\right), \tag{44}$$

where $\delta = \left(x_{j+\frac{1}{2},r}^n - x_{j+\frac{1}{2},\ell}^n\right)/2 = \frac{\Delta t^n}{2}\left(a_{j+\frac{1}{2}}^{n+} - a_{j+\frac{1}{2}}^{n-}\right)$ and the minmod function, defined as

$$\text{minmod}(a,b) = \frac{\text{sign}(a) + \text{sign}(b)}{2} \cdot \min\{|a|,|b|\},$$

is applied in (44) in a componentwise manner.

The resulting fully discrete central-upwind scheme (42), (37)–(40), (43), (44) is very accurate and robust and its numerical dissipation vanishes as $\max_n\{\Delta t^n\} \to 0$ (unlike the numerical dissipation of the Nessyahu–Tadmor and other staggered central schemes). This scheme has been tested on a number of numerical examples in Kurganov and Lin (2007); see also Kurganov et al. (2001); Kurganov and Tadmor (2000), where different, more dissipative versions of the fully discrete central-upwind schemes were derived and studied.

6.1 Semidiscrete Central-Upwind Schemes

A major disadvantage of the fully discrete central-upwind scheme (42), (37)–(40), (43), (44) is its relatively high complexity. This is especially pronounced when the scheme is extended to the two-dimensional (2D) case in a rigorous, genuinely multidimensional—not in a "dimension-by-dimension"—manner; see Kurganov and Lin (2007), Kurganov and Petrova (2001), Kurganov and Petrova (2005) and Kurganov et al. (2016). However, one may pass to a semidiscrete limit as $\max_n\{\Delta t^n\}\to 0$ and derive semidiscrete central-upwind schemes, which are substantially simpler than their fully discrete counterparts and yet accurate and robust.

In order to derive a semidiscrete central-upwind scheme, we need to compute the following limit:

$$\frac{d}{dt}\bar{\boldsymbol{q}}_j(t^n)=\lim_{\Delta t^n\to 0}\frac{\bar{\boldsymbol{q}}_j^{n+1}-\bar{\boldsymbol{q}}_j^n}{\Delta t^n}. \tag{45}$$

After substituting (42) and then (37)–(40), (43) and (44) into (45) (see details in Kurganov and Lin (2007) and also in Kurganov et al. (2001), Kurganov and Petrova (2000) and Kurganov and Tadmor (2000)), we arrive at a particularly simple semidiscrete scheme, which can be written in the flux form as follows:

$$\frac{d}{dt}\bar{\boldsymbol{q}}_j(t)=-\frac{\boldsymbol{H}_{j+\frac{1}{2}}(t)-\boldsymbol{H}_{j-\frac{1}{2}}(t)}{\Delta x} \tag{46}$$

with the numerical fluxes

$$\boldsymbol{H}_{j+\frac{1}{2}}(t)=\frac{a^+_{j+\frac{1}{2}}\boldsymbol{f}\left(\boldsymbol{q}^-_{j+\frac{1}{2}}\right)-a^-_{j+\frac{1}{2}}\boldsymbol{f}\left(\boldsymbol{q}^+_{j+\frac{1}{2}}\right)}{a^+_{j+\frac{1}{2}}-a^-_{j+\frac{1}{2}}}+a^+_{j+\frac{1}{2}}a^-_{j+\frac{1}{2}}\left[\frac{\boldsymbol{q}^+_{j+\frac{1}{2}}-\boldsymbol{q}^-_{j+\frac{1}{2}}}{a^+_{j+\frac{1}{2}}-a^-_{j+\frac{1}{2}}}-\boldsymbol{d}_{j+\frac{1}{2}}\right], \tag{47}$$

where

$$\boldsymbol{d}_{j+\frac{1}{2}}=\frac{1}{2}\lim_{\Delta t^n\to 0}\left\{\Delta t^n(\boldsymbol{q}_x)^{\rm int}_{j+\frac{1}{2}}\right\}={\rm minmod}\left(\frac{\boldsymbol{q}^+_{j+\frac{1}{2}}-\boldsymbol{q}^*_{j+\frac{1}{2}}}{a^+_{j+\frac{1}{2}}-a^-_{j+\frac{1}{2}}},\frac{\boldsymbol{q}^*_{j+\frac{1}{2}}-\boldsymbol{q}^-_{j+\frac{1}{2}}}{a^+_{j+\frac{1}{2}}-a^-_{j+\frac{1}{2}}}\right) \tag{48}$$

is a built-in "anti-diffusion" term and

$$\boldsymbol{q}^*_{j+\frac{1}{2}}=\lim_{\Delta t^n\to 0}\bar{\boldsymbol{q}}^{\rm int}_{j+\frac{1}{2}}=\frac{a^+_{j+\frac{1}{2}}\boldsymbol{q}^+_{j+\frac{1}{2}}-a^-_{j+\frac{1}{2}}\boldsymbol{q}^-_{j+\frac{1}{2}}-\left\{\boldsymbol{f}(\boldsymbol{q}^+_{j+\frac{1}{2}})-\boldsymbol{f}(\boldsymbol{q}^-_{j+\frac{1}{2}})\right\}}{a^+_{j+\frac{1}{2}}-a^-_{j+\frac{1}{2}}} \tag{49}$$

In (47)–(49), the reconstructed point values $\boldsymbol{q}^{\pm}_{j+\frac{1}{2}}$ and the local one-sided speeds of propagation $a^{\pm}_{j+\frac{1}{2}}$ are given by (34) and (35), respectively, but without the upper index n, since these quantities are now computed at some time level t rather than t^n. Note that all of the terms on the RHS of (47) and in (48) and (49) depend on t, but we omit this dependence for the sake of brevity.

Remark 9. The semidiscretization (46)–(49) results in a system of time-dependent ODEs, which should be integrated using a sufficiently accurate and stable ODE solver. In the purely convective and convective-dominated cases, the ODE system is nonstiff and we usually solve it using the three-stage third-order strong stability preserving Runge-Kutta method; see Gottlieb et al. (2011, 2001).

Remark 10. Both the fully and semidiscrete central-upwind schemes belong to the class of Godunov-type Riemann-problem-solver-free central schemes, but since they are constructed using some upwind information (one-sided local speeds), we call them central-upwind schemes.

Remark 11. In the older works on central-upwind schemes (Kurganov et al., 2001; Kurganov and Petrova, 2000, 2001; Kurganov and Tadmor, 2000, 2002), the slopes $(\boldsymbol{q}_x)^{\text{int}}_{j+\frac{1}{2}}$ in (42) were not computed in a sharp way (formula (44) was only proposed in Kurganov and Lin (2007)) and therefore, the built-in anti-diffusion term $\boldsymbol{d}_{j+\frac{1}{2}}$ in (47) was equal to zero.

Remark 12. The first central-upwind scheme, introduced in Kurganov and Tadmor (2000), was obtained by setting the symmetric bounds on the local speeds, namely, by replacing (35) with

$$a^{n\pm}_{j+\frac{1}{2}} = \pm\max\left\{\rho\left(A\left(\bar{\boldsymbol{q}}^{\,n+}_{j+\frac{1}{2}}\right)\right), \rho\left(A\left(\bar{\boldsymbol{q}}^{\,n-}_{j+\frac{1}{2}}\right)\right)\right\}. \tag{50}$$

Remark 13. If the piecewise constant reconstruction (7) and the forward Euler ODE solver are used, the central-upwind schemes from Kurganov and Petrova (2001) and Kurganov and Tadmor (2000) reduce to the first-order Rusanov scheme (11), (16) and (17), while the central-upwind schemes from Kurganov et al. (2001), Kurganov and Petrova (2000) and Kurganov and Tadmor (2002) reduce to the first-order HLL scheme (11), (18) and (19).

Remark 14. If the point values $\left\{\boldsymbol{q}^{\pm}_{j+\frac{1}{2}}\right\}$ are computed using a reconstruction of order r (see Section 5.3), then the semidiscrete central-upwind scheme (46)–(49) will be (formally) rth order accurate.

Remark 15. Semidiscrete central-upwind schemes have been rigorously (using a genuinely multidimensional approach) extended to general 2D hyperbolic systems on a variety of different grids. We refer the reader to Kurganov and Lin (2007), Kurganov et al. (2001), Kurganov and Petrova (2001) and Kurganov and Tadmor (2002) for the central-upwind schemes on the Cartesian meshes. Triangular version of the central-upwind scheme was derived in Kurganov and Petrova (2005). Central-upwind scheme on general quadrilateral grids was introduced in Shirkhani et al. (2016) (see also Kurganov et al. (2016)). Finally, central-upwind schemes on cell-vertex polygonal meshes was developed in Beljadid et al. (2016).

Remark 16. When semidiscrete central-upwind schemes are applied to systems of balance laws

$$\boldsymbol{q}_t + \boldsymbol{f}(\boldsymbol{q})_x = \boldsymbol{S}(x,t,\boldsymbol{q}), \tag{51}$$

the numerical fluxes are still given by (47)–(49) and the only degree of freedom is in the approximation of cell averages of the source term, $\frac{1}{\Delta x}\int_{C_j} \boldsymbol{S}(x,t,\boldsymbol{q})\,dx$, which has to be added to the RHS of (46). In order to construct a reliable and robust method, one has to carefully choose an appropriate quadrature, which respects a delicates balance between the flux and source terms in (51). For instance, semidiscrete central-upwind schemes have been applied to a variety of shallow water models with the source terms describing the bottom topography, friction, and Coriolis forces. Well-balanced central-upwind schemes were developed in, e.g., Beljadid et al. (2016), Bryson et al. (2011), Cheng and Kurganov (2016), Chertock et al. (2015), Chertock et al. (2016), Kurganov and Levy (2002), Kurganov and Petrova (2007) and Shirkhani et al. (2016).

ACKNOWLEDGEMENTS

The author was supported in part by the NSF Grants DMS-1216957 and DMS-1521009.

REFERENCES

Abgrall, R., 1994. On essentially non-oscillatory schemes on unstructured meshes: analysis and implementation. J. Comput. Phys. 114, 45–58.

Arminjon, P., Viallon, M.-C., Madrane, A., 1997. A finite volume extension of the Lax-Friedrichs and Nessyahu-Tadmor schemes for conservation laws on unstructured grids. Int. J. Comput. Fluid Dyn. 9, 1–22.

Beljadid, A., Mohammadian, A., Kurganov, A., 2016. Well-balanced positivity preserving cell-vertex central-upwind scheme for shallow water flows. Comput. Fluids 136, 193–206.

Ben-Artzi, M., Falcovitz, J., 2003. Generalized Riemann Problems in Computational Fluid Dynamics, Cambridge Monographs on Applied and Computational Mathematics, vol. 11. Cambridge University Press, Cambridge, xvi+349. ISBN 0-521-77296-6.

Bianco, F., Puppo, G., Russo, G., 1999. High order central schemes for hyperbolic systems of conservation laws. SIAM J. Sci. Comput 21, 294–322.

Bryson, S., Epshteyn, Y., Kurganov, A., Petrova, G., 2011. Well-balanced positivity preserving central-upwind scheme on triangular grids for the Saint-Venant system. M2AN Math. Model. Numer. Anal. 45 (3), 423–446.

Buffard, T., Gallouët, T., Hérard, J.-M., 2000. A sequel to a rough Godunov scheme: application to real gases. Comput. Fluids 29 (7), 813–847.

Caramana, E.J., Shashkov, M.J., Whalen, P.P., 1998. Formulations of artificial viscosity for multi-dimensional shock wave computations. J. Comput. Phys. 144 (1), 70–97.
Cheng, Y., Kurganov, A., 2016. Moving-water equilibria preserving central-upwind schemes for the shallow water equations. Commun. Math. Sci. 14 (6), 1643–1663.
Chertock, A., Cui, S., Kurganov, A., Wu, T., 2015. Well-balanced positivity preserving central-upwind scheme for the shallow water system with friction terms. Int. J. Numer. Meth. Fluids 78, 355–383.
Chertock, A., Dudzinski, M., Kurganov, A., Lukáčová-Medviďová, M., 2016. Well-balanced schemes for the shallow water equations with Coriolis forces (submitted for publication).
Cockburn, B., Johnson, C., Shu, C.-W., Tadmor, E., 1998. Advanced numerical approximation of nonlinear hyperbolic equations. In: Quarteroni, A. (Ed.), CIME Lecture Notes. Lecture Notes in Mathematics, vol. 1697. Springer-Verlag, Berlin, Heidelberg.
Dafermos, C.M., 2010. Hyperbolic Conservation Laws in Continuum Physics, third ed. Grundlehren der Mathematischen Wissenschaften [Fundamental Principles of Mathematical Sciences], vol. 325. Springer-Verlag, Berlin, xxxvi+708.
Einfeld, B., 1988. On Godunov-type methods for gas dynamics. SIAM J. Numer. Anal. 25, 294–318.
Friedrichs, K.O., 1954. Symmetric hyperbolic linear differential equations. Comm. Pure Appl. Math. 7, 345–392.
Gallouët, T., Hérard, J.-M., Seguin, N., 2002. Some recent finite volume schemes to compute Euler equations using real gas EOS. Int. J. Numer. Methods Fluids 39 (12), 1073–1138.
Godlewski, E., Raviart, P.-A., 1996. Numerical Approximation of Hyperbolic Systems of Conservation Laws. Applied Mathematical Sciences, vol. 118. Springer-Verlag, New York, viii+509.
Godunov, S.K., 1959. A difference method for numerical calculation of discontinuous solutions of the equations of hydrodynamics. Mat. Sb. (N.S.) 47 (89), 271–306.
Gottlieb, S., Shu, C.-W., Tadmor, E., 2001. Strong stability-preserving high-order time discretization methods. SIAM Rev. 43, 89–112.
Gottlieb, S., Ketcheson, D., Shu, C.-W., 2011. Strong Stability Preserving Runge-Kutta and Multistep Time Discretizations. World Scientific Publishing Co. Pte. Ltd., Hackensack, NJ, xii+176.
Guermond, J.-L., Pasquetti, R., 2008. Entropy-based nonlinear viscosity for Fourier approximations of conservation laws. C. R. Math. Acad. Sci. Paris 346 (13–14), 801–806.
Guermond, J.-L., Pasquetti, R., Popov, B., 2011. Entropy viscosity method for nonlinear conservation laws. J. Comput. Phys. 230 (11), 4248–4267.
Harten, A., Lax, P., van Leer, B., 1983. On upstream differencing and Godunov-type schemes for hyperbolic conservation laws. SIAM Rev. 25, 35–61.
Harten, A., Engquist, B., Osher, S., Chakravarthy, S.R., 1987. Uniformly high-order accurate essentially nonoscillatory schemes. III. J. Comput. Phys. 71 (2), 231–303.
Jiang, G.-S., Shu, C.-W., 1996. Efficient implementation of weighted ENO schemes. J. Comput. Phys. 126 (1), 202–228.
Jiang, G.-S., Tadmor, E., 1998. Nonoscillatory central schemes for multidimensional hyperbolic conservation laws. SIAM J. Sci. Comput. 19 (6), 1892–1917. (electronic).
Kröner, D., 1997. Numerical Schemes for Conservation Laws. Wiley-Teubner Series Advances in Numerical Mathematics. John Wiley & Sons Ltd., Chichester, viii+508.
Kurganov, A., Levy, D., 2002. Central-upwind schemes for the Saint-Venant system. M2AN Math. Model. Numer. Anal. 36, 397–425.
Kurganov, A., Lin, C.T., 2007. On the reduction of numerical dissipation in central-upwind schemes. Commun. Comput. Phys. 2, 141–163.

Kurganov, A., Liu, Y., 2012. New adaptive artificial viscosity method for hyperbolic systems of conservation laws. J. Comput. Phys. 231, 8114–8132.

Kurganov, A., Petrova, G., 2000. Central schemes and contact discontinuities. M2AN Math. Model. Numer. Anal. 34 (6), 1259–1275.

Kurganov, A., Petrova, G., 2001. A third-order semi-discrete genuinely multidimensional central scheme for hyperbolic conservation laws and related problems. Numer. Math. 88 (4), 683–729.

Kurganov, A., Petrova, G., 2005. Central-upwind schemes on triangular grids for hyperbolic systems of conservation laws. Numer. Methods Partial Differ. Equ. 21, 536–552.

Kurganov, A., Petrova, G., 2007. A second-order well-balanced positivity preserving central-upwind scheme for the Saint-Venant system. Commun. Math. Sci. 5, 133–160.

Kurganov, A., Petrova, G., 2009. Central-upwind schemes for two-layer shallow equations. SIAM J. Sci. Comput. 31, 1742–1773.

Kurganov, A., Tadmor, E., 2000. New high resolution central schemes for nonlinear conservation laws and convection-diffusion equations. J. Comput. Phys. 160, 241–282.

Kurganov, A., Tadmor, E., 2002. Solution of two-dimensional Riemann problems for gas dynamics without Riemann problem solvers. Numer. Methods Partial Differ. Equ. 18, 584–608.

Kurganov, A., Noelle, S., Petrova, G., 2001. Semi-discrete central-upwind scheme for hyperbolic conservation laws and Hamilton-Jacobi equations. SIAM J. Sci. Comput. 23, 707–740.

Kurganov, A., Petrova, G., Popov, B., 2007. Adaptive semi-discrete central-upwind schemes for nonconvex hyperbolic conservation laws. SIAM J. Sci. Comput. 29, 2381–2401.

Kurganov, A., Prugger, M., Wu, T., 2016. Second-order fully discrete central-upwind scheme for two-dimensional hyperbolic systems of conservation laws (submitted for publication).

Lax, P.D., 1954. Weak solutions of nonlinear hyperbolic equations and their numerical computation. Comm. Pure Appl. Math. 7, 159–193.

LeFloch, P.G., 2002. Hyperbolic Systems of Conservation Laws. Lectures in Mathematics ETH Zürich. Birkhäuser Verlag, Basel. x+294.

LeVeque, R.J., 2002. Finite Volume Methods for Hyperbolic Problems. Cambridge texts in applied mathematics. Cambridge University Press, Cambridge, pp. xx+558.

Levy, D., Puppo, G., Russo, G., 1999. Central WENO schemes for hyperbolic systems of conservation laws. M2AN Math. Model. Numer. Anal. 33 (3), pp. 547–571.

Levy, D., Puppo, G., Russo, G., 2000. Compact central WENO schemes for multidimensional conservation laws. SIAM J. Sci. Comput. 22 (2), 656–672. (electronic).

Levy, D., Puppo, G., Russo, G., 2002. A fourth-order central WENO scheme for multidimensional hyperbolic systems of conservation laws. SIAM J. Sci. Comput. 24 (2), 480–506.

Li, J., Zhang, T., Yang, S., 1998. The Two-dimensional Riemann Problem in Gas Dynamics. Pitman Monographs and Surveys in Pure and Applied Mathematics, vol. 98. Longman, Harlow. x+300. 0-8493-0693-0.

Lie, K.A., Noelle, S., 2003. On the artificial compression method for second-order nonoscillatory central difference schemes for systems of conservation laws. SIAM J. Sci. Comput. 24 (4), 1157–1174.

Liu, Y., 2005. Central schemes on overlapping cells. J. Comput. Phys. 209 (1), 82–104.

Liu, X.D., Osher, S., 1996. Nonoscillatory high order accurate self-similar maximum principle satisfying shock capturing schemes. I. SIAM J. Numer. Anal. 33 (2), 760–779.

Liu, X.D., Tadmor, E., 1998. Third order nonoscillatory central scheme for hyperbolic conservation laws. Numer. Math. 79 (3), 397–425.

Liu, Y., Shu, C.-W., Tadmor, E., Zhang, M., 2007. Non-oscillatory hierarchical reconstruction for central and finite volume schemes. Commun. Comput. Phys. 2 (5), 933–963.

Liu, Y., Shu, C.-W., Xu, Z., 2009. Hierarchical reconstruction with up to second degree remainder for solving nonlinear conservation laws. Nonlinearity 22 (12), 2799–2812.

Liu, X., Mohammadian, A., Kurganov, A., Infante Sedano, J.A., 2015. Well-balanced central scheme for a fully coupled shallow water system modeling flows over erodible bed. J. Comput. Phys. 300, 202–218.

Masella, J.M., Faille, I., Gallouët, T., 1999. On an approximate Godunov scheme. Int. J. Comput. Fluid Dyn. 12 (2), 133–149.

Nessyahu, H., Tadmor, E., 1990. Nonoscillatory central differencing for hyperbolic conservation laws. J. Comput. Phys. 87 (2), 408–463.

Qiu, J., Shu, C.-W., 2002. On the construction, comparison, and local characteristic decomposition for high-order central WENO schemes. J. Comput. Phys. 183 (1), 187–209.

Roe, P.L., 1981. Approximate Riemann solvers, parameter vectors, and difference schemes. J. Comput. Phys. 43 (2), 357–372.

Rusanov, V.V., 1961. The calculation of the interaction of non-stationary shock waves with barriers. Zh. Vychisl. Mat. Mat. Fiz. 1, 267–279.

Serre, D., 1999. Systems of Conservation Laws. 1. Cambridge University Press, Cambridge, xxii+263.

Shchepetkin, A.F., McWilliams, J.C., 1998. Quasi-monotone advection schemes based on explicit locally adaptive dissipation. Mon. Weather Rev. 126, 1541–1580.

Shi, J., Hu, C., Shu, C.-W., 2002. A technique of treating negative weights in weno schemes. J. Comput. Phys. 175 (1), 108–127.

Shirkhani, H., Mohammadian, A., Seidou, O., Kurganov, A., 2016. A well-balanced positivity-preserving central-upwind scheme for shallow water equations on unstructured quadrilateral grids. Comput. Fluids 126, 25–40.

Shu, C.-W., 2003. High-order finite difference and finite volume WENO schemes and discontinuous Galerkin methods for CFD. Int. J. Comput. Fluid Dyn. 17 (2), 107–118.

Shu, C.-W., 2009. High order weighted essentially nonoscillatory schemes for convection dominated problems. SIAM Rev. 51 (1), 82–126.

Smoller, J., 1994. Shock Waves and Reaction-Diffusion Equations, second ed. Grundlehren der Mathematischen Wissenschaften [Fundamental Principles of Mathematical Sciences], vol. 258. Springer-Verlag, New York, xxiv–632.

Sweby, P.K., 1984. High resolution schemes using flux limiters for hyperbolic conservation laws. SIAM J. Numer. Anal. 21 (5), 995–1011.

Tadmor, E., 1984a. The large-time behavior of the scalar, genuinely nonlinear Lax-Friedrichs scheme. Math. Comp. 43 (168), 353–368.

Tadmor, E., 1984b. Numerical viscosity and the entropy condition for conservative difference schemes. Math. Comp. 43 (168), 369–381.

Toro, E.F., 2009. Riemann Solvers and Numerical Methods for Fluid Dynamics: A Practical Introduction, third ed. Springer-Verlag, Berlin, Heidelberg, xx+724.

Toro, E.F., Spruce, M., Speares, W., 1994. Restoration of the contact surface in the HLL-Riemann solver. Shock Waves 4 (1), 25–34.

van Leer, B., 1979. Towards the ultimate conservative difference scheme. V. A second-order sequel to Godunov's method. J. Comput. Phys. 32 (1), 101–136.

von Neumann, J., Richtmyer, R.D., 1950. A method for the numerical calculation of hydrodynamic shocks. J. Appl. Phys. 21, 232–237.

Wilkins, M.L., 1980. Use of artificial viscosity in multidimensional fluid dynamic calculations. J. Comput. Phys. 36, 281–303.

Yang, S., Kurganov, A., Liu, Y., 2015. Well-balanced central schemes on overlapping cells with constant subtraction techniques for the Saint-Venant shallow water system. J. Sci. Comput. 63 (3), 678–698.

Zheng, Y., 2001. Systems of Conservation Laws, Progress in Nonlinear Differential Equations and their Applications, 38, Birkhäuser Boston, Inc., Boston, MA. xvi+317. ISBN 0-8176-4080-0.

Chapter 21

Time Discretization Techniques

S. Gottlieb* and D.I. Ketcheson†
*University of Massachusetts Dartmouth, North Dartmouth, MA, United States
†CEMSE Division, King Abdullah University of Science and Technology (KAUST), Thuwal, Saudi Arabia

Chapter Outline

ABSTRACT

The time discretization of hyperbolic partial differential equations is typically the evolution of a system of ordinary differential equations obtained by spatial discretization of the original problem. Methods for this time evolution include multistep, multistage, or multiderivative methods, as well as a combination of these approaches. The time step constraint is mainly a result of the absolute stability requirement, as well as additional conditions that mimic physical properties of the solution, such as positivity or total variation stability. These conditions may be required for stability when the solution develops shocks or sharp gradients. This chapter contains a review of some of the methods

Handbook of Numerical Analysis, Vol. 17. http://dx.doi.org/10.1016/bs.hna.2016.08.001

historically used for the evolution of hyperbolic PDEs, as well as cutting edge methods that are now commonly used.

Keywords: Runge–Kutta, Multistep, Multiderivative, Strong stability preserving, Deferred correction, Multirate, Parallel

AMS Classification Code: 65

1 OVERVIEW

Hyperbolic partial differential equations (PDEs) involve both spatial and temporal derivatives. While the spatial discretization is the primary focus of many chapters in this book, in this chapter we focus on the approaches to time evolution of hyperbolic PDEs. Most numerical discretizations of hyperbolic problems follow the method of lines approach, in which the conservation law

$$U_t + f(U)_x = 0 \tag{1}$$

is first discretized in space, leading to a system of ordinary differential equations of the form

$$u_t = F(u) \tag{2}$$

where u is a vector that approximates U at the spatial gridpoints. The temporal integration of the initial-value problem (2) is the subject of this chapter. It can be accomplished using, for instance, Runge–Kutta (multistage), multistep, or Taylor series (multiderivative) methods. Combinations of the standard approaches produce methods with multiple stages, steps and/or derivatives.

Methods used in time evolution of hyperbolic PDEs are typically multistep or Runge–Kutta methods, or a combination of these approaches. Recently, multiderivative methods have come into consideration as well. Some general remarks can be made regarding the computational properties of the aforementioned methods. Multistep methods require little computation per step but more storage while multistage methods require more computation per step but can be designed to have lower storage requirements. The computation of additional derivatives may further alleviate the storage requirement, but these derivatives are not always easy or computationally cheap to calculate.

The primary consideration when integrating any initial-value ODE is the computational effort required to advance to a given time; i.e. the step size allowed by a given method normalized by the cost per step. Clearly then, the goal is to permit the largest allowable step-size relative to the number of functional evaluations. The allowable step size is usually determined based on the following three factors:

1. *Absolute stability (linear stability).* Most hyperbolic problems are solved by explicit methods, where the famous Courant–Friedrichs–Lewy (CFL) stability condition requires that the time step size must be taken approximately proportional to the spatial mesh width. Absolute stability typically

requires an even smaller step size than the CFL condition would dictate. For smooth problems, the step size dictated by absolute stability is both necessary and sufficient to guarantee convergence. However, in the presence of shocks or sharp gradients we must require additional nonlinear, noninner product stability properties to ensure convergence.

2. *Robustness (nonlinear stability).* It is often desirable that the numerical approximation produce results that satisfy certain physical or mathematical constraints. We refer to this as *robustness*. For example, in many problems it is required that certain quantities (like density or pressure) remain positive. Furthermore, spurious oscillations should be avoided for both physical and theoretical reasons. Spatial discretizations are often designed to satisfy such physical or mathematical constraints, under certain step-size restrictions which may be tighter than those required for absolute stability. Violation of this step size may lead to nonphysical negative values and oscillations.

   As mentioned above, nonlinear stability conditions become critical for convergence in the presence of shocks or sharp gradients. In space, techniques involving limiters have been introduced to ensure total variation bounds, entropy stability, maximum principle preservation, or positivity. These properties of the spatial discretization can be preserved when used with higher-order strong stability preserving (SSP) time discretizations, discussed in Section 4.

3. *Accuracy.* Given that the temporal step size is generally of the size of the spatial grid, accuracy considerations typically dictate that the time-stepping method used will be approximately the same *order* as the spatial discretization. However, beyond this mild requirement, accuracy of the temporal method is not usually a major concern. Unlike the case for other areas, for hyperbolic problems step-size control based on error estimation is not widely used, and accuracy is not generally considered a limiting factor in choosing a step size for time integration.

Typically, guidelines for choosing a step size that ensures accuracy, stability, and robustness are obtained by analysis of simple (e.g. one-dimensional and/or scalar) problems. Practical step sizes for multidimensional systems in complex geometries may need to be adjusted based on experience.

Although most integrators are designed to run in serial, for certain algorithms this cost can be reduced by taking advantage of concurrency. In some cases memory requirements can also be a limiting factor here as well.

The method of lines approach may be extended to more complex problems. For example, consider a PDE of the form

$$U_t + f(U)_x = g(U)$$

where g may be a (hyperbolic or parabolic) differential operator or a reaction term. If g represents a physical process with much faster dynamics than those of f, or if its discretization is stiff, then a fully explicit discretization will require a time step that is excessively small compared to timescales that are relevant to f. Typical examples include viscous and/or reactive fluid flows. The resulting system of ODEs has the form

$$u_t = F(u) + G(u)$$

and it is often advantageous to treat F and G differently in time, e.g. by using an implicit–explicit (IMEX) method (see Section 6), an exponential integrator or an integrating factor (see Section 7). This is especially advantageous if the stiff term (G) is linear.

This chapter will focus on developing fully discrete methods by the method of lines: discretizing first in space and then applying an initial value ODE solver in time. Although we do not address these in this chapter, it is important to note that there are cases in which going completely outside the method of lines approach leads to innovative approaches. These include the Lax–Wendroff approach (Lax and Wendroff, 1960) as well as the use of Lax–Wendroff in the context of discontinuous Galerkin methods (Qiu et al., 2005). Other notable examples include the arbitrary derivative Riemann problem (ADER) methods (Balsara et al., 2009; Takakura, 2006;Titarev and Toro, 2002, 2005; Toro and Titarev, 2005; Zahran, 2008) and the methods derived in Seal et al. (2016).

2 CLASSICAL METHODS

Once the conservation law has been semidiscretized in space, we obtain a system of ordinary differential equations (ODEs) (2) that can be evolved using any standard approach. In this section, we review the typical approaches for numerical solution of ODEs. The simplest method is the explicit first-order method

$$u^{n+1} = u^n + \Delta t F(u^n)$$

known as the forward Euler method. For hyperbolic conservation laws, the spectrum of the spatial differential operator is purely imaginary; semidiscretizations may have purely imaginary eigenvalues (in the case of centered finite differences or Fourier spectral collocation, for instance) or eigenvalues that lie in the left half-plane near the imaginary axis (in the case of upwind methods). Since the absolute stability region of the forward Euler method intersects the imaginary axis only at the origin, it is typically not a stable choice of time discretization; furthermore, it is only first-order accurate. However, this method is a one-step, one-stage, one-derivative method and so serves as the basis for many other methods, as discussed later in this chapter.

The methods discussed in this section are classical, and so many have been applied by default to hyperbolic PDEs over the years. More recently, the study of the strong stability preserving properties of these classes of methods, discussed in Section 4, suggests which specific methods are best suited for evolution of hyperbolic PDEs.

2.1 Runge–Kutta

The easiest extension of the forward Euler method is known as the improved Euler method, or Heun's method. It is obtained by first using Euler's method and then applying the trapezoidal rule. However, this approach yields a second-order method whose absolute stability region intersects the imaginary axis only at the origin and is therefore is not suitable for evolution of hyperbolic conservation laws.

Using more intermediate stages allows for increases in both the order and the size of the stability region of the method. The three-stage third-order method

$$\begin{aligned}
u^{(1)} &= u^n \\
u^{(2)} &= u^n + \Delta t F(u^{(1)}) \\
u^{(3)} &= u^n + \frac{1}{2}\Delta t\left(F(u^{(1)}) + F(u^{(2)})\right) \\
u^{n+1} &= u^n + \frac{1}{2}\Delta t\left(F(u^{(1)}) + F(u^{(3)})\right).
\end{aligned}$$

and the four-stage fourth-order method

$$\begin{aligned}
u^{(1)} &= u^n \\
u^{(2)} &= u^n + \frac{1}{2}\Delta t F(u^{(1)}) \\
u^{(3)} &= u^n + \frac{1}{2}\Delta t F(u^{(2)}) \\
u^{(4)} &= u^n + \Delta t F(u^{(3)}) \\
u^{n+1} &= u^n + \frac{1}{6}\Delta t\left(F(u^{(1)}) + 2F(u^{(2)}) + 2F(u^{(3)}) + F(u^{(4)})\right)
\end{aligned}$$

have successively larger regions of absolute stability that include the imaginary axis and have been popular for evolution of hyperbolic PDEs (Jameson et al., 1981). While the increase in stages in the above methods leads to better accuracy and larger allowable time step in terms of absolute stability, the computational cost also increases with the number of stages, as one function evaluation is required per stage. However, this increased cost is usually more than offset by the fact that larger time steps can be taken whether accuracy or stability is the limiting factor.

For problems with time-dependent boundary conditions, some caution is needed. The natural method of imposing time-dependent boundary conditions in Runge–Kutta methods reduces the accuracy of the method to first-order locally, and second-order globally, regardless of the spatial operator (Carpenter et al., 1995). One approach to dealing with this is to impose the exact boundary conditions only at each step, not at the intermediate stages. This approach retains accuracy, but requires a much smaller time step to maintain stability. A second approach is to impose boundary conditions derived from the physical boundary condition and its derivatives at each intermediate stage. This approach retains the time step required for stability, but if the hyperbolic PDE is nonlinear this approach only works for Runge–Kutta schemes with order $p \leq 3$.

The Runge–Kutta methods above are all explicit, and the most commonly used methods for hyperbolic problems are all explicit. Implicit methods can be of use when dealing with fast waves that do not need to be accurately resolved. These methods also serve as a basis for implicit–explicit methods described in Section 6.

2.2 Multistep

A second approach to the extension of Euler's method is to add more information about the solution at previous steps. The k-step methods are typically written in the form

$$\sum_{i=0}^{k} \alpha_i u_{n+1-i} = \sum_{i=0}^{k} \beta_i F(u_{n+1-i})$$

where the coefficients α_i and β_i are chosen to optimize accuracy and the desired stability properties. The method is explicit if $\beta_0 = 0$, and implicit otherwise. The most widely used methods in this class are the Adams-type methods, for which $\alpha_j = 0$ for all but $j = 0$ and $j = 1$. The explicit methods of this type are known as Adams–Bashforth methods while the implicit methods are called Adams–Moulton methods. The two-step second-order Adams–Bashforth method is

$$u^{n+1} = u^n + \frac{1}{2}\Delta t(3F(u^n) - F(u^{n-1}),$$

while the three-step third-order method is

$$u^{n+1} = u^n + \frac{1}{12}\Delta t(23F(u^n) - 16F(u^{n-1}) + 5F(u^{n-2}))\cdot$$

The second-order implicit one-step method of the Adams–Moulton type is known as the Crank–Nicolson method

$$u^{n+1} = u^n + \frac{1}{2}\Delta t(F(u^n) + F(u^{n+1}),$$

and is one of the best known methods in its class.

Another approach to multistep methods is to evaluate the function F only at the right endpoint, i.e. $\beta_j = 0$ whenever $j \neq 0$. These are known as backward differentiation formulas (BDF). The second-order BDF method is

$$u^{n+1} = \frac{4}{3}u^n - \frac{1}{3}u^{n-1} + \frac{2}{3}\Delta t F(u^{n+1})\cdot$$

The advantage of multistep methods is that computation cost is steady at one function evaluation per iteration, and does not increase when more steps are used to attain higher order. One disadvantage is that the storage requirement increases with number of stages; for large systems (for instance, those coming from PDEs in three spatial dimensions) this can be prohibitive. Another difficulty is the fact that these methods need starting values for the k initial steps. These starting values can be produced using a one-step method with very small step sizes. If the old values are used only for the function evaluations, as in Adams-type methods, we can use a one-step method of order one less than the multistep method. High-order explicit multistep methods tend to have smaller stability regions than high-order explicit Runge–Kutta methods.

2.3 Multistage Multistep Methods

Combining multiple stages and steps may provide higher-order methods while balancing the computational cost and storage requirements. These methods are sometimes called hybrid methods or pseudo-Runge–Kutta methods, or general linear methods. This class of methods was originally proposed in Butcher (1965), Gear (1965) and Gragg (1964) and has been studied extensively in Butcher (1996, 2006, 2009) and Jackiewicz (2009). These methods have been gaining popularity and have been widely discussed for evolution of hyperbolic PDEs, with a description of an implementation in Vos et al. (2011).

2.4 Taylor Series Methods

Another approach to raising the order of accuracy is the inclusion of higher-order time derivatives. Given the ODE system (2) we use the Taylor series approximation to form a numerical method by replacing the derivatives of u with the derivatives of F. For example, the second-order Taylor series method is

$$u^{n+1} = u^n + \Delta t F(u^n) + \frac{1}{2}\Delta t^2 \dot{F}(u^n)\cdot$$

The challenge in Taylor series methods is the increasingly difficult computation of the derivatives, so these methods are rarely used in practice. However, the idea behind these methods appear in various other approaches, such as the Lax–Wendroff schemes and the multistage multiderivative methods described later.

2.5 Multistage Multiderivative Methods

As seen earlier, to increase the possible order of any method, we can use more stages (e.g. Runge–Kutta methods), more steps (e.g. linear multistep methods) or more derivatives (Taylor series methods). It is also possible to combine these approaches, as we saw in Section 2.3 where methods with multiple steps and stages are discussed. Methods with multiple stages and multiple derivatives have also been used for hyperbolic PDEs. Multistage multiderivative integration methods were first considered in Obreschkoff (1940), Turán (1950) and Stancu and Stroud (1963), and multiderivative time integrators for ordinary differential equations were developed in Shintani (1971, 1972), Kastlunger and Wanner (1972a,b), Mitsui (1982), Ono and Yoshida (2004) and Chan and Tsai (2010). These methods have recently been proposed for use with partial differential equations (PDEs) (Seal et al., 2014; Tsai et al., 2014).

Using the system of ODEs (2) resulting from the spatial discretization of a hyperbolic PDE of the form (1), we can use a two-derivative Taylor series-type method as a building block $u^{n+1} = u^n + \alpha \Delta t F(u^n) + \beta \Delta t^2 \dot{F}(u^n)$ where $\alpha \geq 0$ and $\beta \geq 0$ are coefficients chosen to ensure the desired order. The second-order Taylor series method is obtained through the choice of coefficients $\alpha = 1$ and $\beta = \frac{1}{2}$.

To obtain higher-order explicit methods, we can add more stages:

$$
\begin{aligned}
y^{(i)} &= u^n + \Delta t \sum_{j=1}^{i-1} \left(a_{ij} F(y^{(j)}) + \Delta t \hat{a}_{ij} \dot{F}(y^{(j)}) \right), \quad i = 1, \ldots, s \\
u^{n+1} &= u^n + \Delta t \sum_{j=1}^{s} \left(b_j F(y^{(j)}) + \Delta t \hat{b}_j \dot{F}(y^{(j)}) \right).
\end{aligned} \tag{3}
$$

The coefficients are then selected to attain the desired order, based on the order conditions described in Chan and Tsai (2010) and Gekeler and Widmann (1986).

While it is usually prohibitive to use methods that require more derivatives, the computation of the second derivative is often easily accomplished for hyperbolic PDEs because they require only the readily available computation of the Jacobian. The explicit two-derivative Runge–Kutta methods that have been used with hyperbolic PDEs include the two-stage third-order method (Seal et al., 2014)

$$
\begin{aligned}
u^* &= u^n + \Delta t F(u^n) + \frac{1}{2} \Delta t^2 \dot{F}(u^n) \\
u^{n+1} &= u^n + \frac{2}{3} \Delta t F(u^n) + \frac{1}{3} \Delta t F(u^*) + \frac{1}{6} \Delta t^2 \dot{F}(u^n),
\end{aligned}
$$

the two-stage fourth-order method (Seal et al., 2014; Tsai et al., 2014)

$$
\begin{aligned}
u^* &= u^n + \frac{\Delta t}{2} F(u^n) + \frac{\Delta t^2}{8} \dot{F}(u^n) \\
u^{n+1} &= u^n + \Delta t F(u^n) + \frac{\Delta t^2}{6} (\dot{F}(u^n) + 2\dot{F}(u^*)),
\end{aligned} \tag{4}
$$

and the three-stage fifth-order method (Seal et al., 2014)

$$
\begin{aligned}
u^* &= u^n + \frac{2}{5}\Delta t F(u^n) + \frac{2}{25}\Delta t^2 \dot{F}(u^n) \\
u^{**} &= u^n + \Delta t F(u^n) - \frac{1}{4}\Delta t^2 \dot{F}(u^n) + \frac{3}{4}\Delta t^2 \dot{F}(u^*) \\
u^{n+1} &= u^n + \Delta t F(u^n) + \Delta t^2 \left(\frac{1}{8}\dot{F}(u^n) + \frac{25}{72}\dot{F}(u^*) + \frac{1}{36}\dot{F}(u^{**})\right),
\end{aligned}
$$

Two-derivative Runge–Kutta methods were used with discontinuous Galerkin and finite difference WENO spatial discretizations in Seal et al. (2014) and with finite difference and compact finite difference methods in Tsai et al. (2014). In both these cases, the results were competitive with standard approaches, indicating that these methods may have potential in the evolution of hyperbolic PDEs.

3 DEFERRED CORRECTION METHODS

Spectral deferred correction (SDC) methods (Dutt et al., 2000; Liu et al., 2008; Minion, 2003; Xia et al., 2007) are explicit or implicit one-step methods that can be easily and systematically constructed for any order of accuracy. In this way they seem to be advantageous over Runge–Kutta methods which are more difficult to construct for higher order of accuracy, and multistep methods which need more storage space and are more difficult to restart with a different choice of the timestep Δt. Absolute stability of the deferred correction approach was studied in, e.g. Dutt et al. (2000), Minion (2003) and Xia et al. (2007).

In the deferred correction approach, the interval $[t^n, t^{n+1}]$ (where $t^{n+1} = t^n + \Delta t$) is subdivided into s subintervals, at intermediate values (or nodes) $t^{(m)}$ and associated intermediate time step $\Delta t^{(m)}$. These nodes can be equally spaced or can be the Chebyshev Gauss–Lobatto points. We first compute a first-order approximate solution u_1 on all the nodes, using the forward Euler methods. We then proceed to *correct* this solution for all the nodes in the interval using

$$
u_2^{(m+1)} = u_2^{(m)} + \theta_1 \Delta t^{(m)} (F(u_2^{(m)}) - F(u_1^{(m)})) + I_m^{m+1}(F(u_1)), \tag{5}
$$

for some value $0 \leq \theta_1 \leq 1$ where $I_m^{m+1}(F(u_k))$ is the integral of the s-th degree interpolating polynomial on the $s+1$ points $(t^{(\ell)}, F(u_1^{(\ell)}))_{\ell=0}^{s}$ over the subinterval $[t^{(m)}, t^{(m+1)}]$. This process is repeated until we obtain $u^{n+1} = u_{s+1}^{(s)}$.

The scheme described above with $\theta_k = 1$ is the one discussed in Dutt et al. (2000) and Minion (2003). In Xia et al. (2007), the scheme is also discussed with general $0 \leq \theta_k \leq 1$ to enhance linear stability. The term with the coefficient θ_k does not change the order of accuracy. It can be easily shown that this method is simply a (adaptive-order) Runge–Kutta method, and so inherits all the properties of this class of methods.

The SDC method led to the development of other deferred correction approaches. For example, the multilevel SDC method uses a multilevel coarsening approach that is computationally more efficient (Speck et al., 2015). Another approach is the integral deferred correction method (IDC) in Christlieb et al. (2010c), which uses high-order Runge–Kutta methods in place of the forward Euler method in the above algorithm, which results in superior numerical stability region and enhanced accuracy. A variant of the IDC approach, the revisionist integral deferred correction method (Christlieb et al., 2010a), can be very efficiently implemented on multicore multi-CPU computers and on GPU clusters.

4 STRONG STABILITY PRESERVING METHODS

Given a linear differential equation and a consistent linear numerical method, linear stability is both a necessary and sufficient condition for convergence of the numerical approximation to the solution of a PDE (Strikwerda, 1989). Absolute stability is important for nonlinear problems as well, because if a numerical method is consistent and its *linearization* is L_2 stable and adequately dissipative, then for sufficiently smooth problems the nonlinear approximation is convergent (Strang, 1964). However, when dealing with hyperbolic PDEs, discontinuous solutions often arise. In such cases linear stability theory no longer guarantees convergence. A famous example of this is the linearly stable second-order Lax–Wendroff scheme, which for the nonlinear Burgers' equation becomes L_2 nonlinearly unstable near stagnation points (Majda and Osher, 1978). Some other kind of stability is needed to guarantee convergence for nonlinear problems where the solution is not smooth.

Linear stability can often be proved directly even for complicated multistage or multistep time discretizations, but nonlinear stability is usually more difficult to study. For this reason, most research efforts focus on developing high-order spatial discretizations that have the desired nonlinear stability properties for approximating discontinuous solutions of hyperbolic PDEs, *when coupled with the simple forward Euler time-stepping method* (Cockburn and Shu, 1989; Harten, 1983; Kurganov and Tadmor, 2000; Liu et al., 1994; Osher and Chakravarthy, 1984; Sweby, 1984; Tadmor, 1998). However, this nonlinear stability property may not hold when the spatial discretization is coupled with a linearly stable higher-order time discretization, as is usually needed for practical computation. Strong stability preserving time discretization methods were developed to guarantee that the nonlinear stability properties satisfied by the spatial discretization when coupled with forward Euler integration will be preserved when the same spatial discretization is coupled with these higher-order methods (Shu, 1988; Shu and Osher, 1988).

For hyperbolic PDEs, the relevant nonlinear stability property sometimes takes the form of total variation diminishing (TVD), total variation bounded (TVB) or some nonoscillatory requirement. These properties may be desirable even for a linear problem, where they are not required for convergence. The class of high-order SSP time discretization methods was first developed in conjunction with TVD spatial discretizations (Shu, 1988; Shu and Osher,

1988) and were called TVD time discretizations. However, these time-discretizations preserve the nonlinear stability properties satisfied by the forward Euler method in any norm, seminorm or convex functional. In recognition of their general applicability to any strong stability property, these methods were renamed strong stability preserving (SSP) methods in Gottlieb and Shu (1998). This class of methods has since been further studied in many publications including Gottlieb and Shu (1998), Gottlieb et al. (2001), Ruuth and Spiteri (2002), Shu (2002), Spiteri and Ruuth (2002, 2003), Gottlieb and Gottlieb (2003), Hundsdorfer et al. (2003), Ruuth and Spiteri (2004), Ruuth (2006), Hundsdorfer and Ruuth (2005), Spijker (2008), Ketcheson (2008), Ketcheson et al. (2009) and Liu et al. (2008).

4.1 The SSP Property

The motivation for strong stability preserving methods comes from assuming that when coupled with forward Euler time stepping, the method-of-lines semidiscretization (2) satisfies some desired strong stability property in a given norm, seminorm, or convex functional $\|\cdot\|$, as long as the time step Δt is suitably restricted; i.e. for all u we assume

$$\| u+\Delta t F(u) \| \leq \| u \| \quad \text{for } 0 \leq \Delta t \leq \Delta t_{\mathrm{FE}}. \tag{6}$$

We then wish to find a higher-order time discretization that preserves this strong stability property, perhaps under a different time step restriction. We say that method is *strong stability preserving* (SSP) with SSP coefficient $\mathcal{C}$ if

$$\| u^{n+1} \| \leq \| u^n \| \tag{7}$$

provided only that (6) holds and the time step satisfies

$$\Delta t \leq \mathcal{C} \Delta t_{\mathrm{FE}}. \tag{8}$$

We note that the forward Euler time step Δt_{FE} depends only on the spatial discretization, while the SSP coefficient $\mathcal{C}$ is a property of the time-stepping method.

The SSP property can be ensured if a time-stepping method can be decomposed into convex combinations of forward Euler steps. For explicit Runge–Kutta methods, this decomposition requires rewriting a method in the form

$$u^{(i)} = \sum_{k=0}^{i-1} \left(\alpha_{i,k} u^{(k)} + \Delta t \beta_{i,k} F(u^{(k)}) \right) \qquad 2 \leq i \leq s+1 \tag{9}$$

where $u^{(1)} = u^n$ and $u^{n+1} = u^{(s+1)}$. The form (9), known as the Shu–Osher form, is widely used as it is convenient for SSP analysis, and often results in a low-storage formulation.

Clearly, the stages of the method (9) are simply linear combinations of forward Euler steps with a modified time step $\Delta t \frac{\beta_{i,k}}{\alpha_{i,k}}$. Consistency requires that

$\sum_{k=0}^{i-1}\alpha_{i,k}=1$, so that as long as all the coefficients $\alpha_{i,\,k}$ and $\beta_{i,\,k}$ are nonnegative, we have a convex combination of forward Euler steps, which preserves any strong stability properties under the modified time step restriction

$$\max_{i,k}\frac{\beta_{i,k}}{\alpha_{i,k}}\Delta t \leq \Delta t_{\mathrm{FE}}.$$

Hence if a Runge–Kutta method that can decomposed into this form with $\alpha_{i,\,k} \geq 0$ and $\beta_{i,\,k} \geq 0$, and where if $\alpha_{i,\,k}=0$ then the corresponding $\beta_{i,\,k}=0$, then the method will be SSP with SSP coefficient $\mathcal{C}=\min_{i,k}\frac{\alpha_{i,k}}{\beta_{i,k}}$. The major aim of the research in this field is to design methods that maximize the SSP coefficient, thus allowing the largest possible time step.

A similar property can be shown for multistep methods, but in this case the definition of strong stability becomes $\| u^{n+1} \| \leq \max\left\{\| u^n \|, \| u^{n-1} \|,\ldots, \| u^{n-s+1} \|\right\}$.

The search for methods with large SSP coefficient benefitted greatly from the discovery of connections between SSP theory and the theory of absolute monotonicity of Runge–Kutta methods (Ferracina and Spijker, 2004, 2005; Higueras, 2004, 2005; Kraaijevanger, 1991). The radius of absolute monotonicity was already known to govern the permissible time step for guaranteeing contractivity in arbitrary norms. It turns out that the radius of absolute monotonicity corresponds to the largest SSP coefficient over all possible Shu–Osher forms of a given Runge–Kutta method. With that understanding, the SSP time step restriction is not only sufficient but also necessary for monotonicity. The connections between strong stability and contractivity imply that the optimal contractive methods are also optimal SSP methods. This observation has led to the development of new optimal and efficient SSP methods (Ferracina and Spjker, 2008; Ketcheson, 2008; Ketcheson et al., 2009).

4.2 Optimal Explicit Methods

Optimal explicit SSP Runge–Kutta methods have proven particularly popular for the evolution of hyperbolic PDEs. The three-stage third-order Runge–Kutta scheme SSPRK(3,3)

$$\begin{aligned} u^{(1)} &= u^n + \Delta t F(u^n) \\ u^{(2)} &= \frac{3}{4}u^n + \frac{1}{4}u^{(1)} + \frac{1}{4}\Delta t F(u^{(1)}) \\ u^{n+1} &= \frac{1}{3}u^n + \frac{2}{3}u^{(2)} + \frac{2}{3}\Delta t F(u^{(2)}). \end{aligned} \tag{10}$$

is generally known as the Shu–Osher method and is the most widely used of these methods. This method has an SSP coefficient $\mathcal{C}=1$ and an attractive low-storage formulation. This method was first presented in Shu and Osher

(1988) and shown to be optimal among all three-stage third-order explicit Runge–Kutta methods in Gottlieb and Shu (1998).

The SSP coefficient can be raised if we allow more stages. For example, a four-stage third-order method with $\mathcal{C}=2$ was given in Kraaijevanger (1991), and a five-stage third-order method was given in Spiteri and Ruuth (2002) with $\mathcal{C}=2.65$. Table 1 (top) contains the optimal effective SSP coefficients of explicit Runge–Kutta methods. Of course, we need to include the increased cost incurred by adding more stages, and therefore more function evaluations. To account for this additional cost, we normalize the SSP coefficient by the number of stages and define the *effective SSP coefficient* $\mathcal{C}_{\text{eff}}=\frac{\mathcal{C}}{s}$. Using this measure, we find that the Shu–Osher method above has $\mathcal{C}_{\text{eff}}=\frac{1}{3}$, Kraaijevanger's four-stage method has $\mathcal{C}_{\text{eff}}=\frac{1}{2}$, and the Spiteri and Ruuth five-stage method has $\mathcal{C}_{\text{eff}}\approx 0.53$. So far, the Shu–Osher method nevertheless continues to be more popular.

No four-stage fourth-order Runge–Kutta method with $\mathcal{C}>0$ exists (Gottlieb and Shu, 1998; Kraaijevanger, 1991). If we increase the number of stages, however, we can find good fourth-order SSP explicit Runge–Kutta methods. A numerically optimal five-stage method was found in Kraaijevanger (1991) and again independently in Spiteri and Ruuth (2002), and guaranteed optimal in Ruuth (2006). This method has SSP coefficient $\mathcal{C}=1.508$, and effective SSP coefficient $\mathcal{C}_{\text{eff}}=0.302$. In Ruuth (2006), Ruuth presented ten low-storage methods, of orders $p=3$ and $p=4$ resulting from numerical optimization. Some of these methods are guaranteed optimal, others are the best found in extensive numerical searches.

An alternative that is widely gaining popularity is the low-storage ten-stage fourth-order explicit SSP Runge–Kutta method found in Ketcheson (2008)

$$\beta_{i,i-1}=\begin{cases}\frac{1}{6} & i\in\{1\ldots4,6\ldots9\}\\ \frac{1}{15} & i=5\\ \frac{1}{10} & i=10\end{cases}\qquad \alpha_{i,i-1}=\begin{cases}1 & i\in\{1\ldots4,6\ldots9\}\\ \frac{2}{5} & i=5\\ \frac{3}{5} & i=10\end{cases}\tag{11a}$$

$$\beta_{10,4}=\frac{3}{50}\qquad \alpha_{10,4}=\frac{9}{25}\tag{11b}$$

$$\alpha_{5,0}=\frac{3}{5}\qquad \alpha_{10,0}=\frac{1}{25}.\tag{11c}$$

This method has an SSP coefficient $\mathcal{C}=6$, which is provably optimal. It is also low storage and has attractive rational coefficients.

A potential problem for Runge–Kutta methods with many stages is the amplification of roundoff errors occurring in the intermediate stages. This situation can lead to instabilities. For this reason, we wish to ensure that the SSP

TABLE 1 Effective SSP Coefficients $\mathcal{C}_{\text{eff}}$ of Best Known Explicit (Top) and Implicit (Bottom) Runge–Kutta Methods

	Order		
Stages	*2*	*3*	*4*
1	–	–	–
2	**0.5**	–	–
3	**0.67**	**0.33**	–
4	**0.75**	**0.5**	–
5	**0.8**	**0.53**	0.30
6	**0.83**	**0.59**	0.38
7	**0.86**	**0.61**	0.47
8	**0.88**	**0.64**	0.52
9	**0.89**	**0.67**	0.54
10	**0.9**	**0.68**	**0.60**
11	**0.91**	**0.69**	0.59

	Order				
Stages	*2*	*3*	*4*	*5*	*6*
1	2	–	–	–	–
2	2	1.37	–	–	–
3	2	1.61	0.68	–	–
4	2	1.72	1.11	0.29	
5	2	1.78	1.21	0.64	
6	2	1.82	1.30	0.83	0.030
7	2	1.85	1.31	0.89	0.038
8	2	1.87	1.33	0.94	0.28
9	2	1.89	1.34	0.99	0.63
10	2	1.90	1.36	1.01	0.81
11	2	1.91	1.38	1.03	0.80

A dash indicates that SSP methods of this type cannot exist. A blank space indicates that no SSP methods of this type were found. Bold entries indicate the methods have the optimal SSP coefficient for linear problems.

methods we use are *internally stable*. The methods in this section have been shown to satisfy this property as well (Ketcheson, 2008).

As mentioned above, the optimal contractive methods are also optimal SSP methods. An extensive study of contractive linear multistep methods (Lenferink, 1989, 1991) implies that for explicit s-step linear multistep methods of order p, the SSP coefficient $\mathcal{C} \leq \frac{s-p}{s-1}$ for $s > 1$; while for implicit methods of order $p > 1$, $\mathcal{C} \leq 2$. Existence of SSP linear multistep methods of arbitrarily high order was proved for explicit methods in Németh and Ketcheson (2015). But in practice high-order accurate methods of this type are subject to very small time step restrictions and require very many steps.

4.3 Optimal Implicit Methods

If we have a spatial discretization that satisfies (6) for some norm (or convex functional) then the fully discrete solution will also be strongly stable, in the same norm (or convex functional), for the implicit Euler method, without any time step restriction (Higueras, 2004; Hundsdorfer et al., 2003). However, all Runge–Kutta, linear multistep, and other general linear methods of greater than first order have finite SSP coefficient (Spijker, 1983). The question becomes whether the allowable step size is large enough to offset the additional computational cost of solving an algebraic system at each step.

For linear multistep methods, the results of Lenferink (1991) imply that any linear multistep method of order $p > 1$ has SSP coefficient no greater than $\mathcal{C} = 2$ (Hundsdorfer et al., 2003). Even when special starting procedures are used the step-size restrictions for the implicit multistep methods are hardly better than those of explicit methods. Furthermore, methods of this type with order $p > 1$ are subject to the same maximal SSP coefficient of two (Hundsdorfer and Ruuth, 2005). Thus, although no order barrier on implicit SSP multistep methods exists (Sand, 1986), these step-size restrictions are too severe to make the use of these methods efficient for typical cases.

For implicit Runge–Kutta methods, the size of this finite time step has not been established analytically. However, extensive numerical searches suggest that the time step restrictions for implicit SSP methods are limited to twice the number of stages, so that their *effective SSP coefficient* is bounded by $\mathcal{C}_{\text{eff}} \leq 2$ (Ketcheson et al., 2009). Implicit SSP Runge–Kutta methods of order greater than six do not exist. Methods of up to order five were found in Kraaijevanger (1991). Later, Ferracina and Spjker (2008) showed that optimal singly diagonally implicit SSP Runge–Kutta methods may exist for up to fourth order and found such methods. In this work, they also conjectured the form of optimal second and third-order singly diagonally implicit SSP Runge–Kutta methods with any number of stages. Fully implicit SSP Runge–Kutta methods were investigated via numerical optimization in Ketcheson et al. (2009), and the first sixth-order method was found in that work. Results therein showed that the second- and third-order optimal methods among the class of all fully implicit methods were the same singly diagonally implicit methods found in Ferracina and Spjker (2008). The optimal methods of fourth through sixth order were found to be diagonally implicit.

The results on the SSP coefficient bound for implicit methods have been disappointing. When the time step is limited by a linear stability requirement, or even by a nonlinear stability requirement involving classical stability properties (such as absolute stability of B-stability), there exist some well-known classes of implicit methods that allow the use of arbitrarily large time steps. However, the SSP condition does not allow such large time steps. Table 1 (bottom) contains the optimal effective SSP coefficients of implicit Runge–Kutta methods. Furthermore, this bound of $\mathcal{C}_{\text{eff}} \leq 2$ is observed to be practically relevant in even the simplest applications, indicating that implicit SSP methods are unlikely to be efficient enough to out-perform the explicit methods.

4.4 Optimal SSP Runge–Kutta Methods for Linear Constant Coefficient Problems

Although SSP methods were developed for preserving nonlinear stability properties needed for the stable evolution of nonlinear hyperbolic PDEs with nonsmooth solutions, this property can be useful for linear problems as well. One reason is that the nonlinear stability properties frequently used for ensuring reliable solutions to nonlinear problems mimic the properties of the solution, such as positivity, TVD or TVB, or maximum norm stability. The spatial discretizations that preserve these properties are then also advantageous for linear problems as well, such as Maxwell's equations and linear elasticity. To guarantee that these special properties are preserved when the spatial discretizations are coupled with higher-order time discretizations, we must use SSP methods. Another benefit of SSP methods is their use from the point of view of stability analysis. For example, in Levy and Tadmor (1998), the authors used the energy method to analyze the stability of Runge–Kutta methods for ODEs resulting from coercive approximations such as those in Gottlieb and Tadmor (1991). Using this method it can be proved, for example, that the fourth-order Runge–Kutta method preserved a certain stability property with a CFL number of $\frac{1}{31}$. However, using SSP theory, one easily shows that the same stability property is preserved in the linear case under a CFL number as large as 1.

As we saw earlier, explicit SSP Runge–Kutta methods exist only up to fourth order. However, if we restrict ourselves to solving only linear autonomous problems, the order conditions simplify and this order barrier is lifted: explicit SSP Runge–Kutta methods of any linear order p_{lin} exist. Observe that for linear problems, strong stability (or monotonicity) is equivalent to contractivity. Optimally contractive explicit Runge–Kutta methods were studied by Kraaijevanger in Kraaijevanger (1986), where he gives optimal *linear* methods for many values of s and p, including $1 \leq p_{lin} \leq s \leq 10$, and $p_{lin} \in \{1, 2, 3, 4, s-1, s-2, s-3, s-4\}$ for any s. Ketcheson (2009) proposed an efficient algorithm for finding optimal methods in the class of general linear methods and found methods with very high-order accuracy that use many steps and/or stages. First- and second-order methods that have s stages and methods of order p_{lin} with $s = p_{lin}$ and $s = p_{lin} + 1$ that attain the theoretical bound can be found in Kraaijevanger (1986), Gottlieb and Gottlieb (2003) and Gottlieb et al. (2011).

The family of s-stage, linear order $p_{lin} = s$ methods with $\mathcal{C} = 1$ has the form

$$u^{(i)} = u^{(i-1)} + \Delta t F\left(u^{(i-1)}\right), \quad i = 1, \ldots, s-1$$

$$u^{(s)} = \sum_{k=0}^{s-2} \alpha_k^s u^{(k)} + \alpha_{s-1}^s \left(u^{(s-1)} + \Delta t F\left(u^{(s-1)}\right)\right),$$

where $u^{(0)} = u^n$ and $u^{n+1} = u^{(s)}$. The coefficients α_k^s of the final stage of the s-stage method are constructed iteratively by

$$\alpha_k^s = \frac{1}{k}\alpha_{k-1}^{s-1} \text{ for } k = 1, \ldots, s-2, \quad \alpha_{s-1}^m = \frac{1}{s!}, \quad \alpha_0^s = 1 - \sum_{k=1}^{s-1} \alpha_k^s,$$

starting from the coefficient of the forward Euler method $\alpha_0^1 = 1$.

These linear methods with high linear order p_{lin} have low nonlinear order $p = 2$. In Gottlieb et al. (2015) explicit SSP Runge–Kutta methods that have a high linear order $p_{lin} > 4$ while retaining the highest possible nonlinear order $p = 4$ were constructed. These methods are optimal in terms of SSP coefficient and order for linear problems, without compromising order when applied to nonlinear problems.

It was observed that methods that have nonlinear order $p = 3$ and linear order $p_{lin} = 2, \ldots, 12$ have the same SSP coefficients and the same linear stability regions as the 'linear' methods that have nonlinear order $p = 2$ and linear order $p_{lin} = 2, \ldots, 12$ (see Table 2). The methods that have nonlinear order $p = 4$ and linear order $p_{lin} = 2, \ldots, 12$ have slightly lower SSP coefficients in some cases. However, in many cases the SSP coefficients of the explicit Runge–Kutta methods of nonlinear order $p = 4$ and linear order $p_{lin} > 4$ are identical to the SSP coefficients of the methods with nonlinear order $p = 2$ and linear order $p_{lin} > 4$. While the order barrier of $p \leq 4$ for explicit SSP Runge–Kutta methods indicates the importance of the nonlinear order to the SSP property, we clearly see that the *size* of the SSP coefficient is typically more constrained by the linear order conditions. Thus, if a method of high linear order is desired, one need not sacrifice the higher nonlinear order.

4.5 Optimal Multistep Runge–Kutta Methods

To break the order barrier of SSP Runge–Kutta methods while increasing the SSP coefficient seen in multistep methods, we can combine the two approaches in a multistep Runge–Kutta method of the form:

TABLE 2 SSP Coefficients for Explicit Runge–Kutta Methods With Nonlinear Order $p = 2$ or $p = 3$ and Linear Order $p_{lin} > 4$

s	$p_{lin} = 5$	6	7	8	9	10	11	12
5	1	–	–	–	–	–	–	–
6	2	1	–	–	–	–	–	–
7	2.6506	2	1	–	–	–	–	–
8	3.3733	2.6506	2	1	–	–	–	–
9	4.1	3.3733	2.6506	2	1	–	–	–
10	4.8308	4.1	3.3733	2.6506	2	1	–	–
11	5.5193	4.8308	4.1	3.3733	2.6506	2	1	–
12	6.349	5.5193	4.686	4.1	3.3733	2.6506	2	1

$$y_1^n = u^n \tag{12a}$$

$$y_i^n = \sum_{l=1}^{k} d_{il} u^{n-k+l} + \Delta t \sum_{l=1}^{k-1} \hat{a}_{il} F(u^{n-k+l}) + \Delta t \sum_{j=1}^{i-1} a_{ij} F(y_j^n) \quad 2 \le i \le s \tag{12b}$$

$$u^{n+1} = \sum_{l=1}^{k} \theta_l u^{n-k+l} + \Delta t \sum_{l=1}^{k-1} \hat{b}_l F(u^{n-k+l}) + \Delta t \sum_{j=1}^{s} b_j F(y_j^n) \tag{12c}$$

where u^{n-k+j} are the previous steps and y_j^n the intermediate stages used to compute u^{n+1}. The SSP properties of such multistep Runge–Kutta methods were investigated in Ketcheson et al. (2012) and Bresten et al. (2015). Whereas explicit SSP Runge–Kutta methods have order at most four, explicit SSP two-step Runge–Kutta methods have order at most eight (Ketcheson et al., 2012), and adding more steps breaks this order barrier (Bresten et al., 2015). Furthermore, these methods have significantly larger SSP coefficients than typical linear multistep methods. Any general linear method has $\mathcal{C}_{\text{eff}} \le 1$ (Gottlieb et al., 2011). In Bresten et al. (2015) it was shown that second-order methods with $s > 0$ stage and $k > 1$ steps have

$$\mathcal{C} \le \frac{(k-2)s + \sqrt{(k-2)^2 s^2 + 4s(s-1)(k-1)}}{2(k-1)}. \tag{13}$$

Methods that attain this bound were found in Bresten et al. (2015). This upper bounds applies beyond methods of the form (12) to all explicit multistep multistage methods.

Numerically optimal methods of the form (12) were found in Ketcheson et al. (2012) and Bresten et al. (2015). The true benefit of these methods begins past fourth order, where explicit SSP Runge–Kutta methods are not available. Some useful methods found in this work were the fifth-order methods $(s, k, p) = (3, 4, 5)$ method with $\mathcal{C}_{eff} = 0.33$ and the $(s, k, p) = (6, 3, 5)$ method with $\mathcal{C}_{\text{eff}} = 0.404$. Competitive sixth-order methods found were the $(s, k, p) = (5, 3, 6)$ method with $\mathcal{C}_{\text{eff}} = 0.272$, or the $(s, k, p) = (6, 5, 6)$ method with $\mathcal{C}_{\text{eff}} = 0.345$ if the additional storage cost of five steps is acceptable. Seventh-order methods with little loss in the effective SSP coefficient are the $(s, k, p) = (7, 3, 7)$ method with $\mathcal{C}_{\text{eff}} = 0.243$ or the $(s, k, p) = (7, 4, 7)$ method with $\mathcal{C}_{\text{eff}} = 0.286$. The eighth-order method $(s, k, p) = (9, 5, 8)$ also has a decent effective SSP coefficient of $\mathcal{C}_{\text{eff}} = 0.229$. The ninth- and tenth-order methods do suffer from smaller effective SSP coefficients, but are still competitive with the $(s, k, p) = (9, 4, 9)$ method having $\mathcal{C}_{\text{eff}} = 0.1766$ and by adding many stages but reducing the steps we have the tenth-order method $(s, k, p) = (20, 3, 10)$ with $\mathcal{C}_{\text{eff}} = 0.0917$. These methods are efficient and practical. They have a simple form, larger SSP coefficients than any known methods of the same order of accuracy, and in many cases may be implemented with relatively modest storage requirements. These methods do not require start-up methods with high stage order methods since they do not reuse stages from previous steps. However, to maintain the SSP property, an SSP one-step starting method should be used.

The situation with implicit methods is just as we would expect from the disappointing results on both implicit SSP Runge–Kutta methods and implicit SSP multistep methods: a bound of $\mathcal{C}_{\text{eff}} \leq 2$ is observed, making implicit methods in this class not competitive for most applications.

To date, explicit SSP multistep Runge–Kutta methods have not been widely implemented for solutions of hyperbolic problems. However, they are a promising option when higher-order explicit SSP methods with competitive time steps are needed and may be of use in the future.

4.6 Strong Stability Properties of Multiderivative Methods

Another approach to breaking the order barrier seen in explicit SSP Runge–Kutta methods is to look to multiderivative methods. Furthermore, the recent use of multistage multiderivative methods as time-stepping methods for hyperbolic problems (Seal et al., 2014; Tsai et al., 2014), brings to the forefront the question of strong stability properties for these methods. A first effort in this direction was made for two-derivative methods in Christlieb et al. (2016), and in this section we describe these results.

As usual, we begin with the PDE (1) and spatially discretize it to obtain the system of ODEs (2). The spatial discretization F is designed so that it satisfies the forward Euler condition (6) for the desired convex-functional stability property $\|\cdot\|$. As an additional derivative is used, we also approximate the second derivative in time u_{tt}, represented by $\dot{F}$. We note that $\dot{F}$ can either be obtained from the ODE system (2) by making some approximation to the time derivative of the spatial discretization F, or it can be obtained from the PDE (1) as a direct approximation to $f(U)_{xt}$. To account for the effect of the $\Delta t^2 \dot{F}$ term, we can impose the condition:

$$\textbf{Second-derivative condition} \quad \| u^n + \Delta t^2 \dot{F}(u^n) \| \le \| u^n \| \quad \text{for} \quad \Delta t \le K \Delta t_{FE}, \tag{14}$$

where K is a scaling factor that compares the stability condition of the second-derivative term to that of the forward Euler term.

Given spatial discretizations F and $\dot{F}$ that satisfy (6) and (14), a set of conditions under which the two-derivative multistage method of the form (3) is SSP with coefficient $\mathcal{C}$ was given in Christlieb et al. (2016). These conditions led to the formulation of an optimization problem which enabled to find optimal explicit SSP multistage two-derivative methods of up to order five, thus breaking the SSP order barrier for explicit SSP Runge–Kutta methods.

An interesting method in this class is the unique two-stage fourth-order method (4). While the method is unique, its SSP coefficient $\mathcal{C}$ depend on the value of K in the second-derivative condition. Analytically, we can show the the value of $\mathcal{C}$ is given by the smallest positive root of the polynomial $r^4 + 4K^2r^3 - 12K^2r^2 - 24K^4r + 24K^4$. The optimal methods that breaks the order barrier for explicit Runge–Kutta methods are a family of three-stage fifth-order Runge–Kutta method given in Christlieb et al. (2016).

The SSP properties of multistep multiderivative methods can be studied using different second-derivative conditions as an alternative to (14). This is ongoing work and these methods' utility as time-evolution approaches for hyperbolic PDEs will be determined over the next decade.

4.7 Widespread Applicability of SSP Methods

SSP methods are frequently employed in the solution of hyperbolic PDEs, as they provide a guarantee of provable stability in any norm, while requiring only that the spatial discretization satisfy the desired property when coupled with forward Euler. In addition, since the stability arguments are based only on convex decompositions of high-order methods in terms of the first-order Euler method, *any* convex function will be preserved by SSP high-order time discretizations. This includes, for example, the cell entropy stability property of high-order schemes studied in Osher and Tadmor (1988) and Nessyahu and Tadmor (1990).

SSP time evolution methods have been combined with a wide variety of spatial discretizations such as ENO methods (Baiotti et al., 2005; Caiden et al., 2001; Del Zanna and Bucciantini, 2002), WENO methods (Balbás and Tadmor, 2005; Bassano, 2003; Carrillo et al., 2003; Feng et al., 2004; Labrunie et al., 2004; Pantano et al., 2007; Tanguay and Colonius, 2003; Zhang and MacFayden, 2006), discontinuous Galerkin methods (Cockburn et al., 2004), level set methods (Caiden et al., 2001; Cheng et al., 2003; Cockburn et al., 2005; Enright et al., 2002; Jin et al., 2005; Peng et al., 1999), spectral finite volume methods (Cheruvu et al., 2007; Sun et al., 2006) and spectral difference methods (Wang and Liu, 2005; Wang et al., 2007). Of particular note is the popularity of explicit SSP Runge–Kutta methods within the Runge–Kutta discontinuous Galerkin (RKDG) methods (Cockburn and Shu, 2001). Furthermore, they have been used in a large range of applications, including but not limited to compressible flow (Wang and Liu, 2005), incompressible flow (Patel and Drikakis, 2005), viscous flow (Sun et al., 2006), two-phase flow (Bassano, 2003; Caiden et al., 2001), relativistic flow (Baiotti et al., 2005; Del Zanna and Bucciantini, 2002; Zhang and MacFayden, 2006), cosmological hydrodynamics (Feng et al., 2004), magnetohydrodynamics (Balbás and Tadmor, 2005), radiation hydrodynamics (Mignone, 2005), two-species plasma flow (Labrunie et al., 2004), atmospheric transport (Cheruvu et al., 2007), large-eddy simulation (Pantano et al., 2007), Maxwell's equations (Cockburn et al., 2004), semiconductor devices (Carrillo et al., 2003), lithotripsy (Tanguay and Colonius, 2003), geometrical optics (Cockburn et al., 2005) and Schrodinger equations (Cheng et al., 2003; Jin et al., 2005). These lists are intended only as a sample and are not meant to be exhaustive. We note that all of these examples involve the use of SSP Runge–Kutta methods; it is our impression that SSP multistep methods are not widely used in applications. The value of the newer classes of SSP methods remains an open question.

5 OTHER NUMERICALLY OPTIMIZED METHODS

In Section 4 we reviewed methods that are optimized with respect to the SSP coefficient. For some applications and semidiscretizations, the SSP step size may not be relevant (due to the lack of a forward Euler monotonicity property) or may not be the limiting factor in choosing the time step. In that case it may be useful to optimize other properties of the method, or to simultaneously optimize the SSP coefficient and other properties. In this section we review some other kinds of optimized methods that are relevant to hyperbolic problems.

For discontinuous Galerkin semidiscretizations of hyperbolic problems, it is the absolutely stable step size of the linearized scheme that governs the practically useful time step. In Toulorge and Desmet (2012) and Kubatko et al. (2014), explicit Runge–Kutta methods are optimized primarily for absolute stability for discontinuous Galerkin semidiscretizations, and secondarily for SSP coefficient or low storage. The resulting schemes allow for larger stable time steps. A similar approach was applied to spectral difference semidiscretizations, optimizing also for low storage and small truncation error, in Parsani et al. (2013). This approach was shown to yield efficient integrators for solution of the 2D Euler equations, even on unstructured and highly nonuniform grids. Much earlier, Mead and Renaut (1999) developed optimized Runge–Kutta methods for pseudospectral semidiscretizations, paying attention to both absolute stability and numerical dispersion and dissipation.

Typical improvements in the step size in all these works range from a few percent up to 50–70%. Much larger improvements in absolute stability of explicit methods can be obtained for problems whose eigenvalues have large negative real parts; these typically arise from parabolic terms. Torrilhon and Jeltsch (2007) investigated methods with stability region optimized for hyperbolic–parabolic problems, and a general approach to optimizing stability regions was developed in Ketcheson and Ahmadia (2012).

For linear applications, such as acoustics, strong stability is typically unimportant and control of dispersion or dissipation errors is paramount. A great deal of work has gone into finding optimized explicit embedded Runge–Kutta pairs for such applications; see, e.g. Bogey and Bailly (2004), Berland et al. (2006), Calvo et al. (2003, 2004), Hu et al. (1996), Tselios and Simos (2007).

Finally, we refer the reader to a comprehensive investigation of explicit Runge–Kutta pairs optimized simultaneously for various properties including storage, accuracy, SSP coefficient and absolute stability in Kennedy et al. (2000). The goal of the study was to find efficient methods for compressible Navier–Stokes flows, but several of the methods found are likely to be efficient also for purely hyperbolic problems.

Numerical optimization to determine time integrator coefficients for specific applications is now a widespread practice and will most likely become even more common in the future.

6 IMEX METHODS

Semidiscretizations of purely hyperbolic problems are not usually considered to be stiff, since the time step size dictated by the CFL condition is about the same as the step size needed to accurately capture wave propagation. However, the semidiscretizations of a hypebolic PDE can behave like a stiff problem in cases where there are fast waves that need not be resolved accurately, e.g. acoustic waves in some low mach number flows. Furthermore, problems with hyperbolic terms can be stiff because of the discretization of diffusive terms, or the presence of reaction terms with a fast timescale. Finally, stiffness of the semidiscrete system may be a result of nonuniform or anisotropic spatial grids, or in the case of a highly anisotropic flow.

Perhaps the most prominent example in which stiffness plays a major role is that of the Navier–Stokes equations, which include first-order hyperbolic terms and second-order diffusive terms. The latter lead to stiffness upon semidiscretization. In this and many other examples, the stiff terms are linear whereas some of the nonstiff terms are nonlinear. To make things concrete, in this section we consider the semidiscrete problem

$$u'(t) = F(u) + G(u),$$

where G is stiff and F is not.

In order to deal with stiffness and the attendant step-size restrictions, implicit time discretization can be used. If an implicit method is applied to the full problem—including nonlinear hyperbolic terms—the resulting nonlinear algebraic systems are extremely challenging to solve. Instead, it is common to apply implicit integration only to the stiff terms, while applying explicit integration to the nonstiff terms. Such methods are known as implicit–explicit (IMEX) or semiimplicit[a]. This approach is especially advantageous if the stiff terms are linear (as in diffusion) or if the stiff terms include no spatial derivatives (as in reaction), since then the implicit solves are much cheaper.

The simplest IMEX methods are known as fractional step methods and amount merely to alternating between the solution of $u'(t) = F(u)$ by an explicit method and $u'(t) = G(u)$ by an implicit method. This approach is only first-order accurate; it can be made second-order accurate by using Strang splitting instead. Another common second-order approach is to apply the trapezoidal rule (Crank–Nicolson) to G and three-step Adams–Bashforth to F (Zhong, 1995). To achieve higher order, additive IMEX schemes of both multistep and Runge–Kutta type have been developed. The focus has shifted away from combining existing methods to designing additive pairs from scratch. The literature on such methods (especially of Runge–Kutta type) is now too extensive to review completely here, but we will mention some examples and key developments.

Ascher et al. (1995) studied stability and accuracy of multistep IMEX schemes of up to fourth order for some convection-dominated problems and found that L-stability of the implicit method is important. Optimized second-order multistep IMEX schemes with a SSP explicit part were developed in Gjesdal (2007). IMEX multistep methods based on explicit SSP methods were constructed and compared to IMEX Runge–Kutta methods in Hundsdorfer and Ruuth (2007).

Ascher et al. (1997) studied IMEX Runge–Kutta schemes of up to fourth order, focusing on methods whose implicit part is diagonally implicit and L-stable, showing that such schemes could have larger stability regions than corresponding multistep schemes. Kennedy and Carpenter (2003) developed many IMEX Runge–Kutta schemes of up to fifth order, including error control and dense output and designed for application to convection–diffusion–reaction problems. Typically, the implicit portion of an IMEX Runge–Kutta scheme is diagonally implicit and A-stable or A(α)-stable. Sometimes a Rosenbrock or semiimplicit approach is used in which just a single Newton iteration is applied to the nonlinear algebraic equations at each stage (rather than iterating to convergence) (Zhong, 1995). The explicit part is often designed to be strong stability preserving (Pareschi and Russo, 2005).

[a]The term semiimplicit is also used sometimes to refer to Rosenbrock-type methods.

Additional design considerations that may be considered include L-stability (Pareschi and Russo, 2005), absolute stability near the imaginary axis, and low-storage implementation. For a recent example of designing schemes that possess several desired properties based on a specific field of application (see Higueras et al., 2014).

It is necessary to analyze both parts of an IMEX method together in order to reach correct conclusions about both accuracy and stability. For IMEX Runge–Kutta methods, accuracy of the explicit and implicit parts separately does not guarantee accuracy of the IMEX method; a number of additional order conditions that involve coefficients of both methods must also be satisfied. Stability analysis is generally performed by assuming the eigenvalues of F' are purely imaginary and the eigenvalues of G' are negative real, based on the Navier–Stokes equations as prototype. This leads to a 2-dimensional stability region analogous to the stability region of a traditional method.

7 EXPONENTIAL TIME DIFFERENCING

For some PDEs (including Navier–Stokes and Korteweg–de Vries, for example), semidiscretization results in a system of equations of the form

$$u_t = Lu + N(u)$$

where L is a linear operator and N is nonlinear. We can integrate this *exactly* over one time step with the formula

$$u(t^{n+1}) = e^{L\Delta t}u(t^n) + e^{L\Delta t}\int_0^{\Delta t} e^{-L\tau}N(u(t^n+\tau), t^n+\tau)d\tau \cdot$$

A variety of *exponential time-differencing* (ETD) methods have been developed by approximating the matrix exponential and the integral in this formula (Cox and Matthews, 2002). Such methods are valuable if the linear operator L is stiff, since they are usually subject to a stability condition based only on N. This is often the case when dealing with problems containing both hyperbolic terms and higher-order spatial derivatives (such as Navier–Stokes and Korteweg–de Vries).

A first-order ETD scheme can be derived by assuming $N(u)$ is constant over the interval of interest. The resulting method has the form

$$u(t^{n+1}) = e^{L\Delta t}u(t^n) + L^{-1}\left(e^{L\Delta t} - I\right)N(u(t^n)) \cdot$$

To extend this approach to higher order, we can either include more steps (leading to exponential multistep methods) or more stages (leading to exponential Runge–Kutta methods). Cox and Matthews applied a Runge–Kutta approach to develop higher-order ETD methods. The second-order one-step multistage ETD method takes the form:

$$u^{(1)} = e^{L\Delta t} u(t^n) + L^{-1}\left(e^{L\Delta t} - I\right) N(u(t^n)) \cdot$$
$$u(t^{n+1}) = e^{L\Delta t} u(t^n) + \frac{1}{\Delta t} L^{-2}\left(e^{L\Delta t} - I - \Delta t L\right) N\left(u^1\right) \cdot$$

In Kassam and Trefethen (2005), contour integration was used to derive a fourth-order ETD method that improves upon that given in Cox and Matthews (2002) in terms of stability.

ETD methods have been used widely with local discontinuous Galerkin methods (Cockburn and Shu, 1998), developed for solving nonlinear convection diffusion equations and useful for handling wave equations that contain higher-order derivatives. In general, the presence of higher-order derivatives introduces stiff terms into the discontinuous Galerkin (and local discontinuous Galerkin) discretizations, which require some form of implicit solver to enable reasonable time steps. In Xu and Shu (2005), ETD methods were used to advantage to handle the stiff terms in the local discontinuous Galerkin approximation of a variety of PDEs. The work by Xia et al. (2007) compared the SDC, an additive Runge–Kutta with a diagonally implicit IMEX approach and an exponential time differencing method were applied to solution of a local discontinuous Galerkin method applied to a series of PDEs with higher-order spatial derivatives. This work concluded that all three approaches are suitable for evolving the resulting ODEs, but pointed to the advantages afforded by the ETD methods.

A similar idea to exponential differencing is the integrating factor, or Lawson method approach (Lawson, 1969). To derive these methods, we begin with the equation

$$u_t = Lu + N(u)$$

and apply an integrating factor to the linear parts:

$$u_t - Lu = N(u) \rightarrow \left(e^{-Lt}u\right)_t = e^{-Lt} N(u) \rightarrow v_t = e^{-Lt} N\left(e^{Lt} v\right) \cdot$$

This equation for v is then integrated using a standard Runge–Kutta method. These methods were compared to the ETD Runge–Kutta approach by Cox and Matthews (2002) and found inferior in terms of accuracy; this may be because Lawson methods introduce the fast time-scale directly into the nonlinear terms, which may lead to larger error constants. Kassam and Trefethen (2005) showed that this approach works well for Burgers' equation, but was inferior to the ETD approach for the KdV and Allen–Cahn equations with spectral methods spatial discretizations (Fourier or Chebyshev, as indicated by the boundary conditions). More significantly, the integrating factor method suffered from significant instability for a nondiagonal operator L. The downside of both the ETD and Lawson method approaches is the need to compute the matrix exponential: the cost of this can become prohibitive as the spatial dimension of the problem increases.

8 MULTIRATE TIME STEPPING

When computing on a nonuniform (possibly adaptive) mesh, it may occur that some small cells lead to a CFL restriction that is much more stringent than what would be required over the rest of the mesh. This can also occur if much larger wavespeeds (or other fast dynamics) appear only in a localized region. In such cases it can be useful to apply different time step sizes over different parts of the mesh. This is referred to as *multirate* time stepping. In the most extreme approach, each computational node or cell is allowed to progress using an appropriately chosen step size; this is often referred to as *local time stepping*. More commonly, various regions of the mesh will each advance with their own step size. Typically, the step sizes are chosen in simple integer ratios so that they periodically synchronize with each other; this is known as *subcycling*.

The simplest idea in this vein is to apply, say, a single Runge–Kutta method with one-step size on one part of the grid (Region 1) and a step size (say) half as large over another part of the grid (Region 2). This setting is already enough to illustrate some of the difficulties that appear. A strategy must be chosen for providing the values corresponding to Region 1 near the boundary at the intermediate time for updating Region 2. Depending on the strategy chosen, two things can go wrong. First, the local accuracy of the full discretization at the boundary between regions may be less than the accuracy on a uniform mesh. This order reduction is in a way similar to that appearing when Runge–Kutta methods are applied to problems with time-dependent boundary conditions. Second, even if the method is conservative on a uniform mesh, it may not be conservative at the boundary between regions. Conservation is ensured if both parts of the method share the same weights. If they do not, conservation can still be preserved by using a consistent flux across the boundary. However, this may negatively impact accuracy. For a discussion of these issues with examples, we refer the reader to Hundsdorfer et al. (2014).

A multirate scheme may naturally be viewed as an additive or partitioned Runge–Kutta method. This perspective allows tools developed to understand additive methods to be applied to multirate methods; for an example of this approach see Günther and Sandu (2013).

A number of simple low-order multirate Runge–Kutta methods have been proposed in the literature specifically for hyperbolic problems, some with desirable properies such as strong stability preservation; see, e.g. Tang and Warnecke (2006), Constantinescu and Sandu (2007), Krivodonova (2010), Schlegel et al. (2009), Hundsdorfer et al. (2014) and references therein. For an example of a practical implementation of a multirate scheme in an application setting with parallel load balancing (see Seny et al., 2014).

The application of multirate methods in the presence of spatial adaptive mesh refinement is even more challenging and is an ongoing area of research.

9 PARALLEL IN TIME METHODS

Many problems involving first-order hyperbolic systems are computationally demanding. For instance, direct numerical simulation of three-dimensional turbulent flow requires the resolution of a huge range of scales and is only possible on machines with large numbers of processors. In order to make use of such machines, it is necessary to exploit concurrency in the numerical algorithms employed. The most common way of doing this is by spatial *domain decomposition*, wherein each processor is responsible for a subset of the entire spatial domain. Domain decomposition is particularly effective for hyperbolic problems because the finite speed of propagation and the use of explicit algorithms means that each processor need only communicate information about the solution near the subdomain boundary to its immediate neighbors.

Spatial concurrency does not get around the fact that such computations require a huge number of time steps, due to the fine scales involved and the nature of the CFL condition. If the fastest waves in the problem must be resolved accurately, implicit time stepping also yields no advantage in this regard. It is natural then to ask whether concurrency can be exploited within the time-stepping scheme itself. For typical integrators the answer is no. But specialized methods can be developed with the potential for concurrency. For a recent review of parallel time integration, we refer the reader to Gander (2015).

9.1 Concurrency Across the Method

Perhaps the simplest conceptual approach uses multiple independent function evaluations to achieve high-order accuracy. Algorithmically, such methods can often be expressed as Runge–Kutta methods with some stages that are independent of others. This includes for instance extrapolation methods (Deuflhard, 1985; Ketcheson and bin Waheed, 2014) and certain deferred correction methods (Christlieb et al., 2010b). Essentially, concurrency is used to increase the order of accuracy of the solution. As such, the amount of parallelism that can be achieved is typically limited to a factor equal to the order of accuracy or even less. However, these methods usually achieve excellent parallel efficiency and can be combined with spatial domain decomposition.

9.2 Concurrency Across the Time Domain

Alternatively, one may employ domain decomposition in time, but this is challenging due to causality—later blocks cannot be computed without at least some approximation of the solution from earlier blocks. These methods use a cheap ‘coarse’ integrator first in serial to compute a rough approximation over the full time domain. Then one or more increasingly expensive 'fine' integrators are used to refine the solxution over each subdomain. Examples of such algorithms include Parareal (Dai and Maday, 2013; Lions et al., 2001; Maday, 2010) and PITA (Cortial and Farhat, 2006, 2009; Farhat et al., 2006). Due to the need for an initial coarse serial integration step, the parallel efficiency of these methods is limited. Furthermore, they have largely been developed in the context of dissipative systems; dissipation is helpful because the effect of errors committed in previous temporal subdomains becomes small at later times. It is challenging to apply them to hyperbolic problems because of the lack of dissipation.

One approach that combines concurrency across the domain and across the method is the parallel full approximation scheme in space and time (PFASST) (Emmett and Minion, 2012). This scheme uses the SDC scheme and combines multiple levels of spatial and temporal refinements in a multigrid-like approach to decrease the cost of the coarse propagator relative to the fine propagator.

REFERENCES

Ascher, U.M., Ruuth, S.J., Wetton, B.T., 1995. Implicit-explicit methods for time-dependent partial differential equations. SIAM J. Numer. Anal. 32 (3), 797–823.

Ascher, U.M., Ruuth, S.J., Spiteri, R.J., 1997. Implicit-explicit Runge-Kutta methods for time-dependent partial differential equations. Appl. Numer. Math. 25 (2), 151–167.

Baiotti, L., Hawke, I., Montero, P.J., Loffler, F., Rezzolla, L., Stergioulas, N., Font, J.A., Seidel, E., 2005. Three-dimensional relativistic simulations of rotating neutron-star collapse to a Kerr black hole. Phys. Rev. D 71, 024035.

Balbás, J., Tadmor, E., 2005. A central differencing simulation of the Orszag-Tang Vortex system. IEEE Trans. Plasma Sci. 33 (2), 470–471.

Balsara, D.S., Rumpf, T., Dumbser, M., Munz, C.-D., 2009. Efficient, high accuracy ADER-WENO schemes for hydrodynamics and divergence-free magnetohydrodynamics. J. Comput. Phys. 228 (7), 2480–2516.

Bassano, E., 2003. Numerical simulation of thermo-solutal-capillary migration of a dissolving drop in a cavity. Int. J. Numer. Methods Fluids 41, 765–788.

Berland, J., Bogey, C., Bailly, C., 2006. Low-dissipation and low-dispersion fourth-order Runge-Kutta algorithm. Comput. Fluids 35, 1459–1463.

Bogey, C., Bailly, C., 2004. A family of low dispersive and low dissipative explicit schemes for flow and noise computations. J. Comput. Phys. 194 (1), 194–214.

Bresten, C., Gottlieb, S., Grant, Z., Higgs, D., Ketcheson, D.I., Németh, A., 2015. Strong stability preserving multistep Runge-Kutta methods. Math. Comput. 84, 2743–2761.

Butcher, J.C., 1965. A modified multistep method for the numerical integration of ordinary differential equations. J. Assoc. Comput. Mach. 12 (1), 124–135.

Butcher, J.C., 1996. General linear methods. Comput. Math. Appl. 31 (4–5), 105–112.

Butcher, J., 2006. General linear methods. Acta Numer. 15, 157–256.

Butcher, J., 2009. General linear methods for ordinary differential equations. Math. Comput. Simul. 79 (6), 1834–1845.

Caiden, R., Fedkiw, R.P., Anderson, C., 2001. A numerical method for two-phase flow consisting of separate compressible and incompressible regions. J. Comput. Phys. 166, 1–27.

Calvo, M., Franco, J.M., Rández, L., 2003. Minimum storage Runge-Kutta schemes for computational acoustics. Comput. Math. Appl. 45, 535–545.

Calvo, M., Franco, J.M., Randez, L., 2004. A new minimum storage Runge-Kutta scheme for computational acoustics. J. Comput. Phys. 201 (1), 1–12.

Carpenter, M.H., Gottlieb, D., Abarbanel, S., Don, W.-S., 1995. The theoretical accuracy of Runge-Kutta time discretizations for the initial boundary value problem: a study of the boundary error. SIAM J. Sci. Comput. 16 (6), 1241–1252.

Carrillo, J., Gamba, I.M., Majorana, A., Shu, C.-W., 2003. A WENO-solver for the transients of Boltzmann-Poisson system for semiconductor devices: performance and comparisons with Monte Carlo methods. J. Comput. Phys. 184, 498–525.

Chan, R.P.K., Tsai, A.Y.J., 2010. On explicit two-derivative Runge-Kutta methods. Numer. Algorithms 53, 171–194.

Cheng, L.-T., Liu, H., Osher, S., 2003. Computational high-frequency wave propagation using the level set method, with applications to the semi-classical limit of Schrödinger equations. Comm. Math. Sci. 1, 593–621.

Cheruvu, V., Nair, R.D., Turfo, H.M.A., 2007. A spectral finite volume transport scheme on the cubed-sphere. Appl. Numer. Math. 57, 1021–1032.

Christlieb, A.J., Macdonald, C.B., Ong, B.W., 2010. Parallel high-order integrators. SIAM J. Sci. Comput. 32 (2), 818–835.

Christlieb, A.J., Macdonald, C.B., Ong, B.W., 2010. Parallel high-order integrators. SIAM J. Sci. Comput. 32 (2), 818–835.

Christlieb, A.J., Ong, B., Qiu, J.-M., 2010. Integral deferred correction methods constructed with high order Runge-Kutta integrators. Math. Comput. 79, 761–783.

Christlieb, A., Gottlieb, S., Grant, Z., Seal, D.C., 2016. Explicit strong stability preserving multistage two-derivative time-stepping schemes. J. Sci. Comput. 68, 914–942.

Cockburn, B., Shu, C.-W., 1989. TVB Runge-Kutta local projection discontinuous Galerkin finite element method for conservation laws. II: General framework. Math. Comput. 52, 411–435.

Cockburn, B., Shu, C.-W., 1998. The local discontinuous Galerkin method for time-dependent convection-diffusion systems. SIAM J. Numer. Anal. 35 (6), 2440–2463.

Cockburn, B., Shu, C.-W., 2001. Runge-Kutta discontinuous Galerkin methods for convection-dominated problems. J. Sci. Comput. 16, 173–261.

Cockburn, B., Li, F., Shu, C.-W., 2004. Locally divergence-free discontinuous Galerkin methods for the Maxwell equations. J. Comput. Phys. 194, 588–610.

Cockburn, B., Qian, J., Reitich, F., Wang, J., 2005. An accurate spectral/discontinuous finite-element formulation of a phase-space-based level set approach to geometrical optics. J. Comput. Phys. 208, 175–195.

Constantinescu, E.M., Sandu, A., 2007. Multirate timestepping methods for hyperbolic conservation laws. J. Sci. Comput. 33 (3), 239–278.

Cortial, J., Farhat, C., 2006. A time-parallel implicit methodology for the near-real-time solution of systems of linear oscillators. In: Biegler, L., Ghattas, O., Heinkenschloss, M., Keyes, D., van Bloemen Wanders, B. (Eds.), Real-Time PDE-Constrained Optimization. Springer, pp. 697–724.

Cortial, J., Farhat, C., 2009. A time-parallel implicit method for accelerating the solution of non-linear structural dynamics problems. Int. J. Numer. Methods Eng. 77 (4), 451–470.

Cox, S., Matthews, P., 2002. Exponential time differencing for stiff systems. J. Comput. Phys. 176 (2), 430–455.

Dai, X., Maday, Y., 2013. Stable parareal in time method for first- and second-order hyperbolic systems. SIAM J. Sci. Comput. 35 (1), A52–A78.

Del Zanna, L., Bucciantini, N., 2002. An efficient shock-capturing central-type scheme for multi-dimensional relativistic flows: I. Hydrodynamics. Astron. Astrophys. 390, 1177–1186.

Deuflhard, P., 1985. Recent progress in extrapolation methods for ordinary differential equations. SIAM Rev. 27 (4), 505–535.

Dutt, A., Greengard, L., Rokhlin, V., 2000. Spectral deferred correction methods for ordinary differential equations. BIT Numer. Math. 40, 241–266.

Emmett, M., Minion, M.L., 2012. Toward an efficient parallel in time method for partial differential equations. Commun. Appl. Math. Comput. Sci. 7 (1), 105–132.

Enright, D., Fedkiw, R., Ferziger, J., Mitchell, I., 2002. A hybrid particle level set method for improved interface capturing. J. Comput. Phys. 183, 83–116.

Farhat, C., Cortial, J., Dastillung, C., Bavestrello, H., 2006. Time-parallel implicit integrators for the near-real-time prediction of linear structural dynamic responses. Int. J. Numer. Methods Eng. 67 (5), 697–724.

Feng, L., Shu, C., Zhang, M., 2004. A hybrid cosmological hydrodynamic/*N*-body code based on a weighted essentially nonoscillatory scheme. Astrophys. J. 612 (1), 1–13.

Ferracina, L., Spijker, M.N., 2004. Stepsize restrictions for the total-variation-diminishing property in general Runge-Kutta methods. SIAM J. Numer. Anal. 42, 1073–1093.

Ferracina, L., Spijker, M.N., 2005. An extension and analysis of the Shu-Osher representation of Runge-Kutta methods. Math. Comput. 249, 201–219.

Ferracina, L., Spjker, M., 2008. Strong stability of singly-diagonally-implicit Runge-Kutta methods. Appl. Numer. Math. 58, 1675–1686.

Gander, M.J., 2015. 50 years of time parallel time integration. In: Carraro, Th., Geiger, M., Körkel, S., Rannacher, R. (Eds.), Multiple Shooting and Time Domain Decomposition Methods, Springer International Publishing, Switzerland, pp. 69–113. MuS-TDD, Heidelberg, May 6–8, 2013.

Gear, C., 1965. Hybrid methods for initial value problems in ordinary differential equations. Soc. Ind. Appl. Math. B Numer. Anal. 2 (1), 69–86.

Gekeler, E., Widmann, R., 1986. On the order conditions of Runge-Kutta methods with higher derivatives. Numer. Math. 50 (2), 183–203.

Gjesdal, T., 2007. Implicit-explicit methods based on strong stability preserving multistep time discretizations. Appl. Numer. Math. 57 (8), 911–919.

Gottlieb, S., Gottlieb, L.-A., 2003. Strong stability preserving properties of Runge-Kutta time discretization methods for linear constant coefficient operators. J. Sci. Comput. 18, 83–109.

Gottlieb, S., Shu, C.-W., 1998. Total variation diminishing Runge-Kutta schemes. Math. Comput. 67, 73–85.

Gottlieb, D., Tadmor, E., 1991. The CFL condition for spectral approximations to hyperbolic initial-boundary value problems. Math. Comput. 56, 565–588.

Gottlieb, S., Shu, C.-W., Tadmor, E., 2001. Strong stability preserving high-order time discretization methods. SIAM Rev. 43, 89–112.

Gottlieb, S., Ketcheson, D.I., Shu, C.-W., 2011. Strong Stability Preserving Runge-Kutta and Multistep Time Discretizations. World Scientific Press, New Jersey, London, Singapore.

Gottlieb, S., Grant, Z., Higgs, D., 2015. Optimal explicit strong stability preserving Runge-Kutta methods with high linear order and optimal nonlinear order. Math. Comput. 84, 2743–2761.

Gragg, W.B.H.J.S., 1964. Generalized multistep predictor-corrector methods. J.ACM 11 (2), 188–209.

Günther, M., Sandu, A., 2013. Multirate generalized additive Runge-Kutta methods. Numer. Math, 133 (3), 497–524.

Harten, A., 1983. High resolution schemes for hyperbolic conservation laws. J. Comput. Phys. 49, 357–393.

Higueras, I., 2004. On strong stability preserving time discretization methods. J. Sci. Comput. 21, 193–223.

Higueras, I., 2005. Representations of Runge-Kutta methods and strong stability preserving methods. SIAM J. Numer. Anal. 43, 924–948.

Higueras, I., Happenhofer, N., Koch, O., Kupka, F., 2014. Optimized strong stability preserving IMEX Runge-Kutta methods. J. Comput. Appl. Math. 272, 116–140.

Hu, F., Hussaini, M., Manthey, J., 1996. Low-dissipation and low-dispersion Runge-Kutta schemes for computational acoustics. J. Comput. Phys. 124 (1), 177–191.

Hundsdorfer, W., Ruuth, S.J., 2005. On monotonicity and boundedness properties of linear multistep methods. Math. Comput. 75, 655–672.

Hundsdorfer, W., Ruuth, S.J., 2007. IMEX extensions of linear multistep methods with general monotonicity and boundedness properties. J. Comput. Phys. 225, 2016–2042.

Hundsdorfer, W., Ruuth, S.J., Spiteri, R.J., 2003. Monotonicity-preserving linear multistep methods. SIAM J. Numer. Anal. 41, 605–623.

Hundsdorfer, W., Ketcheson, D.I., Savostianov, I., 2014. Error analysis of explicit partitioned Runge-Kutta schemes for conservation laws. J. Sci. Comput. 63 (3), 633–653.

Jackiewicz, Z., 2009. General Linear Methods for Ordinary Differential Equations. John Wiley and Sons, Ltd. Hoboken, NJ.

Jameson, A., Schmidt, W., Turkel, E., 1981. Numerical solutions of the Euler equations by finite volume methods using Runge-Kutta time-stepping schemes. In: AIAA 14th Fluid and Plasma Dynamic Conference. vol. 81, 1259.

Jin, S., Liu, H., Osher, S., Tsai, Y.-H.R., 2005. Computing multivalued physical observables for the semiclassical limit of the Schrödinger equation. J. Comput. Phys. 205, 222–241.

Kassam, A.-K., Trefethen, L.N., 2005. Fourth-order time-stepping for stiff PDEs. SIAM J. Sci. Comput. 26 (4), 1214–1233.

Kastlunger, K., Wanner, G., 1972. On Turan type implicit Runge-Kutta methods. Computing (Arch. Elektron. Rechnen) 9, 317–325.

Kastlunger, K.H., Wanner, G., 1972. Runge Kutta processes with multiple nodes. Computing (Arch. Elektron. Rechnen) 9, 9–24.

Kennedy, C.A., Carpenter, M.H., 2003. Additive Runge-Kutta schemes for convection-diffusion-reaction equations. Appl. Numer. Math. 44, 139–181.

Kennedy, C.A., Carpenter, M.H., Lewis, R.M., 2000. Low-storage, explicit Runge-Kutta schemes for the compressible Navier-Stokes equations. Appl. Numer. Math. 35 (3), 177–219.

Ketcheson, D.I., 2008. Highly efficient strong stability preserving Runge-Kutta methods with low-storage implementations. SIAM J. Sci. Comput. 30, 2113–2136.

Ketcheson, D.I., 2009. Computation of optimal monotonicity preserving general linear methods. Math. Comput. 78 (267), 1497–1513.

Ketcheson, D.I., Ahmadia, A.J., 2012. Optimal stability polynomials for numerical integration of initial value problems. Commun. Appl. Math. Comput. Sci. 7 (2), 247–271.

Ketcheson, D.I., bin Waheed, U., 2014. A comparison of high order explicit Runge-Kutta, extrapolation, and deferred correction methods in serial and parallel. Comm. App. Math. Comp. Sci. 9 (2), 175–200.

Ketcheson, D.I., Macdonald, C.B., Gottlieb, S., 2009. Optimal implicit strong stability preserving Runge-Kutta methods. Appl. Numer. Math. 59 (2), 373–392.

Ketcheson, D.I., Gottlieb, S., Macdonald, C.B., 2012. Strong stability preserving two-step Runge-Kutta methods. SIAM J. Numer. Anal. 49, 2618–2639.

Kraaijevanger, J.F.B.M., 1986. Absolute monotonicity of polynomials occurring in the numerical solution of initial value problems. Numer. Math. 48 (3), 303–322.

Kraaijevanger, J.F.B.M., 1991. Contractivity of Runge-Kutta methods. BIT Numer. Math. 31, 482–528.

Krivodonova, L., 2010. An efficient local time-stepping scheme for solution of nonlinear conservation laws. J. Comput. Phys. 229 (22), 8537–8551.

Kubatko, E.J., Yeager, B.A., Ketcheson, D.I., 2014. Optimal strong-stability-preserving Runge-Kutta time discretizations for discontinuous Galerkin methods. J. Sci. Comput. 60 (2), 313–344.

Kurganov, A., Tadmor, E., 2000. New high-resolution schemes for nonlinear conservation laws and convection-diffusion equations. J. Comput. Phys. 160, 241–282.

Labrunie, S., Carrillo, J., Bertrand, P., 2004. Numerical study on hydrodynamic and quasi-neutral approximations for collisionless two-species plasmas. J. Comput. Phys. 200, 267–298.

Lawson, J., 1969. Generalized Runge-Kutta processes for stable systems with large Lipschitz constants. SIAM J. Numer. Anal. 4, 372–390.

Lax, P., Wendroff, B., 1960. Systems of conservation laws. Commun. Pure Appl. Math. 13 (2), 217–237.

Lenferink, H.W.J., 1989. Contractivity-preserving explicit linear multistep methods. Numer. Math. 55, 213–223.

Lenferink, H.W.J., 1991. Contractivity-preserving implicit linear multistep methods. Math. Comp. 56, 177–199.

Levy, D., Tadmor, E., 1998. From semi-discrete to fully discrete: stability of Runge-Kutta schemes by the energy method. SIAM Rev. 40, 40–73.

Lions, J.-L., Maday, Y., Turinici, G., 2001. Re'solution dedp par un sche'ma en temps parare'el. C. R. Acad Sci. Paris Ser. I Mathematique 332, 661–668.

Liu, X.-D., Osher, S., Chan, T., 1994. Weighted essentially non-oscillatory schemes. J. Comput. Phys. 115 (1), 200–212.

Liu, Y., Shu, C.-W., Zhang, M., 2008. Strong stability preserving property of the deferred correction time discretization. J. Comput. Math. 26, 633–656.

Maday, Y., 2010. The pararcal in time algorithm. In: Magoulès, F. (Ed.), Sub-Structuring Techniques and Domain Decomposition Methods. Computational Science, Engineering & Technology Series, Saxe-Coburg Publications, United Kingdom, pp. 19–44.

Majda, A., Osher, S., 1978. A systematic approach for correcting nonlinear instabilities. the Lax-Wendroff scheme for scalar conservation laws. Numer. Math. 30, 429–452.

Mead, J.L., Renaut, R.A., 1999. Optimal Runge-Kutta methods for first order pseudospectral operators. J. Comput. Phys. 152 (1), 404–419.

Mignone, A., 2005. The dynamics of radiative shock waves: linear and nonlinear evolution. Astrophys. J. 626, 373–388.

Minion, M., 2003. Semi-implicit spectral deferred correction methods for ordinary differential equations. SIAM J. Numer. Anal. 41, 605–623.

Mitsui, T., 1982. Runge-Kutta type integration formulas including the evaluation of the second derivative. I. Publ. Res. Inst. Math. Sci. 18 (1), 325–364.

Németh, A., Ketcheson, D., 2015. Existence and optimality of strong stability preserving linear multistep methods: a duality-based approach. ArXiv preprint arXiv:1504.03930.

Nessyahu, H., Tadmor, E., 1990. Non-oscillatory central differencing for hyperbolic conservation laws. J. Comput. Phys. 87, 408–463.

Obreschkoff, N., 1940. Neue quadraturformeln. Abh. Preuss. Akad. Wiss. Math.-Nat. Kl. 4 (20), 1–20.

Ono, H., Yoshida, T., 2004. Two-stage explicit Runge-Kutta type methods using derivatives. Jpn. J. Indust. Appl. Math. 21 (3), 361–374.

Osher, S., Chakravarthy, S., 1984. High resolution schemes and the entropy condition. SIAM J. Numer. Anal. 21, 955–984.

Osher, S., Tadmor, E., 1988. On the convergence of difference approximations to scalar conservation laws. Math. Comput. 50, 19–51.

Pantano, C., Deiterding, R., Hill, D., Pullin, D., 2007. A low numerical dissipation patch-based adaptive mesh refinement method for large-eddy simulation of compressible flows. J. Comput. Phys. 221, 63–87.

Pareschi, L., Russo, G., 2005. Implicit-explicit Runge-Kutta schemes and applications to hyperbolic systems with relaxation. J. Sci. Comput. 25 (1–2), 129–155.

Parsani, M., Ketcheson, D.I., Deconinck, W., 2013. Optimized explicit Runge-Kutta schemes for the spectral difference method applied to wave propagation problems. SIAM J. Sci. Comput. 35 (2), A957–A986.

Patel, S., Drikakis, D., 2005. Effects of preconditioning on the accuracy and efficiency of incompressible flows. Int. J. Numer. Methods Fluids 47, 963–970.

Peng, D., Merriman, B., Osher, S., Zhao, H., Kang, M., 1999. A PDE-based fast local level set method. J. Comput. Phys. 155, 410–438.

Qiu, J., Dumbser, M., Shu, C., 2005. The discontinuous Galerkin method with Lax-Wendroff type time discretizations. Comput. Methods Appl. Mech. Eng. 194, 4528–4543.

Speck, R., Ruprecht, D., Emmett, M., Minion, M., Bolten, M., Krause, R., 2015. A multi-level spectral deferred correction method. BIT Numer. Math. 55 (3), 843–867.

Ruuth, S., 2006. Global optimization of explicit strong-stability-preserving Runge-Kutta methods. Math. Comp. 75, 183–207.

Ruuth, S.J., Spiteri, R.J., 2002. Two barriers on strong-stability-preserving time discretization methods. J. Sci. Commun. 17, 211–220.

Ruuth, S.J., Spiteri, R.J., 2004. High-order strong-stability-preserving Runge-Kutta methods with downwind-biased spatial discretizations. SIAM J. Numer. Anal. 42, 974–996.

Sand, J., 1986. Circle contractive linear multistep methods. BIT Numer. Math. 26, 114–122.

Schlegel, M., Knoth, O., Arnold, M., Wolke, R., 2009. Multirate Runge-Kutta schemes for advection equations. J. Comput. Appl. Math. 226, 345–357.

Seal, D.C., Guclu, Y., Christlieb, A.J., 2014. High-order multiderivative time integrators for hyperbolic conservation laws. J. Sci. Comput. 60 (1), 101–140.

Seal, D.C., Tang, Q., Xu, Z., Christlieb, A.J., 2016. An explicit high-order single-stage single-step positivity-preserving finite difference WENO method for the compressible Euler equations. J. Sci. Comput. 68 (1), 171–190.

Seny, B., Lambrechts, J., Toulorge, T., Legat, V., Remacle, J.-F., 2014. An efficient parallel implementation of explicit multirate Runge-Kutta schemes for discontinuous Galerkin computations. J. Comput. Phys. 256, 135–160.

Shintani, H., 1971. On one-step methods utilizing the second derivative. Hiroshima Math. J. 1, 349–372.

Shintani, H., 1972. On explicit one-step methods utilizing the second derivative. Hiroshima Math. J. 2, 353–368.

Shu, C.-W., 1988. Total-variation diminishing time discretizations. SIAM J. Sci. Stat. Comp. 9, 1073–1084.

Shu, C.-W., 2002. A survey of strong stability-preserving high-order time discretization methods. In: Estep, D., Tavener, S. (Eds.), Collected Lectures on the Preservation of Stability under Discretization, SIAM, Philadelphia, PA.

Shu, C.-W., Osher, S., 1988. Efficient implementation of essentially non-oscillatory shock-capturing schemes. J. Comput. Phys. 77, 439–471.

Spijker, M.N., 1983. Contractivity in the numerical solution of initial value problems. Numer. Math. 42 (3), 271–290.

Spijker, M., 2008. Stepsize conditions for general monotonicity in numerical initial value problems. SIAM J. Numer. Anal. 45 (3), 1226–1245.

Spiteri, R.J., Ruuth, S.J., 2002. A new class of optimal high-order strong-stability-preserving time discretization methods. SIAM J. Numer. Anal. 40, 469–491.

Spiteri, R.J., Ruuth, S.J., 2003. Nonlinear evolution using optimal fourth-order strong-stability-preserving Runge-Kutta methods. Math. Comput. Simul. 62, 125–135.

Stancu, D.D., Stroud, A.H., 1963. Quadrature formulas with simple Gaussian nodes and multiple fixed nodes. Math. Comp. 17, 384–394.

Strang, G., 1964. Accurate partial difference methods. II: Nonlinear problems. Numer. Math. 6, 37–46.

Strikwerda, J.C., 1989. Finite Difference Schemes and Partial Differential Equations. Wadsworth and Brooks, California.

Sun, Y., Wang, Z., Liu, Y., 2006. Spectral (finite) volume method for conservation laws on unstructured grids VI: Extension to viscous flow. J. Comput. Phys. 215, 41–58.

Sweby, P., 1984. High resolution schemes using flux limiters for hyperbolic conservation laws. SIAM J. Numer. Anal. 21, 995–1011.

Tadmor, E., 1998. Approximate solutions of nonlinear conservation laws. Advanced Numerical Approximation of Nonlinear Hyperbolic Equations, Springer-Verlag, Berlin, pp. 1–150.

Takakura, Y., 2006. Direct-expansion forms of ADER schemes for conservation laws and their verification. J. Comput. Phys. 219 (2), 855–878.

Tang, H.-Z., Warnecke, G., 2006. High resolution schemes for conservation laws and convection-diffusion equations with varying time and space grids. J. Comput. Math. Int. Ed. 24 (2), 121.

Tanguay, M., Colonius, T., 2003. Progress in modeling and simulation of shock wave lithotripsy (SWL). In: Fifth International Symposium on Cavitation (CAV2003).

Titarev, V.A., Toro, E.F., 2002. ADER: arbitrary high order Godunov approach. J. Sci. Comput. 17 (1), 609–618.

Titarev, V., Toro, E.A., 2005. ADER schemes for three-dimensional non-linear hyperbolic systems. J. Comput. Phys. 204 (2), 715–736.

Toro, E.F., Titarev, V.A., 2005. TVD fluxes for the high-order ADER schemes. J. Sci. Comput. 24 (3), 285–309.

Torrilhon, M., Jeltsch, R., 2007. Essentially optimal explicit Runge-Kutta methods with application to hyperbolic-parabolic equations. Numer. Math. 106 (2), 303–334.

Toulorge, T., Desmet, W., 2012. Optimal Runge-Kutta schemes for discontinuous Galerkin space discretizations applied to wave propagation problems. J. Comput. Phys. 231, 2067–2091.

Tsai, A.Y.J., Chan, R.P.K., Wang, S., 2014. Two-derivative Runge-Kutta methods for PDEs using a novel discretization approach. Numer. Algorithms 65, 687–703.

Tselios, K., Simos, T.E., 2007. Optimized Runge-Kutta methods with minimal dispersion and dissipation for problems arising from computational acoustics. Phys. Lett. A 363, 38–47.

Turán, P., 1950. On the theory of the mechanical quadrature. Acta Sci. Math. (Szeged) 12, 30–37. Leopoldo Fejer et Frederico Riesz LXX annos natis dedicatus, Pars A.

Vos, P.E.J., Eskilsson, C., Bolis, A., Chun, S., Kirby, R.M., Sherwin, S.J., 2011. A generic framework for time-stepping partial differential equations (PDEs): general linear methods, object-oriented implementation and application to fluid problems. Int. J. Comput. Fluid Dyn. 25 (3), 107–125.

Wang, Z., Liu, Y., 2005. The spectral difference method for the 2D Euler equations on unstructured grids. In: 17th AIAA Computational Fluid Dynamics Conference. AIAA.

Wang, Z., Liu, Y., May, G., Jameson, A., 2007. Spectral difference method for unstructured grids II: extension to the Euler equations. J. Sci. Comput. 32 (1), 45–71.

Xia, Y., Xu, Y., Shu, C.W., 2007. Efficient time discretization for local discontinuous Galerkin methods. Discrete Continuous Dyn. Syst. Ser. B 8, 677–693.

Xu, Y., Shu, C.-W., 2005. Local discontinuous Galerkin methods for two classes of two-dimensional nonlinear wave equations. Physica D 208 (1), 21–58.

Zahran, Y.H., 2008. Central ADER schemes for hyperbolic conservation laws. J. Math. Anal. Appl. 346 (1), 120–140.

Zhang, W., MacFayden, A.I., 2006. RAM: a relativistic adaptive mesh refinement hydrodynamics code. Astrophys. J. Suppl. Ser. 164, 255–279.

Zhong, X., 1995. New high-order semi-implicit Runge-Kutta schemes for computing transient nonequilibrium hypersonic flow. AIAA Paper, 95-2007.

Chapter 22

The Fast Sweeping Method for Stationary Hamilton–Jacobi Equations

H. Zhao
University of California, Irvine, CA, United States

Chapter Outline

ABSTRACT

Hamilton–Jacobi equations are a type of nonlinear hyperbolic partial differential equations that arise in many applications such as optimal control, game theory, moving interface problems, geophysics and seismology, computer vision and image processing. The fast sweeping method (FSM) is a simple and efficient iterative numerical method that can achieve optimal complexity for solving stationary convex Hamilton–Jacobi equations. In this chapter, we will give a brief survey on the FSM, its modification/extension to more general hyperbolic problems and its many successful applications in geophysics, optimal control, image processing and computer vision.

Keywords: Hamilton–Jacobi equation, Hyperbolic partial differential equation, Viscosity solution, Monotone upwind scheme, Fast sweeping method, Fast marching method, Gauss–Seidel iteration

AMS Classification Codes: 65N06, 65N12, 65N15, 35L60

Handbook of Numerical Analysis, Vol. 17. http://dx.doi.org/10.1016/bs.hna.2016.09.012

1 INTRODUCTION TO HAMILTON–JACOBI EQUATION

Hamilton–Jacobi equations (HJE) are nonlinear hyperbolic partial differential equations (PDE) typically of the form

$$H(\mathbf{x}, \nabla u) = 0 \quad \mathbf{x} \in \Omega \subset R^d, \tag{1}$$

where $H(\mathbf{x}, \mathbf{p})$ is the Hamiltonian and a boundary condition is prescribed $u = g$ at the noncharacteristic part of the boundary Ω. The characteristic equations for the above HJE are given by

$$\begin{cases} \dfrac{d\mathbf{x}}{dt} &= \nabla_{\mathbf{p}} H, \\ \dfrac{d\mathbf{p}}{dt} &= -\nabla_{\mathbf{x}} H, \\ \dfrac{du}{dt} &= \mathbf{p} \cdot \dfrac{d\mathbf{x}}{dt} = \mathbf{p} \cdot \nabla_{\mathbf{p}} H, \end{cases} \tag{2}$$

with initial conditions $\mathbf{x}(0) = \mathbf{x}_0$, $\mathbf{p}(0) = \mathbf{p}_0$, $u(0) = g(\mathbf{x}_0)$ for $\mathbf{x}_0 \in \partial\Omega$ and $\mathbf{p}_0$ satisfying a consistency condition $H(\mathbf{x}_0, \mathbf{p}_0) = 0$. As is the case for hyperbolic PDEs, information propagates from the boundary to the interior of the domain. Due to the nonlinearity, however, characteristics may intersect and a classical solution of (1) does not exist in general. A proper definition of weak solution, the viscosity solution, is defined (see Crandall and Lions, 1983; Crandall et al., 1984) as: a function $u \in C^{0,1}(\overline{\Omega})$ is a viscosity subsolution (supersolution) if for all $v \in C_0^{\infty}(\Omega)$ such that $u - v$ attains a local maximum (minimum, resp.) at some $\mathbf{x}_0 \in \Omega$, then

$$H(\mathbf{x}_0, \nabla v(\mathbf{x}_0)) \leq 0 \ \ (\geq 0, \ \text{resp.}). \tag{3}$$

and a viscosity solution is both a viscosity subsolution and supersolution. The viscosity solution can also be defined as the limit of the following quasilinear elliptic PDE,

$$H(\mathbf{x}, \nabla u^{\epsilon}) = \epsilon \Delta u^{\epsilon}, \ \ \mathbf{x} \in \Omega \subset R^d, \ \ u^{\epsilon}|_{\partial\Omega} = g, \tag{4}$$

as $\epsilon \to 0$, which is a singular perturbation of the original nonlinear hyperbolic PDE.

However, if the Hamiltonian $H(\mathbf{x}, \mathbf{p})$ is continuous and convex in $\mathbf{p}$, satisfying coercivity condition, $H(\mathbf{x}, \mathbf{p}) \to \infty$ as $|\mathbf{p}| \to \infty$ uniformly for $\mathbf{x} \in \overline{\Omega}$, and compatibility condition, $H(\mathbf{x}, 0) \leq 0$ for all $\mathbf{x} \in \Omega$, the viscosity solution corresponds to an optimal control solution (Lions, 1982) formulated as follows. Define the optical distance $l(\mathbf{x}, \mathbf{y})$ between two points as

$$l(\mathbf{x}, \mathbf{y}) = \inf_{\xi(t)} \left\{ \int_0^1 \rho(\xi(t), -\xi'(t)) dt \middle| \xi \in C^{0,1}([0,1], \overline{\Omega}), \xi(0) = \mathbf{x}, \xi(1) = \mathbf{y} \right\}, \tag{5}$$

where $\xi(t)$ is a path connecting $\mathbf{x}$, $\mathbf{y}$ and $\rho(\mathbf{x}, \mathbf{q}) = \max_{\{\mathbf{p}|H(\mathbf{x},\mathbf{p})=0\}} <\mathbf{p}, \mathbf{q}>$ is the support function for $H(\mathbf{x}, \mathbf{p}) = 0$. If the boundary data g is Lipschitz continuous and satisfies a compatibility condition, $g(\mathbf{x}) - g(\mathbf{y}) \leq l(\mathbf{x}, \mathbf{y})$ for all $\mathbf{x}, \mathbf{y} \in \partial\Omega$, then the viscosity solution agrees with the following optimal control solution:

$$u(\mathbf{x}) = \inf_{\mathbf{y} \in \partial\Omega} \{g(\mathbf{y}) + l(\mathbf{x}, \mathbf{y})\}, \tag{6}$$

In other words, $u(\mathbf{x})$ is the minimal total cost that includes the optical distance along the path and the exit cost at the boundary $\partial\Omega$ given by g among all possible paths starting at $\mathbf{x}$ and exiting Ω at $\mathbf{y}$. Furthermore, if the Hamiltonian H is homogeneous in space, i.e., H does not depend on $\mathbf{x}$, and Ω is convex, then the optical path between two points is a straight line. In particular the optical distance between two points becomes

$$l(\mathbf{x}, \mathbf{y}) = \max_{\{\mathbf{p} | H(\mathbf{p}) = 0\}} (\mathbf{x} - \mathbf{y}) \cdot \mathbf{p} \tag{7}$$

This optimal control interpretation provides a clear picture for the resolution when characteristics defined by (2) in phase space projected to the physical space (local optimal paths) intersect at point $\mathbf{x}$, the path that gives the least cost, which is the global optimal path, defines $u(\mathbf{x})$. Once $u(\mathbf{x})$ is computed, one can find the optimal path starting from any point $\mathbf{x} \in \Omega$ by solving the first ODE of the characteristic equations (2) in the physical space, where $\mathbf{p} = \nabla u$ is available. As we can see later, these nice control interpretations can provide the motivation and construction of fast algorithms for computing the viscosity solution for HJEs with convex Hamiltonians. If the Hamiltonian is nonconvex, mathematically it is not clear how to piece together solutions from all characteristics to give the single valued viscosity solution. However, the definition of viscosity based on the singular perturbation (4) can be used in numerical algorithms by incorporating numerical viscosity.

In many applications, there is a special variable, t (often it is the time), such that the Hamiltonian H is linear in u_t, then the HJE becomes a time-dependent problem with initial and boundary conditions as follows:

$$\begin{cases} u_t(\mathbf{x}, t) + H(\mathbf{x}, t, \nabla u) = 0 & (\mathbf{x}, t) \in \Omega \times (0, +\infty) \\ u(\mathbf{x}, 0) = h(\mathbf{x}) & \mathbf{x} \in \Omega \\ u(\mathbf{x}, t) = g(\mathbf{x}, t) & (\mathbf{x}, t) \in \partial\Omega \times (0, +\infty). \end{cases} \tag{8}$$

In this case, the directions of all characteristics in t always point upward. In other words, the solution at a later time can be determined from the solution at an earlier time. In particular, if the boundary condition does not depend on t, $u(\mathbf{x}, t)$ will converge to viscosity solution of the steady state HJE solution $H(\mathbf{x}, \nabla u) = 0$ with boundary condition $u(\mathbf{x}) = g(\mathbf{x})$, $\mathbf{x} \in \partial\Omega$ as $t \to \infty$. For example, this type of HJEs appear in the level set method (Osher and Sethian, 1988) which is a powerful numerical method for capturing moving interfaces and boundaries.

Here is the outline of the remaining of the chapter. We first give a brief survey of numerical methods for solving HJEs in Section 2. Then we focus on the fast sweeping method (FSM), an efficient iterative method for computing the solutions to HJEs on Eulerian meshes in Section 3.

2 SURVEY OF NUMERICAL METHODS FOR HAMILTON–JACOBI EQUATIONS

There are two important ingredients in designing a numerical method for solving a general HJE (1): a proper discretization that guarantees convergence of the numerical solution to the correct viscosity solution, and a fast solver for a large nonlinear system of equations after discretization. It is well known that a consistent and monotone scheme guarantees the convergence of the numerical solution to the viscosity solution (Barles and Souganidis, 1991; Crandall and Lions, 1984). However, monotone schemes can be of first-order accuracy at most. Since HJEs are nonlinear hyperbolic equations, their solutions have singularities (discontinuities in derivatives) in general. Hence a discretization is desirable to handle discontinuities while achieving higher-order accuracy away from discontinuities. High-order schemes based on essentially nonoscillatory (ENO)/weighted ENO (WENO) schemes (Abgrall, 1996; Osher and Shu, 1991; Qiu and Shu, 2005; Shu, 2007; Zhang and Shu, 2003) and discontinuous Galerkin (DG) method (Cheng and Shu, 2007; Hu and Shu, 1999; Shu, 2013) were developed for time-dependent HJEs (8).

Since the PDE is nonlinear, a system of nonlinear equations has to be solved after discretization. This is the most important part for developing a fast algorithm. In the case of a time-dependent HJE (8), since information propagates along characteristics from earlier time to later time, one can use explicit time marching method from the initial data to solve the discretized system with a constraint on time step due to the Courant–Friedrichs–Lewy (CFL) condition, $\frac{h}{\Delta t} \geq \sup_{\mathbf{x},\mathbf{p}} \| \nabla_p H(\mathbf{x}, \nabla u) \|$, where Δt is the time step and h is the spatial mesh size. However, for an HJE of the general form (1), which is a boundary value problem, it is more challenging to design an efficient method to solve the resulting large system of nonlinear equations. A simple approach is to add a pseudo time and turns the boundary value problem (1) to a time-dependent problem (8) with the same boundary condition and an artificial initial condition. As $t \to \infty$, initial condition will propagate out of the domain and the boundary condition will propagate into the domain, hence the steady state solution is the solution to the original boundary value problem. (This is similar to solving a Laplace equation using heat equation.) This approach can also be viewed as an iterative method using a simple Jacobian iteration. Due to the finite speed of propagation and the CFL condition for time step, the number of iterations has to be of the order $O(\frac{D(\Omega)}{h})$, where $D(\Omega)$ is the diameter of domain Ω, in order for the artificial initial condition to propagate out of the domain and the boundary condition to propagate into and throughout the domain. So with degrees of freedom $O(h^{-d})$ the complexity of such an approach is of the order $O(h^{-(d+1)})$, which is not optimal.

A faster solver for HJEs needs to exploit the hyperbolic nature of the problem, i.e., boundary condition propagates along characteristics to the interior of the domain. Moreover, conflicts have to be settled through an proper mechanism when characteristics meet to render the right viscosity solution. For nonlinear hyperbolic PDEs, such as the boundary value problem (1), the main difficulties come from the facts: (1) the characteristics have all possible directions and are unknown a priori, (2) how to enforce the right condition when characteristics intersect. However, if the Hamiltonian $H(\mathbf{x}, \mathbf{p})$ is convex in $\mathbf{p}$, faster solver of optimal complexity can be designed to solve the discretized nonlinear system by exploiting hyperbolic nature of the problem and using the optimal control interpretation of the viscosity solution. To exploit the hyperbolic nature, one needs to design a consistent monotone upwind discretization of the convex Hamiltonian, which will be discussed in details in Section 3, to enforce the correct causality or information propagation along characteristics. The key observation is that this nonlinear system resulting from a monotone upwind discretization can be put into a triangular system with the right ordering once the characteristics are known. However, as mentioned before, this ordering can be determined only after the solution is known since the ODE for the characteristics and the solution are coupled in the characteristics system (2). Hence, one needs to exploit the ordering and solution simultaneously in a clever way.

There are usually two types of approaches to solve the discretized system efficiently. One is based on iterative method. In particular, the FSM is the most efficient iterative method, which will be explained in more depth in Section 3. The FSM was first proposed in Boúe and Dupuis (1999) to solve a deterministic control problem with quadratic running cost using Markov chain approximation. Convergence was proved and a mesh-size independent small number of iterations was noticed in numerical tests. The FSM was first formulated using monotone upwind scheme for eikonal equation to compute the distance function in Zhao et al. (2000). It was extended to anisotropic eikonal equation in Tsai et al. (2003). A systematic study for the FSM was carried out in Zhao (2005) which showed that the key ingredients for the FSM are (1) a consistent *upwind* monotone scheme that allows information propagates along characteristics efficiently, and (2) Gauss–Seidel iteration with alternating orderings combined with an upwind scheme propagates information along characteristics in all directions in a systematic and effective way. Moreover, optimal complexity for the FSM and a sharp error estimate for distance function were proved in the same work. From an algebraic point of view, a small number of iterations is possible because the system of nonlinear equations resulting from a discretization using an upwind scheme can be put into a triangular form with the right ordering and Gauss–Seidel iteration with alternating orderings can sort out the system effectively. The most important advantages of FSM are its simplicity and optimal complexity $O(h^d)$ for general convex HJEs on structured and nonstructured meshes.

The number of iterations for the FSM, which depends on the Hamiltonian and the dimension d, is finite and is independent of mesh size h. The strategy of FSM as an iterative method can often be modified and extended to more general situations, e.g., high-order methods, a parallel implementation and nonconvex HJEs. However, having a finite number of iterations independent of mesh size may not hold in general. Recently, the FSM has also been extended to computing steady state solutions to hyperbolic conservation laws (Engquist et al., 2013, 2015).

The other approach is noniterative method based on Dijkstra's local one pass algorithm (Dijkstra, 1959), such as the fast marching method (FMM) (Sethian, 1996; Tsitsiklis, 1995). The key assumption for one pass methods is that a right ordering which satisfies the causality principle, i.e., the values at grid points can be determined in an order such that the value at a grid point determined later will not affect the value at a grid determined earlier, can be sorted out on the fly. During the computation, all grid points are typically labelled into three groups: (1) the accepted ones, those grid points whose values are determined (starting from the boundary), (2) the adjacent or considered ones, those grid points that are neighbours to the accepted ones and (3) untouched ones. The key step is how to compute the tentative values for those grid points in the adjacent group and determine which one(s) gets the true value(s) and move them to the accepted group. For simple HJEs, such as the eikonal equation (9) for which the characteristic direction (before characteristics intersect) align with the gradient of the solution, the causality principle simply says that smaller values determine larger values. Hence the FMM based on a monotone upwind scheme accepts the point(s) with minimum tentative value(s) from the adjacent group, which is equivalent to computing the solution for all grid points one by one in ascending order. In this case the complexity of the FMM is $O(h^d|\log h|)$, where the constant depends only on the dimension d and the factor $|\log h|$ comes from sorting the minimum value in the adjacent group in each step. With the same monotone scheme, the solutions from FMM and FSM are exactly the same since they solve the same system of discretized equations. However, for more general convex HJEs, such as anisotropic eikonal equations where the characteristic direction can form a large angle with the gradient of the solution, the original simple FMM may not work. It might happen that a point in the adjacent group cannot be determined from its immediate mesh neighbours in the accepted group. Its modified version, the ordered upwind scheme (OUM) was developed in Sethian and Vladimirsky (2003) to solve more general convex HJEs. The main modification in OUM is that the stencil for the monotone upwind discretization needs to be enlarged on the fly in order to follow the characteristic direction to reach the accepted group to enforce the causality principle correctly. Hence the stencil and discretization cannot be determined and fixed a priori. In other words, the discretized system of nonlinear equations are unknown a priori. The enlargement of the stencils, which depends on the degree of anisotropy, increases both the truncation error and the computation cost. Moreover, it is shown in Cacace et al. (2014) that any local single-pass method may break down for general convex HJEs while iterative methods, such as the FSM, always work.

3 THE FSM

3.1 The FSM on a Rectangular Grid

We start with an introduction to the FSM by solving the following eikonal equation on a rectangular grid in two dimensions:

$$\begin{cases} \| |\nabla u(\mathbf{x})\|_2 = f(\mathbf{x}) & \mathbf{x} \in \Omega \subset R^2 \\ u(\mathbf{x}) = 0 & \mathbf{x} \in \partial\Omega, \end{cases} \tag{9}$$

where $f(\mathbf{x}) > 0$ and $\|\cdot\|$ denotes the Euclidean norm of a vector. One interpretation of the solution u is to regard it as the first arrival time of a wave front initially located at $\partial\Omega$ whose normal velocity is $c(\mathbf{x}) = \frac{1}{f(\mathbf{x})}$. By turning it into a time-dependent problem with an initial function $\tilde{u}_0(\mathbf{x})$ whose zero level set is $\partial\Omega = \{\mathbf{x} \in R^2 | \tilde{u}_0(\mathbf{x}) = 0\}$,

$$\begin{cases} \tilde{u}_t(\mathbf{x},t) - c(\mathbf{x}) \| \nabla \tilde{u}(\mathbf{x},t)\|_2 = 0 & \mathbf{x} \in R^2 \\ \tilde{u}(\mathbf{x},0) = \tilde{u}_0(\mathbf{x}), \end{cases} \tag{10}$$

one gets the level set method that captures the front propagation. In particular, $\{\mathbf{x} \in R^2 | \tilde{u}(\mathbf{x},t) = 0\} = \{\mathbf{x} \in R^2 | u(\mathbf{x}) = t\}, \forall t \geq 0$. As discussed before, the complexity of solving such a time-dependent HJE is not optimal. The FSM, a simple iterative method, can solve the eikonal equation (9) with an optimal complexity.

Let $\Omega^h \subset R^2$ be a rectangular computation domain with a uniform rectangular grid (x_i, y_j), $i = 1, 2, \ldots, I$, $j = 1, 2, \ldots, J$ with a grid size h. Denote the numerical solution as $u^h_{i,j} = u^h(x_i, y_j)$.

(1) Discretization. The upwind monotone scheme at an interior grid point is

$$\begin{aligned} &[(u^h_{i,j} - u^h_{xmin})^+]^2 + [(u^h_{i,j} - u^h_{ymin})^+]^2 = f^2_{i,j} h^2 \\ &i = 2, \ldots, I-1, \quad j = 2, \ldots, J-1, \end{aligned} \tag{11}$$

where $u^h_{xmin} = \min(u^h_{i-1,j}, u^h_{i+1,j})$, $u^h_{ymin} = \min(u^h_{i,j-1}, u^h_{i,j+1})$ and $(x)^+ = \begin{cases} x & x > 0 \\ 0 & x \leq 0 \end{cases}$. One sided difference is used at the boundary of the computational domain. For example, at a left boundary point (x_1, y_j), a one sided difference is used in the x direction,

$$[(u^h_{1,j} - u^h_{2,j})^+]^2 + [(u^h_{1,j} - u^h_{ymin})^+]^2 = f^2_{1,j} h^2.$$

(2) Initialization. Enforce the boundary condition at $\partial\Omega$ by assigning $u_{i,j}^h$ at grid points (x_i, y_j) on or near $\partial\Omega$. These values are fixed in later calculations. Assign ∞ at all other grid points. These values will be updated later.

(3) Gauss–Seidel iterations with alternating sweeping orderings. At each grid whose value is not fixed during the initialization, update its value to be the solution to (11) computed from the current values of its neighbours. The computation domain is swept through by the following four alternating orderings repeatedly till convergence.

$$\begin{array}{ll} (1)\ i=1:I, j=1:J & (2)\ i=I:1, j=1:J \\ (3)\ i=I:1, j=J:1 & (4)\ i=1:I, j=J:1 \end{array}$$

Remark 1. Discretization (11) can be derived as a special case of the general Godunov numerical Hamiltonian proposed in Bardi and Osher (1991) and Rouy and Tourin (1992). It can also be derived as a special case using optimal control formulation (see Remark 3). Assume the discretization at a grid point i has the general form

$$F_i(u_i, u|_{j\in N(i)}) = 0, \tag{12}$$

where F_i denotes a relation among the value at grid i and values at its neighbours $j \in N(i)$. The scheme is monotone if $\frac{\partial F_i}{\partial u_i} \geq 0$ and $\frac{\partial F_i}{\partial u_j} \leq 0, j \in N(i)$. The scheme (11) is consistent (with first-order local truncation error), upwind ($u_{i,\,j}$ depends only on one of its neighbours in x and y direction, respectively) and monotone. Monotonicity guarantees the numerical solution to converge to the viscosity solution of the eikonal equation (Barles and Souganidis, 1991; Crandall and Lions, 1984). However, the upwind nature of the discretization is one of the crucial ingredients that make the FSM work. The discretization can be extended to any dimension and the solution to (11) can be computed explicitly (Zhao, 2005).

The key motivation of the FSM method is that in each sweeping ordering, information along characteristics with certain directions is propagated correctly due to the use of monotone upwind scheme and Gauss–Seidel iteration. So all characteristics can be divided into a finite number of groups according to their directions and each of the group can be covered by one of the sweeping orderings simultaneously. Moreover, values at all grid points are monotonically nonincreasing with iterations. Once the correct value at a grid point is obtained, i.e., the minimum value it can achieve, it will not be changed in later iterations.

In the special case that $f(\mathbf{x}) = 1$, i.e., $u(\mathbf{x})$ is the distance function to $\partial\Omega$, all characteristics are straight lines. It was shown in Zhao (2005) that 2^d iterations are needed for the FSM in d-dimensions to obtain a numerical solution of first-order accuracy. The source singularity produces a numerical error of order $O(h|\log h|)$. Fig. 1A shows that when computing the distance function

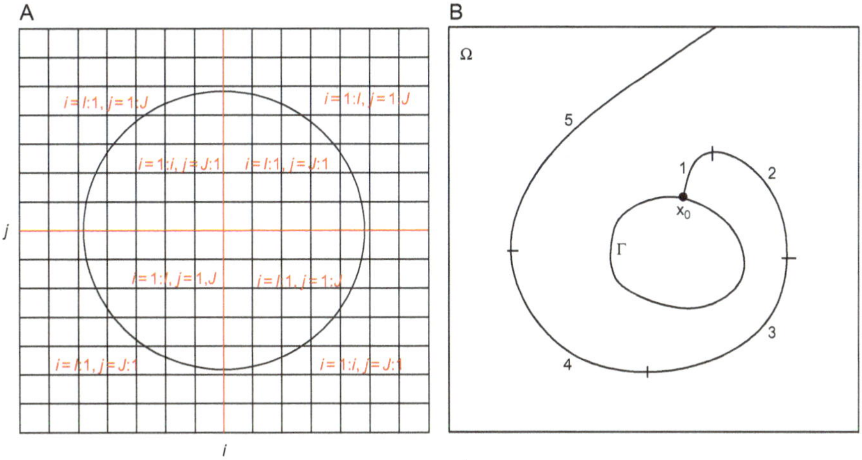

FIG. 1 (A) Distance to a circle. (B) A curved characteristic.

to a circle in 2D, all grid points in the computation domain can be classified into four groups according to the direction of the characteristic at the grid point and all grid points in each group will obtain the correct values simultaneously in one iteration with one of the four orderings. Moreover, once a grid point get its correct value it reaches the minimum and will not be affected in later iterations.

For a general $f(\mathbf{x}) > 0$, the characteristics are curves. However, as shown in Fig. 1B, each characteristic can be segmented into pieces and each piece can be covered by one of the sweeping orderings. The number of iterations is proportional to the maximum number of pieces for all characteristics. It was shown in Zhao (2005) and Qian et al. (2007a) that this number is finite for a bounded domain independent of the mesh size and is proportional to $\frac{D(\Omega)Kf_M}{f_m}$, where $D(\Omega)$ is the diameter of the domain and $K = \sup_{\mathbf{x}\in\Omega}\left|\frac{\nabla f(\mathbf{x})}{f(\mathbf{x})}\right|, f_M = \sup_{\mathbf{x}\in\Omega} f(\mathbf{x}), f_m = \inf_{\mathbf{x}\in\Omega} f(\mathbf{x})$.

Remark 2. The idea of alternating ordering was also used in Danielson's algorithm (Danielsson, 1980) which computes the distance mapping for Manhattan distance on a grid based on a discrete formulation.

3.2 The FSM for General Convex Hamilton–Jacobi Equations and on Triangular Meshes

The FSM can be extended to general convex HJEs and to triangular meshes. We first start with the design of monotone upwind scheme on a triangular mesh using control interpretation (Bornemann and Rasch, 2006). We use a two dimensional setup to illustrate the idea. Given a local triangular mesh, $\Omega^h_{\mathbf{C}}$, around the vertex **C** as shown in Fig. 2A, the numerical Hamiltonian in

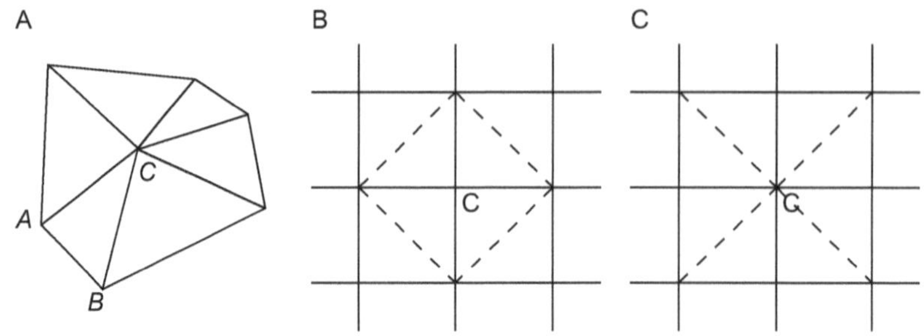

FIG. 2 A local mesh at vertex **C**. (A) Triangula mesh $\Omega^h_{\mathbf{C}}$. (B) 5-point stencil. (C) 9-point stencil.

the local mesh is $H^h_{\mathbf{C}}(\mathbf{p}) = H(\mathbf{C},\mathbf{p})$, which is a local constant approximation of the true Hamiltonian. In this local triangular mesh, the optimal control formula (6) can be used to derive the numerical discretization at **C**,

$$u^h(\mathbf{C}) = \inf_{\mathbf{y}\in\partial\Omega^h_{\mathbf{C}}} \left\{ l^h_{\mathbf{C}}(\mathbf{C},\mathbf{y}) + g^h_{\mathbf{C}}(\mathbf{y}) \right\}. \tag{13}$$

Moreover, since $H^h_{\mathbf{C}}(\mathbf{p})$ is convex in **p** and independent of **x** in this local convex polygon $\Omega^h_{\mathbf{C}}$, $l^h_{\mathbf{C}}(\mathbf{C},\mathbf{y})$ is the optical distance of the straight line connecting C and **y** with the simple form (7) corresponding to the local numerical Hamiltonian $H^h_{\mathbf{C}}(\mathbf{p})$. $g^h_{\mathbf{C}}(\mathbf{y})$ is the piecewise linear interpolation of u^h at vertices along $\partial\Omega^h_{\mathbf{C}}$. For example, for **y** on the line segment $\overline{AB}$, $g^h_{\mathbf{C}}(\mathbf{y})$ is the linear interpolation of $u^h(\mathbf{A})$ and $u^h(\mathbf{B})$. By minimizing over all straight lines starting from vertex **C**, (13) defines $u^h(\mathbf{C})$ as a function of u^h at its neighbouring vertices. In particular, this gives a discretization of a convex HJE that is monotone, since $u^h(\mathbf{C})$ is a nondecreasing function of its neighbouring values, and upwind, since $u^h(\mathbf{C})$ will be determined by one triangle with **C** being one of the vertices.

Remark 3. Finite difference schemes on a rectangular grid can be derived as a special case. For example, a 5-point stencil and 9-point stencil scheme can be derived using the triangular mesh shown in Fig. 2B and C, respectively. 9-point stencil scheme is more accurate since it has a better directional resolution. A PDE-based discretization for convex HJEs on triangular mesh was developed in Qian et al. (2007a) and it was shown to be equivalent to the above control formulation in Luo and Zhao (to appear). A finite difference scheme was derived by Kao et al. (2005) based on discretization of the Legendre transform of the numerical Hamiltonian. Semi-Lagrangian schemes based on discrete dynamic programming principle was developed in Falcone (1994) and Cristiani and Falcone (2007).

In the case of anisotropic eikonal equation of the form,

$$[\nabla u(\mathbf{x}) M(\mathbf{x}) \nabla u(\mathbf{x})]^{\frac{1}{2}} = 1, \tag{14}$$

where $M(\mathbf{x})$ is a $d \times d$ symmetric positive definite matrix and the anisotropy at $\mathbf{x}$ is defined by $\eta(\mathbf{x}) = \sqrt{\frac{\lambda_{\max}(M(\mathbf{x}))}{\lambda_{\min}(M(\mathbf{x}))}}$, explicit formula for the local discretization can be derived and solved (Bornemann and Rasch, 2006; Qian et al., 2007a).

Once the above upwind monotone scheme is applied to all vertices of the mesh, a large system of nonlinear equations is obtained. It turns out that the FSM can be extended to general convex HJEs on unstructured meshes and provide an efficient solver for the corresponding discretized nonlinear systems. The main issue different from a rectangular mesh is how to design a finite number of orderings that can cover characteristics in all directions effectively. One approach was proposed in Qian et al. (2007b) which arranges all vertices according to their distance to three (four) reference points in increasing and decreasing orders and generates 6 (8) alternating orderings in 2D (3D). It was shown that each characteristic curve can be segmented into a finite number of pieces and each piece can be covered by one of the orderings.

The FSM provides a simple, efficient and stable numerical method for computing numerical solutions to general convex HJEs on rectangular or triangular meshes. In particular, the FSM is an iterative method using a discretization scheme fixed a priori with known properties. This is a major difference from a single pass algorithm, e.g., the OUM (Sethian and Vladimirsky, 2003), which needs to modify the stencils of the discretization on the fly in order to be able to compute the correct values at grid points one by one for more general convex HJEs such as the anisotropic eikonal equation (14). This may lead to the deterioration of accuracy as well as breakdown for one pass algorithms as shown in Cacace et al. (2014) even on rectangular grids.

Simple improvements were proposed in Bak et al. (2010) for the implementation of the FSM which can make it even more efficient especially when the Hamiltonian changes rapidly. One improvement, called locking, is to avoid updates at points that are unnecessary by keeping track of grid points that have either already obtained the correct value or do not needed to be updated since none of its neighbours have changed during the current iteration. The second improvement, called the two queue method, keeps all of the unlocked points in a data structure so that the locked points no longer need to be visited at all.

The fast convergence of FSM as an iterative method can be understood and explained using arguments based on characteristics for HJEs. In particular the number of iterations is independent of mesh size. It depends on the Hamiltonian $H(\mathbf{x}, \mathbf{p})$, especially the inhomogeneity in $\mathbf{x}$, since it determines how fast characteristics can turn. On the other hand, it depends very mildly on the anisotropy of the Hamiltonian in $\mathbf{p}$. In a recent work (Luo and Zhao, to appear), the convergence of FSM was studied through the contraction property of the monotone upwind discretization (13) on general meshes. Using examples of convex HJEs with periodic boundary conditions, e.g., HJEs defined on a cylinder, it was shown that the mechanism of convergence for the FSM for hyperbolic problems is different from that of iterative methods for elliptic problems. In particular, it was shown the number of iterations for the FSM can be decreasing as the mesh is refined.

3.3 Extension of the FSM

There are a lot of successful applications and new developments for FSM in the past decade because of its simplicity, efficiency and generality as an iterative method. FSM is well developed and well understood for convex HJEs on general meshes when monotone upwind schemes are used. Since monotone upwind scheme is only first order in local truncation error, several works have been done to extend the FSM to high-order discretization. In Zhang et al. (2006b) the FSM using weighted essentially nonoscillatory (WENO) schemes was proposed. Fifth-order WENO FSM with Richardson Extrapolation and Lax–Wendroff-type procedure for accurate inflow boundary condition treatment was developed in Xiong et al. (2010). Recently several algorithms were developed for the FSM using high-order discontinuous Galerkin methods (Li et al., 2008; Luo, 2013; Wu and Zhang, 2015; Zhang et al., 2011) . Once high-order discretizations are used, loss of monotonicity of the scheme also means loss of the monotonicity of the numerical solution during the sweeping iterations. In order for the iterations to converge, it is important to use the solution from the first-order FSM as a good initial guess and some causality guidance for the higher-order methods. Although finite number of iterations independent of mesh size cannot be guaranteed for these high-order methods, the number of iterations of FSM is still very mildly dependent on the mesh size. The key issue is how to design a good causality criterion for high-order methods so that information propagates along characteristics and handle singularities correctly. A deferred correction on first-order FSM solution based on a second-order compact upwind scheme was developed in Benamou et al. (2010).

A common difficulty in many applications is source singularity, such as travel time computation for point sources in geophysics. Due to the hyperbolic nature of HJEs, numerical errors at sources will propagate through characteristics and contaminate the whole computation domain. The main difficulty for dealing a source singularity is due to the poor resolution in directions for all characteristics emanating from the source on an Eulerian mesh. An FSM method based on factored eikonal equation that can successfully deal with the source singularity was first proposed in Fomel et al. (2009). The key idea is to factor the solution to the eikonal equation as a product of an explicit function, whose singularity at the source agrees with the true solution, and an unknown function. A PDE is derived for the unknown function and an

FSM was developed to solve the new PDE which shares the same causality as the original eikonal equation. Further improvements and extensions and applications to travel time computation in geophysics were developed in later works (Luo et al., 2012, 2014).

Another important and challenging question is how to develop the FSM for nonconvex HJEs. The key ingredients for the success of the FSM for convex HJEs are monotone upwind scheme and Gauss–Seidel iterations with alternating sweeping orderings. Control interpretation of convex HJEs provides a simple causality of the PDE which can be easily incorporated into the discretization and during the iterative procedure. However, for nonconvex HJEs, although there are monotone schemes, e.g., the Lax–Friedrichs scheme, that guarantees the numerical solution to converge to the viscosity solution of the HJE, there is no monotone upwind schemes available. The Lax–Friedrichs scheme can be viewed as a discretization of the singular perturbed HJE (1), which is an elliptic problem, by adding the numerical viscosity. Since the Lax–Friedrichs scheme uses central difference to approximate derivatives and hence is not upwind, finite number of iterations independent of mesh size cannot be expected. However, the strategy of the FSW can still be used to accelerate the convergence of iterative methods for nonconvex HJEs as demonstrated in Kao et al. (2004) and Zhang et al. (2006a).

Since hyperbolic conservation laws is another type of first-order nonlinear hyperbolic PDEs, a natural question is whether the FSM can be designed for steady state hyperbolic conservation laws. For convex HJEs where the control interpretation gives a simple rule for causality/upwind condition along a characteristic and for picking the right value for the viscosity solution when characteristics meet: smaller value rules. The FSM provides a way to propagate information from the boundary to the whole domain correctly and efficiently. However, for hyperbolic conservation laws, there is no such simple rule. In particular, the determination of valid shocks, where propagation of information along characteristic stops, needs to be determined by both the Rankine–Hugoniot condition and the entropy condition. However, these conditions involve derivatives of the solution which is available when the correct solution is known on both sides of the shocks. Due to this difficulty, designing an efficient FSM for hyperbolic conservation laws is more challenging. A relative easy approach (Chen et al., 2013) is using the Lax–Friedrichs scheme to discretize the PDE and applying the FSM as an iterative method to solve the system of nonlinear equations after discretization. This is the same approach as computing the viscosity solution for HJEs by explicitly adding the numerical viscosity. While it was shown that the number of iterations is significantly fewer than other iterative methods, e.g., using time marching method, the number of iterations is still quite significant and increases as the mesh is refined since the Lax–Friedrichs scheme is not upwind. In Chen (2014) and Wu et al. (2016), fixed-point iterations using FSM were applied to high-order WENO schemes for solving steady state hyperbolic conservation laws. Especially, it was found that the FSM technique can improve the linear stability of high-order WENO schemes with forward Euler time discretization. Since time direction accuracy does not contribute to the accuracy of steady state solutions, the forward Euler time discretization is preferred since only one stage is involved in time evolution. However, high-order spatial discretizations may have linear stability issue when it is coupled with the forward Euler, which can be resolved by using the FSM technique. Recently a more elegant approach was proposed for steady state hyperbolic conservation laws in Engquist et al. (2013) and Engquist et al. (2015). The key idea is that the shock location and the solution, a piecewise smooth function, are determined by PDE, the jump conditions across shocks and boundary conditions together, which can be regarded as a hyperbolic free boundary (the shocks) problem. The sweeping strategy gives an upwind way to compute the smooth solution in each region from the given boundary condition all the way to the shock efficiently while the correct jump conditions and entropy conditions at the shocks couple the shock location with the solution simultaneously. The algorithm is efficient and allows the use of higher-order scheme to achieve high-order accuracy in both determination of the shock location and the solution. However, it requires some knowledge of shock structure in two and higher dimensions so that there is a way to update the solution and follow the shock development in a proper direction from certain part of the boundary condition.

To solve large scale problems in practice, parallel implementation is desirable to distribute computation cost and memory. Parallelization of the FSM using monotone upwind scheme was first developed in Zhao (2007) based on domain decomposition. The key points are (1) there is no need of overlap between two adjacent subdomains; (2) communication between two adjacent subdomains at the shared boundary is extremely simple: take the smaller value between the solutions from two subdomains. Both additive and multiplicative versions of domain decomposition method can be used. It can be easily seen that the parallelized FSM produces a monotone converging sequence of solutions and converges in a finite number of iterations. The number of iterations does not depend on the grid size but may depend on the number of subdomains. For the multiplicative version, sweeping through different orderings of the subdomains can accelerate the convergence. The FSM method can be parallelized further by noticing that all grid points can be decomposed into a union of codimension one fronts depending on the ordering. Grid points on the same front are independent of each others and can be updated simultaneously during each sweep. This observation was used to develop another way of parallel implementation of the FSM in Detrixhe et al. (2013). This strategy is desirable for HJEs for which the local solver at each grid point is computationally expensive and for a typical GPU structure consisting of many cores and a shared memory. A hybrid massively parallel FSM combing these two approaches was developed in Detrixhe and Gibou (2016).

Other successful applications of the FSM include a static linear PDE approach for multidimensional extrapolation (Aslam et al., 2014). Arbitrary orders of polynomial extrapolation can be obtained through solutions of a series of static linear PDEs. The FSM of first and second orders were used to solve the PDEs for constant, linear and quadratic extrapolation. An efficient algorithm based on the FSM for computing the Euclidean skeleton of an object directly from a point cloud representation on an underlying grid was developed in Luo et al. (2011a). The key point of this algorithm is to identify those grid points that are (approximately) on the skeleton using the closest point information of a grid point and its neighbours. The information was propagated through the domain using the FSM strategy. The idea of the FSM was used in Gao and Zhao (2009) to develop a fast iterative solver for radiative transport equation. In Luo et al. (2011b), the FSM was successfully developed for the homogenization of a class of HJEs with fast oscillations based on a new approximation for effective Hamiltonians.

ACKNOWLEDGEMENT

The work of H.Z. is partially supported by NSF grant DMS-1418422.

REFERENCES

Abgrall, R., 1996. Numerical discretization of the first-order Hamilton-Jacobi equations on triangular meshes. Comm. Pure Appl. Math. 49, 1339–1377.

Aslam, T., Luo, S., Zhao, H., 2014. A static PDE approach for multi-dimensional extrapolation using fast sweeping methods. SIAM J. Sci. Comput. 36 (6), 2907–2928.

Bak, S., McLaughlin, J., Renzi, D., 2010. Some improvement for the fast sweeping method. SIAM J. Sci. Comput. 32 (5), 2853–2874.

Bardi, M., Osher, S., 1991. The nonconvex multi-dimensional Riemann problem for Hamilton-Jacobi equations. SIAM J. Math. Anal. 22 (2), 344–351.

Barles, G., Souganidis, P.E., 1991. Convergence of approximation schemes for fully nonlinear second order equations. Asymptot. Anal. 4, 271–283.

Benamou, J.D., Luo, S., Zhao, H., 2010. A compact upwind second order scheme for Eikonal equations. J. Comput. Math. 28, 489–516.

Bornemann, F., Rasch, C., 2006. Finite-element discretization of static Hamilton-Jacobi equations based on a local variational principle. Comput. Vis. Sci. 9 (2), 57–69.

Boúe, M., Dupuis, P., 1999. Markov chain approximations for deterministic control problems with affine dynamics and quadratic cost in the control. SIAM J. Numer. Anal. 36 (3), 667–695.

Cacace, S., Cristiani, E., Falcone, M., 2014. Can local single-pass methods solve any stationary Hamilton-Jacobi-Bellman equation. SIAM J. Sci. Comput. 36 (2), 570–587.

Chen, S., 2014. Fixed-point fast sweeping WENO methods for steady state solution of scalar hyperbolic conservation laws. Int. J. Numer. Anal. Model. 11, 117–130.

Chen, W., Chou, C.S., Kao, C.Y., 2013. Lax-Friedrichs fast sweeping methods for steady state problems for hyperbolic conservation laws. J. Comput. Phys. 234, 452–471.

Cheng, Y., Shu, C.-W., 2007. A discontinuous Galerkin finite element method for directly solving the Hamilton-Jacobi equations. J. Comput. Phys. 223, 398–415.

Crandall, M.G., Lions, P.L., 1983. Viscosity solutions of Hamilton-Jacobi equations. Trans. Amer. Math. Soc. 277, 1–42.

Crandall, M.G., Lions, P.L., 1984. Two approximations of solutions of Hamilton-Jacobi equations. Math. Comput. 43, 1–19.

Crandall, M.G., Evans, L.C., Lions, P.L., 1984. Some property of viscosity solutions of Hamilton-Jacobi equations. Trans. Amer. Math. Soc. 282, 487–502.

Cristiani, E., Falcone, M., 2007. Fast semi-Lagrangian schemes for the Eikonal equation and applications. SIAM J. Numer. Anal. 45 (5), 1979–2011.

Danielsson, P., 1980. Euclidean distance mapping. Comput. Graphics Image Process. 14, 227–248.

Detrixhe, M., Gibou, F., 2016. Hybrid massively parallel fast sweeping method for static Hamilton-Jacobi equations. J. Comput. Phys. 322, 199–223.

Detrixhe, M., Min, C., Gibou, F., 2013. A parallel fast sweeping method for the Eikonal equation. J. Comput. Phys. 237 (15), 46–55.

Dijkstra, E.W., 1959. A note on two problems in connexion with graphs. Numer. Math. 1, 269–271.

Engquist, B., Froese, B.D., Tsai, R., 2013. Fast sweeping methods for hyperbolic systems of conservation laws at steady state. J. Comput. Phys. 255, 316–338.

Engquist, B., Froese, B.D., Tsai, R., 2015. Fast sweeping methods for hyperbolic systems of conservation laws at steady state II. J. Comput. Phys. 286, 70–86.

Falcone, M., 1994. The minimum time problem and its applications to front propagation. In: Visintin, A., Buttazzo, G. (Eds.), Motion by Mean Curvature and Related Topics. de Gruyter, Berlin, pp. 70–88.

Fomel, S., Luo, S., Zhao, H., 2009. Fast sweeping method for the factored Eikonal equation. J. Comput. Phys. 228 (17), 6440–6455.

Gao, H., Zhao, H., 2009. A fast forward solver of radiative transfer equation in optical imaging. Transp. Theory Stat. Phys. 38 (3), 149–192.

Hu, C., Shu, C.-W., 1999. A discontinuous Galerkin finite element method for Hamilton-Jacobi equations. SIAM J. Sci. Comput. 21, 666–690.

Kao, C.Y., Osher, S., Qian, J., 2004. Lax-Friedrichs sweeping scheme for static Hamilton-Jacobi equations. J. Comput. Phys. 196 (1), 367–391.

Kao, C.Y., Osher, S., Tsai, R., 2005. Fast sweeping method for static Hamilton-Jacobi equations. SIAM J. Numer. Anal. 42 (6), 2612–2632.

Li, F., Shu, C.-W., Zhang, Y.-T., Zhao, H., 2008. Second order discontinuous Galerkin fast sweeping method for Eikonal equations. J. Comput. Phys. 227 (17), 8191–8208.

Lions, P.L., 1982. Generalized Solutions of Hamilton-Jacobi Equations. Pitman, Boston.

Luo, S., 2013. A uniformly second order fast sweeping method for Eikonal equations. J. Comput. Phys. 241, 104–117.

Luo, S., Zhao, H., to appear. Convergence study for the fast sweeping method, Res. Math. Sci.

Luo, S., Guibas, L.J., Zhao, H., 2011a. Euclidean skeletons using closest points. Inverse Probl. Imaging 5, 95–113.

Luo, S., Yu, Y., Zhao, H., 2011b. A new approximation for effective Hamiltonians for homogenization of a class of Hamilton-Jacobi equations. SIAM J. Multiscale Model. Simul. 9 (2), 711–734.

Luo, S., Qian, J., Zhao, H., 2012. Higher-order schemes for 3-D traveltimes and amplitudes. Geophysics 77 (2), 47–56.

Luo, S., Qian, J., Burridge, R., 2014. High-order factorizations and high order schemes for point-source Eikonal equations. SIAM J. Numer. Anal. 52, 23–44.

Osher, S., Sethian, J.A., 1988. Fronts propagating with curvature dependent speech: algorithms based on Hamilton-Jacobi formulations. J. Comput. Phys. 79 (1), 12–49.

Osher, S., Shu, C.-W., 1991. High-order essentially nonoscillatory schemes for Hamilton-Jacobi equations. SIAM J. Numer. Anal. 28, 907–922.

Qiu, J., Shu, C.-W., 2005. Hermite WENO schemes for Hamilton-Jacobi equations. J. Comput. Phys. 204, 82–99.

Qian, J., Zhang, Y., Zhao, H., 2007a. A fast sweeping method for static convex Hamilton-Jacobi equations. J. Sci. Comput. 31 (1), 237–271.

Qian, J., Zhang, Y., Zhao, H., 2007b. Fast sweeping methods for Eikonal equations on triangulated meshes. SIAM J. Numer. Anal. 45 (1), 83–107.

Rouy, E., Tourin, A., 1992. A viscosity solutions approach to shape-from-shading. SIAM J. Numer. Anal. 29 (3), 867–884.

Sethian, J.A., 1996. A fast marching level set method for monotonically advancing fronts. Proc. Natl. Acad. Sci. USA 93, 1591–1595.

Sethian, J.A., Vladimirsky, A., 2003. Ordered upwind methods for static Hamilton-Jacobi equations: theory and algorithms. SIAM J. Numer. Anal. 41 (1), 325–363.

Shu, C.-W., 2007. High order numerical methods for time dependent Hamilton-Jacobi equations. In: Goh, S.S., Ron, A., Shen, Z. (Eds.), Mathematics and Computation in Imaging Science and Information Processing. Lecture Notes Series, Institute for Mathematical Sciences, National University of Singapore, vol. 11. World Scientific Press, Singapore, pp. 47–91.

Shu, C.-W., 2013. Survey on discontinuous Galerkin methods for Hamilton-Jacobi equations. In: Proceedings of the Eighth International Conference on Scientific Computing and Applications, Contemporary Mathematics, American Mathematical Society. vol. 586. American Mathematical Society, pp. 323–330.

Tsai, Y.R., Cheng, L.T., Osher, S., Zhao, H., 2003. Fast sweeping algorithms for a class of Hamilton-Jacobi equations. SIAM J. Numer. Anal. 41 (2), 673–694.

Tsitsiklis, J.N., 1995. Efficient algorithms for globally optimal trajectories. IEEE Trans. Automat. Control 40, 1528–1538.

Wu, L., Zhang, Y.-T., 2015. A third order fast sweeping method with linear computational complexity for Eikonal equations. J. Sci. Comput. 62, 198–229.

Wu, L., Zhang, Y.-T., Zhang, S., Shu, C.-W., 2016. High order fixed-point sweeping WENO methods for steady state of hyperbolic conservation laws and its convergence study. Commun. Comput. Phys. 20 (4), 835–869.

Xiong, T., Zhang, M., Zhang, Y.-T., Shu, C.-W., 2010. Fast sweeping fifth order WENO scheme for static Hamilton-Jacobi equations with accurate boundary treatment. J. Sci. Comput. 45, 514–536.

Zhang, Y.-T., Shu, C.-W., 2003. High order WENO schemes for Hamilton-Jacobi equations on triangular meshes. SIAM J. Sci. Comput. 24, 1005–1030.

Zhang, Y.-T., Zhao, H., Chen, S., 2006a. Fixed-point iterative sweeping methods for steady-states of Hamilton-Jacobi equations. Methods Appl. Anal. 13, 299–320.

Zhang, Y.-T., Zhao, H., Qian, J., 2006b. High order fast sweeping methods for static Hamilton-Jacobi equations. J. Sci. Comput. 29 (1), 25–56.

Zhang, Y.-T., Chen, S., Li, F., Zhao, H., Shu, C.-W., 2011. Uniformly accurate discontinuous Galerkin fast sweeping methods for Eikonal equations. SIAM J. Sci. Comput. 33 (4), 1873–1896.

Zhao, H., 2005. Fast sweeping method for Eikonal equations. Math. Comput. 74, 603–627.

Zhao, H., 2007. Parallel implementation of fast sweeping method. J. Comput. Math. 25 (4), 421–429.

Zhao, H., Osher, S., Merriman, B., Kang, M., 2000. Implicit and non-parametric shape reconstruction from unorganized points using variational level set method. Comput. Vis. Image Underst. 80 (3), 295–319.

Chapter 23

Numerical Methods for Hamilton–Jacobi Type Equations

M. Falcone* and R. Ferretti†
*Università di Roma "La Sapienza", Roma, Italy
†Università Roma Tre, Roma, Italy

Chapter Outline

ABSTRACT

We give an overview of numerical methods for first-order Hamilton–Jacobi equations. After a short presentation of the theory of viscosity solutions, we show their link with entropy solutions of conservation laws. Then, we review theory and construction of monotone numerical methods in finite difference and semi-Lagrangian form, also providing a numerical test which shows the main features of this class of schemes. Finally, we sketch the main ideas behind high-order methods and more recent developments.

Keywords: Hamilton–Jacobi equations, Viscosity solutions, Numerical methods, Convergence

AMS Classification Codes: 49L25, 65M06, 65M12, 65N06, 49M25

Handbook of Numerical Analysis, Vol. 17. http://dx.doi.org/10.1016/bs.hna.2016.09.018

1 INTRODUCTION AND MOTIVATIONS

The analysis and approximation of first-order partial differential equations of Hamilton–Jacobi (HJ) type have an important role in a number of fields such as fluid dynamics, optimal control and differential games, image processing and material science just to mention a few. In the 1980s, the notion of weak solutions in the viscosity sense, introduced by Crandall and Lions (1983), has had a crucial impact in giving a sound theoretical framework for both the analytical and the numerical study. The goal of this chapter is to sketch this theory and give some introductory material on the construction of approximation schemes for viscosity solutions. Due to space restrictions, we will only provide the main concepts on both theory and numerical methods—for a more complete exposition of the analytical theory, we refer the interested readers to the books by Lions (1982), Barles (1998), Bardi and Capuzzo-Dolcetta (1997) and Evans (2010). A detailed survey on the related applications and numerical methods can be found in the books by Sethian (1996), Osher and Fedkiw (2003) and Falcone and Ferretti (2014) (this latter monograph contains an expanded version of most of the material presented here).

We start by presenting two typical examples of HJ equations, arising in front propagation and optimal control problems.

1.1 Front Propagation via Level Set Method

The evolutive problem related to the *level set formulation of a front propagating in the normal direction with a (known) velocity* $c : \mathbb{R}^d \to \mathbb{R}$ is

$$\begin{cases} v_t + c(x)|Dv| = 0 & (x,t) \in \mathbb{R}^d \times (0,T), \\ v(x,0) = v_0(x) & x \in \mathbb{R}^d \end{cases} \tag{1}$$

where c is typically required to be strictly positive, and $v_0 : \mathbb{R}^d \to \mathbb{R}$ is a proper representation of the initial front Γ_0. For simplicity, assume that Γ_0 is a piecewise smooth surface, boundary of a compact domain Ω_0. Then, the initial condition v_0 must change sign on Γ_0, so that

$$\begin{cases} v_0(x) < 0 & x \in \Omega^0 \\ v_0(x) = 0 & x \in \Gamma_0 \\ v_0(x) > 0 & x \in \mathbb{R}^d \backslash \Omega_0 \end{cases}$$

and, for any $t \geq 0$, the front is identified with the 0-level set of $v(x, t)$. This approach allows for topology changes of the front, and can be extended to more general situations, such as the case of a curvature-related propagation speed (see Osher and Fedkiw, 2003; Sethian, 1996 for an extensive presentation).

1.2 The Infinite Horizon Problem

Consider the controlled system of ordinary differential equations

$$\begin{cases} \dot{y}(t) = f(y(t), \alpha(t)) \\ y(0) = x \in \mathbb{R}^d, \end{cases} \tag{2}$$

where $y \in \mathbb{R}^d$ $\alpha \in \mathcal{A} := \{\alpha : [0, +\infty[\to A, \alpha \text{ measurable}\}$ and $A \subset \mathbb{R}^M$ is compact. Define a cost functional as

$$J_x(\alpha) := \int_0^\infty g(y(s), \alpha(s)) e^{-\lambda s} ds,$$

where $\lambda > 0$ is a discount factor for the costs. The *value function* is defined as $v(x) := \inf_{\alpha \in \mathcal{A}} J_x(\alpha)$, and via the Dynamic Programming principle one can derive the stationary HJ equation

$$\lambda v(x) + \max_{a \in A} \{-f(x,a) \cdot \nabla u(x) - g(x,a)\} = 0, \; x \in \mathbb{R}^d. \tag{3}$$

A classical result shows that, under general assumptions, the value function is the unique viscosity solution of (3). A similar characterization can be obtained for the value function of the *finite horizon problem* of optimal control theory, leading to an evolutive HJ equation (see, e.g., Bardi and Capuzzo-Dolcetta, 1997 for more details).

In what follows, we will mainly treat the case of first-order HJ equations with a convex Hamiltonian H. Nonetheless, nonconvex Hamiltonians may occur in various relevant applications, such as differential games (Fleming and Souganidis, 1989; Isaacs, 1965). On the other hand, stochastic optimal control problems lead to consider second-order HJ equations, for which the books by Kushner and Dupuis (2001) and Fleming and Soner (1993) provide classical references.

2 BASICS ON VISCOSITY SOLUTIONS

Let us start with the stationary model problem of Dirichlet type in an open subset $\Omega \subset \mathbb{R}^d$,

$$\begin{cases} H(x, v, Dv) = 0 & x \in \Omega \\ v(x) = b(x) & x \in \partial\Omega, \end{cases} \tag{4}$$

where b is a given boundary condition, and $H : \Omega \times \mathbb{R} \times \mathbb{R}^d \to \mathbb{R}$ is the *Hamiltonian function* which will be required to satisfy the basic assumptions

(A1) $H(\cdot, \cdot, \cdot)$ is uniformly continuous

(A2) $H(x, v, \cdot)$ is convex

(A3) $H(x, \cdot, p)$ is monotone

In order to show that solutions are not expected to be smooth, we consider the following Dirichlet problem for the *eikonal equation* in one space dimension:

$$\begin{cases} |v_x| = 1 & x \in \Omega = (-1, 1) \\ v(x) = 0 & x = \pm 1 \end{cases} \tag{5}$$

Clearly, $v_1(x) = x$ and $v_2(x) = -x$ satisfy the equation, but not the boundary conditions, and a C^1 solution cannot exist due to Rolle's Theorem. However, both functions $v_3(x) = |x| - 1$ and $v_4(x) = 1 - |x|$ satisfy (almost everywhere) the equation along with the boundary conditions. In fact, there exist infinitely many a.e. solutions of the equation, which may be constructed as piecewise affine functions with slope ± 1, satisfying the boundary conditions. Then, it is clear that the notion of "a.e. solution" is unsuitable for a uniqueness result. One possibility to single out a solution is to perform an *elliptic regularization* of the equation, in the form

$$-\varepsilon v_{xx} + |v_x| = 1, x \in \Omega$$

(with the same boundary conditions as before) and pass to the limit for $\varepsilon \to 0$. This problem has a regular solution $v^\varepsilon \in C^2(-1, 1)$ for every positive ε, and passing to the limit for vanishing ε, we get

$$\lim_{\varepsilon \to 0} v^\varepsilon(x) = \bar{v}(x) = 1 - |x|, \tag{6}$$

which will be defined to be the weak solution of our problem. This is the so-called "vanishing viscosity method", and is the origin for the name "viscosity solution".

What is now considered as the usual definition of viscosity solution makes no longer any reference to a regularization and/or a limit. In what follows, $BUC(\Omega)$ will denote the space of bounded and uniformly continuous functions over the open set Ω.

Definition 1. A function $v \in BUC(\Omega)$ is a viscosity solution of (4) if and only if, for any $\varphi \in C^1(\Omega)$, the following conditions hold:

(i) at every *local maximum* point $x_0 \in \Omega$ for $v - \varphi$,

$$H(x_0, v(x_0), D\varphi(x_0)) \leq 0$$

(ii) at every *local minimum* point $x_0 \in \Omega$ for $v - \varphi$,

$$H(x_0, v(x_0), D\varphi(x_0)) \geq 0$$

We say that v is *a viscosity sub(super)-solution* if *(i)* (resp. *(ii)*) is satisfied. Resuming the previous example, it can be proved that $\bar{v}$ is the unique solution according to this definition.

For solutions defined in this form, some good properties may be proved. First, if v is a classical C^1 solution (i.e., it satisfies the equation pointwise), then it is also a viscosity solution. Vice versa, if v is a regular viscosity solution, then it is also a classical solution. The viscosity solution v is the maximal subsolution and is the vanishing viscosity limit of the elliptic regularization.

The crucial point when dealing with viscosity solutions is to prove uniqueness. This difficulty is typically overcome by using a *comparison principle* (also called *maximum principle*), stating that if u, $v \in BUC(\Omega)$ are respectively a sub- and a super-solution for (4) and $u(x) \le v(x)$ for any $x \in \partial\Omega$, then $u(x) \le v(x)$ for any $x \in \Omega$. Indeed, let u and v be two viscosity solutions of (4). Clearly, they are both sub- and super-solutions, so that we have

$$u(x) \le v(x) \text{ for any } x \in \Omega$$

and, reverting the role of u and v, we also have

$$u(x) \ge v(x) \text{ for any } x \in \Omega$$

which implies $u(x) = v(x)$ in Ω.

A sufficient condition for uniqueness may be given as follows. Let $\varphi : \mathbb{R} \to \mathbb{R}_+$ be continuous and ω be a modulus of continuity. Assume that, for any x, $y \in \Omega$, $u \in [-R, R]$ and $p \in \mathbb{R}^n$,

(A4) $|H(x, u, p) - H(y, u, p)| \le \omega(|x - y|(1+|p|))Q_R(x, y, u, p)$,

with $Q_R(x,y,u,p) := \max(\varphi(H(x,u,p)), \varphi(H(y,u,p)))$. Then, we have the following

Theorem 1. *Let the assumptions (A1)–(A4) be satisfied. Then, the comparison principle holds for* (4), *i.e., the viscosity solution is unique.*

We mention that in general *boundary conditions* should be stated in a suitable weak sense. We refer to Barles (1998) for more details and for other types of boundary conditions (e.g., Neumann and state constraints).

For reader's convenience, we add the definition of viscosity solution, adapted for the evolutive case.

Definition 2. (Evolutive case) A function $v \in BUC(\Omega \times (0, T))$ is a viscosity solution of

$$v_t + H(x, v, Dv) = 0$$

in $\Omega \times (0, T)$ if and only if, for any $\varphi \in C^1(\Omega \times (0, T))$ the following conditions hold:

(i) at every *local maximum* point $(x_0, t_0) \in \Omega \times (0, T)$ for $u - \varphi$

$$\varphi_t(x_0, t_0) + H(x_0, v(x_0, t_0), D\varphi(x_0, t_0)) \le 0;$$

(ii) at every *local minimum* point $(x_0, t_0) \in \Omega \times (0, T)$ for $u - \varphi$

$$\varphi_t(x_0, t_0) + H(x_0, u(x_0, t_0), D\varphi(x_0, t_0)) \ge 0.$$

In some special cases, a representation formula for the viscosity solution can be constructed. This is the case for the problem

$$\begin{cases} v_t + H(Dv) = 0 & (x,t) \in \mathbb{R}^d \times (0,T), \\ v(x,0) = v_0(x) & x \in \mathbb{R}^d \end{cases} \tag{7}$$

where the Hamiltonian $H : \mathbb{R}^d \to \mathbb{R}$ is continuous and convex. Assuming that H is also coercive, i.e.,

$$\lim_{|p| \to +\infty} \frac{H(p)}{|p|} = +\infty,$$

the *Legendre–Fenchel conjugate* of H may be defined, for $p \in \mathbb{R}^d$, as

$$H^*(p) := \sup_{q \in \mathbb{R}^n} \{p \cdot q - H(q)\}.$$

In this particular case, the solution of (7) is given by the Hopf–Lax representation formula as

$$v(x,t) = \inf_{a \in \mathbb{R}^d} \{v_0(x - ta) + tH^*(a)\}. \tag{8}$$

We will see later that this formula can also be used for numerical purposes.

Last, we examine the link between entropy solutions and viscosity solutions. This link is exploited to set up numerical methods originating from conservation laws (Leveque, 1992), and, by a splitting argument, to derive multidimensional schemes (Tourin, 2006). We refer the interested reader to Crandall and Lions (1984), Bardi and Osher (1991) and to Lions and Souganidis (1995). Consider the following two problems: the evolutive HJ equation

$$\begin{cases} v_t + H(v_x) = 0 & (x,t) \in \mathbb{R} \times (0,T), \\ v(x,0) = v_0(x) & x \in \mathbb{R}, \end{cases} \tag{9}$$

and the associated conservation law

$$\begin{cases} u_t + H(u)_x = 0 & (x,t) \in \mathbb{R} \times (0,T), \\ u(x,0) = u_0(x) & x \in \mathbb{R}, \end{cases} \tag{10}$$

and define

$$v_0(x) := \int_{-\infty}^{x} u_0(\xi) d\xi.$$

It can be proved (see Corrias et al., 1995) that, if u is the entropy solution of (10), then

$$v(x,t) = \int_{-\infty}^{x} u(\xi,t) d\xi$$

is the unique viscosity solution of (9), and vice versa. The previous relationship also admits a multidimensional analogue. In fact, the viscosity solution of the Cauchy problem

$$\begin{cases} v_t + H(\nabla v) = 0 & (x,t) \in \mathbb{R}^d \times (0,T), \\ v(x,0) = v_0(x) & x \in \mathbb{R}^d, \end{cases} \tag{11}$$

is equivalent to the entropic solution of the system of conservation laws

$$\begin{cases} \boldsymbol{p}_t + \nabla H(\boldsymbol{p}) = 0 & (x,t) \in \mathbb{R}^d \times (0,T), \\ \boldsymbol{p}(x,0) = \boldsymbol{p}_0(x) = \nabla v_0(x) & x \in \mathbb{R}^d \end{cases} \tag{12}$$

where $\boldsymbol{p} := \nabla v$ (see Jin and Xin, 1998 for a sketch of the proof).

2.1 Convergence Results

To state the main convergence results for the approximation of viscosity solutions, we carry out the discretization in the usual difference scheme framework. Time is discretized with a (fixed) time step Δt, so that $t_k = k\Delta t$; space is discretized with a fixed space step Δx. A generic node will be denoted by $x_j = j\Delta x$, $j \in \mathbb{Z}^d$. We also define $\Delta := (\Delta x, \Delta t)$. In some cases, more general options can be considered, e.g., variable time steps and/or unstructured grids, but we will not treat such situations in detail (see, e.g. Abgrall, 1966).

We write the scheme in compact form as

$$V^{n+1} = S(\Delta; V^n) \tag{13}$$

where S may be defined in terms of its components S_j, $j \in \mathbb{Z}^d$. We denote by v_j^n the desired approximation of $v(x_j, t_n)$, by V^n the set of nodal values for the numerical solution at time t_n, by U (respectively, $U(t)$) that for the exact solution $v(x)$ (resp. $v(x, t)$). We also denote by W and Φ (resp. $W(t)$ and $\Phi(t)$) the sets of nodal values of generic functions $w(x)$ and $\phi(x)$ (resp. $w(x, t)$ and $\phi(x, t)$). In general, we will refer to the set of nodal values as to a (possibly infinite) vector.

In this section we collect two key results, which make use of monotonicity as a stability assumption. The *Crandall–Lions theorem* is inspired by the convergence result of monotone conservative schemes for conservation laws and assumes that the scheme structure parallels that of conservative schemes. The *Barles–Souganidis theorem* is suitable for more general situations, including second-order HJ equations, provided a comparison principle holds, and does not assume any particular structure for the scheme.

We present the result of Crandall–Lions in two space dimensions, the extension to an arbitrary number of dimensions being straightforward. Let us explicitly write the evolutive HJ equation as:

$$v_t + H(v_{x_1}, v_{x_2}) = 0. \tag{14}$$

We define an approximation of the partial derivative v_{x_i} at the point x_j by the right (partial) incremental ratio, that is

$$D_{i,j}[V] = \frac{v_{j+e_i} - v_j}{\Delta x}, \qquad i = 1, 2.$$

In agreement with the definition of schemes in *conservative form* for conservation laws, we define here the class of schemes in *differenced form.*

Definition 3. A scheme S is said to be *in differenced form* if it has the form

$$v_j^{n+1} = S_j(V^n) := v_j^n - \Delta t \mathcal{H}\big(D_{1,j-p}[V^n], \ldots, D_{1,j+q}[V^n]; D_{2,j-p}[V^n], \ldots, D_{2,j+q}[V^n]\big), \tag{15}$$

for two multiindices p and q with positive components and for a Lipschitz continuous function $\mathcal{H}$ (called the *numerical Hamiltonian*).

In practice, (15) defines schemes where the dependence on V^n appears only through its finite differences, computed on a rectangular stencil of points around the node x_j. The differenced form of a scheme lends itself to an easier formulation of the consistency condition, which is given in the following definition.

Definition 4. A scheme in differenced form is *consistent* if, for any $a, b \in \mathbb{R}$,

$$\mathcal{H}(a, \ldots, a; b, \ldots, b) = H(a, b). \tag{16}$$

On the other hand, a *monotone scheme* is defined as follows.

Definition 5. The scheme S is said to be *monotone* if

$$S(\Delta; V) - S(\Delta; W) \geq 0 \tag{17}$$

for any pair of vectors V and W such that $V - W \geq 0$, this inequality to be intended component by component.

In the nonlinear case, we expect that monotonicity may or may not hold depending on the solution propagation speed, which is related to the Lipschitz constant of v_0. We will say that the scheme is *monotone on* $[-R, R]$ if Definition 5 is satisfied for any V and W such that $|D_{i,j}[V]|, |D_{i,j}[W]| \leq R$.

We can now state the Crandall–Lions convergence theorem (see Crandall and Lions, 1984 for the proof).

Theorem 2. Let $H: \mathbb{R}^2 \to \mathbb{R}$ *be continuous, the initial condition* v_0 *be bounded and Lipschitz continuous (with Lipschitz constant* L*) on* $\mathbb{R}^2$ *and* $v_j^0 = v_0(x_j)$. *Let the scheme* (15) *be monotone on* $[-(L+1), L+1]$ *and consistent, for a locally Lipschitz continuous numerical Hamiltonian* $\mathcal{H}$. *Then, there exists a constant* C *such that, for any* $n \leq T/\Delta t$,

$$\left| v_j^n - v(x_j, t_n) \right| \leq C \Delta t^{1/2} \tag{18}$$

for $\Delta t \to 0$, $\Delta x = \lambda \Delta t$.

While it still requires monotonicity, the *Barles–Souganidis convergence theorem* (Barles and Souganidis, 1991) gives a more abstract and general framework for convergence of schemes, including the possibility of treating second-order, degenerate and singular equations. Roughly speaking, this theory states that any monotone, stable and consistent scheme converges to the exact solution *provided there exists a comparison principle for the limiting equation*. The Cauchy problem under consideration is:

$$\begin{cases} v_t + H(x,v,Dv) = 0 & (x,t) \in \mathbb{R}^d \times (0,T), \\ v(x,0) = v_0(x) & x \in \mathbb{R}^d. \end{cases} \tag{19}$$

The function $H : \mathbb{R}^d \times \mathbb{R} \times \mathbb{R}^d \to \mathbb{R}$ is a continuous Hamiltonian for all $w \in \mathbb{R}$, $x,p \in \mathbb{R}^d$. We assume that a *comparison principle* holds true for (19).

Consider a scheme in the general form (13). We assume the following *generalized consistency condition*:

Definition 6. Let $\Delta_m = (\Delta x_m, \Delta t_m)$ be a generic sequence of discretization parameters, (x_{j_m}, t_{n_m}) be a generic sequence of nodes in the space–time grid such that, for $m \to \infty$,

$$(\Delta x_m, \Delta t_m) \to 0 \quad \text{and} \quad (x_{j_m}, t_{n_m}) \to (x,t). \tag{20}$$

Let $\phi \in C^\infty(\mathbb{R}^d \times (0,T])$. Then, the scheme S is said to be *consistent* if

$$\liminf_{m\to\infty} \frac{\phi(x_{j_m}, t_{n_m}) - S_{j_m}(\Delta_m; \Phi(t_{n_m-1}))}{\Delta t_m} \geq \phi_t(x,t) + \underline{H}(x, \phi(x,t), D\phi(x,t)), \tag{21}$$

$$\limsup_{m\to\infty} \frac{\phi(x_{j_m}, t_{n_m}) - S_{j_m}(\Delta_m; \Phi(t_{n_m-1}))}{\Delta t_m} \leq \phi_t(x,t) + \overline{H}(x, \phi(x,t), D\phi(x,t)), \tag{22}$$

where $\underline{H}$ and $\overline{H}$ denote respectively the lower and upper semicontinuous envelopes of H. In (21)–(22), the index of the sequence is m, while j_m and n_m denote the corresponding node indices with respect to the mth space-time grid; we also recall that Φ or $\Phi(t)$ denote the vector of node values for respectively $\phi(x)$ and $\phi(x, t)$. Note that, if a scheme is consistent in the usual sense, it also satisfies (21)–(22).

The standard definition of monotonicity is replaced by the following *generalized monotonicity* assumption.

Definition 7. Let $(\Delta x_m, \Delta t_m)$ and (x_{j_m}, t_{n_m}) be generic sequences satisfying (20). Then, the scheme S is said to be *monotone* (in the generalized sense) if it satisfies the following conditions:

$$\text{if} \quad v_{j_m} \leq \phi_{j_m} \quad \text{then} \quad S_{j_m}(\Delta_m; V) \leq S_{j_m}(\Delta_m; \Phi) + o(\Delta t_m) \tag{23}$$

$$\text{if} \quad \phi_{j_m} \leq v_{j_m} \quad \text{then} \quad S_{j_m}(\Delta_m; \Phi) \leq S_{j_m}(\Delta_m; V) + o(\Delta t_m). \tag{24}$$

for any smooth function $\phi(x)$.

Also in this case, if a scheme is monotone in the sense of Definition 5, then it also satisfies (23)–(24).

Given a numerical solution V^n, we define its piecewise constant (in time) interpolation $v^{\Delta t}$ as:

$$v^{\Delta t}(x,t) = \begin{cases} I[V^n](x) & \text{if } t \in [t_n, t_{n+1}), \\ v_0(x) & \text{if } t \in [0, \Delta t), \end{cases}$$

where $I[V^n](x)$ denotes an interpolation of the node values in V, computed at x. We remark that whenever an interpolation operator is used, as is the case in the definition of $v^{\Delta t}$ or in the semi-Lagrangian (SL) approach, the interpolation operator has to satisfy some monotonicity (or relaxed monotonicity) properties to obtain a monotone scheme.

We can now state the extended version of the convergence result given in (Barles and Souganidis, 1991):

Theorem 3. *Let $v(x, t)$ be the unique viscosity solution of* (19). *Assume that* (21)–(22) and (23)–(24) *hold. Assume in addition that the family $v^{\Delta t}$ is uniformly bounded in L^∞. Then, $v^{\Delta t}(x,t) \to v(x,t)$ locally uniformly on $\mathbb{R}^d \times [0,T]$ as $\Delta \to 0$.*

3 EVOLUTIVE PROBLEMS

To introduce the schemes for time-dependent HJ equations, we refer to the basic problem

$$\begin{cases} v_t(x,t) + H(v_x(x,t)) = 0, & (x,t) \in \mathbb{R} \times [0,T] \\ v(x,0) = v_0(x), & x \in \mathbb{R}. \end{cases} \tag{25}$$

We will make the standing assumption that H is convex and that there exists $\alpha_0 \in \mathbb{R}$ such that

$$\begin{cases} H'(\alpha) \le 0 & \text{if } \alpha \le \alpha_0, \\ H'(\alpha) \ge 0 & \text{if } \alpha \ge \alpha_0. \end{cases} \tag{26}$$

We also define:

$$M_{H'}(L) = \max_{[-L,L]} |H'| \tag{27}$$

i.e., the maximum propagation speed of a solution with Lipschitz constant L.

3.1 Monotone Schemes in Differenced Form

The construction outlined will follow the guidelines of Crandall and Lions (1984). Note that, by construction, schemes in differenced form are necessarily invariant for the addition of constants. Therefore, l^∞ stability follows from monotonicity.

3.1.1 Upwind Discretization

In adapting the upwind scheme to the nonlinear case, it should be taken into consideration that $H'(v_x)$ is the propagation speed of the solution. While it is perfectly clear how to construct an upwind scheme for a speed of constant sign, care should be taken at points where the speed changes sign, in order to obtain a monotone scheme.

3.1.1.1 Construction of the Scheme

The differenced form of the upwind scheme is

$$v_j^{n+1} = v_j^n - \Delta t \mathcal{H}^{Up}\left(D_{j-1}[V^n], D_j[V^n]\right), \tag{28}$$

where the numerical Hamiltonian $\mathcal{H}^{Up}$ is typically defined by

$$\mathcal{H}^{Up}(\alpha,\beta) := \begin{cases} H(\alpha) & \text{if } \alpha,\beta \geq \alpha_0, \\ H(\beta)+H(\alpha)-H(\alpha_0) & \text{if } \alpha \geq \alpha_0, \beta \leq \alpha_0, \\ H(\alpha_0) & \text{if } \alpha \leq \alpha_0, \beta \geq \alpha_0, \\ H(\beta) & \text{if } \alpha,\beta \leq \alpha_0. \end{cases} \tag{29}$$

Note that the situation in which speed changes sign is subject to a different handling, depending on whether characteristics converge or diverge.

3.1.1.2 Consistency

Since the scheme is in differenced form, it actually suffices to apply the consistency condition in Definition 4. If $\alpha = \beta = a$, then the numerical Hamiltonian (29) satisfies

$$\mathcal{H}^{Up}(a,a) = H(a). \tag{30}$$

Note that, in (29), the second and third cases only occur if $a = \alpha_0$.

3.1.1.3 Monotonicity

The partial derivative of the jth component of the scheme is written as

$$\frac{\partial}{\partial v_i} S_j^{Up}(\Delta;V) = \delta_{ij} - \Delta t \left[\frac{\partial \mathcal{H}^{Up}}{\partial \alpha}\frac{\partial D_{j-1}[V]}{\partial v_i} + \frac{\partial \mathcal{H}^{Up}}{\partial \beta}\frac{\partial D_j[V]}{\partial v_i}\right] \tag{31}$$

where α and β are the dummy variables used in the definition (29), and δ_{ij} is the Kronecker symbol. A simple computation shows that, if $i \neq j$, then the monotonicity condition

$$\frac{\partial}{\partial v_i} S_j^{Up}(\Delta;V) \geq 0 \tag{32}$$

is always satisfied, whereas for $i = j$ it is satisfied provided

$$\frac{\Delta t}{\Delta x} \leq \frac{1}{2M_{H'}(L_V)}, \tag{33}$$

where L_V denotes the Lipschitz constant of the sequence V. In contrast to the linear case, this condition is more restrictive than the CFL condition.

3.1.2 *Central Discretization*

Rather than using more general forms of the scheme, we will restrict here to the particular form that directly generalizes the linear Lax–Friedrichs (LF) scheme.

3.1.2.1 Construction of the Scheme

The simplest way to recast LF scheme for the HJ equation is to define it in the form

$$v_j^{n+1} = \frac{v_{j-1}^n + v_{j+1}^n}{2} - \Delta t H\left(D_j^c[V^n]\right), \tag{34}$$

where $D_j^c[V^n]$ is the centred difference at x_j defined by

$$D_j^c[V^n] = \frac{v_{j+1}^n - v_{j-1}^n}{2\Delta x} = \frac{D_{j-1}[V^n] + D_j[V^n]}{2}. \tag{35}$$

This definition of the LF scheme completely parallels the linear case and is also suitable to be treated in the framework of the Crandall–Lions theorem. Indeed, keeping in mind that

$$\frac{v_{j-1}^n + v_{j+1}^n}{2} = v_j^n + \frac{\Delta x}{2}(D_j[V^n] - D_{j-1}[V^n]),$$

(34) can be written in the differenced form

$$v_j^{n+1} = v_j^n - \Delta t \mathcal{H}^{LF}(D_{j-1}[V^n], D_j[V^n]) \tag{36}$$

by setting

$$\mathcal{H}^{LF}(\alpha, \beta) := H\left(\frac{\alpha + \beta}{2}\right) - \frac{\Delta x}{2\Delta t}(\beta - \alpha).$$

3.1.2.2 Consistency

The LF scheme (34) satisfies condition (16), and in fact

$$\mathcal{H}^{LF}(a, a) = H\left(\frac{a + a}{2}\right) = H(a). \tag{37}$$

Consistency is therefore satisfied.

3.1.2.3 Monotonicity

In examining monotonicity, it is convenient to refer to the LF scheme in the form (34). Clearly, the jth component $S_j^{LF}(\Delta;V)$ depends only on the values $v_{j\pm1}$, so that

$$\frac{\partial}{\partial v_i} S_j^{LF}(\Delta;V) = 0 \quad (i \neq j \pm 1)$$

On the other hand, if $i = j \pm 1$, we have

$$\frac{\partial}{\partial v_{j\pm1}} S_j^{LF}(\Delta;V) = \frac{1}{2} \mp \frac{\Delta t}{2\Delta x} H'\left(D_j^c[V]\right).$$

Therefore, if L_V is the Lipschitz constant of the sequence V, the scheme is monotone provided

$$\frac{\Delta t}{\Delta x} \leq \frac{1}{M_{H'}(L_V)}. \tag{38}$$

3.2 SL Discretization

We analyze here the monotone version of the SL scheme, that is, the version obtained with $\mathbb{P}_1$ interpolation (see Falcone and Ferretti, 2014).

3.2.1 Construction of the Scheme

In the SL discretization of the HJ equation, what is really discretized is the representation formula for the solution. In the case of convex HJ equations, the formula under consideration is the Hopf–Lax formula (8). Once rewritten in a single space dimension, and at a point (x_j, t_{n+1}) of the space–time grid, it reads:

$$\begin{aligned} u(x_j, t+\Delta t) &= \min_{a\in\mathbb{R}} \left\{ \Delta t H^*(a) + u(x_j - a\Delta t, t) \right\} \\ &= \Delta t H^*(\bar{a}_j) + u(x_j - \bar{a}_j \Delta t, t), \end{aligned} \tag{39}$$

where $\bar{a}_j$ denotes the minimizer at the node x_j in (39). In the special case of (25), characteristics are straight lines, so that no special care should be taken about the accuracy of time discretization (we will comment on this later). The value $u(x_j - a\Delta t, t)$ should be reconstructed by a monotone space interpolation, for example in piecewise linear ($\mathbb{P}_1$) form. The resulting scheme is therefore:

$$\begin{cases} v_j^{n+1} = \min_{a\in\mathbb{R}} \left\{ \Delta t H^*(\alpha) + I_1[V^n](x_j - \alpha\Delta t) \right\} \\ v_j^0 = u_0(x_j), \end{cases} \tag{40}$$

in which $I_1[V](x)$ denotes the $\mathbb{P}_1$-interpolate of the sequence V, computed at the point x. For the computation of $H^*(a)$ one can use the Fast Legendre–Fenchel transform as in Corrias (1996).

Since the SL scheme is not in differenced form, the convergence analysis is carried out in the framework of the Barles–Souganidis theorem. Given the invariance of the $\mathbb{P}_1$ interpolation for the sum of constants, it suffices to check consistency and monotonicity.

3.2.1.1 Consistency

The scheme is compared with the representation formula (8). Denoting by $\overline{\alpha}_j$ the minimizer at the node x_j in (40), we define

$$S_j^{SL}(\Delta;V) = \Delta t H^*(\overline{\alpha}_j) + I_1[V](x_j - \overline{\alpha}_j \Delta t).$$

Let u be a smooth solution of (25), and U the sequence of its node values. First, recall that

$$|u(x) - I_1[U](x)| \leq C\Delta x^2 \min_{m\in\mathbb{Z}} \frac{x - x_m}{\Delta x}. \tag{41}$$

Writing now $u(x_j, t+\Delta t)$ by means of (39), and using (41), via two unilateral estimates the consistency error is bounded as

$$\begin{aligned} \left|L_j^{SL}(\Delta;t,U(t))\right| &= \left|\frac{1}{\Delta t}\left[u(x_j,t+\Delta t) - S_j^{SL}(\Delta;t,U(t))\right]\right| \\ &\leq C\min\left(\Delta x, \frac{\Delta x^2}{\Delta t}\right), \end{aligned}$$

which implies consistency for any $\Delta x/\Delta t$ relationship. Note that this estimate would suggest that the scheme achieves its best result when going to the final time in a single time step. In practice, when characteristics are not straight lines, errors in characteristics tracking should also be taken into consideration.

3.2.1.2 Monotonicity

First, note that the SL scheme is invariant for the addition of constants since, for a Lagrange interpolation of any order, $I[V+c](x) \equiv I[V](x)+c$.

To check that the SL scheme is monotone, consider two sequences V and W such that $V - W \geq 0$ componentwise. A simple computation gives:

$$S_j^{SL}(\Delta;V) - S_j^{SL}(\Delta;W) \geq I_1[V](x_j - \overline{\alpha}_j\Delta t) - I_1[W](x_j - \overline{\alpha}_j\Delta t),$$

where $\overline{\alpha}_j$ is the minimizer obtained for the sequence V. Since $\mathbb{P}_1$ interpolation is monotone itself (that is, $I_1[V] - I_1[W] \geq 0$), we get

$$S_j^{SL}(\Delta;V) - S_j^{SL}(\Delta;W) \geq 0. \tag{42}$$

Last, since the scheme is invariant for the addition of constants, L^∞ stability is implied by monotonicity.

3.3 Convergence

For the monotone approximations outlined above, it is possible to prove explicit a priori error estimates. More precisely:

- For monotone schemes in differenced form, the Crandall–Lions theorem applies, providing a theoretical convergence rate of order 1/2 under a linear CFL condition. We have therefore (Crandall and Lions, 1984) the following

Theorem 4. *Let H satisfy the basic assumptions,* $u_0 \in W^{1,\infty}$, *u be the solution of* (25) *with L as its Lipschitz constant and* v_j^n *be defined by* (28) *(respectively,* (36)*) with* $v_j^0 = v_0(x_j)$. *Then, for any* $j \in \mathbb{Z}$ *and* $n \in [1, T/\Delta t]$, *and for some positive constant C,*

$$\left| v_j^n - v(x_j, t_n) \right| \leq C \Delta t^{1/2} \tag{43}$$

as $\Delta t \to 0$, *with* $2M_{H'}(L+1)\Delta t \leq \Delta x$ *(resp.,* $M_{H'}(L+1)\Delta t \leq \Delta x$*).*

- For monotone SL schemes, convergence by Barles–Souganidis theorem would not provide an explicit convergence rate. An ad hoc convergence proof leads (Falcone and Ferretti, 2014) to the following

Theorem 5. Let H satisfy the basic assumptions, $u_0 \in W^{1,\infty}$, *u be the solution of* (25) *with L as its Lipschitz constant and* v_j^n *be defined by* (40) *with* $v_j^0 = v_0(x_j)$. *Then, for any* $j \in \mathbb{Z}$ *and* $n \in [1, T/\Delta t]$, *and for some positive constant C,*

$$\left| v_j^n - v(x_j, t_n) \right| \leq C \frac{\Delta x}{\Delta t} \tag{44}$$

as $\Delta x, \Delta t \to 0$, *with* $\Delta x = o(\Delta t)$.

We point out that, as soon as characteristics are no longer straight lines, this convergence estimate becomes

$$\left| v_j^n - v(x_j, t_n) \right| \leq C \left(\Delta t^{\gamma} + \frac{\Delta x}{\Delta t} \right)$$

with γ denoting the order of approximation of characteristics (see Falcone and Ferretti, 1994, 2014).

Last, it is observed in the numerical practice that, if the solution is uniformly semiconcave in $[0, T]$, then the actual convergence rate improves. In fact, in this case, singularities of the gradient are generated by characteristics coming from regular regions of the solution, and this causes the propagation of smaller numerical errors, as shown by the following numerical example.

3.3.1 A Numerical Example

We present a simple numerical example, using the one-dimensional model problem

$$\begin{cases} u_t(x,t) + \frac{1}{2}|u_x(x,t)|^2 = 0 & (x,t) \in (0,1) \times (0,T) \\ u(x,0) = u_0(x) \end{cases} \tag{45}$$

with $T = 0.05$ and two different Lipschitz continuous initial conditions u_0 with bounded support. The first is:

$$u_0(x) = \max\left(1 - 16(x - 0.25)^2, 0\right), \tag{46}$$

whereas the second, obtained by a simple change of sign, is also semiconcave:

$$u_0(x) = -\max\left(1 - 16(x - 0.25)^2, 0\right). \tag{47}$$

Using the initial condition (46), the solution eventually develops a singularity with nonempty superdifferential. After the onset of the singularity, the exact solution reads

$$u(x,t) = \begin{cases} \dfrac{\left(\left|x - \frac{1}{2}\right| - \frac{1}{4}\right)^2}{2t} & \text{if } \dfrac{1}{4} \le x \le \dfrac{3}{4} \\ 0 & \text{else} \end{cases}$$

On the other hand, using the initial condition (47), the solution has the expression

$$u(x,t) = \min\left(\frac{\left|x - \frac{1}{2}\right|^2}{2t + \frac{1}{16}} - 1, 0 \right),$$

and is uniformly semiconcave for $t > 0$. The test is performed with Upwind, LF and SL schemes. In this case, the refinement has been carried out with $\Delta t = \Delta x/40$ for Upwind scheme, $\Delta t = \Delta x/20$ for LF scheme and $\Delta t = 0.01$ (fixed) for SL scheme. Fig. 1 compares exact with numerical solutions for

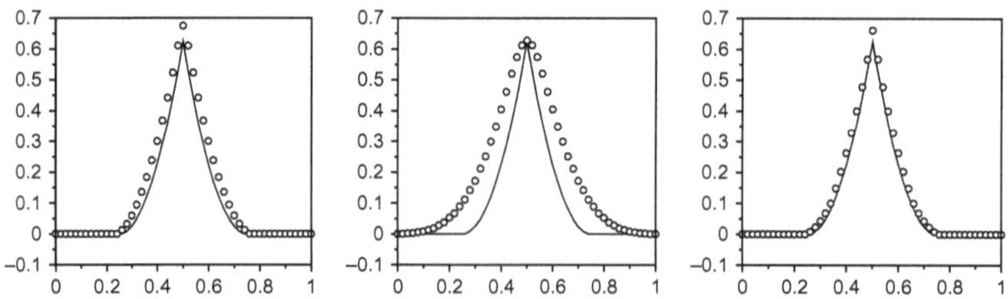

FIG. 1 Numerical results for problem (45)–(46), obtained via Upwind (*left*), Lax–Friedrichs (*centre*) and Semi-Lagrangian (*right*) schemes, 50 nodes.

TABLE 1 Errors in the ∞-Norm for Problem (45), Upwind, Lax–Friedrichs and Semi-Lagrangian Schemes

	$W^{1,\infty}$ Initial Condition			Semiconcave Initial Condition		
n_n	*Upwind*	*LF*	*SL*	*Upwind*	*LF*	*SL*
25	1.13×10^{-1}	2.84×10^{-1}	8.82×10^{-2}	6.42×10^{-2}	3.64×10^{-1}	2.11×10^{-2}
50	1.01×10^{-1}	2.51×10^{-1}	3.53×10^{-2}	3.58×10^{-2}	1.97×10^{-1}	5.02×10^{-3}
100	6.62×10^{-2}	1.89×10^{-1}	1.81×10^{-2}	1.92×10^{-2}	9.76×10^{-2}	1.24×10^{-3}
200	4.08×10^{-2}	1.27×10^{-1}	8.26×10^{-3}	9.97×10^{-3}	4.83×10^{-2}	3.07×10^{-4}
400	2.42×10^{-2}	8.0×10^{-2}	3.94×10^{-3}	5.08×10^{-3}	2.4×10^{-2}	7.63×10^{-5}
Rate	0.56	0.46	1.12	0.91	0.98	2.03

the first initial condition, whereas Table 1 shows numerical errors in the ∞-norm, showing that the theoretical convergence rates are optimal in lack of uniform semiconcavity, but improve in the semiconcave case. Among the schemes in differenced form, the LF scheme has an apparently higher numerical viscosity, but similar convergence rate with respect to the upwind scheme.

4 STATIONARY PROBLEMS

In adapting the various schemes to stationary HJ equations, we refer to the stationary model which in some sense parallels (25), that is

$$\lambda v(x) + H(v_x(x)) = g(x) \quad x \in \mathbb{R}, \tag{48}$$

for $\lambda > 0$. A typical setting to discretize (48) is to consider time-marching schemes, either in differenced form or of SL type. This amounts to look for fixed points of the numerical scheme.

4.1 Discretization in Differenced Form

In differenced form time-marching schemes, the schemes are applied to the evolutive equation

$$v_t + \lambda v + H(v_x) = g,$$

whose solution converges to a regime state satisfying (48). Keeping the scheme in its most general we have

$$\frac{v_j^{n+1} - v_j^n}{\Delta t} + \lambda v_j^n + \mathcal{H}(D_{j-p}[V^n], \ldots, D_{j+q}[V^n]) = g(x_j),$$

which is clearly a consistent scheme. Hence, replacing the time index with an iteration index k, we obtain

$$v_j^{(k+1)} = (1-\lambda\Delta t)v_j^{(k)} - \Delta t\mathcal{H}\left(D_{j-p}\left[V^{(k)}\right],\ldots,D_{j+q}\left[V^{(k)}\right]\right) + \Delta t g(x_j). \quad (49)$$

4.2 SL Discretization

In the case of SL discretization, we apply a generalized form of the Hopf–Lax formula, which applies to the solution of (48), in the form

$$v(x) = \inf_{\alpha\in\mathcal{A}}\left\{\int_0^\tau [g(y_x(s;\alpha)) + H^*(\alpha(s))]e^{-\lambda s}ds + e^{-\lambda\tau}v(y_x(\tau;\alpha))\right\},$$

where $\mathcal{A}$ is the set of measurable functions mapping $[0,+\infty)$ into $\mathbb{R}^M$, and $y_x(s;\alpha)$ satisfies

$$\begin{cases} \dot{y}_x(s;\alpha) = \alpha(s) \\ y_x(0;\alpha) = x. \end{cases}$$

Then, a SL type discretization can be written in iterative form as:

$$v_j^{(k+1)} = \min_{a\in\mathbb{R}}\left\{(1-e^{-\lambda\Delta t})H^*(a) + e^{-\lambda\Delta t}I_1\left[V^{(k)}\right](x_j - a\Delta t)\right\} + \Delta t g(x_j). \quad (50)$$

We mention that a different approximation scheme for stationary HJ equations, also based on a control interpretation, has been proposed in Bornemann and Rasch (2006).

4.3 Convergence and a Priori Error Estimates

It can be easily proved that the right-hand side of both (49) and (50) is a contraction in l^∞ (this follows from monotonicity, as shown in Bokanowsi et al., 2015). Therefore, the iteration converges towards a unique fixed point. Although relatively inefficient (the contraction coefficient is $L_S = 1 - O(\Delta t)$), this procedure is simple and robust. On the other hand, the search for more efficient solvers has motivated a development of Fast Marching and Fast Sweeping methods based on the schemes presented in this chapter (more details can be found in Qian et al., 2007; Sethian, 1996).

With some further work, it could be proved that the adaptation of monotone schemes to the stationary forms (49)–(50) satisfies both Barles–Souganidis and (in the differenced form) Crandall–Lions theory, with the due changes necessary to treat the stationary case. In both cases, the convergence estimates parallel the estimates of the time-dependent case, i.e.,

$$\left|v_j - v(x_j)\right| \leq C\Delta t^{1/2} \quad (51)$$

for the schemes in differenced form, and

$$|v_j - v(x_j)| \leq C\left(\Delta t + \frac{\Delta x}{\Delta t}\right) \tag{52}$$

for the SL scheme. Similar estimates for the nonconvex case can be found in Soravia (1998).

5 HIGH-ORDER APPROXIMATION METHODS

In this section, we sketch some basic ideas about high-order approximations for HJ equations. The topic is undergoing a fast development, and these notes are intended only as a general introduction.

5.1 Theoretical Tools

Out of the framework of monotone schemes, the convergence theory for approximations of HJ equations becomes less classical, and no general recipe has been singled out yet. However, in the last years a couple of techniques have been proved to be viable devices in the convergence analysis of high-order numerical scheme. One is the relaxation of the monotonicity assumption to quasi-monotonicity, the other is semiconcave stability.

5.1.1 ε-Monotonicity

Despite being usually applied to strictly monotone schemes, the Barles–Souganidis theorem allows for some small (more precisely, $o(\Delta t)$) monotonicity defect. Within this margin, it is sometimes possible to prove convergence for quasi-monotone schemes. Notably, this technique has been applied to high-order SL schemes and to filtered schemes (for which the monotonicity defect can be set a priori). Applications of this framework are given in Augoula and Abgrall (2000), Falcone and Ferretti (2014) and Bokanowsi et al. (2015).

5.1.2 Lin–Tadmor Theory

In Lin–Tadmor convergence theory (which derives from the *Lip'*-stability theory for conservation laws, see Lin and Tadmor, 2001), a different concept of stability is singled out, i.e., the concept of *semiconcave stability*:

Definition 8. A family of approximate solutions v^ϵ of (7) is said to be *semiconcave stable* if there exists a function $k(t) \in L^1([0, T])$ such that

$$D^2u^\epsilon(x,t) \leq k(t)I \tag{53}$$

for $t \in [0, T]$ (where I is the identity matrix).

More explicitly, condition (53) means that the matrix

$$D^2u^\epsilon(x,t) - k(t)I$$

is negative semidefinite, that is (since we are dealing with symmetric matrices), that all eigenvalues of D^2u are bounded from above by $k(t)$. In practice, the semiconcave stability is replaced by a bound on the second directional incremental ratios in the form:

$$\frac{v^{\Delta}(x+\delta,t)-2v^{\Delta}(x,t)+v^{\Delta}(x-\delta,t)}{|\delta|^2} \leq k(t). \tag{54}$$

Here, the function $k(t) \in L^1([0, T])$ plays the same role as in the original definition, and δ is a vector whose norm should remain bounded away from zero, and more precisely

$$|\delta| \geq C\Delta x. \tag{55}$$

The core of the theory is an abstract result of convergence for perturbed semiconcave stable solutions.

Theorem 6. *Consider problem* (7) *for a semiconcave initial condition* v_0 *with compact support, and assume the family* v^{ϵ} *is semiconcave stable. Define the truncation error associated to* v^{ϵ} *as*

$$F(x,t) := v_t^{\epsilon}(x,t) + H(\nabla v^{\epsilon}(x,t)). \tag{56}$$

Then, for any $t \in [0, T]$,

$$\| v(t) - v^{\epsilon}(t)\|_{L^1(\mathbb{R}^d)} \leq C_1 \| v_0 - v^{\epsilon}(0)\|_{L^1(\mathbb{R}^d)} + C_2 \| F\|_{L^1(\mathbb{R}^d \times [0,T])}. \tag{57}$$

The second ingredient of the Lin–Tadmor theory, i.e., the L^1 estimation of the truncation error, may be difficult in general, since it requires a reversed approach in measuring the truncation error, that is, by plugging the numerical solution into the exact equation. An easier expression can be derived for Godunov-type schemes, taking into account that in this case numerical errors are generated only by the projection step, and not by the evolution operator, which is in principle exact. A practical application of this theory to a Godunov-type scheme is presented in Lin and Tadmor (2001).

5.2 High-order FD Schemes

The basic strategy for constructing high-order finite difference methods has been first proposed in Osher and Shu (1991), and passes through the intermediate step of a semidiscrete scheme. In a second step, a TVD Runge–Kutta method is applied to the semidiscrete scheme, to obtain a fully discrete approximation.

The semidiscrete scheme is constructed using a monotone numerical Hamiltonian, in the form

$$\dot{v}_j = -\mathcal{H}\left(D_j^-[V], D_j^+[V]\right), \tag{58}$$

in which $\mathcal{H}(\cdot, \cdot)$ is increasing with respect to its first argument, and decreasing with respect to the second. In (58), $D_j^{\pm}[V]$ denote high-order approximations of the right/left derivative at the node x_j, which replace in the numerical Hamiltonian the mere right/left incremental ratios. In the most extensively studied versions of the scheme, these estimates are usually obtained via nonoscillatory (ENO/WENO) techniques (see Bryson and Levy, 2003; Jiang and Peng, 2000; Kurganov and Petrova, 2006; Kurganov et al., 2001; Osher and Shu, 1991).

5.3 High-order SL Schemes

The SL scheme (40) is easily extended to a higher consistency rate by replacing the $\mathbb{P}_1$ space interpolation I_1 with an interpolation of higher accuracy (Carlini et al., 2005; Falcone and Ferretti, 2002). In more general cases, in which characteristics are not straight lines, a more accurate method of characteristics tracking is also desirable (Falcone and Ferretti, 1994).

In some model cases (a single space dimension, no x-dependence of the Hamiltonian) convergence of high-order SL schemes, for both the evolutive and the stationary case, can be proved by showing their quasi-monotonicity. Here, a crucial role is played by the Lipschitz stability of the scheme, along with the inverse CFL condition $\Delta x = O(\Delta t^2)$ (see Bokanowsi et al., 2015; Falcone and Ferretti, 2014; Ferretti, 2002)

5.4 Discontinuous Galerkin

The application to HJ equations of Discontinuous Galerkin (DG) methods uses the relationship with conservation laws. In fact, what is discretized in this case is the CL (or system of CLs, see (12) and Jin and Xin, 1998) associated to the HJ equation. Following Hu and Shu (1999), the approximate solution is constructed in the space

$$V_{\Delta x}^k = \left\{w : w|_{I_j} \in \mathbb{P}_k(I_j)\right\}, \tag{59}$$

in which I_j denotes the j-th element of the computational domain, for example the interval

$$[x_{j-1/2}, x_{j+1/2}] = [x_j - \Delta x/2, x_j + \Delta x/2]$$

in a one-dimensional uniform grid. A DG scheme of order k for the Eq. (25) is defined by looking for the function $w \in V_{\Delta x}^k$ such that

$$\int_{I_j} w_{xt}\phi dx - \int_{I_j} H(w_x)\phi_x dx + \mathcal{H}_{j+1/2}\phi^-_{j+1/2} - \mathcal{H}_{j-1/2}\phi^+_{j-1/2} = 0$$

for all j and all $\phi \in V^{k-1}_{\Delta x}$. In (59), the values $\mathcal{H}_{j-1/2}$ are defined by

$$\mathcal{H}_{j\pm 1/2} := \mathcal{H}\left(w_x\left(x^-_{j\pm 1/2}\right), w_x\left(x^+_{j\pm 1/2}\right)\right),$$

and $\mathcal{H}$ is as usual a monotone numerical Hamiltonian. Note that the outcome of the scheme is the derivative w_x (or, in the multidimensional case, all the partial derivatives). Suitable techniques allow to recover the approximate solution w from this information.

5.5 Filtered Schemes

The general idea behind the construction of filtered scheme is to provide a clever coupling between a monotone and a high-order scheme. The lack of regularity may cause high-order schemes to introduce spurious oscillations, so the idea is to apply the high-order scheme only where the solution is regular enough, this being accomplished by a suitable "filter" function.

The construction of a filtered scheme needs three ingredients: a monotone scheme (denoted by S^M), a high-order scheme (denoted by S^{HO}) and a bounded (not necessarily smooth) *filter function*, $F : \mathbb{R} \to \mathbb{R}$. The filtered scheme S^F is then defined as

$$v_j^{n+1} = S_j^F(V^n) := S_j^M(V^n) + \epsilon \Delta t F\left(\frac{S_j^{HO}(V^n) - S_j^M(V^n)}{\epsilon \Delta t}\right), \tag{60}$$

where $\epsilon = \epsilon(\Delta) > 0$ is a parameter vanishing for Δt, $\Delta x \to 0$, whose choice controls the monotonicity defect of the filtered scheme (more hints on the choice of ϵ can be found in Bokanoswki et al., 2016; Froese and Oberman, 2013). A typical filter function is given by

$$F(x) := \text{sign}(x)\max(1 - ||x| - 1|, 0) = \begin{cases} x & |x| \le 1. \\ 0 & |x| \ge 2. \\ -x+2 & 1 \le x \le 2. \\ -x-2 & -2 \le x \le -1. \end{cases}$$

This definition of the filter function blends the two schemes according to the ratio $\rho = (S^A - S^M)/(\Delta t\epsilon)$. If $|\rho| \le 1$, then $S^F = S^M + \Delta t\epsilon F(\rho) \equiv S^{HO}$, whereas if $|\rho| \ge 2$, then $F(\rho) = 0$ and $S^F \equiv S^M$, i.e., the scheme coincides with the monotone scheme. It can be shown that, for a suitable choice of ϵ, the filtered scheme converges to the viscosity solution. Moreover, a high-order consistency rate can be proved in the regular case, although globally the filtered scheme is not expected to have more than an $O(\sqrt{\Delta x})$ rate of convergence on Lipschitz continuous solutions.

REFERENCES

Abgrall, R., 1996. Numerical discretization of the first-order Hamilton-Jacobi equation on triangular meshes. Commun. Pure Appl. Math. 49, 1339–1373.

Augoula, S., Abgrall, R., 2000. High order numerical discretization for Hamilton-Jacobi equations on triangular meshes. J. Sci. Comput. 15, 197–229.

Bardi, M., Capuzzo-Dolcetta, I., 1997. Optimal Control and Viscosity Solutions of Hamilton-Jacobi-Bellman Equations. Birkhauser, Boston.

Bardi, M., Osher, S., 1991. The non-convex multi-dimensional Riemann problem for Hamilton-Jacobi equations. SIAM J. Math. Anal. 22, 344–351.

Barles, G., 1998. Solutions de viscosité des equations d'Hamilton-Jacobi. Springer-Verlag, Berlin.

Barles, G., Souganidis, P.E., 1991. Convergence of approximation schemes for fully nonlinear second order equations. Asymptot. Anal. 4, 271–283.

Bokanoswki, O., Falcone, M., Sahu, S., 2016. An efficient filtered scheme for some first order Hamilton-Jacobi-Bellman equations. SIAM J. Sci. Comput. 38, 171–195.

Bokanowsi, O., Falcone, M., Ferretti, R., Grüne, L., Kalise, D., Zidani, H., 2015. Value iteration convergence of ε-monotone schemes for stationary Hamilton-Jacobi equations. Discrete Contin. Dyn. Syst. Ser. A 35, 4041–4070.

Bornemann, F., Rasch, C., 2006. Finite-element discretization of static Hamilton-Jacobi equations based on a local variational principle. Comput. Vis. Sci. 9, 57–69.

Bryson, S., Levy, D., 2003. High-order semi-discrete central-upwind schemes for multidimensional Hamilton-Jacobi equations. J. Comput. Phys. 189, 63–87.

Carlini, E., Ferretti, R., Russo, G., 2005. A weighted essentially nonoscillatory, large time-step scheme for Hamilton-Jacobi equations. SIAM J. Sci. Comput. 27, 1071–1091.

Corrias, L., 1996. Fast Legendre-Fenchel transform and applications to Hamilton-Jacobi equations and conservation laws. SIAM J. Numer. Anal. 33, 1534–1558.

Corrias, L., Falcone, M., Natalini, R., 1995. Numerical schemes for conservation laws via Hamilton-Jacobi equations. Math. Comput. 64, 555–580.

Crandall, M.G., Lions, P.L., 1983. Viscosity solutions of Hamilton-Jacobi equations. Trans. Am. Math. Soc. 277, 1–42.

Crandall, M.G., Lions, P.L., 1984. Two approximations of solutions of Hamilton-Jacobi equations. Math. Comput. 43, 1–19.

Evans, L.C., 2010. Partial Differential Equations, second ed. Graduate Studies in Mathematics, vol. 19. American Mathematical Society, Providence.

Falcone, M., Ferretti, R., 1994. Discrete-time high-order schemes for viscosity solutions of Hamilton-Jacobi equations. Numer. Math. 67, 315–344.

Falcone, M., Ferretti, R., 2002. Semi-Lagrangian schemes for Hamilton-Jacobi equations, discrete representation formulae and Godunov methods. J. Comput. Phys. 175, 559–575.

Falcone, M., Ferretti, R., 2014. Semi-Lagrangian Approximation Schemes for Linear and Hamilton-Jacobi Equations. SIAM, Philadelphia.

Ferretti, R., 2002. Convergence of semi-Lagrangian approximations to convex Hamilton-Jacobi equations under (very) large Courant numbers. SIAM J. Numer. Anal. 40, 2240–2253.

Fleming, W.H., Soner, H.M., 1993. Controlled Markov Processes and Viscosity Solutions. Springer-Verlag, New York.

Fleming, W.H., Souganidis, P.E., 1989. On the existence of value function of two-players, zero-sum differential games. Indiana Univ. Math. J. 38, 293–314.

Froese, B.D., Oberman, A.M., 2013. Convergent filtered schemes for the Monge-Ampère partial differential equation. SIAM J. Numer. Anal. 51, 423–444.

Hu, C., Shu, C.W., 1999. A discontinuous Galerkin finite element method for Hamilton-Jacobi equations. SIAM J. Sci. Comput. 21, 666–690.

Isaacs, R., 1965. Differential Games. Wiley, New York.

Jiang, G., Peng, D.P., 2000. Weighted ENO schemes for Hamilton-Jacobi equations. SIAM J. Sci. Comput. 21, 2126–2143.

Jin, S., Xin, Z., 1998. Numerical passage from system of conservation laws to Hamilton-Jacobi equations, and a relaxation scheme. SIAM J. Numer. Anal. 35, 2385–2404.

Kurganov, A., Petrova, G., 2006. Adaptive central-upwind schemes for Hamilton-Jacobi equations with nonconvex Hamiltonians. J. Sci. Comput. 27, 323–333.

Kurganov, A., Noelle, S., Petrova, G., 2001. Semi-discrete central-upwind scheme for hyperbolic conservation laws and Hamilton-Jacobi equations. SIAM J. Sci. Comput. 23, 707–740.

Kushner, H.J., Dupuis, P., 2001. Numerical Methods for Stochastic Control Problems in Continuous Time. Springer-Verlag, Berlin.

Leveque, R.J., 1992. Numerical Methods for Conservation Laws. Birkhäuser, Basel.

Lin, C.T., Tadmor, E., 2001. L^1 stability and error estimates for hamilton-jacobi solutions. Numer. Math. 87, 701–735.

Lions, P.L., 1982. Generalized Solution of Hamilton-Jacobi Equations. Pitman, London.

Lions, P.L., Souganidis, P., 1995. Convergence of MUSCL and filtered schemes for scalar conservation laws and Hamilton-Jacobi equations. Numer. Math. 69, 441–470.

Osher, S., Fedkiw, R.P., 2003. Level Set Methods and Dynamic Implicit Surfaces. Springer-Verlag, New York.

Osher, S., Shu, C.W., 1991. High-order essentially non-oscillatory schemes for Hamilton-Jacobi equations. SIAM J. Numer. Anal. 28, 907–922.

Qian, J., Zhang, Y., Zhao, H., 2007. A fast sweeping method for static convex Hamilton-Jacobi equations. J. Sci. Comput. 31, 237–271.

Sethian, J.A., 1996. Level Set Method. Evolving Interfaces in Geometry, Fluid Mechanics, Computer Vision, and Materials Science. Cambridge University Press, Cambridge.

Soravia, P., 1998. Estimates of convergence of fully discrete schemes for the Isaacs equation of pursuit-evasion differential games via maximum principle. SIAM J. Control Optim. 36, 1–11.

Tourin, A., 2006. Splitting methods for Hamilton-Jacobi equations. Numer. Methods Partial Differ. Equ. 22, 381–396.

Index

Note: Page numbers followed by "*f*" indicate figures, "*t*" indicate tables, "*b*" indicate boxes, and "*np*" indicate footnotes.

F

G

H

M

N

O

U

V

W

FSC
www.fsc.org
MIX
Papier | Fördert gute Waldnutzung
FSC® C083411

Kolibri 360 GmbH
Ferdinand-Jühlke-Straße 7
99095 Erfurt, Deutschland
produktsicherheit@kolibri360.de